INSTRUCTOR EDITION

WELCOME At Thomson Course Technology, our mission is to help people teach and learn about technology. This special Instructor Edition is all about helping YOU teach. We know you have a great deal to manage, and so we provide you with useful tools and resources with every textbook. This Instructor Edition will help you sort through all the material, and will enable you to choose the right combination of tools to help you teach the way you want to teach.

CONTENTS AT A GLANCE

Page IE 2: **CoursePort**

Learn about how you can have your students track and share their results with you for selected activities on the Succeeding with Technology Student Web site.

Page IE 3: **Instructor Resources for Succeeding with Technology, 2nd Edition**

We provide you with an in-depth description of everything this textbook has to offer to help you teach, so you can choose what makes the most sense for your classroom.

Page IE 6: **Succeeding with Technology Web site**

See how you can enhance your course with a variety of online components including Practice Tests, Student Edition Labs, CourseCasts, and more!

Page IE 8: **Succeeding with Technology Online Instructor Community**

This brand new resource provides you with an opportunity to collaborate with the authors and other instructors using the text.

Page IE 9: **Passwords and Security**

Tired of trying to remember passwords for three different systems? This page not only explains the Thomson Course Technology system for protecting sensitive materials such as answer keys and testbanks from students, it also provides you with a handy, single place for you to mark down all of your passwords.

Page IE 10: **Annotated Table of Contents for Succeeding with Technology, 2nd Edition**

The Annotated Table of Contents includes a chapter-by-chapter guide of what each chapter of this book has to offer you and your students as well as additional information to help you teach the course. This TOC provides a chapter-by-chapter explanation of what's available; we'll even make suggestions on how you can implement these materials in your classroom.

TAKE CONTROL OF YOUR CLASSROOM WITH COURSEPORT AND COURSE.COM

THE COURSEPORT STUDENT EXPERIENCE

Students will follow the instructions on the inside front cover of this text to create a CoursePort account and gain access to the Online Companion. Completing the registration process is easy! No keycode is required for students to access the Online Companion for this text!

- Visit http://login.course.com.
- Click ***New User Registration***
- Enter the required, basic information on ***New User Registration*** screen.
- You will then be brought to the ***Choose Your Product*** page.
- Check the box next to the title of the text and upon clicking ***Submit***, you're in!
- At anytime, students can use the ***My Account*** tab to launch their Online Companion or view their own progress in the Universal Gradebook.
- If you choose to track your students' activities in the Universal Gradebook, they will need to ***Join a Class*** from the ***My Account*** tab.

THE INSTRUCTOR EXPERIENCE

As an instructor, you'll also gain access to the Online Companion for this text. But *your* login will provide you with unique resources dedicated to make teaching with this textbook streamlined, powerful, and simple.

Use your Thomson Course Technology username and password to launch CoursePort. If you do not currently have a Thomson Course Technology username and password for www.course.com, contact us:

For Colleges and Universities in the US: Call Thomson Course Technology at **1.800.648.7450** and select option 3 for Support Services.
For Private Career Colleges in the US: Call **1.800.477.3692**
For High Schools: Call Thomson Learning-School at: **1.800.824.5179**
For Corporations, IT Training Centers, and Federal Government Agencies: Call Thomson Course Technology at **1.800.648.7450**

Ensuring students don't get their hands on the answers, test banks, or other resources that we provide you through our Instructor Resources, password protection is at the forefront of our minds at all times. Note that each and every time a caller requests a password, we verify the caller's affiliation with the school that he or she indicates.

My Instructor Login

Username: ______________________________

Password: ______________________________

TAKE ADVANTAGE OF THE POWER AND FLEXIBILITY OF COURSEPORT

- A single password allows you to access the Online Companion for this text, all of the Instructor Resources, and—if you choose—to manage your class with the Universal Gradebook.
- Students can access the Online Companion and track their progress with the activities for this text regardless of how heavily you choose to use CoursePort.
- By setting up a Class Code, you can easily link each student to your class and view results for the individuals or the class as a whole. Customized reporting reveals exactly what content is engaging your students and what is challenging them, so you can adjust your instruction or syllabus accordingly.
- To obtain a Class Code, simply login, click on the Create a Class link and follow the onscreen instructions.

INSTRUCTOR RESOURCES

We know you need more than great textbooks to effectively teach your class. That's why we take the next step in providing you with outstanding Instructor Resources—developed by educators and tested through our rigorous Quality Assurance process. Our goal is to make the teaching and learning experience in your classroom the best it can be. With Course Technology's resources, you'll spend less time prepping, and more time teaching.

INSTRUCTOR RESOURCES CD & WEB SITE: This is the first place to go when preparing for your next class. Both the CD (ISBN: 1-4188-3929-9) and the Web site (www.course.com) contain everything you need to get started. Check the site before each semester for updates to the Instructor Resources for this title.

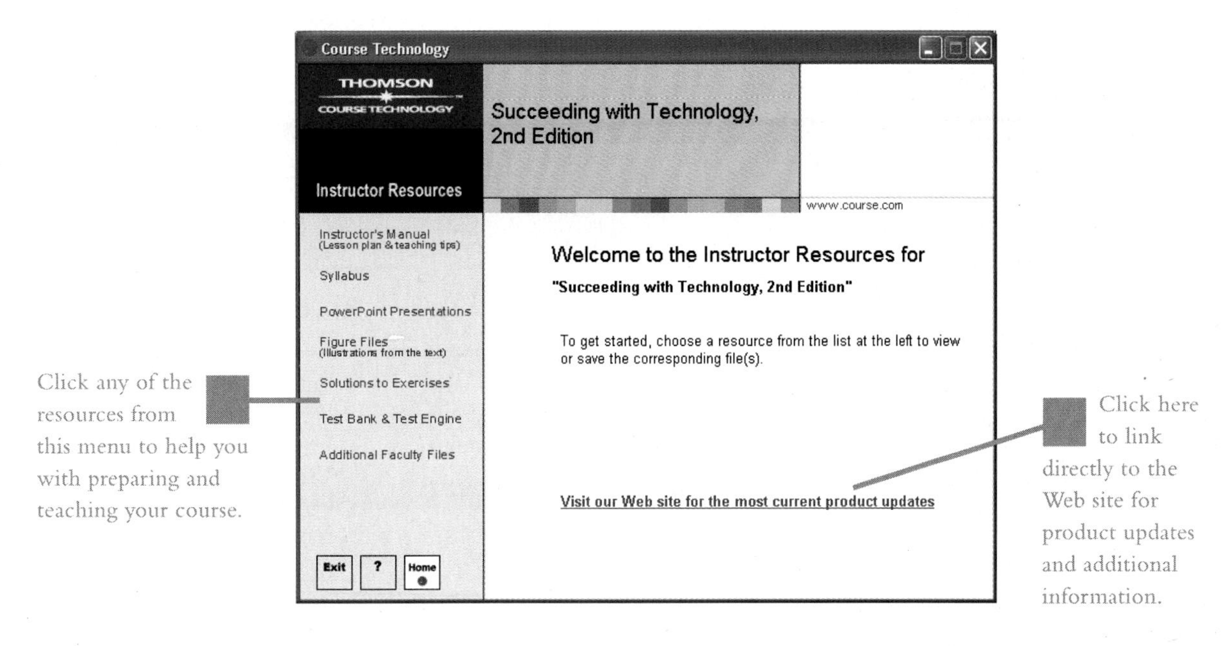

INSTRUCTOR'S MANUAL: Need to compile a lecture on information security for your class that's starting in an hour? Need a Quick Quiz to test your students' understanding of the material? Or looking for ways to challenge your students with group projects? The electronic Instructor's Manual is a great place to look for solutions to all of these dilemmas. Each chapter in the Instructor's Manual is a Microsoft Word document that you can customize easily with your own notes, but it is also filled with great ideas from instructors like you.

EXAMVIEW TESTBANK: ExamView features a user-friendly testing environment that allows you to not only publish traditional paper and LAN-based tests, but also Web-deliverable exams. Utilize the ultra-efficient Quick Test Wizard to create an exam in less than five minutes; or take advantage of the Course Technology question banks; or even customize your own exams from scratch.

WEBCT AND BLACKBOARD: WebCT and Blackboard are the leading distance learning solutions available today. In the past few years, they've also become popular class management platforms. Course Technology has partnered with WebCT and Blackboard to bring you online content that fits into both platforms. *Succeeding with Technology, 2nd Edition* is available with online content in WebCT and Blackboard format, with includes the following components:

- Topic Reviews
- Review Questions
- Case Projects
- PowerPoint Presentations
- Test Banks
- Custom Syllabus
- And More!

POWERPOINT PRESENTATIONS: Delivering engaging and visually impressive lectures is easy with the professionally-designed PowerPoint presentations available for each tutorial in this book. You can edit the files to fit your needs, post them to your network for students to review key concepts, or save them to the Web for your Distance Learning students.

WWW.COURSE.COM

Course Technology is the world's leading Information Technology publisher. Because we focus solely on IT, we have the unique ability to address the needs of customers like you.

Find out about the latest technology trends, products, and courseware solutions on the Course Technology Web site, **www.course.com**.

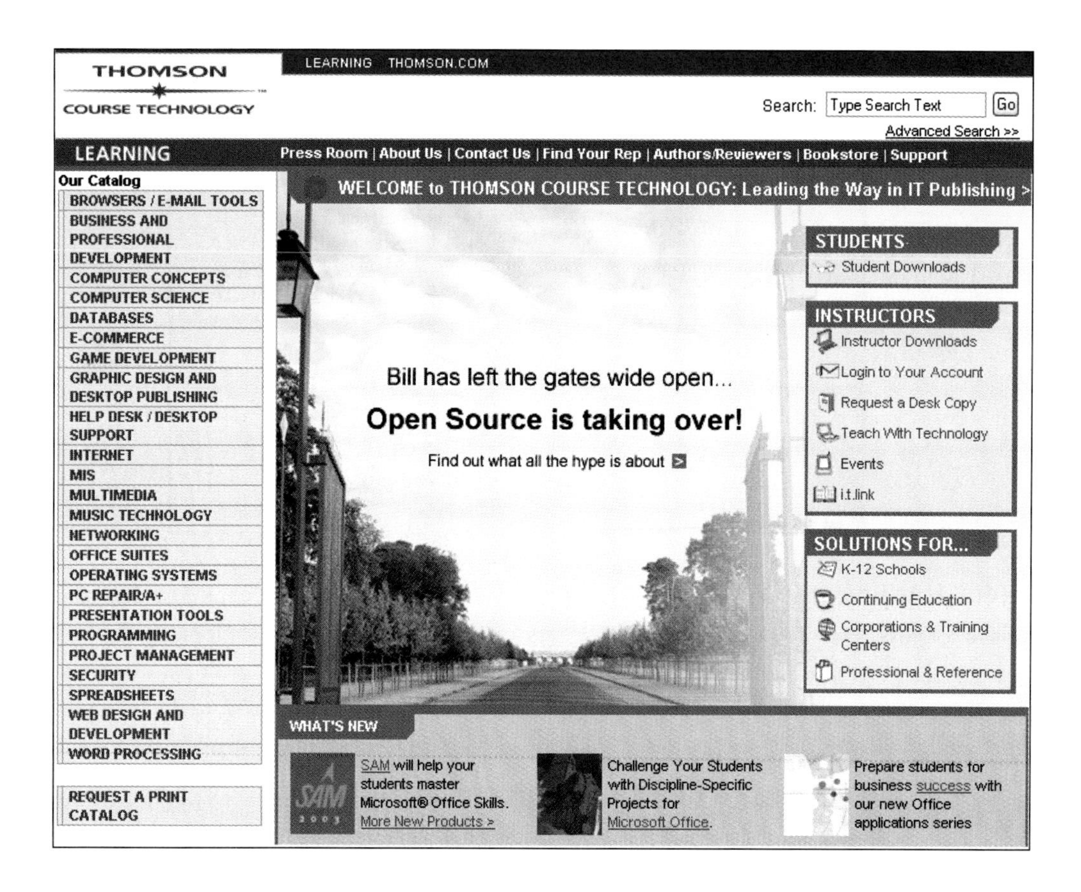

Visit often to:

- Connect with your peers through our Online forums
- Learn about the latest software releases and how they will impact you in the classroom
- Browse our online catalog
- Locate and contact your local sales representative
- Register for the next Conference for Information Technology Educators and our other educational events

SAM COMPUTER CONCEPTS

Add more muscle and flexibility to your "Expand your Knowledge" Student Edition Labs, with SAM (Skills Assessment Manager) Computer Concepts! SAM Computer Concepts adds the power of skill-based assessment and the award-winning SAM classroom administration system to your "Expand Your Knowledge" Student Edition Labs, putting you in control of how you deliver exams and training in your course.

By adding SAM Computer Concepts to your curriculum, you can:

- Reinforce your students' knowledge of key computer concepts with hands-on application exercises.
- Build hands-on computer concepts exams from a test bank of more than 200 skill-based concepts tasks.
- Schedule your students' concepts training and testing exercises with powerful administrative tools.
- Track student exam grades and training progress using more than one dozen student and classroom reports.

Teach your introductory course with the simplicity of a single system! You can now administer your entire Computer Concepts and Microsoft Office course through the SAM platform. For more information on the SAM administration system, SAM Computer Concepts, and other SAM products, please visit **www.course.com/sam**.

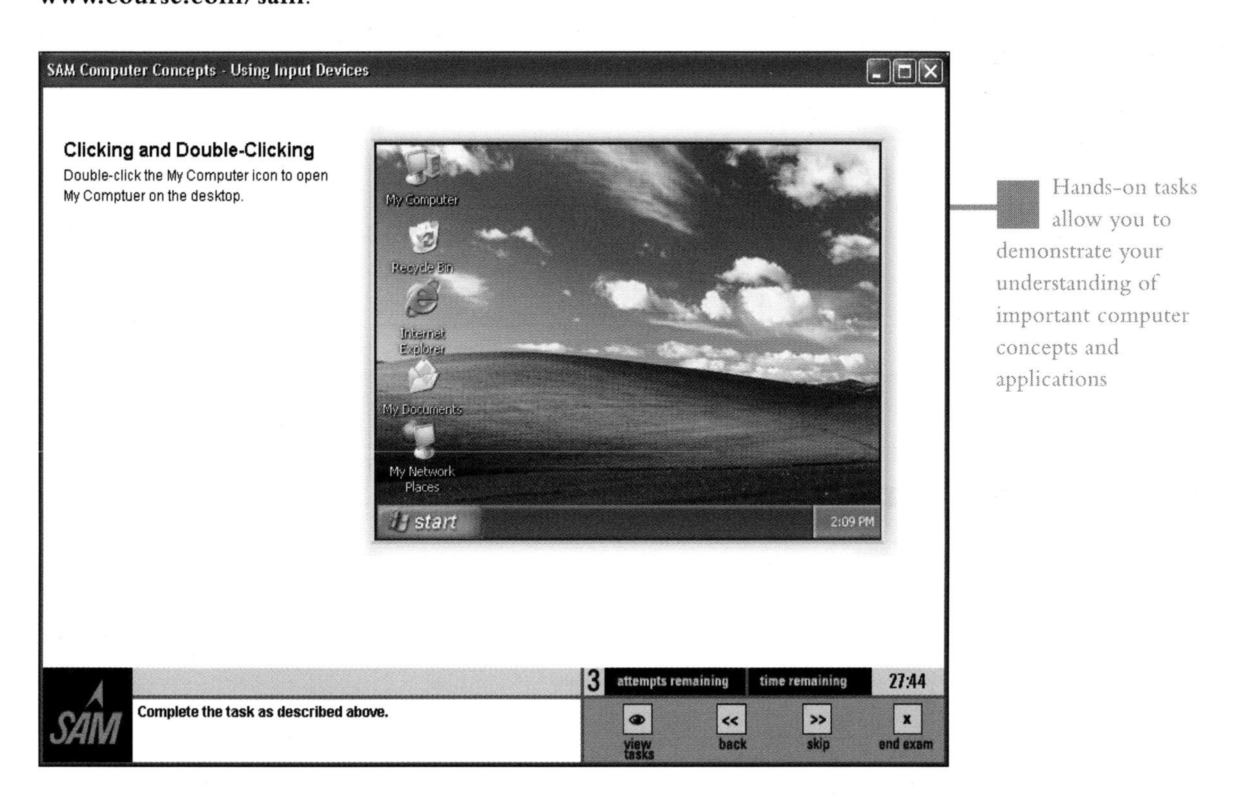

Hands-on tasks allow you to demonstrate your understanding of important computer concepts and applications

STUDENT ONLINE COMPANION

Students deserve a wealth of resources to reinforce their studies. This text's Student Online Companion, is available at **www.course.com/swt2**.

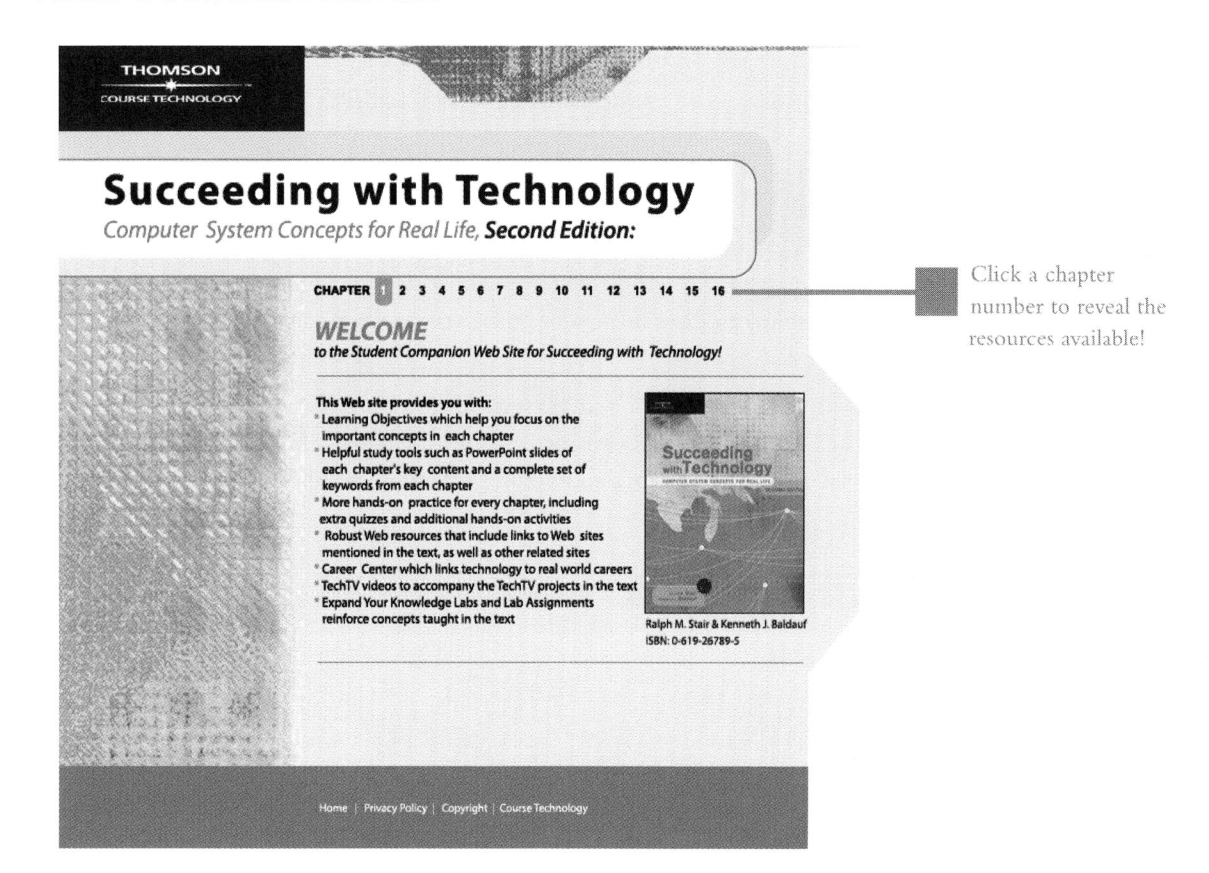

Click a chapter number to reveal the resources available!

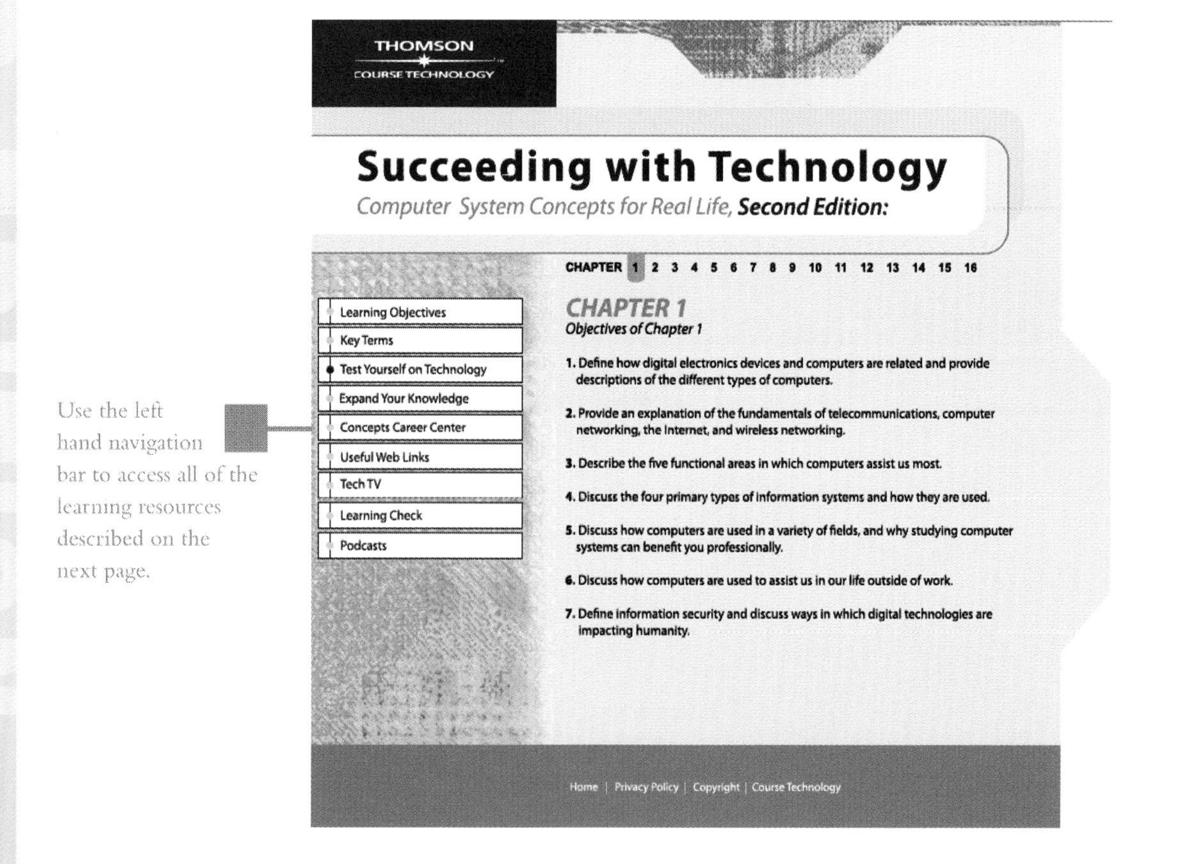

Use the left hand navigation bar to access all of the learning resources described on the next page.

The Student Online Companion features:

- **Learning Objectives:** These provide students with a quick review of the objectives of each chapter.
- **Key Terms:** Students can use the key terms listing as a study tool to test their technology vocabulary from each chapter.
- **Test Yourself on Technology (Practice Tests):** Allow students to test themselves on the key terms and concepts of each chapter and get instant results. As an additional student benefit, the answers include specific page references so they can use it as a study tool.
- **Expand Your Knowledge (Student Edition Labs and Assignments):** These robust lab simulations give students the chance to explore their knowledge on a variety of related subjects. A Lab Assignment to compliment each lab is provided as well. Throughout this book you will see icons, to direct students to the related lab online.

  Student Edition Labs
- **Concepts Career Center:** This online resource gives your students a fun and easy way to learn more about jobs and industries that they may want to explore one day.
- **Useful Web links:** With this resource, students can go directly to the Web pages addressed in the chapters.
- **TechTV:** Help your students stay on top of emerging technologies and technology related issues with our library of TechTV video clips. These videos accompany the TechTV assignments included as an end-of-chapter activity.
- **CourseCast:** Students can download these audio flash cards and chapter review files to their mp3 players and study for exams in the car, on their way to class, at the gym, virtually anywhere while on the go!

ONLINE INSTRUCTOR COMMUNITY

Have you ever wanted to know how an author of the book teaches their course? Have you ever wanted to share your ideas and tips with other instructors or possibly get some ideas and tips from instructors like yourself? We are excited to offer an online community geared specifically for instructors teaching with Succeeding with Technology 2nd Edition.

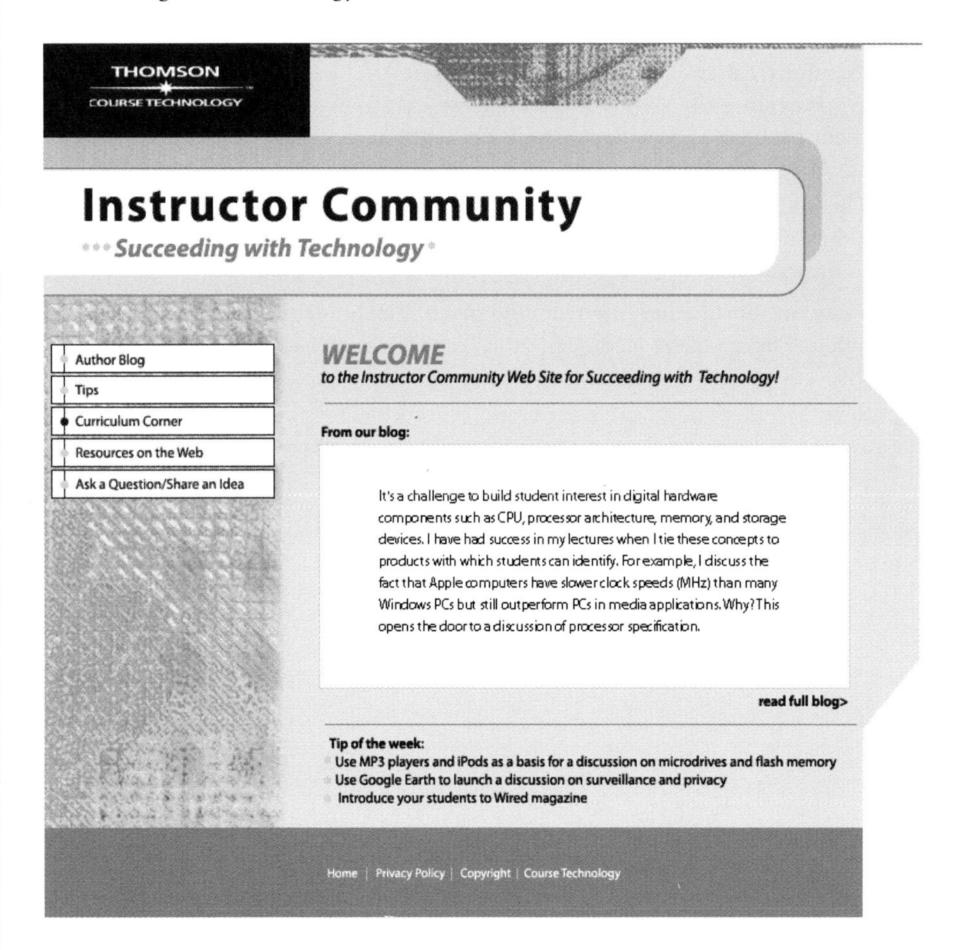

A few of the features you will find on the site are:

- **An author blog:** Ken Baldauf, one of the authors of the text, teaches more than 5,000 students annually and is going to be sharing his experiences, ideas, and tips in a weekly blog. Instructors will have an opportunity to respond to his entries, and ask specific questions about the book and teaching the course.
- **Tips:** A special Tips page will be filled with ideas and tips submitted by instructors just like you.
- **Curriculum Corner:** This section will be filled with teaching files (syllabi, PowerPoint Presentations, projects etc.) from the authors or submitted by instructors like you.
- **Resources on the Web:** This section will be a collection of web resources, focused not on just IT topics, but also interesting teaching and education topics. If you find a site you think is worth sharing, you can submit it for posting!

PASSWORDS AND SECURITY

INSTRUCTOR RESOURCES ON COURSE.COM

Depending on the type of educational institution you are, please call one of the support services teams to obtain your password to our online Instructor Resources.

For Colleges and Universities in the US:
Call Course Technology at **1.800.648.7450** and select option 3 for Support Services.

For High Schools:
Call Thomson Learning—School at **1.800.824.5179**

For Private Career Colleges:
Call Thomson Learning—Career at **1.800.477.3692**

For Corporations, IT Training Centers, and Federal Government Agencies:
Call Course Technology at **1.800.648.7450**

Password protection is of utmost importance at Thomson Course Technology. We monitor each and every caller requesting an instructor resources password by verifying their names with the school the caller indicates as their affiliation. This ensures that students never have access to the answers, test banks, or other resources that we provide for you on course.com.

OTHER RESOURCES AND PASSWORDS

Thomson Course Technology knows that protecting these passwords from students is critical, but keeping track of all the passwords for all the tools you may adopt in a given course (one for your online instructor resources, another for SAM, another for your schools' servers…) can be quite a hassle. For your convenience, we're providing space below for you to keep track of all your login information (either literally or with your own code that students won't understand!) for this course.

RESOURCE	USERNAME	PASSWORD

INSTRUCTOR EDITION

ANNOTATED TABLE OF CONTENTS

Now that you have a better understanding of all the tools that come with *Succeeding with Technology, 2nd Edition*, what about the text itself? What does each chapter have to offer you and your students? This annotated TOC highlights some of the chapter features that you won't want to miss and also gives you some ideas on how to make the classroom experience even more engaging for your students.

We have taken great pains to present the material in a logical and engaging manner that is easy to teach and easy to learn. Here are some suggestions that will assist you in delivering the content of each chapter.

CHAPTER 1: WHY STUDY COMPUTERS AND DIGITAL TECHNOLOGIES?

OVERVIEW

This chapter provides some fundamental concepts and presents a framework for understanding computers and digital technologies. The chapter begins by defining the term computer and relating it to the many digital devices we interact with each day. The chapter then moves to a discussion of computer networks, and the Internet emphasizing the power of network connections in our lives. Next is a discussion of the types of activities for which computers are used; what they do well, and what they don't do so well. This is followed by an overview of the ways information systems are used in today's businesses and organizations. The next sections examine how the power of digital technologies is being harnessed in various industries and careers as well as in our lives outside of work. The final section discusses security, social, and ethical issues surrounding technology. This topics covered in this chapter will set the stage for future chapters and help you unlock the potential of computers to achieve your personal and professional goals.

OBJECTIVES

After completing Chapter 1, the student will be able to:

1. Define how digital electronics devices and computers are related and provide descriptions of the different types of computers.
2. Provide an explanation of the fundamentals of telecommunications, computer networking, the Internet, and wireless networking.
3. Describe the five functional areas in which computers assist us most.
4. Discuss the four primary types of information systems and how they are used.
5. Discuss how computers are used in a variety of fields, and why studying computer systems can benefit you professionally.
6. Discuss how computers are used to assist us in our life outside of work.
7. Define information security and discuss ways in which digital technologies are impacting humanity.

STUDENT PREP

Before you present and discuss this chapter, students should read the chapter and consider the following:

- Which of the electronics devices that you use are digital?
- Out of all the things you do on computers and digital devices, which requires being connected to a network?
- In what ways do digital electronics devices assist you in managing information, communicating with others, being entertained, computing and calculating,
- What are common examples of transaction processing systems, management information systems, and decision support systems that most students interact with?
- In what ways do you think computers and digital electronics devices benefit professionals in your field or areas of interest?
- How important is information security to students like you?

INVITE A GUEST

- Have a salesperson from a local computer store come and demo the latest and greatest mobile digital technologies.
- Have a representative from a company that hires grads from your school discuss the importance of computer literacy for non-technology majors and describe the types of computer skills and knowledge required on the job.

OVERVIEW OF PEDAGOGICAL ELEMENTS

- Technology 360 and Action Plan – Tonya Roberts faced the same dilemma as many college freshman. She had no idea what major to pursue. Although she had discussed career alternatives with her high school counselor, family, and friends, she still had difficulty choosing a major. In this scenario Tonya checks out the career center's resources and finds out how useful computers and digital electronics devices can be.
- Community Technology – Digitizing the World. We have become adept at digitizing things we see and hear and using digital devices to communicate text, numbers, pictures, movies, voice, and music. But what about smell, taste, and touch? Can we digitize information perceived by these senses?
- Job Technology – Information Systems Save Lives. The Chicago Police Department turned to a computer-based information system to provide new methods for targeting crime. They developed a geographic information system (GIS) that presented crime information in geographic context, on a map of the city, in hopes of preventing murders and instances of aggravated battery with firearms.

CHAPTER 2: HARDWARE DESIGNED TO MEET THE NEED

OVERVIEW

How can a person keep up with the ever-increasing momentum of technological innovation? If you tried to learn all the details of every new digital device, you would soon be overwhelmed. It makes better sense to first learn the underlying technology that all these devices share.

From large server computers that support the needs of thousands of users, to the smallest handheld computer, computing devices share certain characteristics. They all manipulate digital data, and they all share fundamental components: processing components, data storage, and input and output capabilities.

This chapter begins by explaining how bits and bytes are used to represent information of value to people. It continues on to discuss hardware concepts in the areas of processing, storage technologies, and input/output techniques and devices. It concludes by applying the concepts learned to the process of selecting and purchasing a computer.

OBJECTIVES

After completing Chapter 2, the student will be able to:

1. Understand how bits and bytes are used to represent information of value to people.
2. Identify the functions of the components of a CPU, the relationship between the CPU and memory, and factors that contribute to processing speed.
3. Identify different types of memory and storage media, and understand the unique properties of each.
4. Identify different types of input and output devices and how they are used to meet a variety of personal and professional needs.
5. Understand the decision-making process involved in purchasing a computer system.

STUDENT PREP

Before you present and discuss this chapter, students should read the chapter and consider the following:

- How is digital convergence exemplified in the digital devices that you use?
- In what ways does processor speed affect the computing activities in which you participate?
- What digital storage devices do you use while at home and away? Keep in mind that even cell phones include digital storage.
- How do the forms of input differ for stationary and portable computers and digital electronics devices?
- What type of computer(s) best suits your needs? Desktop, laptop, tablet, and or handheld?

INVITE A GUEST

- Invite an Apple user or representative to demonstrate the difference between Apple and Windows PCs from a hardware perspective.
- Have a system administrator from your school or a local business discuss the decision process behind deciding how much processing power, memory, and storage is required for mobile computers, desktop computers, and servers.

OVERVIEW OF PEDAGOGICAL ELEMENTS

- Technology 360 and Action Plan – Mark Bates recently graduated from college and started a new job as a graphic artist for Lamar Advertising. Now that he has a steady income, Mark wants to replace his 4-year-old PC with a new computer. This scenario describes the decision process Mark goes through to decide on which computer and computer components are best to suit his needs.
- Community Technology – Hard Drives that Deliver Power to the People. Five years ago most computers and digital media kept users tethered to desks or entertainment centers. New technologies such as high-capacity hard drives have freed us from those shackles and allowed us to take our computing and media wherever we go.
- Job Technology – iPod Goes to College. Duke University in Durham, North Carolina, was the first school to provide all incoming freshmen with their own 20 GB iPod. Later they expanded the program to include some upper-class students. How did they justify this expense? Rather than viewing the iPod as a music player for recreation, they viewed it as a student productivity tool.
- Job Technology – 3D Displays for Architects, Engineers, and Doctors. A German research institute has developed a 3D display technology that does not require the use of special glasses. The Fraunhofer Institute for Telecommunications developed the Free2C 3D display that generates two slightly different images to make objects appear three-dimensional—even when viewed from the side.
- Home Technology – Strategies for Computer Shopping. Purchasing a new computer, whether it is a desktop, notebook, tablet, or handheld, requires an organized process to ensure that you get a computer that is well suited to your needs. This box provides important considerations for new computer purchases.

CHAPTER 3: SOFTWARE SOLUTIONS FOR PERSONAL AND PROFESSIONAL GAIN

OVERVIEW

Software is the key to unlocking the potential of any computer system and developing effective computer applications. Without software, the fastest, most powerful computer would be useless. It can do nothing without instructions to follow and programs to execute. With software, people and organizations can accomplish more in less time.

This chapter covers the basic principles of software, including programming languages, the operating system, application software, and proprietary software. This chapter concludes with some important software issues, such as how to fix software bugs, software licenses, and copyrights.

OBJECTIVES

After completing Chapter 3, the student will be able to:

1. Discuss the importance and types of software.
2. Discuss the functions of some important programming languages.
3. Describe the functions of systems software and operating systems.
4. Describe the support provided by applications software.
5. Discuss how software can be acquired, customized, installed, removed, and managed.

STUDENT PREP

Before you present and discuss this chapter, students should read the chapter and consider the following:

- What five software programs do you use most frequently?
- What software do you interact with that may have been written by your college?
- What software is designed to assist professionals in your chosen career or interest area?
- Are you satisfied with the software that you use, or is there room for improvement? How could your favorite software be improved?
- What are the biggest potential threats or problems to you in using system and application software?

INVITE A GUEST

- Have a computer salesperson or technician demonstrate, compare, and contrast the features of Microsoft Windows, Mac OS, and Linux.
- Have someone from the college or a local business describe how application programs are used to achieve a goal or perform a specific task.

OVERVIEW OF PEDAGOGICAL ELEMENTS

- Technology 360 and Action Plan – Alexandra Pollack is a second-semester freshman in college and working part time to help pay her expenses. In this scenario, she is faced with deciding what software she should purchase to satisfy her personal and college needs.
- Job Technology – Software Reduces Legal Costs. The legal staff at Cisco, a company that sells sophisticated telecommunications systems, uses software to automate the process of applying for patents for its innovative products. The company saves about $3,000 per patent application to the federal government.
- Job Technology – UPS Provides Shipping Management Software to Customers. Quantum View Manage, a Web-based software product from UPS, helps people and businesses track their shipments, display important shipping information, and automatically triggers an invoice when the package is delivered to the customer.
- Job Technology – Telstra Benefits from E-Training for Managers. Telstra turned to IBM Learning Solutions to train their managers at their own pace. Seventy-five percent of the training is online, slashing training time and costs.
- Community Technology – Open Source or Proprietary Software. When the city of Munich, Germany was considering Linux, the open source operating system, a Microsoft representative flew to Germany to offer a very inexpensive version of Windows. The city decided to use Linux to give them more control over their technology destiny.

CHAPTER 4: THE INTERNET AND THE WORLD WIDE WEB

OVERVIEW

The Internet connects people on a global scale and provides a host of communication platforms, information, and services. The Web is an Internet service that provides a user-friendly interface to resources on the Internet. It organizes and presents information on the Internet in a manner that is easy to navigate. More than any other technology, the Web has empowered individuals by providing a public forum to share ideas.

This chapter discusses all of these issues. It begins with an explanation of the underlying technology –software, hardware, and protocols—that support the Internet and Web. It provides a survey of uses of the Internet and Web, and provides a glimpse into the future of these technologies and our use of them.

OBJECTIVES

After completing Chapter 4, the student will be able to:

1. Describe how the Internet developed and how hardware, protocols, and software work together to create the Internet.
2. Explain the underlying structure of the Web and the technologies that support it.
3. Define the categories of information and services that the Internet and Web provide and the forms of communication they support.
4. Explain what Internet2 is and the types of applications it will provide in the future.

STUDENT PREP

Before you present and discuss this chapter, students should read the chapter and consider the following:

- In what manner(s) do you connect to the Internet. Is your connection speed always adequate? How might it be improved?
- What Internet services do you use that are delivered over the Web? What Internet services do you use that are not Web-based?
- How might the Internet be improved to deliver more of the types of services that you desire more efficiently?

INVITE A GUEST

- Invite a representative from an Internet Service Provider to explain the services that the company provides.
- Invite a network researcher with knowledge of new and future Internet technologies to give a presentation.
- Ask your college system administrator or CIO to provide statistics regarding Internet use on campus. How many individuals are logged on to the Internet each evening? What are peak network traffic hours? How is your school handling illegal file sharing?

OVERVIEW OF PEDAGOGICAL ELEMENTS

- Technology 360 and Action Plan – Josh Greene has been using e-mail and the Web for as long as he can remember. His Web activities are so predictable that he imagines he has worn ruts in the paths he repeatedly travels. He has become disenchanted. In this scenario Josh learns new, interesting, and safe Internet activities.
- Home Technology – Selecting an Internet Service Provider. Most colleges offer free Internet access to students. Upon graduating, students often find that monthly Internet access is a considerable expense; sometimes more costly than phone, cable TV, and other utilities. It pays to understand your options and to make sure you are paying only for services that you need. This box provides some useful information.
- Community Technology – Tracking File Swappers by IP Address. In the past few years the Recording Industry Association of America (RIAA) has sued over 4000 individuals for copying and sharing copyrighted music recordings over the Internet. Part of the evidence needed for conviction in such cases is a positive identification of the individual on the computer during the file transfer process. Such identification is only possible by tracking the IP address associated with the offending computer and finding out who was logged on to that computer when the file was transferred. The only entity with this information is the Internet service provider (ISP) providing the user with Internet access.
- Job Technology – Television Marketers Turn To the Web. With the popularity of high-speed Internet connections and video-capable Web browsers, television programmers are planning to make the Web a lot more like TV. One of the first examples of this trend is *www.living.com*, the all-video Web site published by the E.W. Scripps Company and financed by GMC.
- Community Technology – Wikipedia: The Communal Encyclopedia. Wikipedia.com has become one of the most popular sites on the Web for accessing information on any topic. Wikipedia is a unique online encyclopedia created and maintained by the general Web community.

OVERVIEW

In selecting the components of a network, you must consider the speed and capacity of the medium that carries the communication signal. The communications medium works in conjunction with communication devices and software to provide a data communication network. There are numerous types of networks, each supporting the unique needs of their environment. Wireless mobile technologies are making it increasingly common to access networks anywhere, anytime. This chapter examines a variety of network types and their components to provide an understanding of how networks help us be connected and more productive in business, at home, and while traveling.

OBJECTIVES

After completing Chapter 5, the student will be able to:

1. Understand the fundamentals of data communications and the criteria for choosing a communications medium.
2. Explain how networking media, devices, and software work together to provide data-networking services, and describe the benefits of various types of media.
3. List and describe the most popular forms of wireless telecommunications technologies.
4. List the different classifications of computer networks and their defining characteristics, and understand the basics of wireless home networking.

STUDENT PREP

Before you present and discuss this chapter, students should read the chapter and consider the following:

- What are the benefits of connecting computers within a residence?
- How would your life change if you were able to access your computer resources anywhere, anytime?
- What are the benefits of connecting computers in a work place in your chosen career or interest area?
- How are wireless technologies changing the way we live?

INVITE A GUEST

Have a network technician demonstrate a wireless network that uses a wireless access point, and two computing devices. Include discussions on what hardware devices are necessary for such a network, how much they cost, how to set up the network, and what security precautions are necessary. Is wireless Internet available on your campus?

OVERVIEW OF PEDAGOGICAL ELEMENTS

- Technology 360 and Action Plan – Amanda Jackson is a news correspondent recently hired by National Public Radio (NPR). In her previous position with her small town newspaper, she had a desk and PC at which she did most of her work. Her new job entails traveling the world reporting on stories where they occur. Amanda has been given a startup budget to purchase the reporting equipment she requires. In this scenario Amanda learns what networks technologies will best support her new career.
- Job Technology – Dealing with Wireless on Campus. Colleges are racing to provide wireless Internet access on campus to allow students, faculty, administrators, and staff to be more productive and as an incentive in recruiting new students. Many colleges are providing wireless Internet access in classrooms to students who bring their laptop computers. Some teachers are concerned that having Internet access may be distracting to students during lectures.
- Community Technology – PAN Gets More Personal. Prior to Bluetooth, the expression PAN was coined by Thomas Zimmerman, a student at the MIT Media Lab back in 1996 for a unique new networking device he developed and embedded in a pair of gym shoes. Two individuals wearing the customized gym shoes could pass information from one device to the other simply by touching each other.
- Home Technology – Setting up a Wireless Home Network. To set up a wireless home network you need a high-speed Internet connection, a wireless access point/router, and wireless network adapters for all devices that will connect to the network. This box explains how to set it up.

INSTRUCTOR EDITION

OVERVIEW

Digital technology and information systems are tremendously useful in many practical ways. Digital media brings these systems to life with stunning and vivid imagery, powerful sound and music, and realistic, interactive animated 3D environments. If technology were alive, multimedia might be considered its heart and soul. Digital media provides a technical venue for people to express themselves through audio and visual output. This chapter provides an overview of all areas of digital media, including digital music and audio, 2D and 3D digital graphics and animation, digital photography and video, and interactive media such as video games and interactive TV. This chapter examines state-of-the-art media technologies and how they affect us in our personal and professional lives.

OBJECTIVES

After completing Chapter 6, the student will be able to:

1. Understand the uses of digital audio and today's digital music technologies.
2. Describe the many uses of 2D and 3D digital graphics and the technologies behind them.
3. Explain the technologies available to acquire, edit, distribute, and print digital photos, and list new advances in video technologies and distribution.
4. Discuss how interactive media is used to educate and entertain.

STUDENT PREP

Before you present and discuss this chapter, students should read the chapter and consider the following:

- How has technology changed your music purchasing and listening habits over the past five years?
- How has technology changed your photo taking and processing/printing routine over the past five years?
- What impact does digital media have on individuals in your chosen career or interest area?
- How has digital technology affected the motion picture industry?

INVITE A GUEST

Have a professional musician or artist discuss how technology has affected his or her career. Have the guest present his or her opinions on intellectual property and copyrights.

OVERVIEW OF PEDAGOGICAL ELEMENTS

- Technology 360 and Action Plan – Ana Arguello was in her first semester of college. Away from home, on her own for the first time, she had a real sense of freedom—along with a bit of apprehension. Most of her friends had chosen a major and knew what they wanted to do in life. Ana enjoyed so many different things that she had trouble choosing just one area to focus on. She had always been an expressive, artistic person. In this scenario Ana explores a variety of artistic careers and the technology that supports them.
- Home Technology – Selecting a Digital Music Service. When selecting a legal digital music service, you need to consider many options and decide which are valuable to you. Subscription services often provide a free month for users to evaluate the service before having to pay. This box provides important considerations for selecting an online service.
- Community Technology – Independent Film Producer Robert Rodriguez. Independent film producer Robert Rodriguez is known as a troublemaker and rebel in the industry. Armed with an arsenal of digital filmmaking tools and talent, he produces, directs, scores, and embellishes his movies with lavish special effects.
- Community Technology – Google Digitizing the World. Google Earth is an application that combines satellite imagery of the earth with map overlays. The software allows you to type any address, city, state, or country and view high-resolution satellite photos of the location, with roads, landmarks, and other information from maps superimposed on the images, which can be zoomed and panned. The software can be used to get travel directions, find local businesses, measure distances, and a host of other useful applications. Google Earth Pro is used by professionals in commercial and residential real estate, architects, engineers, construction managers, insurance adjusters and investigators, the media, state and local government officials, as well as professionals in defense, intelligence, and homeland security.

OVERVIEW

Everyone and every organization needs a way to store data and to have the ability to convert raw data into important information. Databases serve this function. Without databases, today's businesses could not survive. Databases are also used in medicine, science, engineering, the military, and most other fields. Databases are also useful to individuals for keeping track of items in an apartment, music in a CD collection, and expenses used to prepare a budget and complete a tax return. This chapter describes basic data management concepts.

OBJECTIVES

After completing Chapter 7, the student will be able to:

1. Understand basic data management concepts.
2. Describe database models and characteristics.
3. Discuss the different types of database management systems and their design and use by individuals and organizations.
4. Describe how organizations use database systems to perform routine processing, provide information, and provide decision support, and how they use data warehouses, marts, and mining.
5. Discuss additional database systems, including object-oriented and distributed systems.
6. Describe the role of the database administrator (DBA) and database policies and security practices.

STUDENT PREP

Before you present and discuss this chapter, students should read the chapter and consider the following:

- How might databases be used at your school?
- How might databases be used at Web sites such as amazon.com and Yahoo?
- How might databases be used by individuals in your chosen career or interest area?
- How can databases be a source of concern for privacy advocates?

INVITE A GUEST

- Ask a database administrator to discuss the process of designing a database for use in a company or organization. Have the guest address the following questions: Who takes part in the design process? How is a database administrator made aware of problems in the database? What is a worst case scenario for database failure? Why should individuals in non-technical career paths be interested in databases?
- Ask someone from your college to discuss how databases are being used at your school. What are some of the potential social issues, including privacy and identity theft?

OVERVIEW OF PEDAGOGICAL ELEMENTS

- Technology 360 and Action Plan – Mary Bolger found her grandfather's recipes in a small box in the attic and turned it into a small business. In this box, she must determine what database capabilities she will need to organize her business.
- Job Technology – Hyundai Shoots for #1 with Executive DBMS. Hyundai uses a sophisticated database system, called Enterprise Information and Management System (EIMS), to provide valuable information to its executives and decision makers. The DBMS generates a number of reports, personnel records, organizational charts, sales information, foreign exports, and inventory levels.
- Job Technology – Open-Source Database for Continental. Continental Airlines uses the Linux operating system and the MySQL open-source database to cut costs. MySQL also saves customers time in making or changing reservations by allowing them to complete transactions on the Web instead of through a Continental employee that can take 20 minutes or more.
- Job Technology – Web-based DBMS Empowers Cruise Line Personnel. Holland America Cruise Lines developed a Web-based database to help it increase revenues by $1 million by allowing customers to interact directly with the company online. The company is also using a data warehouse to get access to information to help its managers make better decisions.
- Community Technology – Government Regulations and Database Costs. Laws, such as Sarbanes-Oxley, the USA Patriot Act, and other legal provisions have increased costs to U.S. companies that must comply with these regulations. One FDA rule could cost pharmaceutical companies billions of dollars in new equipment and databases.

INSTRUCTOR EDITION

OVERVIEW

E-commerce has provided a fresh platform for business that has changed the way businesses and consumers think about buying and selling. Increasingly, buyers and sellers are turning to their computers to buy and sell products and are enjoying the benefits. Conducting business online offers convenience and savings to both buyers and sellers. This chapter explores the impact of e-commerce and m-commerce on consumers and businesses, what it takes to set up a successful e-commerce Web site, and the challenges and issues faced by e-commerce participants.

OBJECTIVES

After completing Chapter 8, the student will be able to:

1. Define e-commerce, and understand its role as a transaction processing system.
2. List the three types of e-commerce, and explain how e-commerce supports the stages of the buying process and methods of marketing and selling.
3. Discuss several examples of e-commerce applications and services.
4. Define m-commerce, and describe several m-commerce services.
5. List the components of an e-commerce system, and explain how they function together to provide e-commerce services.

STUDENT PREP

Before you present and discuss this chapter, students should read the chapter and consider the following:

- Has the Web changed the way you shop?
- What types of purchases, if any, are you more likely to make on the Web than at your local merchant?
- How does e-commerce affect individuals in your chosen career or interest area?
- What concerns do you have about conducting business over the Web?

INVITE A GUEST

Find a local merchant who has expanded operations to the Web. Have the merchant discuss how the expansion changed the business. What types of investments were required, and what new clientele resulted?

OVERVIEW OF PEDAGOGICAL ELEMENTS

- Technology 360 and Action Plan – Alejandro Forrero is the youngest of many generations of leather artisans and is currently completing his college education, which he plans on using to benefit the family business. Alejandro is considering the possibility of selling Forrero leather products direct to consumers on the Web. He'll need to do some research and write a proposal that provides details on the costs and benefits of taking the family business online.
- Job Technology – E-Commerce and Customized Products. One way to get customers to shop online is to provide additional services. Several e-tailors are providing their customers with the ability to customize their online purchase.
- Job Technology – Blogs: Benefit or Backfire? Many businesses are experimenting with using blogs to build community on their Web sites with mixed results.

OVERVIEW

Although people can use computer systems to do tasks more efficiently and less expensively, the true potential of computer systems lies in providing information to help people and organizations make better decisions. In this chapter, we discuss the decision making process and explore how computer systems can provide a wealth of information and decision support to individuals and organizations, regardless of their field or career. We also discuss the forms of computer decision making found in artificial intelligence, expert systems, and special-purpose devices.

OBJECTIVES

After completing Chapter 9, the student will be able to:

1. Define the stages of decision making and problem solving.
2. Discuss the use of management information systems in providing reports to help solve structured problems.
3. Describe how decision support systems are used to solve nonprogrammed and unstructured problems.
4. Explain how a group decision support system can help people and organizations collaborate on team projects.
5. Discuss the uses of artificial intelligence and special-purpose systems.

STUDENT PREP

Before you present and discuss this chapter, students should read the chapter and consider the following:

- In what ways does your college registration system assist you in finding courses to take and the sections of each course that have available seats?
- How might the system be improved to assist you in making your course selections?
- Is it possible to automate the duties of your academic advisor?
- How might information systems and decision support systems assist individuals in your chosen career or interest area?
- What types of special-purpose devices could help you at work or at home?

INVITE A GUEST

- Have a local business manager discuss the decision making process in his or her business. Are decisions made by individuals or through group consultation? How do computers assist in the process? What types of information systems are installed in the organization? What is their value to the organization?
- Invite a school administrator to discuss important decisions that he or she has to make and the information that would be helpful in making these important decisions.

OVERVIEW OF PEDAGOGICAL ELEMENTS

- Technology 360 and Action Plan – Maurice's father had more and more real-estate clients who lived hundreds or even thousands of miles from the Duchane's hometown of New Orleans. He utilizes virtual reality software and a group decision support system to create business opportunities over the Internet.
- Job Technology – A "TaylorMade" Information and Decision Support System. TaylorMade, the world's best-selling golf club manufacturers, created an information and decision support system to provide a variety of reports to help its managers control inventory levels and satisfy customer demand. Sales associates use handheld computers to scan in inventory levels at golf shops around the country.
- Job Technology – Ancient Company Profits from Better Decisions and Reports. Kano has been brewing sake in the small town of Mikage, Kobe, Japan, since 1659. Their sake wine, named *Kiku-Masamune* (chrysanthemum sake) became a staple in Japan. The company uses scheduled reports annually, quarterly, monthly, weekly, and daily that provide information on key indicators such as performance by retailer and profits by product and customer. The company uses this information to determine which of its customers require attention and to expand its customer base.

- Job Technology – Amazon Leverages Artificial Intelligence Against Fraud. For large retailers like Amazon.com, losses to credit card fraud are substantial. Amazon.com has 35 million customers who access its products through five Web sites adapted for different countries and languages. The system developed for Amazon uses classic AI techniques, such as neural networks, to analyze patterns in the data to "learn" which patterns represent fraud, much the same way humans learn. The system works on multiple levels. One portion of the system analyzes transactions as they occur, while the credit-card number is being approved. Another subsystem crunches data in Amazon's huge transaction database, looking for fraudulent activities in past transactions.
- Community Technology – Biologically Inspired Algorithms Fight Terrorists and Guide Businesses. This box discusses AI applications, funded by the U.S. government, through the Defense Advanced Research Projects Agency (DARPA), the same organization that produced the Internet. Research in this new intelligence technology is taking place as part of a $54 million program known as Genoa II. The goal of the project is to employ machine intelligence to anticipate future terrorist threats. Researchers hope to make it possible for humans and computers to "think together" in real time to "anticipate and preempt terrorist threats," according to official program documents.

CHAPTER 10: SYSTEMS DEVELOPMENT

OVERVIEW

This chapter covers the systems development process, including systems investigation, analysis, design, implementation, maintenance, and review. In this chapter, students will see how they can be involved in systems development to advance their careers and help their company or organization. They will also see how computer systems professionals, including systems analysts and computer programmers, work together to develop effective computer systems. They will see how they can be involved in the systems development process to get the systems and software they need. This chapter shows how systems development can be used to realize the true potential of computer systems in almost every field or discipline.

OBJECTIVES

After completing Chapter 10, the student will be able to:

1. Describe the participants in and importance of the systems development life cycle.
2. Discuss the use of CASE, project management, and other systems development tools.
3. Understand how systems development projects are initiated.
4. Describe how an existing system can be evaluated.
5. Discuss what is involved in planning a new system.
6. List the steps for placing a new or modified system into operation.
7. Describe the importance of updating and monitoring a new or modified system.

STUDENT PREP

Before you present and discuss this chapter, students should read the chapter and consider the following:

- Consider the components of your school registration system. What types of hardware, software, databases, computer networks, people, and policies support the system?
- What information systems do you interact with over the course of a week?
- Do information systems play a role in your chosen career or interest area? How so or why not?
- What effect can a poorly designed information system have on the organization it was intended to support?
- What types of systems would you like to be developed to help you on the job or in a career?

INVITE A GUEST

- Have a systems analyst discuss his or her responsibilities, challenges, and rewards. Stress the importance of user involvement in the system development process.
- Invite someone from your college or university to discuss the systems that will likely be developed or modified in the next five years.

OVERVIEW OF PEDAGOGICAL ELEMENTS

- Technology 360 and Action Plan – Two years after getting her undergraduate degree in business, Linda Perez has already been promoted to account manager at Mutual Insurance in Tampa, Florida. Linda would like her company to develop an insurance and risk analysis program that would help her and other account managers generate even more business. This opening box and the Action Plan reveal what has to be done.
- Job Technology – A Technology Makeover with Systems Development. Boehringer Ingelheim is among the world's 20 largest pharmaceutical companies with $7.6 billion in revenue and 32,000 employees in 60 nations. Top managers decided to totally revamp the company's computer systems. It took 14 months to roll out the new system, and many employees needed intensive training. In the end, the results were well worth the investment of time and money.
- Community Technology – When Systems Development Projects Fail. Because of errors in a US Airways computer system, prices for some flights were advertised to be as low as $1.86. Flights from the U.S. to Fiji were quoted on an Internet site as $51. Even though the advertised prices were posted on the Internet for less than an hour, the damage was done.
- Community Technology – Finding Trust in Computer Systems. The Royal Bank of Canada was unable to tell its 10 million Canadian customers exactly how much money was in their accounts. Canada's largest bank had a problem that kept tens of millions of transactions, including every direct payroll deposit it handles, from showing up in accounts. Computer system stability, dependability, and security will remain major goals for governments, industries, and technology specialists to work together to achieve.
- Job Technology: Hudson River Park Trust Looks for Help in Systems Development. The Hudson River Park Trust (HRPT) was founded in 1998 by the state of New York to create a five-mile stretch of parkland along Manhattan's West Side. One of the first challenges of the HRPT's staff was to decide how they were going to share information among the hundreds of contractors and suppliers involved in the restoration projects. HRPT used an applications service provider, called Constructware, to save money. Constructware allows the trust and the hundreds of contractors and suppliers it works with to share over the Internet 37,000 blueprints, diagrams, and other documents for various phases of construction.

CHAPTER 11: COMPUTER CRIME AND INFORMATION SECURITY

OVERVIEW

The global information economy depends on computer and information systems to reliably store, process, and transfer information. Threats to system reliability and information integrity undermine our economy and security. Information is money, and thieves are working to divert the flow of information and money to their own pockets leaving many innocent victims in their wake. In this chapter students learn about the value of information, the threats to information security, and ways that they can secure information for themselves, their employer, their country, and the world.

OBJECTIVES

After completing Chapter 11, the student will be able to:

1. Describe the types of information that must be kept secure and the types of threats against them.
2. Describe five methods of keeping a PC safe and secure.
3. Discuss the threats and defenses unique to multiuser networks.
4. Discuss the threats and defenses unique to wireless Wi-Fi networks.
5. Describe the threats posed by hackers, viruses, spyware, frauds, and scams, and methods of defending against them.

STUDENT PREP

Before you present and discuss this chapter, students should read the chapter and consider the following:

- What information stored on your computer is most valuable to you? How are you protecting it?
- Are the passwords you use strong enough?
- Can other users on your network access your PC?
- Is your PC protected from hackers, viruses, and other forms of attack?
- What can you do to help ensure total information security for yourself, your school, and your country?

INVITE A GUEST

- Invite a computer security specialist from the local law enforcement agency or other security organization to discuss safe computing habits.
- Invite a lawyer who specializes in intellectual property law to discuss issues related to digital media and file sharing networks.

OVERVIEW OF PEDAGOGICAL ELEMENTS

- Technology 360 and Action Plan – During the investigation of the crime, the company noted that Jan had used her credit card on several occasions to make online purchases and suggested that perhaps Jan's credit card information had been stolen from her PC. Jan's PC was connected to the campus network which was in turn connected to the Internet. She had no idea how someone could steal information from her PC, or how to prevent it from happening again. In this scenario Jan finds out how to protect her valuable information.
- Job Technology – Corporations Urged to Take Action Against Crime. The U.S. government is pushing CEOs to take responsibility for the nation's critical information infrastructure.
- Home Technology – A PC Security Checklist. Keeping your computer and the information it holds safe and secure requires a two- pronged approach: applying software security tools, and maintaining safe, vigilant behavior. This box shows you how.

CHAPTER 12: DIGITAL SOCIETY, ETHICS, AND GLOBALIZATION

OVERVIEW

Digital technologies have had a profound impact on most aspects of human life. The rapid pace of technological development has given us one of the most fascinating eras in which to live. Technological advances are leading to life-changing scientific breakthroughs, new business management paradigms, and a smaller, more inclusive and connected global society. The application of digital technologies to accomplish more with fewer resources is turning lives upside down in both negative and positive ways. Much of the social impact of these technologies seems to occur with little or no forethought by those responsible for developing and applying the technology. Governments are scrambling to establish laws to minimize negative impacts, while ethicists struggle to apply traditional ethical standards to brand new modes of human interaction. This chapter examines the impact of digital technologies on our sense of community, freedom of speech, privacy, ethics, and globalization.

OBJECTIVES

After completing Chapter 12, the student will be able to:

1. Describe how technology is affecting our definition of community and list some physical and mental health dangers associated with excessive computer use.
2. Describe the negative and positive impact of technology on freedom of speech and list forms of speech and expression that are censored on the Web.
3. Explain the ways in which technology is used to invade personal privacy, and provide examples of laws that protect us from privacy invasion.
4. List ethical issues relative to digital technology that confront individuals in personal and professional life, businesses, and governments.
5. Explain what globalization is, what forces are behind it, and how it is impacting the United States and other nations.

STUDENT PREP

Before you present and discuss this chapter, students should read the chapter and consider the following:

1. Consider the time you spend on cell phone, computer, iPod, video games – in virtual space. How does it balance against the time you spend engaged in real space?
2. How does the ability to express oneself online benefit and endanger individuals looking for information?
3. In what ways is your life less private today than the lives of your parents when they were your age?
4. What costs and benefits are produced for you, your city, state, and country by globalization?

INVITE A GUEST

- Invite an ethicist to discuss issues of ethics in today's digital world.
- Invite a professor from the business or economics department to discuss the impact of globalization on businesses and the national economy.
- Set up a virtual meeting using video conferencing technology to interview someone of interest to your students to further emphasize the capabilities available in our digital world.

OVERVIEW OF PEDAGOGICAL ELEMENTS

- Technology 360 and Action Plan – John Toh considers himself a naturalist. He has one year to go to finish his degree in Civil and Environmental Engineering. John worries about technology's impact on society and the natural order of the world. He is concerned that people are losing touch with their natural environment and even with each other because of the increasing time everyone spends on computers and cell phones. John's concerns about these issues and others are examined in this scenario.
- Job Technology – Can Businesses Sue or Fire Employees over Blogging? Issues regarding freedom for employees have recently come into question, as big corporations have taken legal action against employees for speaking their minds on the Internet.
- Community Technology – Outsourcing U.S. Education. Increasing numbers of students are turning to after-school tutoring services for extra help in problem areas. A new form of tutoring is being offered online. Companies such as Sylvan Learning and Growing Stars offer live online instruction. Students are provided with a starter kit that includes a headset/microphone set, a digital writing pad, and digital pencil.

Dear Readers,

Since the first edition, *Succeeding with Technology* has made its way into the hands of thousands of instructors and students. As technological developments continue to progress at a rapid rate, *Succeeding with Technology, Second Edition* provides you with the latest information and valuable new features to enhance the learning experience. Like the first edition, *Succeeding with Technology, Second Edition* will give you the technological tools you need to help you achieve your personal and professional goals.

The primary objective of *Succeeding with Technology* is to provide instructors and students with the most up-to-date developments, trends, and issues in technology. The use of contemporary examples and references to technology throughout this new edition makes it especially relevant to students' lives. Its clear, straightforward approach helps instructors and students navigate through the most important recent technological developments.

The table of contents and chapter layout have been improved for the Second Edition. Chapter 1 has been streamlined to give students an important introduction to basic concepts and popular uses of technology in various careers and at home, and an overview of the topics discussed in Chapters 2 through 12. The digital media chapter has been moved closer to the beginning of the book to give it more emphasis, and social issues get more in-depth coverage in this edition with two chapters: a new Chapter 11 on information security and computer crime, and Chapter 12 on important and current social and ethical issues of technology.

If you used a previous edition of *Succeeding with Technology,* you will find many familiar pedagogical elements, including updated Job Technology, Community Technology, and Tech Edge boxes. You will also find updated Technology 360 boxes at the beginning of each chapter and the Technology 360 Action Plans at the end of the chapters, plus brand-new Home Technology boxes featuring practical tips for integrating technology at home. Each chapter is full of new material, real-life examples, and references. All end-of-chapter material has been completely updated.

We sincerely appreciate the instructors and students who have used our books in the past. We are committed to developing the best textbook and accompanying materials possible. We are proud of this new edition and hope you enjoy it.

Ralph M. Stair K Baldauf

Succeeding with Technology

COMPUTER SYSTEM CONCEPTS FOR REAL LIFE

SECOND EDITION

RALPH M. Stair
KEN Baldauf

THOMSON
COURSE TECHNOLOGY™

Australia • Canada • Mexico • Singapore • Spain • United Kingdom • United States

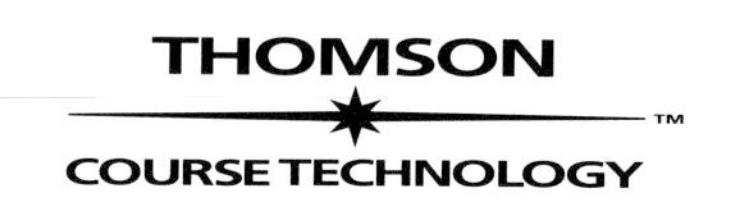

Succeeding with Technology
2nd Edition

by Ralph Stair and Ken Baldauf

Executive Editor:
Rachel Goldberg

Associate Editor:
Amanda Shelton

Senior Product Manager:
Kathy Finnegan

Product Manager:
Brianna Hawes

Associate Product Manager:
Shana Rosenthal

Editorial Assistant:
Janine Tangney

Developmental Editor:
Deb Kaufmann

Marketing Manager:
Joy Stark

Marketing Coordinator:
Melissa Marcoux

Production Editor:
Danielle Chouhan

Designer:
GEX Publishing Services

Cover Designer:
Abby Scholz

Copy Editor:
Mark Goodin

Proofreader:
Katherine Orrino

Indexer:
Joan Green

Compositor:
GEX Publishing Services

COPYRIGHT © 2007 Thomson Course Technology, a division of Thomson Learning™ is a trademark used herein under license.

Printed in the United States of America.

1 2 3 4 5 6 7 8 9 BM 10 09 08 07 06

For more information, contact Thomson Course Technology, 25 Thomson Place, Boston, Massachusetts, 02210

Or find us on the World Wide Web at: www.course.com

ALL RIGHTS RESERVED. No part of this work covered by the copyright hereon may be reproduced or used in any form or by any means—graphic, electronic, or mechanical, including photocopying, recording, taping, Web distribution, or information storage and retrieval systems—without the written permission of the publisher.

For permission to use material from this text or product, submit a request online at www.thomsonrights.com

Any additional questions about permissions can be submitted by e-mail to thomsonrights@thomson.com

Disclaimer
Thomson Course Technology reserves the right to revise this publication and make changes from time to time in its content without notice.

ISBN: 1-4188-3928-0 (Student Edition)
ISBN: 1-4188-3933-7 (Instructor Edition)

For Lila and Leslie
—RMS

For Hannah
—KJB

Brief Contents

Chapter 1	Why Study Computers and Digital Technologies?	2
Chapter 2	Hardware Designed to Meet the Need	56
Chapter 3	Software Solutions for Personal and Professional Gain	110
Chapter 4	The Internet and World Wide Web	174
Chapter 5	Telecommunications, Wireless Technologies, and Computer Networks	230
Chapter 6	Digital Media for Work and Leisure	282
Chapter 7	Database Systems	340
Chapter 8	E-Commerce	388
Chapter 9	Information, Decision Support, Artificial Intelligence, and Special-Purpose Systems	436
Chapter 10	Systems Development	476

Chapter 11 Computer Crime and Information Security 524

Chapter 12 Digital Society, Ethics, and Globalization 580

Glossary 624

Subject Index 648

Career Index 666

Photo Credits 668

Contents

1 Why Study Computers and Digital Technologies? 2

Technology 360° 2
What is a Computer? 5
Digital Technology, 6
Computer Functions, 7
Types of Computers, 8
COMMUNITY TECHNOLOGY — Digitizing the World 8
Mobile Digital Devices, 15
The Power of Connections 17
Telecommunications, 17
The Internet, 18
Wireless Networking, 19
What Can Computers Do? 20
Computing, 21
Automating, 22
Enhancing Communication, 23
Providing Entertainment, 25
Managing Information, 26
Information Systems 27
Types of Information Systems, 28
Using Digital Technologies to Succeed in Your Career 30
Computer-Based Professions, 31
Business and Communications, 32
Science and Mathematics, 33
Engineering, 33
Social Sciences, 33
Fine Arts, 34
Sports, Nutrition, and Exercise, 34
Government and Law, 35
Medicine and Health Care, 36
Criminology, Law Enforcement, and Security, 37
Education and Training, 38
JOB TECHNOLOGY — Information Systems Save Lives 38
Using Digital Technologies to Achieve Personal Goals 39
Personal Finance, 39
Personal Information Management, 40
Personal Research, 40
Personal Relations, 41
Personal Media Center, 41

Information Security and the Social Impact and Implications of Digital Technologies 42
Information Security, 42
The Impact of Digital Technologies on the World, 44
Technology 360° *Action Plan* 47
Summary 48
Test Yourself 50
Key Terms 52
Questions 52
Review Questions, 52
Discussion Questions, 53
Exercises 53
Try It Yourself, 53
Virtual Classroom Activities, 53
Teamwork, 54
TechTV 54
Go Inside Krispy Kreme, 54
Endnotes 54

2 Hardware Designed to Meet the Need 56

Technology 360° 56
The Digital Revolution 58
Representing Characters and Values with Bytes, 59
Bits, Bytes, and People, 60
Integrated Circuits and Processing 61
Integrated Circuits, 62
The Central Processing Unit and Random Access Memory, 63
Storage 70
System Storage, 70
Secondary-Storage Technologies, 73
Optical Storage, 76
Flash Drives and Cards, 78
COMMUNITY TECHNOLOGY — Hard Drives that Deliver Power to the People 78
Evaluating Storage Media: Access Method, Capacity, and Portability, 80
Input, Output, and Expansion 81
Input and Output Concepts, 81
JOB TECHNOLOGY — iPod Goes to College 82
Input Devices, 84
Output Devices, 87
JOB TECHNOLOGY — 3D Displays for Architects, Engineers, and Doctors 90
Special-Purpose I/O Devices, 92
Expansion, 93
Selecting and Purchasing a Computer 95
Researching a Computer Purchase, 96
Computer Vendors, 97
HOME TECHNOLOGY — Strategies for Computer Shopping 99

Technology 360° Action Plan 100

Summary 101
Test Yourself 102
Key Terms 104
Questions 104
Review Questions, 104
Discussion Questions, 104
Exercises 105
Try It Yourself, 105
Virtual Classroom Activities, 106
Teamwork, 107
TechTV 107
Getting Started, 107
Endnotes 107

3 Software Solutions for Personal and Professional Gain 110

Technology 360° 110
An Overview of Software 112
System and Application Software, 112
How Software Works, 113
Programming Languages 114
JOB TECHNOLOGY — Software Reduces Legal Costs 115
Newer Programming Languages, 116
Programming Language Translators, 118
System Software 120
Operating Systems, 120
PC Operating Systems Today, 128
Operating Systems for Servers, Networks, and Large Mainframe Computer Systems, 134
Operating Systems for Handheld Computers and Special-Purpose Devices, 135
Utility Programs, 135
Application Software 139
Productivity Software, 140
Additional Application Software for Individuals, 150
Application Software for Groups and Organizations, 152
Application Software for Information, Decision Support, and Specialized Purposes, 153
JOB TECHNOLOGY — UPS Provides Shipping Management Software to Customers 154
Software Issues and Trends 155
Acquiring Application Software, 155
JOB TECHNOLOGY — Telstra Benefits from E-Training for Managers . 157
Installing New Software, Handling Bugs, and Removing Old or Unwanted Software, 158
Copyrights and Licenses, 161
Shareware, Freeware, Open-Source, and Public Domain Software, 162

COMMUNITY TECHNOLOGY — Open-Source or Proprietary Software? 163
Technology 360° Action Plan 165
Summary 165
Test Yourself 168
Key Terms 169
Questions 169
Review Questions, 169
Discussion Questions, 170
Exercises 170
Try It Yourself, 170
Virtual Classroom Activities, 170
Teamwork, 170
TechTV 171
Software Vending Machines, 171
Endnotes 171

4 The Internet and World Wide Web 174

Technology 360° 174
Internet Technology 176
A Brief History of the Internet, 177
How Does the Internet Work?, 178
HOME TECHNOLOGY — Selecting an Internet Service Provider 182
COMMUNITY TECHNOLOGY — Tracking File Swappers by IP Address 185
Web Technology 187
Web Basics, 188
Web Markup Languages, 189
Web-Authoring Software, 191
Programming the Web, 192
Web Browser Plug-Ins, 194
Internet and Web Applications 195
JOB TECHNOLOGY — Television Marketers Turn to Web 195
Search Engines, Subject Directories, and Portals, 196
Communication and Collaboration, 199
News, 208
Education and Training, 210
E-Commerce, 212
Travel, 212
Employment and Careers, 213
Multimedia and Entertainment, 214
Information, 217
COMMUNITY TECHNOLOGY — Wikipedia: The Communal Encyclopedia 219
The Future Internet 220
Internet2 and Beyond, 220
High-Speed Internet Applications, 221

Technology 360° *Action Plan* 223

Summary 223
Test Yourself 225
Key Terms 226
Questions 226
Review Questions, 226
Discussion Questions, 227
Exercises 227
Try It Yourself, 227
Virtual Classroom Activities, 228
Teamwork, 228
TechTV 228
Google Maps Meets Craigslist, 228
Endnotes 228

5 Telecommunications, Wireless Technologies, and Computer Networks 230

Technology 360° 230
Fundamentals of Telecommunications 232
Telecommunications and Data Communications, 233
Characteristics of Telecommunications, 234
Networking Media, Devices, and Software 236
Networking Media, 236
Networking Devices, 239
Industrial Telecommunications Media and Devices, 242
Networking Software, 244
Wireless Telecommunications Technologies 246
Cell Phone Technologies, 246
JOB TECHNOLOGY — Dealing with Wireless on Campus 249
Pagers, 253
Global Positioning Systems, 253
Wireless Fidelity and WiMax, 256
Bluetooth, 258
Infrared Transmission, 260
Radio Frequency Identification (RFID), 260
Networks and Distributed Computing 262
Computer Networking Concepts, 262
Network Types, 264
COMMUNITY TECHNOLOGY — PAN Gets More Personal 265
Home Networks, 269
HOME TECHNOLOGY — Setting up a Wireless Home Network 270

Technology 360° Action Plan 273

Summary 274
Test Yourself 276
Key Terms 277
Questions 277
Review Questions, 277
Discussion Questions, 277
Exercises 278
Try It Yourself, 278
Virtual Classroom Activities, 278
Teamwork, 279
TechTV 279
Free Wi-Fi, 279
Endnotes 279

6 Digital Media for Work and Leisure 282

Technology 360° 282
Digital Music and Audio 285
Digitizing Music and Audio, 285
Digital Sound for Professionals, 286
Digital Music and Audio Production, 289
Digital Music and Audio Formats, Storage Media, Players, and Software, 294
Digital Music and Audio Distribution, 297

HOME TECHNOLOGY — Selecting a Digital Music Service 299

Digital Graphics 301
Digitizing Graphics, 301
Graphics File Formats, 302

COMMUNITY TECHNOLOGY — Independent Film Producer Robert Rodriguez 303

Uses of Digital Graphics, 304
Vector Graphics Software, 308
Three-Dimensional Modeling Software, 310
Computer Animation, 312

COMMUNITY TECHNOLOGY — Google Digitizing the World 312

Digital Photography and Video 316
Digital Photography, 316
Digital Video, 321
Interactive Media 326
Education and Training, 327
Commercial Applications of Interactive Media, 328
Interactive Video Games, 328
Interactive TV, 331

Technology 360° *Action Plan* 333

Summary 333
Test Yourself 335
Key Terms 336
Questions 337
Review Questions, 337
Discussion Questions, 337
Exercises 337
Try It Yourself, 337
Virtual Classroom Activities, 338
Teamwork, 338
TechTV 339
Army's Virtual World, 339
Endnotes 339

7 Database Systems 340

Technology 360° 340
Basic Data Management Concepts 342
Data Management for Individuals and Organizations, 343
The Hierarchy of Data, 344
Data Entities, Attributes, and Keys, 346
JOB-O TECHNOLOGY — Hyundai Shoots for #1 with Executive DBMS . 347
Simple Approaches to Data Management, 348
The Database Approach to Data Management, 348
Organizing Data in a Database 350
The Relational Database Model, 351
Object-Oriented Databases, 354
Database Characteristics, 355
Database Management Systems 356
Overview of Database Types, 356
Database Design, 361
JOB-O TECHNOLOGY — Open-Source Database for Continental 361
Using Databases with Other Software, 363
Data Accuracy and Integrity, 364
Creating and Modifying a Database, 365
Updating a Database, 367
Manipulating Data and Generating Reports, 367
Database Backup and Recovery, 369
Using Database Systems in Organizations 369
Routine Processing, 370
Information and Decision Support, 370
Data Warehouses, Data Marts, and Data Mining, 371
Database Trends 373
Distributed Databases, 373
Database Systems, the Internet, and Networks, 374
Visual, Audio, Unstructured, and Other Database Systems, 375
JOB-O TECHNOLOGY — Web-Based DBMS Empowers Cruise Line Personnel 376

Managing Databases 377
Database Administration, 377
COMMUNITY ⊙ TECHNOLOGY — Government Regulations and Database Costs 378
Database Use, Policies, and Security, 379
Technology 360° *Action Plan* 380
Summary 380
Test Yourself 382
Key Terms 384
Questions 384
Review Questions, 384
Discussion Questions, 384
Exercises 385
Try It Yourself, 385
Virtual Classroom Activities, 385
Teamwork, 386
TechTV 386
Video: Predicting Huge Surf, 386
Endnotes 386

8 E-Commerce 388

Technology 360° 388
The Roots of E-Commerce 392
E-Commerce History, 392
Transaction Processing, 393
The Transaction Processing Cycle, 394
Different Transaction Processing for Different Needs, 394
Overview of Electronic Commerce 395
Types of E-Commerce, 395
E-Commerce from the Buyer's Perspective, 397
E-Commerce from the Seller's Perspective, 399
Benefits and Challenges of E-Commerce, 400
E-Commerce Applications 403
Retail E-Commerce: Shopping Online, 403
Online Clearinghouses, Web Auctions, and Marketplaces, 404
JOB ⊙ TECHNOLOGY — E-Commerce and Customized Products 404
B2B Global Supply Management and Electronic Exchanges, 405
Marketing, 406
Banking, Finance, and Investment, 409
Mobile Commerce 410
M-Commerce Technology, 411
Types of M-Commerce and Applications, 412
E-Commerce Implementation 414
Infrastructure, 415
Hardware and Networking, 417
Software, 417
Building Traffic, 419
JOB ⊙ TECHNOLOGY — Blogs: Benefit or Backfire? 422

Electronic Payment Systems, 423
International Markets, 425
E-Commerce Security Issues, 426
Technology 360° *Action Plan* 429
Summary 429
Test Yourself 432
Key Terms 433
Questions 433
Review Questions, 433
Discussion Questions, 433
Exercises 434
Try It Yourself, 434
Virtual Classroom Activities, 434
Teamwork, 435
TechTV 435
Finding the Best Deals Online, 435
Endnotes 435

9 Information, Decision Support, Artificial Intelligence, and Special-Purpose Systems 436

Technology 360° 436
Decision Making and Problem Solving 438
Programmed Versus Nonprogrammed Decisions, 439
Optimization and Heuristic Approaches, 440
JOB TECHNOLOGY — A "TaylorMade" Information and Decision Support System 441
Management Information Systems 443
Inputs to a Management Information System, 443
Outputs of a Management Information System, 444
Decision Support Systems 447
JOB TECHNOLOGY — Ancient Company Profits from Better Decisions and Reports 447
Characteristics of a Decision Support System, 448
The Group Decision Support System 451
Characteristics of a GDSS, 452
GDSS Software or Groupware, 452
Artificial Intelligence and Special-Purpose Systems 453
An Overview of Artificial Intelligence, 453
Artificial Intelligence in Perspective, 454
The Difference Between Natural and Artificial Intelligence, 454
Components of Artificial Intelligence, 455
Robotics, 456
Vision Systems, 456
Natural Language Processing, 457
Learning Systems, 457
Neural Networks, 458

Fuzzy Logic, 458
Genetic Algorithms, 458

JOB-TECHNOLOGY — Amazon Leverages Artificial Intelligence against Fraud 459

Intelligent Agents, 460
Expert Systems, 460
Specialized Systems, 462

COMMUNITY-TECHNOLOGY — Biologically Inspired Algorithms Fight Terrorists and Guide Businesses . . 464

Action Plan 467

Summary 467
Test Yourself 469
Key Terms 471
Questions 471
Review Questions, 471
Discussion Questions, 471
Exercises 472
Try It Yourself, 472
Virtual Classroom Activities, 473
Teamwork, 473
TechTV 473
Ben Casnoscha, 473
Endnotes 474

10 Systems Development 476

Technology 360° 476
An Overview of Systems Development 478

JOB-TECHNOLOGY — A Technology Makeover with Systems Development 479

Participants in Systems Development, 481

COMMUNITY-TECHNOLOGY — When Systems Development Projects Fail 481

Why Start a Systems Development Project?, 482
Systems Development Planning, 484
End-User Systems Development, 485
Tools and Techniques for Systems Development 486
Computer-Aided Software Engineering, 486
Flowcharts, 487
Decision Tables, 488
Project Management Tools, 489
Prototyping, 490
Outsourcing, 492
Object-Oriented Systems Development, 493
Systems Investigation 494
Feasibility Analysis, 494
Systems Analysis 495
General Analysis Considerations, 495
Collecting Data, 496

Data Analysis, 497
Requirements Analysis, 497
Systems Design 500
COMMUNITY TECHNOLOGY — Finding Trust in Computer Systems . 500
Generating Systems Design Alternatives, 501
Evaluating and Selecting a Systems Design, 501
The Contract, 502
Systems Implementation 502
Acquiring Hardware, 503
Selecting and Acquiring Software: Make, Buy, or Rent, 503
JOB TECHNOLOGY — Hudson River Park Trust Looks for Help in Systems Development 505
Acquiring Database and Telecommunications Systems, 506
User Preparation, 506
Computer Systems Personnel: Hiring and Training, 506
Site Preparation, 507
Data Preparation, 507
Installation, 508
Testing, 509
Startup, 510
User Acceptance and Documentation, 511
Systems Maintenance and Review 511
Reasons for Maintenance, 511
The Financial Implications of Maintenance, 512
Systems Review, 513
Technology 360° *Action Plan* 514
Summary 514
Test Yourself 517
Key Terms 518
Questions 518
Review Questions, 518
Discussion Questions, 519
Exercises 520
Try It Yourself, 520
Virtual Classroom, 520
Teamwork, 520
TechTV 521
Project Management Using Web-Based Tools, 521
Endnotes 521

11 Computer Crime and Information Security 524

Technology 360° 524
Information Security and Vulnerability 528
What Is at Stake?, 528
JOB TECHNOLOGY — Corporations Urged to Take Action Against Crime 531
Threats to Information Security, 535

Machine-Level Security . . . 539
Passwords, 540
ID Devices and Biometrics, 542
Encrypting Stored Data, 543
Backing Up Data and Systems, 544
System Maintenance, 547
Network Security . . . 548
Multiuser System Considerations, 549
Interior Threats, 551
Security and Usage Policies, 552
Wireless Network Security . . . 553
Threats to Wireless Networks, 553
Securing a Wireless Network, 554
Internet Security . . . 556
Hackers on the Internet, 557
Viruses and Worms, 561
Spyware, Adware, and Zombies, 565
Scams, Spam, Fraud, and Hoaxes, 566
HOME TECHNOLOGY — A PC Security Checklist . . . 570
Technology 360° Action Plan . . . 571
Summary . . . 571
Test Yourself . . . 574
Key Terms . . . 575
Questions . . . 575
Review Questions, 575
Discussion Questions, 576
Exercises . . . 576
Try It Yourself, 576
Virtual Classroom Activities, 576
Teamwork, 577
TechTV . . . 577
Understanding Identity, Spoofing, and Internet Attacks, 577
Endnotes . . . 577

12 **Digital Society, Ethics, and Globalization** 580

Technology 360° . . . 580
Living Online . . . 583
Computers and Community, 584
Health Issues: Keeping a Balance, 585
Freedom of Speech . . . 588
Challenging the Establishment and Traditional Institutions, 589
Laws and Censorship, 590
Privacy Issues . . . 593
JOB TECHNOLOGY — Can Businesses Sue or Fire Employees over Blogging? . . . 593

Personal Information Privacy, 594
Privacy and Government, 595
Surveillance Technologies, 597
Ethics and Social Responsibility 602
Personal Ethical Considerations, 602
Professional Ethical Considerations, 603
Governmental Ethical Considerations, 604
Globalization 608
Outsourcing, 609
Offshoring, 610
Business Challenges in Globalization, 611
COMMUNITY TECHNOLOGY — Outsourcing U.S. Education 611
Conclusion 613
Technology 360° Action Plan 615
Summary 615
Test Yourself 618
Key Terms 619
Questions 619
Review Questions, 619
Discussion Questions, 619
Exercises 620
Try It Yourself, 620
Virtual Classroom Activities, 620
Teamwork, 620
TechTV 621
Gmail Privacy, 621
Endnotes 621

Glossary 624

Subject Index 648

Career Index 666

Photo Credits 668

Preface

You know your computer . . .

now what are you going to do with it?

Most students entering college have already had years of exposure to computers. Elementary and high school students use computers to write papers, create presentations, communicate with each other, conduct research, and entertain themselves. They understand the basics of computer use—how to turn them on and how to run popular applications. But today's technological world requires much more. ***Succeeding with Technology*** teaches students how to apply what they already know and what they learn as well as how to apply technology to their futures.

The creation of *Succeeding with Technology* was guided by the philosophy that for students to prosper, they must grasp the underlying principles of the technologies that have an impact on our lives and understand how those principles are related to real world activities. No one is capable of gaining true understanding by memorizing long lists of technical terms. This textbook won't overwhelm you with descriptions of numerous inconsequential devices. Instead, *Succeeding with Technology* provides straightforward explanations of the principles that guide technological development, without overwhelming the reader with too much detail. An understanding of the concepts provided in this book will translate into a practical understanding of the specific devices and practices in use today and in years to come.

The authors understand that technology in and of itself is not interesting to most people. What most people do find interesting are the exciting ways that technology is being used to improve our day-to-day lives, our professional productivity, society, and the world. *Succeeding with Technology* invests as much effort in showing how technology is used as it does in explaining how technology works. Every concept presented is backed up with practical examples of how it is making an impact on everyday life.

We are proud of the second edition of this unique textbook that takes readers beyond traditional computer competence and fluency, to a deeper understanding of not only how digital technology works, but, more importantly, how it can be harnessed to improve your life.

Welcome to the 2nd Edition of *Succeeding with Technology: Computer System Concepts for Real Life.*

Approach

Succeeding with Technology employs a different approach from many of the other computer concepts books on the market. It is a direct outgrowth of the trends that are causing the introductory computer course to change. From its high-impact graphic design to its content, the unique approach of this textbook is sure to engage and excite readers.

- **Focus on careers.** This textbook contains a wealth of examples of how technology systems are used in different disciplines. Starting with Chapter 1, almost every page of this textbook contains exciting, current examples of how real people and organizations have used technology to achieve success. Through reading the examples, students can evaluate what careers and fields match their interests and talents.
- **Beyond PCs.** We go beyond desktop PCs to provide coverage of the wide variety of computers and computer systems that are in use today. From cell phones to servers, from wireless networks to virtual private networks, we cover all the latest technology that students are likely to encounter at home and at work.
- **Speaks to students with varied skill levels.** Although this textbook assumes that the reader has used a computer, the material is presented in a manner that is sure not to leave anyone behind, while engaging even the most experienced computer user.
- **Less jargon and fewer key terms.** We include everything readers need to know and nothing they don't. Only important terms that students are likely to encounter in the real world appear in bold as key terms.
- **Important social issues explored.** This textbook confronts important controversial issues head on with complete coverage from all perspectives. P2P file sharing, the digital divide, students and plagiarism, violent video games, and many other current issues are explored, providing an excellent launching pad for class discussion.

New to the 2nd Edition

With the 2nd edition of *Succeeding with Technology* we continue to work to minimize the amount of techno-babble that is of no relevance to users of technology while placing emphasis on the application of digital technologies towards the success of professional and personal lives.

- **Up-to-date coverage:** Each chapter of *Succeeding with Technology* has been updated with new and relevant examples and coverage of the latest technologies, with increased coverage of: mobile digital technologies for communication, information access, and entertainment; a wider variety of computer types, computing platforms, and digital electronics devices; a wider variety of careers covered in our examples of digital technologies at work; the importance of information security in everyone's lives; globalization; and digital technologies as they are used outside of work.

- **Streamlined introductory chapter:** Chapter 1 has been reorganized to provide increased coverage of digital technology basics with less depth than was provided in the previous edition. This approach has been adopted in order to engage students sooner.

- ***New* Information Security chapter:** This brand-new Chapter 11 emphasizes the importance of information security in our digital world. The chapter illuminates the student's important role in total information security on multiple levels. Total information security is defined as securing all components of the global digital information infrastructure, from cell phones to PCs to business and government networks to Internet routers and communications satellites.
- ***New* Home Technology boxed feature:** This new box provides students with practical information that will help them in their lives today. For example, there are Home Technology boxes on Strategies for Computer Shopping, Selecting an Internet Service Provider, Setting up a Home Wireless Network, Selecting a Digital Music Service, and more!
- ***New* TechTV content:** The TechTV video content, available through the Succeeding with Technology Web site, for select chapters has been updated to give you cutting edge topics for your students to view, such as Understanding Identity Theft, Project Management Using Web-based Tools, and Google Maps Meets Craigslist.

FEATURES

Succeeding with Technology has a set of features designed to engage interest, show how to solve problems, and demonstrate what people can accomplish with computers.

Chapter Content

Learning Objectives and **contents** show you exactly what subjects will be covered in each chapter. Read this before you dive in so that you know what to expect.

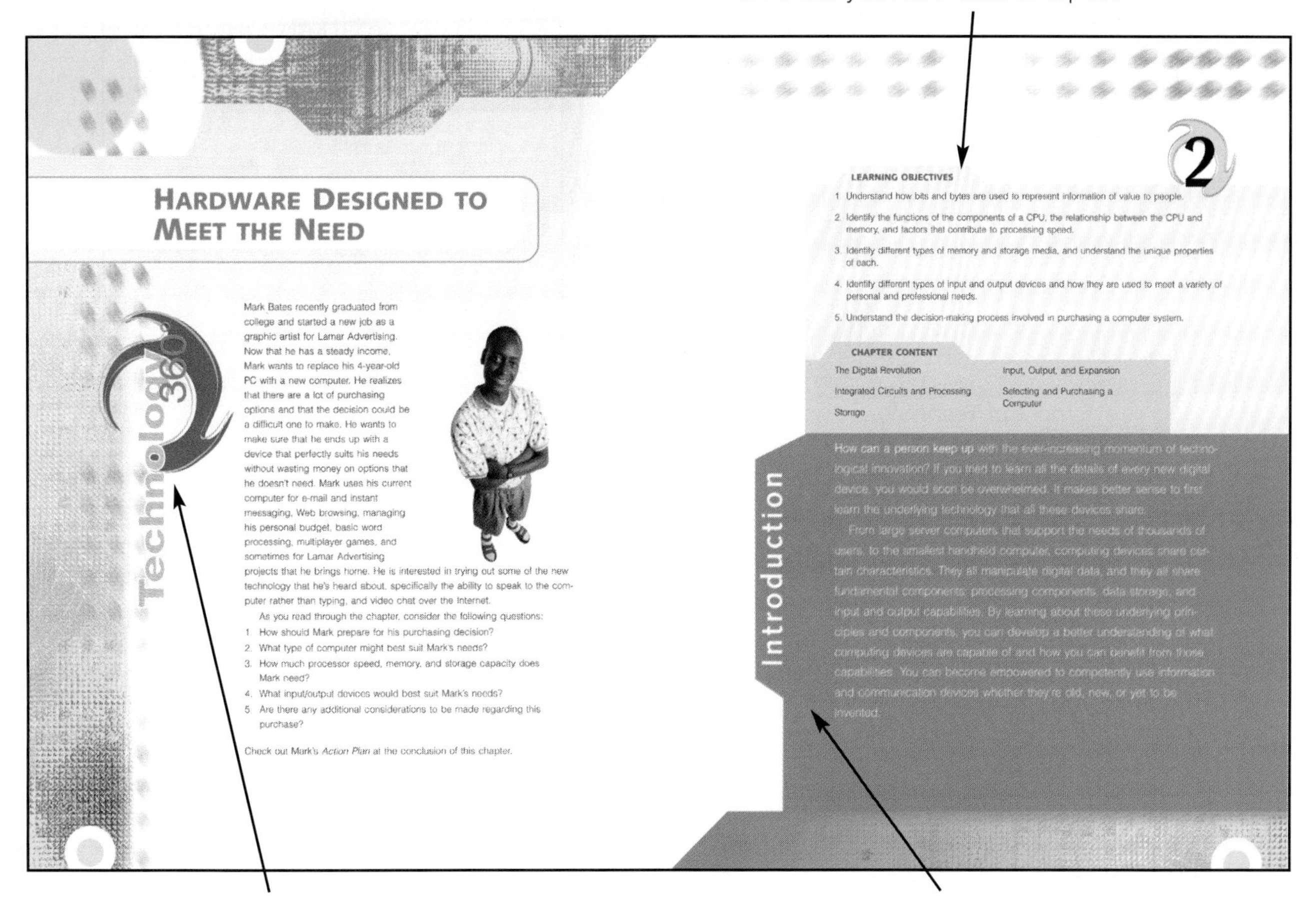

HARDWARE DESIGNED TO MEET THE NEED

Technology360

Mark Bates recently graduated from college and started a new job as a graphic artist for Lamar Advertising. Now that he has a steady income, Mark wants to replace his 4-year-old PC with a new computer. He realizes that there are a lot of purchasing options and that the decision could be a difficult one to make. He wants to make sure that he ends up with a device that perfectly suits his needs without wasting money on options that he doesn't need. Mark uses his current computer for e-mail and instant messaging, Web browsing, managing his personal budget, basic word processing, multiplayer games, and sometimes for Lamar Advertising projects that he brings home. He is interested in trying out some of the new technology that he's heard about, specifically the ability to speak to the computer rather than typing, and video chat over the Internet.

As you read through the chapter, consider the following questions:

1. How should Mark prepare for his purchasing decision?
2. What type of computer might best suit Mark's needs?
3. How much processor speed, memory, and storage capacity does Mark need?
4. What input/output devices would best suit Mark's needs?
5. Are there any additional considerations to be made regarding this purchase?

Check out Mark's *Action Plan* at the conclusion of this chapter.

2

LEARNING OBJECTIVES

1. Understand how bits and bytes are used to represent information of value to people.
2. Identify the functions of the components of a CPU, the relationship between the CPU and memory, and factors that contribute to processing speed.
3. Identify different types of memory and storage media, and understand the unique properties of each.
4. Identify different types of input and output devices and how they are used to meet a variety of personal and professional needs.
5. Understand the decision-making process involved in purchasing a computer system.

CHAPTER CONTENT

The Digital Revolution
Integrated Circuits and Processing
Storage
Input, Output, and Expansion
Selecting and Purchasing a Computer

Introduction

How can a person keep up with the ever-increasing momentum of technological innovation? If you tried to learn all the details of every new digital device, you would soon be overwhelmed. It makes better sense to first learn the underlying technology that all these devices share.

From large server computers that support the needs of thousands of users, to the smallest handheld computer, computing devices share certain characteristics. They all manipulate digital data, and they all share fundamental components: processing components, data storage, and input and output capabilities. By learning about these underlying principles and components, you can develop a better understanding of what computing devices are capable of and how you can benefit from those capabilities. You can become empowered to competently use information and communication devices whether they're old, new, or yet to be invented.

Technology360 scenario discusses people like you who face problems that can be solved with the help of technology. You will learn how they solve their problems using concepts presented in the chapter.

The **Chapter Introduction** welcomes you to the chapter. Learn what is most important in each chapter before you start reading.

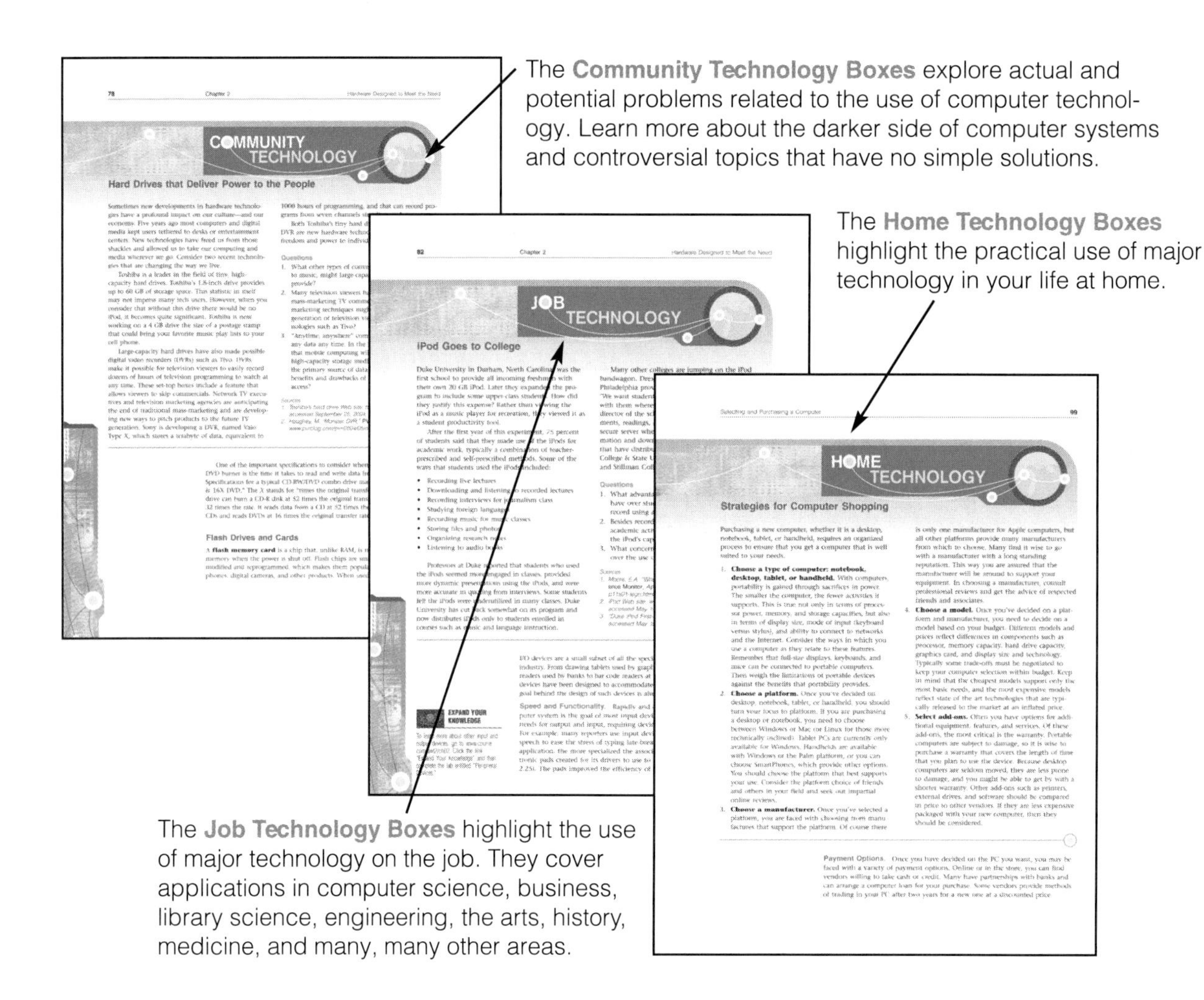

The **Community Technology Boxes** explore actual and potential problems related to the use of computer technology. Learn more about the darker side of computer systems and controversial topics that have no simple solutions.

The **Home Technology Boxes** highlight the practical use of major technology in your life at home.

The **Job Technology Boxes** highlight the use of major technology on the job. They cover applications in computer science, business, library science, engineering, the arts, history, medicine, and many, many other areas.

The Tech Edge elements are brief news-based items that tip readers off to a broad range of the amazing things that are currently happening in our technology-driven culture.

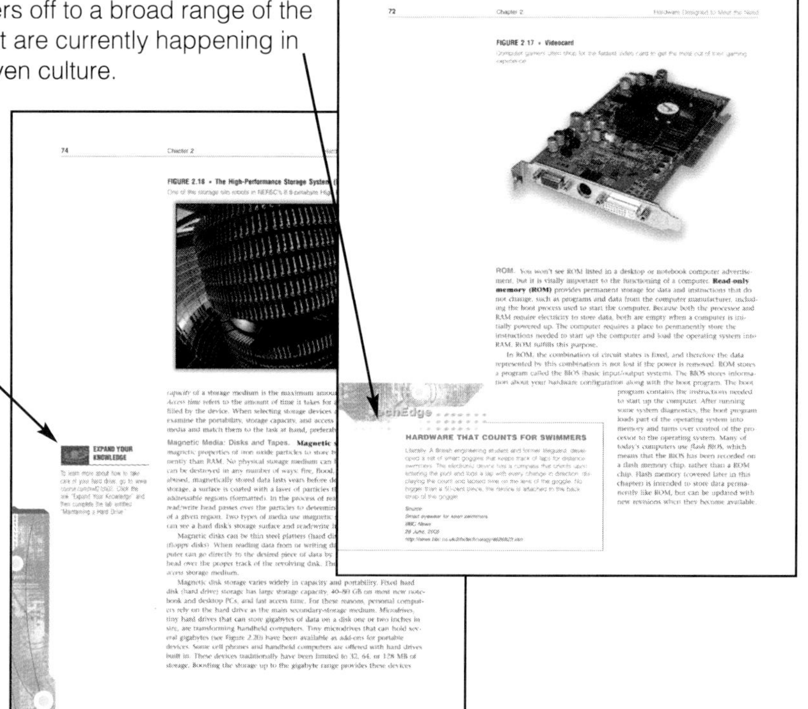

The **Expand Your Knowledge Student EditionLabs,** available on the Succeeding with Technology Web site, help students review and extend their knowledge of the concepts through observations, hands-on simulations, and challenging objective-based questions.

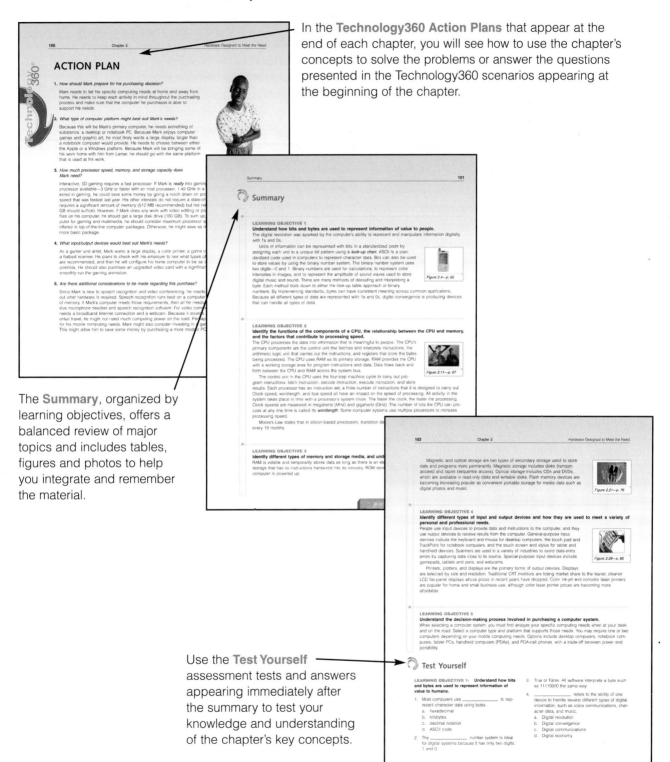

End-of-Chapter

A wealth of end of chapter material will help students retain concepts and use them beyond the course.

In the **Technology360 Action Plans** that appear at the end of each chapter, you will see how to use the chapter's concepts to solve the problems or answer the questions presented in the Technology360 scenarios appearing at the beginning of the chapter.

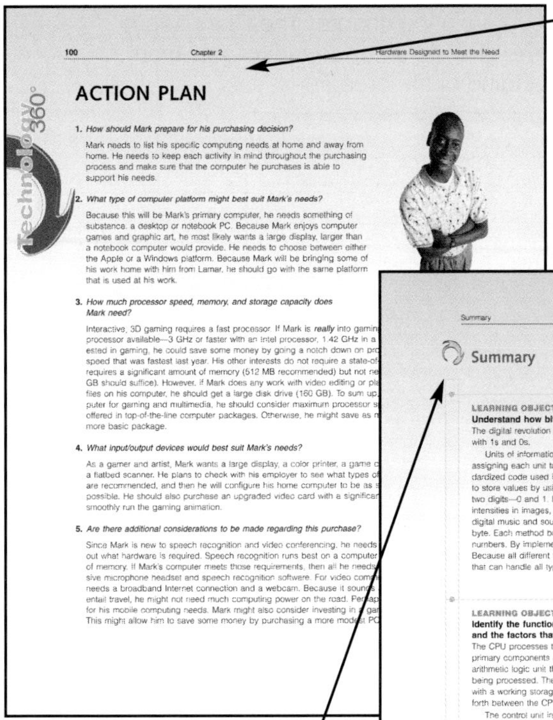
100 Chapter 2 Hardware Designed to Meet the Need

ACTION PLAN

1. *How should Mark prepare for his purchasing decision?*
Mark needs to list his specific computing needs at home and away from home. He needs to keep each activity in mind throughout the purchasing process and make sure that the computer he purchases is able to support his needs.

2. *What type of computer platform might best suit Mark's needs?*
Because this will be Mark's primary computer, he needs something of substance, a desktop or notebook PC. Because Mark enjoys computer games and graphic art, he most likely wants a large display, larger than a notebook computer would provide. He needs to choose between either the Apple or a Windows platform. Because Mark will be bringing some of his work home with him from Lamar, he should go with the same platform that is used at his work.

3. *How much processor speed, memory, and storage capacity does Mark need?*
Interactive, 3D gaming requires a fast processor. If Mark is *really* into gamin[g] processor available—3 GHz or faster with an Intel processor, 1.42 GHz in a ... ested in gaming, he could save some money by going a notch down on pr[ocessor] speed that was fastest last year. His other interests do not require a state-of-... requires a significant amount of memory (512 MB recommended) but not ne[cessarily] GB should suffice). However, if Mark does any work with video editing or pl[ans] files on his computer, he should get a large disk drive (160 GB). To sum up, ... puter for gaming and multimedia, he should consider maximum processor s[peed] offered in top-of-the-line computer packages. Otherwise, he might save as m[uch] more basic package.

4. *What input/output devices would best suit Mark's needs?*
As a gamer and artist, Mark wants a large display, a color printer, a game c[ontroller] a flatbed scanner. He plans to check with his employer to see what types of ... are recommended, and then he will configure his home computer to be as ... possible. He should also purchase an upgraded video card with a significan[t] ... smoothly run the gaming animation.

5. *Are there additional considerations to be made regarding this purchase?*
Since Mark is new to speech recognition and video conferencing, he needs ... out what hardware is required. Speech recognition runs best on a computer ... of memory. If Mark's computer meets those requirements, then all he needs ... sive microphone headset and speech recognition software. For video con[ferencing he] needs a broadband Internet connection and a webcam. Because it soun[ds] ... entail travel, he might not need much computing power on the road. Pe[rhaps] ... for his mobile computing needs. Mark might also consider investing in ... This might allow him to save some money by purchasing a more mod[est] PC ...

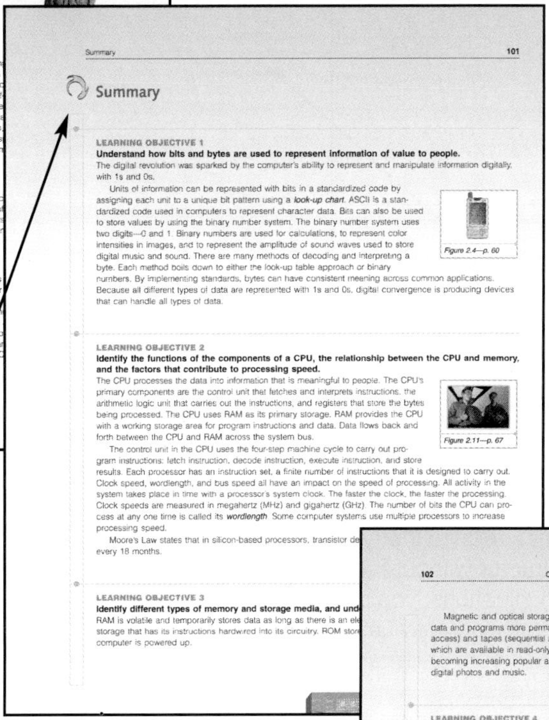
Summary 101

Summary

LEARNING OBJECTIVE 1
Understand how bits and bytes are used to represent information of value to people.
The digital revolution was sparked by the computer's ability to represent and manipulate information digitally, with 1s and 0s.
Units of information can be represented with bits in a standardized code by assigning each unit to a unique bit pattern using a *look-up chart*. ASCII is a standardized code used in computers to represent character data. Bits can also be used to store values by using the binary number system. The binary number system uses two digits—0 and 1. Binary numbers are used for calculations, to represent color intensities in images, and to represent the amplitude of sound waves used to store digital music and sound. There are many methods of decoding and interpreting a byte. Each method boils down to either the look-up table approach or binary numbers. By implementing standards, bytes can have consistent meaning across common applications. Because all different types of data are represented with 1s and 0s, digital convergence is producing devices that can handle all types of data.

Figure 2.4—p. 60

LEARNING OBJECTIVE 2
Identify the functions of the components of a CPU, the relationship between the CPU and memory, and the factors that contribute to processing speed.
The CPU processes the data into information that is meaningful to people. The CPU's primary components are the control unit that fetches and interprets instructions, the arithmetic logic unit that carries out the instructions, and registers that store the bytes being processed. The CPU uses RAM as its primary storage. RAM provides the CPU with a working storage area for program instructions and data. Data flows back and forth between the CPU and RAM across the system bus.
The control unit in the CPU uses the four-step machine cycle to carry out program instructions: fetch instruction, decode instruction, execute instruction, and store results. Each processor has an instruction set, a finite number of instructions that it is designed to carry out. Clock speed, wordlength, and bus speed all have an impact on the speed of processing. All activity in the system takes place in time with a processor's system clock. The faster the clock, the faster the processing. Clock speeds are measured in megahertz (MHz) and gigahertz (GHz). The number of bits the CPU can process at any one time is called its *wordlength*. Some computer systems use multiple processors to increase processing speed.
Moore's Law states that in silicon-based processors, transistor de[nsity] ... every 18 months.

Figure 2.11—p. 67

LEARNING OBJECTIVE 3
Identify different types of memory and storage media, and und[erstand] ...
RAM is volatile and temporarily stores data as long as there is an el[ectric] ... storage that has its instructions hardwired into its circuitry. ROM stor[age] ... computer is powered up.

The **Summary**, organized by learning objectives, offers a balanced review of major topics and includes tables, figures and photos to help you integrate and remember the material.

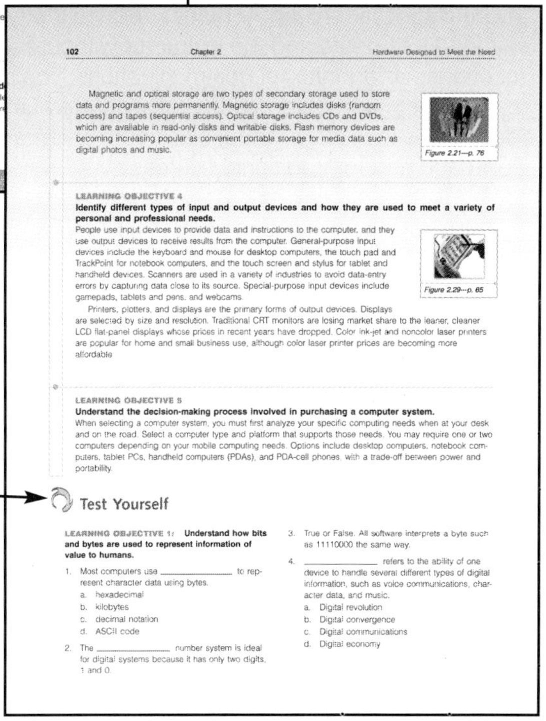
102 Chapter 2 Hardware Designed to Meet the Need

Magnetic and optical storage are two types of secondary storage used to store data and programs more permanently. Magnetic storage includes disks (random access) and tapes (sequential access). Optical storage includes CDs and DVDs, which are available in read-only disks and writable disks. Flash memory devices are becoming increasingly popular as convenient portable storage for media data such as digital photos and music.

Figure 2.21—p. 76

LEARNING OBJECTIVE 4
Identify different types of input and output devices and how they are used to meet a variety of personal and professional needs.
People use input devices to provide data and instructions to the computer, and they use output devices to receive results from the computer. General-purpose input devices include the keyboard and mouse for desktop computers, the touch pad and TrackPoint for notebook computers, and the touch screen and stylus for tablet and handheld devices. Scanners are used in a variety of industries to avoid data-entry errors by capturing data close to its source. Special-purpose input devices include gamepads, tablets and pens, and webcams.
Printers, plotters, and displays are the primary forms of output devices. Displays are selected by size and resolution. Traditional CRT monitors are losing market share to the leaner, cleaner LCD flat-panel displays whose prices in recent years have dropped. Color ink-jet and noncolor laser printers are popular for home and small business use, although color laser printer prices are becoming more affordable.

Figure 2.29—p. 85

LEARNING OBJECTIVE 5
Understand the decision-making process involved in purchasing a computer system.
When selecting a computer system, you must first analyze your specific computing needs when at your desk and on the road. Select a computer type and platform that supports those needs. You may require one or two computers depending on your mobile computing needs. Options include desktop computers, notebook computers, tablet PCs, handheld computers (PDAs), and PDA-cell phones, with a trade-off between power and portability.

Test Yourself

LEARNING OBJECTIVE 1: **Understand how bits and bytes are used to represent information of value to humans.**

1. Most computers use ______________ to represent character data using bytes.
   a. hexadecimal
   b. kilobytes
   c. decimal notation
   d. ASCII code
2. The ______________ number system is ideal for digital systems because it has only two digits, 1 and 0.
3. True or False. All software interprets a byte such as 11110000 the same way.
4. ______________ refers to the ability of one device to handle several different types of digital information, such as voice communications, character data, and music.
   a. Digital revolution
   b. Digital convergence
   c. Digital communications
   d. Digital economy

Use the **Test Yourself** assessment tests and answers appearing immediately after the summary to test your knowledge and understanding of the chapter's key concepts.

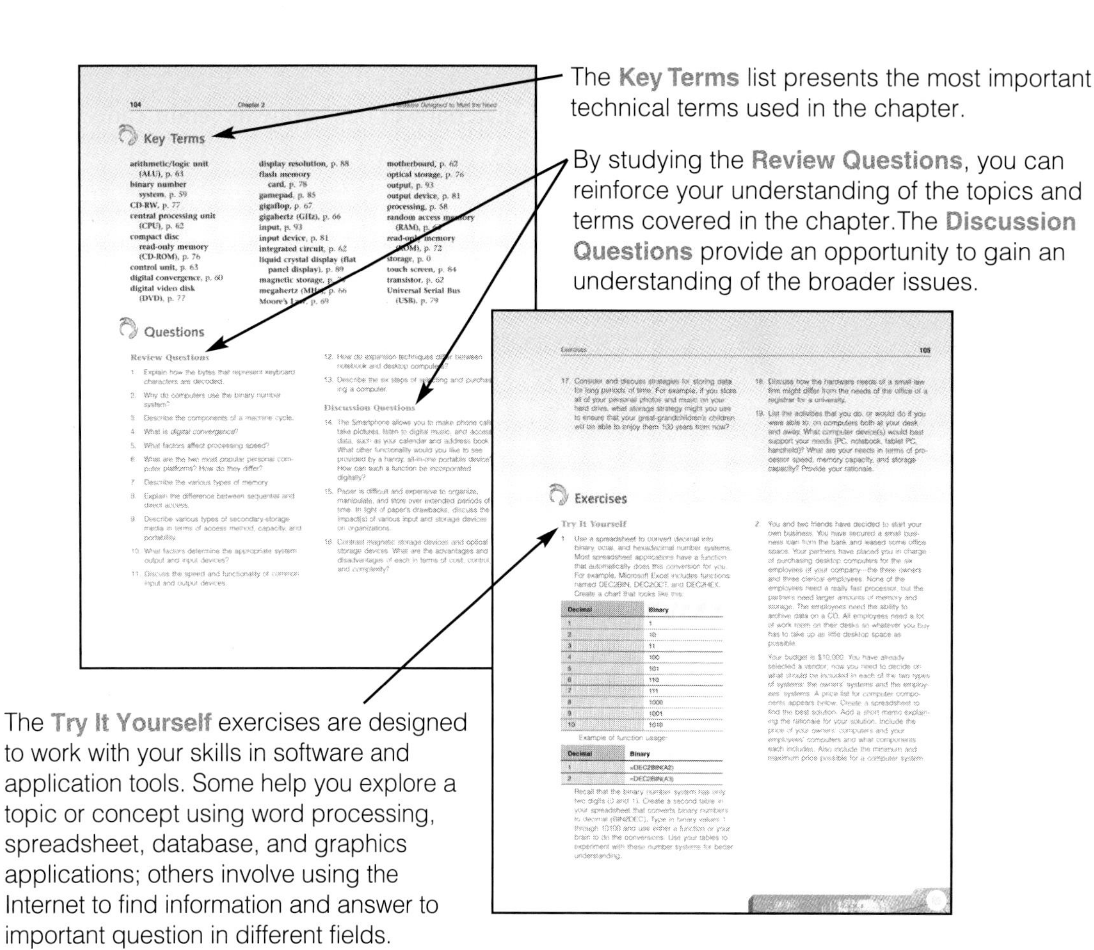

The **Key Terms** list presents the most important technical terms used in the chapter.

By studying the **Review Questions**, you can reinforce your understanding of the topics and terms covered in the chapter.The **Discussion Questions** provide an opportunity to gain an understanding of the broader issues.

The **Try It Yourself** exercises are designed to work with your skills in software and application tools. Some help you explore a topic or concept using word processing, spreadsheet, database, and graphics applications; others involve using the Internet to find information and answer to important question in different fields.

The **Virtual Classroom Activities** require that you interact with other students and the instructor without face-to-face meetings. Using Internet communications, you can develop joint projects with other students.These activities are valuable for courses that use a virtual classroom or distance learning approach and for traditional courses.

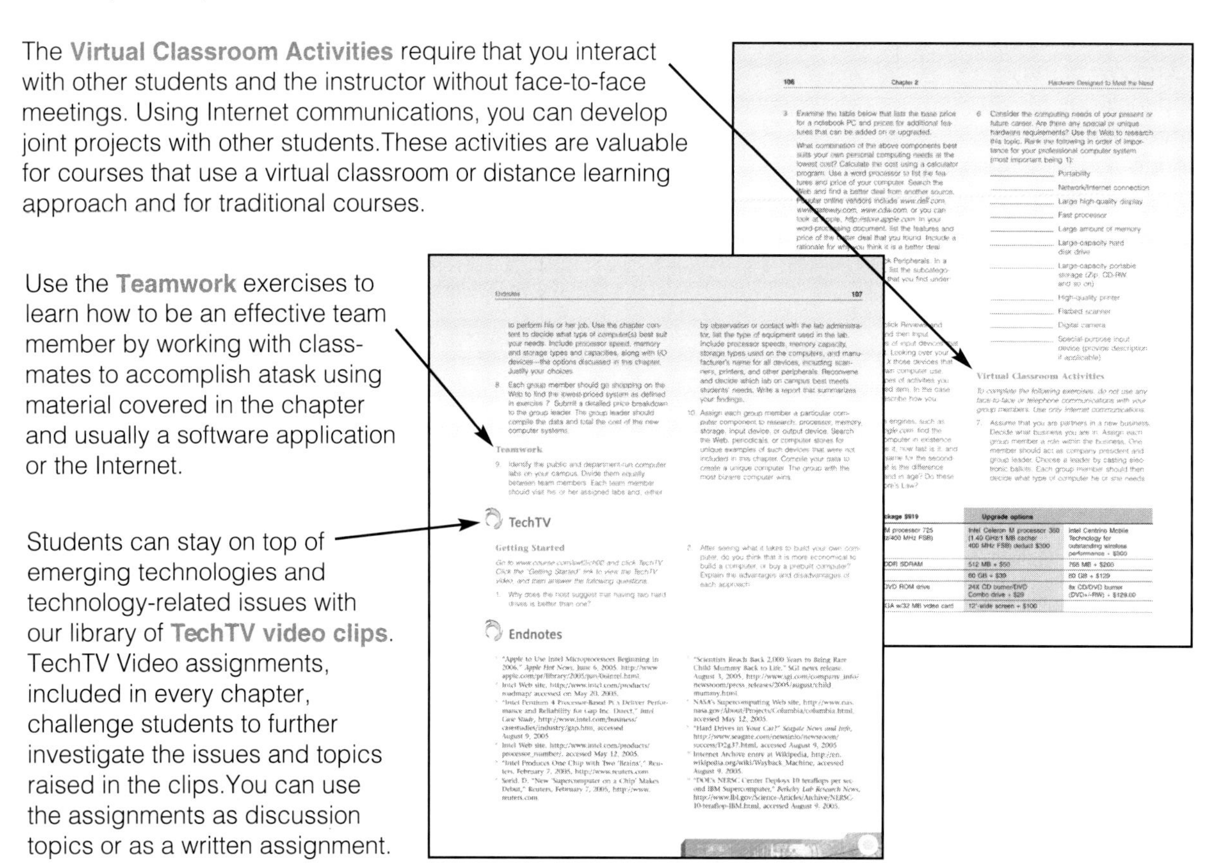

Use the **Teamwork** exercises to learn how to be an effective team member by working with class-mates to accomplish atask using material covered in the chapter and usually a software application or the Internet.

Students can stay on top of emerging technologies and technology-related issues with our library of **TechTV video clips**. TechTV Video assignments, included in every chapter, challenge students to further investigate the issues and topics raised in the clips.You can use the assignments as discussion topics or as a written assignment.

STUDENT ONLINE COMPANION

We have created an exciting online companion for you to use as you work through this book, which will help you learn and practice new concepts every step of the way. You can find this online companion at **www.course.com/swt2**

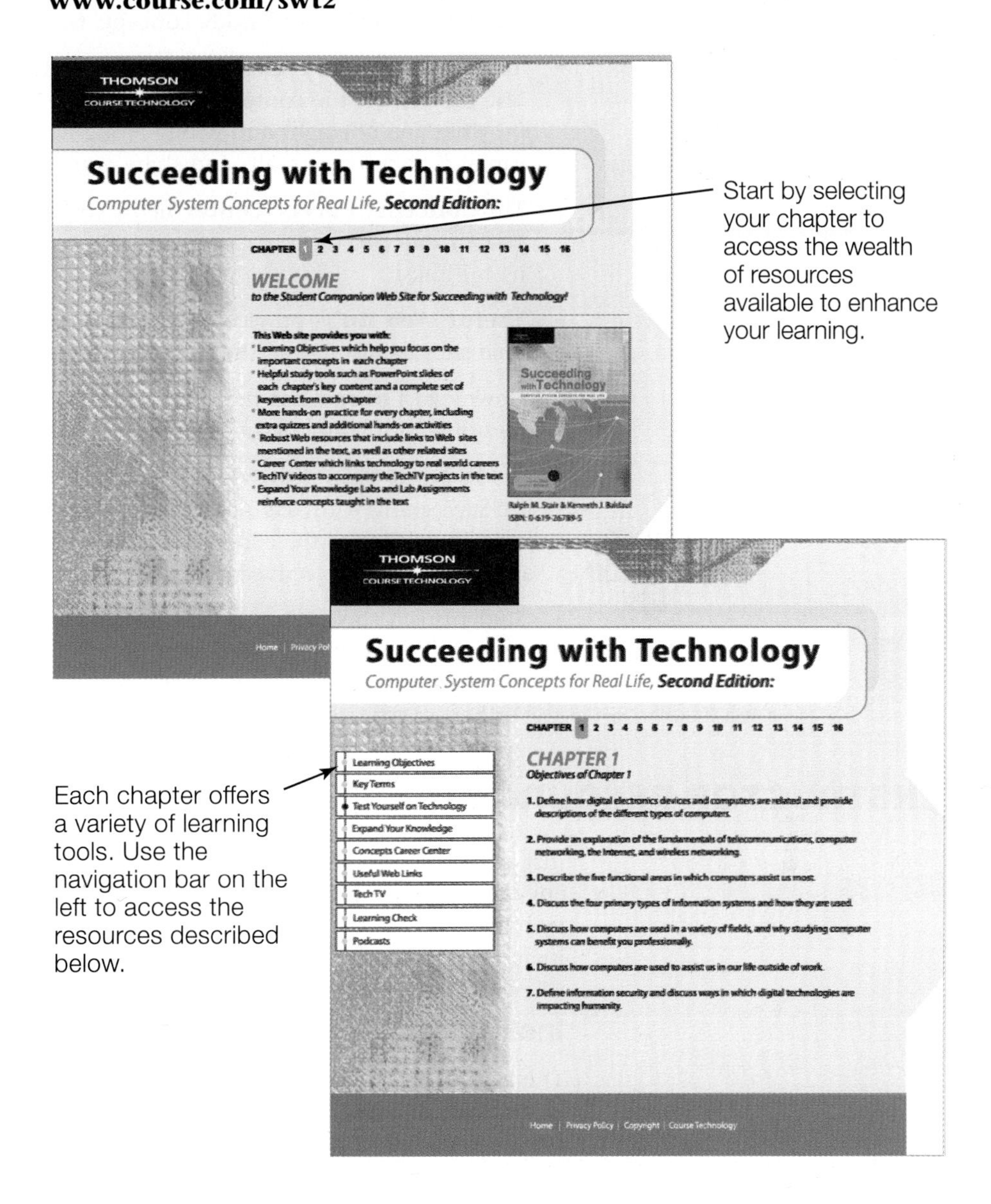

The Succeeding with Technology Student Online Companion contains the following features:

- The home page for each chapter lists the **learning objectives** for the chapter. Review these to get an overview of what is covered.
- The **key terms** from each chapter are provided along with their definitions; use these to review and study concepts from the chapter.
- **Test Yourself on Technology** are **Practice Tests** created to allow you to test yourself on the content in each chapter and immediately get feedback on what you got right and wrong along with a customized study plan. You can also track your results and share your results with your instructor.
- The **Concepts Career Center** helps you access information that will give you a chance to explore careers in technology-related fields and other useful links!
- **Useful Web links** provide you access to the home pages of the primary Web sites mentioned in the chapters and more!
- Download **CourseCast** files to your MP3 player and study on the go! These files offer you audio flashcards, concepts review, and more!

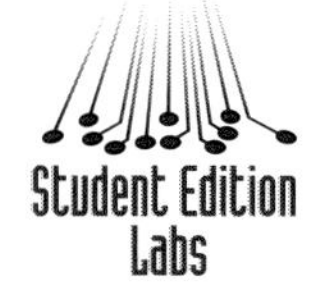

- The **Expand Your Knowledge Student Edition Labs**, available through the Succeeding with Technology Web site, help students review and extend their knowledge of the concepts through observations, hands-on simulations, and challenging objective-based questions.
- Students can stay on top of emerging technologies and technology-related issues with our updated library of online **TechTV video clips**. The **TechTV assignments**, included at the end of every chapter, challenge students to further investigate the issues and topics raised in the videos.

INSTRUCTOR RESOURCES

Just as ***Succeeding with Technology*** goes beyond computer literacy and fluency to provide students with the information they need to be successful with technology, we also take the instructor resources to the next level. The package includes:

Instructor's Manual

The Instructor's Manual provides materials to help instructors make their classes informative and interesting. The manual offers several approaches to teaching the material with a sample syllabus and comments on different components. It also suggests alternative course outlines and ideas for projects. For each chapter, the manual includes a chapter outline, learning objectives, lecture notes (including discussion topics) and teaching tips.

ExamView® Test Bank

This objective-based test generator lets the instructor create paper, LAN, or Web based tests from test banks, containing 150 questions, specifically designed for this text. Use the exams as is, or modify to create your own unique tests for your course.

PowerPoint Presentations

Microsoft PowerPoint slides are included for each chapter. Instructors can you use the slides in a variety of ways, including as teaching aids during classroom presentations or as printed handouts for classroom distribution. Instructors can add their own slides for additional topics introduced in class.

Solutions

Solutions for all of the end-of-chapter questions and activities can be found on the Instructor Resources CD or at **www.course.com**.

Figure Files

Figure files for all images used in the textbook are provided on the Instructor CD. Instructors can use these files to create their own presentations or customize the ones that accompany the book.

DISTANCE LEARNING CONTENT

Thomson Course Technology is proud to present online courses in WebCT and Blackboard to provide the most complete and dynamic learning experience possible. For more information on how to bring distance learning to your course, instructors should contact their Thomson Course Technology sales representative.

SAM COMPUTER CONCEPTS

If your instructor has chosen to use SAM Computer Concepts (assessment and training software) in your course, you will have access to hands-on assessment and interactive training simulations that reinforce lessons presented in this text.

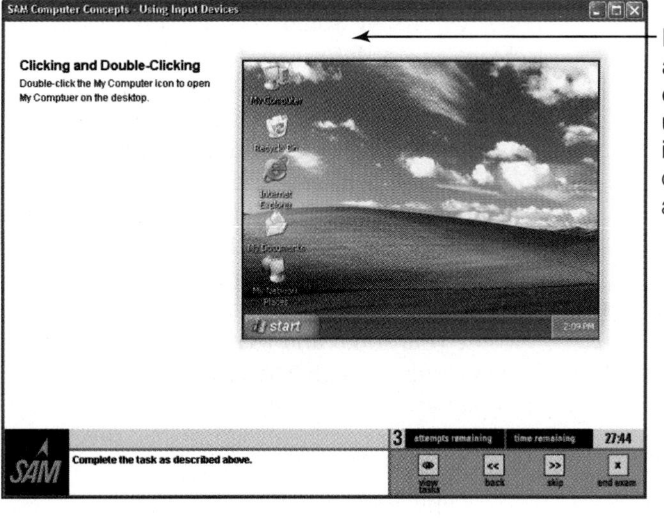

Hands-on tasks allow you to demonstrate your understanding of important computer concepts and applications

ACKNOWLEDGMENTS

Developing any book is a difficult undertaking. We would like to thank our teammates at Thomson Course Technology for their dedication and hard work. Special thanks to Brianna Hawes, Product Manager, and Rachel Goldberg, Executive Editor. We would also like to thank Kristen Duerr, Senior Vice President Computing Technology and Information Systems. We would like to acknowledge and thank Deb Kaufmann, our Developmental Editor. She deserves special recognition for her tireless effort and help in all stages of this project. Danielle Chouhan guided the book through the production process. Christina Micek helped with the photos and illustrations. We would also like to thank Robin Ireland, who researched and wrote the Tech Edge elements that appear in the margins of every chapter.

We greatly appreciate the perceptive feedback from all the reviewers who worked so hard to assist us on both the First and Second editions, including:

Lancie Affonso, College of Charleston
Gregg W Asher, Minnesota State University, Mankato
Jim P. Borden, Villanova University
Joseph DeLibero, Arizona State University
Nichol W. Free, Computer Learning Network
Alla Grinberg, Montgomery College
J. Scott Hilberg, Towson University
Martha Lindberg, Minnesota State University, Mankato
Bill Littlefield, Indiana University—Kelley School of Business
Joan Lumpkin, Wright State University
Cathy Radziemski, Arizona State University
Richard Schwartz, Macomb Community College
Elizabeth Spooner, Holmes Community College
DeLyse Totten, Portland Community College
Therese Viscelli, Georgia State University
Amy B. Woszczynski, Kennesaw State University

Ralph Stair would like to thank the faculty and staff of the Department of Management Information Systems, College of Business Administration, at Florida State University for their support and encouragement. He would also like to thank his family, Lila and Leslie, for their support.

Ken Baldauf would like to thank the faculty and staff of the Computer Science department at Florida State University for their support of this project, and his family for their support and patience throughout the writing process.

We are committed to listening to our adopters and readers and to developing creative solutions to meet their needs. We strongly encourage your participation in helping us provide the freshest, most relevant information possible. We welcome your input and feedback. If you have any questions or comments, please contact us through Thomson Course Technology or your local representative, via e-mail at CT.Succeeding@course.com, via the Internet at www.course.com, or address your comments, criticisms, suggestions, and ideas to:

Ralph Stair
Ken Baldauf
Thomson Course Technology
25 Thomson Place
Boston, MA 02210

Ralph M. Stair K Baldauf

THE AUTHOR TEAM

Succeeding with Technology brings together an experienced author team. **Ralph Stair** is Professor Emeritus in the College of Business at Florida State University, and has spent about 20 years teaching introductory computer courses. While at Florida State University, he developed market-leading information systems textbooks (Principles of Information Systems and Fundamentals of Information Systems) that were among the first to bring important upper-level information systems content to the introductory computer course used primarily in business schools. These popular textbooks are used around the world. Ralph Stair enjoys listening to people who use his textbooks and developing the best textbooks possible. The success of these textbooks allowed him to retire early from Florida State University to devote more time to research and writing. *Succeeding with Technology* is an outgrowth of his devotion to showing students how they can succeed with technology.

Teaching over 5,000 Florida State University computer literacy students annually, **Ken Baldauf** brings additional practical experience and insight to *Succeeding with Technology*. With a background in computer science, Ken started out teaching computer programming and Web development classes. He developed an interest in the impact of computers on society; in particular, how technology supports individuals in leading more productive and fulfilling lives. This interest led Ken to head the Computer Literacy program at Florida State University where he has been instrumental in guiding computer literacy standards. For the past eight years Ken has supervised and taught computer literacy classes including a general Computer Literacy class, a Computer Literacy II class, and a Computer Literacy class for business majors (see *http://lit.cs.fsu.edu* for details). Dissatisfied with traditional Computer Concepts textbooks, Ken took on *Succeeding with Technology* to create a next-generation textbook that students would read and appreciate. For more about Ken Baldauf visit his Web site at www.kenbaldauf.com.

WHY STUDY COMPUTERS AND DIGITAL TECHNOLOGIES?

Tonya Roberts faced the same dilemma as many college freshmen: she had no idea what major to pursue. A fellow student told Tonya about the career center at the university and the computer systems available for use there. Tonya decided to speak with a career counselor. She loaded her favorite music mix onto her MP3 player, donned her headphones, threw her cell phone in her backpack, and took off for the career center. At the center, a counselor showed Tonya how to use computerized questionnaires to relate her aptitudes, interests, and work environment preferences to various careers. Simply by typing in responses on the questionnaires Tonya was able to generate lists of potential careers, including what the job involved, salaries earned, where positions occurred, education required, current job market statistics, and other important information. Rather than waste paper with a printout, Tonya connected her MP3 player to the computer's USB port and saved the career information on her device.

Back at her apartment, Tonya transferred the information to her PC and studied her possible futures. The career center's Web site provided links to several online career and job placement services. Even though it would be a long time before Tonya was ready to start looking for work, she thought it might be useful to see what the job market looked like and read some job descriptions. She learned that there are all kinds of jobs out there for individuals with the right credentials and found several that captured her interest. The job postings she found acted as an incentive throughout her college education.

In her senior year Tonya revisited the career center, this time to arrange for job interviews. She had come a long way since her first visit. She had the knowledge required to begin work that she gained from her studies and a very insightful internship. Even though she was not earning a "tech degree," she had learned alot of specialized computer skills required in her field.

As you read through this chapter, consider the following questions:

1. What types of computers and digital devices played a role in Tonya's search for the ideal career and job?
2. How did telecommunications, the Internet, and computer networks assist Tonya in her quest?
3. What types of computer-based information systems might have been used by Tonya and the career center in looking for a career and job?
4. Why are computers and digital technologies important to Tonya's career and personal life?
5. What security and ethical issues might be involved in the use of databases at the career center?

Check out Tonya's *Action Plan* at the conclusion of this chapter.

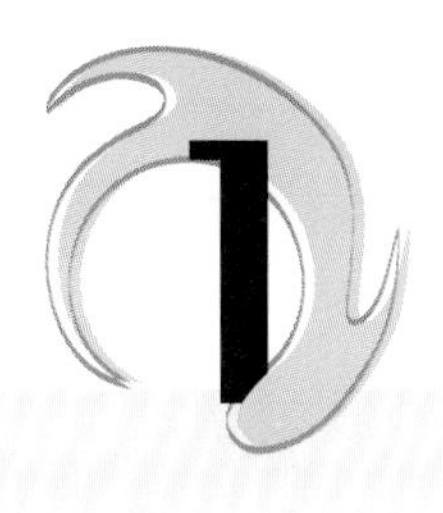

LEARNING OBJECTIVES

1. Define how digital electronics devices and computers are related, and provide descriptions of the different types of computers.
2. Provide an explanation of the fundamentals of telecommunications, computer networking, the Internet, and wireless networking.
3. Describe the five functional areas in which computers assist people most.
4. Discuss the four primary types of information systems and how they are used.
5. Discuss how computers are used in a variety of fields, and why studying computer systems can benefit you professionally.
6. Discuss how computers are used to assist people in their life outside of work.
7. Define information security, and discuss ways in which digital technologies are impacting humanity.

CHAPTER CONTENT

What Is a Computer?

The Power of Connections

What Can Computers Do?

Information Systems

Using Digital Technologies to Succeed in Your Career

Using Digital Technologies to Achieve Personal Goals

Information Security and the Social Impact and Implications of Digital Technologies

Introduction

Why study computers and digital technologies? As you will see throughout this book, understanding computers and digital technologies can help you achieve your personal and professional goals. You can use computers to obtain and play the latest music, keep track of your expenses and develop a monthly budget, and obtain information about almost any topic. Today, computers are used in all career areas from anthropology to zoology. Computers and digital technologies are indispensable in business, engineering, science, the fine arts, the military, and all other fields. In this book, you will see hundreds of examples of how individuals and organizations have used computers to achieve their goals.

This book introduces you to essential concepts in digital technologies that every student should know prior to starting a career. To realize the vast potential of computer systems, you will learn the latest about hardware, software, and the Internet. You will also learn about telecommunications and network systems, databases, electronic commerce and transaction processing, and information and decision support systems. Important information on information security, computer systems development, multimedia, and ethical and societal issues will show you how computer systems can be acquired and controlled to maximize their potential.

This society thrives on information. People depend on computer-based systems to create, store, process, access, and distribute information. Information is knowledge, and knowledge is power! This is true for all ages in all walks of life. Students use computers to research homework topics, communicate with friends, and acquire the latest music; investors use computers to make multimillion-dollar decisions; financial institutions employ computers to transfer billions of dollars around the world electronically; and chemical engineers use computers to store and process information about chemical reaction rates to make stronger and lighter plastics.

Employers expect graduating college students to be as fluent with computers and digital technology as they are with the English language. Table 1.1 illustrates how many job ads, even in nontechnical fields, specify computer skills. Employers advertising professional positions that require a bachelor's degree assume that applicants will have an understanding of digital technologies and their uses. Applicants who don't are simply squeezed out of the running. The more you know about computers and technology the more marketable you are—no matter what your career. Employers recognize that computers can assist in acquiring important job skills and knowledge, communicating effectively, and working efficiently. Whatever your career area, you must excel not only in your area of expertise but also in digital information technologies in order to be successful. A *knowledge worker* is a professional who makes use of information and knowledge. Will your future career require you to be a knowledge worker?

TABLE 1.1 • Monster.com job ads

Today's employers expect applicants to have a thorough understanding of digital technologies and their uses.

Employer	Position	Computer Skills Requirements
Nike	*Visual merchandising specialist*	"Computer skills, including Microsoft Word, Excel, and Outlook"
Levi Strauss & Co.	*Manager, graphic design*	"Current knowledge of industry trends, graphic design technology, and innovative printing techniques"
GE Healthcare	*Human resources administrative assistant*	"Experience with MS Word, Excel, PowerPoint, and a database is required"
Skyhawks Sports Academy	*Professional coach*	"Internet and computer skills are a must!"
Walt Disney World Resort	*Designer*	"Minimum 2 years experience working with Microsoft applications (Word, Excel, PowerPoint, and Outlook), Minimum 3 years AutoCAD (latest version) experience/ proficiency"

Computer literacy is a working understanding of the fundamentals of computers and their uses. Originally computer literacy was focused on desktop computer skills. Today typical computing experience extends far beyond the desktop to numerous mobile and networked devices. Rather than accessing digital information and services through one desktop location, today people interact with a wide range of digital devices from many locations (see Figure 1.1). This increased interaction with and reliance on digital technologies has made understanding them all the more important. Today you need much more than just a fundamental understanding; you need to become a master of technology.

Some technology educators feel that the phrase *computer literacy* isn't sufficient to describe the mastery of technology that is required today. Other titles

FIGURE 1.1 • The digital world

Digital technologies are an integral part of everyone's lives that require a high degree of understanding and knowledge.

have been suggested such as *information technology literacy, computer competency, computer fluency,* even *information and communications technology literacy.* Whatever you call it, everyone agrees that a deep and broad understanding of digital technologies and their use is a valuable asset towards success in careers and life. Knowing how to use certain software applications is not enough; you need to understand how to apply your knowledge of computers and digital technology to real-life problems. This textbook provides computer literacy by examining all types of digital technologies and showing how they are being used by individuals to enhance their professional and personal lives.

Digital technologies have forever changed society, businesses, organizations, and personal lives. This chapter provides some fundamental concepts and presents a framework for understanding computers and digital technologies. This understanding will set the stage for future chapters and help you unlock the potential of computers to achieve your personal and professional goals.

WHAT IS A COMPUTER?

What is a computer? This may seem like a silly question in today's computer-saturated world, but if you think about it, there are so many different types of digital devices in use today that it has become difficult to define what is and isn't a computer! Although it might be safe to assume that most devices that have a keyboard and mouse are considered computers, what about other digital devices like the iPod, cell phone, and digital camera? Here's a brief definition: a **computer** is a digital electronics device that combines hardware and software to accept the input of data, process and store the data, and produce some useful output. To understand this definition, and to help define what is and is not a computer, you must first understand the basics of digital technology and the general functions of a computer.

Digital Technology

The purpose of the textbook, summed up by its title, is to provide information that can help students succeed in reaching their personal and professional goals by using the strengths of technology. **Technology** refers to tools, materials, and processes that help solve human problems. Many of today's technologies fall under the classification of digital electronics. These are the technologies that you will study. Today's well-equipped knowledge worker, such as the one shown in Figure 1.2, makes use of many digital technologies. Each serves a specific function that meets a unique need in a particular environment. Personal computers, cell phones, digital cameras, and digital music players like the iPod are all considered digital electronics devices. What does it mean to be digital?

FIGURE 1.2 • Knowledge worker

Today's workers employ many different types of digital electronics devices for maximum productivity.

A **digital electronics device** is any device that stores and processes bits electronically. A **bit** (short for *binary digit*) represents data using technologies that can be set to one of two states, such as on or off, charged or not charged. Each state is assigned a 1 or a 0, the only two possible values for binary digits. For example, *on* might be assigned a 1, and *off* a 0. A string of bits could be notated as 10011011. This is the essence of a digital device—the ability to represent, process, transfer, and store data and information as 1s and 0s.

A group of eight bits is called a **byte**. Bytes can represent all types of useful data and information, such as characters, words, or sounds. Bytes can be grouped together to create an electronic **file,** a named collection of instructions or data stored in the computer or digital device. **Data** refers to the items stored on a digital electronics device including numbers (values), characters (letters), sounds (such as your voice over a cell phone network), music (CDs and MP3s), or graphics (photos, drawings, and movies). Anything that can be expressed and recorded can be represented as 1s and 0s and stored as files in digital devices. The process of transforming nondigital information, such as things you experience with your senses—written or spoken words, music, artwork, photographs, movies, even tastes and fragrances—to 1s and 0s is called **digitization**.

When the bits and bytes are processed to a format that is useful to people—statistics in a graph, the results of a Web search, music in your headphones, photos on a display—it is called information. **Information** is data organized and presented in a manner that has additional value beyond the value of the data itself. The primary purpose of digital electronics devices is to process digital data (such as sales figures) into information (such as a graph comparing sales by month).

At the heart of all digital electronics devices is a microprocessor. A **microprocessor**, sometimes called a *chip* or just a *processor*, combines microscopic electronic components on a single integrated circuit that processes bits according to software instructions.

Computers, computer networks, and other digital electronics devices store, transfer, and process bits and bytes in vast quantities. It is typical to refer to the power of digital technologies in terms of thousands, millions, billions, even trillions of bits and bytes. For example, today's typical PC can store billions of bytes on a hard drive, and millions of bytes in memory. High-speed Internet connections are able to send and receive billions of bits per second. Table 1.2 shows common prefixes used to express multiples of bits, bytes, and other digital metrics, and Table 1.3 shows the size in bytes of some common digital items.

Bytes are typically represented with an uppercase *B* and bits with a lower case *b*. For example, *KB* stands for kilobyte and *Kb* for kilobits. These are important metrics to understand if you want to measure the power and effectiveness of digital technologies.

TABLE 1.2 • Prefixes for digital technology metrics

Prefix	Value	Amount
Kilo	1,000	Thousand
Mega	1,000,000	Million
Giga	1,000,000,000	Billion
Tera	1,000,000,000,000	Trillion
Peta	1,000,000,000,000,000	Quadrillion
Exa	1,000,000,000,000,000,000	Quintillion

TABLE 1.3 • Storage capacity examples

Example storage sizes	
Text file	1 KB
Spreadsheet file	50 KB
High-resolution digital photo	1.2 MB
MP3 digital music file	5 MB
Music file on CD	64 MB
Microsoft Office software	640 MB
The Matrix motion picture	7.38 GB
Sears inventory and customer database	55 TB
CERN database (nuclear and particle physics research)	20 PB
Google database	30 PB

Computer Functions

Now that you understand what a digital electronics device is, what about the rest of the definition of a computer? Recall that a computer is a digital electronics device that combines hardware and software to accept the input of data, process and store the data, and produce some useful output (Figure 1.3). **Hardware** refers to the tangible components of a computer system or digital device. **Software** comprises the electronic instructions that govern the computer system's functioning.

FIGURE 1.3 • Computer functions

A computer is used to input, process, and store data and provide useful output.

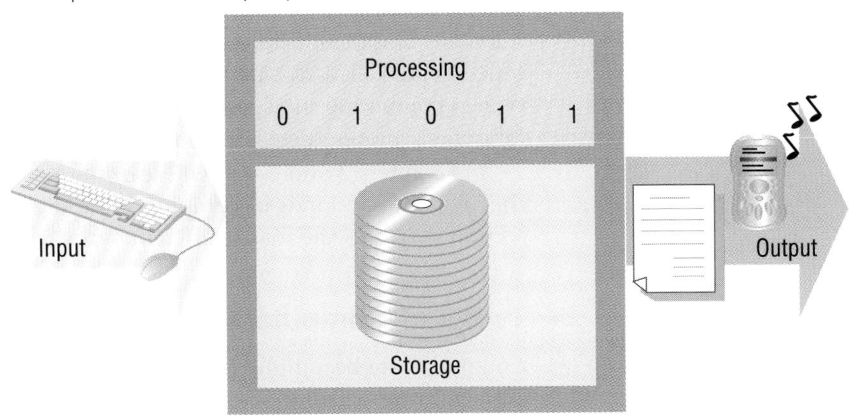

Strictly speaking, this definition covers not only desktop, notebook, and handheld computers, but also today's popular digital electronics devices. Cell phones, digital cameras, handheld gaming devices, and digital music players all process input into some form of output—voice, photos, video, music—based on program instructions or software. So, although nearly all digital electronics devices fall

Digitizing the World

This chapter makes the statement that "anything that can be expressed and recorded can be represented as 1s and 0s and stored as files in digital devices." It is interesting to consider the limits of such a statement. Is there anything that cannot be digitized?

People have become adept at digitizing things that are seen and heard, and use digital devices to communicate text, numbers, pictures, movies, voice, and music. But what about smell, taste, and touch? Can information perceived by these senses be digitized?

Of these three, smell and taste are the most likely to turn up on PCs first. Trisenx, a company based in Savannah, Georgia, has developed digital technologies that allow users to generate aromas and flavors from their PC. Aromas are produced by a digital device they call the Scent Dome, and flavors can be printed onto thick fiber sheets for licking. Once the technology catches on, scents and flavors can embellish interactive games and be attached to e-mail and Web sites. Imagine the aroma of a florist's shop wafting from your computer when you visit *www.ftd.com*. Or imagine smelling the tar as you chase a creature through the tar pits of an imaginary world on your PS2 game console.

Questions:

1. List six creative examples, other than those provided here, of uses for digital aromas.
2. How might digital aromas be abused by marketing companies and hackers?
3. How might it be possible to communicate information to all five senses without the use of headphones, displays, or boxes like the Scent Dome?

Sources

1. *Trisenx Web site, http://www.trisenx.com/intro.html, accessed on April 10, 2005.*
2. *Sood, V. "Digital Scents May Promote Products on Net,"* The India Tribune Online, *March 11, 2002, http://www.tribuneindia.com/2002/20020311/login/main2.htm, accessed on April 24, 2005.*
3. *"How the Net Has Become Multi-Sensory," Computer Research & Technology Web site, http://www.crt.net.au/etopics/sens.htm, accessed on April 24, 2005.*

under the definition of a computer, in casual conversation the term *computer* is generally used to refer to *general-purpose computers*—those that can run any of hundreds of different software applications. Computers designed for one particular function are called *special-purpose computers* and are typically referred to by the type of computing they perform—digital music player, digital video recorder, digital camera, motion detector, controller, blood pressure monitor, and so on.

In Chapters 2 and 3 you will learn more about hardware and software as they relate to all kinds of computers and digital electronics devices. This next section describes the many different types of computers in use today.

Types of Computers

Computers are becoming increasingly pervasive. As you walk across campus you are bound to see some students working on notebook computers, others talking or sending text messages on their cell phones, and still others walking to the beat of their favorite music playing on an iPod. You may pass an ATM machine or two or kiosks that provide special services such as campus maps or access to college records. Maybe you take advantage of a break and play a quick round of your favorite video game on your cell phone. Computers of one sort or another are within reach wherever you go.

FIGURE 1.4 • Apple iMac

All-in-one PCs like the iMac sandwich the system unit and display together in one case.

FIGURE 1.5 • Apple PowerBook

Today's notebook computers have power that rivals larger desktop PCs.

Personal Computers. The type of digital device that most people associate with the word *computer* is called a personal computer. A **personal computer**, or **PC**, is a general-purpose computer designed to accommodate the many needs of an individual. Personal computers come in a wide variety of types and styles from a wide variety of manufacturers. The most traditional and powerful type of PC is the desktop computer. *Desktop computers,* such as the Dell Dimension or Apple iMac, are designed to be stationary and used at a desk. They typically include a tower case (also called the system unit and houses the circuitry), a display, keyboard, mouse, speakers, and a printer. Some of today's "all-in-one" PCs sandwich the system unit and display into one case to save desktop space (see Figure 1.4).

Notebook computers, also called *laptop computers,* such as the IBM Thinkpad or Apple iBook are also considered personal computers and provide near-desktop power in a portable case. Notebook computers have gotten quite powerful in recent years. David Herbruck and Beat Baudenbacher started their company Loyalkaspar (*www.loyalkaspar.com*), a high-tech New York design boutique, with two Apple PowerBooks (see Figure 1.5), and an iPod.[1] They didn't have enough money for office space so they rented two desks. By the end of their first two years in business, they had developed a reputation for innovative work in high-definition motion graphics. They are well known in the world of HDTV (high-definition television) and have hired six additional graphic artists to keep up with demand. They have their own office and their computing power has grown to include a bank of six Power Mac G5 desktop computers. Examples such as this illustrate what amazing things can be accomplished with nothing more than imagination, a notebook computer, and valued computer skills.

Tablet PCs are portable personal computers, similar to a notebook computer in size, that provide a touch-sensitive display on which you can write and draw. A special electronic stylus is used to write or draw on the screen, as well as to select menu items and manipulate the cursor. All that is written and drawn on the display is interpreted, processed, and stored digitally. Tablet PCs are useful in environments where you have to access your computer while on the go and on your feet (see Figure 1.6).

The convertible model tablet PC converts between notebook PC and tablet by allowing the display to be rotated. Convertible tablets are useful in that they provide both keyboard and stylus methods of input. A slate model tablet PC does not provide a keyboard. Slates are lighter than convertible models and allow you to connect to a larger display, keyboard, and mouse through a docking station.

The Boeing Company is investigating how tablet PCs might assist its 167,000 employees. In a pilot study Boeing provided tablet PCs to selected users in representative work environments.[2] Those who participated in the pilot study were

FIGURE 1.6 • Tablet PC

Tablet PCs are ideal for environments where you have to access your computer while on your feet.

thrilled to be able to work on their computers wherever they went without having to continually return to their desk. The tablet PCs enhanced their time spent in meetings and freed them up to move around and work with others. All participants reported an increase in productivity.

Tablet PCs are ideal solutions for certain computing environments. However, since tablets are relatively new technology, they are typically more expensive than notebooks and provide fewer expansion and upgrade options.

Handheld computers, also called *PDA*s (for personal digital assistant), are personal computers that are only slightly larger than traditional cell phones. Like tablet PCs, handheld computers use a touch-sensitive display and stylus for input. Handwriting recognition software translates written characters to editable word processing characters. Handhelds have traditionally been used for personal information management (PIM) such as maintaining to-do lists, storing appointments on a personal calendar, and storing contacts in an address book.

Today's handhelds do a lot more than just PIM. They can run many of the same software applications as a desktop PC, but typically with fewer features. For example there are versions of Microsoft Word, Excel, Access, and PowerPoint for handheld computers. Handheld computers equipped with wireless networking technologies allow users to access e-mail, the Web, and private networks. Handheld PCs are used to read *e-books,* digital versions of books and other publications. A handheld computer can function as an MP3 digital music player, digital photo library, video game arcade, and a digital video player. *Smart phones* are handheld computers that include cell phone capabilities. A high-end smart phone can offer the benefits of several digital devices in one: a PDA, a cell phone, a digital music player, and a portable video player (Figure 1.7). Typical PDA applications are listed in Table 1.4.

The data stored on a PDA is typically synchronized with the data stored on a desktop or notebook PC. This way files that are edited on one device or the other are the same across all. The PDA connects to the other PC through a *docking station* or *cradle*—a small stand for a handheld device that is used to recharge its battery and to connect to a PC. Upon connecting, synchronization software automatically is engaged to transfer and update files. Synchronization can also take place without the cradle using wireless networking technologies.

The Los Angeles-based law firm of Keesal, Young & Logan found that the notebook computers they gave their legal staff apparently didn't meet their needs because they were left unused in their offices 80 percent of the time.[3] After a careful examination of the needs of the staff they discovered that handheld computers provided the best usability for lawyers who were often doing their computing from the client's location. After switching to smart phones, lawyers with the firm were able to check their e-mail; access mission-critical applications such as a client information application, case management software, and workload management software; and

POCKET-SIZED OFFICE

Rather than tote around a four-to-eight pound laptop, more professionals are using USB keychain drives to carry files they need with them. The problem? Having the right application upon arrival to access those files. In answer to that growing growl of frustration, two flash drive manufacturers have joined forces to create one with "brains built in." Now execs can carry system preferences, any must-have applications, e-mail, browser and bookmarks, and calendar with them wherever they go.

Source:
U3 Sees USB Revolution in Ability to Put a World on a Keychain Drive
By Janet Rae-Dupree
Silicon Valley/San Jose Business Journal
http://sanjose.bizjournals.com/sanjose/stories/2005/06/13/story5.html
June 10, 2005

FIGURE 1.7 • Smart phones

Smartphones such as the Treo from Palm offer mobility and compact computing.

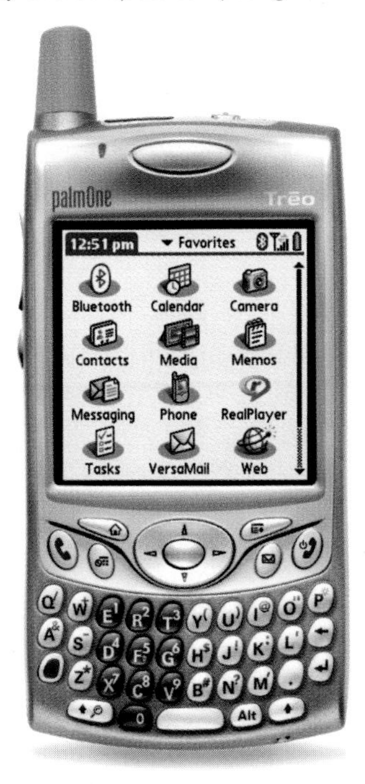

TABLE 1.4 • Typical PDA applications

PDA applications
• Calculator
• Calendar
• Contacts
• Database with forms for data entry
• E-book reader
• E-mail/messaging
• Games
• Maps/Global Positioning System
• Media player for music and video
• PDF viewer
• Photo viewer
• Presentation software
• Spreadsheet
• Synchronization
• To-do list
• Voice recorder
• Web browser
• Word processor/notes

make phone calls, all with a device that fits in a pocket. Different types of PCs cater to different needs and environment. It is important for professionals to take advantage of the unique capabilities of each of the various types of PCs.

Computer Platforms. A computer's type, processor, and operating system define its **computing platform**. When describing a computer's platform in terms of hardware, one would list the type of computer—handheld PC, desktop PC, server, and so on, along with a description of the unique hardware features that make up the computer including processor type and other internal component specifications. Because operating systems are designed for a specific type of hardware platform, the operating system itself is often used to define a computing platform. For example, Microsoft Windows runs on Intel and Intel-compatible processors. Defining a platform as Windows implies Intel. The computing platform is important for two reasons: (1) it defines the user's experience and interaction with the computer, and (2) it provides specifications for software developers. Typically, software is written to run on one particular platform.

There are many competing platforms for all types of computer systems. Table 1.5 lists some of the most popular personal computing platforms for desktops, notebooks, tablets, and handhelds. Varying computer platforms support varying needs, and each has a unique look and feel. Two popular smart phone platforms are shown in Figure 1.8, and another (Palm) was shown in Figure 1.7. The two most popular personal computer platforms are IBM-compatible, commonly referred to as Windows, and Apple. Hundreds of manufacturers make Windows computers. Only one company makes Apple computers.

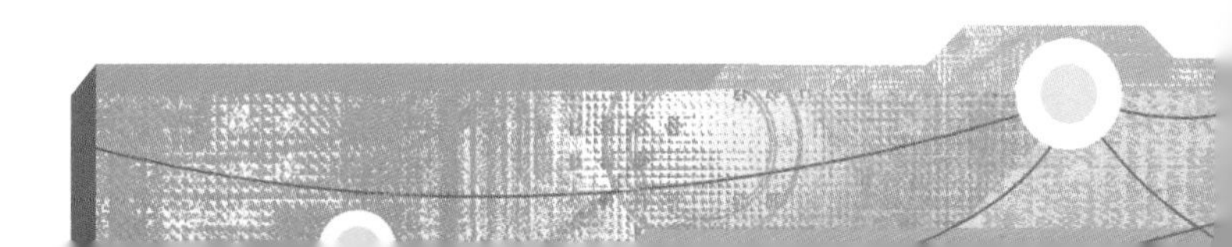

TABLE 1.5 • Popular personal computer platforms

	Desktop/notebook	Tablet	Handheld/smart phone
Microsoft Windows	X	X	X
Apple OS	X		
Linux	X	X	X
Symbian			X
Palm			X

FIGURE 1.8 • Smart phone platforms

Smart phones running operating systems by Symbian (left) and Microsoft (right) provide different features and experiences for the user.

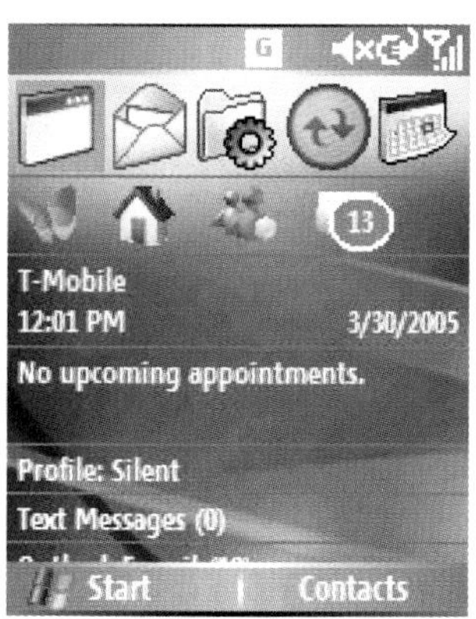

Servers. The larger computers that power today's network and Internet services are called servers. **Servers** are powerful general-purpose computers that provide information services to numerous users over a computer network. Servers often have multiple processors and large amounts of storage to support many simultaneous users. In most cases, servers run 24 hours a day, 365 days a year, in order to provide services at any time. Although servers are general-purpose computers that can run a variety of software applications, most are dedicated to specific duties. For example, a *Web server* is responsible for serving up Web pages over the Internet. A *file server* may store and deliver files to employees' desktop computers over a company's private network. An *e-mail server* handles sending and receiving e-mail messages. A *print server* manages the printing requests for a printer shared by multiple users on a network. Large businesses and organizations typically have many types of servers providing a wide assortment of services to their employees and the public.

Servers come in a wide variety of sizes with varying capacities, from serving dozens to thousands of users at once (see Figure 1.9). A *midrange server* has the capacity to service dozens or even hundreds of users at a time. One of the world's leading newspapers, the *Chicago Tribune*, has consolidated all of its data onto several midrange servers.[4] Its five critical areas—editorial pagination, classified ad production, circulation/customer care, data warehousing, and clip art archival—are accessed by staff throughout the company from central servers over the company network.

A *mainframe server* can service hundreds or thousands of users at a time over a computer network. Mainframe servers, often simply called *mainframes*, have been used in companies and organizations since the late 1950s. Prior to the

FIGURE 1.9 • Mainframe server

Large servers can support thousands of users simultaneously.

introduction of PCs in the early 1980s, employees used *terminals*, desktop computers with a keyboard and display, but little else, to connect to a mainframe and access data. Many mainframes have been retired in favor of smaller, faster, and more efficient servers, but they are still required for the large data environments found in government institutions and large corporations. For example, Deutsche Bahn, the company that manages the German railway system, uses a mainframe computer to manage and distribute data from over 5000 databases, containing 6.5 terabytes of data, to its 55,000 employees around the country.[5]

Supercomputers. **Supercomputers**, the most powerful computers manufactured, harness the strength of hundreds or even thousands of processors simultaneously to accomplish very difficult tasks. IBM's Blue Gene/L computer, shown in Figure 1.10, with 32,768 processors, is being used to solve problems for the U.S. Department of Energy's National Nuclear Security Agency. The Earth Simulator (*www.es.jamstec.go.jp*) uses 5120 processors to simulate the Earth's climate in an effort to determine what steps are necessary to ensure a healthy future for the planet and its inhabitants. Disney uses a supercomputer with 900 processors to create its feature animations.

Supercomputers have helped to answer many of life's most perplexing questions and solve some important problems. The human genome (the structure of human DNA) was decoded using supercomputers. Many of today's most useful

FIGURE 1.10 • Supercomputers

Supercomputers such as this BlueGene/L employ thousands of processors to answer life's most difficult questions.

pharmaceuticals were developed with the assistance of supercomputers. Supercomputers, through highly advanced simulations, put an end to nuclear testing in developed countries. There are still many difficult questions and problems that may only be solved using future generations of supercomputers.

Special-Purpose Computers. Special-purpose computers are manufactured to serve a specific purpose, such as game consoles, digital media players, digital cameras, and smart home appliances.

You may have noticed an increasing number of public-access computers, called kiosks, in public areas and stores. A **kiosk** is a computer station that provides the public with specific and useful information and services. Usually equipped with a touch screen, kiosks provide everything from online store catalogues, such as at JC Penney's and Staples, to maps and exhibit information, such as in cities, on campuses, and in museums. The ATM where you can withdraw money from your bank account is a kiosk. Kiosks that allow customers to print photos from their digital cameras, such as the one in Figure 1.11, have become very popular. There are even church kiosks that accept donations and respond with a blessing.

Special-purpose computers also work behind the scenes in the devices that support professionals and control the automated equipment that is increasingly prevalent in everyone's daily lives. **Embedded computers**, sometimes called *microcontrollers*, are special-purpose computers (typically an entire computer on one chip) that are embedded in electrical and mechanical devices in order to control them. Automatic doors; washing machines; elevators; automobile systems such as fuel injection, braking, and airbag deployment; microwave ovens; copy machines; telephones; and numerous other gadgets encountered everyday are controlled by embedded computers. In fact, there are many more embedded computers in use today than general-purpose computers. A home may have only one or two PCs, but it probably contains dozens of embedded computers.

FIGURE 1.11 • Kiosks

Kodak provides kiosks that produce prints from digital cameras.

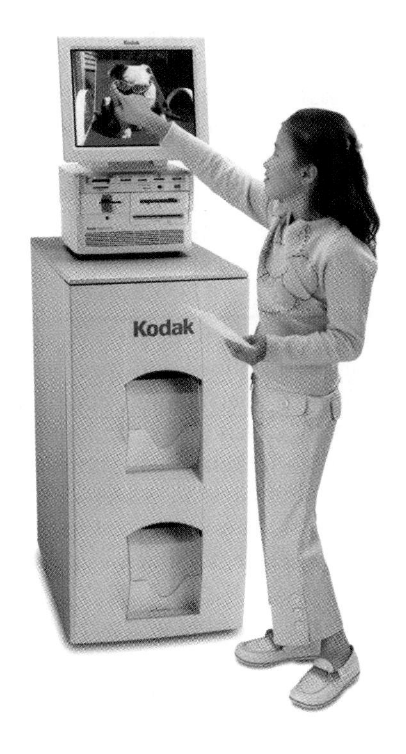

Embedded computers control special-purpose devices used in nearly all professions. Embedded computers can be found at the heart of systems used by security experts such as airport screening devices, metal detectors, and bomb detection devices, as well as the radar guns used to catch speeders. The manufacturing industry makes use of all types of automated systems controlled by embedded computers to manufacture products. Embedded computers are essential to the specialized tools used by doctors, nurses, and medical technicians to examine and diagnose patients, such as in MRI and CAT scan medical imaging machines (see Figure 1.12).

Mobile Digital Devices

The largest area of growth in technology has been in mobile devices. Today, people are doing their work on the go more than ever before. Working life has been transformed from one in which individuals work independently at a desk, to one in which teamwork and mobility are the norm. The "desk" has to be wherever you are. The trend of digital electronic devices to become smaller and increasingly powerful has fully supported the move to an increasingly mobile workforce. The power of a 1990s computer that filled an entire desktop now fits in a six-ounce handheld computer.

FIGURE 1.12 • Embedded computer

Embedded computers are used in a multitude of special-purpose devices used by professionals.

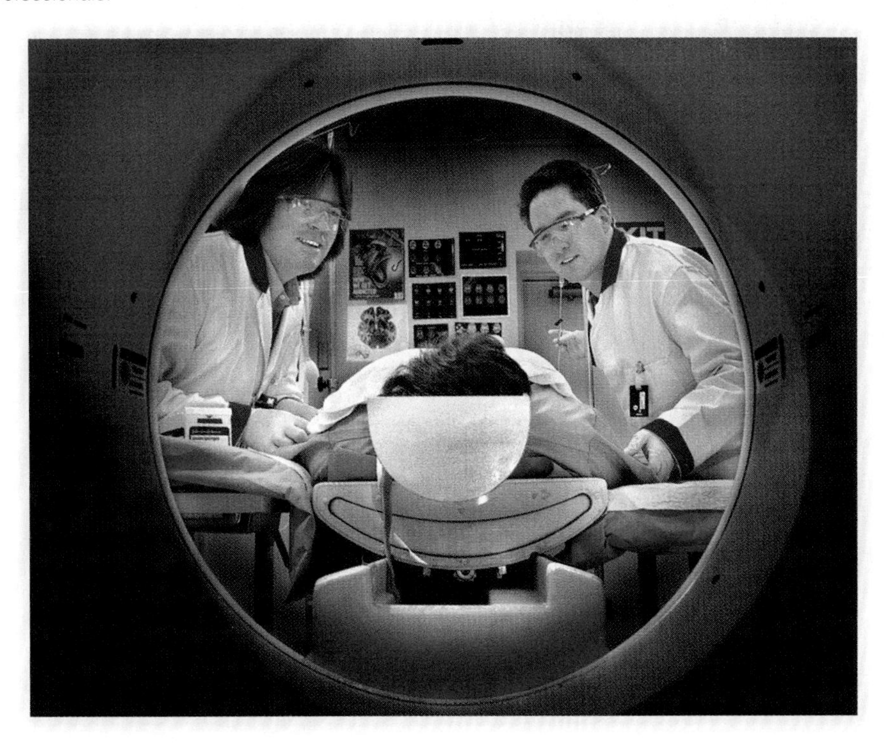

Mobile Computing. Among the types of computers defined in the previous section are several different mobile computers: notebooks, tablets, and handheld PDAs. Every year since 2003 notebook PCs have consistently outsold desktop PCs. Consumers love the freedom that notebooks, tablets, and handhelds provide, especially those equipped with wireless networking technology. As wireless networks grow to cover larger areas, you can expect to see continuing increases in mobile computer sales. Many colleges have wireless campuses where students can use mobile computers to access the school network and the Internet from anywhere on campus. Many cities are working to provide free wireless Internet access in business districts and airports. Services such as Boingo and Wayport provide subscribers with wireless Internet access at thousands of locations around the world. The phenomenal rate of growth in mobile computing is driven by wireless network technologies.

Mobile Communications. Cell phones and cell phone services are among the fastest growing of digital markets. The ability to phone and speak with individuals from almost any location has transformed society. Online communications and associations are beginning to supersede here-and-now communications and associations. Today's cell phones, however, do far more than provide mobile voice communications. *Text messaging,* also known as *Short Message Service (SMS)* and *texting,* involves using a cell phone to send short text messages to other cell phone users. Due to its low price and convenience, texting has become as popular a use of cell phones as voice communications (Figure 1.13).

FIGURE 1.13 • It's a cell world after all

Cell phone communications are changing the way people relate with each other.

Cell phone handsets are available in a wide range of styles with varying levels of capabilities. The most powerful cell phones (referred to earlier as smart phones) double as handheld computers. *Third generation (3G) cell phones* offer high-speed Internet access. For a monthly fee you can connect your notebook computer to the Internet through your cell phone at speeds comparable to high-speed home connections. Other new services for cell phones include music and video clip downloads.

Mobile Media. Recent years have witnessed an explosion of handheld media devices (see Figure 1.14). The launch of the Apple iPod in 2003 brought portable digital music into the mainstream. While portable *MP3 players*—handheld devices that play music stored in the digital MP3 format—have been available for some time, it was the iPod's slick design and marketing that brought it to the attention of the general public. In 2004 several companies, such as Creative and Samsung, hoped to take advantage of public enthusiasm for mobile media by introducing *portable media center* devices that play not only digital music but also play digital movies. Movies, television shows, and other video clips can be downloaded from the Web or transferred from DVD to the devices and displayed on the 3.8" screen. In 2005 the introduction of Sony's PlayStation Portable (PSP) brought handheld gaming to a new level with a high-quality widescreen display for realistic interactive 3-D graphics.

In answer to the question "What is a computer?" you have learned that all digital electronics devices can be classified as computers. Some are general-purpose computers and others are special-purpose computers. Throughout this text

FIGURE 1.14 • The mobile media boom

This may be considered the decade of mobile media, with the introduction of portable media devices such as the iPod, Zen, and PlayStation Portable.

2003 2004 2005

you will learn about all different types of computers and digital devices and how they are used at home, at work, and on the road. From supercomputers to desktop computers, from servers to cell phones, from kiosks to handheld games, you will see how computers provide digital services for every environment and need. You will see that the ability to connect to information sources, individuals, and services from any location is of particular value and importance.

THE POWER OF CONNECTIONS

It is commonly understood that people are able to accomplish more and produce better solutions when they work together rather than individually. "There is strength in numbers," "Two heads are better than one," and other common expressions underscore this widely held belief. Today's businesses and organizations put great emphasis on teamwork. Today's digital networks provide the technical foundation to support the communication that is at the heart of teamwork.

Telecommunications

Telecommunications are communications that take place electronically over a distance. Forms of telecommunication include telephone systems, radio, television, and computer networks. Telecommunications components include a transmitter that sends a signal over a medium, such as fiber-optic cables or through the air, to a receiver (Figure 1.15).

FIGURE 1.15 • Telecommunications

Telecommunications involves a transmitter sending a signal to a receiver over a medium electronically.

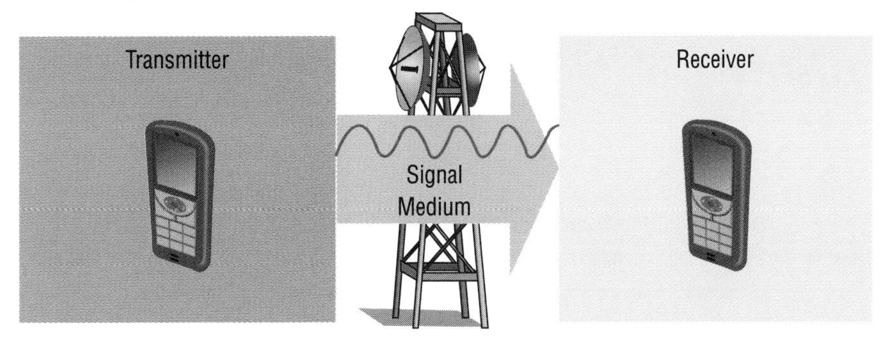

The *telecommunications industry*, often called the *telecom industry*, focuses on electronic voice and data communications. Telecom giants such as Sprint and MCI offer voice, data, and Internet services for residential customers and businesses. Telecom companies maintain the large networks over which telephone and Internet communications travel (see Figure 1.16).

FIGURE 1.16 • The global network

Global telecommunication networks span the United States and the world, providing voice and data communications and carrying Internet traffic.

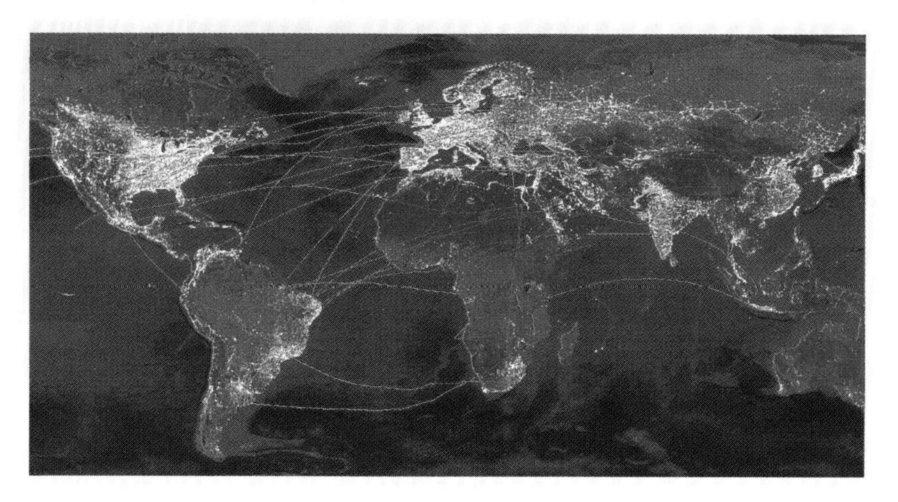

A **computer network** is a telecommunications network that connects two or more computers for the purpose of sharing data, hardware, and software resources. Computer networks can range in size from two computers to thousands of computers or in the case of the Internet millions of computers. For computers to communicate—to send and receive data—they must follow the same set of communication rules called protocols. **Protocols** are rules that allow two or more computers to communicate over a network.

The Internet

Multiple computer networks joined together to form larger networks are called *internetworks*. In 1969 research commenced to build what would eventually become the largest internetwork in the world. That research provided the protocols and technologies that govern today's Internet. The **Internet** is the world's largest public computer network; a network of networks that provides a vast array of services to individuals, businesses, and organizations around the world. The Internet population is rapidly approaching one and a quarter

UNPLUG TO CONNECT

"Techno-rebels" in Portland, Oregon, want to build community by developing a citywide network of hosts to provide free wireless access. Their goal is to unplug the entire city so they can remain connected even if the Internet goes down. Mostly computer programmers who want to share their knowledge and spark ideas off one another, their efforts have been supported by a combination of grant money and altruism. Cities all across the United States are following suit.

Source:
Techno-Rebels Spread Wireless Network Vision
By Elizabeth Armstrong Moore
The Christian Science Monitor
http://www.csmonitor.com/2005/0615/p01s03-ussc.html
June 15, 2005

billion. People rely on the Internet for news and information, communication, education, entertainment, commerce, arranging travel, finding jobs, and many other important activities.

The *Web*, short for **World Wide Web**, is an Internet service that provides convenient access to information through hyperlinks. A *hyperlink* is an object in a Web document that can be clicked to access related information. Developed and released to the public in the early 1990s, the Web opened the Internet to the general public. Because of its ingenious and easy-to-use design, the Web has become the primary tool for accessing most Internet information and services.

Wireless Networking

Wireless is the technology buzzword of the decade. **Wireless networking** uses radio signals rather than cables to connect computers and digital devices to computer networks and through those networks to the Internet. The increasingly mobile workforce is benefiting greatly from wireless networking technologies that allow workers to remain connected to business networks and the Internet while on the go. Just as the cell phone expanded people's telephone capabilities, wireless networking is expanding their computing capabilities.

Wireless networks are being installed in businesses, homes, and on campuses in order to allow individuals within range to access information and services stored on networks. **Wi-Fi**, short for *wireless fidelity*, is a popular wireless networking standard that connects computers to other computers and to computer networks (see Figure 1.17). When you see wireless Internet access advertised in coffee shops, restaurants, book stores, hotels, airports, and other public places, chance are they are using Wi-Fi. Many notebook computers come equipped with Wi-Fi capability. Wi-Fi is popular in homes where it allows residents to share an

FIGURE 1.17 • Wi-Fi

Wi-Fi wireless networking technology is used at many colleges to give students access to the Internet anywhere on campus.

Internet connection, printers, and files between several computers. It is also being used in homes to transmit music and movies between computers and televisions and stereos. Wi-Fi is popular in businesses as it is easy to install and provides employees with the ability to take their computers with them anywhere on the premises without losing a network connection.

Wireless technologies, telecommunications, the Internet, and the Web are all important factors in using technology to succeed. You will find examples throughout this book showing how people in various careers are using the benefits of telecommunications to accomplish their personal and professional goals.

WHAT CAN COMPUTERS DO?

Although computers and digital electronics devices seem to provide almost limitless services, there are some activities for which they excel and others for which they are ill equipped. For example, computers are excellent at carrying out well-defined, repetitive tasks accurately and quickly. Computers have no problem carrying out long complicated calculations at lightning speed. A computer can sort through millions or billions of data records in a matter of seconds to find those that match a keyword or some other criteria—a task that might take a human a lifetime. Fortunately, the things that computers do well—working with large amounts of data and repetitive tasks—are things that people find difficult and monotonous, and the creative and interpersonal endeavors that people find most engaging, computers are ill suited to perform. For example, a computer could write a poem or compose a song, but such would be found lacking in substance. Designing a new marketing campaign, choosing an employee for promotion, listening to the complaints of a patient and interpreting their meaning, and many other professional activities require human creativity and intuition. It is helpful as computer users to consider the areas in which computers can assist people most: computation, automation, communications, digital media and entertainment, and information management (see Figure 1.18).

FIGURE 1.18 • Computer strengths

Computers are most effectively used when performing these five activities.

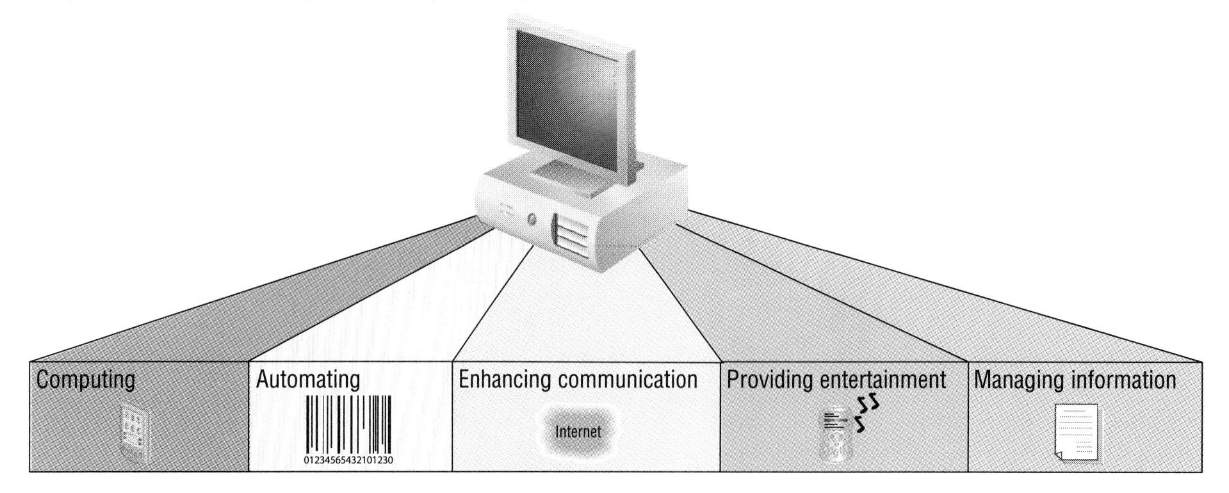

Computing

Computers were invented to compute. While the word *compute* can be interpreted in different ways, it is used here to mean calculating a solution to a mathematical problem. This is the number-crunching aspect of computers. Early computers calculated missile trajectories during World War II. After the war, they were put to work for the U.S. government calculating census data and in private industry calculating financial data. Over the years, computation has remained paramount in the value that computers provide. Computers can crunch endless streams of numbers in a fraction of the time humans can without ever tiring. And they do it faster and better every year. With each new and faster supercomputer, new answers to age-old questions are discovered.

Computational science is an area of computer science that applies the combined power of computer hardware and software to solving difficult problems in various scientific disciplines. For example, at Virginia Tech they have built one of the world's most powerful supercomputers, named systemX, by linking together 1100 dual-processor Apple computers (see Figure 1.19). Meteorologists are using the supercomputer to crunch numbers and simulate the structure of tornadoes so they can better understand them. Biologists are using the supercomputer to examine the many roles water plays in the human body.[6] Scientists in all areas rely on computers to do the mathematical grunt work, and simulate natural phenomena that are difficult or impossible to study in real life.

FIGURE 1.19 • Virginia Tech's systemX

Supercomputers like systemX carry out calculations and simulations that would otherwise be beyond human capabilities.

People rely on computers to find solutions to problems that are beyond the abilities of the human brain, and yet the human brain functions in a manner that computers find impossible to fully duplicate. If the speed of the human brain were measured and represented as a processor speed, it has been estimated that it can carry out 100 trillion operations (thoughts) per second. The world's fastest supercomputer, IBM's Blue Gene/L, has a theoretical peak performance of 360 trillion operations per second. Still, no one has suggested that this machine can think like a human.

Artificial intelligence (**AI**) is an area of computer science that deals with simulating human thought and behavior in computers. Most AI experts are not interested in creating computers that think like humans. People already know how to make more humans; who would want to make computers into humans? It makes more sense to use those things that computers do best to complement human intelligence, to extend natural human abilities, and take over and automate activities that people find tedious, dangerous, and difficult.

Automating

Automation involves utilizing computers to control otherwise human actions and activities. For example, an area of AI called *computer vision* uses video cameras as eyes for a computer system that can tirelessly "watch" objects or areas and accurately interpret what it is "seeing." Vision systems are used in security applications and quality control. *Expert systems*, another branch of AI, automates tasks that are carried out by human experts—tasks that can be well defined and are typically tedious, monotonous, or hazardous to the human expert. *Natural language processing* is a branch of AI that empowers computers with the ability to understand spoken words and provides more convenient ways for people to interact with computers. Throughout this textbook you will find many examples of ways in which the computational strength of computers is assisting in human progress and development.

Another area in which computers are helpful is through the automation of physical tasks. *Robotics* is a branch of AI that empowers computers to control mechanical devices to perform tasks that require a high degree of precision or are otherwise tedious, monotonous, or hazardous for humans. Robotics is mostly known for its use in the manufacturing process, such as robotic arms that perform welding on an assembly line. However, when more closely examined it is clear that robotics is an increasing presence in people's everyday lives. A visit to a public restroom reveals automation that flushes toilets, turns on faucets, and dispenses paper towels. Robotic automation makes it possible for airplanes to fly and rockets to navigate in outer space. In New York, fully automated trains are being used on a 24-station line connecting Manhattan and Brooklyn. A lone train operator sits in the front car watching the controls.[7]

At Sonoma State University, the Jean and Charles Schulz Information and Technology Center uses an automated retrieval system (ARS) to retrieve books from the library shelves. Students key in the code of the book they want retrieved and in under a minute the book is delivered.[8] Increasingly, such systems are being used to store and retrieve warehouse inventory (see Figure 1.20).

FIGURE 1.20 • Automated retrieval system

Automated retrieval systems are able to store and retrieve inventory with the press of a button.

Automation allows people to exceed their natural abilities and empowers those with disabilities. New *smart homes* allow residents to open and close curtains, turn on sprinkler systems, control media throughout the house, and adjust environmental controls from any Internet-connected computer or wall-mounted display (see Figure 1.21).

Enhancing Communication

Computer systems control, support, or provide many forms of communication. Obvious examples are computer-based forms of communication such as e-mail, instant messaging, and Web logs, more commonly known as *blogs*. Even the Web itself is a form of communication. What may be less apparent is that computers and computer networks are increasingly supporting phone-based voice communications as well.

FIGURE 1.21 • Smart home controls

Home lighting is controlled from a central Smart Home control panel, using wireless connections to light switches.

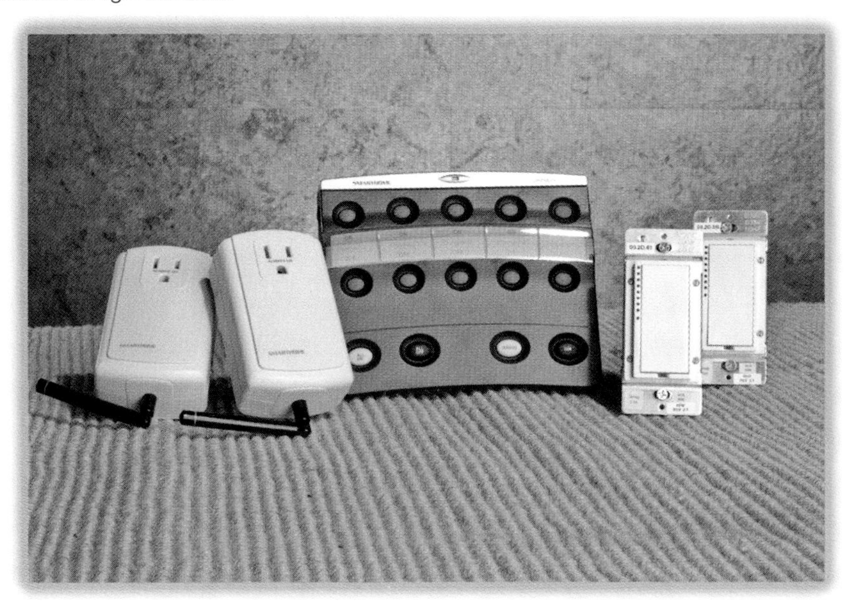

Most phone networks are phasing out traditional phone technologies in favor of digital computer technologies. In 2003 Sprint began moving its entire phone network to the digital technologies used by the Internet.

Businesses and individuals are also exploring using digital phone services to save money and enjoy better service. **Voice over Internet Protocol**, more commonly known as **VoIP** (pronounced *voip*) is a popular technology that allows phone conversations to travel over the Internet or other data networks.

Residential VoIP services, referred to as *broadband phones*, are becoming popular with home users who are interested in low phone bills and many additional features. Residential VoIP subscribers must have a high-speed Internet connection in their home. They connect their standard phone into a device that digitizes the signal to travel over the Internet. Residential VoIP customers can accept calls made to their phone number and make calls from any Internet-connected location to any phone number.

Phone networks and data networks are steadily merging into a single entity that utilizes the efficient technology of the Internet. As communication networks and services become fully integrated, increasing numbers of people will enjoy the benefits offered to VoIP customers as a standard service from their communications provider.

Radio and television are transforming into computer-driven digital media as well. Digital cable and digital radio services offer audiences an increased variety of programming with crystal clear connections. Print media is also going digital. More people read the *New York Times* online than the printed version of the paper. Society is increasingly experiencing the world through computers. People are interacting with each other increasingly through digital connections. Communication is definitely one of today's most important uses of computers and digital technologies.

Providing Entertainment

Digital media refers to music, video, photographs, graphic art, animation, and 3D graphics stored and processed in a digital format. Digital technologies have vastly expanded the creative toolkits of artists of every genre. Music is recorded in a manner that is truer to the quality of the original performance. Three-dimensional animation provides more fuel for the imagination than traditional 2D cartoons. Games that feature 3D graphics provide players with a more lifelike experience as they maneuver within virtual worlds.

Computer-generated special effects make it difficult to tell what is real and what is fabricated in today's motion pictures. The fight scene in *The Matrix Reloaded* where Keanu Reeves fights dozens of agents drove the best minds in the special effects industry to invent previously unheard of techniques. Special effects guru John Gaeta input gigabytes of images of the actor into the computer system and was able to create a 3D computer-generated Keanu Reeves that looked identical to the real thing. It wasn't the actor, or even a stunt double performing those miraculous feats—it was a virtual Keanu. This new form of manipulating virtual movie stars is one form of a technology called *image-based rendering.*

Digital technologies not only enrich people's entertainment experiences but also have revolutionized entertainment distribution (see Figure 1.22). The Internet has opened new marketing and delivery mechanisms that make media easier than ever to acquire. Portable digital media devices have made it possible to enjoy music, photos, and entertainment wherever you go. Digital technologies

FIGURE 1.22 • MSN video download service

Microsoft's video download service allows subscribers to download video clips to play in Windows Media Player on a PC or portable media device.

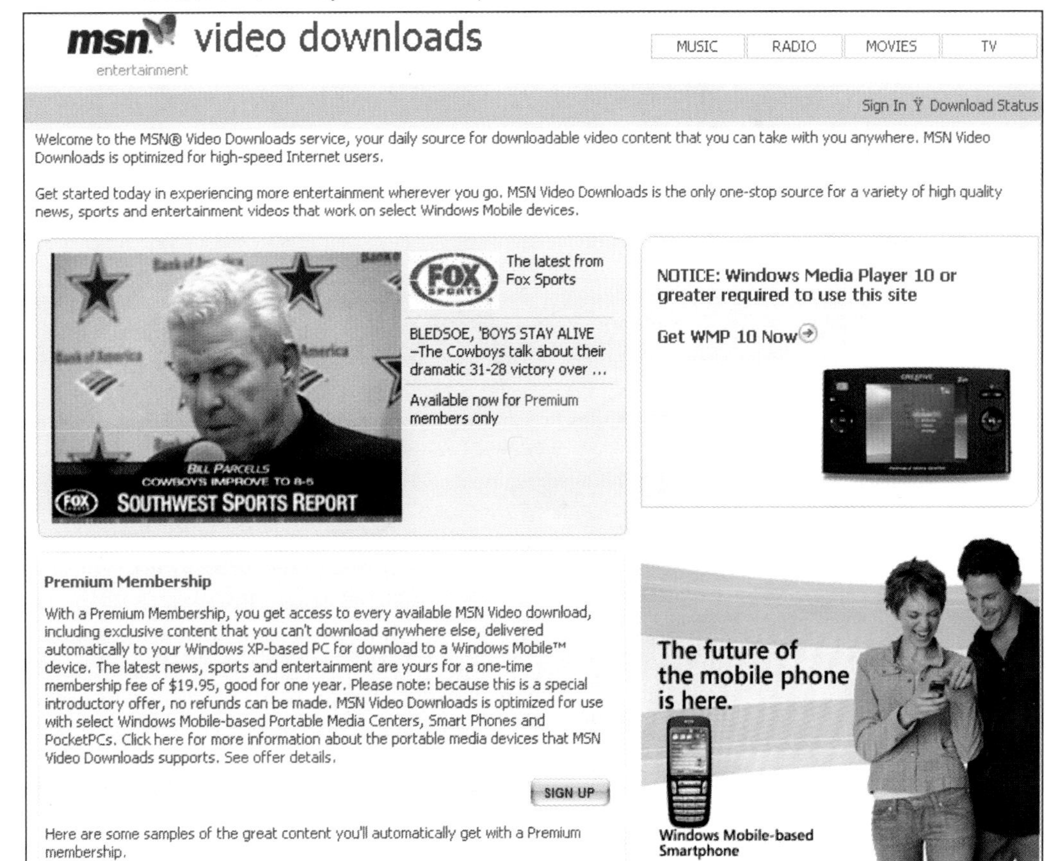

have improved the quality and increased the quantity of creative works in this connected society. The improved circulation of media products can only be considered a positive influence on society from both provider and consumer points of view. Suppliers and distributors of media are working to develop new business models that allow for the free flow of media content while financially supporting those that create and manage the media.

Managing Information

Perhaps the greatest impact of computers on society is in information storage, management, retrieval, and distribution. As documents, expressive and creative works, and other forms of information have been digitized, the resulting amount of stored information has become overwhelming.

Information overload is the common term used to describe a state in which the amount of information available overpowers one's ability to manage and use it. An information system, or more specifically a **computer-based information system** (**CBIS**), makes use of computer hardware and software, databases, telecommunications, people, and procedures to manage and distribute digital information (see Figure 1.23). A **database** is a collection of data stored on a computer, organized to meet users' needs. Databases are a key component to managing information, and information systems are the primary defense against information overload.

FIGURE 1.23 • Computer-based information systems

Many interrelated components make up a CBIS.

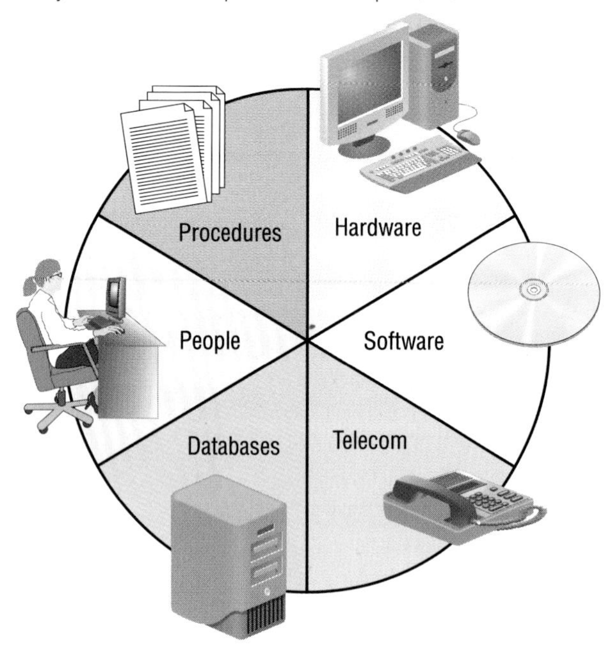

Individuals use information systems to manage their contacts, e-mail, and online documents. Businesses use information systems to analyze business data to determine courses of action that might provide greater profits. The Web might be considered the world's largest information system, providing a convenient framework for publishing and finding information on the Internet. Consider how Yahoo.com makes use of the six components of a computer-based information system. Yahoo.com uses:

- Hardware: Web servers and user PCs
- Software: Web browser and server software
- Databases: To store links to Web pages that are searchable either through keywords or by clicking through menus of topics
- Telecommunications: The Internet
- People: Who manage and use the system
- Procedures: Web protocols manage and deliver information to users in an efficient manner

Information management plays such a crucial role in the interactions with computers that the next section of this chapter looks at it in more detail.

Consider how *you* use computers. You will discover that all of your uses of computers fall under one of the above categories: computation, automation, communications, digital media and entertainment, or information management. To use computers to maximize your success in your chosen career and personal endeavors you must consider how to maximize the way that computers are used in these five areas and work to ensure that you are taking advantage of the latest and greatest technologies. This book is created to do exactly that.

INFORMATION SYSTEMS

The term *information system* is a large blanket that covers a wide variety of computer applications. It would be helpful to pause and consider some of the common terminology associated with information technology. Earlier this chapter defined a digital technology device and a computer. The expression *computer system* is typically used to describe multiple computers working together over a network towards a common goal. An information system was formally defined in the previous section and can be considered more simply as a computer system designed to manage information. The expression **information technology (IT)** can be defined as issues related to the components of an information system. Table 1.6 is provided to assist you in understanding the subtle differences between these commonly used expressions.

TABLE 1.6 • Important distinctions between technical terminologies

Term	Definition
Digital electronics device	Any device that stores and processes bits electronically
Computer	A digital electronics device that utilizes hardware to accept the input of data, and software, or a computer program, to process and store the data and produce some useful output
Computer system	Multiple computers working together over a network towards a common goal
Information system	A computer system that makes use of hardware, software, databases, telecommunications, people, and procedures to manage and distribute digital information
Information technology (IT)	Issues related to the components of an information system

If you think of it, information management is the most common use of personal computers. A word processor assists people in managing the information in a document. E-mail systems, to-do lists, address books, and the Web are all information management systems. Music, movies, and photographs can be considered information, so any applications used to catalogue and play these media files can be considered information management software. When it comes down to personal computer use, a computer's information management abilities are typically used much more than its computational abilities.

The use of information systems escalates vastly in businesses and the work place. No matter what the apparent purpose of any particular business or career, you can be sure that information management plays a large role in its daily activities. Information systems are used to store and manage patient records. Many doctors rely on information systems running on handheld PDAs to assist them in finding appropriate medication for treating their patients. In the real estate industry, huge databases manage information on available properties. Scientists share research data stored and managed by information systems accessible over the Internet. Photographers make use of information systems to store and catalog digital photographs. Whatever profession you name, you can find an information system that assists those professionals in managing their information and in being more efficient and effective in their work (see Figure 1.24).

Although businesses make heavy use of common software packages such as word processing, spreadsheet, and database software, it is more common to find custom-designed software managing most of a company's information. Software that is prepackaged and sold to the public is called *off-the-shelf software*. Software that is created to address the needs of an individual entity is called *custom-designed software*. Because most businesses have unique information management needs they must use some form of custom-designed software. Information systems have

FIGURE 1.24 • Information systems in action

Most professionals depend on information systems to organize information and provide them with valuable insight.

become a major weapon in competition between rival businesses. Today's businesses attempt to develop optimal information processing in order to streamline their work and gain an advantage over their competitors. Businesses compete not only by the quality of their products but also with the quality of their information management systems.

The information systems that people use everyday at work and at home are not haphazardly designed and created. **Systems development** is the activity of creating new or modifying existing systems. *Systems analysts* are information professionals responsible for designing information systems. There are specific and formal types of information systems with which all students should be familiar prior to beginning a professional life.

Types of Information Systems

There are many formally defined information systems that people and businesses use to manage information. The most basic simply records useful data, and the most advanced serve as intelligent advisors. This section introduces you to the four most common types of information system from the most basic to the most intelligent.

Transaction Processing Systems. A *transaction* represents an exchange, such as buying medical supplies at a hospital, downloading music on the Internet, or paying employees. A **transaction processing system (TPS)** is an information system used to support and record transactions. Almost all organizations process transactions to some extent. Consider all the transactions a typical college or university processes. Paying faculty and staff, processing student grades, paying for phone service and electric utilities, and keeping track of gifts and donations from alumni and others are just a few transactions that colleges and universities must process. Travel agencies, a small beauty shop, a hockey team, charitable groups, career counselors, museums, and local nightclubs all have to process transactions.

When transactions are carried out online, over the Internet or using other telecommunications and network systems it is called **e-commerce** (see Figure 1.25). The term *e-commerce* (short for electronic commerce) was coined in the 1990s when businesses began moving in droves to the Web. E-commerce has transformed the marketplace and forever changed the manner in which businesses operate and people shop.

Management Information Systems. A **management information system (MIS)** is an information system used to provide useful information to decision makers, usually in the form of a report. You may use an MIS on your school Web site when selecting classes for a new semester. The MIS searches a database and provides a list of available classes with open seats. The list of classes provided is considered the MIS report. It provides you with useful information on which to base your decision.

The focus of an MIS is on operational efficiency. In a business setting, management information systems typically provide preplanned reports generated with data and information from the transaction processing system for marketing, production, finance, and other functional areas of business. MIS is used in nonbusiness environments as well. Scientists use MISs to examine data generated

FIGURE 1.25 • E-commerce

E-commerce supports and manages transactions electronically over a network.

in experiments (see Figure 1.26). MISs are the primary tool for making sense out of mountains of data. They are a necessity in all aspects of today's society.

Decision Support Systems. A **decision support system (DSS)** is an information system used to support problem-specific decision making. A DSS is like an MIS that is programmed to tackle specific complicated problems. It does more than just search and sort through database records. It applies appropriate formulas and logic to the input to provide informed solutions. The focus of a DSS is on effective decision making. Whereas an MIS helps an organization "do things right," a DSS helps a person "do the right thing." The overall emphasis is to support rather than replace decision making.

Online dating services use decision support systems to match up partners. The DSS uses personality profile data supplied by the customers as input. The DSS software applies scientifically proven mathematical formulas to the data to determine which customers have compatible attributes that are important to relationships. The result is a list of possible matches from which the customer can select.

FIGURE 1.26 • NASA and MIS

NASA scientists use management information systems to examine the data sent from Mars Rovers.

There is one type of DSS designed to support group decision making. A *group decision support system (GDSS)*, sometimes called a computerized collaborative work system, provides tools for groups to reach consensus and design the most effective solutions with minimum time and effort (see Figure 1.27).

FIGURE 1.27 • Decision support systems

Decision support systems assist individuals and groups to come up with logically sound solutions.

Expert Systems. The most advanced information systems come from a branch of artificial intelligence. An **expert system (ES)** is an information system that can make suggestions and reach conclusions in one particular area of expertise much the same way that a human expert can. Expert systems go beyond decision support systems by applying AI techniques that allow the software to draw conclusions from incomplete data. Expert systems often work with probabilities. An expert system may inform you that there is an 80 percent probability that you suffer from attention deficit disorder and suggest that you see a doctor.

These computer systems are like human specialists with many years of experience in a field. In fact, they are developed in part through extensive interviewing and observation of such experts. For example, a college advisor might be intensely interviewed in order to design an online advising service for students. The expert system would be provided with the same input a college advisor receives—a student's chosen major and minor degree programs, the classes the student has successfully completed, the classes required for the degree program, the classes available in the upcoming semester, and the prerequisite requirements of each class. The system may prompt the student for additional input regarding student's preferences for elective classes. The system would then produce an ideal schedule of classes in which the student may enroll. The student can have the schedule reviewed and approved by a human advisor.

USING DIGITAL TECHNOLOGIES TO SUCCEED IN YOUR CAREER

As you will see throughout this book, people and organizations use digital technologies to succeed in the workplace and in life. As you read about the use of technology in various fields, think of how you can use these examples to help you achieve your own personal goals and career aspirations. You will find that an example of how computers are used in one field can often be directly applied in other fields.

Of course, it would be impossible to cover every career or industry. This section gives you a sampling of what is possible, beginning with the obvious: computer-based professions.

Computer-Based Professions

People who work in a computer-based profession are involved in designing and building hardware, software, database systems, telecommunications, and Internet systems. *Software engineers*, for example, design and develop new programming applications. *Computer scientists* use computers to help with the software design process. They are also responsible for conducting state-of-the-art research into computing topics such as artificial intelligence, robotics, and electrical circuits that could profoundly change people's lives in the future (Figure 1.28).

Inside the computer industry, computer professionals work in areas such as design, manufacturing, sales, and services, often with a specific major product line. From a wide array of products, computer-systems professionals may choose to specialize in equipment designed for use within certain areas such as networks and telecommunications, multimedia systems, expert systems, or imaging technology. Others may choose to work in certain industry segments such as education, manufacturing, business, laboratories, engineering, and many other areas.

FIGURE 1.28 • Computer science

Computer scientists conduct research to improve all aspects of computer systems.

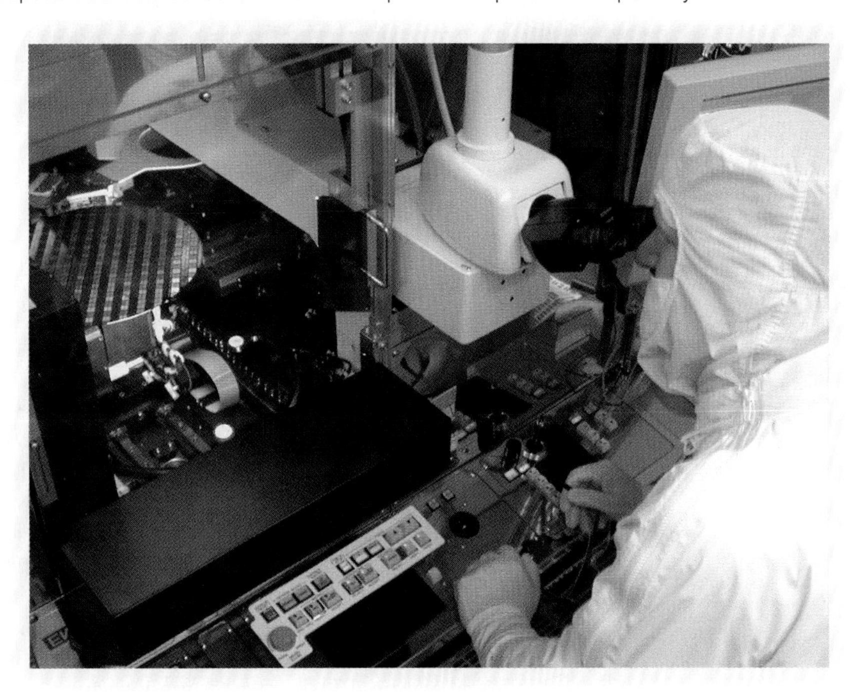

In businesses, computer personnel typically work in a computer department that employs a chief information officer (CIO), computer programmers, systems analysts, computer operators, and a number of other computer personnel. Computer personnel have a variety of job titles and responsibilities. The chief technology officer (CTO), for example, typically works under a CIO and specializes in hardware and related equipment and technology. In addition to technical skills, computer personnel also need skills in written and verbal communication, an understanding of organizations and how they operate, and the ability to work with people (users).

Business and Communications

A variety of career opportunities exist in business and communications, marketing, sales, accounting, finance, organizational behavior, human resources, leisure and hospitality, and information systems. Computers are essential in producing business documents and reports, including payroll checks, inventory reports, tax documents, and many others.

Most businesses succeed in using technology to increase revenues or reduce costs. Here are just a few examples. In factories, computers are being used to design and manufacture products, using *computer-assisted design (CAD)* and *computer-assisted manufacturing (CAM)*. Increasingly, companies are involving customers in the design and engineering process.[9] Toyota, the Japanese automaker, is launching a manufacturing system to allow the company to build almost any type of vehicle in any of its plants.[10] The new CAM system is faster and less expensive. According to one company spokesperson, "The new line represents 50 percent less investment than the one it replaces and lets us add a different car type to the line at 70 percent lower costs than before."

FIGURE 1.29 • Computers in business

Computers are an essential element in most businesses.

In banking, computers are used to instantly move billions of dollars from one institution to another, using *electronic funds transfer (EFT)*. Computers are used in all aspect of retail. Products usually have *Universal Product Codes (UPCs)* placed on them that can be read by scanners. Computer systems are also used to check for customer credit, handle returns, and provide a vast array of reports to managers. Marketing and advertising companies use computers to collect and analyze vast stores of consumer data. The information is used to provide high-quality products and services to customers. Insurance companies use technology to process the majority of their insurance applications and claims electronically. The National Insurance Crime Bureau, a nonprofit organization supported by roughly 1000 property and casualty insurance companies, uses computers to join forces with special investigation units and law enforcement agencies as well as to conduct online fraud-fighting training to investigate and prevent these types of crimes.

Computers allow people to work virtually anywhere at any time. A manager, for example, can spend about 10 percent of her time at her office in Michigan. The rest of her time is spent on the move in airports, hotels, and cars visiting customers and other sales representatives. Work can also be done at vacation homes or on a boat in the Bahamas by using mobile computing systems and connections to the Internet.

Science and Mathematics

Chemistry, biology, mathematics, statistics, astronomy, physics, meteorology, environmental sciences, oceanography, sports science, and military science are just a few fields in science where computers are used. Because older computers didn't have sufficient power or speed, data collected decades ago from the Apollo missions to the moon are now just being analyzed by today's computers.[11] The new data is revealing stunning information about the surface of the moon and ancient moon quakes. Scientists are using computers to analyze string theory equations, which attempts to explain both large-scale physics and the physics of subatomic particles, in which a particle's location in space is not certain at any point in time, but is based on mathematical probability instead. Scientists are also using computers to forecast weather, hunt for hurricanes, analyze the environmental impact of forest fires, and make detailed maps using *geographic information systems (GISs)*.

Advanced machines and computers have enabled medical research scientists to use protein sequencers and synthesizers to map the entire human genome. Scientists have used computers to advance their knowledge so rapidly that people have a hard time keeping up (Figure 1.30). New technology moves from theory to truth in every area everyday.

FIGURE 1.30 • Computers in science

This computer-generated view of a nanotube allows researchers to develop the tiniest of machines that may ultimately change the world.

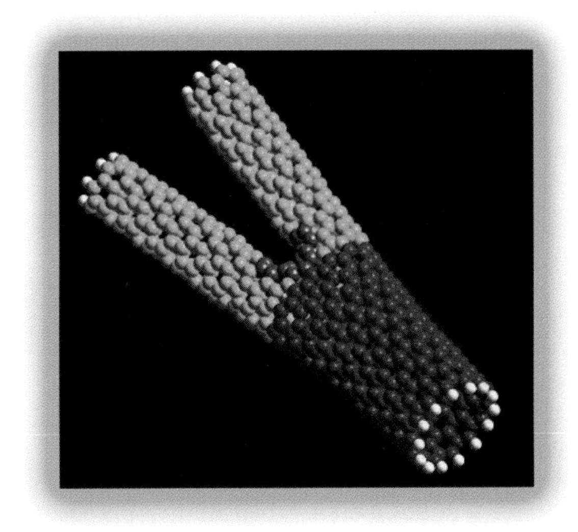

Engineering

Engineering careers include aeronautical, mechanical, electrical, chemical, civil, industrial, and environmental engineering. Engineers use computers in design and operations (see Figure 1.31). In chemical engineering, for example, engineers use computers to design petroleum refinery operations to produce a variety of gasoline and diesel fuels using minimal energy requirements. Computers are also used to monitor and control petroleum refinery operations to make sure that sophisticated refinery towers and systems are efficiently and safely operating. Some universities use computers to design new aircraft and space vehicles. Computers are also used to make complex thermodynamics, power consumption, circuits and signal, and reactor design calculations, to name a few.

Social Sciences

The social sciences include economics, geography, psychology, political science, sociology, and urban planning. In economics, computers are used to determine leading economic indicators, the Consumer Price Index used to monitor inflation, and a number of other indicators to monitor and control the economy. In geography, computers are used to map geographic areas. In political science, computers are used in a wide variety of areas. The results of political surveys and

FIGURE 1.31 • Computers in engineering

The design of bridges and highways is enhanced through the use of computers.

polls are reported in the popular news almost daily. Computers are used in public elections. Aniers, which is a suburb of Geneva, was the first city in Switzerland to use the Internet in a public, binding vote.

Fine Arts

The use of computers in the arts surprises many people. Professionals in fine arts that might appear to be beyond the use of computers are making use of computers as much as other professionals (see Figure 1.32). Film studies, visual arts, theater, literature, dance, photography, and music all benefit from computers.

Computers have been widely used for producing music using *Musical Instrument Digital Interface (MIDI)*, special effects in films such as *The Incredibles* and *Shark Tales*, digital photography, and the ever-popular video games. Computers make it possible to film a scene, change it within the computer, and scan the altered images back onto a file to create an image undetectable by the human eye. In dance, computers are used to capture motion and choreograph complex dance movements. In graphic arts, sophisticated software such as Adobe Creative Suite can be used to design, develop, print, and place beautiful advertising, brochures, posters, prints, and videos on the Internet. Artists are now developing digital art that can be viewed on TV screens, HDTV sets, and computer monitors when these devices are not used for their normal purposes.[12] This type of digital art is often purchased on DVDs or over the Internet.

Sports, Nutrition, and Exercise

Computers have been used in all aspects of sports, nutrition, and exercise (see Figure 1.33). National Football League teams, for example, use computers to provide instant feedback. Computers are programmed to diagram and analyze offensive and defensive plays of teams and their opponents. The computer analyzes an opposing team's play from the past few games and predicts what the opponent will do in specific situations. The computer has also been used to design football equipment that reduces the chance of permanent paralysis or brain

FIGURE 1.32 • Computers in fine arts

Software such as Adobe Creative Suite empowers graphic artists to create effective images efficiently.

FIGURE 1.33 • Computers in sports

Subtle details detected by the computer aid coaches, medical experts, and sports equipment companies in achieving their unique goals.

damage to players by analyzing films during which an injury takes place and producing graphs showing the force in pounds absorbed at points of the player's body at certain moments in time. Yet another use of the computer in sports is biomechanics, a familiar concept in Russian and East German athletics since the early 1960s. It has been used in the United States as well to coach its Olympic athletes. A computer programmed to watch athletes draws attention to factors too subtle to be detected by the human coach and shows athletes how to improve their techniques. Diet and nutrition for athletes and others can be analyzed by software to reduce weight and improve performance.

Government and Law

Since the 1950s computers have been used in compiling the U.S. Census. Massive databases are part of most of the government's operations, such as the Internal Revenue Service. Other governmental agencies, including the United States Postal Service (USPS), the Census Bureau, the Department of Homeland Security, and a variety of state and local agencies would have difficulty operating without computers (Figure 1.34). Pentagon and military activities including security and nuclear defense depend on computers. Computers are also helping the military transform from "command and control" to "sense and respond" to meet contemporary military challenges.[13] The space program would still be in the realm of science fiction without computers.

Computer technology is improving legal practice in several major areas. In one area, computers provide information on a firm's attorneys and staff, clients,

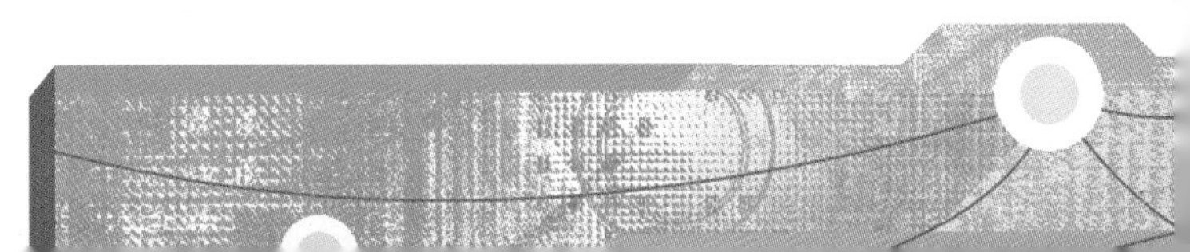

FIGURE 1.34 • Computers in government

A census-taker for the U.S. Census Bureau uses a hand-held computer to collect data for the 2006 Census Test.

billing, customer relationship management, and news items. iManage developed WorkSite to provide document management as well as collaboration with others using a private network and the Internet. WorkSite presents relevant information on each client such as word processing documents, e-mail, billing information, and images in a single view accessible to everyone on the team. Millions of publications and records of interest to legal professionals and researchers are becoming increasingly digitized and available over the Internet through services such as Lexis-Nexis. Computers are also used in the courtroom for evidence presentation and courtroom communications. Support services such as document imaging and management, graphics and animation for litigation, and electronic evidence management are all possible using the computer.

Medicine and Health Care

Careers in medicine and health care include nursing, nutrition, exercise physiology, social work, psychology, family therapy, and medicine. Computers produce medical records and reports to hospital administrators, insurance companies, and government agencies. Electronic health records (EHRs) can improve health care and reduce health care costs by billions of dollars annually.[14] President Bush appointed a national coordinator for health information technology and by an executive order called for broad implementation of EHRs by 2014.

Medical informatics is an area of medicine involved with storing medical records in a digital format on a national database. Computerized physician order entry (CPOE) enables physicians to place orders on the computer and notify nurses, laboratory technicians, pharmacists, clerks in accounting, and anyone else involved immediately when the order is entered (Figure 1.35). When integrated with patient monitors and electronic messaging, physicians can receive and send messages and orders via e-mail.

Some medical computers help doctors diagnose diseases and prescribe treatment. Specialized medical expert systems, software that is programmed to act like a team of expert doctors, can convert a patient's symptoms and problems into likely diseases, and include an estimate of the chances that the patient has each disease. Some medical expert systems also have the ability to suggest treatment options. In some cases, expert systems software explores possible diseases and treatments that might have been overlooked by even the best doctor.

Magnetic resonance imaging (MRI) machines are used in most hospitals today. Surgeons use the three-dimensional images produced by computer graphics to help them operate. They also use software to locate and remove some of the causes of seizures. Tests that detect cancer or measure chemical levels are performed using microchip-loaded probes that are threaded into the body via a catheter. Computer-aided surgery can be used to perform precise surgical operations from remote locations. Electronic sensors detect chemical changes in the lens of the eye that precede cataracts; this new treatment enables the use of drug intervention that may eliminate surgery. Specialists use video linkups to communicate treatment to emergency medical technicians. Dentists use intraoral video cameras to record and project images that are then logged as digital files into a database.

FIGURE 1.35 • Computers in medicine

Computerized physician order entry makes communication among medical professionals much faster and more accurate.

Criminology, Law Enforcement, and Security

Computers are used extensively in criminology, law enforcement, and security. Computers provide crime fighters with invaluable information on criminals, stolen vehicles, and missing persons. The Missing Child Act authorized the creation of a database to help local and state law enforcement authorities locate and identify the thousands of children reported missing each year. The creation of a database of unidentified dead bodies has helped eliminate the uselessness of a family spending its life savings to search for a missing child whose body had already turned up in another state.

Instead of a polygraph, computers can be used to "read" a voice, to detect the stress produced by lying, and to produce a voiceprint, which is as unique as fingerprints. Computers can be used to capture facial thermograms (systems of blood vessels), which, like fingerprints, are distinct and unique in each individual, and can be read using an infrared camera, a computer, and a database. Responding to 911 calls, helping to find missing children, locating missing motor vehicles, and analyzing crime scene information are just a few additional applications of computers.

Security and law enforcement agencies can search through a company's computer files, documents, and e-mails to match potential criminals to crimes and produce proof for convictions in such cases as sexual harassment, pornography, computer fraud, and industrial espionage. Crime investigators, for example, can use computers to track down crime syndicates who steal more than one-third of the software used in Western Europe. A detective can use software to analyze high-volume crimes and identify burglars most likely to have committed them with a high degree of accuracy. With the increase in computer attacks, there are new and exciting careers in security and fraud detection and prevention. The University of Denver, for example, offers a master's program in cyber security.[15]

Information Systems Save Lives

The homicide rate for Chicago was highest in the nation in 2003. The Chicago Police Department turned to a computer-based information system to provide new methods for targeting crime. They developed a geographic information system (GIS) that presented crime information in geographic context, on a map of the city, in hopes of preventing murders and instances of aggravated battery with firearms. The primary use of the information provided by the system was to assist officers in making decisions about which areas of the city required additional police power.

Within the first six months of using the new GIS there was an 18 percent drop in murders compared to the same time period a year earlier. Evidently applying additional officers to the worst areas of the city was having an effect. At the close of 2004, Chicago's homicide rate was down 26 percent from the 2003 rate. Much of their success they contribute to their "cops on the dots" strategy, which floods high-crime areas identified by the GIS with previously desk-bound officers.

The amount of statistical crime-related data that the department collected was too massive for human interpretation. A traditional information system would make it possible to call up detailed information for a given locale, but still made it difficult to get the "big picture" of crime in the city. The geographic information system provided the data in a more human-friendly manner that provided the key to solving the problem. The human mind is exceptional at taking in and interpreting visual information from images and maps. Computers are exceptional at sorting through massive amounts of data to find useful information. The Chicago Police Department's GIS makes use of the best qualities of both humans and computers.

Joe Kezon, the manager of the GIS project, summed up the project in these words: "The commander of the Deployment Operations Center saw the importance of having the ability to do some mapping and analysis that would allow them to make key judgments of where they should create police deployment areas...Our desire is to ensure zero tolerance in crime areas, and the technology enables us to determine where those areas are."

Questions:

1. Why did a GIS provide the perfect solution to the problem in Chicago?
2. What other industry might profit from the use of a GIS and why?
3. Why is it important for professionals in law enforcement to keep up with current technologies?

Sources

1. *Chen, A. "GIS Fights Crime in Chicago,"* eWeek Online, *May 31, 2004, http://www.eweek.com.*
2. *Associated Press. "Chicago Topped Nation in Homicides in 2003,"* USAToday Online, *January 1, 2004, http://www.usatoday.com.*
3. *Brett, M. "Flooding the Block,"* The Chicago Reporter, *January 2005, http://www.chicagoreporter.com/2005/1-2005/Departments/current.htm.*

Like many other universities, the University of Denver is hoping to get accreditation from the National Security Agency for the new degree program that specializes in security.

Education and Training

Computers are used in most aspects of education and training. Computer-aided instruction (CAI) can deliver course content and measure student performance. Distance learning is used to deliver courses and instruction to and from remote locations. Instructors and students can be located around the world. The Internet is increasingly being used for training and assessment. People with vision disabilities can use special hardware and computer monitors. Many software packages have built-in help facilities and educational tools.

USING DIGITAL TECHNOLOGIES TO ACHIEVE PERSONAL GOALS

A successful career can be rewarding in many ways. It makes you feel valuable and gives a boost to your self esteem, improving your general view of the world. If you are fortunate, it may also give a boost to your bank account. Most people would agree that life is also full of many successes that may have nothing to do with one's career or monetary rewards. Many of the best things in life occur outside the workplace. Computers and digital technologies are playing an increasing role in assisting people in achieving personal goals and leading more rewarding, productive, and fulfilling lives outside of work.

Although businesses and organizations often use custom designed software to achieve their goals, individuals depend on off-the-shelf software to support their personal needs. This section will focus on software and technologies designed for home and personal use.

Personal Finance

Personal finance software such as Quicken (see Figure 1.36) can assist individuals with living within a budget, saving for the future, and investing. Web sites such as The Motley Fool (*www.fool.com*) provide practical advice and tutorials on topics such as getting out of debt and investing wisely. Spreadsheet software and online tools allow users to run what-if scenarios to determine how changes in their budget today will influence their wealth in the future.

FIGURE 1.36 • Personal financial management

Software such as Quicken can help individuals take control of their finances to more easily and rapidly meet their financial goals.

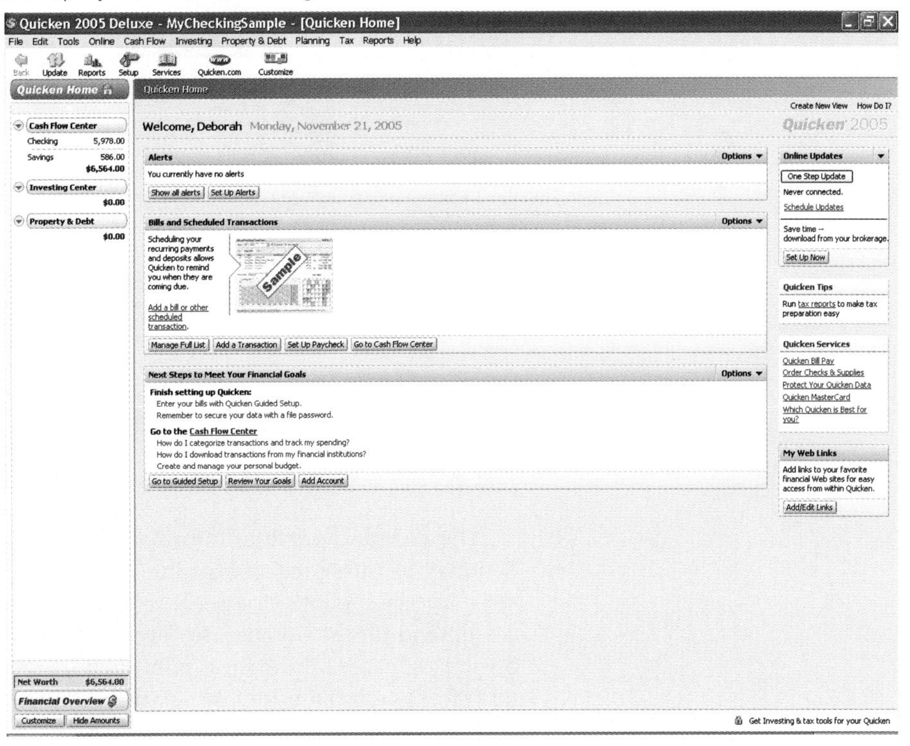

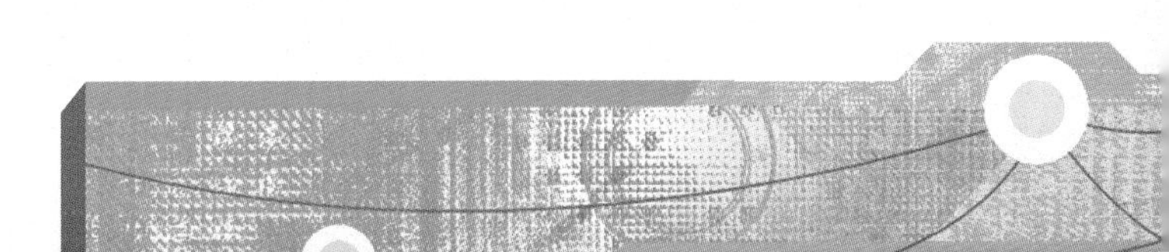

Personal Information Management

PIM software is the "Swiss army knife" of personal use software. It provides a personal calendar to keep track of where you need to be and when. You can set an alarm to remind you of an appointment. PIM also stores your personal to-do lists and address book. Some PIM software comes with personal journal software so that you can record brilliant ideas or your life story.

Personal Research

The Web is increasingly used as the first and primary information source for personal questions and needs. People are relying more heavily on the Internet to provide information during important life changes—pivotal points in their lives such as starting a new career, planning a marriage, buying a home, or having a baby (see Figure 1.37).[16] The Web helps to uncover options and learn strategies for choosing the correct path.

FIGURE 1.37 • Personal research

More than ever, people are using the Internet to help them research important events in their lives.

The Web also provides information from experts in areas of personal interest. You may be interested in learning more about Einstein or finding out what it takes to earn a pilot's license. It is hard to come up with a topic that won't draw a thousand hits at Google. However, it is important to develop the skills for discerning valuable information published by credible experts from rubbish and misinformation published by self-proclaimed experts.

Digital technologies have made personal research so much more convenient than it used to be. In so doing it has empowered people in many ways. In the pre-PC days, a person struck by an idea or question would need to go to the library to gain more information on the topic. Perhaps the person would have been fortunate enough to have a set of encyclopedias at home. In either case, most ideas and questions would probably evaporate by the time the person

gained access to more information on the topic. Today, when you are struck by an idea or question, you can simply look up information online.

Handheld and other computing devices are making it possible to research topics whenever and wherever a thought may strike. Consider the power of this spontaneous information access. For the first time you can progress from a germ of an idea, to research, to new discovery—wherever the idea strikes, with no time wasted. How many brilliant inventions, theorems, and discoveries may be hastened by humankind's ability to research information spontaneously? How many potentially brilliant discoveries have fallen by the wayside in the past due to the delay in the flow of information?

Personal Relations

Many people are turning to the Web to make new friends and look up old friends. The facebook (*www.facebook.com*) has become a popular social venue for college students wanting to meet others with similar interests. Online matchmaker eHarmony.com claims to have created matches that have resulted in thousands of marriages that are going strong (see Figure 1.38). Classmates.com maintains lists and contact information of millions of individuals based on the school they went to and the year they graduated, making it easy to get in touch with old friends. Genealogy.com can map your family tree if you provide your parents' names. Digital technology is making it easier than ever before to create personal connections.

FIGURE 1.38 • eHarmony.com

Online matchmakers like eHarmony.com pair up couples based on detailed personality profiles.

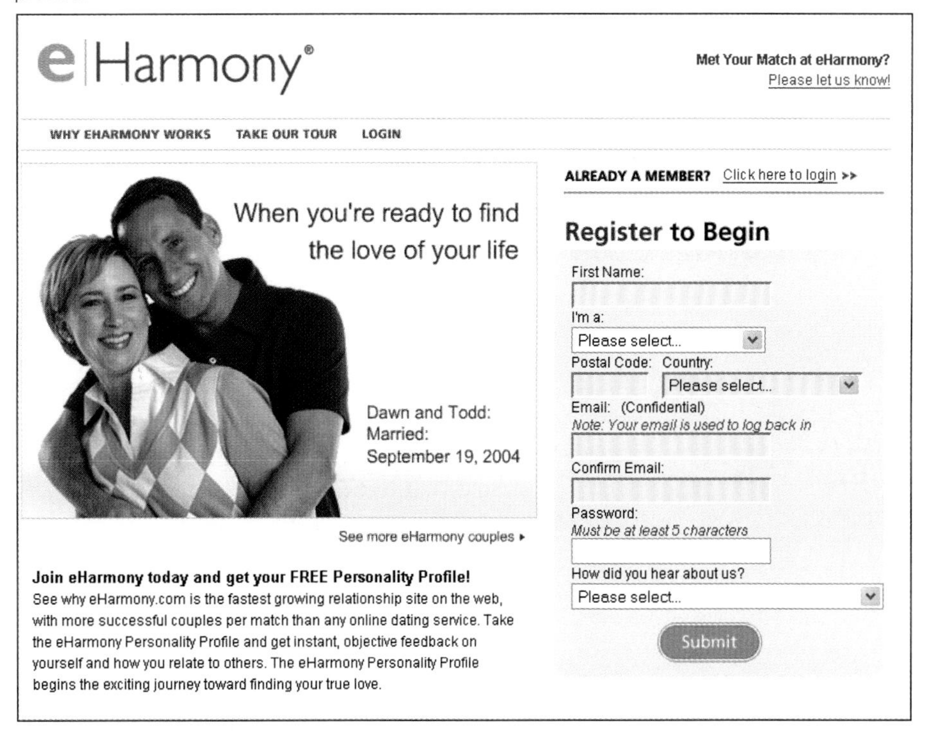

Personal Media Center

A new market in media management software is emerging that assists individuals in managing personal digital photos, music collections, and video. Some of this new software is designed to run on PCs that connect to entertainment systems to control digital television and stereo systems (see Figure 1.39).

FIGURE 1.39 • Personal media center

Windows XP Media Center edition is designed to connect to your entertainment center and control and organize your television viewing, as well as your music, video, and photograph collections.

There is a wide assortment of other personal software applications that include recipe management software, health and nutrition software, home inventory software, kids and parenting software, and assorted packages that support varying hobbies and interests.

INFORMATION SECURITY AND THE SOCIAL IMPACT AND IMPLICATIONS OF DIGITAL TECHNOLOGIES

Digital technologies help people to be more productive and successful. Nearly everyone agrees that technology has raised the standard of living and will continue to provide increasing benefits. But as it is with all major social changes, there are negative and even dangerous effects of technology on society. Society's increasing dependence on digital technology makes society more vulnerable should that technology fail. There are many ways in which digital lifestyles are vulnerable and open to abuse. This section provides an overview and some fundamental concepts on this topic.

Information Security

Information security refers to the protection of information systems and the information they manage against unauthorized access, use, manipulation, or destruction, and against the denial of service to authorized users. Information security is a growing concern as increasing amounts of important and private information is stored digitally on systems connected to public networks and wireless private networks. From bank account and credit account access codes, to personal medical records, to secret business strategies, to national defense initiatives, to e-mail sent to a friend, people trust information systems to keep their most valuable and secret information safe and secure. Unfortunately, there are active forces determined to steal and corrupt that information.

FIGURE 1.40 • Busted!

Kevin Mitnick spent five years behind bars for hacking into private corporate networks and stealing proprietary information.

The word *hacker* traditionally meant a particularly brilliant computer expert. Over time the meaning has changed considerably. Today, according to popular media usage, a **hacker** is an individual who subverts computer security without authorization. This term is used broadly in the news media to label individuals who use technology for terrorism, vandalism, credit card fraud, identity theft, intellectual property theft, and many other forms of crime (see Figure 1.40). It should be noted that many computer experts consider themselves to be hackers even though they do not break or attempt to break any laws. Some prefer to use the term *cracker*, for criminal hacker, and *attacker*, or *intruder*, to identify hackers that break the law.

Hackers are able to access supposedly secure networks and computers, often using Internet connections, by taking advantage of vulnerabilities in software. Hackers can also intercept information as it travels over a network or the Internet. Wireless networks are especially vulnerable to hackers as no physical connection is required to access the network.

Computers and computer systems are also vulnerable to viruses and spyware—software developed by hackers that can be inadvertently contracted from e-mail attachments, Web sites, or files downloaded over the Internet. Viruses and spyware can corrupt your computer files, causing your computer to malfunction. They can open "back door" access to your system to be used by hackers to exploit your system resources. They can access your private data and deliver it to others over the Internet. They can also turn your computer into a "zombie" computer that spreads viruses, spyware, and spam to others over the Internet without you suspecting a thing.

Hackers, viruses, and spyware can mean big trouble to businesses that depend on a healthy technology infrastructure. A company such as Amazon.com or eBay could lose millions if their servers are brought down for even a short time.

Keeping networks and the information they store secure takes effort on many levels. Important safeguards must be implemented by individuals, businesses, and governments in order to achieve a secure information infrastructure. Information security is so important that recently the United Nations has taken an interest in managing it. Information security is also a high priority for the U.S. Department of Homeland Security (DHS). The *National Strategy to Secure Cyberspace* was developed as a framework for protecting cyberspace (the Internet and associated networking infrastructures), which is recognized as essential to the U.S. economy, security, and way of life. The document states that its purpose is to "engage and empower Americans to secure the portions of cyberspace that they own, operate, control, or with which they interact. Securing cyberspace is a difficult strategic challenge that requires coordinated and focused effort from our entire society: the federal government, state and local governments, the private sector, and the American people."

LIFE-SAVING COMPUTATIONS

A steering wheel finger scanner developed by two UK firms fires sound pulses through a driver's finger. The amount of time they take to pass through measures the degree of bone density. The data is then passed to an onboard computer that configures the seatbelt to provide just the right amount of resistance to protect the driver without breaking bones or crushing the chest. Manufacturers expect to save 20 percent more lives with the technology.

Source
Finger Scanner Fine-Tunes Car Safety Settings
By Will Knight
NewScientist.com
http://www.newscientist.com/article.ns?id=dn7520
June 14, 2005

The Impact of Digital Technologies on the World

To *pervade* means to become diffused throughout every part of something. Society is entering the age of pervasive computing. *Pervasive computing* implies that computing and information technologies are diffused throughout every part of the environment. Rather than being obvious and "in your face," the pervasive computing environment is ubiquitous. *Ubiquitous computing* suggests that technology is becoming so much a part of the environment that people don't even notice it. Mark Weiser, who coined the expression *ubiquitous computing*, refers to this as the age of "calm technology," when technology "recedes into the background of our lives."[17] A pervasive and ubiquitous computing environment is one in which you can access digital information and media at any time and in any place with very little effort (see Figure 1.41).

FIGURE 1.41 • Pervasive computing

Increasingly, environments support access to information and media at any time and any place with little effort.

Although this is true in much of the developed world, there are societies and portions of societies that, due to poverty, geographical isolation, or lack of technical infrastructure, have been unable to enjoy the benefits of digital technologies. The **digital divide** is a title used for the social and economic gap between those who have access to computers and information technologies and those that do not. Access to information is empowering. It provides individuals, businesses, and cultures with a distinct advantage over those that do not have it. The digital divide is an issue between developed countries and third-world countries. It is also an issue between wealthy and poor populations within a country. These social divisions often run along lines of race, class, and culture.

Much work has been done in recent years to narrow the digital divide within the United States. Portions of the No Child Left Behind Act[18] and programs such as National Educational Technology Standards (NETS, *http://cnets.iste.org*) have brought Internet connections to 99 percent of public schools with an average student to computer ratio of 5:1. Numerous programs have been established to help narrow the divide between varying income classes, races, and international communities (see Figure 1.42).

Those who enjoy the benefits of technology are being confronted by many technology-related ethical issues. For example, the Internet empowers freedom of expression by providing a means to distribute and access digital information and media, such as art, photographs, music, and movies, worldwide with little effort. This has led to rampant violations of copyright laws that protect the intellectual properties of the creator and/or owners of the media and information. The Internet population is left confused as to what is ethically right and wrong when it comes to sharing intellectual property. In the meantime, the recording industry and motion picture industry are pushing to implement technologies that would

FIGURE 1.42 • The digital divide

Efforts are underway to widen access to computers and technology in the United States as well as in other less developed countries.

control what could be shared over the Internet, but which could also dramatically reduce freedom of expression, impede the development of culture, and stifle the free flow of information.

Ethical questions are also raised over what information should be restricted on the Internet. Should hate groups be allowed to promote their philosophies? Should scientific research that provides information that could pose a danger to national or international security be available online? How can parents protect children from viewing adult content? Who should decide what content is *adult*? These are only a few of the issues with which governments are grappling, sometimes reaching differing conclusions.

Technology provides convenience and valuable services, but these often come at a price of some degree of privacy. The more that an information system knows about you, the better it can provide customized services that meet your unique needs. In many cases you provide information about yourself in order to enjoy the benefits of the service. But much of the information collected and stored about you is done without your knowledge. This *invisible information gathering* makes some people nervous and concerned about their rights to privacy.

Most commercial Web sites track your movement within the pages of the Web site to determine which topics or products are of most interest to you. The next time you visit, those items may be the first thing you see. If you provide the company with information by filling out an online form, you may start receiving e-mail about the items that the company feels interest you most. Many businesses sell customer information to other interested parties.

Digital information is easy to collect, distribute, combine, and analyze. Some companies are in the business of aggregating information from numerous public and private databases to develop profiles of individuals that can be quite revealing. A company by the name of CheckPoint, for example, combines data from credit bureaus, local, state, and federal agencies, telephone records, private corporations, and other sources to build a detailed account of a person's life. CheckPoint has billions of records in its system that it sells to marketing companies, businesses, and government agencies.

Technology-influenced social change is occurring at a frantic pace. In many ways it is like riding a runaway rollercoaster. There is no turning back, and at times it feels as though individuals have little or no control over their technological future. But there is much at stake, and it is important that those involved have a say in what they find acceptable and unacceptable.

This chapter has introduced the basic concepts and terms that are necessary for you to understand in order to move forward in learning about computers, information systems, and other topics in digital electronics. This chapter has also provided an overview of the topics that are covered in the chapters that follow so that they will not seem brand new as you begin each chapter. Take some time to review the keywords for this chapter and work through the exercises that follow so that you have a basic understanding on which to build.

THE BETTER TO SEE YOU WITH, MY DEAR

Intel Research sees the "next wave" in computing will be systems that gather and manage information for individuals. On the burner now is a system of sensors for baby's mattress and a wall-mounted camera that send info to Mom's (or Pop's) PC. It monitors movement, heart rate, and temperature, and can be edited to a home video. Also in the works, a teddy bear with a hidden video camera to catch those candid shots grandparents love.

Source
These walls (and teddy bears) have eyes
By Michael Kanellos
Staff Writer, CNET News.com
http://news.com.com/These+walls+and+teddy+bears+have+eyes/2100-1040_3-5738029.html?tag=nefd.pop
June 9, 2005

ACTION PLAN

Remember Tonya Roberts from the beginning of this chapter? She was the student that was having difficulty deciding on a major. Here are answers to the questions asked about Tonya's situation.

1. **What types of computers and digital devices played a role in Tonya's search for the ideal career and job?**
   - Tonya's MP3 player provided portable storage so that she could transfer files from the career center's system to her own PC.
   - Tonya's PC provided her access to the career center files and Web resources for researching careers and job ads.
   - Tonya's cell phone provided a way for the career center to contact her when an interview slot was available.
   - The career center used PCs running special career center software for students to use.
   - The career center probably also used a server on which it could maintain a database of companies and applicants.
   - The career center used a Web server to provide additional career information to students over the Web.

2. **How did telecommunications, the Internet, and computer networks assist Tonya in her quest?**
   - It is likely that the career center maintained a network of computers from which students accessed career information from a central server.
   - The Internet provided Tonya with information from several Web resources.
   - A digital cell phone network provided a means for the career center to contact Tonya at any time.

3. **What types of computer-based information systems might have been used by Tonya and the career center in looking for a career and job?**
   - Depending on the complexity of the software, either a decision support system or expert system was used to analyze Tonya's personality profile and suggest careers.
   - A management information system is used to manage the database of employer and applicant information.

4. **Why are computers and digital technologies important to Tonya's career and personal life?**
   - Because most information is or is becoming digital, the more Tonya knows about the systems that manage digital information, the more she can take advantage of information to obtain knowledge and enjoy entertainment and services that provide a more successful and fulfilling life.

5. **What security and ethical issues could be involved in the use of databases at the career center?**
   - The information in the database contains personal student information. Individuals in the career center need to judge with whom they can safely share this information. Students should be made aware of who has access to the information. The career center must also take all possible actions to protect their network and database from hackers, viruses, and spyware.

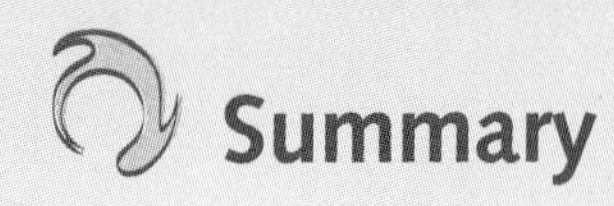

Summary

LEARNING OBJECTIVE 1

Define how digital electronics devices and computers are related, and provide descriptions of the different types of computers.

A digital electronics device is one that stores and processes data as bits, 1s and 0s. A computer is a digital electronics device that combines hardware and software to accept the input of data, process and store the data, and produce some useful information as output. There are general-purpose computers, such as PCs, servers, and supercomputers and there are special-purpose computers such as MP3 players, digital cameras, and scores of other digital devices custom designed to carry out a particular task. Personal computers include desktop, notebook, tablet, and handheld computers. Mobile digital devices, such as mobile computers, communications devices, and media devices are ideal for the increasingly mobile workforce and population.

Figure 1.2—p. 6

LEARNING OBJECTIVE 2

Provide an explanation of the fundamentals of telecommunications, computer networking, the Internet, and wireless networking.

Telecommunications are forms of communication that take place electronically over a distance. Forms of telecommunication include telephone systems, radio, television, and computer networks. A computer network is a telecommunications network that connects two or more computers for the purpose of sharing data, hardware, and software resources. The Internet is the world's largest public computer network; it is a network of networks that provides a vast array of services to individuals, businesses, and organizations around the world. The Web is an application that makes use of the Internet to deliver information and services through a convenient interface that utilizes hyperlinks. Wireless networking is very popular because it uses radio signals rather than cables to connect computers and digital devices to computer networks and the Internet.

Figure 1.17—p. 19

LEARNING OBJECTIVE 3

Describe the five functional areas in which computers assist people most.

The things that computers do well—working with large amounts of data and repetitive tasks—are things that people find difficult and monotonous, and the things that people find most engaging—creative endeavors—computers are ill suited to perform. Computers excel in five functional areas: computation, automation, communications, digital media and entertainment, and information management. Computers are good at computing, that is to say they calculate solutions to mathematical problems. Computers provide automation services such as those provided by robotics systems. Computers support most of today's electronic forms of communication. Much of today's entertainment is provided by computers in the form of digital music, movies, and interactive games. Finally, perhaps the most useful of computer's skills is managing the massive amounts of digital information in existence.

Figure 1.19—p. 21

LEARNING OBJECTIVE 4

Discuss the four primary types of information systems and how they are used.

The four primary types of information systems are transaction processing systems (TPS), management information systems (MIS), decision support systems (DSS), and expert systems (ES). A transaction processing system is an information system used to support and record transactions such as paying for products or paying an employee. When transactions are carried out electronically online, over the Internet or using other telecommunications and network systems it is called *e-commerce*. A management information system is an information system used to provide useful information to decision makers usually in the form of a report. A decision support system is an information system used to support problem-specific decision making. An expert system (ES) is an information system that can make suggestions and reach conclusions in one particular area of expertise much the same way that a human expert can.

Figure 1.26—p. 29

LEARNING OBJECTIVE 5

Discuss how computers are used in a variety of fields, and why studying computer systems can benefit you professionally.

Computers play an important role in all professions including those in business, communications, science, math, engineering, social sciences, fine arts, sports, nutrition, government, law, medicine, health care, criminology, law enforcement, security, education, and training.

Figure 1.30—p. 33

LEARNING OBJECTIVE 6

Discuss how computers are used to assist people in their life outside of work.

Computers and digital technologies are playing an increasing role in assisting people to achieve personal goals and leading more rewarding, productive, and fulfilling lives outside of work. Applications of particular usefulness include personal finance, personal information management, personal research, personal relations, personal media, and entertainment.

Figure 1.39—p. 42

LEARNING OBJECTIVE 7

Define information security, and discuss ways in which digital technologies are impacting humanity.

Information security refers to the protection of information systems and the information they manage against unauthorized access, use, manipulation, or destruction, and against the denial of service to authorized users. Information security has become a growing concern as increasing amounts of important and private information is stored digitally on systems connected to public networks and wireless

Figure 1.40—p. 43

private networks. A hacker is an individual who subverts computer security without authorization. Computers and computer systems are also vulnerable to viruses and spyware that can be inadvertently contracted from e-mail attachments, Web sites, or files downloaded over the Internet. Keeping networks and the information they store secure takes effort on many levels. Important safeguards must be implemented by individuals, businesses, and governments in order to achieve a secure information infrastructure.

You live in an increasingly pervasive and ubiquitous computing environment, one in which you can access digital information and media at any time and in any place with little effort. The digital divide refers to the social and economic gap between those who have access to computers and information technologies and those that do not. Access to information is power. Those who enjoy the benefits of technology are being confronted by many technology-related ethical issues. Pervasive digital communications are influencing the manner in which relationships and communities are defined, established, and maintained. Convenience and valuable services often come at a price of some degree of privacy. Much of the information collected and stored about individuals is done without their knowledge.

Test Yourself

LEARNING OBJECTIVE 1. Define how digital electronics devices and computers are related, and provide descriptions of the different types of computers.

1. The process of transforming nondigital information to 1s and 0s is called ________________ .
   a. processing
   b. technology
   c. digitization
   d. transliteration
2. ________________ is data organized and presented in a manner that has additional value beyond the value of the data itself.
   a. Information
   b. Software
   c. Input
   d. A database
3. A ________________ is useful in environments where you have to access a computer with a full-size display while on the go and on your feet.
4. ________________ are powerful general-purpose computers that provide information services to numerous users over a computer network.
5. True or False: Today, people are working on the go more than ever before.

LEARNING OBJECTIVE 2. Provide an explanation of the fundamentals of telecommunications, computer networking, the Internet, and wireless networking.

6. Communications that take place electronically over a distance are called ________________ .
7. ________________ are rules that are implemented in network software and hardware to establish connections between two or more computers to allow them to communicate.
   a. Network architectures
   b. Internetworks
   c. Access points
   d. Protocols
8. ________________ is a popular wireless networking standard that sends and receives data between network access points and computers.
9. True or False: The Internet and the Web are essentially the same thing.

LEARNING OBJECTIVE 3. Describe the five functional areas in which computers assist people most.

10. ________________ is an area of computer science that deals with simulating human thought and behavior in computers.
11. True or False: People are better than computers at working with large amounts of data.

12. _______________ is a popular technology that allows phone conversations to travel over the Internet or other data networks.
    a. MIS
    b. VoIP
    c. Robotics
    d. Smart Home

13. Which of the following is not one of the five functional areas in which computers excel?
    a. automation
    b. digital media and entertainment
    c. information management
    d. creative endeavors

LEARNING OBJECTIVE 4. Discuss the four primary types of information systems and how they are used.

14. A(n) _______________ is an information system used to support problem-specific decision making.
    a. transaction processing system (TPS)
    b. management information system (MIS)
    c. decision support system (DSS)
    d. expert system (ES)

15. A(n) _______________ is an information system used to support and record transactions.
    a. transaction processing system (TPS)
    b. management information system (MIS)
    c. decision support system (DSS)
    d. expert system (ES)

16. A(n) _______________ is an information system that can make suggestions and reach conclusions in one particular area of expertise much the same way that a human expert can.
    a. transaction processing system (TPS)
    b. management information system (MIS)
    c. decision support system (DSS)
    d. expert system (ES)

17. A(n) _______________ is an information system used to provide useful information to decision makers usually in the form of a report.
    a. transaction processing system (TPS)
    b. management information system (MIS)
    c. decision support system (DSS)
    d. expert system (ES)

18. _______________ are information professionals responsible for designing information systems.

LEARNING OBJECTIVE 5. Discuss how computers are used in a variety of fields, and why studying computer systems can benefit you professionally.

19. In businesses, computer personnel typically work in a computer department that may employ all the following except a _______________.
    a. chief information officer (CIO)
    b. chief executive officer (CEO)
    c. systems analyst
    d. computer operator

20. True or False: Computer systems have not been used to any extent in the fine arts.

21. True or False: The uses of computer systems in almost every field are almost limitless.

LEARNING OBJECTIVE 6. Discuss how computers are used to assist people in their life outside of work.

22. True or False: Computers are more useful at work than they are outside of work.

23. Individuals typically depend on _______________ software to support their personal needs.
    a. custom-designed
    b. expensive
    c. off-the-shelf
    d. client/server

24. _______________ software typically includes a personal calendar, to-do list, and an address book.

LEARNING OBJECTIVE 7. Define information security, and discuss ways in which digital technologies are impacting humanity.

25. _______________ refers to the protection of information systems and the information they manage against unauthorized access, use, manipulation, or destruction, and against the denial of service to authorized users.

26. Which of the following can be accomplished by a spyware program?
    a. Corrupt your computer files.
    b. Crash your hard drive.
    c. Block access to hackers.
    d. Fry your motherboard.

27. The ______________ is a title used for the social and economic gap between those who have access to computers and information technologies and those who do not.

Test Yourself Solutions: **1.** c. digitization, **2.** a. information, **3.** Tablet PC, **4.** Servers, **5.** True, **6.** telecommunications, **7.** d. protocols, **8.**Wi-Fi, **9.** False, **10.** artificial intelligence, **11.** False, **12.** b. VoIP, **13.** d. creative endeavors, **14.** c. DSS, **15.** a. transaction processing system (TPS), **16.** d. expert system (ES), **17.** b. management information system (MIS), **18.** Systems analysts, **19.** b. chief executive officer (CEO), **20.** False, **21.** True, **22.** False, **23.** c. off-the-shelf, **24.** personal information management (PIM), **25.** Information security, **26.** a. corrupt your computer files, **27.** digital divide

Key Terms

artificial intelligence (AI), p. 22
bit, p. 6
byte, p. 6
computer, p. 5
computer literacy, p. 4
computer network, p. 18
computer-based information system (CBIS), p. 26
computing platform, p. 11
data, p. 6
database, p. 26
decision support system (DSS), p. 29
digital divide, p. 44
digital electronics device, p. 6
digitization, p. 6
e-commerce, p. 28
embedded computer, p. 14
expert system (ES), p. 30
file, p. 6
hacker, p. 43
hardware, p. 7
information, p. 6
information security, p. 42
information technology, p. 27
Internet, p. 18
kiosk, p. 14
management information system (MIS), p. 28
microprocessor, p. 6
personal computer (PC), p. 9
protocols, p. 18
servers, p. 12
software, p. 7
supercomputer, p. 13
systems development, p. 28
technology, p. 6
telecommunications, p. 17
transaction processing system (TPS), p. 28
Voice over Internet Protocol (VoIP), p. 24
Wi-Fi, p. 19
wireless networking, p. 19
World Wide Web, p. 19

Questions

Review Questions

1. How many bits are in a byte?
2. List the six prefixes for digital technology metrics in order starting with *kilo*.
3. How does information differ from data?
4. What is a computer?
5. List four different types of PCs.
6. Name three types of special-purpose servers.
7. Describe how five career areas can benefit from computer systems.
8. List four examples of computer kiosks.
9. Name three different types of mobile digital devices.
10. List the five things that computers are good at, and give examples of each.
11. What is an information system? Describe the components of an information system.
12. Describe the various types of electronic commerce.
13. What is the difference between a transaction processing system and a management information system?
14. What is the difference between a decision support system and an expert system?
15. Why is it important to study information systems?
16. Name three threats to information security that may endanger Internet users.

Discussion Questions

17. List the digital devices that you own and use and a description of what essential valuable service each one provides. How much do these devices affect your life?
18. Think about yesterday and the day before. Consider your activities. List all the computers and digital devices you came in contact with including items such as ATMs and digital cable television. Compare your list to others in your class. List five overall benefits that digital technologies have provided you and the businesses that run them.
19. Stand near a busy part of campus during a between-class rush and calculate the percentage of students using cell phones. Discuss how the use of cell phones is changing the social environment of your campus.
20. List the top 10 ways you use computers and categorize them into the five functional areas of computer use: computation, automation, communication, entertainment, and information management. What area of computer functionality do you use most?
21. Describe an information system provided by your school that is frequently used by students. What benefits does it provide to students and the school? Do you consider it well designed? Why or why not?
22. Go to one or more stores that sell computer hardware. Give an example of a digital device that is not currently available that you would consider buying if it were available.
23. What career area or field is of interest to you? Describe how you could use one or more computer systems to advance your career in this area or field.
24. If you own a computer, what tools and techniques have you implemented to protect it from hackers, viruses, and spyware?
25. What social and ethical issues about computers and digital technology concern you the most? Which one is the most dangerous to you and society?

Exercises

Try It Yourself

1. Prepare a data disk or USB thumb drive to use for the lab and Web exercises throughout the text. Create one folder for each chapter in the textbook (you should have 12 folders when you are done). As you work through the Try It Yourself exercises and complete other work using the computer, save your assignments for each chapter in the appropriate folder. Add a text file to the root directory of the disk or drive that includes your name and contact information. Name the file OWNER_INFO.
2. At one or more local computer stores, research your ideal personal computer system, including all hardware, software, databases, and so on. Using a spreadsheet, list the cost of each item and compute the total cost for the entire computer system.
3. Using the Internet, research a career area that interests you. (You can use a search engine, such as Yahoo.com or Google.com.) Using your word processor, prepare a report describing the number and types of computer-related occupations that are available in that career area. In addition, note how many other occupations require some computer systems technology and skills.

Virtual Classroom Activities

For the following exercises, do not use face-to-face or telephone communications with your group members. Use only Internet communications.

4. Use the Internet to research distance learning. What are the advantages of distance learning for a student? What are the disadvantages?
5. With a group or team, investigate invasion of privacy issues, including identity theft. What can you do to avoid invasion of your privacy? What new laws should be passed to protect people from the invasion of privacy?

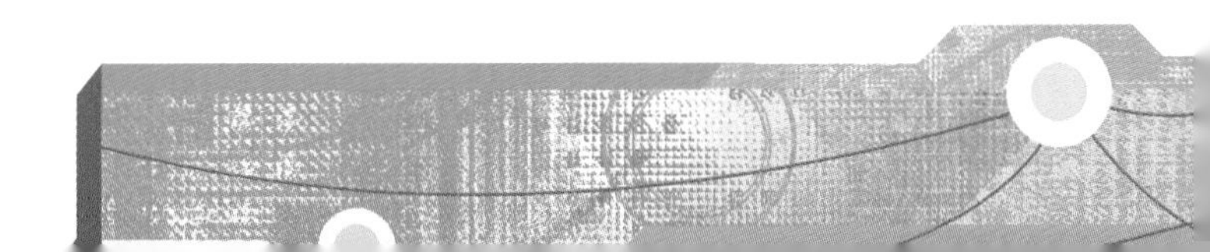

Teamwork

6. Your first task is this: One member of the group must send an e-mail message to a second member of the group, giving an opinion on a current event. The second person should state an opinion in response and forward that message, including the first message, on to the third member of the group. Continue with this process until the last member of the group forwards the entire string of messages back to its originator. Print this final message and submit it to your instructor. It should contain the names and comments of each member of the group.

7. Within your team, brainstorm about the characteristics of a good group or team member. Develop a contract to be used by your team to ensure that all members of the team will work hard to complete all teamwork assignments. Note that you can revisit this document and modify it if necessary. If you do, have all members initial the changes.

8. Your team should explore how a computer system can be used to obtain a competitive advantage in two or more career areas. You can use the Internet to search for ideas. Use your word-processing program to write a report on what you found.

TechTV

Go Inside Krispy Kreme

Go to *www.course.com/swt2/ch01* and click TechTV. Click the Go Inside Krispy Kreme link to view the TechTV video, and then answer the following questions.

1. How have computer systems assisted the many Krispy Kreme franchises across the country to provide consistent products and services for their customers?

2. What role does Web-based multimedia play in providing answers for Krispy Kreme's managers?

3. How have Krispy Kreme's computer and information systems improved Krispy Kreme employees' productivity and the quality of Krispy Kreme's products and services?

Endnotes

[1] Apple Web site, "Finding a Home in HD Animation," http://www.apple.com/pro/video/herbruck-baudenbacher/, accessed on April 3, 2005.

[2] Microsoft Web site, "The Boeing Company Puts the Tablet PC Through Its Paces," http://www.microsoft.com/resources/casestudies/CaseStudy.asp?CaseStudyID=13593, accessed on April 3, 2005.

[3] PalmOne Web site, "Leading Litigation Firm Keesal, Young & Logan, Stays In Sync With PalmOne and PensEra," http://solutions.palmone.com/regac/success_stories/enterprise/enterprise_details.jsp?storyId=776, accessed on April 3, 2005.

[4] Sun Microsystems Web site, "Sun Systems Help Bring Additional Functionality and Reliability to the Chicago Tribune's Newspaper Production Systems," http://www.sun.com/success-servers/pdfs/chicagotribune.pdf, accessed on April 3, 2005.

[5] IBM Web site, "German Railways Run Full Speed Ahead," http://www-1.ibm.com/servers/eserver/zseries/feature020205/, accessed on April 3, 2005.

[6] Herper, M. "Apple's Superfast Computer," *Forbes*, October 26, 2005, http://www.forbes.com/technology/enterprisetech/2004/10/26/cx_mh_1026aapl.html

[7] Glanville, J. "NYC Subway Gets a Computerized Facelift," *Associated Press*, April 11, 2005.

[8] Project ARS Web site, http://mike.passwall.com/ars, accessed on April 22, 2005.

[9] Prahalad, C. K. et al., "Adding Customers to the Design Team," *Business Week,* March 1, 2004, p. 22.

[10] Brown, S. "Toyota's Global Body Shop," *Fortune,* February 9, 2004, p. 120B.

[11] Chang, K. "70's Apollo Data Yields New Information," *The Rocky Mountain News,* February 15, 2005, p. 34A.

[12] Petersen, A. "Art for When There's Nothing on TV," *Wall Street Journal,* February 16, 2005, p. D1.

[13] Lin, G. et al, "New Model for the Military," *OR/MS Today,* January 2005, p. 26.

[14] Gilhooly, K. "For Better Healthcare," *Computerworld,* January 31, 2005, p. 23.

[15] Fillion, R. "DU Program Seeks NSA's Nod," *The Rocky Mountain News,* June 3, 2004, p. 3B.

[16] Lewin, J. "Consumers Relying on Web During Major Life Changes," *ITworld.com*, April 19, 2005, http://www.itworld.com.

[17] Ubiquitous Computing Web site, http://www.ubiq.com/hypertext/weiser/UbiHome.html, accessed on April 22, 2005.

[18] No Child Left Behind, Part B Grants for Education Technology, http://www.whitehouse.gov/news/reports/no-child-left-behind.html#9, accessed on October 7, 2005.

HARDWARE DESIGNED TO MEET THE NEED

Mark Bates recently graduated from college and started a new job as a graphic artist for Lamar Advertising. Now that he has a steady income, Mark wants to replace his 4-year-old PC with a new computer. He realizes that there are a lot of purchasing options and that the decision could be a difficult one to make. He wants to make sure that he ends up with a device that perfectly suits his needs without wasting money on options that he doesn't need. Mark uses his current computer for e-mail and instant messaging, Web browsing, managing his personal budget, basic word processing, multiplayer games, and sometimes for Lamar Advertising projects that he brings home. He is interested in trying out some of the new technology that he's heard about, specifically the ability to speak to the computer rather than typing, and video chat over the Internet.

As you read through the chapter, consider the following questions:

1. How should Mark prepare for his purchasing decision?
2. What type of computer might best suit Mark's needs?
3. How much processor speed, memory, and storage capacity does Mark need?
4. What input/output devices would best suit Mark's needs?
5. Are there any additional considerations to be made regarding this purchase?

Check out Mark's *Action Plan* at the conclusion of this chapter.

2

LEARNING OBJECTIVES

1. Understand how bits and bytes are used to represent information of value to people.
2. Identify the functions of the components of a CPU, the relationship between the CPU and memory, and factors that contribute to processing speed.
3. Identify different types of memory and storage media, and understand the unique properties of each.
4. Identify different types of input and output devices and how they are used to meet a variety of personal and professional needs.
5. Understand the decision-making process involved in purchasing a computer system.

CHAPTER CONTENT

The Digital Revolution

Integrated Circuits and Processing

Storage

Input, Output, and Expansion

Selecting and Purchasing a Computer

Introduction

How can a person keep up with the ever-increasing momentum of technological innovation? If you tried to learn all the details of every new digital device, you would soon be overwhelmed. It makes better sense to first learn the underlying technology that all these devices share.

From large server computers that support the needs of thousands of users, to the smallest handheld computer, computing devices share certain characteristics. They all manipulate digital data, and they all share fundamental components: processing components, data storage, and input and output capabilities. By learning about these underlying principles and components, you can develop a better understanding of what computing devices are capable of and how you can benefit from those capabilities. You can become empowered to competently use information and communication devices whether they're old, new, or yet to be invented.

In Chapter 1 you learned that a computer is a digital electronics device that supports four activities:

- Input: Capturing and gathering raw data
- Processing: Converting or changing raw data into useful outputs
- Storage: Maintaining data within the system on a temporary or permanent basis
- Output: Producing the results of the processing in a manner that is discernable to human senses or used as input into another system

Although it is easy to see how PCs and other types of computers fit this definition, you might sometimes overlook other electronic devices that also meet these criteria. For example, a digital cell phone accepts your voice and keypad entry as input; stores and processes sound-related data, phone numbers, and other data; and outputs an electronic signal and text to the display. Other examples of computers include special-purpose devices such as video game consoles, digital music players, and digital video recorders like TiVo.

This chapter looks in detail at hardware that supports the four activities of computer systems, starting with the most elementary concepts that are shared by all computing devices.

THE DIGITAL REVOLUTION

The world is in the midst of a digital revolution (Figure 2.1), and most experts agree that civilization has only seen the tip of the iceberg. It is impossible to guess how your life will be affected by advances in digital technologies over the next five or 10 years. By understanding the underlying principles of digital electronics, you can be better prepared to thrive in that future. To understand the importance of digitization, it is essential to learn more about how computers use bytes to represent data and information.

FIGURE 2.1 • Digital electronics

Notebook computers, cell phones, MP3 players, and a wide array of other devices are all based on the same digital technology.

Representing Characters and Values with Bytes

In Chapter 1 you learned that digital devices store and process data as *bits* and that in order to be useful, bits are typically organized into groups of eight called *bytes*. Bytes are treated differently depending on their purpose, just as characters are treated differently depending on how they are used. For example, when you see the symbols 322-2413 you treat them differently depending on whether it is a friend's phone number or a math problem, subtracting one value from another.

The same is true with bytes. When bytes are used to represent a finite set of data such as the letters in the alphabet, the 52 cards in a deck, or the model numbers of vehicles manufactured by Ford, a table can be used to associate each unit of data with a unique bit pattern. This approach to data representation can be considered a "look-up table" approach. For example, in an alphabet look-up table, you might look up 10001111 to find that it represents the uppercase letter G. A playing card look-up table might show that 10001111 is the queen of hearts. Bit patterns representing the characters on computer keyboards have been defined in this manner. In the early days of computing, the computer industry agreed on a code for representing keyboard text characters and named it the *American Standard Code for Information Interchange (ASCII)*. Figure 2.2 shows a portion of the ASCII chart.

FIGURE 2.2 • Look-up table

Look-up tables such as the ASCII chart are used to associate data with bytes.

ASCII Chart			
A	01000001	P	01010000
B	01000010	Q	01010001
C	01000011	R	01010010
D	01000100	S	01010011
E	01000101	T	01010100
F	01000110	U	01010101
G	01000111	V	01010110
H	01001000	W	01010111
I	01001001	X	01011000
J	01001010	Y	01011001
K	01001011	Z	01011010
L	01001100		
M	01001101		
N	01001110		
O	01001111		

The look-up table approach to data representation has become standardized for various applications. A *standard* is an agreed upon way of doing something within an industry. Standards are very important to digital technologies because they allow devices from different manufacturers to work together, share data, and function in essentially the same way. The ASCII standard allows digital devices to share keyboard-entered data. The MP3 standard allows a digitized song to be played on numerous different devices. There are standards for every popular look-up table data representation scheme.

The **binary number system** uses only two values, 0 and 1, and is used by computers and digital devices to represent and process data. There are an infinite amount of number systems, all equal in power, and all supporting the same mathematical operations as our decimal number system. Table 2.1 shows the decimal number 238 as it is represented by four popular number systems. The decimal number system, also called base 10, was adopted by humans because of its ease of use for our 10-fingered species. The binary, or base 2 system was adopted for computer use because of its ease of use to a machine that uses two-state bits.

TABLE 2.1 • Number systems

Decimal (base 10)	Binary (base 2)	Octal (base 8)	Hexadecimal (base 16)
238	11101110	356	EE

Because bytes can be used to represent values with binary numbers, computers are able to do some very useful things. Besides their use in performing mathematical calculations, binary numbers are the key ingredient for digitizing sound, music, photographs, drawings, paintings, animation, and movies. Your favorite movie on DVD and songs on a CD or MP3 are nothing more than long lists of binary numbers interpreted by your media player as colors to display or sound waves to play.

As you can see, there are many methods of decoding and interpreting a byte. One particular bit pattern, say 11110000, could be interpreted differently by an e-mail system, a word processor, a media player, and a calculator (see Figure 2.3) Each method boils down to either the look-up table approach or binary numbers. By implementing standards, bytes can have consistent meaning across common applications.

EXPAND YOUR KNOWLEDGE

To learn more about binary numbers, go to www.course.com/swt2/ch02. Click the link "Expand Your Knowledge" and then complete the lab entitled "Binary Numbers."

FIGURE 2.3 • Digital data representation

The same byte can represent different things to different software and digital devices: a pixel in a photo, a moment of sound in a song, or shapes in a drawing.

1 1 1 1 0 0 0 0

FIGURE 2.4 • Digital convergence in your hand

The HP iPAQ Pocket PC combines a digital cell phone, digital camera, and handheld computer in one device.

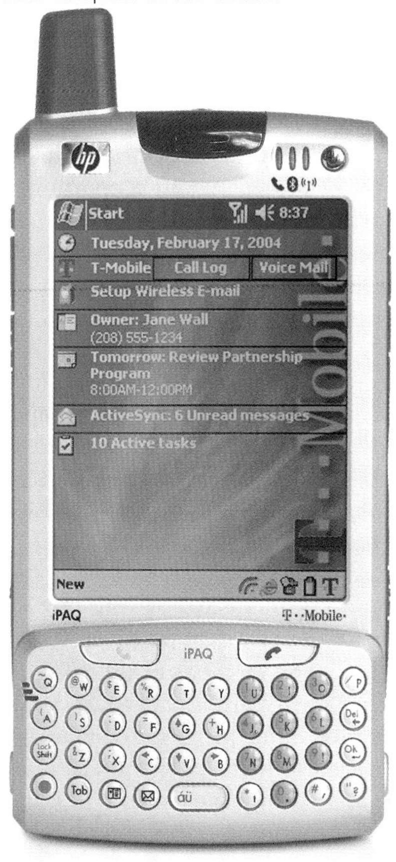

Bits, Bytes, and People

Digitization is the great equalizer of information. Whether the information is headline news, a text message, music, photos, videos, or voice, it is all just a bunch of bits to the machine. This is the underlying principle behind digital convergence. **Digital convergence** is the trend to merge multiple digital services into one device. At home, products such as Microsoft Windows XP Media Center Edition turns a PC into a media, information, and communication center, used for audio and video entertainment, computing, information access, and voice, video, and text communications. At work, many businesses are combining phone networks with data networks to carry voice, video, and text communications and information over the same network lines. On the road, digital convergence takes the form of Smartphones and other multipurpose handheld devices that merge cell phone, PDA (personal digital assistant), MP3 player, digital video player, and digital camera functionality into a single portable device, as shown in Figure 2.4.

The digitization of information grants users power over that information. Learning the basic principles of digitization and binary data representation allows you to move beyond just knowing what buttons to push, to a deeper understanding of what is happening inside the machine and in your culture.

INTEGRATED CIRCUITS AND PROCESSING

The ability to digitize information is at the heart of the digital revolution; however, digitized information is valueless without the ability to process it into useful forms (see Figure 2.5). Processors are important components of all digital devices. CD and MP3 players process digitized music into analog sound; a digital cell phone processes voice sound waves into digital signals and back again; a digital cable TV receiver processes digital video data into images on your television screen; and, of course, computers process data into useful information and services.

FIGURE 2.5 • Neuroimaging on a Mac

Dr. Nouchine Hadjikhani finds the processing power of the Apple Mac G5 essential for viewing MRI images of the brain to unravel the mystery of migraines.

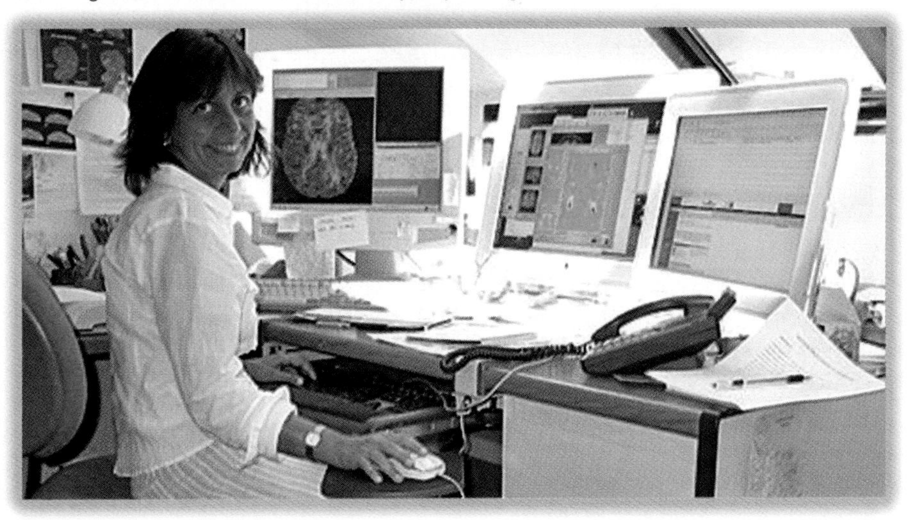

The quality of a digital electronics device is typically a reflection of the speed of its processor. This is especially true in computers: increasing processor speeds provide faster and more robust services. At home, faster processors support increased media applications such as flight simulators and 3D interactive gaming. In the professional world, faster processors are allowing graphic artists to reach new heights in visual effects, while accountants and financial analysts are able to run calculations on increasing amounts of data in decreasing amounts of time. Faster processors are providing additional features in all types of digital devices. Many cell phones now have the processing power to provide speech recognition features that allow you to speak a name rather than dial the number. Handwriting recognition is becoming increasingly reliable in handheld and tablet PCs because of more powerful processors (see Figure 2.6). Smaller, faster processors are providing capabilities in handheld devices that rival desktop computers.

This section provides a look at the technology used in today's processors, as well as how to measure the quality of a processor in order to make wise purchasing decisions. It begins with a look at important components that support the act of processing in integrated circuits, and how a special integrated circuit, called the central processing unit, works with other integrated circuits, called memory, to process data in computers and other digital devices.

FIGURE 2.6 • Processor power

Increases in processor power are providing higher-quality services such as handwriting recognition.

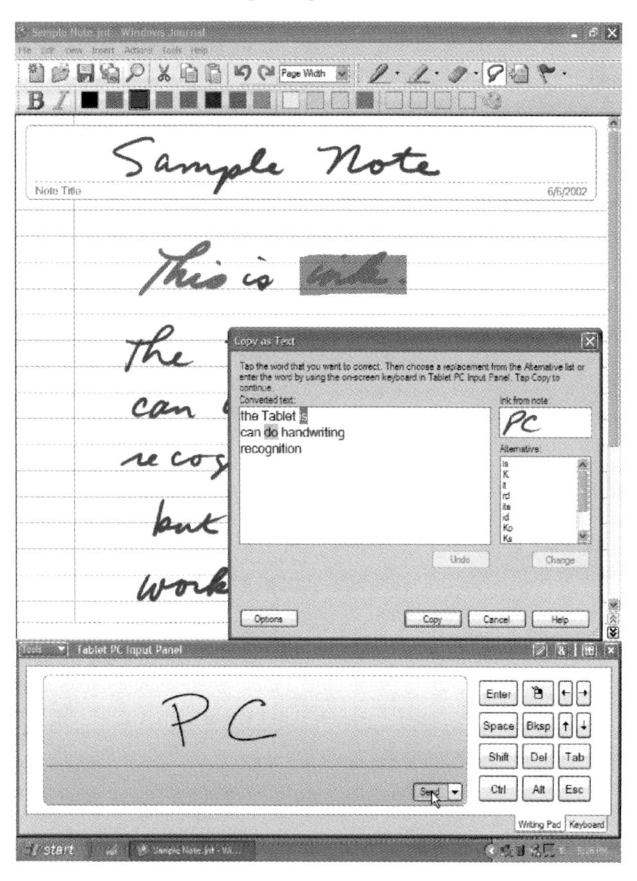

Integrated Circuits

Computers use digital switches not only to store bits and bytes, but also to process them. Chapter 1 showed how any object having two distinct states can be used as a digital switch. Over the years computers have used different devices as digital switches, progressing from physical switches and relays, to vacuum tubes, to transistors. The **transistor** is an electronics component composed typically of silicon, that opens or closes a circuit to alter the flow of electricity to store and manipulate bits. Invented in 1947 at Bell Labs, transistors have become the key ingredient of all digital circuits, including those used in computers. When electricity is flowing through a transistor it represents a 1; when it is not flowing it represents a 0. By combining transistors and using the output from one or more transistors as the input to others, computers control the flow of electricity in a manner that represents mathematical and logical operations.

In the late 1950s, Jack Kilby of Texas Instruments and Robert Noyce of Fairchild Semiconductor developed a method to integrate multiple transistors into a single module called an **integrated circuit**. Integrated circuits, also called *chips*, are used to store and process bits and bytes in today's computers. A group of integrated circuits that work together to perform the processing in a computer system is called the **central processing unit (CPU)**. Today's technology is able to pack all the CPU circuits onto a single module smaller than the size of your smallest fingernail, called a *microprocessor*. The microprocessor and other supportive chips are housed on circuit boards where embedded pathways electronically join the chips together. The primary circuit board of a computing device is called the **motherboard** and the pathways are called *buses* (see Figure 2.7).

FIGURE 2.7 • Motherboard

All computing devices, such as this Nokia N-Gage cell phone/video game, house their chips on a circuit board known as a motherboard.

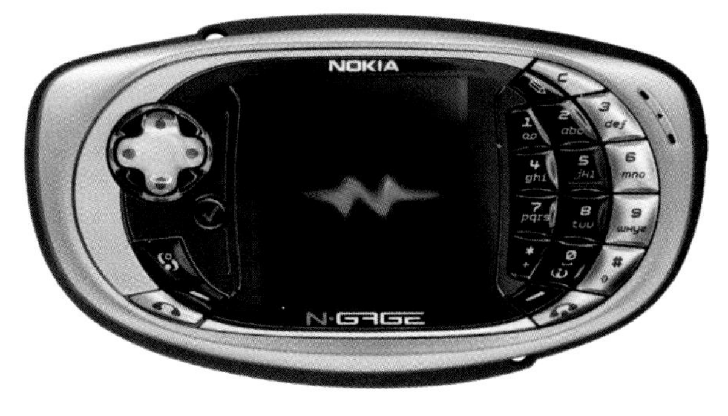

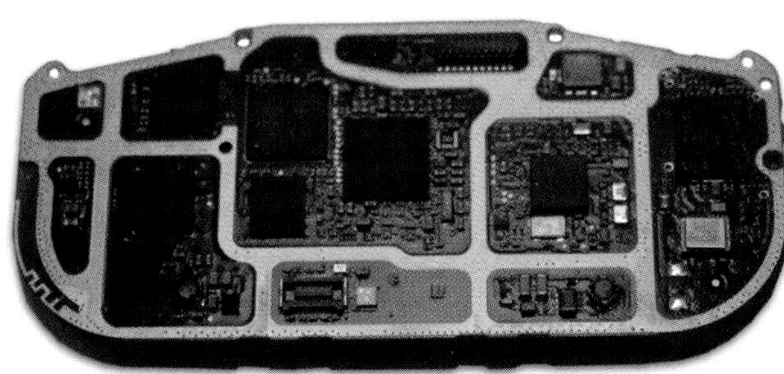

The Central Processing Unit and Random Access Memory

EXPAND YOUR KNOWLEDGE

To learn more about the motherboard and its various components, go to www.course.com/swt2/ch02. Click the link "Expand Your Knowledge" and then complete the lab entitled "Understanding the Motherboard."

Processing is the act of manipulating data in a manner defined by programmed instructions. Computer programs contain lists of instructions for the processor to carry out. A processor is engineered to carry out a specific and finite number of instructions called its *instruction set*. When a computer runs a program, the processor progresses through the program's sequence of instructions, carrying out each instruction with specially designed circuitry (Figure 2.8), and jumping to various subsets of instructions as a user interacts with the program.

A CPU consists of three primary components: the arithmetic/logic unit, the control unit, and registers. The **arithmetic/logic unit** (**ALU**) contains the circuitry to carry out instructions, such as mathematical and logical operations. The **control unit** sequentially accesses program instructions, decodes them, and coordinates the flow of data in and out of the ALU, the registers, RAM, and other system components such as secondary storage, input, and output devices. *Registers* hold the bytes currently being processed.

FIGURE 2.8 • Intel Pentium 4

Central processing units such as the Pentium 4 processor have microscopic architectures more complex than the world's largest cities.

The CPU works closely with random access memory. **Random access memory** (**RAM**) is temporary, or *volatile*, memory that stores bytes of data and program instructions for the processor to access. RAM capacities in today's new PCs typically range from 256 MB to 512 MB with high-end PCs exceeding 1 GB. Data flows back and forth between the CPU and RAM across the system bus. The *system bus* consists of electronic pathways between the CPU and RAM capable of transporting several bytes at once. Typically the system bus connects to a *chipset* that ties together several bus systems sending and receiving bytes from memory, input and output devices, storage, networks, and other motherboard components. Figure 2.9 shows how the system bus connects the CPU to the chipset, and through it to RAM and other components in the Intel Pentium 4 architecture.

To understand the function of processing and the interplay between the CPU and RAM, you need to examine the way a typical computer executes a program instruction.

FIGURE 2.9 • CPU, bus, and RAM

In the Pentium 4 architecture, the system bus connects the processor to a chipset that passes data back and forth with RAM over dual channels.

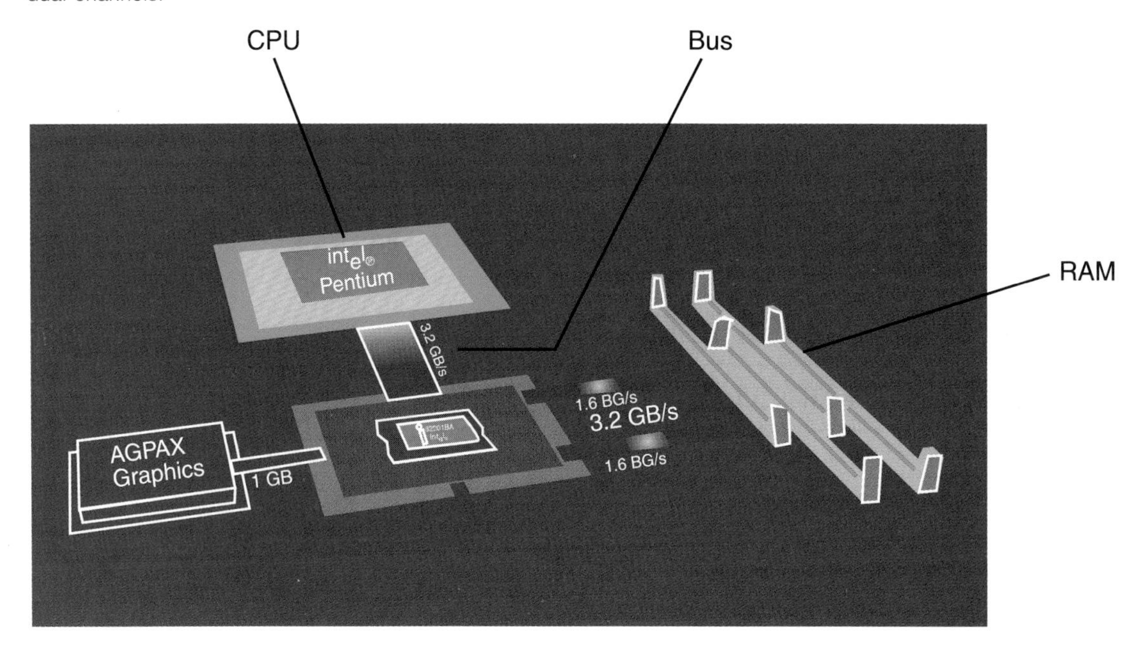

The Machine Cycle. The execution of an instruction is a step-by-step process that involves two phases: the instruction phase and the execution phase. These two phases together make up the *machine cycle*. Figure 2.10 shows the four steps in the machine cycle. In the instruction phase of the machine cycle, the computer carries out the following steps:

Instruction Phase

Step 1: Fetch instruction. The instruction to be executed is accessed from RAM by the control unit. The control unit stores the RAM address of the currently executing instruction.

Step 2: Decode instruction. The instruction is decoded, relevant data is moved from RAM to the CPU registers, and the stored address of the current instruction is incremented to prepare for the next fetch.

In the execution phase of the machine cycle, the computer carries out Steps 3 and 4:

Execution Phase

Step 3: Execute the instruction. The ALU does what it is instructed to do. This could involve making either an arithmetic computation or a logical comparison.

Step 4: Store results. The results are stored in registers or RAM.

Processors perform a variety of instructions on different types of data. For example, the processor in a CD player reads bytes of digitized sound from a CD, and processes the digits into a form you can listen to—music. A digital camera accepts input from a camera lens and outputs a list of color codes to a file. An e-mail program, translates keyboard data into ASCII characters, and forwards the collection of characters to an e-mail server for processing. The efficiency of the manner in which a processor makes use of the machine cycle directly impacts the quality of performance of a digital device.

CPU Characteristics. The first consideration in selecting a computer is typically its speed—how quickly it can carry out such tasks as loading a program, opening a file, and writing to a CD. Processor manufacturers design processors

FIGURE 2.10 • Execution of an instruction

The machine cycle consists of four steps: fetch, decode, execute, and store.

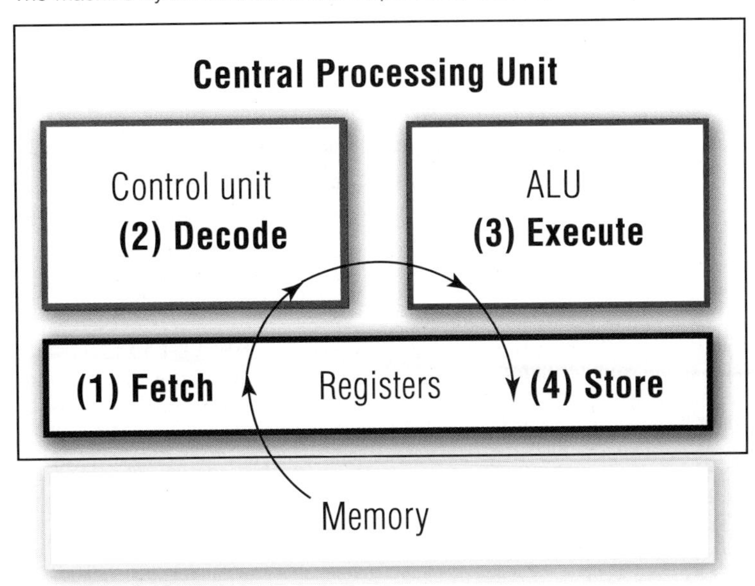

to meet numerous needs and budgets. Other manufactures have similar varieties of processors. For example, IBM and Apple produce PowerPC processors of varying speeds to support different Apple computers sold at varying prices. Recently Apple stunned the tech world by deciding to switch from PowerPC processors to Intel processors in their computers.[1] Table 2.2 shows several types of processors offered by Intel.

TABLE 2.2 • Intel manufactures processors for a variety of computer types and uses.[2]

Processor	Computer type	Description
Intel Celeron	Desktop and notebook PCs	Manufactured for users on a limited budget; doesn't support high-end graphics applications
Intel Pentium	All	Standard processor for typical users
Intel Pentium Extreme	Desktop PCs	For PowerUsers who need the fastest processor for handling advanced graphics
Intel Pentium M and Centrino chipset	Notebook PCs	Provides fast processing, long battery life, and wireless network technology
Intel Itanium	Servers	Designed for multiple processor computing
Intel Xeon	Workstations	Designed for high-speed processing on special-purpose computers

Managers of Gap Inc. Direct, the online division of Gap Inc., the company that owns Gap, Banana Republic, and Old Navy brand names, chose computers with Pentium 4 processors because of their ability to carry out multiple tasks simultaneously. The switch to Pentium 4 helped employees to work on multiple chores at the same time while mission-critical applications were running in the background. The faster, more powerful processor boosted productivity across the entire organization.[3]

One component that contributes to the speed at which a processor can carry out an instruction is the system clock. Each CPU contains a *system clock* that produces a series of electronic pulses at a predetermined rate called the *clock speed*. These pulses are used to synchronize processing activities. Just as a person's heartbeat circulates blood through the body, the system clock distributes bits and bytes through the components of a computer system. Clock speeds in today's digital devices are measured in **megahertz (MHz)**, millions of cycles per second, or **gigahertz (GHz)**, billions of cycles per second. It is this specification that is typically at the top of the list in computer advertisements.

Faster clock speeds generate more heat in a system and require larger cooling systems. For this reason, you'll find that smaller devices generally have a lower maximum available clock speed, as illustrated in Table 2.3.

TABLE 2.3 • Clock speed comparisons

Machine type	Top clock speeds
Handheld computer	624 MHz
Notebook PC	2.6 GHz
Desktop PC	3.73 GHz

Although clock speed has a direct effect on overall system performance, it is not the only contributing factor. In fact clock speeds can be deceptive. An Apple computer with a 2.7 GHz G5 processor can outperform a Dell computer with a faster 3.4 GHz Pentium 4 processor. A notebook computer with a 1.6 GHz Pentium M processor can equal or surpass a Dell notebook with a faster 3 GHz Pentium 4 processor in performance. In today's computers, it is not so much how fast the system clock ticks, but how much the processor can do with each tick of its clock.

Rather than emphasizing clock speed when marketing its processors, Intel has switched to processor code numbers. A notebook computer from 2003 might be advertised as having a Pentium III 1.33 GHz processor. A notebook today might be advertised as having a Pentium M Processor 725. The 725 combines four performance specifications that contribute to system performance. Table 2.4 lists these factors.

TABLE 2.4 • Computer performance factors[4]

Architecture	Basic design of a microprocessor; may include process technology and/or other architectural enhancements
Cache (MB/KB)	A temporary storage area for frequently accessed or recently accessed data; having certain data stored in a cache speeds up the operation of the computer. Cache size is measured in megabytes (MB) or kilobytes (KB).
Clock speed (GHz/MHz)	Speed of the processor's internal clock, which dictates how fast the processor can process data; clock speed is usually measured in GHz (gigahertz, or billions of pulses per second)
Front side bus (GHz/MHz)	The connecting path between the processor and other key components such as the memory controller hub; FSB speed is measured in GHz or MHz.

Cache (pronounced cash) *memory* is a type of high-speed memory that a processor can access more rapidly than RAM. Cache memory functions somewhat like a notebook used to record phone numbers. Although a person's private notebook may contain only 1 percent of all the numbers in the local phone directory, the chance that the person's next call will be to a number in his or her notebook is high. Cache memory works on the same principle: a cache controller makes "intelligent guesses" as to what program instructions and data are needed next and stores them in the nearby cache for quick retrieval.

Three levels of cache are used in today's personal computers: L1, L2, and L3. The levels indicate the cache's closeness to the CPU. L1 is stored on the same chip as the microprocessor; L2 and L3 are on separate chips. Considerably more expensive than RAM, cache memory is provided in much smaller capacities.

Cache sizes vary from processor to processor. A Pentium processor typically has a 512 KB L1 cache. The PowerPC G5 processor uses a more complicated arrangement that includes a 512 KB L2 cache, plus a 64 KB L1 instruction cache and 32 KB L1 data cache. The larger the cache, the faster the processing.

Another important factor that affects a computer's performance is the processor's wordlength. *Wordlength* is the number of bits that a CPU can process at once. A processor with a 32-bit wordlength has the capacity to be twice as fast as a processor with a 16-bit wordlength. Today's personal computers typically use wordlengths of 32 or 64 bits. Intel Pentium 4 processors have a 32-bit wordlength, and Apple G5 processors have a wordlength of 64 bits. The larger the wordlength, the more powerful the computer.

The latest technique in chip design is referred to as *multicore technology* and refers to housing more than one processor on a chip. Intel has designed a *dual-core processor* that uses two processors on one chip that work together to provide twice the speed of traditional single-core chips[5] (see Figure 2.11). Sony has also released a dual-core chip called *Cell* designed for use in the PlayStation 3 video game system.[6] Dual-core and multicore chips require software to be specially designed to take advantage of their unique architecture.

FIGURE 2.11 • Dual-core processors

A researcher displays a wafer that contains over a hundred of Intel's new dual-core processors.

By now you've gathered that judging a computer's quality and speed by its specifications can be complicated. The truest measure of a processor's performance is the amount of time it takes to execute an instruction. This measure is called *MIPS* for millions of instructions per second; a more precise measurement is called *FLOPS* for floating-point operations per second. Today's personal computers carry out billions of instructions per second, or operate in the **gigaflop** range. Supercomputers run in the *teraflop* (trillions) range. For example, IBM's ASCI White computer assists the U.S. government in simulating a nuclear detonation at 12.3 teraflops—12.3 trillion floating-point operations per second. It has been estimated that a human brain's probable processing power is around 100 teraflops, roughly 100 trillion calculations per second. The fastest supercomputer in 2005, the IBM BlueGene/L, was clocked at 70.72 teraflops with near-future generations expected to top 360 teraflops.

This section has covered a considerable amount of information about microprocessors. However, when you decide it is time to buy a new personal computer, you do not typically begin by comparing processors. You typically begin with more general considerations such as whether to get a notebook computer or a desktop computer. Should you get an Apple or a PC? When you decide on a computer type and platform, you have a narrower range of processors from which to choose. Chapter 1 discussed the many types of computer systems available. It is now time to look at the concept of the computer platform.

Multiprocessing and Parallel Processing. Some computing tasks require more powerful computers. Individuals pursuing careers in motion picture special effects, animation, or other demanding graphics production areas can look forward to working on workstations with multiple processors to assist in graphic imaging. Scientists running experiments with large quantities of data also require extra processing power. In Chapter 1, you learned about midrange and mainframe servers that support the computing needs of an entire organization. These computers also use multiprocessing.

As the name implies, *multiprocessing* is processing that occurs using more than one processing unit. The purpose: to increase productivity and performance. One form of multiprocessing involves using coprocessors. Typically used in larger workstations, *coprocessors* are special-purpose processors that speed processing by executing specific types of instructions, while the CPU works on another processing activity. Each type of coprocessor performs a specific function. For example, a math coprocessor chip is used to speed mathematical calculations, and a graphics coprocessor chip decreases the time it takes to manipulate graphics. A popular workstation for motion picture special effects specialists and animators is the Octane2 workstation from Silicon Graphics, Inc. (*www.sgi.com/workstations/*). It includes two CPUs and two graphics coprocessors: a vertex processing engine and an image and texture engine. A machine of this magnitude is able to render a 3D scene in minutes, when the same job might take hours on a typical PC. Using the Silicon Graphics Prism Visualization system, Stanford researchers made headlines by creating a realistic 3D model of a mummified child who lived around AD 25 (see Figure 2.12). The model allowed the researchers to determine the age, and sex of the child as well as reconstructing and viewing the child's face.[7]

FIGURE 2.12 • Multiprocessing brings mummy to life

Using multiprocessing, researchers were able to create a 3D model of a child who lived around AD 25.

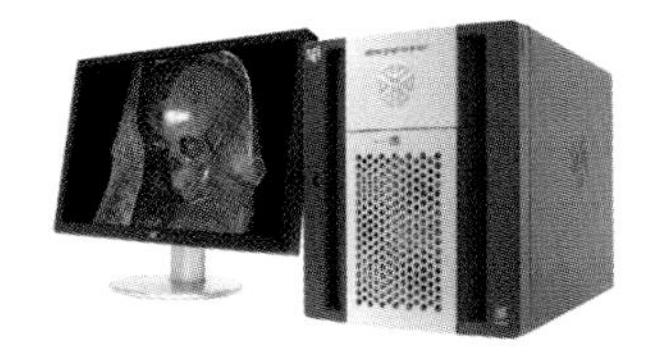

Another form of multiprocessing is called *parallel processing*, which speeds processing by linking several microprocessors to operate at the same time, or in parallel. The challenge of multiprocessing is not in connecting the processors, but making them work effectively as a unified set. Accomplishing this difficult task requires special software that can allocate, monitor, and control multiple processing jobs at the same time. A new technique called *grid computing* or *clustering* allows processors from different computers to work together over a network on complex problems. Some businesses form large parallel processing grids out of the networked PCs that remain behind when the employees go home for the evening. Other organizations such as SETI (*www.seti.org*) allow Internet users to donate their PC's processor power to problems too large for supercomputers such as scanning the heavens for extraterrestrial life (*http://setiathome.ssl.berkeley.edu*).

Massively parallel processing (MPP), used in supercomputers, involves using hundreds or thousands of processors operating together. For example, NASA's most powerful supercomputer, named Columbia (see Figure 2.13), is revitalizing the aerospace agency's high-end computing infrastructure by applying the power of 10,160 processors to make breakthrough scientific discoveries that support agency missions.[8]

FIGURE 2.13 • Supercomputer

NASA's Columbia MPP supercomputer uses 10,160 processors to assist NASA scientists in making breakthrough scientific discoveries.

Physical Characteristics of the CPU. CPU speed is also limited by physical constraints. Most CPUs are collections of digital circuits imprinted on silicon wafers, or chips, each no bigger than the tip of a pencil eraser. To turn a digital circuit within the CPU on or off, electrical current must flow through a medium (usually silicon) from point A to point B. The time it takes for the current to travel between points can be decreased by reducing the distance between the points.

Gordon Moore, cofounder of Intel, observed in 1965 that the continued advances in technological innovation made it possible to reduce the size of transistors, doubling their density on the chip every 18 months. He predicted that this trend would continue. This prediction has come to be known as **Moore's Law**, and over the years it has proved true (see Figure 2.14). When transistor densities increase, so does the speed of the processor, since the electrons have shorter distances to travel. One interpretation of Moore's Law assumes that if transistor densities double, processor speeds will also double every 18 months. This has also proven true. This is helpful information as it allows you to predict how fast computers will be in years to come. It is anticipated that Moore's Law will continue to hold true for the next decade; however, at some point Moore's Law will fail due to the inherent physical limitations of silicon. The high-tech industry is researching alternatives to the silicon-based chip. Table 2.5 lists some active areas of research in processing technologies.

FIGURE 2.14 • Moore's Law

Moore's Law predicts that transistor densities in new silicon-based chips will double every 18 months.

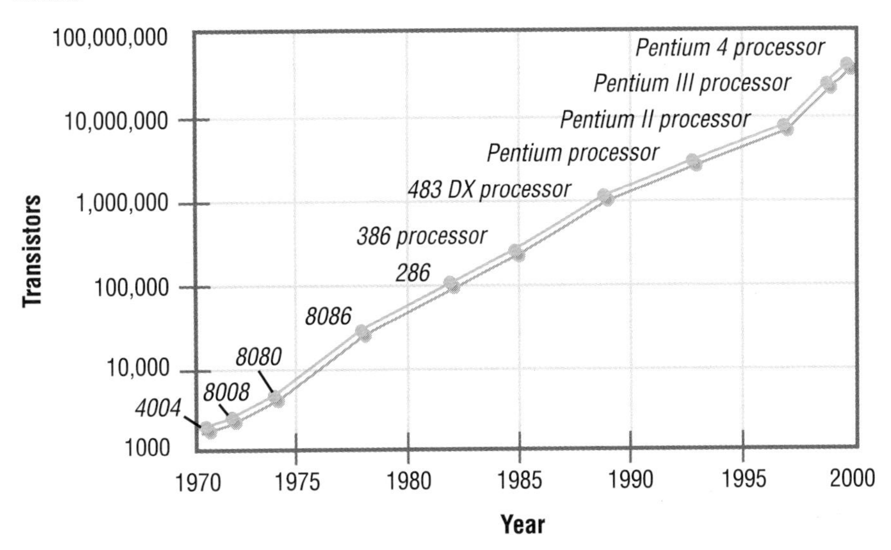

TABLE 2.5 • Active areas of research in processing technologies

Technology	Description
Strained silicon	IBM has perfected a way to stretch silicon to speed the flow of electrons through the transistors on the chip; electrons experience less resistance and flow up to 70 percent faster, which can lead to chips that are up to 35 percent faster—without having to shrink the size of transistors.
Silicon germanium	Silicon germanium includes layers of germanium, a substance that is like silicon, transistors with these layers can switch faster and perform better; IBM has been successful in creating a silicon germanium chip that runs at more than 110 gigahertz. The chip will be used to speed up communications systems.
Optical processing	Some companies are experimenting with chips called *optical processors* that use light waves instead of electrical current. The primary advantage of optical processors is their speed. It has been estimated that optical processors have the potential of being 500 times faster than traditional electronic circuits.
Quantum computing	Quantum computing proposes the manipulation of quantum states to perform computations far faster than is possible on any conventional computer. Understanding quantum computing requires a fundamental understanding of the theoretical principles of quantum mechanics. In brief, a quantum computer, doesn't use bits, but rather a fundamental unit of information called a quantum bit or *qubit*. The qubit displays properties in adherence to the laws of quantum mechanics, which differ radically from the laws of classical physics.
DNA computing	Israeli scientists have devised a computer that can perform 330 trillion operations per second, more than 100,000 times the speed of the fastest PC. The secret: it runs on DNA.

STORAGE

There have been tremendous developments in storage technologies in recent years. People are able to store more data in smaller devices and media for considerably less money than ever before. The large quantities of data that used to be chained to desktop computers can now be easily taken anywhere or copied and shared with others. Storage technologies are affecting our personal and professional lives and society in general. The use of CD-Rs to store music has had a tremendous impact on the music industry. Recordable DVDs are having the same impact on the motion picture industry. Being aware of the storage options available provides opportunities that can improve your quality of life.

Storage has been defined as the ability to maintain data within the system temporarily or permanently. The previous section discussed one form of temporary storage called RAM, sometimes called *memory* or *primary storage*. This section provides a more detailed look at the physical form of RAM, along with other storage used by computer systems. This section also looks at *secondary-storage devices* that are used to store data more permanently than RAM, as when the computer is turned off.

System Storage

System storage is storage that is used by a computer system for standard operations. There are several forms of system storage including RAM, cache, video RAM, ROM, and CMOS. Of these, RAM has the largest capacity. The following section looks at the physical circuit boards that make up RAM. You should find this information helpful when and if you decide to upgrade the memory on your computer.

RAM SIMMs. You have seen how RAM works hand-in-hand with the CPU to carry out program instructions. RAM exists as a set of chips grouped together on a circuit board called a *single in-line memory module*, or *SIMM*. RAM SIMMs are inserted into slots in the motherboard near the processor. A new computer typically comes with two or four RAM slots, half of which are occupied with SIMMs, and half left available for future expansion as shown in Figure 2.15.

FIGURE 2.15 • Motherboard RAM

RAM SIMMs are inserted into slots in the motherboard near the processor.

There are many different types of RAM: DRAM, SDRAM, RD-RAM, DDR-SDRAM, FPMRAM, EDO-RAM, BEDO-RAM—the list goes on and on. Each type of RAM reflects a manufacturer's effort to use a new technology to get data to the processor more quickly. Today's desktop and notebook PCs typically use either DDR-SDRAM, or DDR-II SDRAM. DDR-II doubles the speed of DDR which makes it an ideal partner for today's fastest processors. When upgrading RAM, you should consult your computer documentation or contact the manufacturer to find out what type of RAM your computer uses.

When shopping for a new computer, processor specifications are typically listed first, and then RAM specifications (see Figure 2.16). Most new computers come with 256 MB of RAM. You can usually upgrade to 512 MB for as little as $40. Besides processor and RAM specifications, a computer shopper is likely to run into another confusing specification: video memory. The following section describes how video memory affects system performance.

FIGURE 2.16 • Computer ad

Computer ads like this Web ad typically list processor, storage, and display specifications.

iBook G4 Notebook PowerPC G4 1.2GHz Processor, 256MB RAM, 30GB Hard Drive, 12.1-inch XGA TFT Display, 8X DVD / 24X16X24 CD-RW Combo Drive, Mac OS X Tiger v10.4 Product Number: 316976 Mfr. Part #: M9623LL/A Brand: Apple « Visit their Showcase	$899.99	Add To Cart (Delivery Only) Free Shipping	Compare » «
VAIO S460/B Notebook Intel Pentium M Processor 740, 1.73GHz, with Centrino Technology, 512MB RAM, 80GB Hard Drive, 13.3-inch WXGA TFT Display, 4X DVD+/-RW Drive, Windows XP Home Edition Product Number: 319318 Mfr. Part #: VGNS460/B Brand: Sony « Visit their Showcase	Was: ~~$1,799.99~~ $1,549.99 SAVE $250 after: $250.00 mail-in rebate(s)	Add To Cart (Delivery Only) Free Shipping	Compare » «
3360-EG1 Notebook Intel Pentium M Processor 725, 1.6GHz, with Centrino Technology, 512MB RAM, 60GB Hard Drive, 12.1-inch XGA TFT Display, 4X DVD+/-RW Drive, Windows XP Professional Product Number: 51100010 Mfr. Part #: AV3360-EG1 Brand: Averatec	$1,281.99	Sold Out!	Compare » «
1020-ED1 Notebook Intel Celeron M Processor, 1GHz, 512MB RAM, 60GB Hard Drive, 10.6-inch WXGA TFT Display, 8X DVD / 24X24X24 CD-RW Combo Drive, Windows XP Home Edition Product Number: 319848 Mfr. Part #: 1020ED1 Brand: Averatec	Was: ~~$1,149.99~~ $899.99 SAVE $250 after: $50.00 instant savings $200.00 mail-in rebate(s)	Add To Cart (Delivery Only) Free Shipping	Compare » «

Video RAM. *Video RAM* (VRAM) is used to store image data for a computer display in order to speed the processing and display of video images. Video RAM acts as a *buffer* or intermediate storage area between the microprocessor and the display. When images are to be sent to the display, they are first read by the processor from RAM and then written to video RAM. From video RAM, the data is converted to signals that are sent to the display. Most of today's PCs come equipped with 64 MB of VRAM; however, users who work with video, such as video producers and gamers, need at least twice this amount. VRAM is typically not stored directly on the motherboard, but rather on a video circuit board called a video card (Figure 2.17) that is plugged into the motherboard.

FIGURE 2.17 • Videocard

Computer gamers often shop for the fastest video card to get the most out of their gaming experience.

ROM. You won't see ROM listed in a desktop or notebook computer advertisement, but it is vitally important to the functioning of a computer. **Read-only memory (ROM)** provides permanent storage for data and instructions that do not change, such as programs and data from the computer manufacturer, including the boot process used to start the computer. Because both the processor and RAM require electricity to store data, both are empty when a computer is initially powered up. The computer requires a place to permanently store the instructions needed to start up the computer and load the operating system into RAM. ROM fulfills this purpose.

In ROM, the combination of circuit states is fixed, and therefore the data represented by this combination is not lost if the power is removed. ROM stores a program called the BIOS (basic input/output system). The BIOS stores information about your hardware configuration along with the boot program. The boot program contains the instructions needed to start up the computer. After running some system diagnostics, the boot program loads part of the operating system into memory and turns over control of the processor to the operating system. Many of today's computers use *flash BIOS,* which means that the BIOS has been recorded on a flash memory chip, rather than a ROM chip. Flash memory (covered later in this chapter) is intended to store data permanently like ROM, but can be updated with new revisions when they become available.

HARDWARE THAT COUNTS FOR SWIMMERS

Literally. A British engineering student and former lifeguard, developed a set of smart goggles that keeps track of laps for distance swimmers. The electronic device has a compass that orients upon entering the pool and logs a lap with every change in direction, displaying the count and lapsed time on the lens of the goggle. No bigger than a 50-cent piece, the device is attached to the back strap of the goggle.

Source:
Smart eyewear for keen swimmers
BBC News
28 June, 2005
http://news.bbc.co.uk/2/hi/technology/4626823.stm

Because handheld computers typically don't include a hard disk drive, they use ROM to not only store the BIOS but also to store the operating system and applications that are included with the device.

CMOS Memory. *CMOS memory* (pronounced *see-moss*, short for complementary metal-oxide semiconductor) provides semipermanent storage for system configuration information that may change. CMOS is unique in that it uses a battery to stay powered up. CMOS is the reason that a PC is able to maintain the correct time and date even when it is not plugged in. In addition to keeping time, CMOS stores information about a computer system's disk drive configuration, startup procedures, and low-level operating system settings. On PCs that run Windows, you can view CMOS settings by pressing a designated key such as F2 during the startup procedure when prompted. CMOS provides a service somewhere between RAM and ROM in that it is permanent like ROM and doesn't go away when the computer is turned off, but its contents can be changed like RAM.

EXPAND YOUR KNOWLEDGE

To learn more about managing your files, go to www.course.com/swt2/ch02. Click the link "Expand Your Knowledge" and then complete the lab entitled "Managing Your Files."

Secondary-Storage Technologies

In addition to the system storage used in the operation of the computer system, there is also *secondary storage,* used to store data and software more permanently without the need for electricity. Hard drives, CDs, DVDs, USB flash drives, and portable MP3 players are all forms of secondary storage. They allow you to store large quantities of data on your computer even when the computer has no power. Secondary storage also allows you to take data with you wherever you go. Data storage has become ubiquitous—available nearly everywhere. You may have a secondary-storage device on you now. Market researcher IDC estimates that hard drives in cars will soar from 520,000 units in 2003 to nearly 6 million units in 2008.[9] Car hard drives will store sophisticated navigation systems, entertainment, and automotive diagnostic systems.

It is not unusual for students to carry millions or even billions of bytes around with them on USB drives dangling from key chains and on MP3 players storing music and other kinds of computer files. At the same time, Internet storage facilities store gigabytes of data for you to access over any Internet-connected device. Soon you may not need to tote data around at all.

The need to store increasing amounts of data for scientific and archival purpose is barely met by today's storage technologies. A project known as the Internet Archive has been storing snapshots of the entire Web each year since 1996. By going to the Internet Archive Web site (*www.archive.org/*), and using the "Wayback Machine" you can call up any Web site and see its contents at any time since 1996. The archive's total collection of Web sites in 2003 was around 100 terabytes (that's 1000 gigabytes) of data. In 2004 the Internet Archive contained approximately 1 petabyte (a million gigabytes) of data, and in 2005 2.5 petabytes.[10]

The National Energy Research Scientific Computing Center (NERSC) in Berkeley, California, supports a great deal of research done for the U.S. Department of Energy. As one of the largest facilities in the world devoted to providing computational resources and expertise for scientific research, NERSC includes a large supercomputer and a huge amount of storage. Its High-Performance Storage System (HPSS) stores 8.8 petabytes of scientific data[11] (see Figure 2.18.) From huge data banks, to tiny USB flash drives, secondary storage affects everyone on personal and societal levels.

When discussing secondary storage, the term *storage device* refers to the device itself, the drive or circuitry that reads and writes data. *Storage media* refers to the objects that hold the data, such as disks. Some storage devices, such as flash memory drives, combine the device and media in one module. The *storage*

FIGURE 2.18 • The High-Performance Storage System (HPSS) at NERSC

One of the storage silo robots in NERSC's 8.8-petabyte High Performance Storage System.

capacity of a storage medium is the maximum amount of bytes that it can hold. *Access time* refers to the amount of time it takes for a request for data to be fulfilled by the device. When selecting storage devices and media, you should examine the portability, storage capacity, and access time associated with the media and match them to the task at hand, preferably at the least cost.

EXPAND YOUR KNOWLEDGE

To learn more about how to take care of your hard drive, go to www.course.com/swt2/ch02. Click the link "Expand Your Knowledge" and then complete the lab entitled "Maintaining a Hard Drive."

Magnetic Media: Disks and Tapes. **Magnetic storage** devices use the magnetic properties of iron oxide particles to store bits and bytes more permanently than RAM. No physical storage medium can be genuinely permanent. It can be destroyed in any number of ways: fire, flood, sledge hammer. But if not abused, magnetically stored data lasts years before deteriorating. In magnetic storage, a surface is coated with a layer of particles that are organized into addressable regions (formatted). In the process of reading and writing data, a read/write head passes over the particles to determine, or set, the magnetic state of a given region. Two types of media use magnetic storage: disks and tapes. You can see a hard disk's storage surface and read/write heads in Figure 2.19.

Magnetic disks can be thin steel platters (hard disks) or Mylar plastic film (floppy disks). When reading data from or writing data onto a disk, the computer can go directly to the desired piece of data by positioning the read/write head over the proper track of the revolving disk. Thus, the disk is called a *direct access* storage medium.

Magnetic disk storage varies widely in capacity and portability. Fixed hard disk (hard drive) storage has large storage capacity, 40–80 GB on most new notebook and desktop PCs, and fast access time. For these reasons, personal computers rely on the hard drive as the main secondary-storage medium. *Microdrives,* tiny hard drives that can store gigabytes of data on a disk one or two inches in size, are transforming handheld computers. Tiny microdrives that can hold several gigabytes (see Figure 2.20) have been available as add-ons for portable devices. Some cell phones and handheld computers are offered with hard drives built in. These devices traditionally have been limited to 32, 64, or 128 MB of storage. Boosting the storage up to the gigabyte range provides these devices

FIGURE 2.19 • Hard disk drive

Hard disk drives store data on multiple stacked disks called *platters* that are read by read/write heads.

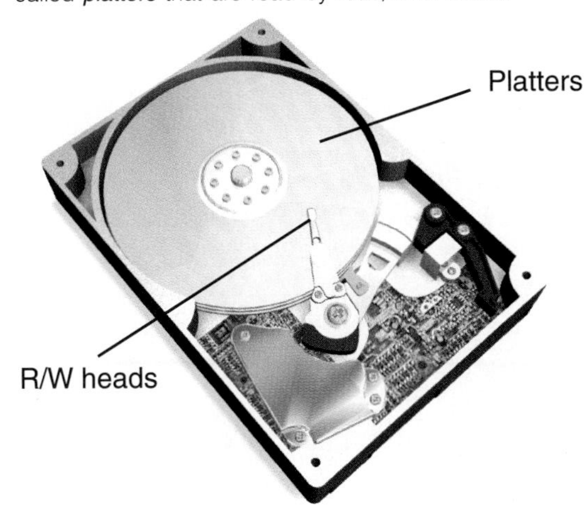

with the capability to store hours of music, video (TV shows, movies, or personal video), hundreds of photos, data files, and software. The new LifeDrive handheld from PalmOne is the first mainstream handheld PC to include a 4 GB hard drive as standard equipment.[12] Many more such devices are sure to follow. Adding a hard drive is a major step in bringing full desktop capabilities to handheld devices. The 20 GB and 40 GB Toshiba microdrives are what made the first iPods so popular.

Floppy disks are a portable, low-capacity (1.44 MB) storage medium. The most popular form of portable storage in the 1990s, floppies are being phased out in favor of higher capacity, less expensive media. Once standard on new PCs, floppy drives are now an optional feature that is increasingly being turned down by computer shoppers. To understand why, consider the fact that a 256 MB flash drive holds 177 floppy disks worth of data, a CD holds 486 floppy disks worth, and the microdrive in Figure 2.20 holds over 4000 floppies worth of data. Clearly floppy disks are old technology.

High-capacity diskettes, such as the Iomega Zip disk, and the Imation Superdisk allow you to store up to 83 times as much data as on a standard floppy disk in about the same amount of space and cost a bit more than floppy disks. The release of writable CDs and flash drives has significantly and negatively affected the market for all portable magnetic disk storage.

FIGURE 2.20 • Microdrive

This tiny microdrive can hold about 1200 MP3 songs or several full-length motion pictures.

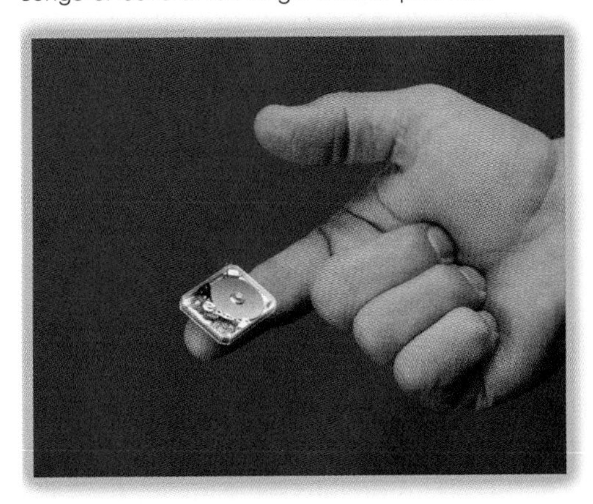

Magnetic tape is a storage medium used by businesses and organizations that need to store and back up large quantities of data (Figure 2.21). Similar to the kind of tape found in audio or video cassettes, magnetic tape is Mylar film coated with iron oxide particles. Magnetic tape is an example of a *sequential access* storage medium because data is written and read in sequential order from the beginning of the tape to the end. Although access is slower, magnetic tape is usually less expensive than disk storage. For applications that require access to very large amounts of data in a set order, sequential access is ideal. For example, government agencies, such as the U.S. Census Bureau, and large insurance corporations maintain too much data to store on regular hard drives. Magnetic tape provides a medium that can handle these large data sets. When these organizations conclude work on a data set, the tape is removed and stored, and a new tape takes its place. Magnetic tapes and tape cartridges are also used to back up disk drives and to store data off-site for recovery in case of disaster. Increases in hard disk drive capacities and advances in disk storage technologies are causing many businesses to turn from tapes to disks for their high-capacity data storage needs.

Business and organizations often provide large quantities of storage to employees over a network. Server computers can provide a central store for important corporate data for employees to share. Arrays of disks can be formed and used in groups to handle terabytes of data. A relatively new technology called a *storage area network* or *SAN*, links together many storage devices over a network and treats them as one large disk. Hackensack University Medical Center, in Hackensack New Jersey, used a SAN to assist in going fully digital—paperless and filmless. The hospital stores digital images of X-rays, ultrasounds, and

FIGURE 2.21 • Magnetic tape storage

Frank Elliott, IBM vice president for tape storage, holds IBM's latest 1 terabyte tape cartridge that holds the equivalent of the 1500 CDs that surround him.

MRIs, prescription medicine orders, medical test results, patient's medical histories, medical references, and of course patient billing and other business related-data all on a huge, high-speed SAN, that interconnects different kinds of data storage devices with associated data servers.

In order to safeguard the valuable data stored on their computer systems, some businesses implement RAID. *RAID* or *redundant array of independent disks* uses a second system of disks to maintain a backup copy of the data stored on the primary disks. If the original drives or data become damaged, the secondary disks can take over with little loss of time or work. RAID can turn a tragedy into only a momentary inconvenience and safeguards companies from the risk of down time.

Optical Storage

Optical storage media, such as CDs and DVDs, store bits by using an optical laser to burn pits into the surface of a highly reflective disk surface. A pit represents a 0, and the lack of a pit represents a 1. The 1s and 0s are read from the disk surface by using a low-power laser that measures the difference in reflected light caused by the pits (or lack thereof) on the disk. Audio CDs that store music, data CDs that store software, and DVDs that store motion pictures all use the same fundamental technology. The advantages of optical storage over portable magnetic storage are in capacity and longevity. Optical media can store thousands of times more data than a floppy disk in nearly the same amount of disk space. Data stored on magnetic media may begin to deteriorate in less than 10 years, but data on a CD could last more than 100 years.

The first optical media to be mass-marketed to the general public was the CD-ROM. A **compact disk read-only memory (CD-ROM)**, commonly referred to as a CD, is an optical media that stores up to 700 MB of data. Just like the ROM on your motherboard, once data has been recorded on a CD, it cannot be modified—the disk is read-only. Originally designed to store music, the CD soon replaced the cassette tape as the most popular form of music distribution, and migrated to more general data storage uses in the computer marketplace. The CD ideally satisfied the increasing storage demands of large software programs. Software that formerly required dozens of floppy disks could be distributed on one inexpensive and convenient CD.

A **digital video disk (DVD)** stores over 4.7 GB of data in a fashion similar to CDs except that DVDs are able to write and read much smaller pits on the disk surface (see Figure 2.22) and can sometimes write to and read from multiple disk layers. Unlike CDs, DVDs can store an entire digitized motion picture. An additional benefit of DVD drives is that they are backward-compatible with CD-ROMs, meaning that they can play CDs as well as DVDs. *Backward-compatible* is an expression used to indicate that a new version of some technology still supports the specifications of the old version.

FIGURE 2.22 • CD, DVD, and Blue

CDs, DVDs, and new Blu-ray (BD) and Blue-laser disks (BDs) all use optical technologies with varying levels of precision to store increasing amounts of data in the same amount of space.

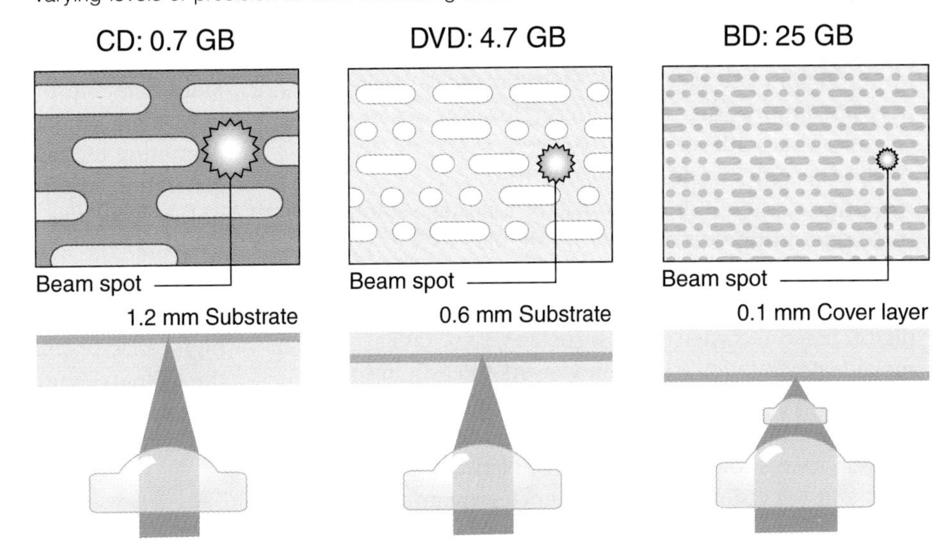

Optical disk capacities will continue to increase, and access times will continue to decrease. Proof can be found in the new blue-laser-based optical disk system that stores 50 GB on a standard size optical disk. Toshiba's Blue-laser system and the Blu-ray system developed by the Blu-ray Disc Association (BDA) (*www.blu-ray.com*) are able to store more data than DVDs because of the shorter wavelength of blue light. Newer versions of the technology have extended the capacity to 100 GB—enough capacity to store 100 hours of video on one optical disk.

The popularity of the CD and DVD has soared since the introduction of recordable optical disks. The process of writing to an optical disk is sometimes called *burning*. A number of different CD and DVD burners are available today at reasonable prices. Different manufacturers support different standards and formats in efforts to win customers over to their products. Manufacturers use *R* to indicate that a media is recordable; that is, it can be written to only once. *RW*, for rewritable, is used to indicate that a disk can be rewritten numerous times just as you would a hard drive. Currently, the most popular format for writable CDs is **CD-RW.** The most popular DVD burners are combination drives that support the two most popular DVD standards, +RW and –RW. CD-RW and DVD drives are also being combined and have become popular as standard equipment on new PCs.

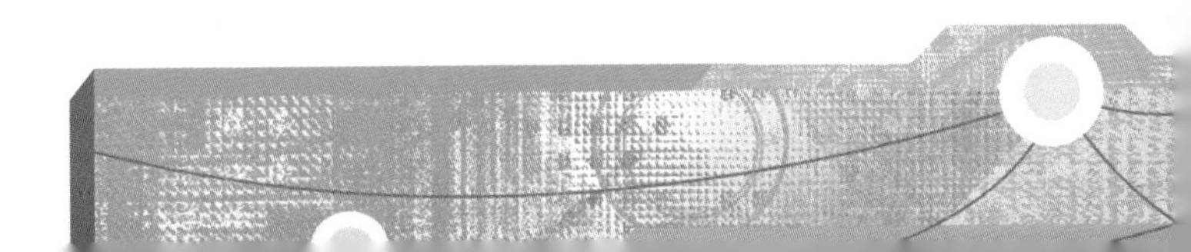

Hard Drives that Deliver Power to the People

Sometimes new developments in hardware technologies have a profound impact on our culture—and our economy. Five years ago most computers and digital media kept users tethered to desks or entertainment centers. New technologies have freed us from those shackles and allowed us to take our computing and media wherever we go. Consider two recent technologies that are changing the way we live.

Toshiba is a leader in the field of tiny, high-capacity hard drives. Toshiba's 1.8-inch drive provides up to 60 GB of storage space. This statistic in itself may not impress many tech users. However, when you consider that without this drive there would be no iPod, it becomes quite significant. Toshiba is now working on a 4 GB drive the size of a postage stamp that could bring your favorite music play lists to your cell phone.

Large-capacity hard drives have also made possible digital video recorders (DVRs) such as Tivo. DVRs make it possible for television viewers to easily record dozens of hours of television programming to watch at any time. These set-top boxes include a feature that allows viewers to skip commercials. Network TV executives and television marketing agencies are anticipating the end of traditional mass-marketing and are developing new ways to pitch products to the future TV generation. Sony is developing a DVR, named Vaio Type X, which stores a terabyte of data, equivalent to 1000 hours of programming, and that can record programs from seven channels simultaneously.

Both Toshiba's tiny hard drives and Sony's mega DVR are new hardware technologies that provide more freedom and power to individuals.

Questions

1. What other types of conveniences, besides listening to music, might large-capacity, portable, minidrives provide?
2. Many television viewers have lost tolerance for mass-marketing TV commercials. What television marketing techniques might appeal better to a new generation of television viewers using new technologies such as Tivo?
3. "Anytime, anywhere" computing requires access to any data any time. In the future, do you predict that mobile computing will make use of portable high-capacity storage media or wireless networks as the primary source of data access? What are the benefits and drawbacks of both methods of data access?

Sources

1. *Toshiba's hard drive Web site, http://sdd.toshiba.com, accessed September 25, 2004.*
2. *Haughey, M. "Monster DVR,"* PVRBlog, *May 10, 2004, www.pvrblog.com/pvr/2004/05/monster_dvr.html.*

One of the important specifications to consider when purchasing a CD or DVD burner is the time it takes to read and write data from and to the disk. Specifications for a typical CD-RW/DVD combo drive may read "52x32x52 CD-RW & 16X DVD." The *X* stands for "times the original transfer rate" of a disc. This drive can burn a CD-R disk at 52 times the original transfer rate and a CD-RW at 32 times the rate. It reads data from a CD at 52 times the original transfer rate of CDs and reads DVDs at 16 times the original transfer rate of DVDs.

Flash Drives and Cards

A **flash memory card** is a chip that, unlike RAM, is nonvolatile and keeps its memory when the power is shut off. Flash chips are small and can be easily modified and reprogrammed, which makes them popular in computers, cellular phones, digital cameras, and other products. When used in media devices such as digital cameras, camcorders, and portable MP3 players, flash memory cards are sometimes referred to as *media cards*. They serve media devices well, since they are small with high capacity. Compared to other types of secondary storage, flash memory can be accessed more quickly, consumes less power, and is smaller in size. The primary disadvantage is cost. A flash chip can cost almost three times more per megabyte than a traditional hard disk. Nonetheless, the market for flash chips has exploded in recent years.

USB flash drives, also called *thumb drives*, are small flash memory modules about the size of your thumb or smaller, shown in Figure 2.23, that conveniently plug into the USB port of a PC or other digital electronics device to provide convenient, portable, high-capacity storage. **Universal Serial Bus (USB)** is a standard that allows a wide array of devices to connect to a computer through a common port. USB drives have all but replaced floppy disks on many college campuses. Although they are called *drives*, they contain no mechanical parts as do magnetic or optical drives. They use flash memory technology to store anywhere from 16 MB to multiple gigabytes of data, ranging in price from $10 to over $1000. USB drives attach to the computer using the USB port found on most computers. Once attached, the operating system recognizes the drive and assigns it a drive letter, such as the G: drive. Saving a file to the G: drive stores it on the USB flash drive. When done, you simply remove the flash drive from the PC and take it with you to use later on some other PC. Although USB

FIGURE 2.23 • Flash drives

Flash drives attach to a USB port and come in a variety of shapes and styles. Some can store more than a gigabyte of data.

flash drives do not come close to CD-RWs and DVD+RWs in terms of cost per MB, the convenience they offer make them well worth the expense. Flash drives come in a variety of shapes and styles and some are attached to key rings to double as key fobs.

Evaluating Storage Media: Access Method, Capacity, and Portability

As with other computer system components, the access methods, storage capacities, and portability you require of secondary-storage media are determined by your objectives. An objective of a credit-card company's computer system, for example, might be to rapidly retrieve stored customer data in order to approve customer purchases. A fast access method is critical to the success of the system. In a hospital setting in which physicians are evaluating patients, then portability, capacity, and privacy might be major considerations in selecting and using secondary-storage media and devices.

Storage media that provide faster access methods are generally more expensive than media that provide slower access. The cost of additional storage capacity and portability varies widely, but is also a factor to consider. For example, you can burn over 4 GB of data to a DVD that costs less than one dollar; 4-GB flash drives are just entering the market at $1700. The flash drive offers benefits of speed, flexibility, and portability. In addition to cost, you may also need to address security issues. How should the secondary-storage devices be controlled so as to allow only authorized people access to important data and programs?

Clearly the trend in storage media, as in most areas of technology, is toward smaller, more powerful (higher-capacity), and less expensive devices, such as the use of iPods as storage devices. The magnetic hard drive still reigns supreme as the choice for day-to-day storage because of its large capacity and low price. Writable optical disks are rapidly taking over the portable data storage market due to their high capacity and low price. Computer vendors are phasing out floppy drives, leaving them off new computers unless specifically requested by the customer, and CD-RW/DVD drives have become standard. Table 2.6 and Figure 2.24 compare and contrast the differences in storage media.

TABLE 2.6 • Comparing data storage devices and media

Storage device	Avg. cost of device	Capacity	Cost of media (risk/tape)	Avg. cost per MB	Technology	Portable
Floppy disk	$20	1.44 MB	$0.22 in bulk	$0.15	Magnetic	Yes
Iomega ZIP disk	$100	250 MB	$11.00	$0.04	Magnetic	Yes
CD-RW	$40	700 MB	$1.00 in bulk	$0.0014	Optical	Yes
DVD+RW	$90	4.7 GB	$1.00 in bulk	$0.0002	Optical	Yes
Flash drive	N/A	16 MB–2 GB	$12–$200	$.75	Solid State	Yes
Hard drive	N/A	20 GB–400 GB	$30–$300	$0.0007	Magnetic	No
Tape	$700	2 GB–400 GB	$2–$35 in bulk	$0.0002	Magnetic (sequential)	No

FIGURE 2.24 • Optical versus magnetic portable storage

The dramatic increase in capacity and the lower cost of optical media over magnetic storage make optical media the preferred choice for today's computer users.

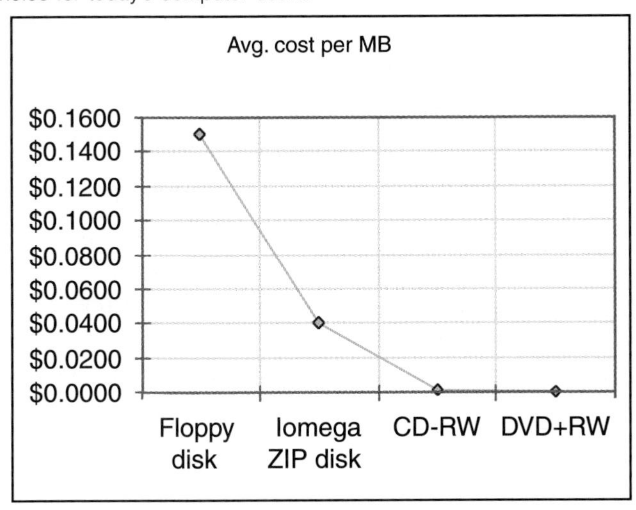

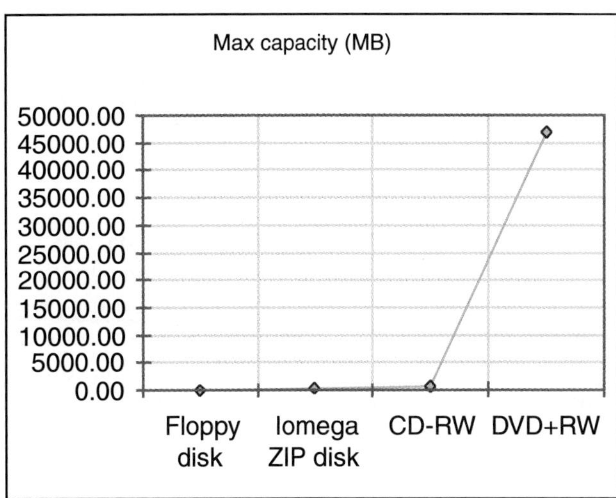

INPUT, OUTPUT, AND EXPANSION

Users interact with computers through input and output (I/O) devices. Of all the computer hardware components, I/O devices have the most direct impact on a user's computing experience. To accommodate a wide variety of data and the many environments in which data is processed, there are literally hundreds of different input devices on the market. By learning about input devices, you also learn what computers are capable of. Output devices connect directly with our senses. Although most output from a computer is visual, much is auditory, and some exotic devices even affects our other senses. This section explores input and output and the different peripheral devices that users add to their computers to expand their functionality.

Input and Output Concepts

An **input device** assists in capturing and entering raw data into the computer system. Successful input devices must be easy to learn to use and effortless to manipulate. An **output device** allows you to observe the results of computer processing with one or more of your senses. Personal computers typically use only a handful of I/O devices. There is a good chance that you have used a keyboard, mouse, display monitor, and printer for much of your life. Using these devices is second nature to many people. New personal computing devices and product designs are often challenging to learn new techniques for input. Touch pads and TrackPoints are easy to become accustomed to on notebook computers. Touch screens, stylus, and handwriting recognition may be a bit more challenging on handheld computers. These basic

DON'T TOUCH THAT DIAL

A problem for doctors in the operating theater has always been an inability to operate electronic equipment because it can't be sterilized. A hologram embedded with sensors now allows the surgeon to activate the dial without touching it. The hologram of the relevant control panel floats in the air somewhere within easy reach of the surgeon's hand. When they "touch" it, it relays a message to whatever electronic equipment it is connected to.

Source:
Something in the air: Holographic controls
By Patricia Resende
Mass High Tech
July 11, 2005
www.masshightech.com/displayarticledetail.asp?art_id=69090&cat_id=95

iPod Goes to College

Duke University in Durham, North Carolina, was the first school to provide all incoming freshmen with their own 20 GB iPod. Later they expanded the program to include some upper-class students. How did they justify this expense? Rather than viewing the iPod as a music player for recreation, they viewed it as a student productivity tool.

After the first year of this experiment, 75 percent of students said that they made use of the iPods for academic work, typically a combination of teacher-prescribed and self-prescribed methods. Some of the ways that students used the iPods included:

- Recording live lectures
- Downloading and listening to recorded lectures
- Recording interviews for journalism class
- Studying foreign languages
- Recording music for music classes
- Storing files and photos
- Organizing research notes
- Listening to audio books

Professors at Duke reported that students who used the iPods seemed more engaged in classes, provided more dynamic presentations using the iPods, and were more accurate in quoting from interviews. Some students felt the iPods were underutilized in many classes. Duke University has cut back somewhat on its program and now distributes iPods only to students enrolled in courses such as music and language instruction.

Many other colleges are jumping on the iPod bandwagon. Drexel University's School of Education in Philadelphia provides iPods to faculty and freshmen. "We want students to be able to take the professor with them wherever they go," says William Lynch, director of the school. Professors upload class assignments, readings, audio files, and other material to a secure server where the students can access the information and download it to their iPods. Other colleges that have distributed iPods to students include Georgia College & State University in Milledgeville, Georgia, and Stillman College in Tuscaloosa, Alabama.

Questions

1. What advantages do students who use an iPod have over students who take handwritten notes or record using a tape recorder?
2. Besides recording lectures, what types of classes or academic activities might be particularly suited for the iPod's capabilities? Why?
3. What concerns might teachers and students have over the use of the iPod in class?

Sources

1. *Moore, E.A. "When iPod Goes Collegiate,"* The Christian Science Monitor, *April 19, 2005, www.csmonitor.com/2005/0419/p11s01-legn.html.*
2. *iPod Web site, www.apple.com/ipod/musicandmore.html, accessed May 19, 2005.*
3. *"Duke iPod First-Year Experience," www.duke.edu/ipod/, accessed May 19, 2005.*

I/O devices are a small subset of all the specially designed input devices used in industry. From drawing tablets used by graphic artists to magnetic ink character readers used by banks to bar code readers at the supermarket, a variety of input devices have been designed to accommodate the unique needs of professionals. The goal behind the design of such devices is always speed and functionality.

EXPAND YOUR KNOWLEDGE

To learn more about other input and output devices, go to www.course.com/swt2/ch02. Click the link "Expand Your Knowledge" and then complete the lab entitled "Peripheral Devices."

Speed and Functionality. Rapidly and accurately getting data into a computer system is the goal of most input devices. Some activities have very specific needs for output and input, requiring devices that perform specific functions. For example, many reporters use input devices that record and transcribe human speech to ease the stress of typing late-breaking stories. The more specialized the application, the more specialized the associated I/O device. UPS had special electronic pads created for its drivers to use to collect customer signatures (see Figure 2.25). The pads improved the efficiency of UPS' recordkeeping.

FIGURE 2.25 • UPS signature device

UPS drivers collect customers' signatures by having them sign an electronic pad that stores the signature electronically.

The Nature of Data. *Human-readable* data can be directly read and understood by humans. A sheet of paper containing lists of customers is an example of human-readable data. By contrast, *machine-readable data* is read by computer devices. Customer data that is stored on a disk is an example of machine-readable data. Note that it is possible for data to be both human-readable and machine-readable. For example, both human beings and computer system input devices can read the magnetic ink on bank checks.

Source Data Automation. Regardless of how data gets into the computer, it is important that it be captured near its source. *Source data automation* involves automating data entry where the data is created, thus ensuring accuracy and timeliness. Source-data automation is used by librarians who use scanners to check out and check in library materials. The moment a book is scanned at check out, its status changes in the online card catalogue to "checked out." Some rental-car companies use automated scanners that scan vehicles as they exit and enter the rental lot. The scanner collects data on the vehicle, including the mileage and gas gauge data, date, and time, and prepares a customer invoice before the customer returns to the counter.

As you enter into the era of pervasive and ubiquitous computing, everyday objects may serve as I/O devices. Microsoft researchers are working on technologies that turn any surface into a touch sensitive display (Figure 2.26). Imagine checking text messages on your kitchen table as you eat breakfast, and viewing your photo album on the kitchen wall!

FIGURE 2.26 • New forms of I/O

Microsoft's surface computing project uses combinations of sensors, cameras, and projectors to turn various surfaces, such as kitchen tables, desks, counters, or walls, into computing interfaces.

FIGURE 2.27 • Thumb navigation

The iPod Click Wheel is a special-purpose input device that allows for easy navigating of the device's menu system.

Input Devices

Input devices can be classified as either general purpose or special purpose. A *general-purpose I/O device* is designed for use in a variety of environments. This category of I/O devices includes keyboards and displays. A *special-purpose I/O device* is designed for one unique purpose. An example of a special purpose I/O device is the pill-sized camera from Given Imaging that, when swallowed, records images of the stomach and the small intestine as it passes through the digestive system. Another example of a special-purpose input device is Apple's patented Click Wheel (Figure 2.27) that is used to navigate the menu system of the iPod. Apple advertises that "without lifting that trusty thumb of yours from the wheel, you can easily select playlists, scroll through thousands of songs and start the music playing."[13]

Personal Computer Input Devices. Today's PCs are multimedia devices that can input (and output) many kinds of data, such as text, audio, and video. Various input devices are used to capture these types of data, including keyboards, mice, microphones, digital cameras, and scanners.

A computer keyboard and a computer mouse are the most common input devices used for entering data such as characters, text, and basic commands. A number of companies are developing new keyboards that are more comfortable, adjustable, and faster to use. For example, Microsoft's *ergonomic keyboard* is designed in such a way as to reduce the stress on your wrists common with traditional keyboards.

EXPAND YOUR KNOWLEDGE

To learn more about input devices, go to www.course.com/swt2/ch02. Click the link "Expand Your Knowledge" and then complete the lab entitled "Using Input Devices."

A computer mouse is used to direct the computer's activities by selecting and manipulating symbols, icons, menus, or commands on the screen. Different types of mice are available, including corded and cordless, one-, two-, or three-button, with scroll wheel and without. Some users prefer a trackball over a mouse. A *trackball* sits stationary and allows you to control the mouse pointer by rolling a mounted ball. If you've ever tried to draw pictures using a mouse, you know how frustrating it can be. *Graphics tablets* allow you to draw with a penlike device on a tablet to create drawings on your display (Figure 2.28).

FIGURE 2.28 • Graphics tablet

Graphic artists and photographers use graphics tablets for maximum control of a digital brush, pen, or pencil.

Mobil Input Devices. Other methods for entering data are tailored for mobile computing. Notebook computers integrate the mouse either as a touch-sensitive pad below the spacebar (called a *touch pad*) or a rubber nub, similar to a pencil's eraser, in the center of the keyboard (called a *TrackPoint*). By moving your finger across the pad, or applying directional pressure to the nub, you direct the mouse cursor on the screen.

Smaller mobile devices are doing away with the keyboard and mouse altogether. Handheld computers and tablet PCs use a **touch screen** that allows you to select items on the screen by touching them with your finger or a *stylus*—a short penlike device without ink (see Figure 2.29). These devices translate characters written on the screen with a stylus into ASCII characters that can be stored and edited in a word-processing document. There are different handwriting recognition systems that make use of varying styles of writing. Some require the user to use a specific alphabet of characters that the system can recognize. For example, the alphabet used for Graffiti, the first handwriting recognition system designed by Palm, is shown in Figure 2.30.

FIGURE 2.29 • Stylus and touch screen

Most handheld computers rely on stylus and touch screen for input.

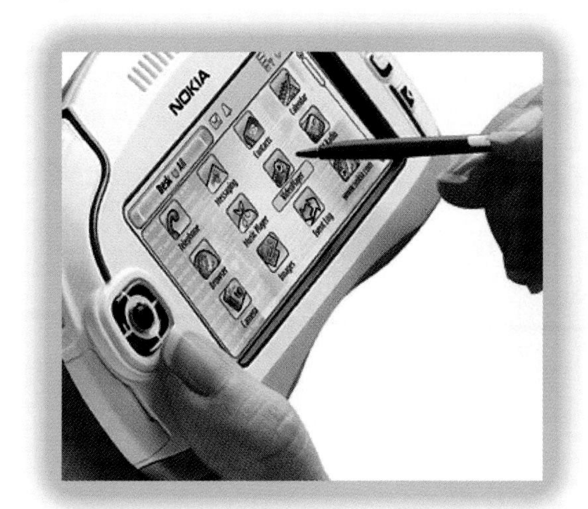

Microphone Input Devices. Microphones can take human speech as input, digitize the sound waves, and use *speech recognition software* to translate the input into dictated text that appears on the screen or into commands. On the factory floor, equipment operators can use speech recognition to give basic commands to machines while using their hands to perform other operations. Speech recognition software can be used to take dictation, translating spoken words into ASCII text. *Voice recognition,* a similar technology, can be used by security systems to allow only authorized personnel into restricted areas.

Gaming Devices. Gamers enjoy specialized input devices that let them quickly react to game action. Most gamers prefer to use a **gamepad** device to control game characters and objects (see Figure 2.31); some games, such as flight simulators, are easier to navigate using a *joystick,* a device resembling a stick shift.

FIGURE 2.30 • Graffiti

Palm invented Graffiti, a standard for handwritten characters for easy handwriting recognition.

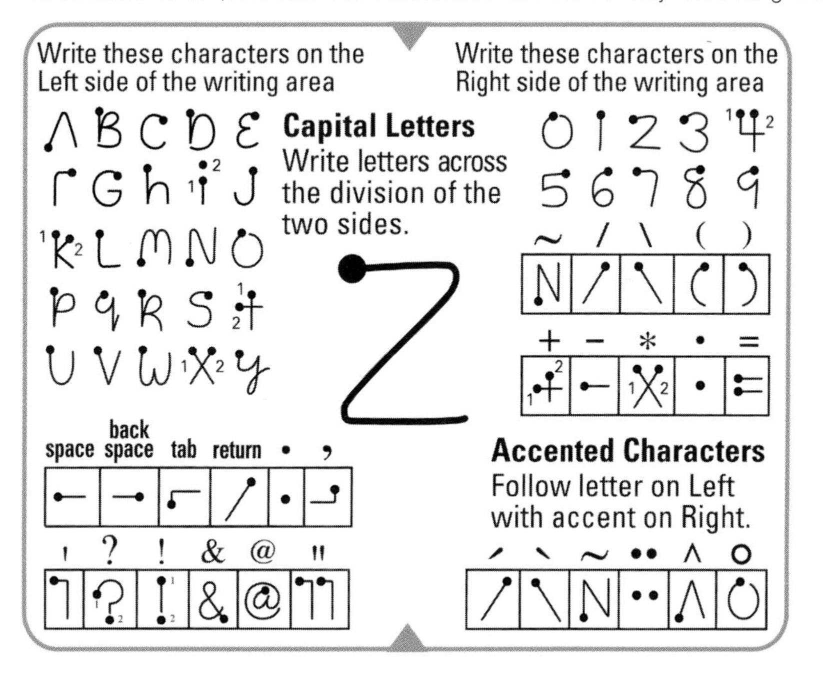

EXPAND YOUR KNOWLEDGE

To learn more about the mouse, keyboard, and other input devices, go to www.course.com/swt2/ch02. Click the link "Expand Your Knowledge" and then complete the lab entitled "Using Input Devices."

Digital Cameras. Many people are switching from film to digits as they discover the convenience of digital photography. A *digital camera* captures images through the camera's lens and stores them digitally rather than on film. The quality of a digital camera is typically judged on how many megapixels (millions of pixels) can be captured in an image.A *pixel* is one of many tiny dots that make up a picture in the computer's memory. A traditional inexpensive film-based camera produces 1.2 megapixel images. Today, you can buy a 2-megapixel digital camera for under $100, and a professional-grade 8.2-megapixel camera could cost as much as $1500. As with most technology, prices in the digital camera market are rapidly dropping. Most cameras capture images on a flash

FIGURE 2.31 • Gamepad

Game enthusiasts use a gamepad device to control game characters in a virtual world.

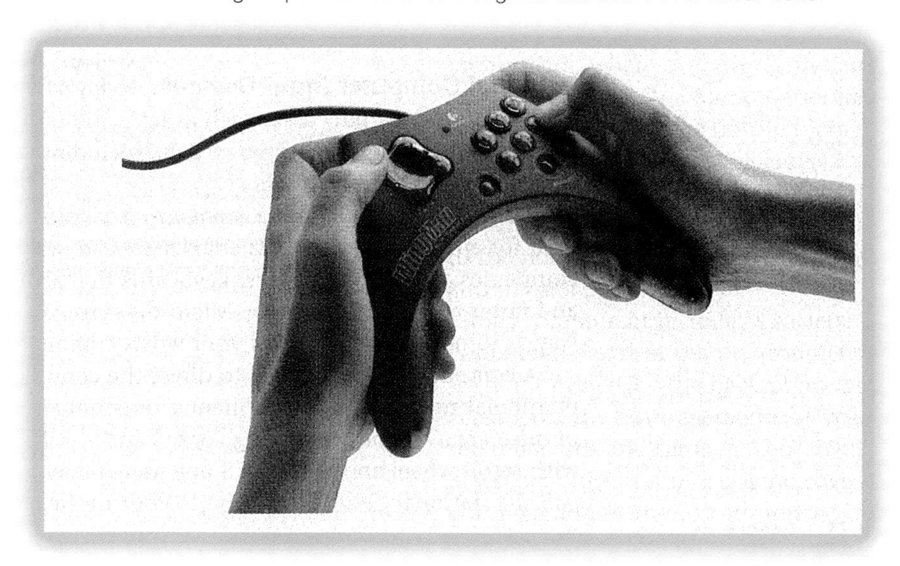

memory card that you can download to your PC using a USB cable. Some computers and printers have flash memory card readers that allow you to download pictures directly to your PC or printer. The images taken by camera phones are also increasing in size and quality. There are several 5-megapixel cell phones on the market, and Samsung has recently released a 7-megapixel model in Korea.[14]

Many digital cameras, in addition to taking still images, allow you to capture short video recordings. If you are interested in longer video, you can purchase a camcorder. A bit more costly, *digital camcorders* allow you to take full-length digital video that you can watch on your TV, download to your computer, or transfer to CD, DVD, or VCR tape. *Webcams* provide a lower-priced video camera for use as a computer input device. They are ideal for video conferencing over the Web, as shown in Figure 2.32.

Scanning Devices. You can input both image and character data using a scanning device. Both *page scanners* and *handheld scanners* can convert monochrome or color pictures, forms, text, and other images into digital images. It has been estimated that U.S. enterprises generate over one billion pieces of paper daily. To cut down on the high cost of using and processing paper, many companies look to scanning devices to help them manage their documents. Combined with *character recognition software*, a scanner can transform document images into editable word-processing documents.

The field of biometrics uses a variety of scanning devices. *Biometrics* is the study of measurable biological characteristics. Biometrics is becoming increasingly useful in the area of security as a tool for confirming a person's identity. Scanners and software can verify an individual's identity by examining biological traits, such as retinal or iris patterns, fingerprints, or facial features. First Financial Credit Union is using one such device to verify customer identity. The bank has installed kiosks at remote locations where customers are able to access all the services that a branch office has to offer, including applications for new accounts and loans. The kiosks save the bank the cost of opening new branches. Fingerprint scans assure the bank that the customers making the transactions are who they say they are.

FIGURE 2.32 • Webcam

Webcams connect to your computer for video conferencing over the Internet.

Businesses and organizations use a number of special-purpose scanners and optical readers to collect data. A *magnetic ink character recognition* (*MICR*) device reads special magnetic-ink characters such as those written on the bottom of checks. *Optical mark recognition* (*OMR*) readers read "bubbled-in" forms commonly used in examinations and polling. *Optical character recognition* (*OCR*) readers read hand-printed characters. *Point-of-sale (POS) devices* are terminals, or I/O devices connected to larger systems, with scanners that read codes on retail items and enter the item number into a computer system.

Output Devices

Computer output consists of the results of processing produced in a manner that is observable by human senses or that can be used as input into another system. Output can be visual—on a display or printed page—or audio—through speakers; in the case of virtual reality systems, output can even be tactile and olfactory to recreate real environments. Output from one system can be fed into another system as input. For example, consider the system that controls natural gas distribution throughout the United Kingdom. One computer system monitors the pressure in the natural gas pipes that crisscross the country. The output from that system is fed into the computer system that regulates the flow of gas to maintain consistent and safe levels.

TechEdge

HELPING-HAND HARDWARE

Her vocabulary isn't much—just 260 words—and she only has a "command grammar" of 75 phrases, but Clarissa aids astronauts as they perform any of the 12,000 procedures assigned. The first voice-activated computer in space, she asks them what they need and then reads them the instructions as they ask for them, allowing them to keep their eyes and hands on what they are doing. And she isn't going to take over the space station anytime soon.

Source:
Space station gets HAL-like computer
By Maggie McKee
NewScientist.com news service
June 2005
www.newscientistspace.com/article.ns?id=dn7584

Display Monitors. Remarkable progress has been made with display screens, including those used with personal computers. With today's wide selection of monitors and displays, price and overall quality can vary tremendously.

The first consideration when selecting a display is typically size. Display size is measured diagonally and ranges from 13, 14, and 15 inches for notebook computers to 17, 19, and 21 inches for desktop displays. There are of course extremes. For example, Dell offers a Media Center PC with a 24" widescreen display that supports high-definition TV. Apple offers a line of widescreen "Cinema Displays" the largest of which is 30 inches and holds more than 4 million pixels (Figure 2.33). If you use a lot of graphics applications, games, video, artwork, or like to multitask with many windows open on the screen at once, you probably want to go with a larger display size. If you mostly deal in text, e-mail, and Web pages, and wish to conserve desktop space or go mobile, a smaller display size should suit your needs. Widescreen displays have become popular both on notebook computers and desktops as increasing numbers of users are viewing movies on their PCs.

FIGURE 2.33 • Apple's cinema displays

Widescreen displays have become popular on desktop and notebook systems for viewing widescreen high-definition movies.

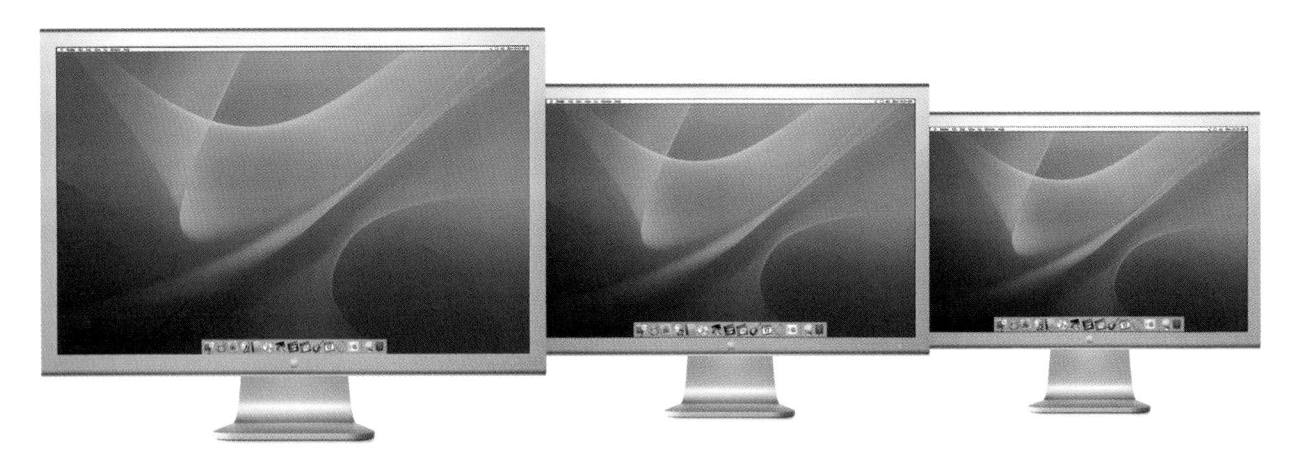

The quality of a screen is often measured in pixels. **Display resolution** is a measure, in width by height, of the number of pixels on the screen. *Dot pitch* refers to the measure of space between pixels. The lower the dot pitch, the better the image quality. The higher the resolution, the higher the level of detail and image quality. A screen with a 1024 × 768 resolution (786,432 pixels) has a higher level of sharpness than one with a resolution of 640 × 350 (224,000 pixels). The higher the resolution, the smaller the pixels and images they create. If you want to fit more on your display, you could do so by choosing a higher resolution. On the other hand, if you found that the font or images on your display are to small to see clearly, you could choose a lower resolution that would increase the size of the objects on the display.

TechEdge

NEW COP ON THE BEAT

Cross the optical character recognition system developed to read addresses for the Italian Post Office with four-eyed hardware and you get the newest member of the Los Angeles County Sheriff's Department. Called the *Mobil Plate Hunter 900*, it is bolted to the light bar of a patrol car, where it scans 500 to 800 license plates an hour looking for stolen cars. Happily, it is as accurate as it is fast, with less than 1 in 100,000 wrong.

Source:
Grand Theft Auto Meets Robocop
By Cyrus Farivar
Wired News
www.wirednews.com/news/autotech/0,2554,67864,00.html?tw=wn_tophead_5
June 17, 2005

Knowledge of display technology has become useful when shopping for televisions. Today's display technologies offer many options for televisions as well as computers. *High definition TV* (HDTV) uses a resolution that is twice that of traditional television displays for sharper, crisper images. HDTV uses a widescreen format, which means it uses the same height and width ratio used in movie theaters. Standard television displays and traditional computer displays use a 4:3 ratio representing four inches of width for every three inches of height. Widescreen displays use a ratio of 16:9. As PCs and entertainment media continue to merge it becomes useful to understand the display options of both.

A **liquid crystal display (LCD),** or **flat panel display**, is a thin, flat display that uses liquid crystals—an organic, oillike material—placed between two pieces of glass to form characters and graphic images on a backlit screen. Once used primarily as a laptop display, LCD displays are quickly displacing the old style CRT (cathode-ray tube) displays for desktop systems. The two primary choices in LCD screens are passive-matrix and active-matrix LCD displays. Passive-matrix displays are typically dimmer, slower, but less expensive. Active-matrix displays are bright, clear, and have wider viewing angles than passive-matrix displays.

You have probably heard of plasma display televisions. A *plasma display* is a flat panel display that uses plasma gas between two flat panels to excite phosphors and create light. The use of plasma in displays became popular in the early part of the decade because of its ability to be used to create large, flat, thin televisions at a reasonable cost. LCD has lower limits on size, and, at the time, was more expensive to manufacture. Over time, as LCD production has become less costly, its use in televisions has grown to favorably compete with plasma televisions; especially for televisions under 36 inches in size.

FIGURE 2.34 • The Free2C 3D kiosk

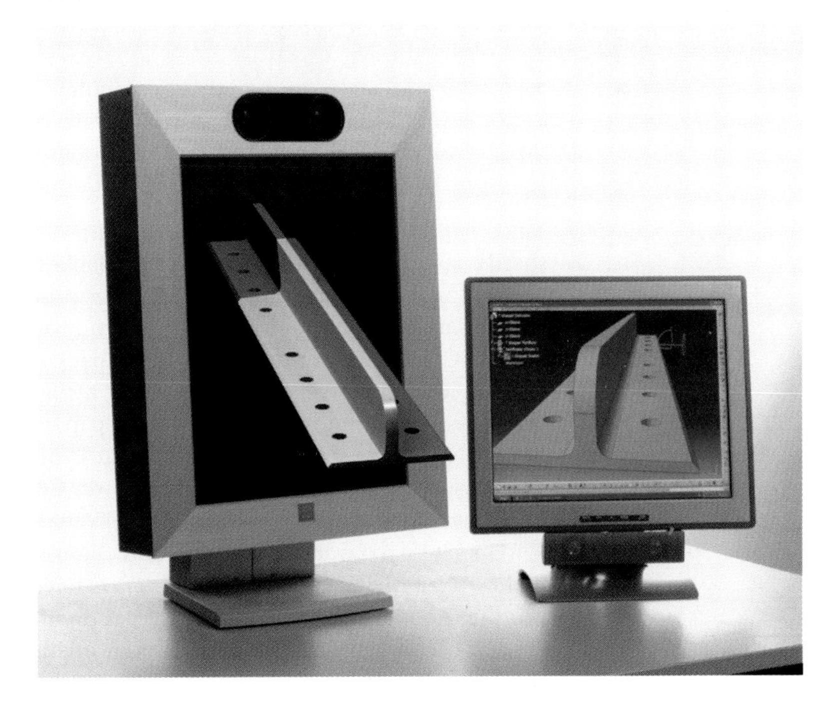

LCD projectors are designed for projecting presentations from your computer onto a larger screen. Typically costing thousands of dollars, these small portable devices are a must-have item for people in businesses that require them to make

3D Displays for Architects, Engineers, and Doctors

A German research institute has developed a 3D display technology that does not require the use of special glasses. The Fraunhofer Institute for Telecommunications developed the Free2C 3D display that generates two slightly different images to make objects appear three-dimensional—even when viewed from the side (Figure 2.34). A lens in front of the display directs one beam of light to the left eye of the viewer and a second beam to the right. A camera at the top of the display tracks the viewer's movement and locks onto right and left eyes to provide the 3D effect from any angle. Other cameras and infrared sensors can be used to manipulate objects on the screen with basic hand gestures.

Doctors at the German University of Teubingen have used the 3D display to train for minimally invasive surgery. In addition to doctors, the device is also being marketed to architects and engineers to assist in visualizing their designs. One implementation packages the 3D display in a kiosk for use in high-tech showrooms.

Questions

1. How might a 3D display be used on your home computer?
2. What uses might professionals in your chosen career area find for a 3D display?
3. What are the limitations of this technology?

Sources

1. *"New 3D Screen Requires No Special Goggles," Reuters, March 10, 2005, www.reuters.com.*
2. *The Free2C 3D Display Web page, www.hhi.fraunhofer.de/english/im/products/Cebit/free2C/, accessed May 19, 2005.*

top, notebook, and even handheld computers. Some people use LCD projectors at home to project HDTV onto a white wall. One drawback is that LCD projection only works well in a semidarkened room.

Printers and Plotters. One of the most useful and popular forms of output is called *hard copy*, paper output from a printer. A variety of printers with different speeds, features, and capabilities are available. The two most popular types of printers are color ink-jet printers and laser printers. A *laser printer* uses techniques similar to those of photocopiers to provide the highest-quality printed output. Color laser printers are rather expensive, so many home users settle for either a less expensive, noncolor laser or a color ink-jet printer. An *ink-jet printer* sprays droplets from ink cartridges onto paper to create pixels. Although ink-jet printers create good looking hard copy, it is not quite as polished as laser printers provide. Also, ink may run if it gets wet, so use care when printing addresses onto envelopes and hope they aren't delivered on a rainy day.

The latest trend in home printers is the photo printer (Figure 2.35). Sparked by low-price digital cameras, *photo printers* have become a popular method for printing photo-quality images on special photo-quality paper. Some photo printers don't require a PC. Just connect your digital camera or media card directly to the printer, use the onboard display to edit the photo if necessary, and print it directly to professional grade photo paper.

Small business owners or people that work from home can benefit from multifunction printers. *Multifunction printers* combine the functionality of a printer,

FIGURE 2.35 • Photo printer

Some photo printers allow you to print directly from the camera or media card without the need for a PC.

fax machine, copy machine, and digital scanner in one device—digital convergence applied to printing technologies. Although you would expect such a device to be very expensive, many can be purchased for less than $200.

The speed of a printer is typically measured by the number of *pages printed per minute* (*ppm*). The quality of resolution of printers is similar to the resolution of display screens. A printer's output resolution depends on the number of dots printed per inch. A printer with a 600 *dot-per-inch* (*dpi*) resolution prints more clearly than one with a 300-dpi resolution. When shopping for a printer, consider the quality of the output (judged by resolution specs and personal evaluation of printed copy), speed of the printer (ppm), price of printer, and how quickly the printer consumes toner or ink along with the price of refills. The initial investment in a printer can end up paling in comparison to the price of toner or ink cartridge refills over time.

Two other older printer technologies can still be found in some workplaces. *Impact printers* use a forcible impact to transfer ink to the paper in order to print characters. *Dot-matrix* printers use impact as well, but are capable of printing characters and graphical images using a matrix of small printing pins.

Plotters are a type of hard-copy output device used for printing large graphic designs. Businesses typically use plotters to generate paper or acetate blueprints and schematics, or print drawings of buildings or new products onto paper or transparencies.

Computer Sound Systems. Most of today's desktop personal computer systems include at least low-quality speakers to output sound. Sound is used by the computer operating system and other software to cue the user to certain events, for example, the flourish of music that plays when Microsoft Windows starts up and shuts down. Other event-driven sounds are included to draw the user's attention to important information; for example, the beep that sounds when a warning dialog box is displayed, or the "You've Got Mail!" exclamation that is familiar even to those who don't use AOL.

Sound systems also support entertainment applications such as CD/MP3 music players and DVD movie players. The sound systems on most notebook and hand-held computers are not of sufficient quality to do justice to such media; so users of such systems typically turn to headphones for higher-quality sound.

Multimedia and gaming enthusiasts often purchase more expensive sound systems for their computers. For example, a surround sound system with a sub-woofer can provide additional realism to games in a virtual reality simulation (Figure 2.36).

FIGURE 2.36 • Surround sound speaker system

This immersive surround sound system from Logitech helps gamers feel as though they are right in the middle of the action.

A computer's ability to output sound is particularly important to individuals who have limited vision. Screen-reader programs such as JAWS, by the Freedom Scientific Corporation, read aloud the text displayed on thc screen.

Special-Purpose I/O Devices

Many special-purpose input/output devices are designed to support scientific and medical research. For example, a new type of digital movie camera that uses ultrafast laser pulses is able to record things faster and smaller than ever. It is providing scientists with striking, 3D color movies of atoms, molecules, and living cells in action. Neurobiologists use the camera to watch the brain think, develop, age, and deal with disease. They can see neurons grow inside the living brain. Such close examination of the workings of the brain are leading to new understandings. For example, a recent study of monkeys controlling a robotic arm using only brain signals found that the monkeys did more than just control the arm; their brain structures adapted to treat the arm as if it were their own appendage. This discovery has profound implications for the potential clinical success of brain-operated devices for humans with disabilities.[15]

Chemists are using another new input device to track the motions of the particles—electrons, protons and neutrons—that make up an atom. To do so, they use an advanced laser strobe light that slices time into the shortest bits yet achieved—*attoseconds*—a billionth of a billionth of a second.[16]

Computer scientists and musicians at the MIT Media Lab are experimenting with special input devices that allow children to compose and perform music without any musical instrument skills. The input devices include a drawing pad that plays music based on children's drawings; soft, squeezable, colorful instruments that let kids mold, transform, and explore musical material; and a variety of other devices that release the creativity in children. The output is beautiful and interesting music that reflects the child's creativity.[17]

FIGURE 2.37 • Wearable PC

Wearable PCs such as this one for divers often use head-mounted displays (attached here to the diving mask) and special hands-free or one-handed input devices.

The area of virtual reality has produced a number of unique and interesting I/O devices. For example, the *virtual reality headset* can project output in the form of three-dimensional color images. Spatial sensors in the headset act as input devices, and when you move your head, images and sounds in your headset change. Virtual reality devices already allow architects to design and "walk through" buildings before they begin construction. They allow physicians to practice surgery though virtual operations and pilots to simulate flights without ever leaving the ground.

Wearable PCs have small system units that can clip to a belt, or fit in a pack, head-mounted displays that only partially obscure vision, and hands-free or one-handed input devices. They are used in industry for individuals who need to have access to data while doing physical work such as repairing a jet engine, or inspecting a city's underground utility tunnels. Figure 2.37 shows an underwater research scientist using a wearable PC designed for use underwater.

Expansion

Most computers provide users with the means to add devices and expand their computers' functionality. A desktop computer user might wish to add a scanner, a notebook user might want to add a webcam for video conferencing, and a handheld computer user might want to add a folding keyboard for more convenient data entry. This section describes the most common methods of expanding a computer system.

Desktop Computer Expansion. One fact that has traditionally complicated system expansion is that peripheral devices use a variety of cables and connectors. Desktop computers provide standard *ports* (sockets) for display, keyboard, printer, and mouse connectors, but have not easily accommodated the many other special-purpose cables and connectors. Those days are over. Today's computers come with a number of USB ports and all kinds of devices that use them (see Figure 2.38). It is not unusual to find six or more USB ports on a new computer, into which you can plug the keyboard, mouse, and additional devices of your choice, such as handheld computers, digital cameras, portable MP3 players, network devices, joysticks, memory modules, and many others, all using a common connector design. USB provides not only a connection to the computer for data transfer, but also a power line that can be used to power a number of useful and entertaining gadgets.

More specialized peripheral devices may come with their own circuit board, called an *expansion board* or *expansion card*, to be installed in your computer.

FIGURE 2.38 • Universal Serial Bus

Many of today's peripheral devices connect to a computer using the USB port.

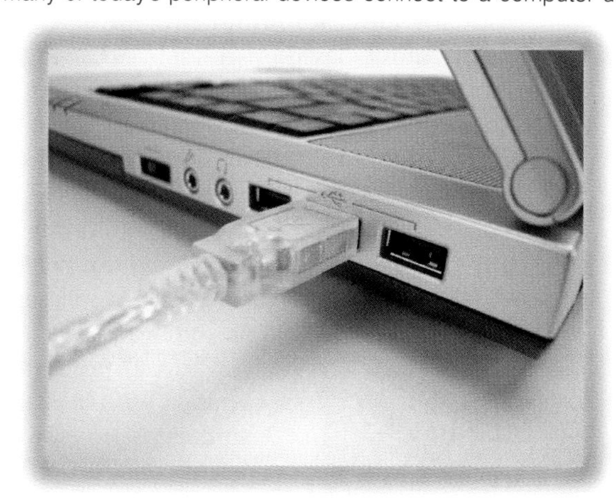

Installation of these devices is not as convenient as simply plugging in a USB connector, but typically it is easy enough for most computer users to handle. The method of installing them varies from machine to machine, so users should consult their owner's manual for specific instructions. After removing the cover from your computer, you see a bank of *expansion slots* on the computer's motherboard (see Figure 2.39); some slots may already contain cards. By viewing the width of the card you wish to install, you should be able to determine which slot can accommodate the card. After installing the card according to instructions, the port on the card is exposed at the back of the computer for use.

FIGURE 2.39 • Expansion slots

More specialized peripheral devices may come with an expansion card to be installed in the expansion slots on the motherboard of your computer.

FireWire (also known as IEEE 1394) competes with USB as a standard for connecting devices to PCs. FireWire was developed by Apple in the 1990s, but now can also be found on Windows PCs. Videographers, musicians, and media hobbyists appreciate the speed at which FireWire transfers data between the computer and devices such as video cameras, digital music players, recording equipment, and other peripheral devices. Although FireWire typically out performes the original USB standard, the latest version of USB, USB 2, competes more closely with FireWire in terms of data transfer speed. Today, more peripheral devices support USB than FireWire. Apple computers include several USB and FireWire ports as standard equipment. Windows PCs do not typically include FireWire support, but users can add a FireWire expansion card.

Bob D'Amico, chief photographer for ABC TV, says that FireWire and Apple computers allow him and his crew to take digital photos with nearly the speed and ease of a film camera. Using PowerBook computers connected via FireWire to their Hasselblad H1 digital cameras, D'Amico and his crew are able to snap off a shot every 1.2 seconds for up to 30 frames stored directly to the PowerBooks.[18]

Mobile Computer Expansion. Like desktop PCs, notebook computers include USB ports for convenient expansion. Notebook computers also provide *PCMCIA slots* that accept *PCMCIA cards*, shown in Figure 2.40, usually called *PC Cards*. PC Cards support a number of devices. For example, wireless network adapters and additional storage devices can be purchased as PC Cards for notebook computers. Notebook computers also include ports to add a standard keyboard, mouse, printer, and display or LCD projector.

FIGURE 2.40 • PC Card

Notebook computers offer expansion opportunities using PC Cards.

Handheld computers can also accommodate additional devices. For instance, you can connect a global positioning system (GPS) to a handheld computer, which when combined with GPS software can display your location on a map. Other options for handheld computers include wireless Internet access, webcams, and LCD projectors. With the addition of a PCMCIA adapter, a handheld can support all the peripherals available to notebook computers.

SELECTING AND PURCHASING A COMPUTER

Putting together a complete computer system is more involved than simply connecting computer devices. Take a moment to think about how car manufacturers build a car. They avoid installing a transmission incapable of delivering the engine's full power to the wheels. Instead, car manufacturers match the components to the intended use of the vehicle. Racing cars, for example, require special types of engines, transmissions, and tires. To select the right transmission for a racing car, you must consider not only how much of the engine's power can

FIGURE 2.41 • Handheld computers in action

Wyndham Resorts decided that handheld computers were the best way to improve its check-in process.

be delivered to the wheels (efficiency and effectiveness), but also how expensive the transmission is (cost), how reliable it is (control), and how many gears it has (complexity). Similarly, you must assemble computer systems so that they are effective, efficient, and well balanced.

You should always base a computer system purchase on a careful study of the needs that it will address. Wyndham Resorts had issues with getting customers through the check-in and checkout process efficiently. To address the problem they equipped their hospitality staff with handheld computers (see Figure 2.41) and sent them out to check customers in at the curb or even when they arrived at the airport. Other resorts have installed kiosks that allow customers to check in and out without the assistance of staff. Kiosks and handheld computers ideally address the specified problem—checking customers in and out as quickly and accurately as possible.

Researching a Computer Purchase

Businesses such as Wyndham Resorts employ IT (information technology) professionals to select and purchase computer systems to meet their business needs. For your own computer purchases you need to function as the IT professional. IT professionals typically begin the process of designing a computer system by studying the needs of the organization. You too, should begin by considering the ways in which you plan to use your computer. Make a list of your typical computing activities and the software that you use. Include software that you are interested in using in the future at work and at home. This list should guide you through this complex decision-making process.

This chapter has provided you with the technical knowledge that is required for making a wise computer purchase. Even armed with this understanding, purchasing a new computer system requires a considerable amount of research. The computer market changes so rapidly that no matter how much you know about computers, it takes some time to catch up with the current state of technology and the market. Fortunately, many free resources are available to assist you in learning as much as you need to know.

Web sites such as *www.cnet.com* and *www.zdnet.com* can help you decide on a computer type and platform. Browsing around your local computer store is another method of investigating the market. Magazines such as *Computer Shopper* and *MacWorld* can provide additional information. Many computer industry magazines have information online as well.Visiting manufacturers' Web sites can be very helpful. Viewing the PCs at *www.gateway.com* and *www.dell.com* can give you

NECESSITY PARENTS INVENTION

Using recycled parts, creative engineering, and offbeat (and dust-tolerant) power sources, India's engineers are coming up with innovations for affordable computing. Stripped-down cheap computers, domain computers that focus on fixed functions, automatic teller kiosks that double as Internet access for a village, computers with flash memory chips for OS and applications instead of a hard drive, and even communal arrangements like "computer grids" are a few of the solutions for bringing technology to underdeveloped areas.

Source:
India's Tech Rennaissance: The $100 computer is key to India's tech fortunes
By Michael Kanellos
CNET News.com
http://news.com.com/Indias+renaissance+The+100+computer/2009-1041_3-5752054.html?tag=nefd.lede
June 29, 2005 4:00 AM PT

a good idea of the state of the market in terms of PC computer specifications and price. Check out *www.apple.com* for information on Apple desktop and notebook computers. Computer retailers such as *www.cdw.com* and *www.mobileplanet.com* provide a convenient format for shopping across a variety of computer types and manufacturers.

Computer Vendors

Computer systems can be purchased online, over the phone, or in a local computer store. They can be purchased new or used, leased, or paid for in installments. The method you use to purchase a computer depends on your personal preferences and financial situation.

Online Vendors. Apple and many of the Windows PC manufacturers provide online stores that allow customers to configure and purchase computers (see Figure 2.42). Buying directly from the manufacturer can sometimes save you money. The main benefit though, is that it allows you to custom configure your PC. Online customers can choose processor, memory capacity, hard disk capacity, secondary-storage devices, display type and size, and numerous other options. PCs purchased direct from the manufacturer, are usually delivered in under two weeks. Users can choose service options that provide in home service should any trouble occur with the PC. Some vendors provide overnight pickup and delivery service that supports repairs by mail for notebook PCs in two working days.

FIGURE 2.42 • The Apple Store

Apple and most PC manufacturers provide online stores that allow customers to purchase directly from the manufacturer.

Online computer retailers such as Computer Discount Warehouse (*www.cdw.com*) sell computers from varying manufacturers (see Figure 2.43). Customers can compare packages from varying manufacturers to find the best configuration and price. Often these sites provide filters that allow you to sort by price, processor speed, popularity, and other specifications. Since these computers are already configured and ready to go, they offer little in the way of customization but very speedy delivery. Computers are covered by the manufacturer's warranty. Online computer retailers provide a good way to scout out computer manufacturers, compare prices, and narrow the field.

FIGURE 2.43 • Online computer retailers

Online computer retailers such as Computer Discount Warehouse sell computers from many manufacturers and include tools that allow you to sort by price, brand, and sometimes popularity.

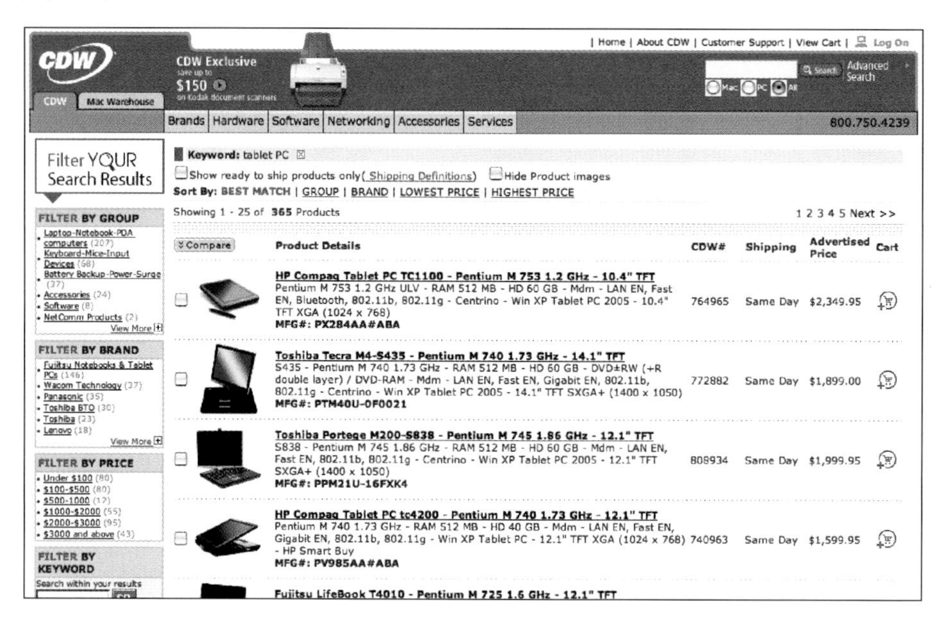

Computer auction sites such as eBay provide yet another option for online computer shopping. Secondhand computers and other bargains found on eBay typically do not include a warranty and thus incur a risk. Because computers lose value quickly, any computer older than four years old is probably not worth anything on the market. Before purchasing a used computer, compare it in price and quality to new PCs. With PC prices dropping dramatically every year, there is little market for used computers.

Local Vendors. Your local computer store offers you the advantage of being able to try out the PC you are considering prior to the purchase. You may also benefit from the advice of a qualified salesperson. What's more, if you get your computer home and find some trouble with it, you may be able to return it to the store for another one. Local stores may also provide special sale prices, package deals, or other incentives that make this option worth your while. Local merchants understand that there are good deals online and often try to match them. For this reason, it is a good idea to shop online prior to visiting a store.

The downside of purchasing locally is that you have little or no opportunity to customize your PC—what you see is what you get. One should also be sure that you are not sold a PC that the store is looking to get rid of, but doesn't really match your needs.

Strategies for Computer Shopping

Purchasing a new computer, whether it is a desktop, notebook, tablet, or handheld, requires an organized process to ensure that you get a computer that is well suited to your needs.

1. **Choose a type of computer: notebook, desktop, tablet, or handheld.** With computers, portability is gained through sacrifices in power. The smaller the computer, the fewer activities it supports. This is true not only in terms of processor power, memory, and storage capacities, but also in terms of display size, mode of input (keyboard versus stylus), and ability to connect to networks and the Internet. Consider the ways in which you use a computer as they relate to these features. Remember that full-size displays, keyboards, and mice can be connected to portable computers. Then weigh the limitations of portable devices against the benefits that portability provides.
2. **Choose a platform.** Once you've decided on desktop, notebook, tablet, or handheld, you should turn your focus to platform. If you are purchasing a desktop or notebook, you need to choose between Windows or Mac (or Linux for those more technically inclined). Tablet PCs are currently only available for Windows. Handhelds are available with Windows or the Palm platform, or you can choose SmartPhones, which provide other options. You should choose the platform that best supports your use. Consider the platform choice of friends and others in your field and seek out impartial online reviews.
3. **Choose a manufacturer.** Once you've selected a platform, you are faced with choosing from manufactures that support the platform. Of course there is only one manufacturer for Apple computers, but all other platforms provide many manufacturers from which to choose. Many find it wise to go with a manufacturer with a long-standing reputation. This way you are assured that the manufacturer will be around to support your equipment. In choosing a manufacturer, consult professional reviews and get the advice of respected friends and associates.
4. **Choose a model.** Once you've decided on a platform and manufacturer, you need to decide on a model based on your budget. Different models and prices reflect differences in components such as processor, memory capacity, hard drive capacity, graphics card, and display size and technology. Typically some trade-offs must be negotiated to keep your computer selection within budget. Keep in mind that the cheapest models support only the most basic needs, and the most expensive models reflect state of the art technologies that are typically released to the market at an inflated price.
5. **Select add-ons.** Often you have options for additional equipment, features, and services. Of these add-ons, the most critical is the warranty. Portable computers are subject to damage, so it is wise to purchase a warranty that covers the length of time that you plan to use the device. Because desktop computers are seldom moved, they are less prone to damage, and you might be able to get by with a shorter warranty. Other add-ons such as printers, external drives, and software should be compared in price to other vendors. If they are less expensive packaged with your new computer, then they should be considered.

Payment Options. Once you have decided on the PC you want, you may be faced with a variety of payment options. Online or in the store, you can find vendors willing to take cash or credit. Many have partnerships with banks and can arrange a computer loan for your purchase. Some vendors provide methods of trading in your PC after two years for a new one at a discounted price.

ACTION PLAN

1. *How should Mark prepare for his purchasing decision?*

   Mark needs to list his specific computing needs at home and away from home. He needs to keep each activity in mind throughout the purchasing process and make sure that the computer he purchases is able to support his needs.

2. *What type of computer platform might best suit Mark's needs?*

   Because this will be Mark's primary computer, he needs something of substance: a desktop or notebook PC. Because Mark enjoys computer games and graphic art, he most likely wants a large display, larger than a notebook computer would provide. He needs to choose between either the Apple or a Windows platform. Because Mark will be bringing some of his work home with him from Lamar, he should go with the same platform that is used at his work.

3. *How much processor speed, memory, and storage capacity does Mark need?*

   Interactive, 3D gaming requires a fast processor. If Mark is ***really*** into gaming he should go with the fastest processor available—3 GHz or faster with an Intel processor, 1.42 GHz in a Mac. If he is only mildly interested in gaming, he could save some money by going a notch down on processor speed to perhaps the speed that was fastest last year. His other interests do not require a state-of-the-art processor. Gaming also requires a significant amount of memory (512 MB recommended) but not necessarily a lot of storage (40 GB should suffice). However, if Mark does any work with video editing or plans to store his music or video files on his computer, he should get a large disk drive (160 GB). To sum up, if Mark will be using his computer for gaming and multimedia, he should consider maximum processor speed, memory, and storage offered in top-of-the-line computer packages. Otherwise, he might save as much as $800 by going for a more basic package.

4. *What input/output devices would best suit Mark's needs?*

   As a gamer and artist, Mark wants a large display, a color printer, a game controller, a drawing tablet, and a flatbed scanner. He plans to check with his employer to see what types of displays and drawing tablets are recommended, and then he will configure his home computer to be as similar to the one at work as possible. He should also purchase an upgraded video card with a significant amount of memory to smoothly run the gaming animation.

5. *Are there additional considerations to be made regarding this purchase?*

   Since Mark is new to speech recognition and video conferencing, he needs to do some research to find out what hardware is required. Speech recognition runs best on a computer with a fast processor and lots of memory. If Mark's computer meets those requirements, then all he needs to do is purchase an inexpensive microphone headset and speech recognition software. For video communication over the Internet, he needs a broadband Internet connection and a webcam. Because it sounds as though Mark's job does not entail travel, he might not need much computing power on the road. Perhaps a PDA cell phone will suffice for his mobile computing needs. Mark might also consider investing in a game system for his television. This might allow him to save some money by purchasing a more modest PC.

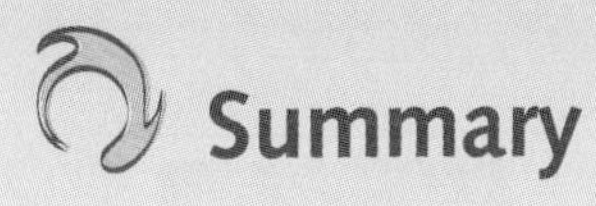

Summary

LEARNING OBJECTIVE 1

Understand how bits and bytes are used to represent information of value to people.

The digital revolution was sparked by the computer's ability to represent and manipulate information digitally, with 1s and 0s.

Units of information can be represented with bits in a standardized code by assigning each unit to a unique bit pattern using a *look-up chart*. ASCII is a standardized code used in computers to represent character data. Bits can also be used to store values by using the binary number system. The binary number system uses two digits—0 and 1. Binary numbers are used for calculations, to represent color intensities in images, and to represent the amplitude of sound waves used to store digital music and sound. There are many methods of decoding and interpreting a byte. Each method boils down to either the look-up table approach or binary numbers. By implementing standards, bytes can have consistent meaning across common applications. Because all different types of data are represented with 1s and 0s, digital convergence is producing devices that can handle all types of data.

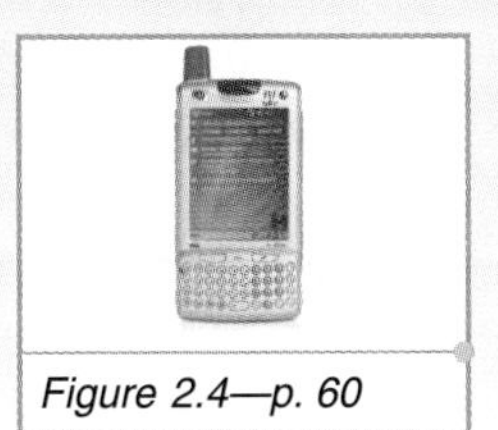

Figure 2.4—p. 60

LEARNING OBJECTIVE 2

Identify the functions of the components of a CPU, the relationship between the CPU and memory, and the factors that contribute to processing speed.

The CPU processes the data into information that is meaningful to people. The CPU's primary components are the control unit that fetches and interprets instructions, the arithmetic logic unit that carries out the instructions, and registers that store the bytes being processed. The CPU uses RAM as its primary storage. RAM provides the CPU with a working storage area for program instructions and data. Data flows back and forth between the CPU and RAM across the system bus.

The control unit in the CPU uses the four-step machine cycle to carry out program instructions: fetch instruction, decode instruction, execute instruction, and store results. Each processor has an instruction set, a finite number of instructions that it is designed to carry out. Clock speed, wordlength, and bus speed all have an impact on the speed of processing. All activity in the system takes place in time with a processor's system clock. The faster the clock, the faster the processing. Clock speeds are measured in megahertz (MHz) and gigahertz (GHz). The number of bits the CPU can process at any one time is called its ***wordlength***. Some computer systems use multiple processors to increase processing speed.

Moore's Law states that in silicon-based processors, transistor densities in an integrated circuit will double every 18 months.

Figure 2.11—p. 67

LEARNING OBJECTIVE 3

Identify different types of memory and storage media, and understand the unique properties of each.

RAM is volatile and temporarily stores data as long as there is an electrical current. ROM is more permanent storage that has its instructions hardwired into its circuitry. ROM stores the boot process that runs when the computer is powered up.

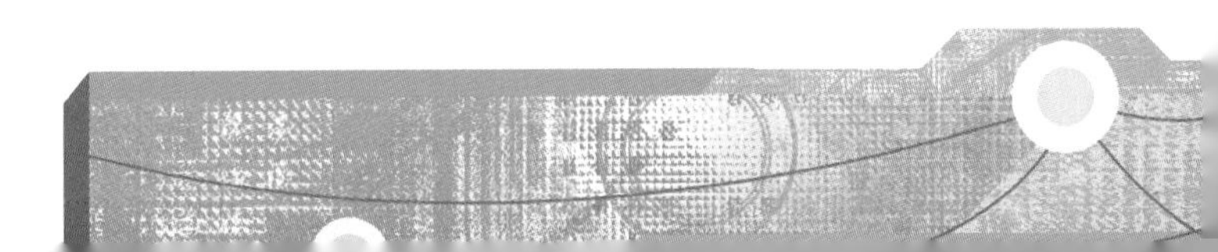

Magnetic and optical storage are two types of secondary storage used to store data and programs more permanently. Magnetic storage includes disks (random access) and tapes (sequential access). Optical storage includes CDs and DVDs, which are available in read-only disks and writable disks. Flash memory devices are becoming increasing popular as convenient portable storage for media data such as digital photos and music.

Figure 2.21—p. 76

LEARNING OBJECTIVE 4

Identify different types of input and output devices and how they are used to meet a variety of personal and professional needs.

People use input devices to provide data and instructions to the computer, and they use output devices to receive results from the computer. General-purpose input devices include the keyboard and mouse for desktop computers, the touch pad and TrackPoint for notebook computers, and the touch screen and stylus for tablet and handheld devices. Scanners are used in a variety of industries to avoid data-entry errors by capturing data close to its source. Special-purpose input devices include gamepads, tablets and pens, and webcams.

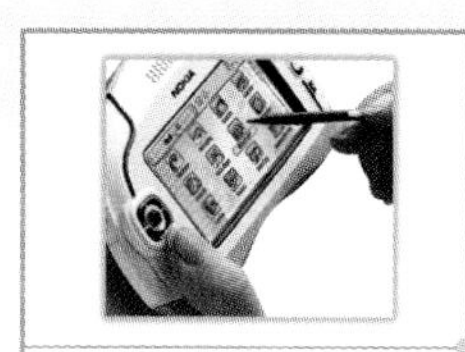

Figure 2.29—p. 85

Printers, plotters, and displays are the primary forms of output devices. Displays are selected by size and resolution. Traditional CRT monitors are losing market share to the leaner, cleaner LCD flat-panel displays whose prices in recent years have dropped. Color ink-jet and noncolor laser printers are popular for home and small business use, although color laser printer prices are becoming more affordable.

LEARNING OBJECTIVE 5

Understand the decision-making process involved in purchasing a computer system.

When selecting a computer system, you must first analyze your specific computing needs when at your desk and on the road. Select a computer type and platform that supports those needs. You may require one or two computers depending on your mobile computing needs. Options include desktop computers, notebook computers, tablet PCs, handheld computers (PDAs), and PDA-cell phones, with a trade-off between power and portability.

Test Yourself

LEARNING OBJECTIVE 1: Understand how bits and bytes are used to represent information of value to humans.

1. Most computers use ________________ to represent character data using bytes.
   a. hexadecimal
   b. kilobytes
   c. decimal notation
   d. ASCII code
2. The ________________ number system is ideal for digital systems because it has only two digits, 1 and 0.
3. True or False: All software interprets a byte such as 11110000 the same way.
4. ________________ refers to the ability of one device to handle several different types of digital information, such as voice communications, character data, and music.
   a. Digital revolution
   b. Digital convergence
   c. Digital communications
   d. Digital economy

LEARNING OBJECTIVE 2: Identify the functions of the components of a CPU, the relationship between the CPU and memory, and the factors that contribute to processing speed.

5. In the CPU, the ________________ performs mathematical calculations, and the ________________ sequentially accesses program instructions.

6. True or False: Data flows back and forth between the CPU and memory across the system bus.

7. Which of the following does not contribute to processor speed?
   a. clock speed
   b. ROM
   c. front side bus (FSB) speed
   d. wordlength

LEARNING OBJECTIVE 3: Identify different types of memory and storage media, and understand the unique properties of each.

8. When a computer is first powered on, the processor is fed instructions from ________________ .
   a. register storage
   b. RAM
   c. ROM
   d. secondary storage

9. True or False: The two types of optical storage are random access and sequential access.

10. Notebook computers use ________________ cards as one method of connecting external devices.

LEARNING OBJECTIVE 4: Identify different types of input and output devices and how they are used to meet a variety of personal and professional needs.

11. ________________ involves automating data entry where the data is created, thus ensuring accuracy and timeliness.
    a. Keyboard input
    b. Source data automation
    c. Secure data entry
    d. Special-purpose device input

12. Most gamers prefer to use a ________________ device to control game characters.

13. True or False: LCD displays are becoming more popular than the old-fashioned CRT displays.

LEARNING OBJECTIVE 5: Understand the decision-making process involved in purchasing a computer system.

14. What is the first thing you should do when considering a computer purchase?
    a. Choose a manufacturer.
    b. Consider a platform.
    c. Decide between a desktop, notebook, tablet, or handheld.
    d. Examine your usage and needs.

15. The most important computer system component that you purchase is the ________________ .

16. The first consideration in selecting a computer system is ________________ .
    a. platform
    b. processor
    c. portability
    d. manufacturer

Test Yourself Solutions: **1.** d. ASCII code, **2.** binary, **3.** False, **4.** b. Digital convergence, **5.** ALU, control unit, **6.** True, **7.** b. ROM, **8.** c. ROM, **9.** False, **10.** PC, **11.** b. Source data automation, **12.** gamepad, **13.** True, **14.** d. Examine your usage and needs, **15.** warranty, **16.** c. portability.

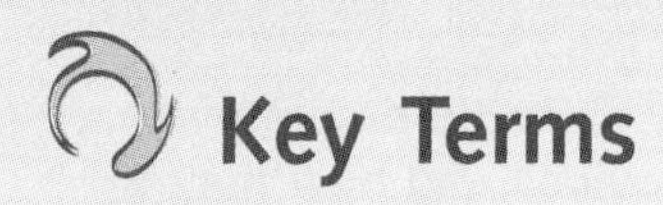

Key Terms

arithmetic/logic unit (ALU), p. 63
binary number system, p. 59
CD-RW, p. 77
central processing unit (CPU), p. 62
compact disc read-only memory (CD-ROM), p. 76
control unit, p. 63
digital convergence, p. 60
digital video disk (DVD), p. 77
display resolution, p. 88
flash memory card, p. 78
gamepad, p. 85
gigaflop, p. 67
gigahertz (GHz), p. 66
input device, p. 81
integrated circuit, p. 62
liquid crystal display (flat panel display), p. 89
magnetic storage, p. 74
megahertz (MHz), p. 66
Moore's Law, p. 69
motherboard, p. 62
optical storage, p. 76
output device, p. 81
random access memory (RAM), p. 63
read-only memory (ROM), p. 72
touch screen, p. 84
transistor, p. 62
Universal Serial Bus (USB), p. 79

Questions

Review Questions

1. Explain how the bytes that represent keyboard characters are decoded.
2. Why do computers use the binary number system?
3. Describe the components of a machine cycle.
4. What is *digital convergence*?
5. What factors affect processing speed?
6. What are the two most popular personal computer platforms? How do they differ?
7. Describe the various types of memory.
8. Explain the difference between sequential and direct access.
9. Describe various types of secondary-storage media in terms of access method, capacity, and portability.
10. What factors determine the appropriate system output and input devices?
11. Discuss the speed and functionality of common input and output devices.
12. How do expansion techniques differ between notebook and desktop computers?
13. Describe the six steps of selecting and purchasing a computer.

Discussion Questions

14. The Smartphone allows you to make phone calls, take pictures, listen to digital music, and access data, such as your calendar and address book. What other functionality would you like to see provided by a handy, all-in-one portable device? How can such a function be incorporated digitally?
15. Paper is difficult and expensive to organize, manipulate, and store over extended periods of time. In light of paper's drawbacks, discuss the impact(s) of various input and storage devices on organizations.
16. Contrast magnetic storage devices and optical storage devices. What are the advantages and disadvantages of each in terms of cost, control, and complexity?

17. Consider and discuss strategies for storing data for long periods of time. For example, if you store all of your personal photos and music on your hard drive, what storage strategy might you use to ensure that your great-grandchildren's children will be able to enjoy them 100 years from now?

18. Discuss how the hardware needs of a small law firm might differ from the needs of the office of a registrar for a university.

19. List the activities that you do, or would do if you were able to, on computers both at your desk and away. What computer device(s) would best support your needs (PC, notebook, tablet PC, handheld)? What are your needs in terms of processor speed, memory capacity, and storage capacity? Provide your rationale.

Exercises

Try It Yourself

1. Use a spreadsheet to convert decimal into binary, octal, and hexadecimal number systems. Most spreadsheet applications have a function that automatically does this conversion for you. For example, Microsoft Excel includes functions named DEC2BIN, DEC2OCT, and DEC2HEX. Create a chart that looks like this:

Decimal	Binary
1	1
2	10
3	11
4	100
5	101
6	110
7	111
8	1000
9	1001
10	1010

Example of function usage:

Decimal	Binary
1	=DEC2BIN(A2)
2	=DEC2BIN(A3)

Recall that the binary number system has only two digits (0 and 1). Create a second table in your spreadsheet that converts binary numbers to decimal (BIN2DEC). Type in binary values 1 through 10100 and use either a function or your brain to do the conversions. Use your tables to experiment with these number systems for better understanding.

2. You and two friends have decided to start your own business. You have secured a small business loan from the bank and leased some office space. Your partners have placed you in charge of purchasing desktop computers for the six employees of your company—the three owners and three clerical employees. None of the employees need a really fast processor, but the partners need larger amounts of memory and storage. The employees need the ability to archive data on a CD. All employees need a lot of work room on their desks so whatever you buy has to take up as little desktop space as possible.

Your budget is $10,000. You have already selected a vendor; now you need to decide on what should be included in each of the two types of systems: the owners' systems and the employees' systems. A price list for computer components appears below. Create a spreadsheet to find the best solution. Add a short memo explaining the rationale for your solution. Include the price of your owners' computers and your employees' computers and what components each includes. Also include the minimum and maximum price possible for a computer system.

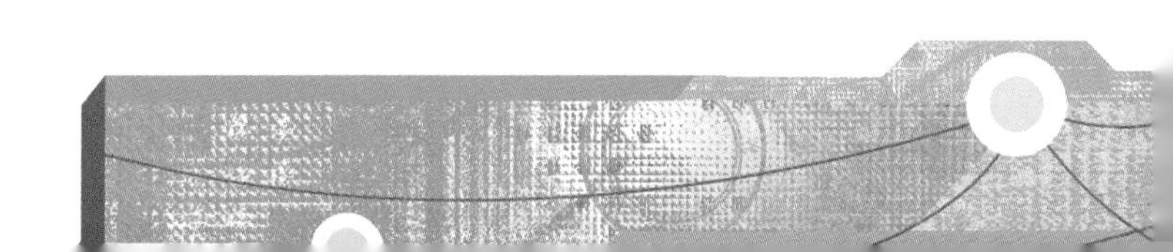

3. Examine the table below that lists the base price for a notebook PC and prices for additional features that can be added on or upgraded.

   What combination of the above components best suits your own personal computing needs at the lowest cost? Calculate the cost using a calculator program. Use a word processor to list the features and price of your computer. Search the Web and find a better deal from another source. Popular online vendors include *www.dell.com*, *www.gateway.com*, *www.cdw.com*, or you can look at Apple, *http://store.apple.com*. In your word-processing document, list the features and price of the better deal that you found. Include a rationale for why you think it is a better deal.

4. Visit *www.cnet.com* and click Peripherals. In a word-processing document, list the subcategories of input/output devices that you find under:

   - Monitors
   - Printers
   - Scanners

   Now visit *www.zdnet.com*, click Reviews and Prices, then All Products, and then Input Devices. Add the categories of input devices that are listed to your document. Looking over your complete list, mark with an *X* those devices that would benefit you in your own computer use. Add a description of the types of activities you do that requires each marked item. In the case of monitors and printers, describe how you decided on the type.

5. Using common Web search engines, such as *www.ask.com* and *www.google.com*, find the name of the fastest supercomputer in existence. What is its name, who made it, how fast is it, and what is it used for? Do the same for the second-fastest supercomputer. What is the difference between the two in speed and in age? Do these two computers support Moore's Law?

6. Consider the computing needs of your present or future career. Are there any special or unique hardware requirements? Use the Web to research this topic. Rank the following in order of importance for your professional computer system (most important being 1):

   ________________ Portability

   ________________ Network/Internet connection

   ________________ Large high-quality display

   ________________ Fast processor

   ________________ Large amount of memory

   ________________ Large-capacity hard disk drive

   ________________ Large-capacity portable storage (Zip, CD-RW, and so on)

   ________________ High-quality printer

   ________________ Flatbed scanner

   ________________ Digital camera

   ________________ Special-purpose input device (provide description if applicable)

Virtual Classroom Activities

To complete the following exercises, do not use any face-to-face or telephone communications with your group members. Use only Internet communications.

7. Assume that you are partners in a new business. Decide what business you are in. Assign each group member a role within the business. One member should act as company president and group leader. Choose a leader by casting electronic ballots. Each group member should then decide what type of computer he or she needs

	Base package $919	Upgrade options	
Processor speed	Pentium M processor 725 (1.60 GHz/400 MHz FSB)	Intel Celeron M processor 360 (1.40 GHz/1 MB cache/ 400 MHz FSB) deduct $300	Intel Centrino Mobile Technology for outstanding wireless performance + $300
Memory	256 MB DDR SDRAM	512 MB + $50	768 MB + $200
Hard drive storage	40 GB	60 GB + $39	80 GB + $129
Portable storage	8X Max DVD ROM drive	24X CD burner/DVD Combo drive + $29	8x CD/DVD burner (DVD+/-RW) + $129.00
Display	14.1-in XGA w/32 MB video card	12"-wide screen + $100	

to perform his or her job. Use the chapter content to decide what type of computer(s) best suit your needs. Include processor speed, memory and storage types and capacities, along with I/O devices—the options discussed in this chapter. Justify your choices.

8. Each group member should go shopping on the Web to find the lowest-priced system as defined in exercise 7. Submit a detailed price breakdown to the group leader. The group leader should compile the data and total the cost of the new computer systems.

Teamwork

9. Identify the public and department-run computer labs on your campus. Divide them equally between team members. Each team member should visit his or her assigned labs and, either by observation or contact with the lab administrator, list the type of equipment used in the lab. Include processor speeds, memory capacity, storage types used on the computers, and manufacturer's name for all devices, including scanners, printers, and other peripherals. Reconvene and decide which lab on campus best meets students' needs. Write a report that summarizes your findings.

10. Assign each group member a particular computer component to research: processor, memory, storage, input device, or output device. Search the Web, periodicals, or computer stores for unique examples of such devices that were not included in this chapter. Compile your data to create a unique computer. The group with the most bizarre computer wins.

TechTV

Getting Started

Go to www.course.com/swt2/ch02 and click TechTV. Click the "Getting Started" link to view the TechTV video, and then answer the following questions.

1. Why does the host suggest that having two hard drives is better than one?
2. After seeing what it takes to build your own computer, do you think that it is more economical to build a computer, or buy a prebuilt computer? Explain the advantages and disadvantages of each approach.

Endnotes

[1] "Apple to Use Intel Microprocessors Beginning in 2006," *Apple Hot News*, June 6, 2005, http://www.apple.com/pr/library/2005/jun/06intel.html.

[2] Intel Web site, http://www.intel.com/products/roadmap/ accessed on May 20, 2005.

[3] "Intel Pentium 4 Processor-Based PCs Deliver Performance and Reliability for Gap Inc. Direct," *Intel Case Study*, http://www.intel.com/business/casestudies/industry/gap.htm, accessed August 9, 2005.

[4] Intel Web site, http://www.intel.com/products/processor_number/, accessed May 12, 2005.

[5] "Intel Produces One Chip with Two 'Brains'," Reuters, February 7, 2005, http://www.reuters.com.

[6] Sorid, D. "New 'Supercomputer on a Chip' Makes Debut," Reuters, February 7, 2005, http://www.reuters.com.

[7] "Scientists Reach Back 2,000 Years to Bring Rare Child Mummy Back to Life," SGI news release, August 3, 2005, http://www.sgi.com/company_info/newsroom/press_releases/2005/august/child_mummy.html.

[8] NASA's Supercomputing Web site, http://www.nas.nasa.gov/About/Projects/Columbia/columbia.html, accessed May 12, 2005.

[9] "Hard Drives in Your Car?" *Seagate News and Info*, http://www.seagate.com/newsinfo/newsroom/success/D2g37.html, accessed August 9, 2005

[10] Internet Archive entry at Wikipedia, http://en.wikipedia.org/wiki/Wayback_Machine, accessed August 9, 2005.

[11] "DOE's NERSC Center Deploys 10 teraflops per second IBM Supercomputer," *Berkeley Lab Research News*, http://www.lbl.gov/Science-Articles/Archive/NERSC-10-teraflop-IBM.html, accessed August 9, 2005.

[12] Pogue, D. "A New Spin on a Palmtop (or Inside It)," *New York Times*, May 19, 2005, accessed at http://www.nytimes.com.

[13] iPod Web site, at http://www.apple.com/ipod/, accessed May 20, 2005.

[14] "Samsung SCH-V770, 7-megapixel camera phone," *Mobile Tracker*, March 9, 2005, http://www.mobiletracker.net/archives/2005/03/09/samsung-v770-megapixel.

[15] "Monkeys adapt robot arm as their own," *Science Letter*, June 7, 2005, http://www.lexis-nexis.com.

[16] Boyd, R.S. "Forget Nano; Focus on Atto: Ultra-Fast Cameras May Let Scientists Watch as Atoms Form Chemical Bonds," *Charleston Gazette* (West Virginia), May 22, 2005, http://www.lexis-nexis.com.

[17] "Film Festival to Host Toy Symphony Performance" *Orlando Business Journal*, February 12, 2004, http://orlando.bizjournals.com/orlando/stories/2004/02/09/daily33.html.

[18] Gutoff, B. "24 Years as ABC's Chief Photographer," Apple Pro Case Study, http://www.apple.com/pro/photo/damico/, accessed August 9, 2005.

SOFTWARE SOLUTIONS FOR PERSONAL AND PROFESSIONAL GAIN

Alexandra Pollack is a second-semester freshman in college and working part-time to help pay her expenses. She needs a new computer system, has saved for almost a year, and wants to make sure that she gets what she needs for the rest of college and her part-time job. One of her instructors advised her to decide what software she needs before she buys a new computer system.

Alex is majoring in communications and is interested in going to law school after she graduates. She is currently working as a clerk at a local law firm. The firm allows her to do some of her work from her dorm room, which is convenient for Alex. Working from her dorm room saves her a 25-minute drive and the cost of the expensive clothes she would need to work at the law firm's office. The law firm told her that she needs something called Office XP to do her work from home. A friend recommended that she consider getting Office 2003 instead.

Alex also enjoys music. She would like to be able to download music from the Internet and play the music on a portable device. She currently has an old CD player.

As you read through the chapter, consider the following questions:

1. What software does Alex need for her work for the law firm?
2. What do you recommend for downloading and playing music?
3. What type of operating system does she need and why?

Check out Alex's *Action Plan* at the conclusion of this chapter.

3

LEARNING OBJECTIVES

1. Discuss the importance and types of software
2. Discuss the functions of some popular programming languages
3. Describe the functions of system software and operating systems
4. Describe the support provided by application software
5. Discuss how software can be acquired, customized, installed, removed, and managed

CHAPTER CONTENT

An Overview of Software

Programming Languages

System Software

Application Software

Software Issues and Trends

Introduction

Software is the key to unlocking the potential of any computer system and developing effective computer applications. Without software, the fastest, most powerful computer is useless. It can do nothing without instructions to follow and programs to execute. With software, people and organizations can accomplish more in less time.

All software is developed using a programming language. Programming languages have gone through a number of generations, and with today's languages, writing computer programs is easier and faster than it was 40 years ago. Programming languages are used to develop two basic types of software—system software and application software. System software makes computers run more efficiently, whereas application software helps people, groups, and organizations achieve their goals. Both types of software are discussed in this chapter. This chapter concludes with some important software issues, such as how to fix software bugs, software licenses, open-source software, and copyrights.

In the early days of computing, when computer hardware was expensive, software costs were a comparatively small percentage of total computer system costs. The situation has dramatically changed today. Software can be 75 percent or more of the total cost of a particular computer system because (1) advances in hardware technology have resulted in dramatically reduced hardware costs; (2) increasingly complex software requires more time to develop, hence, it is more costly; and (3) salaries for individuals who develop software have increased because of the increased demand for their skills. In the future, as shown in Figure 3.1, software will constitute an even greater portion of the cost of the overall computer system. The critical functions software serves, however, make it a worthy investment.

FIGURE 3.1 • The importance of software

Organizations have greatly increased their expenditures on software as compared to hardware since the 1950s.

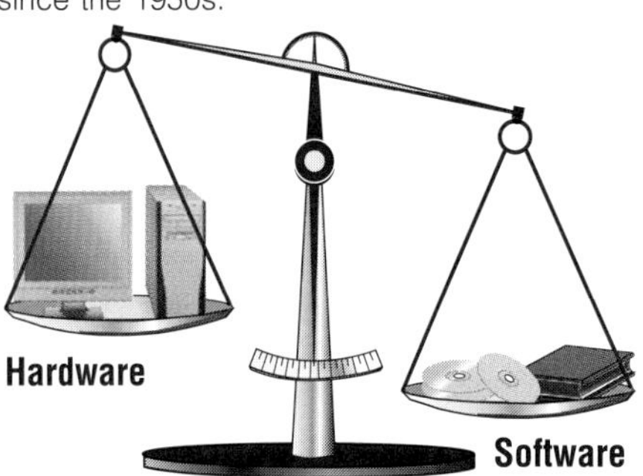

AN OVERVIEW OF SOFTWARE

As you learned in Chapter 1, *software* consists of the electronic instructions and programs that govern a computer system's functioning and that control the workings of the computer hardware. Computer programs are sets of instructions or statements to the computer. Ultimately, these statements direct the circuitry within the hardware to operate in a certain fashion. Computer programmers are people who write or create these instructions or statements that become complete programs. Within the computer industry, programmers are called *software engineers*. Software also often includes documentation that assists the user in learning how to use the software. Today, documentation is usually included in *Help* features or menus within the software, on the Internet, or in manuals produced by the software manufacturer and others.

System and Application Software

There are two basic types of software: system software and application software. System software is the set of programs that coordinates the activities of the hardware and various computer programs. System software is written for a specific set of hardware, most particularly the CPU. The particular hardware configuration of the computer system combined with the particular system software in use is known as the *computer platform*. For example, a PC running the Windows operating systems is a popular computer platform. The other type of software, application software, consists of programs written to solve problems and help people and organizations achieve their goals. Application software applies the power of the computer to give individuals, groups, and organizations the ability to solve problems and perform specific activities or tasks. The drug company Pfizer, for example, has developed application software to detect the early signs of Parkinson's disease.[1] The new software detects slight trembling in speech patterns not detectable by the human ear that predict the disease.

How Software Works

Software usually consists of a number of files, ranging from a few to dozens or more. Files can have an extension at the end of the file name. An *extension* identifies the type of file and is placed after the file name, following a period. At least one of the files is an executable program file with an .exe extension. Other files can store data. A text file typically has a .txt extension. Many word-processing files have a .doc extension. For example, the .doc extension in the file Report1.doc indicates that the Report1 file is a Microsoft Word document, and a .wpd extension indicates a WordPerfect document.

There are many ways to start or run executable files. The most common ways are to click an icon on the desktop. For example, in the Windows XP operating system, you can click the Start button, click All Programs, then locate and select the program. You can also use the run command to launch an application. Click the Start button and then the Run icon to display the Run dialog box. You can then enter the location and name of the program to run. See Figure 3.2. For example, you might have to manually install a new application program by running a program called *setup* from a CD disk on the D drive. Typing D:\setup in the Run dialog box and clicking OK will run the setup program. You can also click the Browse button in the Run dialog box to locate the file.

FIGURE 3.2 • Using the Run dialog box to start a program

Clicking the Start button and then the Run icon is one way of executing a program.

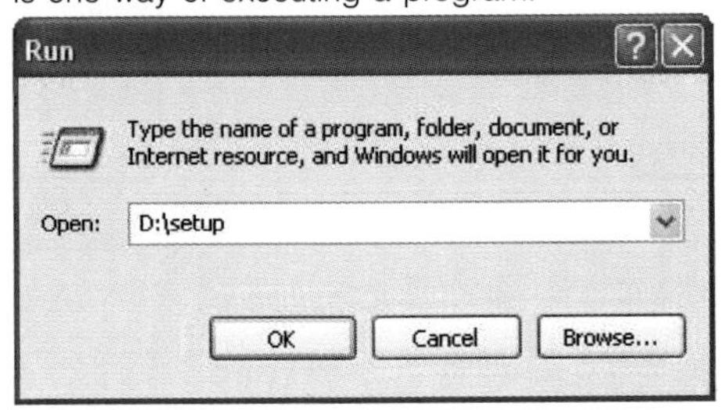

When a program such as a word processor is started, instructions from the program are copied from the computer's disk drive, where they are permanently stored, into the computer's memory. Once in memory, the instructions are transferred to the processor and executed. As you create a document—for example, a term paper including text, bold terms, footnotes, and other features—the word-processing software obeys your commands and does what you wish (see Figure 3.3). When you are finished, you can close the program and stop it from running in memory in a variety of ways, such as clicking icons to close their programs, selecting commands from a menu, or typing in commands such as CLOSE or EXIT. When you close the program, it is removed from the computer's memory. The computer's memory also contains part of the operating system and data that may be needed by the program.

TechEdge

YOU ARE HERE

Students in the UK developed a location-aware blog tool as the winning entry in a competition sponsored by Microsoft. Called OneReach, the software uses a Smartphone with GPS to allow globe-trotting folks to find information specific to where they are. They can append to the existing posts—thereby providing the most up-to-date info imaginable on conditions, sights, or tribulations—and can create a map that allows their near and dear to follow their footsteps.

Source:
Blog tool wins innovation prize
BBC News
http://news.bbc.co.uk/2/hi/technology/4118422.stm
22 June, 2005

FIGURE 3.3 • Using word-processing software

Word-processing programs have many formatting features, such as bold, italics, and footnotes.

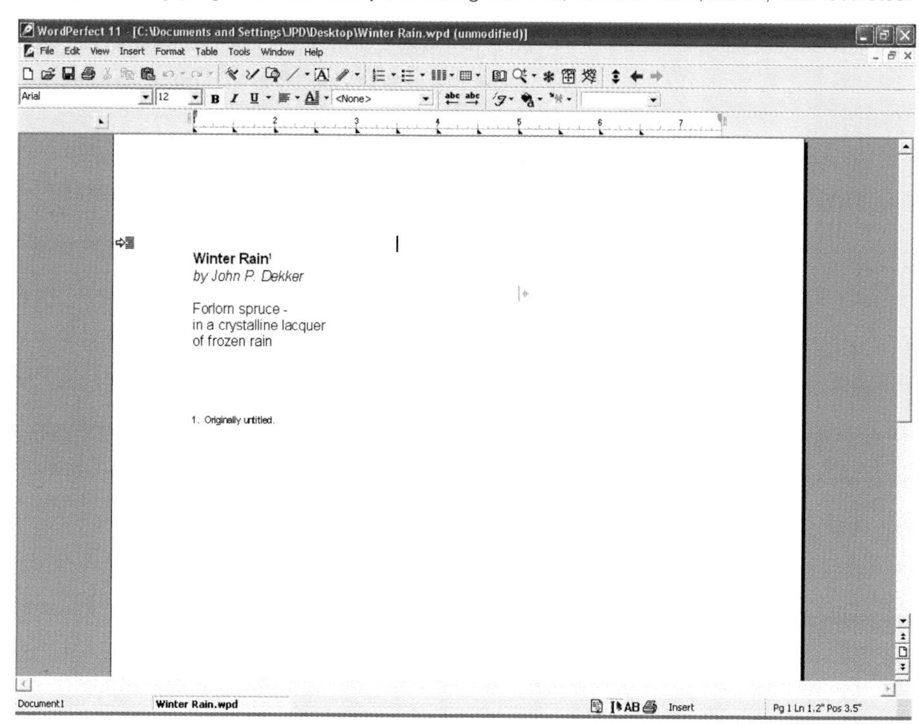

PROGRAMMING LANGUAGES

All software, both system software and application software, is written in coding schemes called programming languages. A programming language, the primary tool of computer programmers, provides commands for writing software that is translated to the detailed step-by-step instructions executed by the processor to achieve an objective or solve a problem. It is impossible to overemphasize the role of programming languages and software in problem solving, decision making, and attaining individual and organizational goals. Because programming languages are becoming easier to learn and use, noncomputer people are increasingly writing their own programs. Students have written their own programs to help them complete homework assignments or class projects. Engineers have developed programs to design bridges, and corporate employees have written programs to help them on the job. Some larger programs, however, can be difficult and complex to develop even with newer programming languages.

In the very early years of computing, writing computer programs was quite complex. The various switches and circuits that composed the computer hardware had to be manually set; that is, data was represented by physically switching various circuits on or off. In modern computers, however, the CPU works in conjunction with various software programs to control the digital circuitry. For this to occur, the CPU must receive signals from the computer program that it can convert into actions (the switching of the circuits on and off). Since these signals are not the actions themselves, but merely instructions that these actions should occur, they are called program code. *Program code* is the set of instructions that signal the CPU to perform circuit-switching operations. In the simplest coding schemes, a line of code typically contains a single instruction such

Software Reduces Legal Costs

Getting expert legal advice and services can be expensive. Some lawyers charge $500 or more per hour for their services. In addition, many large organizations have a complete legal staff that can cost the organization hundreds of thousands of dollars or more each year. To reduce high legal costs, some organizations, such as Cisco Systems, Inc., are now developing specialized software.

Cisco manufactures and sells sophisticated telecommunications systems. According to Mark Chandler, general counsel for the company, "Legal services need to be delivered more simply and more conveniently. Technology lets you commoditize, fix fees, and make competitive bids. All that leads to cost savings." The company is now using software it developed to automate the patent process for its thousands of employees. The old manual system required a tremendous amount of paperwork and a complicated review process. Now, about 1,000 Cisco inventors can submit patent applications to a review board and eventually the U.S. Patent and Trademark Office using the software. Cisco estimates that it saves about $3,000 per patent application, which amounts to savings of about $2.5 million per year. It cost Cisco under $1 million to develop the patent application software. According to Chandler, "It is a gift that keeps on giving."

Other companies are also starting to use legal software. DuPont, the large chemical company, estimates that it saves about $5 million per year using a collaborative software package called Edge that the company developed. Edge allows DuPont's corporate attorneys and others to analyze legal documents and have online meetings about cases and other legal issues the company faces.

Questions:

1. Are there any disadvantages of using legal software?
2. What types of industries or businesses might benefit the most from legal software?

Sources:

1. *Grimes, A. "On the Case,"* The Wall Street Journal, *March 21, 2005, p. R10.*
2. *Tuttle, R. "Cisco Systems Presses for Change in Space Program COTS Buys,"* Aerospace Daily, *April 13, 2005, p. 3.*

as, "Retrieve the data in memory address X." As discussed in Chapter 2, the instruction is then decoded during the instruction phase of the machine cycle.

As mentioned earlier, the primary tool of a computer programmer is a programming language—sets of symbols and rules used to write program code. Like writing a report or a paper in English, writing a computer program in a programming language requires that the programmer follow a set of rules. Each programming language uses a set of symbols that have special meanings, much as English uses the Roman alphabet. Each language also has its own set of rules for how the symbols should be combined into statements capable of conveying meaningful instructions to the CPU. To some extent, programming involves translating what a user wants to accomplish into a logic that the computer can understand and execute.

A *programming language standard,* also called *syntax,* is a set of rules that describes how programming statements and commands should be written. A rule stating "Variable names must start with a letter" is an example. A variable is an item that can take on different values. Program variable names such as SALES, PAYRATE, and TOTAL follow the above rule because they start with a letter, whereas variables such as %INTEREST, $TOTAL, and #POUNDS do not.

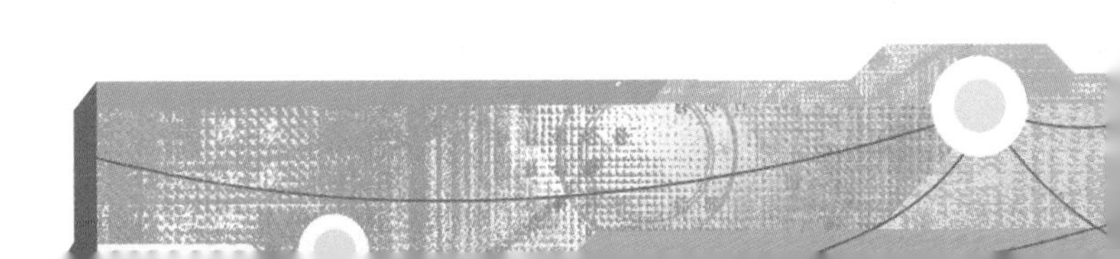

Programming languages have evolved since the early days of computing, and they continue to evolve. As shown in Table 3.1, the evolution of programming languages can be thought of in terms of generations of languages.

TABLE 3.1 • The evolution of programming languages

Generation	Language	Approximate development date	Sample statement or action
First	Machine language	1940s	00010101
Second	Assembly language	1950s	MVC
Third	High-level language	1960s	balance = balance + deposit
Fourth	Query and database languages	1970s	PRINT EMPLOYEE NUMBER IF GROSS PAY > 1000
Beyond fourth	Natural and intelligent languages	1980s	IF certain medical conditions exist, THEN a specific diagnosis is made

EXPAND YOUR KNOWLEDGE

To learn more about visual programming, go to www.course.com/swt2/ch03. Click the link "Expand Your Knowledge" and then complete the lab entitled "Visual Programming."

Newer Programming Languages

Newer languages are more difficult to classify. Languages beyond the fourth generation include artificial intelligence languages, visual languages, and object-oriented languages.

Programming languages used to create artificial intelligence or expert systems applications are often called *artificial intelligence languages*. They have also been called *fifth-generation languages (5GLs)* by some people. Artificial intelligence and expert systems are discussed later in the book. Fifth-generation languages are sometimes called *natural languages* because they use English-like syntax. They allow programmers to communicate with the computer by using normal sentences, as in the example shown in Figure 3.4. For example, a computer programmed in a fifth-generation language can understand such queries as "How many athletic shoes did our company sell last month?" Fifth-generation languages have the potential to predict the weather, diagnose potential diseases given a patient's symptoms and current condition, and determine where to explore for oil and natural gas.

FIGURE 3.5 • An example of a natural-language program

```
GIVE ME A SORTED LIST OF
ALL SALES REPRESENTATIVES
LIVING IN DENVER AND
EARNING OVER $47,500
```

Visual languages use a graphical or visual interface for program development. Prior to visual languages, programmers were required to describe the windows, buttons, text boxes, and menus that they were creating for an application by using programming language commands. With visual languages, the programmer drags and drops graphical objects such as buttons and menus onto the application form. Then, using a programming language, the programmer defines the capabilities of those objects in a separate code window. *Visual Basic* was one of the first visual programming languages. Other languages with visual development interfaces include *Visual Basic.NET* and *Visual C++.NET*. Visual Basic and Visual Basic.NET (see Figure 3.5) can be used to develop applications that run under the Windows operating system.

FIGURE 3.5 • An example of a Visual Basic.NET program

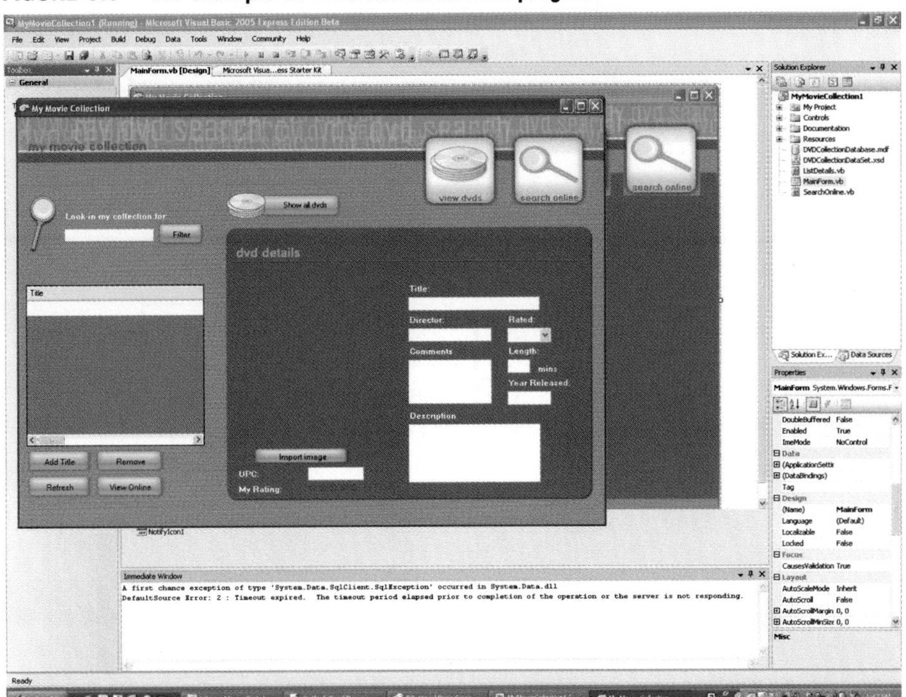

C++ is a powerful and flexible programming language used mostly by computer systems professionals to develop software packages, including operating systems. *Java* is a programming language developed by Sun Microsystems to create programs that can be run on any operating system and on the Internet. Java, which is discussed in Chapter 4, can be used to develop complete applications or smaller applications, called *Java Applets*. The New York Stock Exchange, for example, is developing a new trading system based on Java.[2] Some believe that the new system, that permits trading using wireless handheld devices, could eventually mean the end of the "buy" and "sell" shouts that come from the trading floor. Visual Basic.NET, C++, and Java are also examples of objected-oriented languages, which are discussed next.

Object-oriented programming languages, such as Visual Basic.NET, C++, and Java, allow the creation and interaction of programming objects. In object-oriented programming, data, instructions, and other programming procedures are grouped together. The items in such a group are called an *object*.

Building programs and applications using object-oriented programming languages is like constructing a building using prefabricated modules or parts. The object containing the data, instructions, and procedures is a programming building block. Unlike prefabricated building modules, however, the same objects (modules or parts) can be used repeatedly—millions of times if needed. Object-oriented programming is often used to develop custom software packages for companies in avionics, financial services, health care, and other industries.

This ability to reuse programming code is one of the primary advantages of object-oriented programming (see Figure 3.6). The instruction code within an object can be reused in different programs for a variety of applications, just as the same basic type of prefabricated door can be used in two different houses. Thus, a sorting routine developed for a payroll application could be used in both a billing program and an inventory control program. By reusing program code, programmers are able to write programs for specific application problems more quickly.

FIGURE 3.6 • Object-oriented programming

By combining existing program objects with new ones, programmers can easily and efficiently develop new object-oriented programs to accomplish organizational and personal goals.

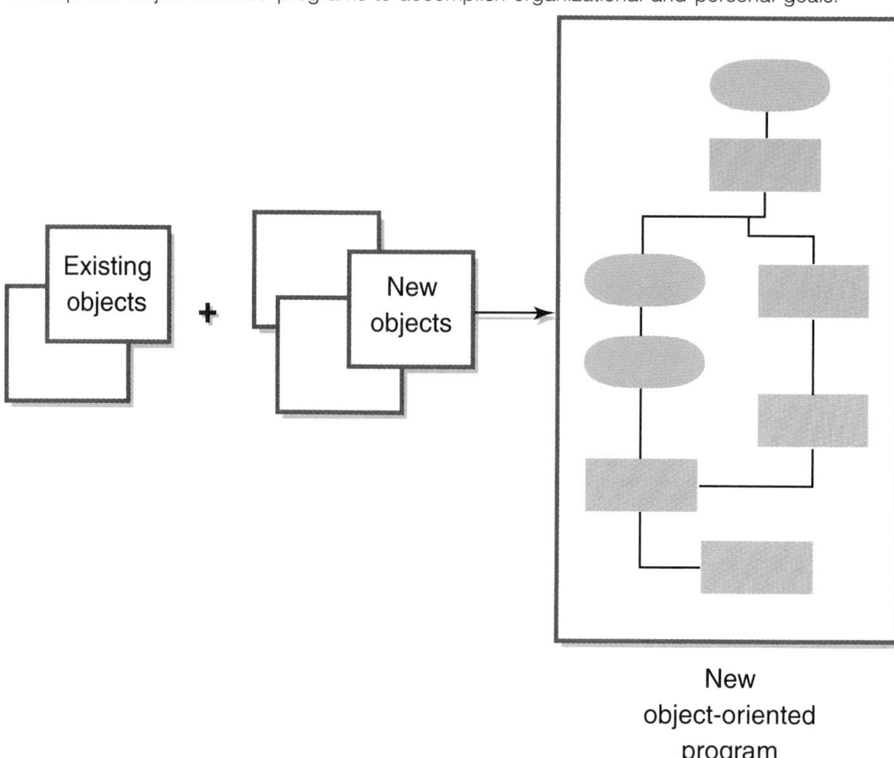

Programming Language Translators

Except for the first generation of programming languages, every programming instruction must be converted into machine language to be executed by the CPU. This translation is done by system software called a *language translator*. A language translator converts a statement from a high-level programming language, such as the statement READ SALES, into machine language. The high-level program code is referred to as the *source code*; the machine-language code is referred to as the *object code*. Two types of translators are interpreters and compilers.

An *interpreter* is a language translator that converts each statement in a programming language, such as a Java program, into machine language and executes the statement, one at a time. An interpreter does not produce a complete machine-language program. After the statement executes, the machine-language statement is discarded, the process continues for the next statement, and so on (see Figure 3.7). Many of the programs that are accessed in Web pages are interpreted. This allows them to run on your computer without being installed.

FIGURE 3.7 • How an interpreter works

An interpreter sequentially translates each program statement or instruction into machine language. The statement is executed, and another statement is then translated. An interpreter does not produce a complete machine-language program.

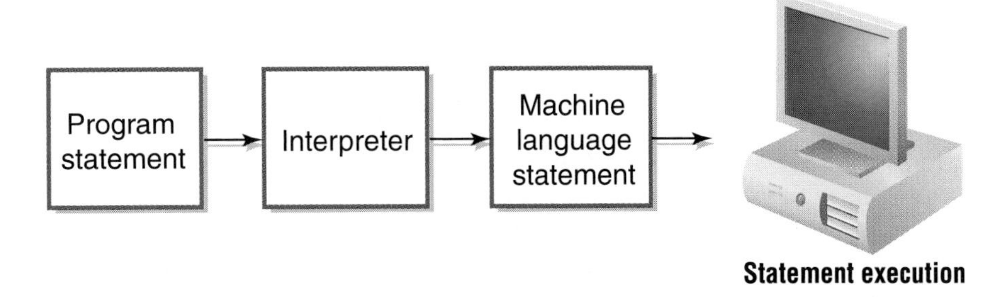

A *compiler* is a language translator that translates a complete program, such as a C++ program, into a complete machine-language, executable file (see Figure 3.8). Once the compiler has translated a complete program into a machine-language program, the machine-language program can be run on the computer as many times as needed. The executable (.exe) files on your computer that contain your favorite programs were created by a compiler. With a compiler, program execution is a two-stage process. First, the compiler translates the program into a machine-language program; second, the machine-language program is executed. Commercial software vendors compile the software and sell the resulting executable files. Some compilers go beyond language translation to utilize libraries and bring in additional code needed to perform functions, such as sorting, computing averages, and other tasks. This is why the software you buy typically includes one executable file and many other supporting files.

FIGURE 3.8 • How a compiler works

A compiler translates a complete program into a complete machine-language program (Stage 1). Once this is done, the machine-language program can be executed in its entirety (Stage 2).

Stage 1: Convert to machine-language program

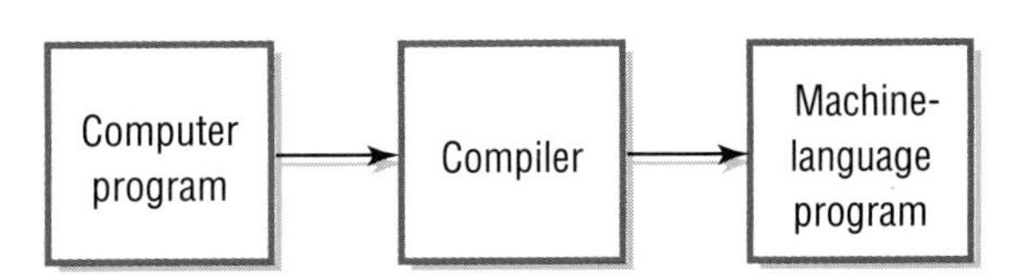

Stage 2: Execute machine-language program

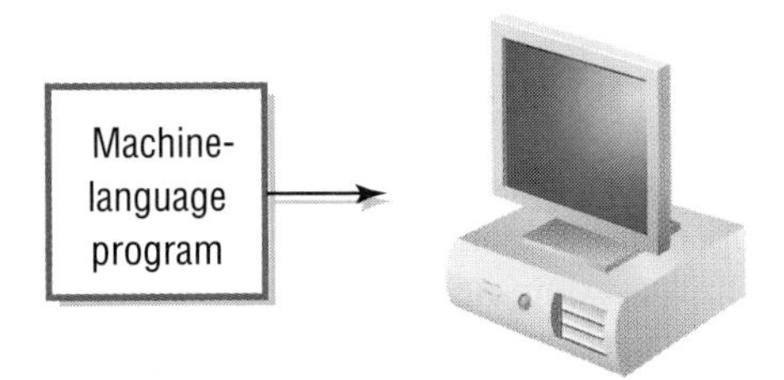

Because compiled programs run faster than programs that have to be translated line by line by an interpreter, compilers are usually preferred for frequently run programs. Compiled programs are also often usable on different computer platforms or operating systems. If a program is infrequently used, an interpreter may be a satisfactory language translator.

SYSTEM SOFTWARE

Controlling the operations of computer hardware is one of the most critical functions of system software. System software includes operating systems and utility programs that interact with the computer hardware and application software programs, creating a layer of insulation between the two.

Operating Systems

An operating system (OS) is a set of computer programs that runs or controls the computer hardware and acts as an interface with both application programs and users (see Figure 3.9). Operating systems can control one computer or multiple computers, or they can allow multiple users to interact with one computer.

FIGURE 3.9 • Operating systems

An operating system (OS) is a set of computer programs that runs or controls the computer hardware and acts as an interface between application programs and users.

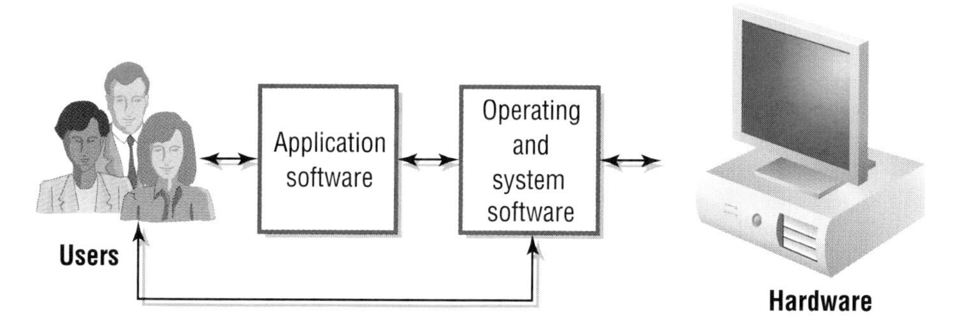

The various combinations of operating systems, computers, and users include:

- A single computer with a single user: This is typical of a personal computer or a handheld computer that allows one user at a time, such as Apple OS X.
- A single computer with multiple users: This is typical of larger mainframe or server computers that can accommodate hundreds or thousands of people, all using the computer at the same time, such as z/OS for large IBM computers.
- Multiple computers: This is typical of a network of computers, such as a home network that has several computers attached or a large computer network with hundreds of computers attached around the world. Some PC operating systems, such as Windows, allow multiple computers to be connected in a network.
- Special-purpose computers: This is typical of a number of special-purpose operating systems that control sophisticated military aircraft, the space shuttle, some home appliances, and a variety of other special-purpose computers.

Although most computers use just one operating system, it is possible to have two or more operating systems stored on a single computer by *partitioning* a hard disk to contain different operating systems. This can be useful when you need to run different programs or applications that require different operating systems on the same computer. When the computer starts, you are given the opportunity to choose which operating system to use.

The operating system plays a central role in the functioning of the complete computer system and is usually stored on the hard drive. For smaller computers however, the operating system can be stored on a computer chip. After a computer system is started, or booted up, much of the operating system is transferred to memory. Once in memory, the operating system instructions are executed, or run. The operating system's collection of programs performs a variety of activities and functions, discussed below.

Nearly all operating systems perform certain common tasks, such as controlling computer hardware, managing memory, managing the processor(s), controlling input and output devices, storing and manipulating files, and providing a user interface. See Figure 3.10. Most operating systems today also provide networking features.

FIGURE 3.10 • Most operating systems perform similar functions

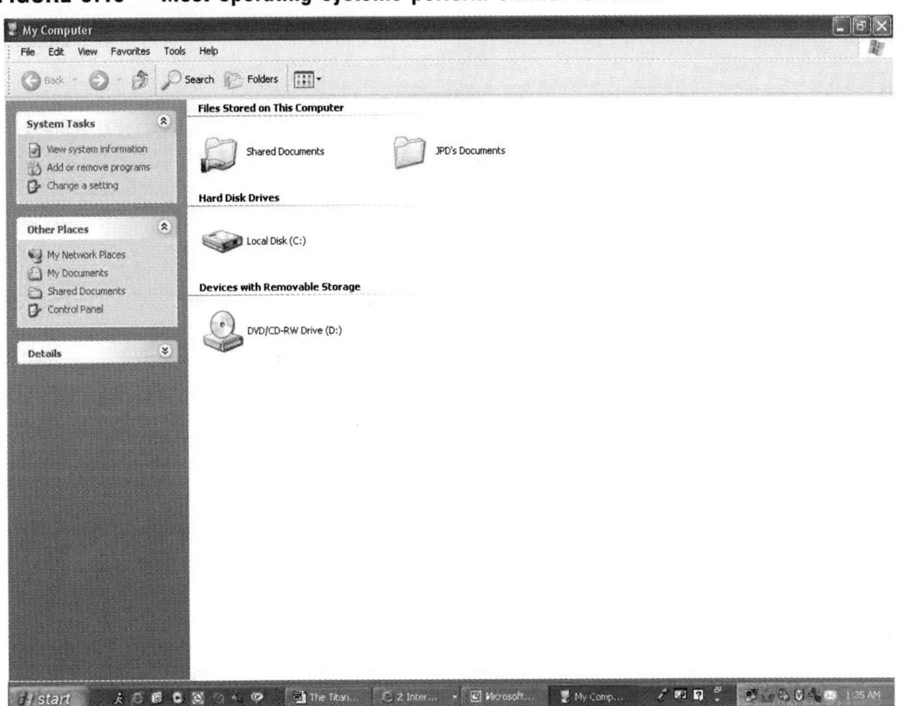

Providing a User Interface. One of the most important functions of any operating system is providing a user interface. A user interface allows one or more people to have access to, and command of, the computer system. The first user interfaces for mainframe and personal computer systems were command based (see Figure 3.11). A **command-based user interface** requires that text commands be typed at a prompt in order to perform basic tasks. For example, the command ERASE FILE1 would cause the computer to erase or delete a file named FILE1. RENAME and COPY are other examples of commands used to rename or copy files from one location to another. Most operating systems provide access to a *command prompt* to allow system administrators and advanced users to interact with the operating system through specific commands. However, the default user interface today is the graphical user interface.

FIGURE 3.11 • Command-based interface

A command-based user interface uses text commands to get the computer to perform basic or administrative activities.

```
Command Prompt
01/30/2002  10:57 PM    <DIR>          MSD33BK
01/30/2002  10:54 PM            67,594 msd33bk.exe
03/19/2002  01:27 PM    <DIR>          TPBDEMO
03/10/2002  12:11 PM    <DIR>          Completed Projects
04/24/2002  12:23 AM           869,876 GTTeleCH08 AF files.zip
06/15/2002  09:52 PM                23 NAVOPTX.DAT
05/21/2002  04:43 PM           360,620 Career Corner HI-RES Comp SalesCbsStkMkt
I-528-0178.JPG
06/03/2003  03:26 PM           481,440 transition.exe
06/15/2002  10:30 PM    <DIR>          Documents and Settings
05/17/2002  04:38 PM            35,019 Fig 1-01a1 HI-RES Microsoft Pocket PC ema
il.jpg
07/01/2002  01:33 PM    <DIR>          COLLWIN
07/05/2002  01:50 PM    <DIR>          TEMP
07/28/2002  02:14 PM               217 search.log
              18 File(s)      3,545,316 bytes
              27 Dir(s)  13,381,861,376 bytes free

C:\>cd\windows

C:\WINDOWS>cd\Documents and Settings

C:\Documents and Settings>dir /p
 Volume in drive C has no label.
 Volume Serial Number is 131E-12E9

 Directory of C:\Documents and Settings

06/15/2002  10:30 PM    <DIR>          .
06/15/2002  10:30 PM    <DIR>          ..
06/15/2002  10:30 PM    <DIR>          All Users
06/15/2002  10:41 PM    <DIR>          default
               0 File(s)              0 bytes
               4 Dir(s)  13,381,861,376 bytes free

C:\Documents and Settings>_
```

A graphical user interface (GUI) uses pictures or icons on the screen and menus to send commands to the computer system. Many people find that GUIs are easier to learn and use because complicated commands are not required. With a mouse, the user can highlight and click a command to perform various operating system functions (see Figure 3.12).

FIGURE 3.12 • A graphical user interface (GUI)

A graphical user interface (GUI) uses pictures or icons on the screen and menus to send commands to the computer system.

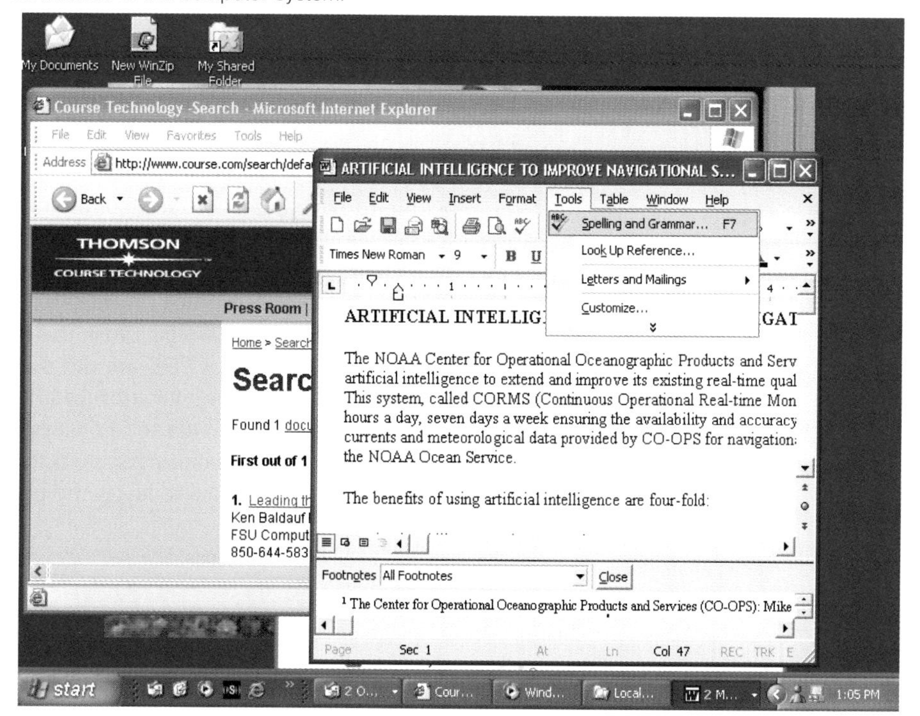

One of the first graphical user interfaces was on the Apple Macintosh computer, developed in the 1980s. Today, the most widely used graphical user interface is Windows by Microsoft. As the name suggests, Windows is based on the use of a window, or a portion of the display screen dedicated to a specific application. The screen can display several windows at once. For this reason, all GUI environments are sometimes referred to as "window" environments, even though they are not Microsoft products. The popularity of Microsoft Windows is in part due to the many advantages of using any graphical user interface, as listed in Table 3.2.

TABLE 3.2 • Advantages of a graphical user interface

Advantage	Description
Performing tasks in a GUI environment is intuitive	To open a file, you click a file icon or symbol; to delete a file, you drag it to a wastebasket icon or press the Delete key after the file has been highlighted.
GUI applications can be easy to use	Besides being intuitive, some software companies, such as Microsoft, include assistance features, such as wizards, that help with the creation of tables and forms; detailed, technical manuals describing complex commands are usually not needed.
The application interfaces are consistent	Once you learn the basics for one application, such as a word-processing program, the same basic commands and approaches work with other applications, such as spreadsheets or database programs.
You can view several running applications simultaneously	Windows can be sized to view the contents of several files and programs at the same time; this comes in handy when you need to view one document while typing in another, or when you need to copy and paste between applications.
The applications are flexible	You can use either a mouse or the keyboard; in addition, you can save files in different ways, in different formats, and to different folders or directories by using different save options.
Mistakes can be corrected easily	If you make a mistake, graphical user interfaces allow you to cancel or undo what you have done; most menu boxes have a cancel option.
Confirmation is requested as a safeguard	GUIs often ask you to confirm important operations, such as saving or deleting a file; most have an OK or Yes option for you to check or click before an important operation is carried out.

All operating systems are either proprietary or portable (sometimes called *generic*). A *proprietary operating system* is one developed by a vendor for use with specific computer hardware. Many cell phones use proprietary operating systems designed specifically for the handset. For example, T-Mobile's Sidekick II cell phone uses a proprietary operating system designed to support the phone's unique hardware features: camera, large display, and full keyboard. The iPod digital music player also uses a proprietary operating system designed for managing and playing music on that particular device. A *portable operating system* is one that can function with different hardware configurations. Microsoft Windows CE and Palm OS are examples of portable operating systems. They can run on many different devices from many different manufacturers as long as the device is designed to support that operating system. For example, Palm OS can be found on a number of handheld computers and cell phones.

Portable operating systems play an important role in interoperability, the ability of devices and programs from different vendors to operate together. Because organizations often upgrade their computer hardware and software to support new activities, portability can be an important consideration in selecting an operating system.

EXPAND YOUR KNOWLEDGE

To learn more about the Windows operating system, go to www.course.com/swt2/ch03. Click the link "Expand Your Knowledge" and then complete the lab entitled "Using Windows."

Controlling Common Computer Hardware Functions. Look back at Figure 3.9, and notice that the application program can only communicate with the system hardware through the operating system. Because of this, the OS serves as a buffer or interface between the application program and the hardware. Operating systems control many common hardware functions. For example:

- Starting the computer (often called the *boot process* or **booting**): A *cold boot* is performed by pushing a button or switch when the computer is not on. A *warm boot* is performed while the computer is currently running. For example, many computers have a *restart* procedure or button that performs a warm boot.
- Performing various tests: During the boot process, a *power-on self test (POST)* is performed to make sure the computer components are working correctly. The *basic input/output system (BIOS)* is activated from one or more computer chips to perform additional testing and to control various input/output devices, such as keyboards, display screens, disk drives, and various ports. Note that the computer can also be started from an operating system stored on a CD or floppy disk, called a *recovery disk.* This can be done if there is a problem with the computer's hard disk and the computer is not booting correctly when turned on. The recovery disk should be easily accessible and stored in a secure location.
- Inputting data from a keyboard, a mouse, or some other input device
- Reading data from and writing data to disk drives or other secondary storage devices
- Outputting or writing information to the computer screen, printer, or other output device: Because printers and similar devices are much slower than the computer's processor, operating systems often use *spooling,* whereby data is written from the processor to the hard disk or memory first in a buffer or waiting area. The output is then transferred to the slower printer, freeing the processor for other tasks.
- Formatting disks: Different operating systems often require different disk formats. Before some disks, such as floppy disks, can be used on a computer, they must be formatted. *Format* is a common function of an operating system.
- Configuring hardware devices: If a new printer is added, for example, the operating system can add controls for thc new printer to the computer using software called a *device driver.* Drivers are also available for external hard disks, scanners, and many other devices. If the needed drivers are not included with the hardware device on a CD-ROM, they are often available through the operating system or the Web site of the operating system or device vendor. The *www.microsoft.com/technet* Web site, for example, contains drivers for Windows operating systems.

The operating system's control of hardware is often rather subtle and is intended to be transparent to the user. *Transparent* means that an activity or task happens automatically and does not have to be considered, understood, or known by the user.

Suppose that a computer manufacturer designs new hardware that can operate much faster than before. Further suppose that this new hardware functions differently from the old hardware, requiring different machine code to perform certain tasks. If operating systems did not exist, all of the application software would have to be rewritten to take advantage of the new, faster hardware. Fortunately, because many application programs usually share a single OS, only the interface with the OS needs to be rewritten to convert the same set of commands on the application side to the new group of instructions needed on the hardware side. Having an OS layer allows software engineers to design many thousands of applications and use these applications on different types of hardware by adapting the OS.

Managing Memory. Operating systems control how memory is accessed and attempt to maximize the use of memory and storage. These functions allow the computer to efficiently and effectively store and retrieve data and instructions, and supply them to the CPU, thus speeding processing.

The memory-management portion of an OS converts a logical request for data or instructions into the physical location where the data or instructions are stored. A *logical view* of data is the way a programmer or user thinks about data. With a logical view, the programmer or user doesn't have to know where the data is physically stored in the computer system. However, a computer understands only the *physical view* of data, which includes the specific location of the data in storage or memory and the techniques needed to access the data. You can think of these concepts as logical versus physical access. For example, the current price of product GIZMO might always be found in the logical location Gizmo$. If the CPU needed to fetch the price of GIZMO as part of a program instruction, the memory-management program of the operating system would translate the logical location Gizmo$ into an actual physical location in memory or secondary storage, as shown in Figure 3.13.

FIGURE 3.13 • The operating system controlling physical access to data

The user prompts the application software for specific data. The operating system translates this prompt into instructions for the hardware to find the data the user requested. Having successfully completed this task, the operating system then relays the data back to the user via the application software.

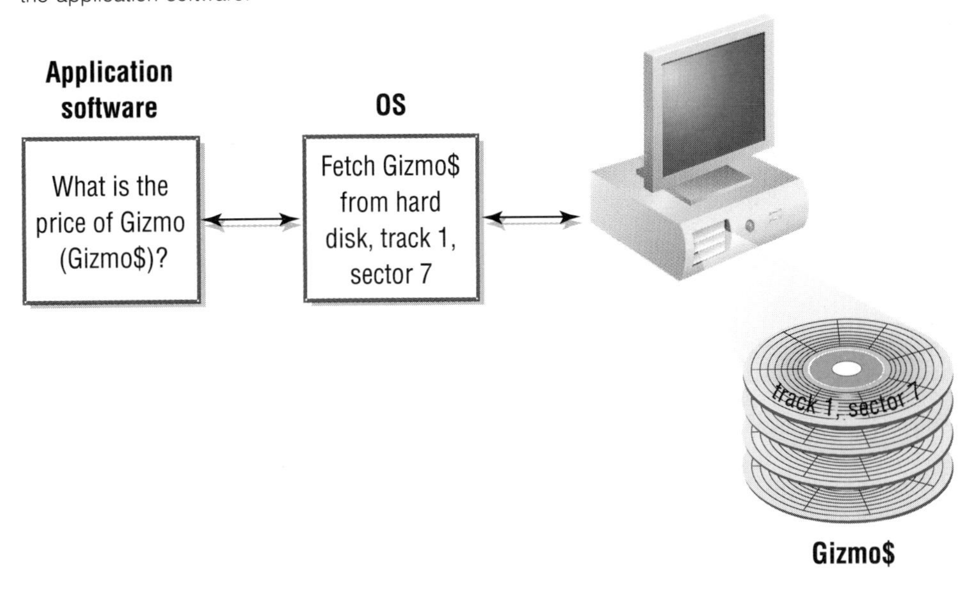

Many operating systems use *virtual memory (VM)*, or *virtual storage*, which allows users to store and retrieve more data without physically increasing the actual storage capacity of memory. Virtual memory extends standard memory by treating a portion of the disk drive as memory. Most computer operating systems make use of VM to allow users to run more applications simultaneously and to maximize the efficiency of memory management for best performance.

Managing Processors. The OS controls the operation of all processors within the computer system. As discussed in Chapter 2, the processor must retrieve each instruction, decode it, and then execute it. In addition, most operating systems permit several programs to be running at the same time, each requiring processor resources. For example, you may be entering text through a keyboard using a word-processing program, be connected to the Internet and playing the latest songs by using a media player, have your appointments and calendar program open and running, and be printing the results from a tax preparation program, all at the same time. The operating system makes sure that all these programs access the computer's processor(s) in an efficient and effective manner. The operating system also makes sure that one program doesn't interfere with the operation of another program.

Today's operating systems permit one user to run several programs or tasks at the same time, which involves multitasking, and allow several users to use the same computer at the same time, which is called time sharing. Multitasking and time sharing work by assigning each of the billions of cycles per second of the system clock to different tasks or users. The speed at which the processor switches attention between multiple tasks or users makes it appear as though there are several processors, each devoted to a specific job.

Multitasking involves running more than one application at the same time. See Figure 3.14. For example, a national sales manager might want to run an inventory control program, a spreadsheet program, and a word-processing program at the same time. With multitasking, he can run all three programs and share data and results among them. Spreadsheet results can be inserted into the inventory control program. Important tables and analysis from the inventory control program can then be inserted directly into the word-processing program. *Time sharing* allows more than one person to use a computer, typically a server computer, at the same time. For example, 15 research assistants in a school district can be entering survey results into a computer system at the same time. In another case, thousands of people may be simultaneously using an online computer service to get stock quotes and valuable news. Web servers provide time sharing for users requesting access to a Web page.

FIGURE 3.14 • Multitasking

With multitasking, you can run several programs simultaneously and share data among them.

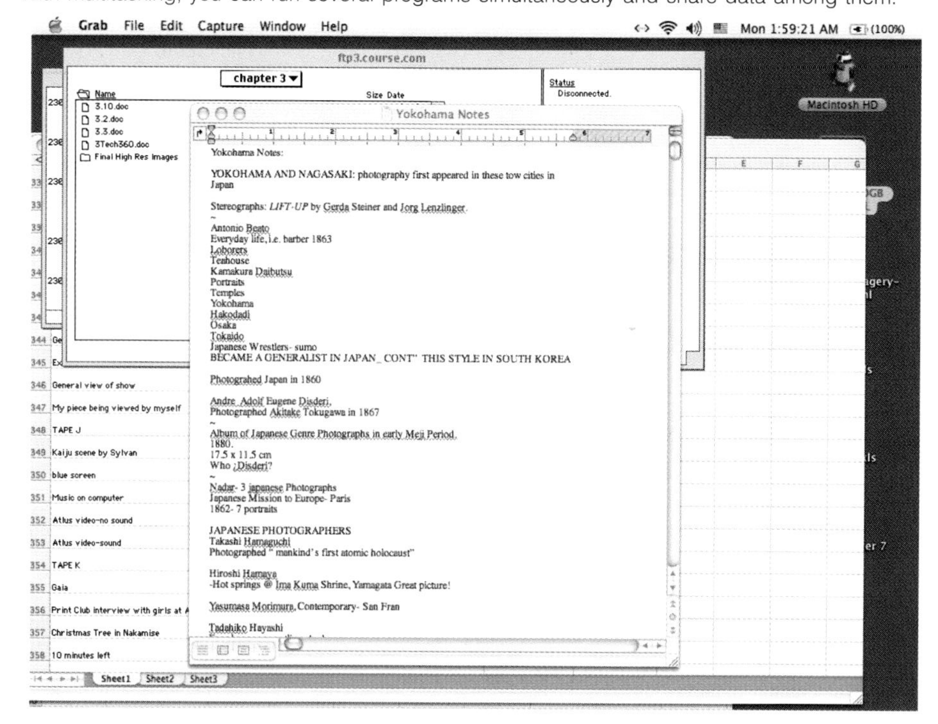

Managing Input, Output, Storage, and Peripherals. Just as the OS manages the hardware inside the computer, it also manages and coordinates the use of input and output devices, storage devices, and other peripheral equipment. Today's computers have keyboards, printers, display screens, and hard disks. Many have CD-ROM and DVD drives, joysticks, and a variety of other peripheral equipment. The operating system must manage all of the devices.

Operating systems also allow you to install new hardware and software easily. A Microsoft Windows feature called *plug and play (PnP)* allows you to attach a new hardware device and have it automatically installed and configured by the operating system. After PnP installs and configures the new device, it is ready to use.

Managing Files. All computers store and manipulate files that can contain data, instructions, or both. Operating systems organize files into folders, sometimes referred to as directories. Folders can hold one or more files. Organizing files into folders or subfolders makes it much easier to locate files, compared to having all files in one large folder or directory. Files can be organized alphabetically, by size, by type, or even by the date they were created or last modified. In addition, operating systems allow files to be copied from hard disks to portable media such as USB drives and CD-Rs. This feature is essential for keeping accurate and current backup files. File management can protect certain files from unwanted users; one approach is to use secure passwords and identification numbers. OS file-management features also allow you to search for files in various folders or directories using keywords or even partial words.

But how do operating systems know where data and files are stored on the computer? The operating system provides a *file system*, a way of organizing how data and files are physically stored and how they are logically manipulated. For example, older Windows operating systems used the *file allocation table (FAT)* file system, and newer Windows operating systems such as Windows XP generally use *New Technology File System (NTFS)*. NTFS is a more secure and stable file system than FAT and supports better data recovery if there are problems. The Linux and UNIX operating systems use several different file systems such as ext3 or ufs.

Each OS file system has conventions that specify how files can be named and organized (see Table 3.3). In Windows, for example, file names can be 256 characters long, both numbers and spaces can be included in the file name, and upper- and lowercase letters can be used. Windows and Linux don't permit spaces in file names, and certain characters, such as \, *, >, < cannot be used. File systems also specify how files can be organized in folders or subfolders.

TABLE 3.3 • Conventions or rules for file names

Convention (Rule)	Windows	Macintosh	Linux
Length in characters	256	256	255
Case sensitive?	No	Yes	Yes
Can numbers be used?	Yes	Yes	Yes
Can spaces be used	Yes	Yes	No

Managing Network Functions. Many of today's operating systems, including ones for personal computers, allow multiple computers to be connected together, sharing disk space, printers, and other computer resources. From smaller PCs and Macintosh computers to huge networks of large mainframe computers, network management features of today's operating systems allow computers to work efficiently and effectively together. Mac OS, Windows XP, Windows Server, Novell NetWare, UNIX, and Linux are a few examples of operating systems that include network management capabilities. These operating systems are covered later in this chapter and in more detail in Chapter 5.

Other Operating System Functions. Operating systems perform a number of other important functions (see Figure 3.15). Windows operating systems store important information on hardware devices, software settings, and user preferences in a database called the *registry*. The registry is automatically updated when new hardware or software is installed, or user preferences are entered. The registry can also be manually changed by skilled computer technicians. Because changes to the registry can dramatically alter or damage how a computer operates, manually changing it by unskilled people is not advised.

You can also modify operating systems to make them more user friendly. For example, visually or physically impaired people can use the Microsoft Windows accessibility options to make using the computer easier. You can alter screen and mouse operations, change printer and modem operations, change the volume and tone of sound, and adjust how battery power is managed and used by laptop computers.

Operating systems typically come with additional software that is not part of the operating system, including limited word processors, graphics programs, and games, such as Solitaire, Minesweeper, and Pinball. Some operating systems also come with Internet browsers and media players. Media Player by Microsoft, for example, can record TV programs, play music, and organize photos. The software also acts as a TV tuner. Microsoft is also developing software and operating systems to deliver TV programs over cable or the Internet using the *Internet Protocol Television (IPTV)* standard.[3] According to an analyst from Forrester Research, "Microsoft's platform is the most promising. There will be a lot of Microsoft software on set-top boxes five years from now."

FIGURE 3.15 • Other operating system functions

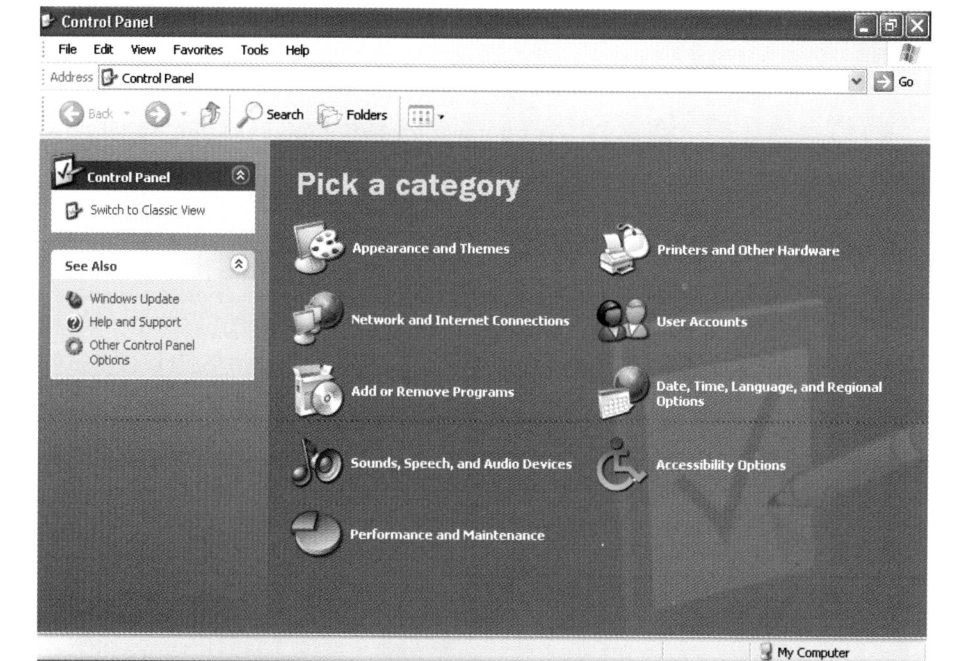

PC Operating Systems Today

Today's PC operating systems are fast and powerful, compared to older operating systems. They define and determine our computing experience. Older operating systems were command based and used such instructions as COPY, RENAME, and FORMAT. These older operating systems include Control Program for Microcomputers (CP/M) developed in the 1970s and Microsoft Disk Operating System (MS-DOS, often called just DOS) developed by Microsoft in the 1980s. Microsoft introduced the first of its Windows family of GUI operating systems in the late 1980s, and its descendents continue to be the most popular operating systems today. Other popular OSs for PCs today are the Apple Macintosh Mac OS, UNIX, and Linux.

Windows. Windows was originally designed to provide a GUI that ran "on top of" DOS; that is, it originally required DOS to run. Windows versions 1, 2, and 3 were DOS based. Newer versions of Windows, such as Windows XP and 2003 can access DOS or DOS-like functions but do not require DOS to run. Significant advances have been made over the years. With Windows 95 for instance, a taskbar was added at the bottom of the screen to keep track of all open applications and make it easier to switch between them, and a file manager was added to browse files and folders. The ability to network several computers together was added as a feature of the Windows 3.11 (Windows for Workgroups) operating system and has been improved and extended in all subsequent Windows operating systems. Windows NT was designed to allow personal computers and server computers to be connected in a network. Windows NT, Windows 2000, and Windows 2003 are often referred to as *network operating systems (NOS)* because they include security and network management features for networks and the servers that handle them. These and other network operating systems are discussed later in the section, "Operating Systems for Servers, Networks, and Large Computer Systems."

The current version of Windows for PCs is Windows XP (see Figure 3.16). It comes in different editions, including Home, Professional, Media Center, and others. Media Center edition enhances Windows XP Professional with additional music, video, and multimedia features. Windows Tablet PC Edition also enhances Windows XP with additional features to support the use of tablet PCs. With Windows XP, two or more users can be logged on at the same time. XP also supports security features for wireless networks.

FIGURE 3.16 • The Windows XP operating system

The Windows XP operating system comes in Home, Professional, Media Center, and other editions.

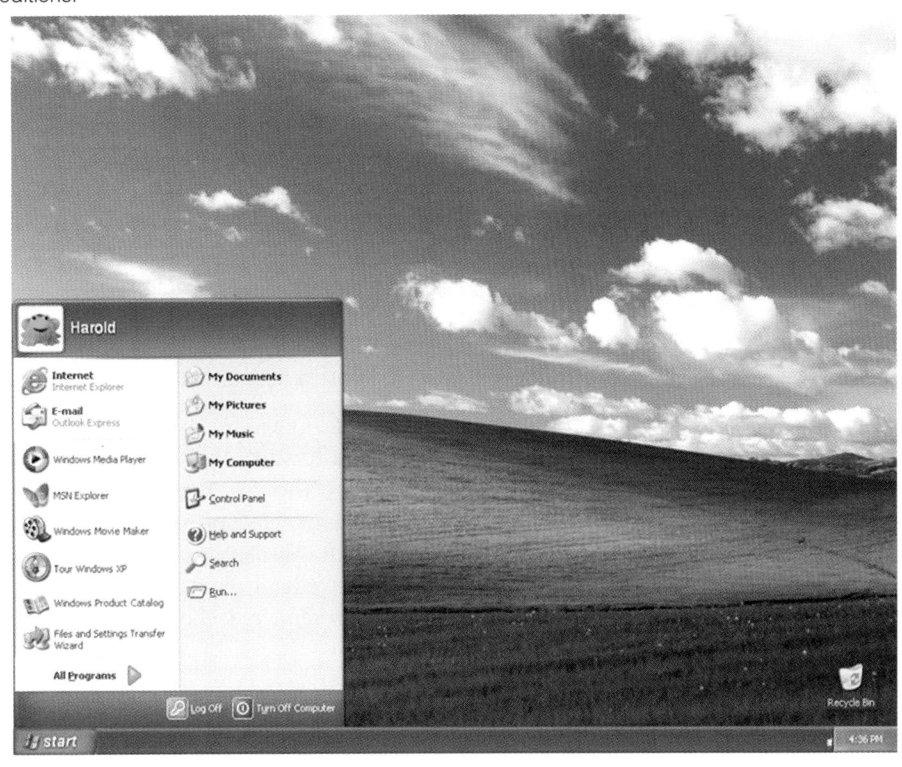

The next major revision of the Windows operating system (called Vista) is expected in 2005 or 2006. Vista will have many new features.[4] For example, InfoCard will allow people to disclose the information they choose to businesses and others on the Web, including name, credit card information, phone numbers, gender, and so on. The new version of Internet Explorer will have new features to help block attacks from the Internet. With the new operating system, Microsoft hopes to give people more choice about protecting their data and information and how they use their computer system. In 2004, Microsoft had about 93 percent of the PC operating system market.[5] Apple had 3 percent of the market, Linux had 3 percent, and other companies account for about 1 percent of the PC operating system market. See Figure 3.17.

FIGURE 3.17 • Microsoft leads the PC operating system market

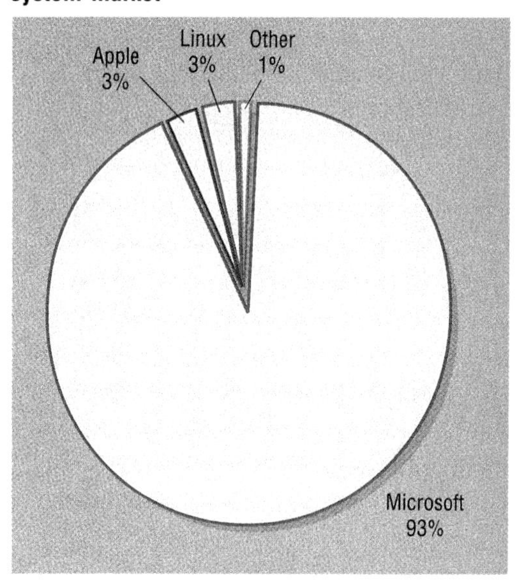

Apple Computer Operating Systems. Although IBM-compatible system platforms traditionally use Windows operating systems and Intel chips (and are often called *Wintel* for this reason), Apple computers typically use a proprietary Apple operating system (see Figure 3.18) and Motorola processors. Like Windows, Apple operating systems have gone through revisions and changes to make them more useful. Easy-to-use graphical user interfaces have been a hallmark of Apple systems from early on; their newer operating systems such as Mac OS X continue to be easy to learn and use. Simplicity does not mean that Apple's latest operating system is not powerful, however. Macintosh operating systems offer many outstanding graphics capabilities, virtual memory, multitasking functions, and the ability to run older "classic" Apple applications. They also have a toolbox of features that can be accessed by clicking icons, including database access. Tiger, the version of OS X released in 2005, has many new features, including support for 64-bit computing.[6] The Spotlight feature makes OS X the first operating system to offer an integrated search feature to find files and folders. Spotlight allows you to conveniently look inside documents, photos, calendars, e-mails, and other files and folders. Dashboard is another feature of Tiger that displays frequently used programs, like calculators, dictionaries, and calendars. Using Dashboard, you can see frequently used programs and execute them.

The Apple platform is preferred by many professionals who require high-speed computing for processor-intensive tasks, such as graphics and mathematical computation. Apple computers make use of a unique hardware architecture coupled with a powerful operating system. Mac OS X is based on the UNIX operating system, which is considered by many computer technicians as the most secure and stable system available. This combination of power and simplicity has won Apple a very loyal following among its small market share.

UNIX. *UNIX* is a powerful operating system developed by AT&T for networked workstations and servers. At the time UNIX was developed in the 1970s, AT&T was not permitted to market the operating system due to federal regulations that prohibited the giant telecommunications company from competing in the computer marketplace. All of this changed in the 1980s, when AT&T was broken up and many of the federal regulations preventing it from entering the computer marketplace were removed. Since then, UNIX has increased in popularity.

FIGURE 3.18 • The Mac OS X operating system

The Mac OS X operating system offers outstanding graphics abilities.

UNIX is popular in research and scientific computing environments. It is a portable operating system and popular for Web servers. Unlike Windows or Mac OS X, UNIX can be used on many computer system types and platforms from personal computers to mainframes. UNIX benefits organizations using both small and large computer systems, because it is compatible with different types of hardware and users have to learn only one operating system.

While UNIX is traditionally a command-based operating system (see Figure 3.19), most users interact with a UNIX computer through a GUI interface supported by the X Window System (see Figure 3.20). UNIX and the X Window System offer users a high degree of customization. Each UNIX X Window desktop can look as unique as each of the users that use and design them. Several different graphical desktop environments can be used under X Window. With its support for 32- and 64-bit processing and high-level computing, UNIX is favored by users who require computational power. UNIX users are required to be technically savvy in order to install, configure, and function in their computing environment, unlike their Windows and Mac OS counterparts. For those with the qualifications, UNIX provides more features and flexibility than its competitors.

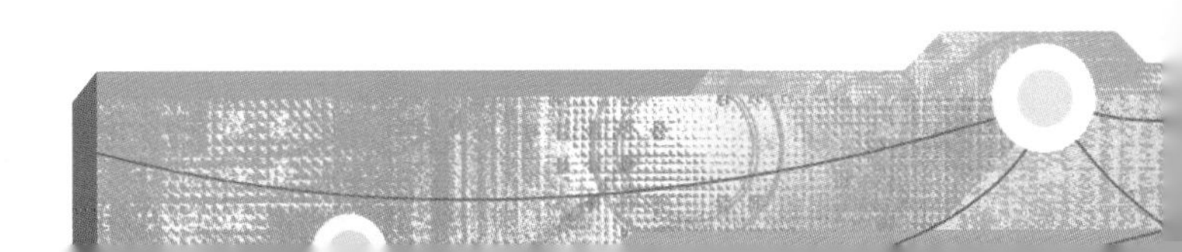

FIGURE 3.19 • The UNIX operating system

UNIX is a command-based operating system, which can be more difficult to learn than GUI operating systems.

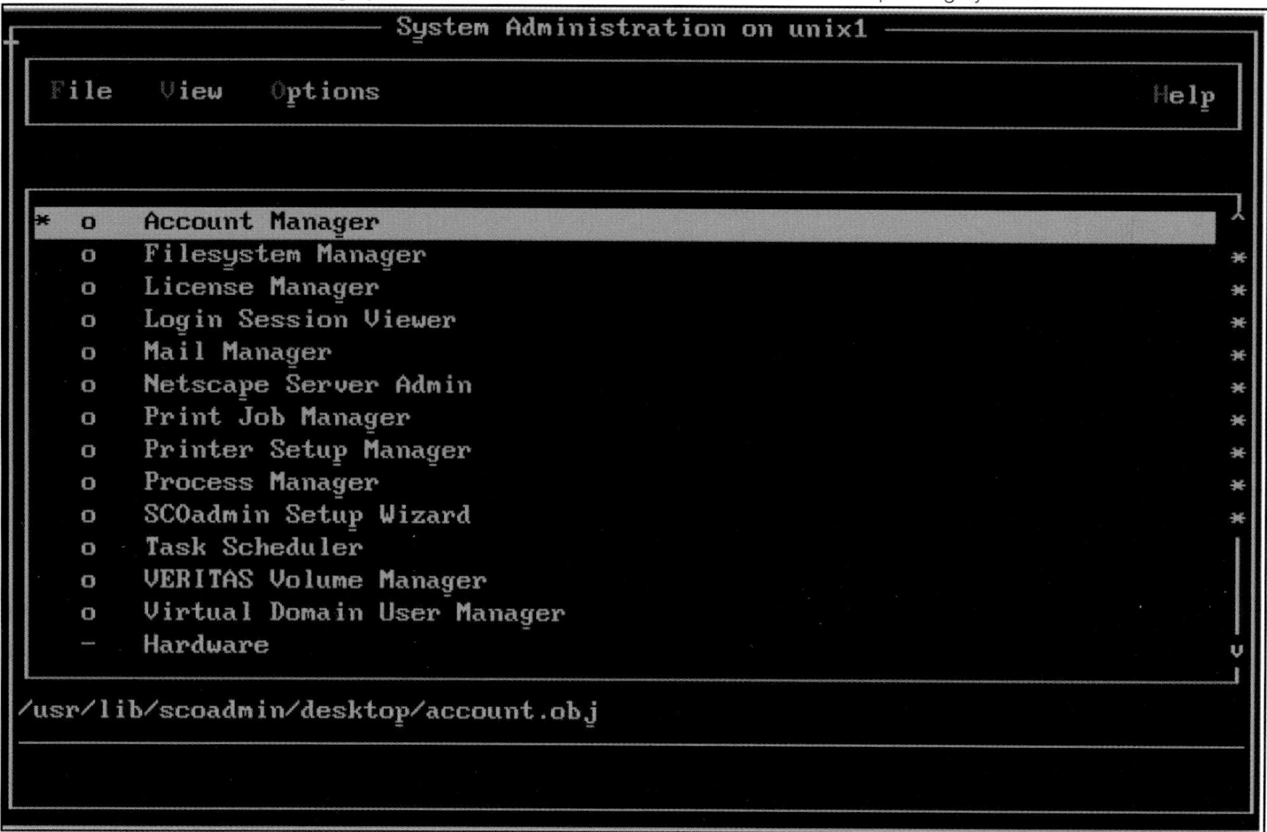

FIGURE 3.20 • X Window

X Window provides high-powered, highly customizable GUIs that run on UNIX and Linux operating systems. This is the GNOME desktop.

Linux. The *Linux* operating system was developed by Linus Torvalds in 1991. Based on the UNIX operating system, Linux was designed to provide a powerful, free, UNIX-like operating system that would run on desktop PCs. As with UNIX, Linux can be used as a command-line operating system or through a GUI such as GNOME or Lindows (see Figure 3.21). Today, it is being used on computers from small personal computers to large mainframe systems. Unlike many other operating systems, Linux is an **open-source software** package, which means that users have access to the source code. The operating system is available free to users under a *General Public License (GPL)* arrangement. Some vendors also offer commercial versions of Linux that include more features and offer user support. The free or low cost of Linux is a big advantage compared to other operating systems such as Windows and Mac OS X. People can not only get copies of Linux free of charge or at low cost, but they can also develop utilities, applications, and enhancements to the operating system. Grocer Hannaford Brothers, for example, used the Linux operating system in about a hundred of its supermarkets.[7] The company expects to save about 30 percent on its software costs by using Linux compared to using Windows or another nonopen-source operating system. In addition, several large computer vendors, including IBM, Hewlett-Packard, and Intel, support the Linux operating system.[8] IBM, for example, has more than 500 programmers working with Linux.

FIGURE 3.21 • Lindows

Lindows is a GUI that runs on Linux and closely resembles Microsoft Windows.

Linux and Linux-compatible applications are available over the Internet and from other sources, including Red Hat Linux and Caldera OpenLinux. Many individuals and organizations are starting to use Linux. A large travel and airline ticketing company, for example, can use Linux to run its Internet site. University researchers can use the Linux operating system along with other application software to help analyze the impact of Atlantic currents on climate.

Some people believe that operating systems like Linux and UNIX are more secure and less vulnerable to hackers than Windows. A survey revealed that many CIOs are considering switching to Linux and open-source software because of security concerns with Microsoft software.[9] A number of companies, including Intel and IBM, support Linux and open-source software and are contributing funds to defend Linux against copyright-infringement lawsuits.[10]

Operating Systems for Servers, Networks, and Large Mainframe Computer Systems

Some operating systems are designed specifically for larger computer systems or computer systems that require a server. These systems operate over a network, so network management functions are an important element in these operating systems. Windows Server 2003, UNIX, Linux, Novell NetWare, and Sun Microsystems Solaris are examples of network operating systems. See Figure 3.22. For example, Microsoft Windows Server 2003 can be used to coordinate large data centers. The operating system also works with other Microsoft applications and operating systems (such as Windows XP Professional) on client computers. It can be used to prevent unauthorized disclosure of information by blocking text and e-mails from being copied, printed, or forwarded to other people. Sun Microsystems hopes that its open-source Solaris will attract developers to make the software even better.[11] Small businesses, for example, often use network operating systems to run networks and perform critical business tasks.[12]

FIGURE 3.22 • Windows Server operating system

The Microsoft Windows Server 2003 interface is similar to that of Windows XP, but Windows Server provides functions to manage multiple users and networks that XP lacks.

Large mainframe computer manufacturers typically provide proprietary operating systems with their specific hardware. z/OS is a 64-bit operating system for IBM's large mainframe computers. MPE/iX is an operating system used by Hewlett-Packard's mainframe computers. Enterprise Systems Architecture/370 (ESA/370) and Multiple Virtual Storage/Enterprise Systems Architecture (MVS/ESA) are older operating systems used on IBM mainframe computers.

Operating Systems for Handheld Computers and Special-Purpose Devices

Operating systems for handheld devices and other small devices that contain computers are called *embedded operating systems* because they are typically embedded in a computer chip. The popular Palm personal digital assistant (PDA) uses the Palm OS. Today, the company has two major businesses. *PalmOne* is a hardware company that provides a line of innovative handheld computers, and *PalmSource* is a software company that supports handheld computers and smart phones by other companies, including Handspring, Aceeca, and others. Software applications include business, multimedia, games, productivity, reference and education, hobbies and entertainment, travel, sports, utilities, and a variety of wireless applications.

Many Macintosh users prefer Palm operating systems for PDAs because they are easy to integrate with Mac programs and applications.[13] Such flexibility has enabled the Palm OS to remain relatively easy to use while adding some expandability and capability as a general-purpose computing platform. An office supplies store, for example, can use wireless devices containing the Palm OS to scan shipments, create an electronic manifest as deliveries are loaded on a truck, and capture customer signatures electronically. Customers can track the status of their orders via its Web site. In 2004, Palm had about 51 percent of the market for PDA and handheld operating systems.[14] Microsoft had about 32 percent of the market, Linux had 3 percent, and other companies accounted for about 14 percent of the market for PDA operating systems. See Figure 3.23.

FIGURE 3.23 • Market share for PDA operating systems

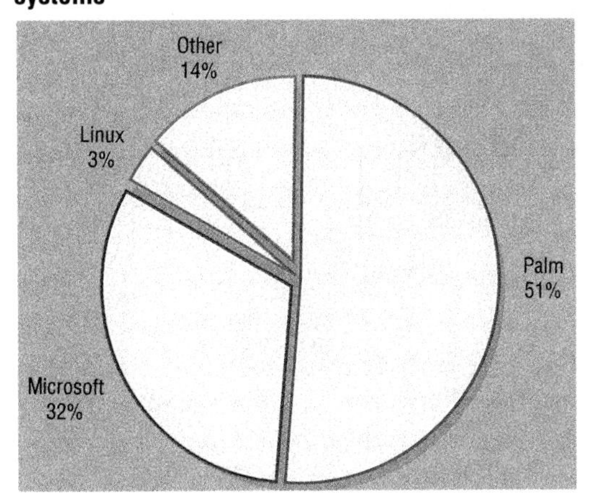

Microsoft has developed a number of operating systems for small computers and devices, including Pocket PC, Handheld PC, Windows Embedded, and Windows Mobile. These operating systems allow handheld devices to be synchronized with PCs using cradles, cables, and wireless connections.

Many other devices, including cell phones, also use embedded operating systems. For example, the Symbian operating system is used in many Nokia, Motorola, and Sony cell phones. Psion also makes cell phone operating systems for Nokia, Ericsson, and Motorola phones. In addition, there are operating systems for special-purpose devices, such as computers on the space shuttle, in military weapons, and in some home appliances. Microsoft makes embedded operating systems for industrial devices, ATMs, set-top boxes (see Figure 3.24), medical devices, and automobile devices. Windows Automotive is built on the Windows CE operating system and is specifically designed to offer voice recognition, Internet access, and other Windows features for vehicles.

Utility Programs

Another type of system software is a utility program. A utility program is any system software besides operating systems that assist in maintaining, managing, and protecting computer system resources. In fact, as operating systems have developed, they have incorporated features that were originally found in separate utility programs, such as disk maintenance utilities. Utilities are used to merge and sort sets of data, keep track of computer jobs being run, and perform other important routine tasks. Utility programs often come installed on computer systems; a number of utility programs can also be purchased. One utility program is even used to make computer systems run better and longer without

FIGURE 3.24 • Embedded operating systems

Many special-purpose devices have embedded operating systems.

problems. Computer scientists at IBM and other companies are developing software to predict when a computer might malfunction and to take corrective action. The software can also help a network of computers share processing jobs to get more done in less time. Here is a brief sampling of some popular types of utility programs:

- Virus detection and recovery: Computer viruses from the Internet and other sources can be a nuisance or can completely disable a computer. Virus detection and recovery software can be installed to constantly monitor your computer for possible viruses. If a virus is found, the software can eliminate the virus or "clean" the virus from the computer system, as shown in Figure 3.25. It is essential to keep virus detection and recovery software current, and most packages provide an automatic or manual update feature that checks for the latest updates when the computer is connected to the Internet. Symantec and McAfee are examples of companies that make virus detection and recovery software. Microsoft offers a free malicious software removal tool on its Web site, *www.microsoft.com.*[15] Microsoft has also acquired several antivirus companies and is expected to release antivirus software in the future.[16]
- File compression: File compression programs can reduce the amount of disk space required to store a file and reduce the time it takes to transfer a file over the Internet. WinZip (*www.winzip.com*) is a popular Windows file compression program that compresses files or groups of files into zip files. A zip file has a .zip extension and can be easily unzipped to the original file. *MP3 (Motion Pictures Experts Group-Layer 3)* is a popular file

FIGURE 3.25 • Virus detection and recovery software

Virus detection and recovery software is an important utility for anyone accessing the Internet and sharing files.

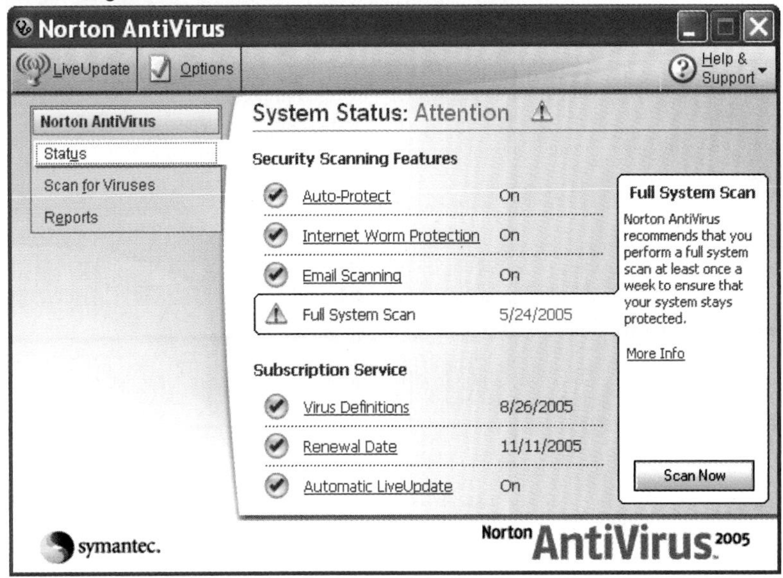

compression format used to store, transfer, and play music. It can compress files to be 10 to 20 times smaller and maintain a sound quality that approaches that found on CDs. Software, such as iTunes from Apple, can be used to store, organize and play MP3 music files.

- Spam and pop-up-ad guards: Getting unwanted e-mail (spam) and having annoying and unwanted ads pop up on your display screen while you are on the Internet can be frustrating and a big waste of time. There are a number of utility programs that can be installed to help block unwanted e-mail spam and pop-up ads, including Cloudmark SpamNet, IhateSpam, Spamnix, McAfee SpamKiller, and Ad-aware.
- Hardware and disk utilities: Hardware utilities can check the status of all parts of the PC including hard disks, memory, modems, speakers, and printers.[17] Disk utilities check the hard disk's boot sector, file allocation tables, and directories, and analyze them to ensure the hard disk hasn't been tampered with. Disk utilities can also optimize the placement of files on a crowded disk. A number of utility programs diagnose and fix hard disk problems. This can involve scanning a disk for problems, fixing any problems, and defragmenting disks. *Defragmenting* involves rearranging files on a hard disk to increase the speed of executing programs and retrieving data stored on the hard disk. These types of disk diagnostic programs are available with many operating systems. (See Figure 3.26.) They can also be purchased from companies that make utility programs, such as Symantec.
- Backup utilities: A number of utilities can help you back up data and programs from a hard disk onto tape, CD, or DVD drive, or other devices. These utilities can back up all data and programs or selected data and programs that may have changed since the last backup. BackUp MyPC, by StompSoft, is an example of a backup utility for Windows PCs. Many operating systems come with built-in backup utilities. Windows System Restore feature, for example, has the ability to take a "snapshot" of the OS (restore point) and return files to their state at that point.

FIGURE 3.26 • Disk defragmenter

This Windows XP utility can help optimize disk performance.

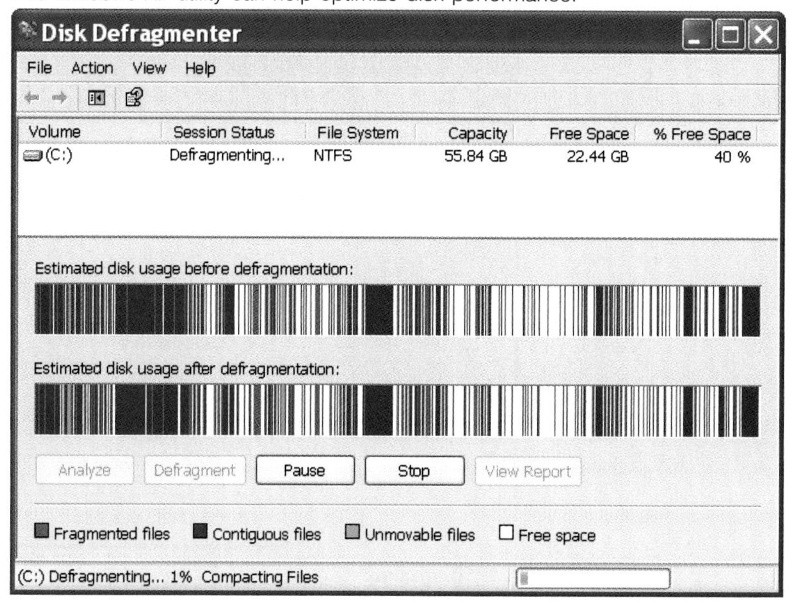

- File transfer utilities: People who transfer large files over the Internet, such as Web developers, commonly use FTP (file transfer protocol) or SFTP (secure file transfer protocol) to transfer files they have created on a PC to a server for others to access. Utilities such as WS_FTP provide a graphical interface to make transfers faster and easier. Transferring programs and files from an old computer to a new one is usually time consuming. Some people reinstall all programs and then copy data to CDs or USB drives to transfer to the new machines. Utilities such as PC Relocator or Move2Mac automate this process. After installing utility software on the old and new machine and connecting both machines with a cable, the software automatically transfers programs and data files to your new machine, saving you time and the inconvenience of manually transferring your programs and files.
- Search utilities: A number of companies have introduced search software that allows you to search the files on your hard disk and locate important information. The software maintains a database of keywords from the content of your e-mail, word-processing documents, spreadsheets, and other documents to quickly track down information. Most of these desktop search tools are free and available from a number of popular Internet sites.[18] Yahoo! Desktop Search, Ask Jeeves Desktop Search, Google Desktop, and MSN Deskbar are examples (see Figure 3.27).
- Other utilities: Utility programs exist for almost every conceivable task or function. Screensavers, for example, can be an important way of securing your computer while you're away from your desk by configuring your screensaver with a password. Utilities called Windows cleaners, for example, clean out unnecessary entries in the Windows registry to help Windows run more efficiently. Microsoft Windows Rights Management Services can be used with Microsoft Office programs to manage and protect important corporate documents.[19] Widgit Software has developed an important software utility that helps people with visual disabilities use the Internet.[20] The software converts icons and symbols into plain text that can be easily seen. There is even a software utility

FIGURE 3.27 • Desktop search

With a desktop search tool (the Google tool is highlighted in this figure), you can search through files on your computer.

that allows a manager to see every keystroke a worker makes on a computer system.[21] Monitoring software can catalogue the Internet sites that employees visit and the time that employees are working at their computer. Microsoft makes a utility, called Virtual PC for Mac that allows Windows programs to be run on Apple Macintosh computers.

APPLICATION SOFTWARE

As discussed earlier in this chapter, the primary function of *application software* is to apply the power of the computer to give people, groups, and organizations the ability to solve problems and perform specific activities or tasks. When someone wants the computer to do something, one or more application programs are used. The application programs then interact with systems software. System software then directs the computer hardware to perform the necessary tasks. This concept was illustrated in Figure 3.9.

Of all the types of software, application software has the greatest potential to help an individual, a group, or an organization achieve its goals. Individuals can use application software to advance their career, to communicate with others, to access a vast array of knowledge, or to serve up entertainment. Groups and organizations use application software to help people make better decisions, reduce costs, improve service, or increase revenues. Web browsers and related Internet software will be fully explored later in the book. This next section looks at some uses of application software in more detail.

Productivity Software

Productivity software is any software designed to help individuals be more productive. Over the years this category of software has included the most popular personal computer applications: word processors, spreadsheets, database-management systems, presentation graphics software, and personal information management software. Productivity software is used at home for personal tasks and at work for professional tasks. For example, a spreadsheet application might be used at home to set up a personal budget or in a business to design a corporate budget.

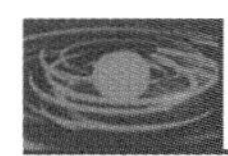

EXPAND YOUR KNOWLEDGE

To learn how to get the most out of your word processing application, go to www.course.com/swt2/ch03. Click the "Expand Your Knowledge" link and then complete the lab entitled, "Word Processing."

Word-Processing Applications. Word processing is perhaps the most highly used application software for individuals. *Word-processing software* allows users to create formatted text documents varying in complexity from simple to-do lists to professional magazine layouts. Microsoft Word is the most popular word-processing program and is available for both PCs and Macintosh computers (see Figure 3.28). Corel's WordPerfect, Lotus' WordPro, and Sun Microsystems' Write, are other examples of word processors.

The features available on today's word processors are stunning. All of the common features you would expect are included, such as easy text entry and formatting, the capability to develop attractive tables, check spelling and grammar, and generate footnotes and endnotes. You can create automatically numbered or bulleted lists in most word-processing programs, and in many you can insert photos, graphics, and drawings. Today's word-processing programs also have sophisticated document processing features such as generating a table of contents at the beginning of the text and an index at the end.

Word-processing programs can be used with a team or group of people collaborating on a project. The authors and editors who developed this book, for example, used the Track Changes and Reviewing features of Microsoft Word to track and make changes to chapter files. You can insert comments in, or make revisions to a document that a coworker can review and either accept or reject.

EXPAND YOUR KNOWLEDGE

To learn how to get the most out of your spreadsheet application, go to www.course.com/swt2/ch03. Click the "Expand Your Knowledge" link and then complete the lab entitled "Spreadsheets."

Spreadsheet Applications. *Spreadsheet software* supports complicated numerical analysis and calculations, and allows users to perform *"what if" analysis* on financial and other data. A spreadsheet contains rows that are numbered and columns that are lettered. Cell D10, for example, is found in row 10 and column D.

FIGURE 3.28 • A word-processing program

Word processing is perhaps the most highly used application software for individuals.

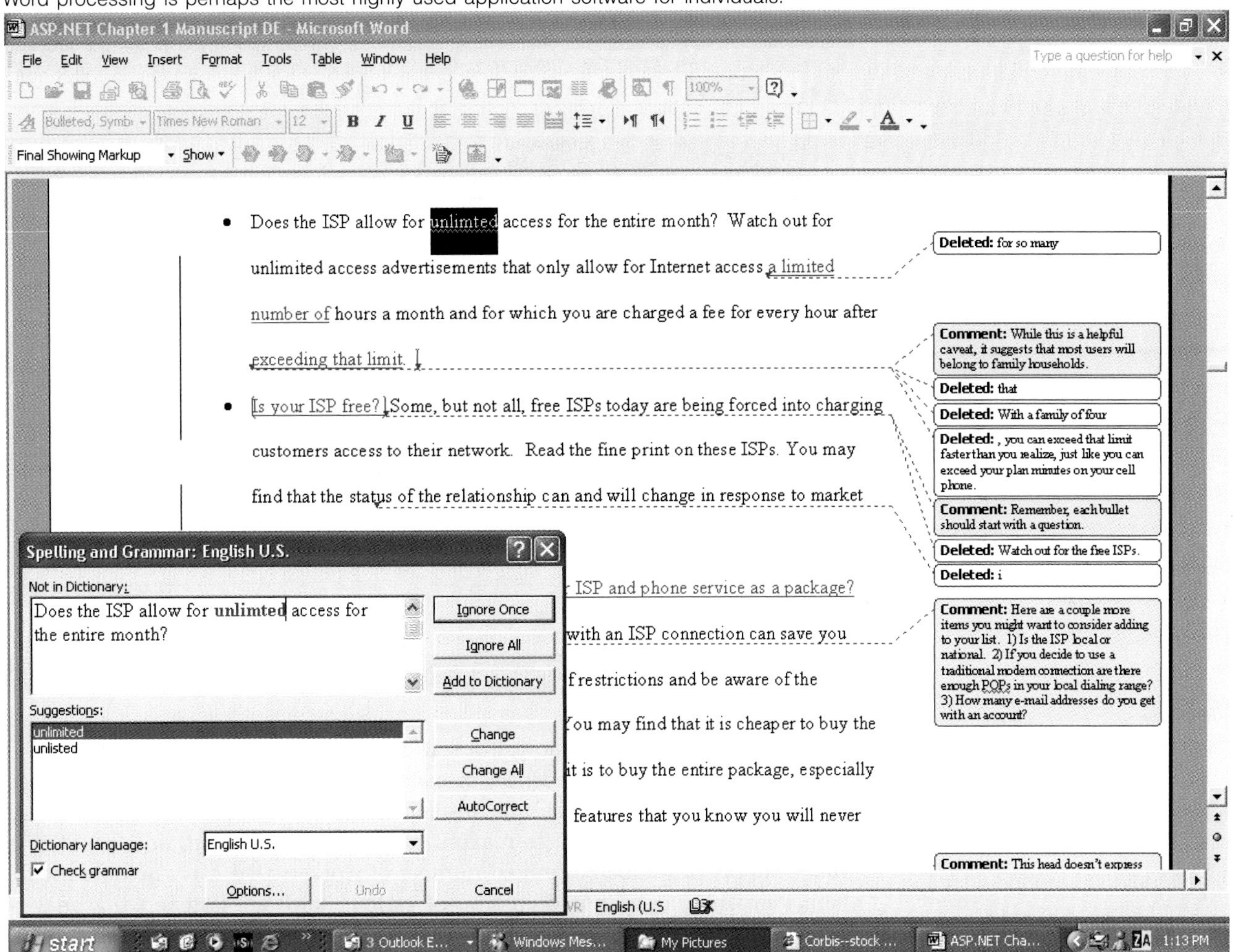

You can enter text, numbers, or complex formulas into a spreadsheet cell (see Figure 3.29). For example, you can have a spreadsheet automatically get the total or average of a column or row of numbers. If you have a budget spreadsheet containing your actual expenses for the last several months, you can change one number and immediately see the impact on your average monthly expenditures, your total for the month, and any other calculations that you have entered into your spreadsheet. In general, if you change a value in a spreadsheet, all the formulas based on the value are changed immediately and automatically. The authors and publisher used a spreadsheet to plan and monitor important deadlines to publish this book in a timely fashion.

Spreadsheet programs have many built-in functions for science and engineering, statistics, and business. The science and engineering functions include sine, cosine, tangent, degrees, maximum, minimum, logarithms, radians, square root, and exponents. The statistical functions include correlation, statistical testing, probability, variance, frequency, mean, median, mode, and much more. The business functions include depreciation, present value, internal rate of return, and the monthly payment on a loan, to name a few.

EXPAND YOUR KNOWLEDGE

To learn more about advanced spreadsheet features, go to www.course.com/swt2/ch03. Click the "Expand Your Knowledge" link and then complete the lab entitled "Advanced Spreadsheets."

FIGURE 3.29 • A spreadsheet program

Spreadsheet applications are excellent for making calculations, as seen in this application.

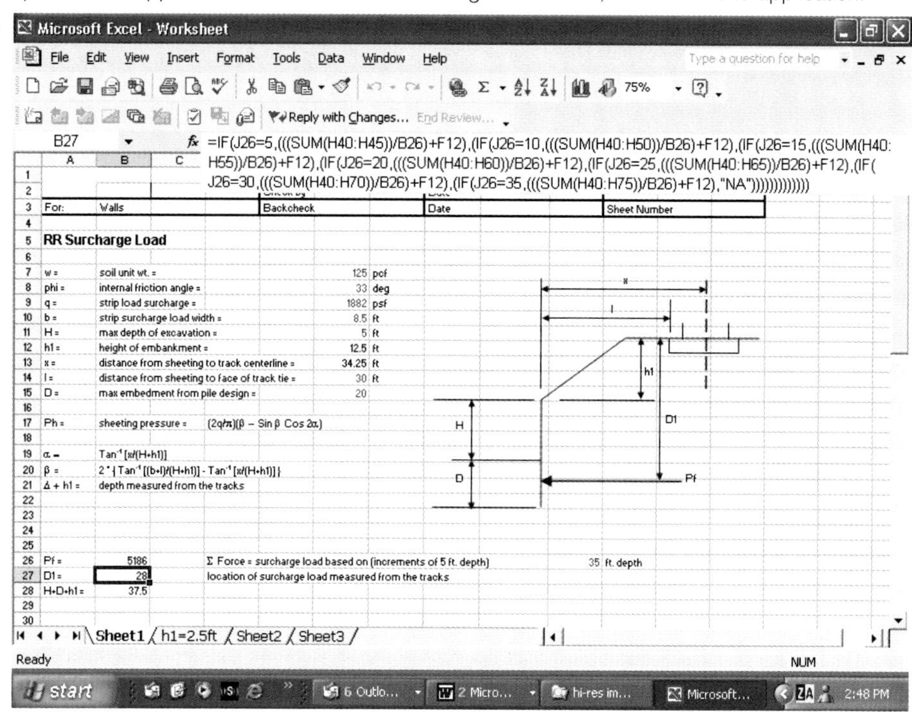

Optimization is another powerful feature of many spreadsheet programs. *Optimization* allows the spreadsheet to maximize or minimize a quantity subject to certain constraints. For example, a small furniture manufacturer that produces chairs and tables might want to maximize its profits. The constraints could be a limited supply of lumber, a limited number of workers that can assemble the chairs and tables, or a limited amount of various hardware fasteners that may be required. Using an optimization feature, such as Solver in Microsoft Excel, the spreadsheet can determine what number of chairs and tables to produce with labor and materials constraints in order to maximize profits. (See Figure 3.30.) As another example, a company that produces dog food might want to minimize its costs, while meeting certain nutritional standards. Minimizing costs becomes the objective, while the nutritional standards are the constraints. Again, an optimization feature can determine the needed blend of dog food ingredients to minimize costs, while meeting the nutritional requirements of the dog food. These are just a few examples of the use of optimization in spreadsheets. Because of the power and popularity of spreadsheet optimization, some colleges and universities offer complete courses based on Solver in the Microsoft Excel spreadsheet. Most of these courses are in engineering and business schools.

"BABEL FISH" TECHNOLOGY

A quick look at the hard drive of any computer reveals dozens of applications, all of which speak nearly as many different languages. A new technology developed by Cisco Systems gives networks the ability to "speak" the languages of the software they support. Previously, different middleware products performed this function. By better linking with the applications directly, the network can work not only smarter and faster, but with greater security as well.

Update: Cisco Networks to Work Closer with Applications
by Duncan Martell
ComputerWorld
http://www.computerworld.com/networkingtopics/networking/story/0,10801,102652,00.html
June 21, 2005

Presentation Graphics. In addition to the limited graphics programs that come with most operating systems, such as painting and drawing programs, there are

FIGURE 3.30 • Optimization

Some spreadsheets offer optimization tools that can perform complex computations based on certain constraints.

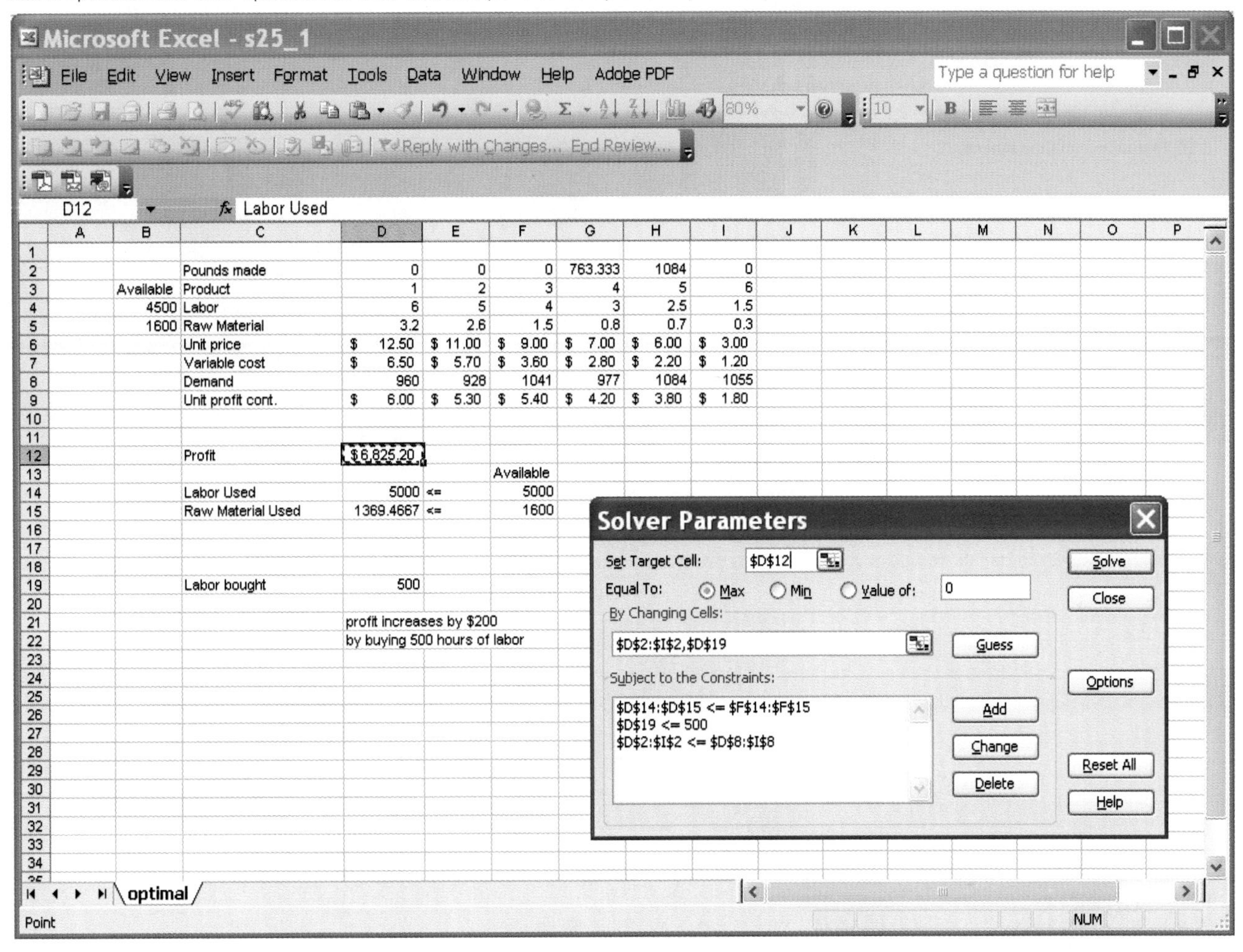

powerful presentation graphics programs that can be used for a variety of purposes (see Figure 3.31). *Presentation graphics software* supports formal presentations by providing graphic "slides" that can be used to accompany and embellish a live presentation or to present the material without the use of a human presenter. Presentation graphics programs are almost essential for making presentations to professional groups and audiences that can vary from a few people to thousands of people. Physicians and medical personnel use presentation graphics to show the results of medical research at conferences. Forest-service consultants use presentation graphics to describe new forest management programs, and businesses almost always use presentation graphics to present financial results or new initiatives to executives and managers. Because of their popularity, many colleges and departments require students to become proficient with presentation graphics programs. Presentation graphics programs are becoming popular with a wide variety of people. Fourth-grade students, for example, have used presentation graphics software to make presentations on sixteenth-century explorers in a history class.

Most presentation graphics programs, such as Freelance Graphics by Lotus, and PowerPoint by Microsoft, consist of a series of slides. Each slide can be displayed on a computer screen, printed as a handout, or (more commonly) projected onto a large viewing screen for audiences. Powerful built-in features allow you to develop attractive slides and complete presentations. You can select a

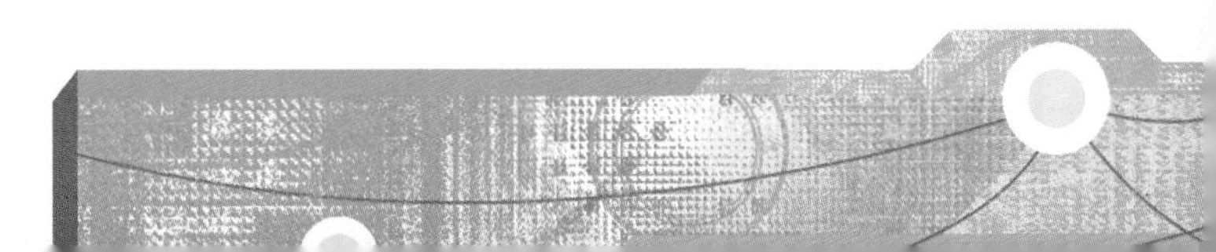

FIGURE 3.31 • A presentation graphics program

Presentation graphics programs are almost essential for making presentations to professional groups and audiences that can vary from a few people to thousands of people.

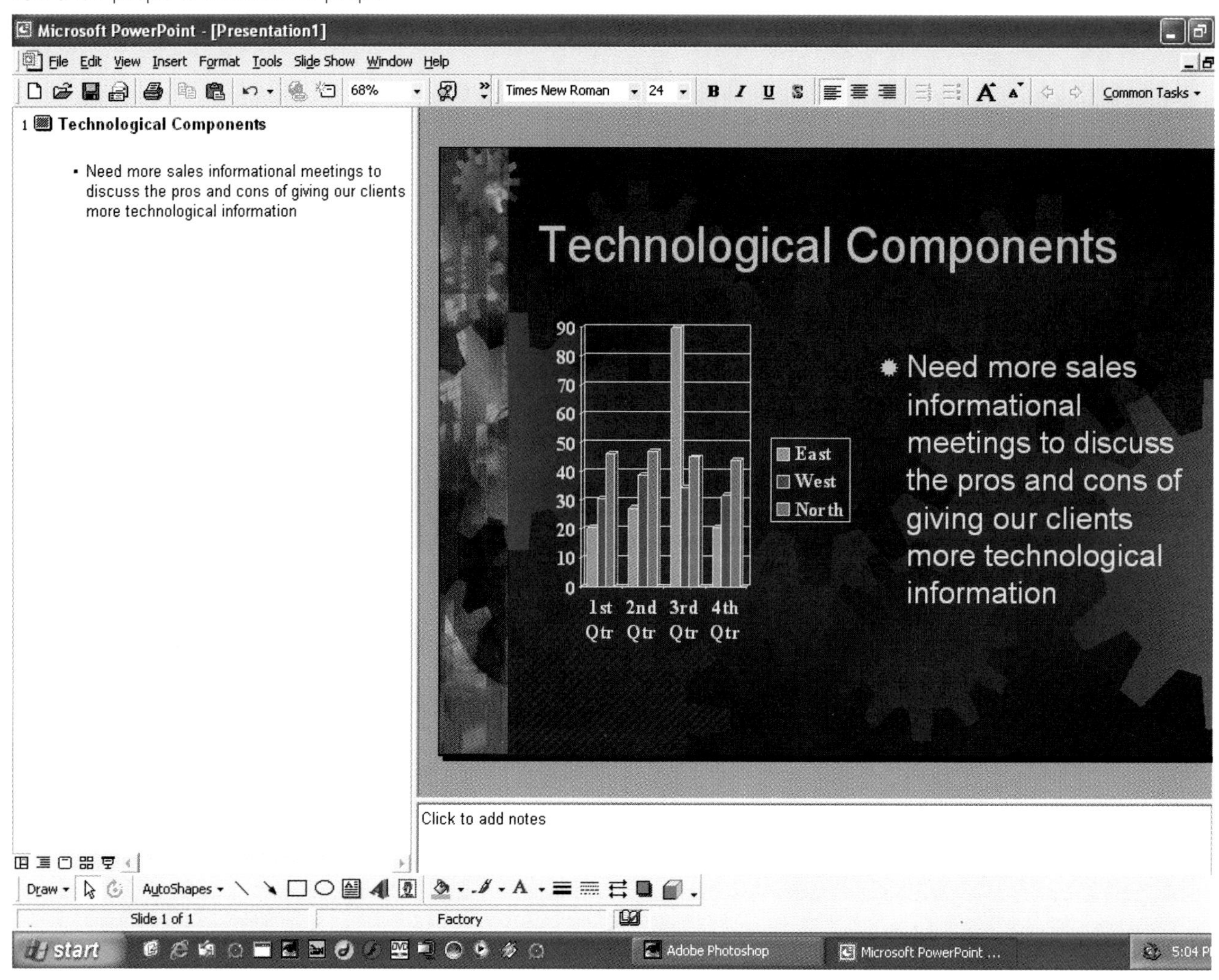

template for a type of presentation, such as recommending a strategy for managers, communicating bad news to a sales force, giving a training presentation, or facilitating a brainstorming session. The presentation graphics program takes you through the presentation step by step, including applying color and attractive formatting. Of course, you can also custom design your own presentation.

Electronic presentations are increasingly being used to provide training over the Web, and to provide information at kiosks in public areas and industry trade shows. Individuals can progress through the presentation themselves by clicking icons on screen, or the presentation can be set to progress automatically at specified time intervals.

EXPAND YOUR KNOWLEDGE

To learn how to get the most out of your presentation graphics application, go to www.course.com/swt2/ch03. Click the "Expand Your Knowledge" link and then complete the lab entitled "Presentation Software."

There are many outstanding features of presentation graphics programs. All types of charts, drawings, and formatting are available. Most presentation graphics programs come with many pieces of *clip art*, such as drawings and photos of people meeting, medical equipment, telecommunications equipment, entertainment media, and much more. This clip art can be inserted into any slide. You can purchase additional clip art from other computer vendors and insert items into a slide. In addition, you can insert movie and sound clips into a presentation. Presentation graphics programs also have collaboration features that allow you to work with several people or groups around the world on a presentation.

Database-Management Programs. *Database-management software* is used to store, manipulate, and manage data in order to find and present useful information. Databases can be used to store large tables of information (see Figure 3.32). Each table can be related. For example, a company can create and store tables that contain customer information, inventory information, employee information, and much more. Assume a customer decides to place an order for two music CDs of a new artist, one for herself and one for a friend, by calling a sales rep she has worked with in the past. The sales rep can record the customer number, the inventory number for the new music CD, and the quantity, two CDs in this case. The database-management software does the rest. First, the database takes the customer number and goes to the customer table to retrieve the name, address, and credit card information. With the employee number of the sales rep from the customer table, the database-management software goes to the employee table to determine any sales commissions that should be paid to the sales rep. Next, the database-management system takes the inventory number for the music CD and goes to the inventory table to get all the information about the music CD and make sure that there are at least two CDs in stock. Once all of this is done, the order can be processed and the two music CDs sent to the customer. All of this is done as a result of giving the customer number, inventory number, and quantity to the database-management software. Before database-management software existed, processing an order was done manually, requiring a lot of time and having the potential for making many mistakes.

In addition to order processing, database-management software can be used to perform all business functions for a small business, including payroll, inventory control, order processing, bill paying, and producing tax returns. Database-management systems can also be used to track and analyze stock and bond prices, analyze weather data to make forecasts for the next several days, and summarize medical research results. At home, you can use database-management software to keep a record of expenses, a list of what is in your apartment or home, or a list of the members of a student government organization.

Personal Information Management. *Personal information management (PIM) software* helps individuals store useful information, such as to-do lists, appointment calendars, and contact lists. In addition, information in a PIM can be linked. For example, you can link an appointment with a sales manager that appears in the calendar with information on the sales manager in the address book. When you click the appointment in the calendar, information on the sales manager from the address book is automatically opened and displayed on the computer screen.

FIGURE 3.32 • A database program

Database-management software can be used to store large tables of information and produce important documents and reports.

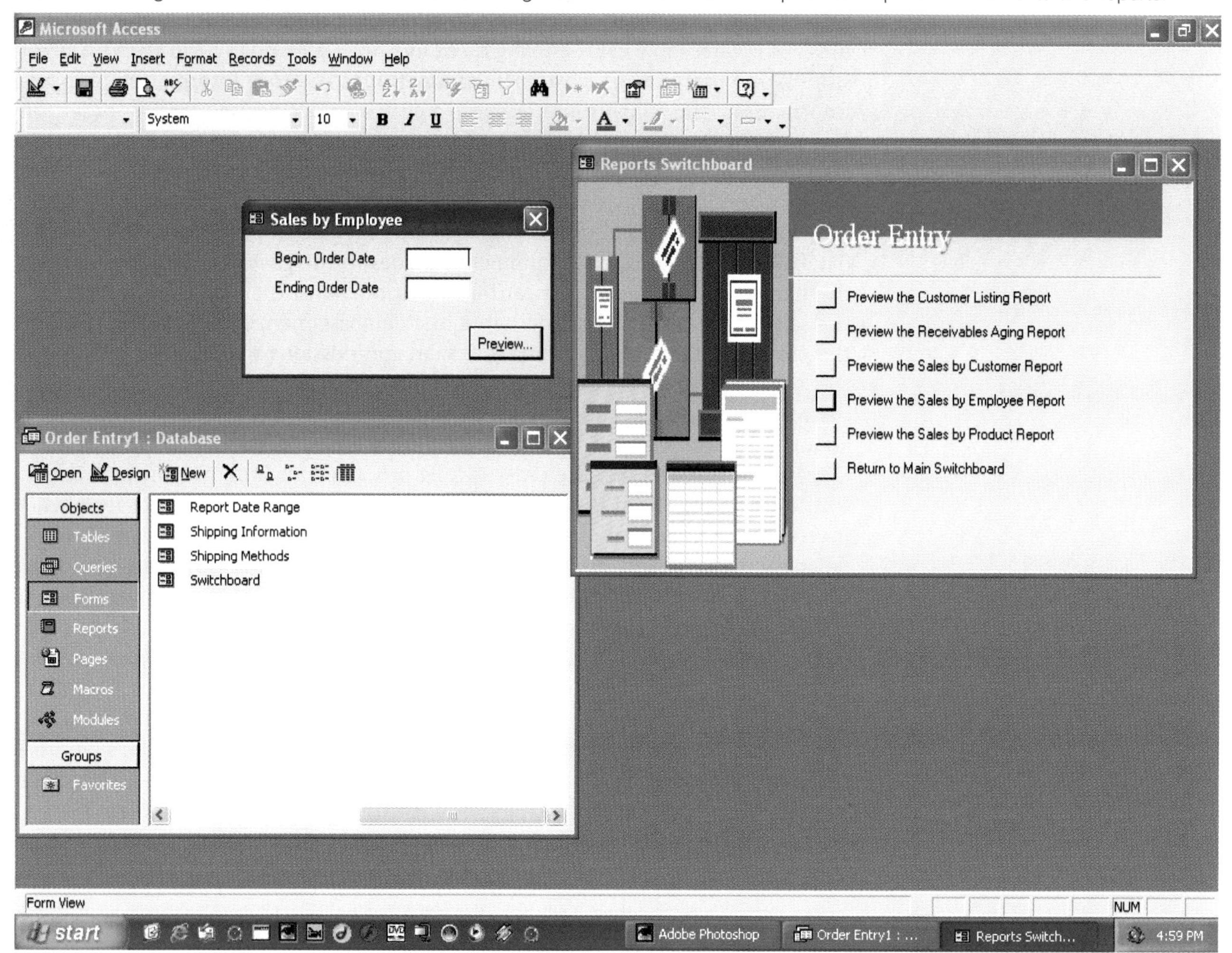

Personal information managers, such as Microsoft Outlook, can be used on handheld computers, laptops, or large-scale computers (see Figure 3.33).[22] For example, you can take a handheld computer with you, make changes while traveling, and have the data automatically uploaded and synchronized with your main computer. This can be done with a *docking station* for your handheld computer. Of course, any changes you make to your PIM on your main computer can be automatically synchronized and transferred to a portable computer or handheld device that you take with you.

FIGURE 3.33 • A personal information manager

Personal Information Managers (PIMs) help individuals, groups, and organizations store useful information, such as a list of tasks to complete or a list of names and addresses.

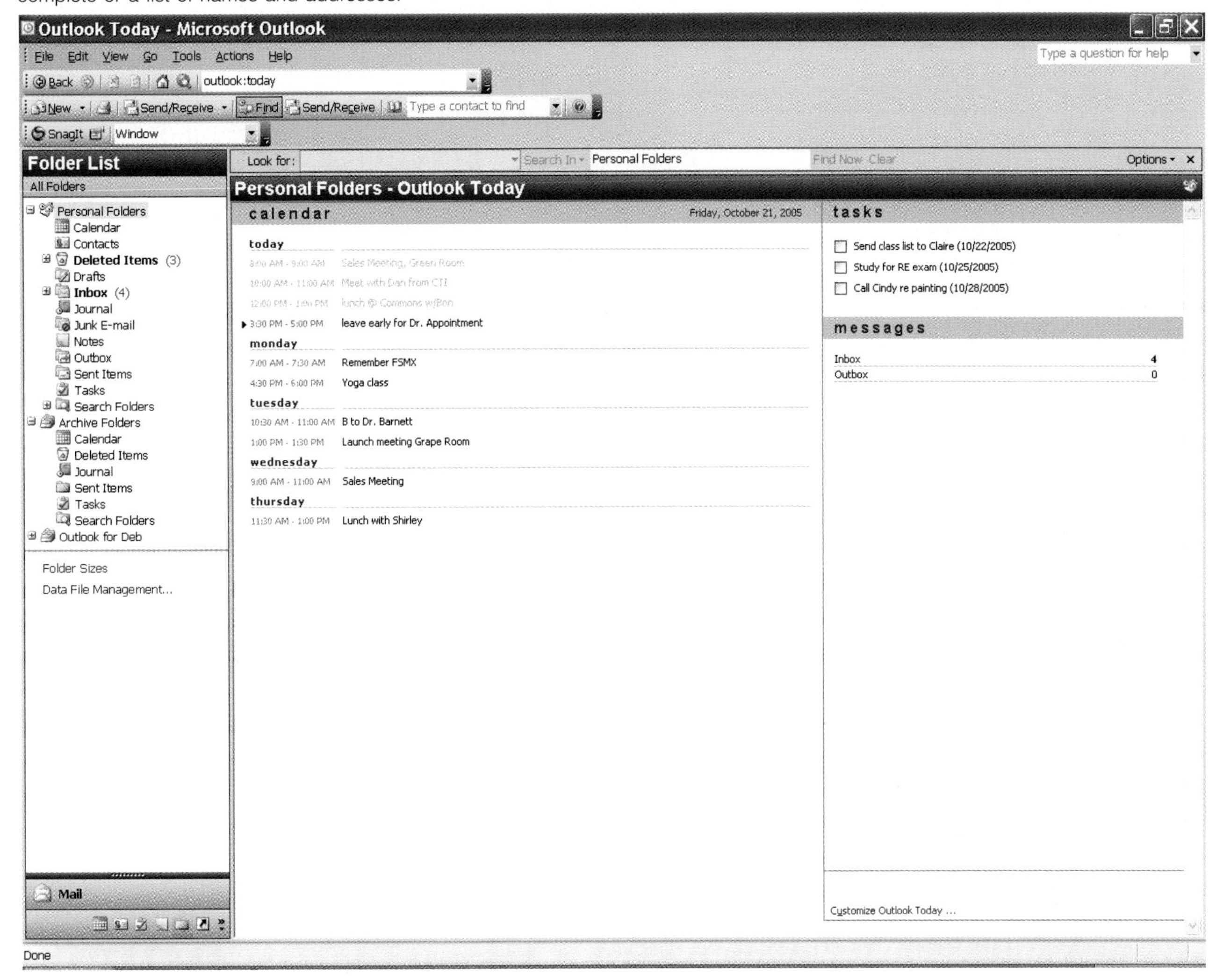

Software Suites. A **software suite** is a collection of application software packages sold together. Software suites can include word processors, spreadsheets, presentation graphics, database-management systems, personal information managers, and more (see Figure 3.34). It is even possible to select which software packages are part of the suite.

Microsoft Office, a software suite from Microsoft, has a number of versions depending on the market and the year, that often includes its word processor (Word), spreadsheet (Excel), database (Access), presentation graphics (PowerPoint), and personal information manager (Outlook). Microsoft Office is available for both PCs and Apple Macintosh computers, and is available in different packages. The Enterprise Edition includes InfoPath, which allows people to create Web pages; Business Contact Manager; OneNote; for taking notes and drawing; and a desktop publishing tool for publishing reports, books, and other documents.

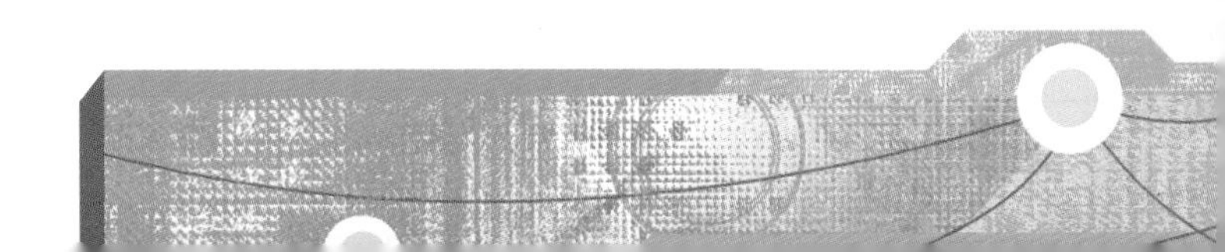

FIGURE 3.34 • An example of a suite

Microsoft Office includes a word processor (Word), spreadsheet (Excel), database (Access), presentation graphics (PowerPoint), and personal information manager (Outlook; not shown here).

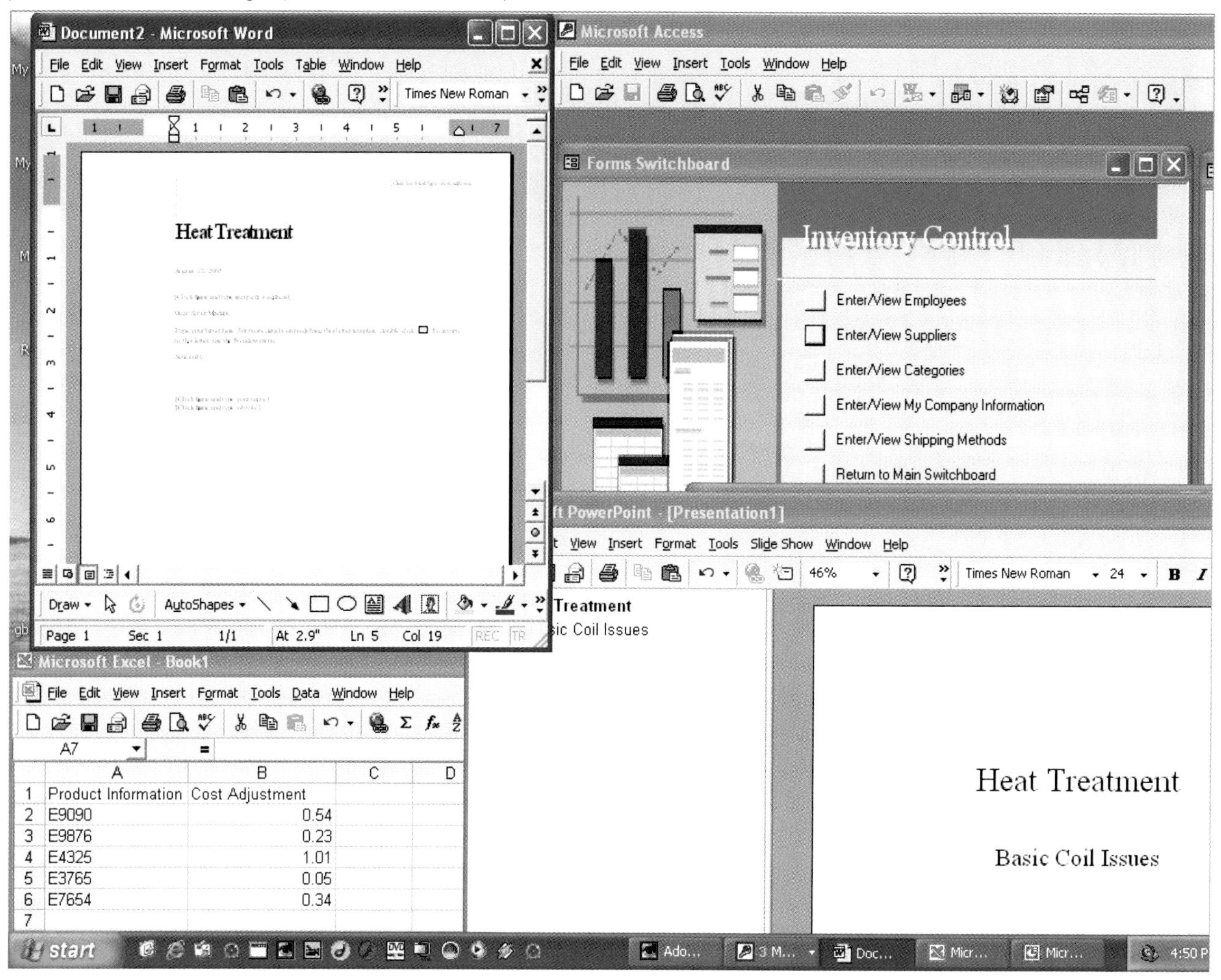

Different versions of many software suites are often designed to work together. In other words, you can work in Office 2003 and share your work with others who are using an older version of Office. Often, a newer version of a software package can automatically read files from an older version of the same software, a feature called *backward compatibility*.

Corel WordPerfect Office, Lotus SmartSuite, and Sun Microsystems StarOffice are other examples of general-purpose software suites for personal computer users. The Free Software Foundation offers software similar to Sun Microsystems's StarOffice that includes word processing, spreadsheet, database, presentation graphics, and e-mail applications for the Linux operating system.[23] OpenOffice.org is an open-source suite for Linux.[24] Each of these software suites includes a spreadsheet program, word processor, database program, and graphics package with the ability to move documents, data, and diagrams among them. Thus, a user can create a spreadsheet and then cut and paste that spreadsheet into a document created using the word-processing application. See Table 3.4 for some popular suites and their functions.

TABLE 3.4 • Popular suites and their functions

Personal productivity function	Microsoft Office	Lotus SmartSuite Millennium Edition	Corel WordPerfect Office	Sun Microsystems StarOffice
Word processor	Word	WordPro	WordPerfect	Writer
Spreadsheet	Excel	Lotus 1.2.3	Quattro Pro	Calc
Presentations	PowerPoint	Freelance Graphics	Presentations	Impress
Database	Access	Lotus Approach	Paradox	Base
Organizer	Outlook	Organizer	Address Book	Not available

There are a number of advantages to using a software suite. The software has been designed to work similarly, so that once you learn the basics for one application, the other applications are easier to learn and use. With newer versions of some suites, files can be used on both PCs and Macintosh computers without a separate translation program. For example, you can create Microsoft Office documents using a Dell (IBM-compatible) computer with Windows XP, transfer the files over the Internet, and open the same files on a Macintosh computer without any conversion or translation.

Microsoft applications can work together through object linking and embedding. Object linking and embedding (OLE), is a technology that allows Microsoft Office users to copy and paste data between office applications and link the data so that when it is changed in the source document it is automatically updated in the target document. With this feature, you can *cut* or *copy* a figure, drawing, chart, table, or other item and *paste* it into another application. For example, you can copy a figure from a presentation graphics program and insert it into a word-processing or spreadsheet program.

Buying software in a bundled suite is cost-effective: together the programs usually sell for less money than they would cost individually. However, since not all the applications may be needed or desired, some people prefer to buy the applications individually.

Integrated Software Packages. Rather than bundling separate programs into a suite, *integrated software packages* provide the functionality of different types of software all in one program. For example, Microsoft Works provides basic word-processing, spreadsheet, database, address book, and other applications features all in one program. Although each application is not as powerful as stand-alone software or comparable applications in software suites, integrated software packages offer basic capabilities for less money. While a suite of applications like Microsoft Office that includes several stand-alone applications may cost several hundred dollars, an integrated package like Works cost as little as $100.

Mobile Software Packages. Tablet and handheld PCs are typically sold with software already installed that takes advantage of the device's unique features. For example, handheld PCs come with PIM software, since it is the most popular software used on handhelds. They also typically include "light" versions of word processing and spreadsheet software, a note-taking program for jotting off quick notes, a calculator, a photo viewer, and media player.

Microsoft Windows XP Tablet PC Edition includes software called OneNote that takes advantage of the tablet's pen and paper style interface (see Figure 3.35). OneNote is a productivity tool that allows users to combine handwritten notes, diagrams, and other images with typed notes on the same page. You can save the pages and easily search for information within the pages. For instance a search for *Becky* might bring up a handwritten note you made to yourself to meet Becky at the mall, an e-mail you sent to Becky last week, and a photo of Becky. OneNote is being marketed to college students as a useful tool for taking class notes. Using your tablet PC, you can take handwritten notes in class, and copy and paste passages and images from the Web right into your notes.

FIGURE 3.35 • Mobile productivity tools

People on the go can be productive with tablet PCs and software like Microsoft OneNote.

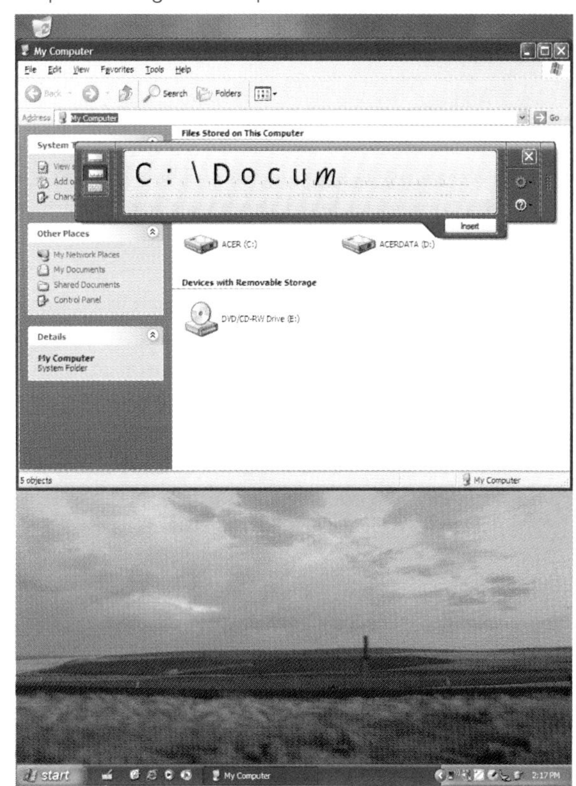

Additional Application Software for Individuals

There are a number of other interesting and powerful application software tools for individuals. GarageBand, for example, is Macintosh software that allows people to create their own music the way a professional does, and it can sound like a small orchestra.[25] Pro Tools is another software program used to edit digital music.[26] Some software contains multimedia features, discussed in detail in Chapter 6. In some cases, the features and capabilities of these application software tools can more than justify the cost of an entire computer system. Some of these programs are listed in Table 3.5, and you can see examples of some of them in Figure 3.36.

FIGURE 3.36 • Additional application software

Examples of specific software applications—video editing, music, and statistical software—are shown here.

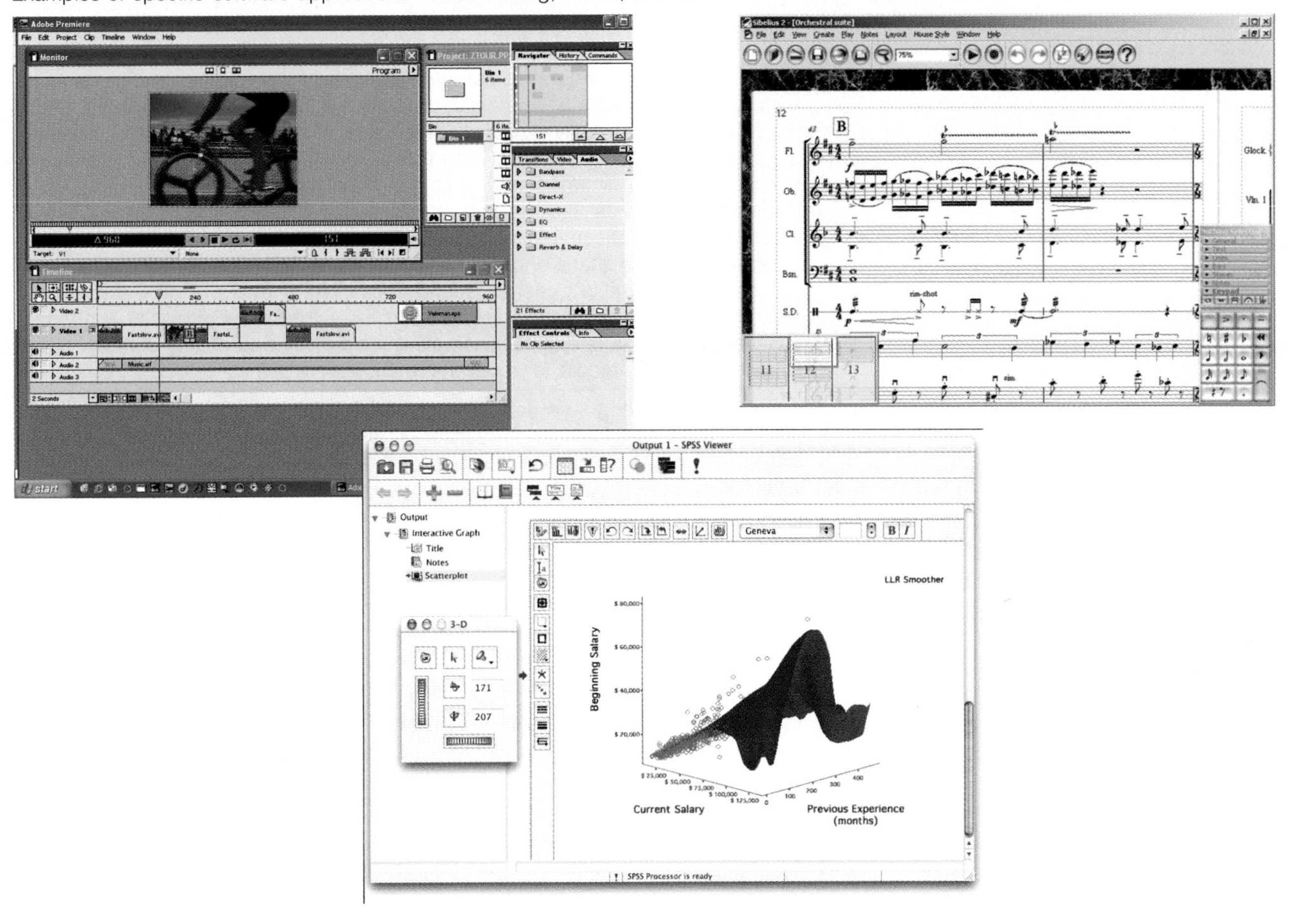

TABLE 3.5 • Examples of additional application software

Type of software	Explanation
Project management	Used to plan, schedule, allocate, and control people and resources (money, time, and technology) needed to complete a project according to a schedule; project by Microsoft is an example of a project-management software package.
Financial management and tax preparation	Provides income and expense tracking and reporting to monitor and plan budgets. Some programs have investment-portfolio-management features. Tax-preparation software allows you to prepare federal and state returns. For the 2005 tax year, the use of some popular tax software is free for those that go to the IRS Web site and file electronically.[30] Many tax-preparation programs can get data from financial-management software packages. Quicken and Money are examples of financial-management packages.
Web authoring	Used to create attractive Web pages and links; HomeSite and FrontPage are examples of Web-authoring tools.
Music	Creates, stores, and compresses music. Sibelius is a score-writing software product available on PCs and Apple Macintosh computers and used around the world by music teachers. Starclass is another music-teaching software package that includes 180 lesson plans. MP3 is a commonly used music and sound compression standard that is used on many portable music players. MIDI is software that can be used to create and synthesize sound and music. See Chapter 6 on multimedia for more information.

TABLE 3.5 • Examples of additional application software (continued)

Type of software	Explanation
Photo and video editing	Used to store, edit, and manipulate digital photographs and video clips. Moving Picture, for example, is a film-editing program that has been used by ABC, CBS, FOX, NBC, PBS, A&E, and other broadcasters. Even large software companies are now investigating software to enhance a filmmaker's ability to store and manipulate digital images and sounds. Microsoft, for example, may have invested as much as $500 million on research for new media software. Video editors are also becoming more powerful. Adobe Premiere has a variety of editing tools, audio filters, and over 300 video templates that can be customized. The software can also generate sound tracks with a range of different tempos and modes. Other video-editing software includes Pinnacle Edition DV and Video Toaster. Sony Vega is another example of powerful video-editing software. See Chapter 6 on multimedia for more information.
Educational and reference	A number of exciting software packages have been developed for training and distance learning. University professors often believe that colleges and universities must invest in distance learning for their students. Some universities offer complete degree programs using this type of software over the Internet. Blackboard and WebCT software helps instructors deliver and administer courses over the Internet. This publisher and others provide content for Blackboard and WebCT for many of its books.
Desktop publishing	Used to create high-quality printed output, including text and graphics. Various styles of pages can be laid out; art and text files from other programs can also be integrated into published pages. A number of the programs in Adobe Creative Suite can be used in desktop publishing.[1]
Computer-aided design (CAD)	Engineers, architects, and designers often use CAD to design and develop buildings, electrical systems, plumbing systems, and more; Autosketch, CorelCAD, and AutoCad are examples of CAD software.
Statistical	Performs a wide array of statistical tests. Colleges and universities often have a number of courses in statistics that use this type of application software. Two popular applications in the social sciences are SPSS and SAS.
Entertainment, games, and leisure	Games and leisure software can be used by itself or with other people while connected to the Internet. The games include adventure, sports, simulation, and strategy games. Sony Computer Entertainment, for example, uses a software program, called Virtually Human, to create "humanlike" characters. The software is also used in medicine, dance, and drama to analyze human movement.

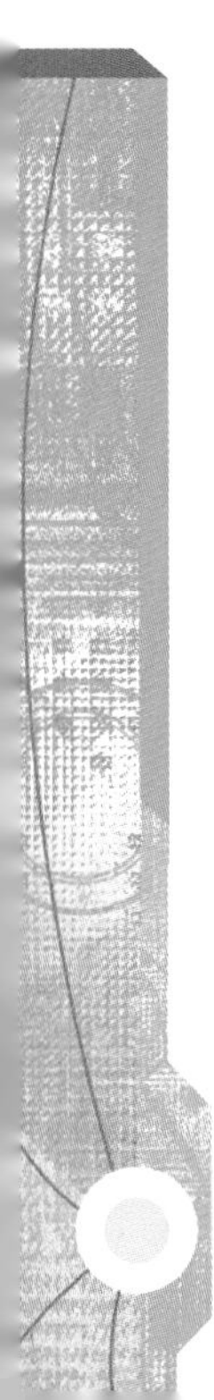

Application Software for Groups and Organizations

Application software is indispensable for groups and organizations. This software can be used to process routine transactions, provide information to help people make better decisions, and perform a number of specialized functions to handle unique but important tasks. Application software is used in all areas of businesses. Philips has developed application software that can send video clips over mobile phones.[27] The software is being used by the British Broadcasting Corporation for breaking news stories or as a backup to broadcast-quality video. Primerica Life Insurance uses an application software package to capture and submit insurance policy applications using small handheld computers.[28] According to Tom Swift, senior VP of field technology, "With 30,000 to 35,000 life insurance applications being processed each month, the potential cost savings are significant." Other application programs can complete sales orders, control inventory, pay bills, write paychecks to employees, and provide financial and marketing information to managers and sales representatives. Xerox, for example, uses NewLeads software to locate potential customers from conferences and trade shows.[29] According to a Xerox representative, "When the

TABLE 3.6 • Examples of application software for routine business activities

Accounts payable
Accounts receivable
Asset management
Billing
Cash-flow analysis
Check processing
General ledger
Human resources management
Inventory control
Invoicing
Order entry
Payroll
Purchasing
Receiving
Sales ordering
Scheduling
Shipping

show's over, all you've got left are your leads." The new software helps Xerox turn curious visitors to their trade shows into real customers. Most of the computerized business jobs and activities discussed in this book involve the use of application software.

Application software is used in almost every field. In fine arts, application software is being used to enhance art and film. IBM, for example, has developed software to help TV production companies convert thousands of TV episodes of the popular shows from the old analog film format to a digital format, saving the production company both time and money. Some movie theaters are converting hundreds of movie screens to a digital cinema system that makes film distribution and playback easier.

Routine Transaction Processing Software. Software that performs routine functions that benefit the entire organization can be developed or purchased. One of the first transaction processing software packages was a payroll program for Lyons Bakeries in England, developed in 1954 on the Leo 1 computer.[32] A fast-food chain might develop a materials ordering and distribution program to make sure that each fast-food franchise gets the necessary raw materials and supplies during the week. This materials ordering and distribution program can be developed internally using staff and resources in the company's Information Technology (IT) Department or purchased from an external software company. Local, state, and national governments also need routine transaction processing software. A European lottery, for example, allows the sale of lottery tickets on the Internet and using mobile phones.

As discussed in Chapter 1, transaction processing software is often a component of larger systems called *transaction processing systems (TPSs)*. A few examples are shown in Table 3.6. Some computer vendors, such as SAP, package these important applications into a unified package called *enterprise resource planning (ERP)* packages.

Transaction processing software is also used by organizations to provide valuable services to individuals. One of the world's most popular Web sites, eBay, owes its success to transaction processing software that supports online transactions between individuals. When an eBay buyer purchases an item such as a Spiderman collectable action figure from an eBay seller, the software transfers funds from the buyer's bank account to the seller's.

TechEdge

EASIER TO SEE THAN SAY

People recognize a face more accurately than they can describe it. This proves problematic when it comes time to describe an assailant. New identification software offers the victim a gallery of 70 faces to peruse. They pick the six closest, and the program "breeds" another 70 faces from those six. After five or six such evolutionary rounds, with each generation further refining the features, the person says, "That's him/her!" and the police grab their keys.

The Darwinian Police Sketch
By David Kohn
Popular Science
http://www.popsci.com/popsci/computers/article/0,20967,1068182,00.html
July 2005

Application Software for Information, Decision Support, and Specialized Purposes

As discussed in Chapter 1, routine transaction activities can store and generate a vast amount of data that can be transformed into useful information to help people or groups make better decisions. Although these systems are popular in businesses and corporations, they are also used in many other areas. Voice stress software is

UPS Provides Shipping Management Software to Customers

Because of the popularity of the Web, and thanks to recent innovations in software development tools, many of today's software programs are written to run on Web servers using Web browsers as the user interface. Businesses are particularly fond of Web-delivered services because they provide universal accessibility and are easy to deploy and maintain. United Parcel Service (UPS) provides many Web-based software applications for its customers, the most recent of which is called Quantum View Manage.

Quantum View Manage is a component of the larger Quantum View system, which includes three Web-based applications that help businesses control their supply chains. Quantum View Manage allows UPS small package shippers to track package movements within their own supply chains. It is designed for customers who ship, on average, 30 or more packages per week and wish to track packages coming in and going out.

Quantum View Manage offers more convenience than traditional tracking software by putting shipping status information at users' fingertips. UPS spokeswoman Laurie Mallis says that one of the primary benefits of the software is that customers don't have to dig to find or enter tracking information. The software maintains and displays all shipping information for an organization in a number of useful report formats. Customers can configure the software to view shipment information for multiple accounts, so they can see when a package is processed, where it is in transit, when it arrives, or whether and why it is delayed, according to Mallis.

The software provides many features for ultimate customer flexibility. It shows both an inbound view of packages being shipped by your vendors and suppliers to designated locations and an outbound view of packages billed to your UPS account. "Customers can customize the information that they're receiving, because different departments within different companies have different needs," Mallis explains. For example, a customer service department might want to know whether a package was shipped or, if there's a delay, what the cause is. Shippers can be notified via e-mail, so they can see problems before their customers do, enabling better service. In addition, a shipper's finance department could use the confirmed delivery notice to automatically trigger an invoice and then view all outstanding COD orders. Some customers use Quantum View Manage to help manage inventory.

Questions

1. What advantage does Web-based software provide to businesses?
2. What risks are involved for companies that rely on Web-based software such as Quantum View Manage in managing their own businesses?

Sources:
1. *Schofield, A. "UPS Plans Distribution Facility At Singapore Airport FTZ,"* Aviation Daily, *April 22, 2005, p. 9.*
2. *Rosencrance, L. "UPS Launches Quantum View Manage,"* Computerworld, *February 4, 2004, www.computerworld.com.*
3. *"UPS Opens Supply Chain 'Window' with Quantum View Manage,"* Business Wire, *February 4, 2004, www.lexisnexis.com.*

being developed to help detect fraud in the insurance industry.[33] Pilot programs have been very successful in detecting people who try to make false claims or in detecting scam insurance companies. Another insurance industry software package called Works4you assists Scottish Life employees in managing their insurance needs and benefits, including retirement programs.[34] Employers can use decision support software to test job performance of current or potential employees.

Physicians also use software to make better decisions. Cancer is a major killer, second only to heart attacks for some age groups. People diagnosed with cancer each year undergo radiation, where X-ray beams are shot into the body to kill the cancer cells. Sophisticated software is now being used to increase the cure rate. This type of software analyzes hundreds of scans of the cancer tumor to create a 3D view of the tumor. The program can then consider thousands of angles and doses of radiation to determine the best radiation program. The software analysis takes only minutes, but the results can save years or decades of life for the patient. Table 3.7 presents some additional examples of applications for information, decision support, and specialized purposes.

TABLE 3.7 • Examples of application software for information, decision support, and specialized purposes

Application	Description
Management reporting software	A variety of reports can be produced from the data generated and stored from routine processing, including reports that are produced on a schedule, reports of exceptional or critical situations requiring immediate attention, and reports that are produced only when requested or demanded.
Groupware	Software to help groups of people work together more efficiently and effectively is often referred to as *groupware*. Groupware can support a team of managers working on the same production problem, letting them share their ideas and work via connected computer systems. Lotus Notes is a popular group software product by IBM that helps people in distant locations work together by sharing schedules, notes, discussions, documents, and actual work on projects.
Decision support software	Special decision support systems (DSSs) can be developed to perform sophisticated qualitative and quantitative analysis for individual decision makers and managers. Software to help decision makers determine the best location for a new warehouse is an example.
Executive support software	This software is designed to support top-level executives and decision makers, such as generals in the military or presidents of companies.
Expert systems software	This software is programmed to act like an expert in a field such as medicine. Expert system software can be used to diagnose complex medical conditions in patients.
Artificial intelligence software	This software is programmed to have humanlike intelligence. The software is used in robots, to simulate human thinking, to learn from new experiences, to handle imprecise and fuzzy information, and to perform a number of other specialized functions.

SOFTWARE ISSUES AND TRENDS

As software increases in importance, software issues become more significant. Knowing how to acquire, install, and control the use of software are important skills, as are knowing when to purchase new software and remove old software. In addition, vendors are doing what they can to protect their software from being copied or duplicated by individual users and other software companies. Today, companies can copyright some software and programs, but this protection is limited. Another approach is to use existing patent laws. These and related issues are explored in this section.

Acquiring Application Software

Individuals, groups, and organizations can either develop or customize a program for a specific application (called *proprietary software*), or purchase and use an existing software program (sometimes called *off-the-shelf software*). It is also possible to modify some off-the-shelf programs, giving a blend of off-the-shelf and customized approaches.

Customized Application Software. One option in acquiring application software is to make or customize software to deliver a specific problem solution.

A publishing company, for example, might want to develop customized software to deliver course material over the Internet. In some cases, this will result in in-house software development, during which the organization's computer personnel are responsible for all aspects of developing the necessary programs. In other cases, customized software can be obtained from external vendors. For example, a third-party software firm, often called a value-added software vendor, may develop or modify a software program to meet the needs of a particular industry or company. A specific software program developed for a particular company or organization is called *contract software*. See Figure 3.37. Some of the advantages inherent in in-house developed software include:

- The software usually meets user requirements: With customized software, you can get exactly what you need in terms of features, reports, and so on. Being involved in the development offers a level of control over the results. With off-the-shelf software, an organization might need to pay for features that are not required and never used. In addition, the software may not have all the required features, and ultimately require modification or customization.
- The software has more flexibility: Customized software enables an organization to have more flexibility in making modifications and changes.

Along with these advantages, organizations must consider an important disadvantage of developing in-house customized software. If an organization chooses in-house development, it must acknowledge the time it takes to develop the required features. It can also be very expensive to develop in-house customized software, and getting a software package that meets organizational goals is not always accomplished.

FIGURE 3.37 • Custom software

Developing custom software allows organizations to create applications tailored specifically to their users and tasks.

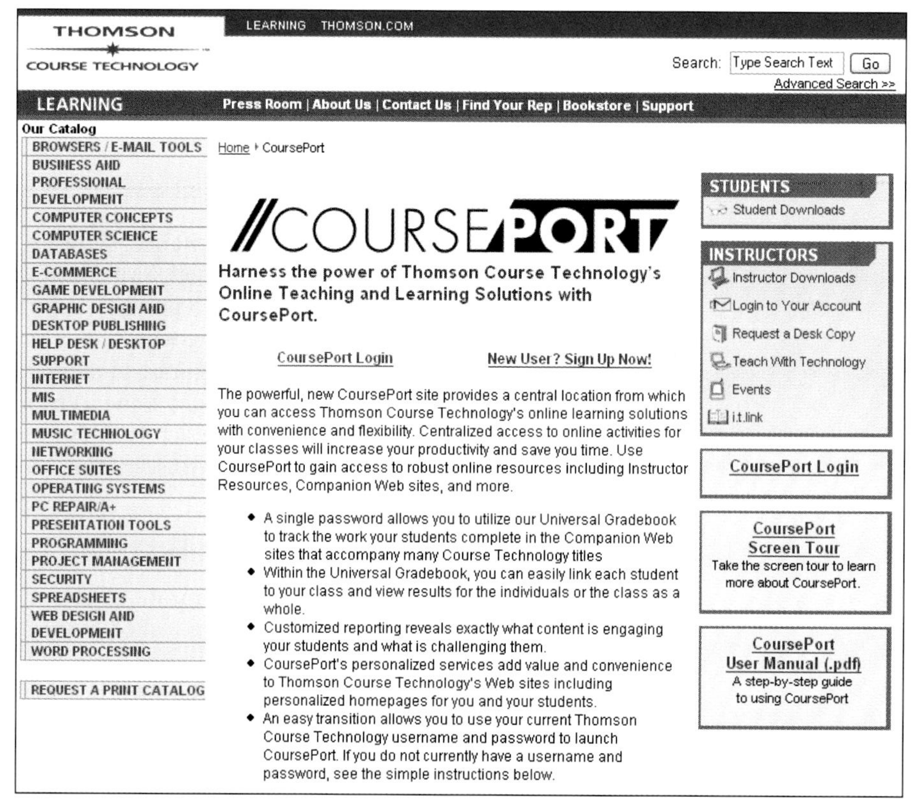

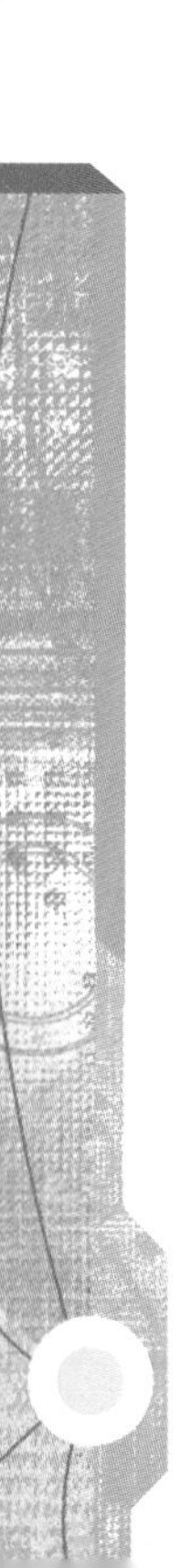

Telstra Benefits from E-Training for Managers

Companies often look to outsiders for help in areas beyond their expertise. Australian telecommunications giant Telstra turned to IBM Learning Solutions for management skills training. In a program Telstra named Frontline Management Foundation Program (FLM), managers from across the organization receive state-of-the-art management training, primarily through online resources.

Managers progress through the course at their own pace, investing only a few hours each week. The course contains a variety of engaging online learning modules covering a range of topics such as leadership, management fundamentals, staffing, teamwork, and coaching. One module titled "The Team Building Process" presents a tutorial on identifying the symptoms of ineffective teams. Managers read about symptoms of an ineffective team and see examples of each symptom. They answer questions as the tutorial progresses from screen to screen to check their understanding. The tutorial concludes with a video of an actual team meeting, and managers must evaluate the team's effectiveness in an online questionnaire. Some questionnaires are graded automatically; others are submitted to a live e-facilitator to assist the student. Each module contains several lessons, and each concludes with an online examination. Students may retake any module at any time until they pass.

Because managers can do most of their class work in the office, at home, or anywhere they have access to the Internet, the training is part of their normal daily routine. Previous classroom training programs required Telstra managers to take four weeks off work for group training sessions. By delivering 75 percent of the course online, the new e-training system requires only four days of classroom work, allowing managers to remain on the job while in training.

After the online training and the classroom workshops, the program progresses to a third phase, in which managers apply their knowledge to their own jobs. Even after completing the course, managers can return to the training software at any time to brush up on their skills.

Telstra plans to train 3300 managers in four years with the system. To date, the program has received highly satisfactory feedback from individual trainees, as well as evidence of a change in management approaches among managers who have taken the course.

Questions

1. As a manager in training, would you prefer to attend a weeks-long training seminar or work at your own pace through an e-training system? Why?
2. As a trainer, would you prefer to teach students through face-to-face meetings in multiweek seminars, or would you prefer to act as an e-facilitator guiding students through the e-training system and assisting them with trouble areas when needed. Why?

Sources:

1. *Taylor, M. "Telstra Employees Want Continued CSS Access,"* Super Review, *April 2005.*
2. *"IBM Success Stories: Telstra," IBM Web site, www.ibm.com, accessed on February 21, 2004.*
3. *"IBM Helps Telstra to Teach Staff," News.com.au, www.news.com.au, accessed on February 22, 2004.*
4. *Telstra Web site, www.telstra.com, accessed on February 21, 2004.*
5. *IBM Australia Web site, www.ibm.com/au/, accessed on February 21, 2004.*
6. *IBM Learning Solutions, www.306.ibm.com/services/learning, accessed on February 21, 2004.*

Off-the-Shelf Application Software. Software can also be purchased, leased, or rented from a company that develops programs and sells them to many computer users and organizations. Software programs developed for a general market are called general software programs or *off-the-shelf software packages*, because these programs can literally be purchased in packages "off the shelf" in a store or over the Internet.

Many individuals and organizations use off-the-shelf software. Most individuals and many organizations, for example, purchase software to perform word-processing, database functions, spreadsheet analysis, and other common activities. Purchasing, leasing, or renting off-the-shelf software offers a number of advantages, including:

- Lower costs: The software company is able to spread the software development costs over a large number of customers, hence reducing the cost any one customer must pay.
- Less risk: With an existing software program, you can analyze the features and performance of the package. If software is to be customized, there is more risk concerning the features and performance of the software that has yet to be developed.
- High quality: Many off-the-shelf packages are of high quality. In addition, a large number of customer or firms have tested the software and helped identify many of its bugs. There is, however, never an iron clad guarantee of high quality, particularly if the intended application is rather unique.
- Less time: Off-the-shelf software can often be installed quickly. It can take years or more to develop and install customized software.
- Fewer resources needed: If software is to be developed or customized in-house, it can take a complete staff of computer personnel. These additional resources are not needed if an organization uses off-the-shelf software.

Combining Customized and Off-the-Shelf Application Software. In some cases, individuals, groups, or organizations use a blend of external and internal software development. That is, off-the-shelf software packages are modified or customized by in-house or external personnel. A manufacturing company, for example, might want to acquire a powerful inventory control program to help reduce inventory costs. The company might also need to include specific inventory calculations for a few important inventory items. The best approach in this case might be to buy or lease off-the-shelf inventory software and to ask the software company to make the necessary changes to the inventory programs, or the buyers or leasors can ask permission to make the changes themselves to the software.

GETTING TO KNOW YOU

Casinos not only want to know who loses, but how much and why. Of special interest are big blackjack betters with poor judgment. Using pattern-recognition software that identifies line patterns on chip edges and playing card faces, and a small camera mounted on the "shoe," they can track every detail of play. Now, when the casino comps the large loser the standard 25% of their loss, they no longer have to estimate it—saving thousands of dollars.

The Digital Pit Boss
By David Talbot
TechnologyReview.com
http://www.technologyreview.com/articles/05/08/issue/brief_boss.asp?p=1
August 2005

Installing New Software, Handling Bugs, and Removing Old or Unwanted Software

Software companies revise their programs and sell new versions every few years or more often. In some cases, the revised software can offer a number of new and valuable enhancements. In other cases, the software may use complex program code that offers little in terms of additional capabilities. Furthermore, revised software can contain bugs or errors. Deciding whether or not to purchase the newest software can be a problem for organizations and individuals with a large investment in software.

Should the newest version be purchased when it is released? Some organizations and individuals do not always get the latest software upgrades or versions, unless there are significant improvements or capabilities that they need. Instead, they may upgrade to newer software only when there are vital new features. Upgrades, however, are typically inevitable, as software vendors eventually discontinue support for older, outdated versions. Also, individuals often feel pressure to upgrade in order to stay compatible with others with whom they may share files that have decided to upgrade. While most new versions of software attempt to be backward compatible and support files made with older versions, this usually only extends to one or two previous versions.

The decision to upgrade or not and when, relies heavily on the software itself. Some basic software utility programs may carry out a simple function that will support your needs for the next 10 years without the need for upgrades. Other applications such as Web browsers should be upgraded whenever a new version is available in order to access the latest Web technologies and stay up to date with security features. For expensive popular software such as productivity software and operating systems, many individuals and organizations feel it wise to wait a year after the initial release prior to upgrading. This provides the software enough time to be market tested to determine the value of upgrading.

Installing New Software. Installing new software is usually straightforward (see Figure 3.38). Software for personal computers typically comes on CDs or is downloaded from the Web. Here is how you can install most software:

- Check the documentation to make sure that your computer meets the storage, processor, and memory requirements to run the software. This information, often called *system requirements,* is usually printed on the outside of the box that contains the software or on the Web page from which you download the software. Be warned! With almost all store-bought software, if you open the box, you cannot return it.
- Place the CD that contains the software in the computer's CD drive. In most cases, the software installation process will start immediately. If not, the instructions that come with the software will lead you through what is needed, depending on your computer and its operating system. Software that is downloaded from the Web will include instructions for installation at the Web site or in an associated README text file. Oftentimes, simply running the file you download will begin the installation process.
- Once the installation process has begun, follow the instructions as they appear on the screen. You will likely be given choices during the installation. Most software products have recommended default installation settings.
- Once the software is installed, you might have to register or activate the software before it can be used. This is usually done on the Internet.
- Check for updates. Some software, such as tax-preparation programs, offer updates to make sure you are using the latest version. In other cases, the Web site for the software company will post any bugs or problems with fixes or patches that you can install on your computer to eliminate potential problems. Some software packages have a large number of patches, which can take time to install. Updates are usually obtained over the Internet.

FIGURE 3.38 • Installing new software

These screens show three points in the process of installing new software.

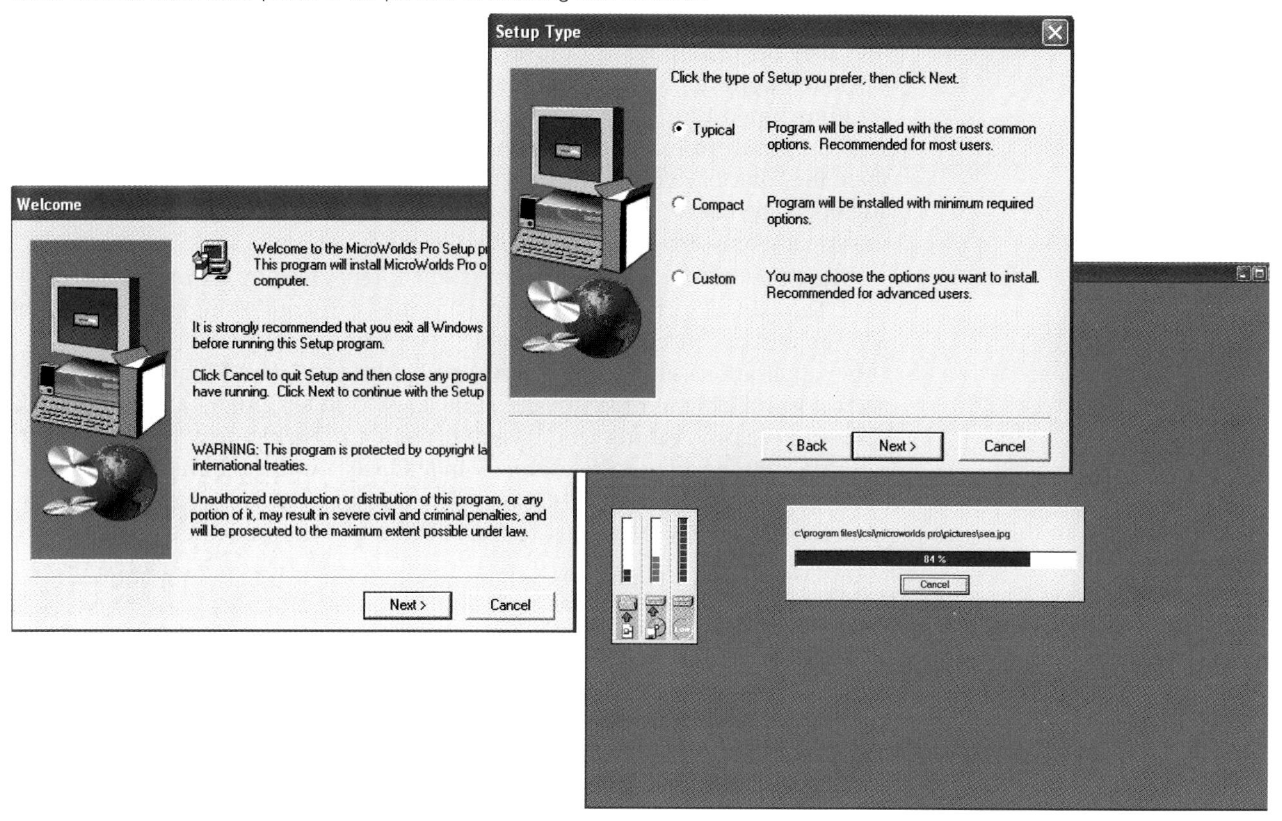

Handling Software Bugs. Software often has *bugs* consisting of one or more defects or problems that prevent the software from working as intended or working at all. Software can contain thousands or millions of lines of computer code, so it can be very difficult to remove all bugs before it is made available to the public. Software companies often provide patches and updates to overcome software bugs. A software patch for Windows XP, called Service Pack 2, is available to make the XP operating system more secure and less venerable to virus and worm attacks. Service Pack 2 also includes a number of tweaks and tools to enhance Windows XP functionality.

Here are some things you can do to overcome pesky software bugs:

- Be careful buying or acquiring the latest software before it has been completely tested and used by others. Some people would rather get software that is a year old or older to make sure that most errors have been found and fixed.
- After you install the software, check the *readme* files that may be included with the software. These files often contain last-minute updates or disclosures, including bugs and how to deal with them.
- Register the software with the software maker. Software companies will often alert you if there is a problem or bug and give you steps to follow to eliminate it.
- Check the Web site of the software vendor often. There can be updates that eliminate any known bugs. In addition, vendors often list bugs that have been found and offer patches or fixes that can be downloaded over

the Internet. Many software applications also offer an option to manually or automatically check for updates over the Internet from within the software.

- Check with popular PC magazines and journals. Popular PC magazines and journals often have articles on software bugs and possible solutions.
- If all else fails, carefully document exactly what happened when you found the bug and then contact the software vendor for a solution.

EXPAND YOUR KNOWLEDGE

To learn more about installing and uninstalling software, go to www.course.com/swt2/ch03. Click the link "Expand Your Knowledge" and then complete the lab entitled "Installing and Uninstalling Software."

Removing (Uninstalling) Software. Most operating systems have an Add/Remove Program feature to assist you in removing software. See Figure 3.39. Note that this feature does not work with all software and that this feature does not always remove all elements of the program or software package. Some utility software packages, such as Norton SystemWorks and McAfee QuickClean for Microsoft Windows PCs, can help eliminate unwanted elements of software you have removed. Trying to remove program files and icons manually can be a problem with some programs and operating systems, which could make it difficult to use or boot the computer in the future. When possible, it is best to remove software using an add/remove software utility

FIGURE 3.39 • Removing software

Most operating systems allow you to remove (uninstall) software.

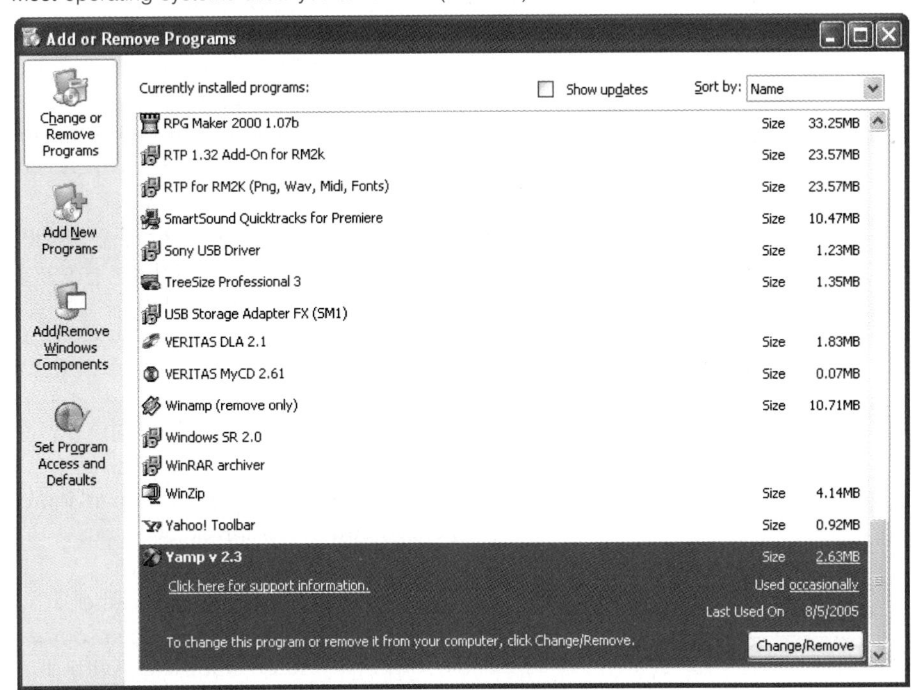

Copyrights and Licenses

Most software products are protected by law using copyright and licensing provisions. In most cases, you don't own the software and can't share it with others. You only have the right to use it on your computer. Copyright and licensing provisions can vary. In some cases, you are given unlimited use of software on one or two computers. This is typical with many applications developed for personal computers. In other cases, you pay as you go. The more computers you use the software on, the more you pay the software vendor. This approach is becoming popular with software placed on networks or larger computers. Most of these protections prevent you from copying software and giving it to others without restrictions. Some software now requires that you *register* or *activate* it before it can be fully used. Registration and activation sometimes put unknown software on your hard disk that can monitor the activities and changes to your computer system. As discussed next, some software doesn't have restrictive copyright or licensing agreements.

Shareware, Freeware, Open-Source, and Public Domain Software

Many software users are doing what they can to minimize software costs. Some are turning to shareware and freeware—software that can be less expensive or free. Shareware is software distributed under a "try before you buy" business model. Typically the user is given 30 days to use and evaluate the software for free. After 30 days, a registration fee is required for continued use of the product. Most shareware software is relatively inexpensive with registration fees in the range of $15 to $80. Some fairly expensive software is distributed as shareware as well. For example, Adobe Photoshop can be downloaded and used for free for 30 days, but costs $599 to register.

Freeware is software that is made available to the public for free. There are a number of reasons why software developers might give away their product. Some wish to build customer interest and name recognition. Others simply don't need the money and wish to make a valuable donation to society. Still others, such as those associated with the Free Software Foundation (*www.fsf.org/*), believe that all software should be free. Some freeware is placed in the *public domain* where anyone can use the software free of charge. Creative works that reach the end of their term of copyright revert to the public domain. Some examples of freeware include:

- Thunderbird: An e-mail and newsgroup software program
- Gaim: An instant messaging software program that runs on Windows and Linux operating systems
- Adobe Reader: A free reader program used to read Adobe PDF files
- AVG Anti-Virus: A free antivirus program that requires registration on the *www.stop-sign.com* Web site
- WinPatrol: Free software to tell you when a spyware program tries to install itself on your computer
- OpenOffice.org: A free and open-source suite of word processing, spreadsheet, and presentation programs
- Irfan-View: A free photo editor
- Free games: Download free games at *www.download-free-games.com*

As noted earlier in the chapter, open-source software makes the source code available to people and organizations. With this type of software, you can make changes to the software or develop your own software that integrates with the open-source software. Weather.com, for example, uses open-source software to cut costs by 50 percent or more compared to traditional software.[35] According to Dan Agronow, CIO for Weather.com, "Where it makes sense, we will always look at open-source alternatives." Not using open-source software can be very expensive. The managers of the "Big Dig" construction project in Boston experienced the high cost of not using open-source software. The construction project hired one company to complete the first phase of developing an integrated software application to analyze highway traffic and help provide security for the project. Unfortunately, the software was not developed using open-source software. When a second company was hired to complete the project, the first company refused to turn over the source code. What were the costs of not using open-source software? The auditor of the state of Massachusetts estimates that it could be as high at $10 million, when legal

Open-Source or Proprietary Software?

When the city of Munich, Germany, decided to switch its 14,000 computers used by local government employees from Microsoft Windows to Linux, an open-source operating system, Microsoft's CEO, Steve Ballmer, flew to Munich to lobby the mayor. In the negotiations, Ballmer even went as far as matching Linux's price—essentially giving the city Microsoft Windows for free—and still the mayor turned him down. The rationale? City officials said the decision was a matter of principle: the municipality wanted to control its technological destiny. It did not wish to place the functioning of government in the hands of a commercial vendor whose primary accountability is to its shareholders rather than to citizens.

Governments and companies around the world are turning to open-source software in increasing numbers. China is working to adopt Linux to become self-sufficient and secure. India, Japan, and South Korea are also aggressively pursuing open-source alternatives to Microsoft software. Ford Motor Company is switching to Linux for much of its server computing.

The one major drawback of open-source software—and the major challenge for the Linux operating system—is the lack of application programs for personal computing. Since Microsoft holds a monopoly in the desktop PC operating system market, the vast majority of software is written for Microsoft Windows. The software developed for Linux is much smaller in quantity, variety, and some may argue, in quality. While an inexpensive alternative to Microsoft Office called Star Office was written to run on Linux, and a Windows-like interface, called Lindows for Linux is available, most users are more comfortable with the familiar Microsoft software. Because a solid business model for open-source development is still in the blueprint stages, developers have little incentive to create software that will be given away for free. Companies like IBM, however, are finding success in distributing Linux to their customers, and profiting from designing, training, and supporting those Linux-based computer systems.

Questions

1. What are the advantages of open-source software?
2. How can software companies like Microsoft compete against open-source software?

Sources:

1. *Songini, M. "Software Ownership Battle Adds $10M to Cost of Big Dig,"* Computerworld, *February 28, 2005, p. 16.*
2. *"Microsoft at the Power Point," Economist.com, September 11, 2003, http://www.economist.com.*
3. *Lettice, J. "Motor Giant Ford to Move to Linux,"* The Register, *September 9, 2003, http://www.theregister.com.*
4. *Linux Online, http://www.linux.org/, accessed February 8, 2004.*
5. *Open Source Initiative Web site, http://www.opensource.org/, accessed February 8, 2004.*
6. *Sun Microsystems Star Office Web site, http://wwws.sun.com/software/star/staroffice/6.0/, accessed February 8, 2004.*

and project delays are included. [36] Some companies are also starting to reveal their source code. IBM recently made about $10 million worth of software patents available to open-source software developers at no charge.[37]

Initiative is a nonprofit corporation dedicated to the development and promotion of open-source software (see *www.opensource.org* for more information on the group's efforts). Table 3.8 gives examples of open-source software.

TABLE 3.8 • Examples of open-source software

Software type	Open-source example
Operating system	Linux
Application software	OpenOffice.org
Database software	MySQL
Internet browser	Mozilla Firefox
Instant messaging	Jabber

Not everyone agrees that open-source software saves money. A number of businesses and organizations have compared the total cost of ownership of licensed and open-source systems only to find that open source wasn't saving them as much money as they had hoped. While open-source systems can be obtained for next to nothing, the up-front costs are only a small piece of the total cost of ownership that accrues over the years that the system is in place.[38] Some claim that open-source systems contain many hidden costs, particularly for user support or if there are problems with the software. Microsoft technology specialist Quazi Zaman asks, "With open source, who's going to support the hundreds of thousands of users?" Licensed software comes with guarantees and support services that open-source software does not. Still, many businesses appreciate the additional freedom that open-source provides.

ACTION PLAN

Remember Alex from the beginning of the chapter? Here are answers to her questions.

1. What does Alex need for her work for the law firm?

Microsoft Office is a software suite that includes a number of popular application software packages, such as a word processor, spreadsheet program, database program, and several other packages. Alex should get Office XP or the newer Office 2003, which can work with files from the older Office XP.

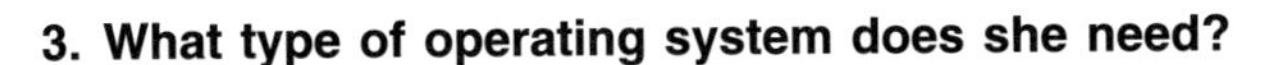

2. What do you recommend for legally downloading and playing music?

To download and play music, Alex needs an Internet connection and the ability to play MP3 files. There are a number of excellent music players that can store and play MP3 files.

3. What type of operating system does she need?

Although Microsoft Office works on either Windows or a Macintosh computer, Alex may want to use the same operating system that she uses at work. Other software packages, however, do not run on both operating systems. Some packages run only on Windows, whereas others run only on a Macintosh computer. Because the type of operating system determine what kind of computer she needs (PC or Apple Macintosh), the operating system she selects is an important decision. Alex should talk to her instructors in her Communications Department at school and her boss at work to get their recommendations and find out which operating systems are used at school and on the job.

Summary

LEARNING OBJECTIVE 1

Discuss the importance and types of software.

Software consists of programs that control the workings of the computer hardware, along with the documentation to explain the programs. Computer programs are sets of instructions to the computer that direct the circuitry within the hardware to operate in a certain fashion. Computer programmers write or create these instructions, which become complete programs. Program documentation is the collection of narrative descriptions designed to assist in the program's use and implementation. There are two main categories of software: system software, which coordinates the activities of hardware and various computer programs; and application software, which is designed to solve user problems and perform specific activities and tasks.

LEARNING OBJECTIVE 2

Discuss the functions of some popular programming languages.

All software programs are written in coding schemes called programming languages, which provide instructions to a computer to perform some processing activity. Programming languages are sets of program code containing instructions that signal the CPU to perform circuit-switching operations. The process of coding these instructions is called ***programming***. A programming language standard is a set of rules that describes how programming statements and commands should be written. An interpreter is a language translator that converts each statement in a program into machine language and executes the statement. A compiler is a language translator that translates a complete program into a complete machine-language program. Compiled programs run faster than interpreted ones.

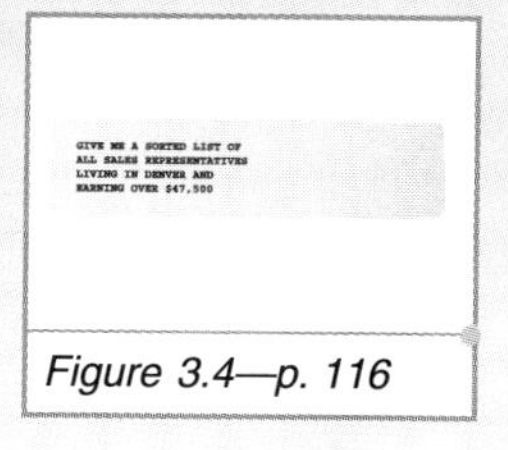

Figure 3.4—p. 116

Programming languages have gone through several generations, with each generation making writing software easier and faster. Artificial intelligence and expert-system programming use fifth-generation languages (5GL). 5GLs are sometimes called ***natural languages*** because they use even more English-like syntax than their predecessors.

Object-oriented programming languages use groups of related data, instructions, and procedures called ***objects***, which serve as reusable modules in various programs. These languages can reduce program development and testing time.

LEARNING OBJECTIVE 3

Describe the functions of system software and operating systems.

System software is a collection of programs that interacts with hardware and application software, creating a layer of communication between them. System software includes operating systems, language-translation programs, and utility programs.

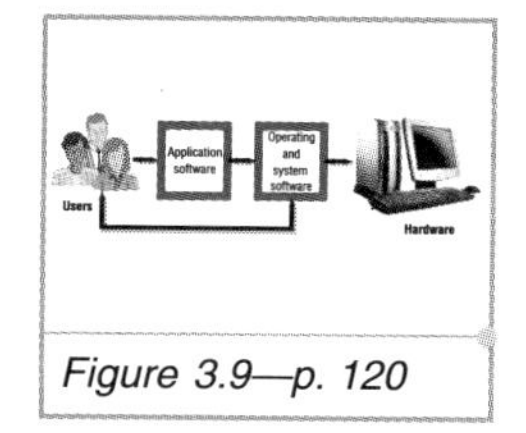

Figure 3.9—p. 120

An operating system (OS) is a set of computer programs that runs or controls the computer hardware and acts as an interface with application programs. OSs convert an instruction from an application into a set of instructions needed by the hardware. The OS also serves as a buffer between application programs and hardware, giving hardware independence.

A computer's operating system performs a number of important functions including memory management, processor management, task management, file management, and network management. An operating system manages memory by converting logical requests into physical locations and by placing data in the best storage space, perhaps in expanded or virtual memory. The OS manages the computer's processor or processors, and controls input, output, storage, and peripheral devices. It allocates computer resources through multitasking, and time sharing. Multitasking involves running more than one application at the same time. Time sharing allows more than one person to use a computer system at the same time. File management involves organizing files into folders or directories. Different operating systems have different file-management conventions. Network management allows multiple computers to work together.

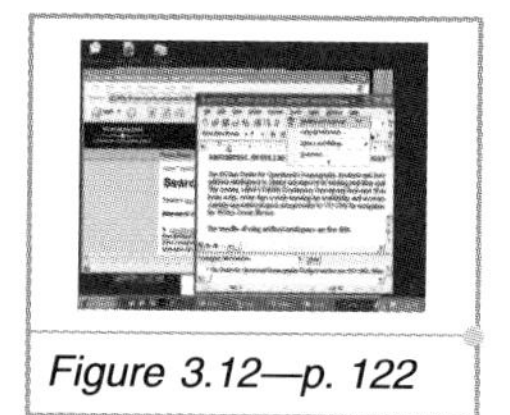
Figure 3.12—p. 122

An OS also provides a user interface. A command-based user interface uses text commands to send instructions; a graphical user interface (GUI), such as Windows, uses icons and menus. In addition, there are a number of other important operating system features, such as plug and play (PnP).

Microsoft Windows is currently the most popular PC operating system. Mac OS X is a proprietary operating system for Apple Macintosh computers. UNIX is the leading portable operating system, usable on many computer system types and platforms. Linux is a UNIX-based operating system available free to users under a General Public License (GPL) arrangement. Server or network operating systems such as Windows Server, UNIX, and Sun Solaris are becoming widespread among networked organizations. Linux can also be used as a network OS. In addition, there are embedded operating systems used in handheld and special-purpose devices, as well as proprietary operating systems for mainframes.

Utility programs include virus detection and recovery, file compression, spam and pop-up guards, hardware and disk utilities, backup functions, and file transfer.

LEARNING OBJECTIVE 4

Describe the support provided by application software.

Organizations can customize application software, buy existing programs off the shelf, or use a combination of customized and off-the-shelf application software. Each approach has advantages and disadvantages.

User software or personal productivity tools are general-purpose programs, including word-processing, spreadsheet, and database-management programs, as well as presentation graphics programs and personal information managers. A software suite is a collection of application software packages bundled together. Some companies produce application software that contains several programs in one integrated package. Additional application software for individuals includes project management, financial management, tax preparation, Web authoring, photo and video editing, educational packages, desktop publishing, and many others.

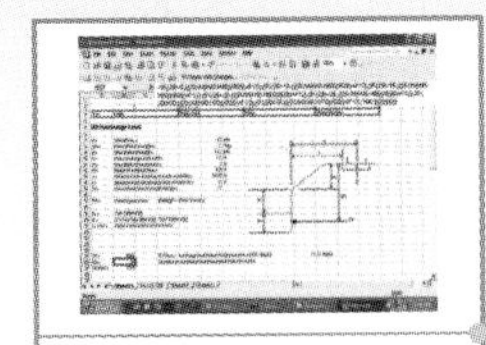

Figure 3.29—p. 142

Software that helps groups work together is often called ***groupware***. Routine transaction processing software includes accounts receivable, accounts payable, asset management, and many other applications. Software can also be purchased or developed for information, decision support, and specialized purposes.

LEARNING OBJECTIVE 5

Discuss how software can be acquired, customized, installed, removed, and managed.

Software issues and trends include acquiring software, installing new software, handling software bugs, removing software, copyrights and licenses, and the use of shareware, freeware, and open-source software. Individuals, groups, and organizations can either develop or customize a program for a specific application (called ***proprietary software***), or purchase and use an existing software program (sometimes called ***off-the-shelf software***). It is also possible to modify some off-the-shelf programs, giving a blend of off-the-shelf and customized approaches. Software bugs can be a problem, particularly with new versions of software. Being careful when buying software, checking readme files, registering software, checking the vendor's Web site, checking popular PC magazines for fixes to software bugs, and contacting the software manufacturer are some of the ways to help find and solve software bugs.

Most software products are protected by copyright law or licensing provisions. Some software companies give you unlimited use of software on one or two computers. People and organizations use shareware and freeware software to reduce software costs. Shareware may not be as powerful or as expensive as professional software, but some people get what they need at a good price. Open-source software makes its source code available to people and organizations. With this type of software, you can make changes to the software or develop your own software that integrates with the open-source software.

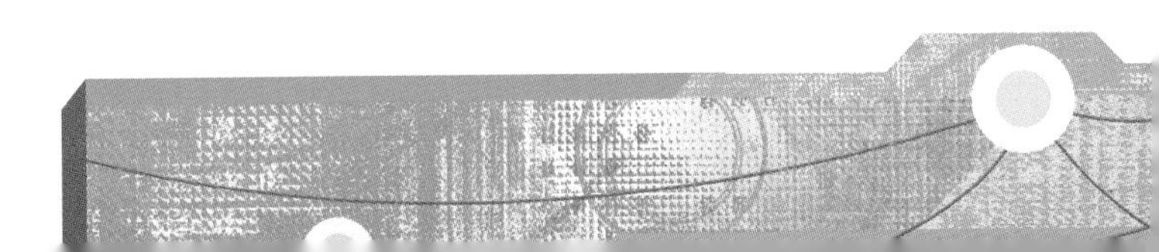

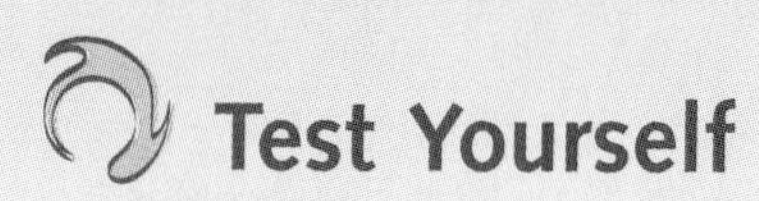

Test Yourself

LEARNING OBJECTIVE 1: Discuss the importance and types of software.

1. _______________ are sets of instructions or statements to the computer that direct its operation.
2. True or False: A word processing program is an example of system software.

LEARNING OBJECTIVE 2: Discuss the functions of some popular programming languages.

3. True or False: An artificial intelligence programming language is also called a first-generation language.
4. C++ is an example of a:
   a. first-generation language
   b. second-generation language
   c. third-generation language
   d. object-oriented language
5. A compiler is an example of a _______________ .
6. True or False: With a visual language, the programmer can drag and drop graphical objects, such as buttons and menus, from a tool box on to the application form.

LEARNING OBJECTIVE 3: Describe the functions of system software and operating systems.

7. _______________ is a collection of programs that interact with the computer and application hardware, creating a layer of insulation between the two.
8. Which of the following is not a typical function of an operating system?
   a. control multiple processors
   b. control special-purpose computers
   c. manage the word-processing function
   d. manage the memory function
9. _______________ allocates computer resources to make the best use of system assets.
10. True or False: Multitasking involves running more than one application at the same time.
11. _______________ allows more than one person to use a computer system at the same time.
12. Windows is an example of:
    a. a graphical user interface
    b. a command-based operating system
    c. a proprietary operating system
    d. a language translator
13. True or False: Mac OS X was one of the first operating systems for personal computers.
14. A(n) _______________ is a language translator that converts each statement in a program into machine language and executes the statement.

LEARNING OBJECTIVE 4: Describe the support provided by application software.

15. True or False: Database programs can contain optimization features to minimize or maximize a quantity subject to constraints.
16. What type of application software would a physician use in a professional meeting or conference to present the results of a research study?
    a. word processing
    b. spreadsheet
    c. database
    d. presentation graphics
17. A(n)_______________ helps individuals, groups, and organizations store useful information, including a list of tasks to complete, a list of names and addresses, and an appointment calendar.
18. Applications can work together through the use of
    a. compilers
    b. interpreters
    c. object linking and embedding
    d. groupware
19. Routine processing systems or applications are often called _______________ .

LEARNING OBJECTIVE 5: Discuss how software can be acquired, customized, installed, removed, and managed.

20. True or False: Individuals, groups, and organizations can develop or customize a program for a specific application. This is called proprietary software.

21. One way to help find and eliminate software bugs is to:
    a. Check the OLE file.
    b. Check the readme file.
    c. Use the Add/Remove feature.
    d. Use the object-oriented error-checking feature.
22. True or False: Most software is protected by copyright law or licensing provisions.
23. _______________ software makes the source code available to users.

Test Yourself Solutions: **1.** Computer programs, **2.** False, **3.** False, **4.** d. object-oriented language, **5.** programming language translator, **6.** True, **7.** System software, **8.** c. manage the word-processing function, **9.** Task management, **10.** True, **11.** Time sharing, **12.** a. a graphical user interface, **13.** False, **14.** interpreter, **15.** False, **16.** d. presentation graphics, **17.** personal information manager (PIM), **18.** c. object linking and embedding, **19.** transaction processing systems (TPS), **20.** True, **21.** b. Check the readme file, **22.** True, **23.** Open-source

Key Terms

application software, p. 112
booting, p. 124
command-based user interface, p. 121
compiler, p. 119
computer program, p. 112
computer programmer, p. 112
freeware, p. 162
graphical user interface (GUI), p. 122
interpreter, p. 118
object linking and embedding (OLE), p. 149
object-oriented programming, p. 117
open-source software, p. 133
operating system (OS), p. 120
productivity software, p. 140
programming language, p. 114
shareware, p. 162
software suite, p. 147
system software, p. 120
utility program, p. 135

Questions

Review Questions

1. Why does software represent such a significant amount of the total cost of a computer system?
2. What is a program? How does it differ from software?
3. Explain the difference between system and application software.
4. Draw a diagram that outlines the relationship among hardware, system software, and application software.
5. What is open-source software?
6. Give several examples of visual programming languages.
7. What are object-oriented programming languages? Why might they be used?
8. What is an operating system? What functions does it perform?
9. What are the advantages of using a graphical user interface (GUI)?
10. What is a compiler? What is an interpreter?
11. Describe four utility programs. How might each be used?
12. What is a software suite? What is an integrated software package?
13. Discuss the advantages and disadvantages of customized software versus off-the-shelf software.
14. What is a software license?
15. What steps would you take to find and eliminate a software bug or error?

Discussion Questions

16. If planning for a computer system begins with software, what do you start with: programming languages, system software, or application software? Why?
17. You have to develop a graphical program that contains menus and buttons. What programming language would you use? Why?
18. What trends are occurring in the development of operating systems for personal computers? What might operating systems be like in five or ten years?
19. When would you use an integrated software package versus a software suite?
20. Pick a career you would like to pursue. What types of application software might you find useful?
21. What advantage does an organization have if users build their own applications? Should users and computer system programmers choose the same development languages? Explain why or why not.

Exercises

Try It Yourself

1. Using a word-processing program, create a document that describes the top five application software packages for a career area of your choice. Give a brief description of each application software package and a description of how it might benefit you professionally.
2. Describe the types of software you would like to use at home for entertainment. Use a database program to develop a table listing different operating systems that would be most appropriate for this software, along with their benefits, disadvantages, and costs.
3. Use the Internet to find information on objected-oriented programming languages. You may be asked to send e-mail to your instructor about what you found.

Virtual Classroom Activities

For the following exercises, do not use face-to-face or telephone communications with your group members. Use only Internet communications.

4. Software is a key component of any distance learning system. Using the Internet or other sources, retrieve information on two or more distance learning systems. You may be asked to write a report or send your instructor e-mail about what you found. What operating systems would you need to run the software?
5. Groupware, software used to support group decision making, can be critical in making decisions in a group setting. Explore the use of groupware and group decision support systems software. You may be asked to write a report or send your instructor e-mail about what you found.
6. Interview one or more managers and ask them what their policy is about updating their application and system software.

Teamwork

7. In a group of three or four classmates, interview three programmers or programmer/analysts from different businesses to determine what programming languages they are using to develop applications. How did these software developers choose the languages they are using? What do they like about their own language and what could be improved? Which language would the programmers choose for software development if the decision were completely their own choice? It is possible that the programmers may each choose a different language. Considering the information obtained from the programmers, select one of these languages and briefly present your selection and the rationale for the selection to the class.

8. Open-source software is popular with many people. Have each team member investigate an open-source software product, such as a word processor, database program, spreadsheet program. Each team member should write a brief description of the software, including the features of the software versus the features of a normal commercial software package. The team should then write a description of the advantages and disadvantages of open-source software compared to commercial software.

9. Have your team determine what types of embedded software packages are being used by classmates, friends, and family members.

TechTV

Software Vending Machines

Go to www.course.com/swt2/ch03 and click TechTV. Click the "Software Vending Machines" link to view the TechTV video, and then answer the following questions.

1. How does the "software vending machine" approach work?

2. Can the consumer get discounted prices and refunds using this approach?

Endnotes

[1] Hensley, S. "New Test May Detect Parkinson's Early," *The Wall Street Journal,* January 7, 2005, p. B1.

[2] Mearian, L. "IBM Builds Java System for NYSE," *Computerworld,* December 20, 2004, p. 13.

[3] Greene, J. et al. "Microsoft May Be a TV Star Yet," *Business Week,* February 7, 2005, p. 78.

[4] Guth, R. "Microsoft Tests Software," *The Wall Street Journal,* March 28, 2005, p. B1.

[5] Spanbauer, S. "After Antitrust," *PC World,* May 2004, p. 30.

[6] Mossberg, W. "Tiger Leaps Out Front," *The Wall Street Journal,* April 28, 2005, p. B1.

[7] Hoffman, T. "Grocer Rings Up Savings," *Computerworld,* January 31, 2005, p. 10.

[8] Hamm, S. "Linux, Inc." *Business Week,* January 31, 2005, p. 60.

[9] Staff. "CIOs Eyeing Open-Source Software," *Information Security,* January 2004, p. 22.

[10] Bank, D. "Intel, IBM to Back New Linux Fund," *The Wall Street Journal,* January 12, 2004, p. B3.

[11] Thibodeau, P. "Sun Begins Its Release of Open-Source Solaris Code," *Computerworld,* January 31, 2005, p. 6.

[12] Janowske, D. et al. "Taking Care of Small Business," *PC Magazine,* February 3, 2004, p. 12.

[13] Drier, T. "Mac On The Go," *PC Magazine,* February 17, 2004, p. 35.

[14] Spanbauer, S. "After Antitrust," *PC World,* May 2004, p. 30.

[15] Rupley, S. "Bad Code?" *PC Magazine,* March 8, 2005, p. 19.

[16] Guth, R. "Aiming to Fix Flaws, Microsoft Buys Another Antivirus Firm," *The Wall Street Journal,* February 9, 2005, p. B1.

[17] Spector, L. "The Trouble-Free PC," *PC World,* February 2004, p. 74.

[18] Rubenking, N. "Yahoo! Enters Desktop Search Fray," *PC Magazine,* February 22, 2005, p. 31.

[19] Dragan, R. "Maintain Control of Business Content," *PC Magazine,* February 17, 2004, p. 42.

[20] Staff. "Web Access Software," *Telecomworldwire,* January 12, 2004.

[21] Richmond, R. "Do You Know Where Your Workers Are?" *The Wall Street Journal,* January 12, 2004, p. R1.

[22] Dragan, R. "Microsoft Outlook 2003," *PC Magazine,* February 17, 2004, p. 64.

[23] Mendelson, E. "StarOffice 7 Makes a Run at Office," *PC Magazine,* February 3, 2004, p. 40.

[24] Albro, E. "The Linux Experiment," *PC World,* February 2004, p. 105.

[25] Mossberg, W. "How to Become a Rock Star," *The Wall Street Journal,* February 4, 2004, p. D1.

[26] Brown, M. "Stars for an Hour," *Rocky Mountain News,* March 6, 2004, p. 1D.

[27] Deibel, M. "Filing Free," *Rocky Mountain News,* February 28, 2005, p. 1B.

[28] Stafford, A. et al., "Adobe's Creative Evolution," *PC World,* May 2005, p. 50.

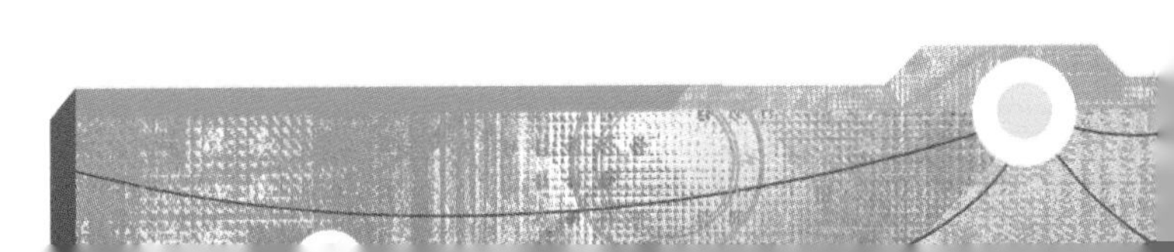

[29] Staff. "Philips Software for Mobiles in Use by the BBC," *New Media Age,* January 29, 2004, p. 10.

[30] Rodgers, J. "Field-Force Mobility," *Insurance & Technology,* February 1, 2004, p. 17.

[31] Staff. "Tricks of the Trade," *Sales & Marketing Management,* January 2004, p. 27.

[32] Aeberhard, J. "Fifty Years of Business Software," *Computing,* January 29, 2004, p. 8.

[33] Staff. "Fraud Software Set to Take Off," *Post Magazine,* January 22, 2004, p. 2.

[34] Staff. "Comprehensive Software Package Launched," *Money Management,* January 1, 2004.

[35] King, J. "A Sunny Forecast for Open Source," *Computerworld,* April 26, 2004, p. 19.

[36] Songini, M., "Software Ownership Battle Adds $10M to Cost of Big Dig," *Computerworld,* February 28, 2005, p. 16.

[37] Hamm, S. "One Way to Hammer at Windows," *Business Week,* January 24, 2005, p. 36.

[38] Alan J. "The Real Cost of Open Source," *FCW.com,* May 3, 2005, www.fcw.com.

THE INTERNET AND WORLD WIDE WEB

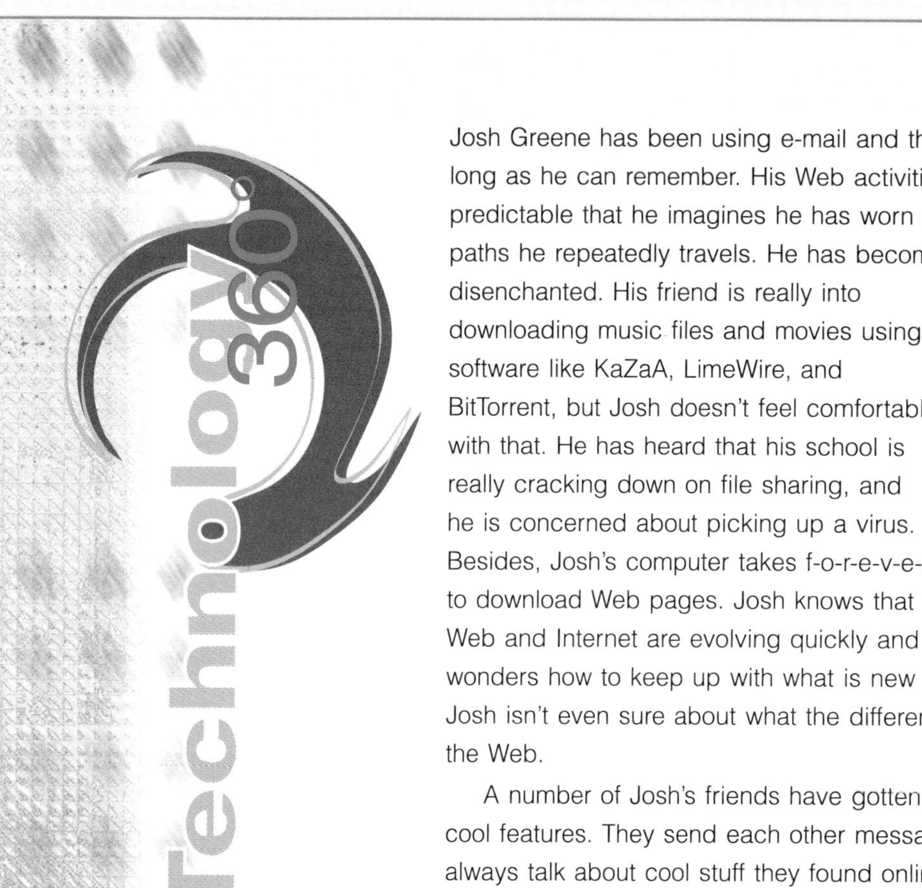

Josh Greene has been using e-mail and the Web for as long as he can remember. His Web activities are so predictable that he imagines he has worn ruts in the paths he repeatedly travels. He has become disenchanted. His friend is really into downloading music files and movies using software like KaZaA, LimeWire, and BitTorrent, but Josh doesn't feel comfortable with that. He has heard that his school is really cracking down on file sharing, and he is concerned about picking up a virus. Besides, Josh's computer takes f-o-r-e-v-e-r to download Web pages. Josh knows that the Web and Internet are evolving quickly and wonders how to keep up with what is new and exciting. Come to think of it, Josh isn't even sure about what the difference is between the Internet and the Web.

A number of Josh's friends have gotten new cell phones with all kinds of cool features. They send each other messages throughout the day. They always talk about cool stuff they found online and spend some evenings partying—on their computers! Josh's roommate is traveling to visit an online friend over the weekend. Josh is feeling a bit left out. Like it or not, Josh will be spending *his* evening online as he has a big research project due tomorrow. He is unable to get to the library so he plans on doing his research on the Internet but isn't sure where to start.

As you read through the chapter, consider the following questions:

1. What online activities can Josh use to enrich his quality of life?
2. Which online activities are legal and safe, and which are not?
3. How should Josh conduct his research on his computer?

Check out Josh's *Action Plan* at the conclusion of this chapter.

4

LEARNING OBJECTIVES

1. Describe how the Internet developed and how hardware, protocols, and software work together to create the Internet.
2. Explain the underlying structure of the Web and the technologies that support it.
3. Define the categories of information and services that the Internet and Web provide and the forms of communication they support.
4. Explain what Internet2 is and the types of applications it will provide in the future.

CHAPTER CONTENT

Internet Technology

Web Technology

Internet and Web Applications

The Future Internet

Introduction

You've learned how hardware and software work together to assist people in achieving their goals. This in itself is highly valuable. However, the true power of computing lies in connecting computing devices and users through the interconnected networks that make up the Internet. The Internet connects people on a global scale and provides a host of communication platforms, information, and services.

The Web is an Internet service that provides a user-friendly interface to resources on the Internet. It organizes and presents information on the Internet in a manner that is easy to navigate. More than any other technology, the Web has empowered individuals by providing a public forum to share ideas. The Internet has leveled the playing field between small and large organizations and provided opportunities for some organizations that would otherwise never stand a chance of competing. The Web, at any given time, can be viewed as a snapshot of the human condition. Anything of personal or professional interest to any person is represented here: the good, the bad, and the ugly. Thus, it has its share of controversy.

For students and professionals, the Web has become a primary source of information in support of scholarly research. Because of the vast amount of information and the lack of quality control, researchers must learn unique methods for sorting the wheat from the chaff. This chapter addresses all of these issues and much more!

The Internet provides a platform on which hundreds of millions of people combine and share knowledge and views. If two heads are better than one, hundreds of millions of heads sharing knowledge and unique perceptions is a powerful thing indeed. The Internet brings the power of global community to computing.

The Internet provides the technology and physical connections between devices to support many applications, the most popular of which are e-mail and the Web. E-mail provides a convenient and low-cost form of communication over the Internet, while the Web provides a convenient method of sharing information and services over the Internet. The Web is not simply another name for the Internet, but rather one of many applications and services available on the Internet. The Web, e-mail, instant messaging, P2P file sharing, Internet radio, and Google Earth are all examples of applications that run on the Internet. People are using the Internet and its applications as a resource for communicating, learning, entertainment, and making new friends. You can use it professionally to collaborate on projects (see Figure 4.1). Businesses invest in the Internet to leverage competitive advantages. Products are marketed and sold over the Internet. Great works of creativity are published and distributed over the Internet. The Internet is highly regarded and a part of many of our daily lives, but what exactly is it?

FIGURE 4.1 • Video conferencing from 36,000 feet (1200 m) above the Pacific Ocean

A CalTech researcher participates in a video/audio collaborative session over the Internet with colleagues in Slovakia and Switzerland while aboard his international flight.

INTERNET TECHNOLOGY

The Internet is relatively young and still in its early stages of development. New ways of using the Internet are being developed and introduced every day. This section provides a brief overview of the origins of the Internet so that you can observe how it has come to be so influential. You learn about the principles that govern the Internet so that you can be more knowledgeable about its capabilities and limitations.

A Brief History of the Internet

As defined in Chapter 1, the Internet is a global, public, network of computer networks. A *computer network* is a collection of computing devices connected together to share resources such as files, software, processors, storage, and printers. There are millions of privately owned networks around the world. Joining networks together into larger networks so that users on different networks can communicate and share data creates an *internetwork*. Today's Internet joins together networks of over 300 million computers, or *Internet hosts*, to create the world's largest internetwork (see Figure 4.2). To fully understand the Internet, you must examine its origins.

FIGURE 4.2 • The Internet

The Internet provides high-speed information and communication thoroughfares between individuals and organizations around the globe.

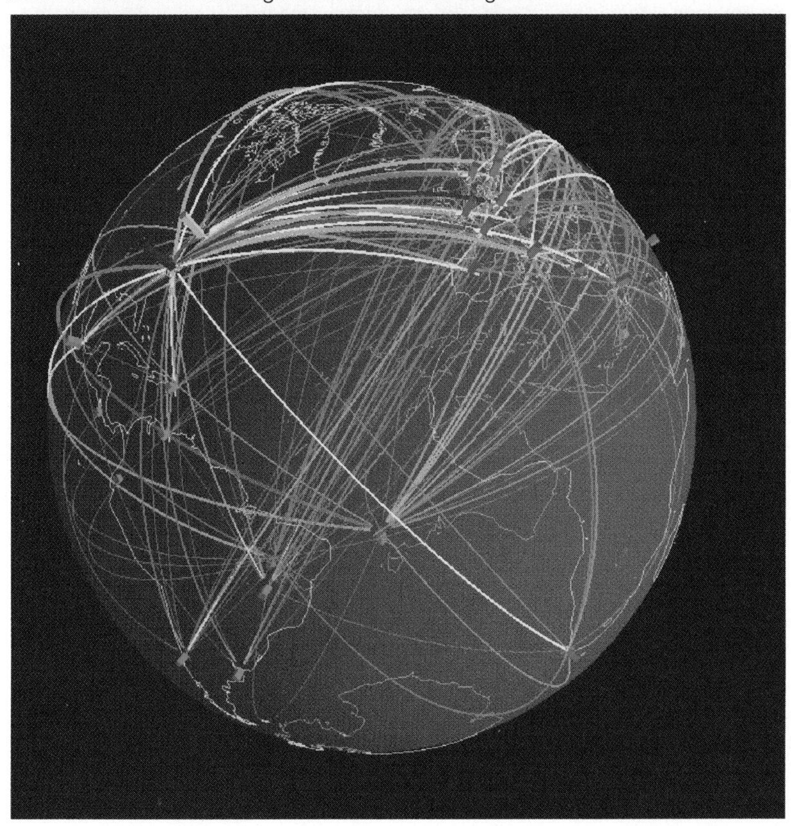

In 1957, computing was done primarily on large mainframe computers accessed from within an organization through a network of terminals. Government agencies, universities, businesses, and other large organizations all used this type of networking environment. In that same year, the U.S.S.R. surprised the world by launching Sputnik, the first artificial earth satellite. The United States viewed the launch as a challenge. The following year President Eisenhower reacted by forming two government agencies under the Department of Defense (DoD) to advance space technologies, weapons, and communication systems: the Advanced Research Projects Agency (ARPA) and the National Aeronautics and Space Administration (NASA). Many amazing achievements were to come from these organizations, but among those that had the most social impact were placing the first man on the moon and the Internet.

In 1969, ARPA commissioned ARPANET for research into networking. Its initial goal was to establish closer communications for research by connecting the computer networks of four research institutions: the University of California at Los Angeles, Stanford, the University of California at Santa Barbara, and the University of Utah. The goal was accomplished within the year. Once the groundwork was laid for successful communications between networks, ARPANET began growing exponentially. Table 4.1 summarizes the early milestones in the development of the Internet that led to the introduction of the World Wide Web at the end of 1993.

With the birth of the Web, the Internet exploded, with a 341,634 percent annual growth rate in Internet hosts in following years. Internet service providers began sprouting up all over the world to provide Internet service to businesses and homes. With increasing amounts of consumers flocking to the Web, businesses recognized it as a powerful new marketing and sales tool. The Internet's focus shifted from supporting solely academic and government interests to supporting public and commercial interests. Today there are over 1 billion Internet

TABLE 4.1 • Early Internet development milestones[1]

Year	Internet hosts	Internet milestones
1970	13	The first cross-country link was installed by AT&T to connect networks across the country
1973	35	ARPANET went international as it expanded overseas to University College in London, England, and the Royal Radar Establishment in Norway
1984	1024	ARPANET was divided into two subnetworks: MILNET, for military needs, and ARPANET, for research
1990	313,000	The ARPANET project was officially concluded, and "the Internet" was turned over to the public, to be managed by the Internet Society (ISOC)
1991	617,000	The Commercial Internet Exchange (CIX) Association was established to allow businesses to connect to the Internet
1993	1,500,000	The first Web browser, Mosaic, was released to Internet users to an unprecedented enthusiastic reception

users—over 15 percent of the world population (see *www.internetworldstats.com/stats.htm* for the most current statistics). Among those that are not connected are populations that cannot afford computers and connections. This important social issue is referred to as the *digital divide*. There are also cultures and individuals who resist technology because of personal or religious philosophies.

How Does the Internet Work?

EXPAND YOUR KNOWLEDGE

To learn more about how the Internet works, go to www.course.com/swt2/ch04. Click the link "Expand Your Knowledge", and then complete the lab entitled "Connecting to the Internet".

The Internet is a combination of hardware, protocols, and software. The hardware provides the physical cables and devices that control and carry Internet data. *Protocols* are the rules that are implemented in network software and hardware to establish connections between two or more computers to allow them to communicate. Software allows users to interact with the Internet to access information and services. The following sections provide detailed explanations of the Internet's hardware, protocols, and software.

Internet Hardware. The hardware over which Internet traffic flows includes the Internet backbone, routers, and the computers that request and serve up information and services. The **Internet backbone**, the main Internet pathways and connections, is made up of the many national and international communication networks that are owned by major telecom companies, such as MCI, AT&T, and Sprint—the same companies and networks that provide telephone service. These companies agreed to connect their networks so that users on all the networks could share information over the Internet. The cables, switching stations, communication towers, and satellites that make up these telecommunication networks provide the hardware over which Internet traffic travels. These large telecom companies are called *network service providers* (*NSPs*). Examples include MCI, Sprint, British Telecom, and AT&T. The combined backbones of these and other NSPs make up today's Internet backbone. The complexity of mapping the physical connections of the Internet has become an overwhelming and daunting task. Some researchers use visualization software tools such as Walrus (*www.caida.org/tools/visualization/walrus/*) to create graphical representations of the Internet based on statistical data (see Figure 4.3).

Network service providers enable Internet users to connect to their networks through utility stations called *points of presence* (PoPs). The visualization of the Internet shown in Figure 4.3 illustrates the hub-and-spokes nature of the Internet. The PoPs act as the hubs, and the connections they provide, the spokes. The PoPs are connected together by the Internet backbone. Points of

FIGURE 4.3 • Visualizing the Internet

Using visualization tools scientists are able to create theoretical 3D models of the complex linked connections that make up the Internet.

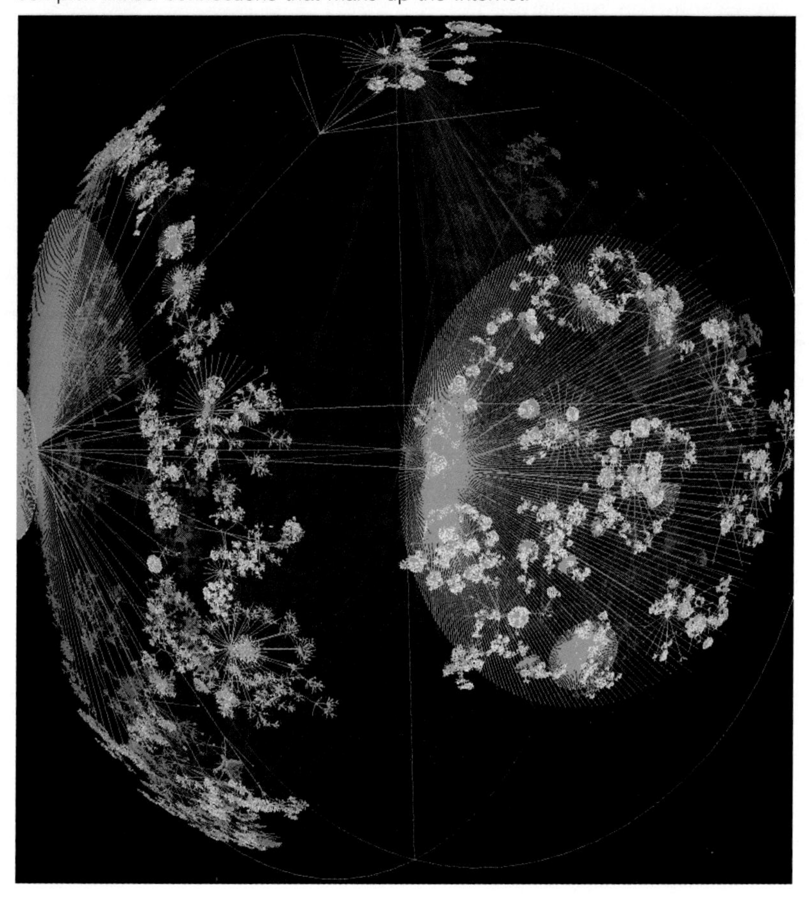

presence include networking hardware that allows individuals, companies, and service providers such as Comcast and America Online, to connect to the Internet backbone. MCI, for example, has more than 4500 PoPs throughout North America, Europe, and the Pacific Rim, and incorporates more than 3.2 million user connections.

Companies that provide users with access to the Internet through these PoPs are called **Internet service providers (ISPs)**. There are hundreds of ISPs from local to international levels. ISPs work as liaisons between Internet users and the telecommunications companies that own the backbones. They charge a monthly fee to Internet users and provide devices by which the user can connect to the Internet. Many companies, organizations, and institutions, work directly with the network service providers to connect their own organizations' networks directly to a PoP.

The Internet uses routers to make sure that information sent is directed to the intended recipient. **Routers** are special-purpose computing devices—typically small to large boxes with network ports—that manage network traffic by evaluating messages and routing them over the fastest path to their destination. Routers are typically located at network junctions, where one network is joined to another network. An e-mail message sent from New York to Los Angeles passes from router to router until it reaches its destination. Such an e-mail message might pass through as many as 20 routers along the way, but would reach its destination within a fraction of a second (see Figure 4.4).

AN END TO OVERDUE BOOKS?

Whether Google is a technology company or a media company, two things are clear: it knows content is king on the Net, and it dreams big. With its Digital Library project, Google has partnered with the libraries of Oxford, Stanford, Harvard, and the University of Michigan, as well as the New York Public Library for an impressive goal. Within 10 years they hope to scan every book published and make it available—and searchable, of course—online.

Source:
The world's most intriguing company
By Gregory M. Lamb
Christian Science Monitor
http://www.csmonitor.com/2005/0627/p13s01-stin.html
June 27, 2005

Accessing the Internet. Businesses, organizations, and individuals purchase access to the Internet from ISPs. Large businesses and institutions such as your college purchase industrial-strength connections to the Internet from telecom companies to connect their existing computer network to the Internet. Individuals and small businesses have several options for connecting to the Internet including a dial-up connection, cable, DSL (digital subscriber line), and digital satellite service.

A **dial-up connection** is a low-speed Internet service that utilizes the customer's phone line for data transfer rates as high as 56 kilobits per second (Kbps), provided

FIGURE 4.4 • Internet routers in action

Software from Visualware (http://visualroute.visualware.com/) allows you to witness routers in action. This Internet message sent from Ashburn, Virginia, to Burbank, California, required the assistance of 16 routers along the way and arrived at its destination 70 microseconds after it was sent.

Analysis: 'www.ucla.edu' was found in 15 hops (TTL=49).

Hop	%Loss	IP Address	Node Name	Location	Tzone	ms
0		161.58.180.113	WIN10115.visualware.com	*		
1		161.58.176.129	-			0
2		161.58.156.140	-			0
3		129.250.28.206	xe-1-2-0-3.r20.asbnva01.us.bb.verio.net	Ashburn, VA, USA	-05:00	0
4		129.250.2.83	p16-7-0-0.r02.asbnva01.us.bb.verio.net	Ashburn, VA, USA	-05:00	0
5		205.171.1.145	dcp-brdr-02.inet.qwest.net			0
6		205.171.251.33	-			0
7		205.171.8.221	dca-core-02.inet.qwest.net			1
8		205.171.205.25	bur-core-01.inet.qwest.net	Burbank, CA, USA	-08:00	69
9		205.171.13.174	bux-edge-01.inet.qwest.net			69
10		63.145.160.158				69
11		137.164.24.22	dc-ucla--lax-ispa-ge.cenic.net			96
12		169.232.6.169	border--border.backbone.ucla.net			70
13		169.232.6.145	border--core.backbone.ucla.net			70
14		169.232.51.14	core--csb1.backbone.ucla.net			70
15		169.232.33.135	www.ucla.edu			70

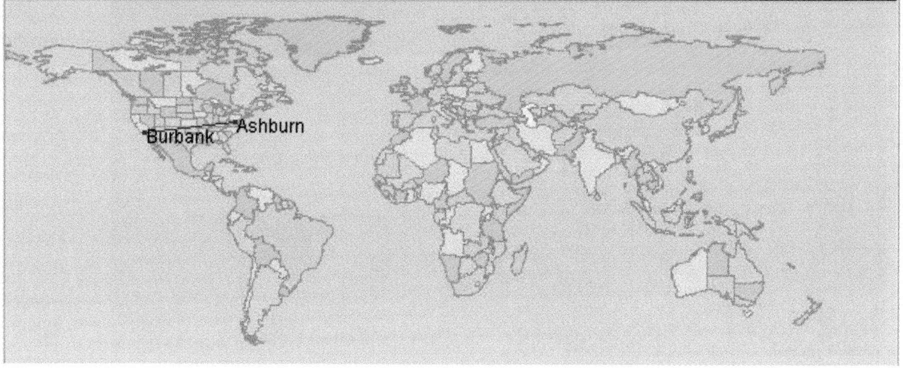

by any one of hundreds of Internet service providers. Customers are typically provided with a local or toll-free phone number along with a username and password. The computer's modem dials the number, establishes a connection to the ISP, and logs on. Dial-up connections tie up the phone line during use and cannot be used simultaneously with voice communications.

Most dial-up service providers advertise an "accelerated" feature that boosts speeds up to five times that of a standard dial-up connection (to 280 Kbps). The technology used is referred to in the industry as Web accelerator or Internet accelerator technology. It makes use of compression techniques to reduce the size of Web pages, images, and other files so that they travel more quickly over the network. While acceleration is rarely five times faster, it does increase dial-up performance considerably.

A **cable modem connection** is a high-speed Internet service, with data transfer rates as high as 6 Mbps, provided by cable television service providers. Some providers offer different plans with varying prices and connection speeds.

When using a cable service, the Internet signal is carried along the same cable as the television signals. A cable TV receiver receives the frequencies reserved for television, and a cable modem receives the frequencies used for Internet. A cable splitter is used for customers who use the cable for both television and the Internet. The cable modem connects to a computer using its Ethernet connection, and can be connected to several computers using a router or switch. Cable modems can be installed by professionals or for users who have existing cable service, a self-installation kit can be purchased.

Cable modem network lines are shared by neighbors, so if many neighbors access the Internet simultaneously, they might experience slower throughput. Some concerns have been raised over the ability of the cable to support the Internet needs of everyone in a given neighborhood.

A **DSL (digital subscriber line) connection** uses the customer's phone line, but there is no dialing up and users can use the Internet and talk on the phone simultaneously. DSL provides high-speed Internet access with data transfer rates up to 3 Mbps. Some DSL providers offer different service packages for varying prices and connection speeds. DSL service is similar to a cable modem, except that here your telephone line is split to carry signals to both a DSL modem and your telephone. Like a cable modem, and unlike a dial-up connection, a DSL line provides an always-on connection. As with a cable modem, a DSL modem is connected to a computer through an Ethernet card or to a router to share the signal between multiple computers. DSL is slower than cable service, but because it is a dedicated line rather than a shared line, you don't compete with neighbors for bandwidth.

FIGURE 4.5 • ISP speed comparison

Today's broadband Internet access options offer connection speeds over 100 times faster than traditional dial-up connections.

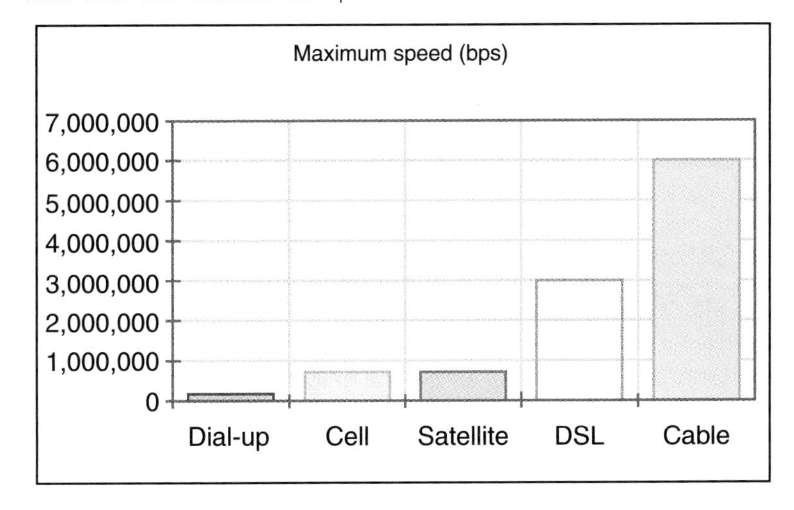

A *digital satellite service (DSS)* connection is a wireless high-speed Internet service, with data transfer rates of around 700 Kbps, provided to your home by companies such as EarthLink and DirecWay. Satellite service is faster than dial-up service but much slower than either cable modem or DSL (see Figure 4.5). For this reason, it is typically not used in areas where either DSL or cable is available. Satellite Internet is installed by professionals in much the same manner as cable modems and DSL. Satellite is the only service with substantial setup and equipment costs; users can expect to spend around $600 up front for installation. DSL and cable typically have promotions that provide free installation and setup. Although DSS offers an always-on connection, users may lose their signal during stormy weather.

Many cell phone companies now support and promote Internet access over cellular networks. At speeds as fast as 700 Kbps, connecting your notebook PC to the Internet either through a line to your cell phone or by using a wireless network adapter is a viable but expensive option (see Table 4.2).

TABLE 4.2 • Internet service fees and speeds (2005–2006)

	Average monthly fee	Max speed (bps)	Up-front cost
Dial-up	$ 15.00	56,000	$ -
Cell	$ 79.99	700,000	$ -
Satellite	$ 59.95	700,000	$ 600.00
DSL	$ 44.95	3,000,000	$ -
Cable	$ 57.95	6,000,000	$ -

Selecting an Internet Service Provider

Most colleges offer free Internet access to students. Upon graduating, students often find that monthly Internet access is a considerable expense; sometimes more costly than phone, cable TV, and other utilities. It pays to understand your options and to make sure you are paying only for services that you need.

This chapter provides general information on speed and cost of different types of services. Not all services are available in all locations. There are several Web sites that provide information on which services are available in your area based on your phone number and address. One such service is available at *www.cnet.com* by clicking "Internet services."

Once you are aware of your options, you should consider the ways that you use the Internet. If you plan to use your Internet connection for accessing media such as video, music, and video games, or if you would like to use a Web cam for video conferencing, you will appreciate a connection faster than 2 Mbps, which means either cable or DSL. If all you plan on doing is to check your e-mail, you might be satisfied with a bargain basement $9.00 per month dial-up connection.

Varying services offer different features. For example, commercial information services such as MSN and AOL provide software packages that include common tools for Internet use such as chat, instant messaging, Web browsing, shopping, news, e-mail, and others. Other Internet service providers provide only an Internet connection, relying on users to install their own Internet utilities such as a Web browser and instant messaging client. The difference in these types of dial-up services can be as much as $15 per month. Some of the more common features to be considered when selecting a service are listed below:

- Web hosting: The ability to post a Web page
- Accelerated dial-up access (5 times faster)
- E-mail account(s)
- Unlimited access/time
- Nationwide access for connecting to the Internet while on the road
- Tech support
- Security, virus, scam, and spam protection
- Pop-up blocker
- Parental controls
- Photo storage and sharing
- File storage and sharing

Choosing an ISP is a matter of balancing your budget against your requirements. The advice of friends can be very helpful. Be cautious of signing up for lengthy contracts, and take advantage of promotions that allow you to try a service for free.

Many other forms of Internet access are emerging as efforts continue to provide access to everyone, anywhere, anytime. Although not yet widely offered, the power companies of the world are poised to offer broadband Internet over the power grids through technology called *broadband over power lines* or *BPL*. The advantage of BPL is that the power grid is the most pervasive network in the world, reaching many places where telephone and cable lines do not. Connecting a computer to the Internet would be as easy as plugging a BPL modem into a wall outlet. BPL access will be provided at speeds of up to 1 Mbps. IBM and CenterPoint Energy are conducting a pilot project providing 220 homes in southwest Houston with broadband over power line (BPL).[2]

Other wireless technologies such as WiMax promise to extend high-speed wireless access to the Internet over large geographic regions. Wireless networking is discussed in detail in the next chapter.

The Internet is becoming increasingly available in locations away from work and home through pay-as-needed services. Many bookstores, coffee shops, airports, and hotels offer Internet access at a per-hour service charge. For example, at this writing, Starbucks charges $6 for the first hour of Internet access and 10 cents per minute thereafter. Many hotels charge $10–$20 per day for guest access to the Internet. Boingo (*www.boingo.com*) is a service that provides wireless Internet access from 20,000 locations around the world for $21.95 per month. United Airlines and other airlines are rolling out in-flight wireless Internet service for passengers. Lufthansa offers Internet access on many of its international flights for a flat fee of $29.95 for an entire flight or $9.95 for a half hour.[3]

In summary, the physical building blocks of the Internet include the high-speed telecommunications networks that make up the Internet backbone, the network access points at which they connect, the routers that manage the traffic traveling over the Internet, and the cable, DSL, dial-up, or other modems or devices that connect to individual computers. The Internet could not exist without this hardware. Equally important are the rules that govern the format of Internet traffic, which are called the protocols.

Internet Protocols. Whether negotiating peace between nations at war, merging corporate infrastructures, or attempting to connect different types of networks, you must begin by striking common ground, and establishing policies and procedures. In networking, policies and procedures for finding common ground for communications between two devices are defined by protocols. The protocols for the Internet are the *Transmission Control Protocol* (TCP) and *Internet Protocol* (IP). Together these two protocols are commonly referred to as **TCP/IP**.

Data is transported over the Internet in packets. A data *packet* is a small group of bytes that includes the data being sent and a header containing information about the data, such as its destination, origin, size, and identification number (see Figure 4.6). The Internet is a *packet-switching* network. Internet applications, such as e-mail, divide up information, such as an e-mail message, into small packets in order to make efficient use of the network. Upon arriving at their destination, the packets are reconstructed into the original message.

FIGURE 4.6 • Data packets

E-mail, Web pages, and all Internet data is broken up into consistent-sized data packets for efficient travel over the Internet.

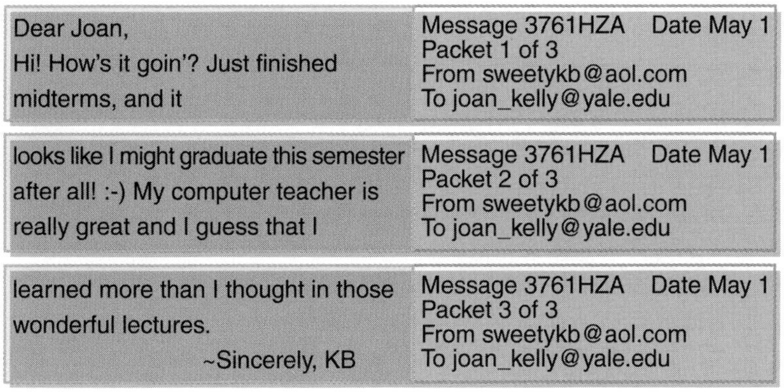

Data	Header
Dear Joan, Hi! How's it goin'? Just finished midterms, and it	Message 3761HZA Date May 1 Packet 1 of 3 From sweetykb@aol.com To joan_kelly@yale.edu
looks like I might graduate this semester after all! :-) My computer teacher is really great and I guess that I	Message 3761HZA Date May 1 Packet 2 of 3 From sweetykb@aol.com To joan_kelly@yale.edu
learned more than I thought in those wonderful lectures. ~Sincerely, KB	Message 3761HZA Date May 1 Packet 3 of 3 From sweetykb@aol.com To joan_kelly@yale.edu

The Internet Protocol (the IP in TCP/IP) defines the format and addressing scheme used for the packets. Routers on the Internet use the information in the packet header to direct the packet to its destination. The Transaction Control Protocol (the TCP in TCP/IP) enables two hosts to establish a connection and exchange streams of data. TCP guarantees delivery of data and also guarantees that packets are delivered in the same order in which they were sent.

The Internet Protocol requires that all devices connected to the Internet have a unique IP address. An **IP address** is a unique identifier for Internet hosts consisting of four numbers (0 to 255) separated by periods, such as 64.233.161.104. IP addresses can be *static*, permanently assigned to a particular computer, or *dynamic*, assigned to computers as needed. Computers that provide services, such as *www.yahoo.com* use static addresses, while computers that dial up to connect to the Internet are assigned temporary IP addresses for the time that they are connected. The over 300 million Internet hosts cited earlier is actually a count of the amount of assigned IP addresses. By using dynamic IP addresses, the Internet hosts can serve the over 888 million users.

Because people are more comfortable dealing with names than numbers, IP addresses are assigned associated English names called *domain names*. For example 64.233.161.104 is also known as *www.google.com*. Domain names and IP addresses are managed by the Internet Corporation for Assigned Names and Numbers (ICANN) and can be purchased from accredited registrars (*www.icann.org/registrars/accredited-list.html*). You can visit a registrar such as *www.enames.org* to find out if a particular name is available (you might find it interesting to see if your own name is registered), and for a small yearly fee you can claim an unregistered name for your own. Registering a domain name does not provide you with your own Web site, only with ownership of the name that you can associate with a Web site you have created and stored on a Web server with an assigned IP address.

The Internet uses the *Domain Name System (DNS)* to translate domain names into IP addresses. A database of addresses and names are stored on DNS servers. Internet services such as e-mail and the Web access DNS servers to translate the English addresses of people and Web sites to numeric IP addresses.

Many protocols are used on the Internet. Each service offered on the Internet—e-mail, the Web, instant messaging—has its own governing protocols. These protocols govern Internet communications and work much like TCP/IP does. Although you don't see these protocols at work, they are essential in all Internet communications.

FIGURE 4.7 • Client/server networking

Client/server technology is the basis of Internet services such as e-mail and the Web and uses server computers to distribute data to client applications such as Internet Explorer.

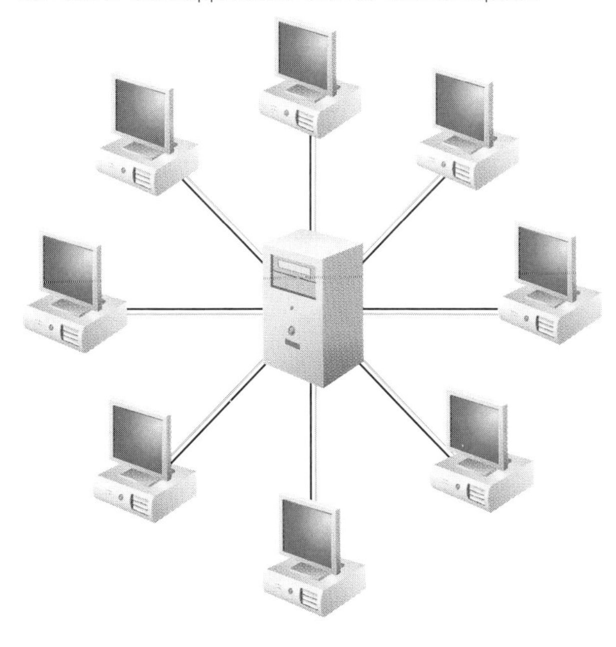

Internet Software. Most computers on the Internet communicate using client/server relationships. **Client/server** describes a relationship between two computer programs in which one program, the *client*, makes a service request from another program, the *server*, which provides the service (see Figure 4.7). A Web browser such as Internet Explorer, for example, is a client that requests a Web page from a Web server, such as *www.monster.com*. An e-mail program, such as Outlook or Eudora, is a client that connects to an e-mail server to retrieve e-mail messages. Instant messaging (IM) clients connect to IM servers to connect and communicate with other users on the Internet, just as Chat clients connect to Chat servers for the same purpose.

Server computers are typically powerful computers that can accommodate many simultaneous user requests. They run 24 hours a day, seven days a week to provide Internet services such as Web pages and e-mail service. The service performed by the server is defined by the type of server software it runs. A server running Web server software replies to Web page requests, and a server running e-mail server software governs the distribution of e-mail to and from the network. Different types of server software respond to requests that arrive at different ports on the server computer. In the client/server context, a **port** is a logical addresses used by clients and servers that is associated with a specific service. For example, a server receives Web page requests on port 80 and domain name system requests on port 53; KaZaA, the popular file-sharing application, uses port 1214. The client application requesting the service includes the correct port number in the outgoing packet headers. The server software accepting requests on that port number can be assured of the nature of the request.

Tracking File Swappers by IP Address

In the past few years the Recording Industry Association of America (RIAA) has sued over 4000 individuals for copying and sharing copyrighted music recordings over the Internet. Part of the evidence needed for conviction in such cases is a positive identification of the individual on the computer during the file transfer process. Such identification is only possible by tracking the IP address associated with the offending computer and finding out who was logged on to that computer when the file was transferred. The only entity with this information is the Internet service provider (ISP) providing the user with Internet access.

Each time your computer connects to the Internet, your ISP assigns it an IP number. Your ISP knows that it is your computer because you connect using your account username and password. Your ISP keeps records of which users are assigned which IP address at any given time. Your ISP may be a commercial company such as AOL, MSN, or Comcast. Or it might be your college or employer. Either way, by use of username, IP address, and port numbers, your activities on the Internet can be tracked.

In most cases, this confidential information is protected by the ISP. This is only smart business because any commercial ISP that disregards customer privacy would soon go out of business. Colleges also place a high value on the privacy of students and faculty both for ethical and economic reasons. However, the RIAA has placed extreme pressure on ISPs to turn over user information on individuals found illegally providing and accessing copyrighted recordings.

In 2003 the RIAA, using a controversial subpoena provision introduced by the 1998 Digital Millennium Copyright Act (DMCA), demanded that Verizon Internet Services reveal the identity of a Verizon subscriber who allegedly used KaZaA peer-to-peer software to share music online. Verizon fought the RIAA through several court battles and appeals to finally win the case in October 2004. The U.S. Court of Appeals ruled that the DMCA doesn't authorize the issuing of subpoenas to ISPs that, "act as mere conduits for the unauthorized transmission of copyrighted works," says a Bureau of National Affairs report.

Such was not the case in the UK where the High Court recently granted the British Phonographic Industry (BPI) an order under which six UK ISPs had to supply the names and addresses of 31 individuals alleged to "have uploaded large numbers of music files on to peer-to-peer filesharing networks," as a BPI press release puts it.

Most colleges complain of being harassed by the RIAA on a daily basis over students distributing music from college networks. Many colleges have closed down file-sharing network access, and are knocking on dorm doors giving students stern warnings and in the worst cases revoking student's access to the Internet altogether.

Questions

1. Do you think that ISPs should be required by law to hold private the records of customers? Why or why not?
2. Is there *any* situation where you feel that it would be ethically correct for an ISP to turn over the records of a customer? When? Why?
3. What steps might colleges take to please both students and the RIAA?

Sources

1. *Beale, M. "MPAA, RIAA Warn College of File Sharers,"* The Dartmouth, *May 16, 2005, www.thedartmouth.com/article.php?aid=2005051601040.*
2. *"RIAA v. Verizon Case Archive," Electronic Frontier Foundation Web site, , at http://www.eff.org/legal/cases/RIAA_v_Verizon/, accessed on June 5, 2005.*
3. *Haines, L. "High Court Orders ISPs to Name File-Sharers,"* The Register, *March 11, 2005, www.theregister.co.uk.*
4. *"Personal Data Apparently Given in File-Sharing Case,"* Reuters, *April 14, 2005, www.reuters.com.*

Some colleges are able to keep students from using controversial file-sharing software by closing off access to specific ports on the campus network. Likewise a user's network activity can be tracked by checking the username for an IP address at a given time, and the port activity associated with that address. Recently, the Record Industry Association of America has applied pressure to get Internet service providers to supply just such information as evidence to use in court in file-sharing copyright infringement cases.

FIGURE 4.8 • Peer-to-peer (P2P) networking

P2P networking allows users to link their computers over the Internet without using a server.

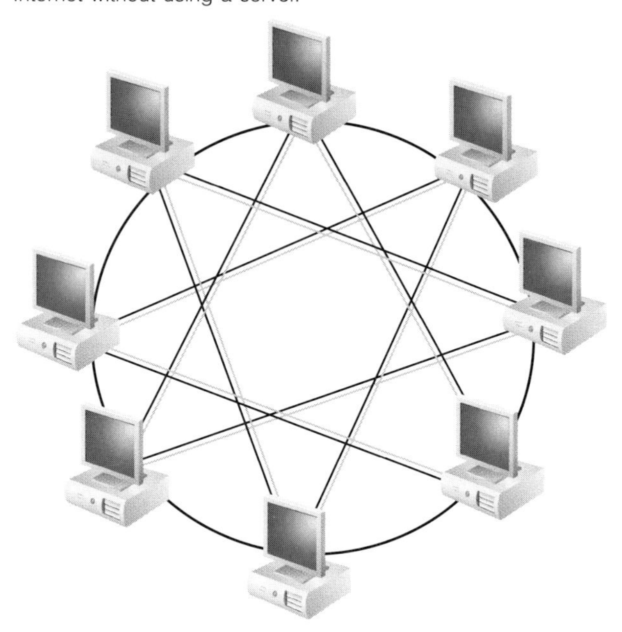

In some cases a computer may act as both a client and a server, as happens in P2P networking. **Peer-to-peer (P2P)** networks do not utilize a central server, but facilitate communications directly between clients (see Figure 4.8). Participants running P2P software make a portion of their file system available to other participants to access directly. In this relationship, an Internet user's personal computer acts as both server (as other users access files) and client (as the user accesses other's files). The Gnutella file-sharing system, which is at the heart of music-sharing services such as KaZaA, makes use of P2P networking. The original Napster used a client/server system, which left the company liable for illegal file-sharing activities. Since P2P does not use a central server, it makes it more difficult to hold the software company liable for what users do with the software. However, by connecting users directly without a central server, users sacrifice speed and some level of safety. Providing access to your hard drive to thousands of strangers presents opportunities to hackers, viruses, and spyware that they would ordinarily not have. Also, since the files being swapped through these systems are unchecked and unregulated, many have been found to contain viruses and other security threats.

A Layered System. The Internet can be viewed conceptually as a multilayer system. You've examined three layers that involve hardware, protocols, and software. Experts refer to the software portion of the Internet as the *application layer*. The protocol portion of the Internet, where client software communicates requests to servers using the packet-switching rules of TCP/IP, is called the *transport layer*. The hardware associated with the Internet is referred to as the *physical layer*. This general three-layer conceptual view of Internet technology is illustrated in Figure 4.9. Network specialists find it useful to use a more detailed, seven-layer model called the *Open System Interconnection (OSI) model*. The OSI model provides network technicians and administrators with a deeper understanding of networking technology for designing and troubleshooting networks. For those of you just interested in using networks, however, the more general view provided here is enough to assist you in practical issues, such as selecting an Internet service provider, working with Internet applications, protecting ourselves from hackers, and other issues that will be discussed as you progress through this book.

FIGURE 4.9 • Conceptual layers of the Internet

The Internet can be viewed conceptually as a multilayer system.

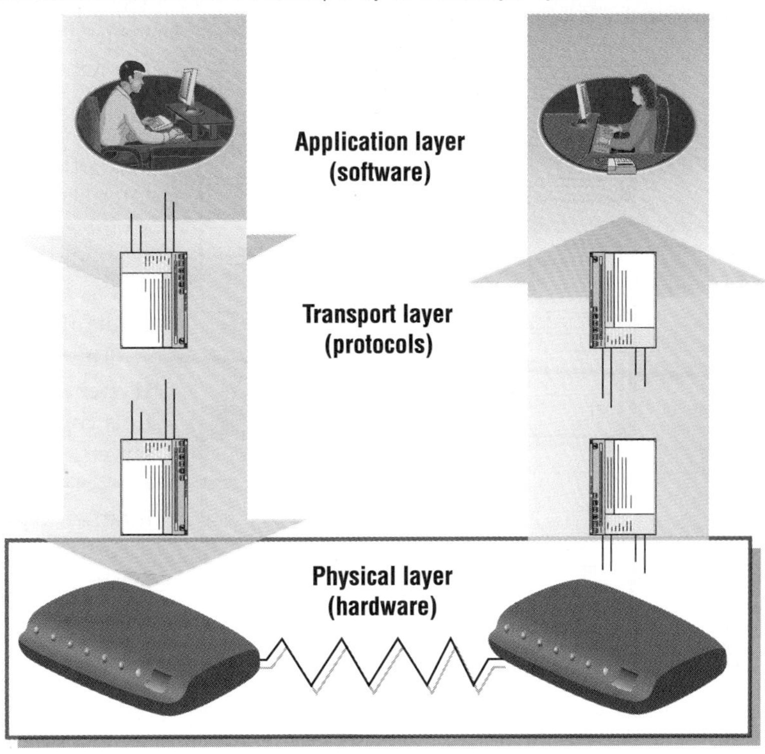

WEB TECHNOLOGY

The World Wide Web was developed by Tim Berners-Lee between 1989 and 1991 in his research at CERN, the European Organization for Nuclear Research (in French, *La Conseil Européen pour la Recherche Nucléaire*) in Geneva, and released to the public in the form of the Mosaic Web browser in 1993. What he originally conceived of as an organizational tool to help keep track of his own personal documents has grown into an organizational tool that helps hundreds of millions of users share and access information on the Internet using an easy-to-use graphical interface. Recall that the Web is defined as an Internet service that provides convenient access to information through hyperlinks.

The process of "linking together" documents from diverse sources requires three components:

1. A defined system for linking the documents
2. Protocols that allow different computers to communicate
3. Tools to assist in creating the documents and the links between them

Tim Berners-Lee came up with all three: hyperlinks for linking documents, the Hypertext Transfer Protocol (HTTP) along with server and client software to manage communications between different computers, and HTML, a language for creating and linking documents.

Over the past decade many new technologies have been developed that work with Berners-Lee's original Web technologies to deliver richer Web content—animation, video, 3D views of objects and locations, music, and computer programs. This section covers the Web from the basics to state-of-the-art technologies.

FIGURE 4.10 • Hyperlinks

By relating documents to each other using hyperlinks, you form a web of interrelated information that is logically arranged and easy to navigate.

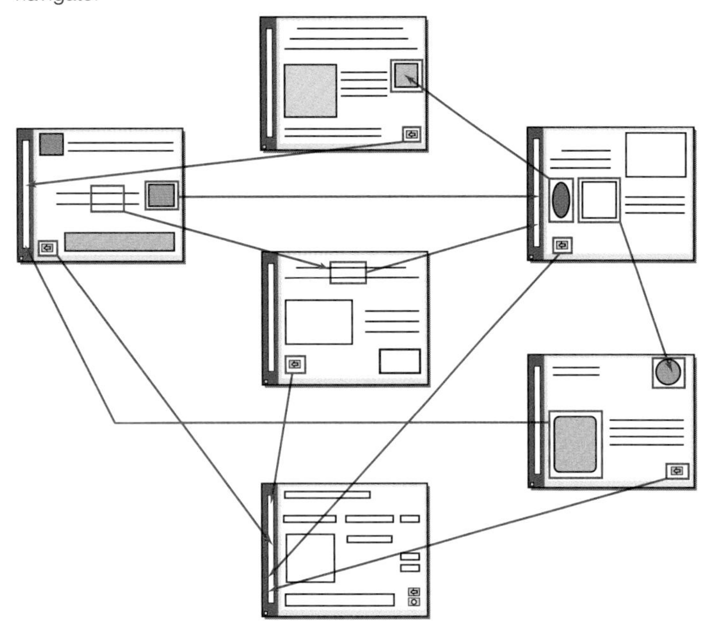

Web Basics

The cornerstone of Tim Berners-Lee's World Wide Web is the hyperlink. A **hyperlink** is an element in an electronic document—a word, phrase or image—that when clicked, opens a related document. By relating documents to each other using hyperlinks, you form a web of interrelated information that is logically arranged and easy to explore (see Figure 4.10). Some differentiate between *hypertext*, text that acts as a link, and *hypermedia*, pictures or other media that act as links.

The protocol of the Web is called HTTP. The **Hypertext Transfer Protocol (HTTP)** controls communication between Web clients and servers. A Web client, usually called a **Web browser**, takes the form of software such as Internet Explorer, Netscape, and Firefox, and is used to request Web pages from Web servers. A **Web server** stores and delivers Web pages and other Web services such as interactive Web content.

Web browsers are available for many different computing platforms. Desktop computers, notebook computers, and handheld computers all include Web browsers. Some cell phones now come equipped with Web browsers. Many of the most popular Web sites are providing content in a stripped down form for convenient navigating on a small PDA or cell phone display (see Figure 4.11). The Web is also available on televisions by using special Internet-connected set-top boxes. In the world of pervasive computing, Web interfaces are cropping up on all kinds of objects—from refrigerators to automobiles.

FIGURE 4.11 • The handheld Web

Many Web publishers offer handheld versions of their sites.

All Web pages are identified and accessed using a URL. A **Uniform Resource Locator**, more commonly called a **URL,** acts as a Web page address, incorporating the domain name of the Web server, and the location of the Web page file on the server. Figure 4.12 identifies the different components of a URL.

The final portion of the domain name, .com, .edu, and so on, is called the *top-level domain* (*TLD*). Top-level domains classify Internet locations by type or, in the case of international Web sites, by location. There are hundreds of TLDs. The most common are listed in Table 4.3. A complete list can be found at *www.icann.org/tlds*.

The .name TLD has been provided for use with personal Web sites. For example, Jerry Jawalski could purchase the name jerry.jawalski.name from an accredited registrar, and lay claim to the e-mail address jerry@jawalski.name, and the URL *http://www.jerry.jawalski.name* for as little as $6 per year. Other incentives to promote the .name TLD are listed at *www.nic.name*.

FIGURE 4.12 • Components of a URL

The URL *http://www.course.com/cca/ch1/index.html* consists of several components.

Protocol	Web server	Domain name	Location on server	Requested file
http://	www.	course.com/	cca/ch1/	index.html

TABLE 4.3 • Popular top-level domains

TLD	Description
.com	Commercial business
.biz	Commercial business
.edu	Educational institution
.org	Nonprofit organization
.net	Networking service
.gov	Government agency
.name	Personal Web site

Web Markup Languages

A *markup language* is used to describe how information is to be displayed. It typically combines the information, such as text and images, along with additional instructions for formatting. The primary markup language that is used to specify the formatting of a Web page is called **Hypertext Markup Language** (**HTML**). Web pages are sometimes called HTML documents. HTML uses tags to describe the formatting of a page. An *HTML tag* is a specific command inside angle brackets (< >) that tells a Web browser how to display items on a page. Figure 4.13 illustrates HTML code and how the browser interprets it. Note that there are opening and closing versions of HTML tags. The closing tag is indicated by a forward slash. HTML commands operate on whatever is between the opening and closing tag. Thus, <b>HTML</b> tells the browser to display the letters *HTML* in a bold font. Viewing the figure, can you guess what effect the <em> command has?

FIGURE 4.13 • Hypertext Markup Language (HTML)

The HTML code at the top is read by a Web browser, which displays the content on the bottom.

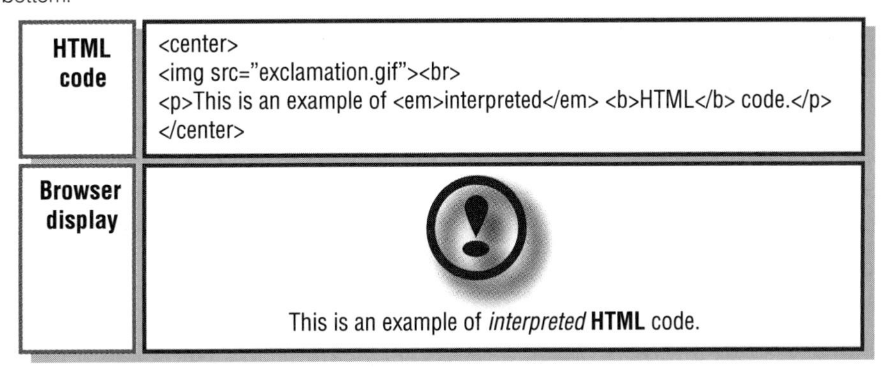

HTML code
<center>
<img src="exclamation.gif">
<p>This is an example of <em>interpreted</em> <b>HTML</b> code.</p>
</center>
Browser display
This is an example of interpreted HTML code.

A number of newer Web markup languages are increasing in popularity. One that is changing the landscape of the Web and other aspects of computing is the **Extensible Markup Language** (**XML**). Although HTML provides a method of describing the format of a Web page, XML provides a method for describing and classifying the content of data in a Web page. The use of XML is leading to what researchers have labeled the Semantic Web (see *www.w3.org/2001/sw*). The *Semantic Web* is an extension of the current Web and provides a common framework that supports the sharing and reuse of data across application, enterprise, and community boundaries. Compare the HTML code and XML code in Figure 4.14. Although the XML code is simplified for the purpose of this example, it clearly illustrates how data is classified in XML code. In contrast, the HTML code says only how the data should look, and says nothing about the purpose of the data.

FIGURE 4.14 • XML

XML provides a method of describing or classifying data in a Web page.

Web content	HTML code	XML code
Reebok® Classic Ace Tennis Shoe $49.95 Soft leather tennis shoe. Lightweight EVA molded midsole. Rubber outsole. China.	<strong><font face="Verdana, Arial, Helvetica, sans-serif">Reebok® Classic Ace Tennis Shoe </font></strong> <strong>$49.95</strong> <table width="100%" border="1"><tr><td>Soft leather tennis shoe. Lightweight EVA molded midsole. Rubber outsole. China. </td></tr></table>	**<product type="shoes">** **<name>** Reebok Classis Ace Tennis Shoe **</name>** **<price>$49.95</price>** **<description>** Soft leather tennis shoe. Lightweight EVA molded midsole. Rubber outsole. China. **</description>** **</product>**

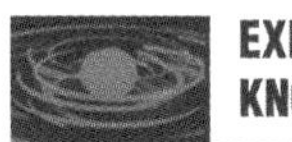

EXPAND YOUR KNOWLEDGE

To learn more about how to create Web pages, go to www.course.com/swt2/ch04. Click the link "Expand Your Knowledge", and then complete the lab entitled "Creating Web Pages".

XML provides several advantages to both Web content publishers and viewers. Publishers are able to use XML to separate data from Web page formatting. This is possible because XML Web content is implemented using several files: one that defines the structure of the data (product, name, price, and description), another that provides the actual data (Reebok, $49.95, and so on), and a third that defines the format of the presentation of the data in a Web browser. This method of organization offers great convenience to organizations that may change the content of a Web page frequently. For example, a news organization can use the same layout for their Web page, but change the data from day to day to reflect the latest news.

The structured data approach of XML provides convenience for Web browsing as well. Web searches become much easier when Web site content is classified and defined using XML tags. Imagine that the HTML and XML code in Figure 4.14 included descriptions of many tennis shoes. It would be much easier for a search engine to find a match for "price < $60" in the XML code than in the HTML code.

XML has gained such approval from Web developers that the World Wide Web Consortium (W3C), the organization that develops and approves Web standards, is endorsing a successor to HTML called XHTML. *XHTML* embodies the best of both HTML and XML in one markup language. It is anticipated that XHTML will gradually replace HTML as the primary markup language of the Web.

XML has been adopted by Microsoft in its Office suite to standardize data descriptions across applications. Because data is described the same with XML in Word, Excel, Access, and PowerPoint, as well as other Microsoft applications, it can be easily shared between those applications and published to the Web (see Figure 4.15). Microsoft is replacing its proprietary file formats (.doc, .xls, .mdb, and .ppt) with standardized compressed .xml files.[4] Because many software vendors support the XML standard of data representation, data can now easily be shared across software from different vendors, and even across a variety of computing platforms. A global standard for data representation is the first step to taking control of information and overcoming information overload.

FIGURE 4.15 • A standard for data representation

XML provides a standard for data representation so that data from different data sources can be shared for varying uses.

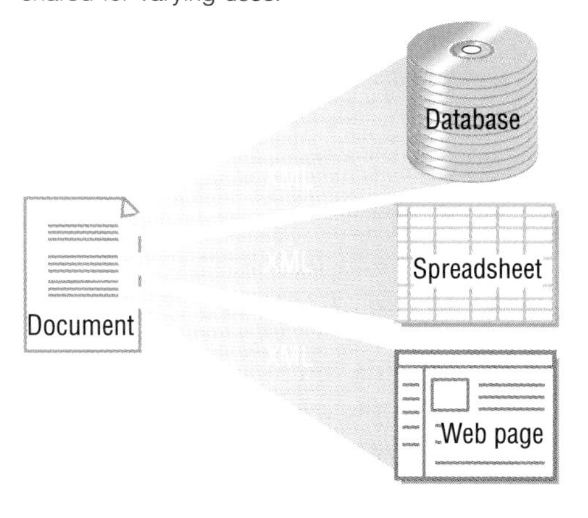

XML is at the heart of a new and useful technology called Web services. A *Web service* uses XML and other standards and protocols to support application-to-application communication over the Internet. For example, a calendar application running on your PC might use Web service technology to communicate with your doctor's office scheduling software to negotiate a time for your next medical exam. Your calendar application, working with your doctor's scheduling application, would provide a list of times and dates for you to choose from for your next checkup. The Web service keeps you from waiting on hold and eliminates the need to hire someone to manage scheduling for the doctor. XML and Web service technology is driving many of today's most useful Internet applications.

Web-Authoring Software

EXPAND YOUR KNOWLEDGE

To learn more about designing Web pages, go to www.course.com/swt2/ch04. Click the link "Expand Your Knowledge", and then complete the lab entitled "Web Design Principles".

When the Web was new, creating Web sites was the domain of techies who took the time to learn HTML, and those that hired them. Today, anyone who can use a word processor can create professional-quality Web sites with little time and effort. All it takes is something worth writing about, a good sense of visual design, and some Web authoring software. **Web authoring software** allows you to create HTML documents using word-processor-like software. Rather than having to type out HTML tags to create Web page formatting, the author defines the formatting using standard menu commands in a what-you-see-is-what-you-get (WYSIWYG) editor (Figure 4.16). *Wysiwyg* (pronounced wizzie-wig) implies that the Web page you design with the Web authoring software will look the same when published on the Web. When you save your Web page, the software creates the HTML file with the appropriate tags.

FIGURE 4.16 • WYSIWYG

In a WYSIWYG editor, such as Macromedia's Dreamweaver, the Web page you design in the Web-authoring software will look the same when published on the Web.

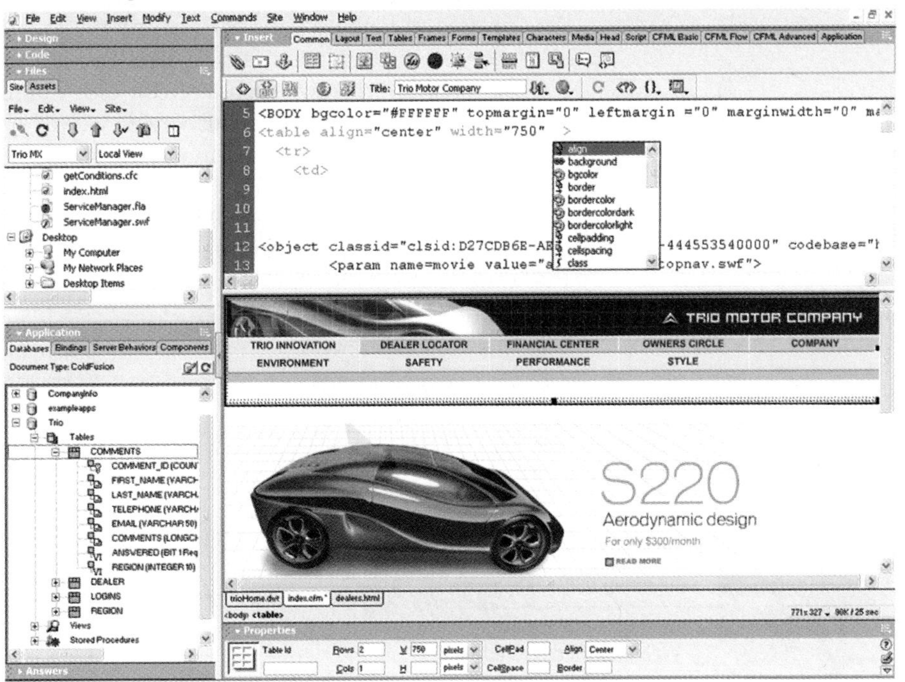

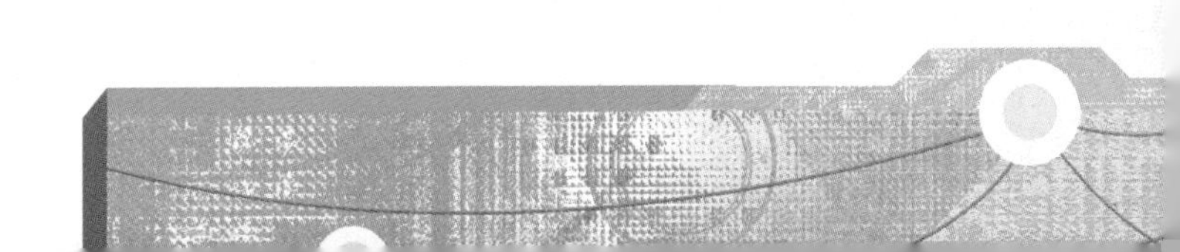

Wysiwyg editors are a great convenience for quickly creating Web pages. However, these editors fall somewhat short of automating all processes involved in Web production. Understanding HTML tags is still a valuable skill to possess in Web development.

Although most people who use the Web do not have a Web site, it is likely that the percentage of people who do publish to the Web will continue to increase. Many nontechnical employees are now responsible for creating or maintaining content on professional Web sites. Outside of work, the Web is an ideal way to share files with friends, family, and strangers. You can use your Web site to share photos with family and friends, share your corporate mission statement with customers, or share your political philosophies with the world.

Programming the Web

While creating *static* Web pages, those that simply present unchanging information, has become easy, today's professional-quality Web sites have evolved well beyond these basics. Today's most popular Web sites are *dynamic*, providing users with an interactive experience tailored to the individual.

Have you ever had a Web page greet you by name? Many of today's commercial Web sites are created *on-the-fly* when a new user accesses it. For example, if you have made purchases at Amazon.com you may notice that each time you visit Amazon.com you are greeted by name and presented with advertisements for products similar to those you have purchased before. Amazon custom-creates a Web page just for you. But, how does the Amazon server know who you are, and what you like? The answer involves cookies.

Today's Web browsers allow Web servers to store small text files on your computer called **cookies**. When you revisit a Web site, even though your computer may have a different IP address than the last time you visited, the Web site can recognize you by reading its associated cookie file from your PC (see Figure 4.17). Cookie files can contain data or perhaps a unique identifier that can be used to look up your personal profile in a database on the server. Information accessed through the use of cookies can include your name, address, and other personal information such as credit card numbers, items you've previously purchased at the Web site, items you've viewed on the Web site, and the amount of time that you viewed each item. This information is obtained through tracking your activities at the Web site and from data from the forms you fill out and submitted at the site. By collecting this information over time, a detailed customer profile can be developed and used to cater to your individual tastes each time you visit the Web site.

FIGURE 4.17 • Browsers, cookies, servers, and software

Some Web servers use cookie files on your computer to customize the Web pages you view.

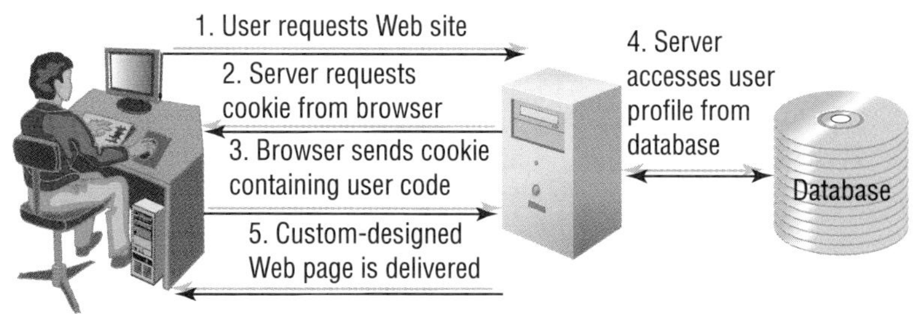
1. User requests Web site
2. Server requests cookie from browser
3. Browser sends cookie containing user code
4. Server accesses user profile from database
5. Custom-designed Web page is delivered
Database

Some worry that cookies may be a threat to security and privacy. Today's browsers provide means with which to heighten security and restrict the use of cookies. Users also have the power to delete cookie files from their systems. These files are so common, however, that nearly every commercial Web site uses them, and most users accept them as a fact of Internet life.

Collecting and interpreting customer information requires the use of software designed to interact with Web browsers and servers. The software typically runs on the server using information collected from Web-based forms and mouse clicks as input. Some Web software is designed to run independently on the client computer. Whether on the server side or the client side, Web software plays a large role in how you use the Web today:

- Search engines use software to query databases with user-provided keywords
- Shopping on the Web involves shopping cart software and software that manages the transaction
- Software manages Web cookies and creates dynamic Web pages custom designed to the visitor
- Online services such as online dating, games, animated cartoons, special-purpose calculators, local weather, and many others make use of Web-driven software

The three most prevalent programming languages used to implement this type of interactivity are Java, JavaScript, and ActiveX. *Java* is an object-oriented programming language that allows software engineers to create programs that run on any computer platform (see Figure 4.18). *JavaScript* is similar to Java in that it provides functionality in Web pages through programming code that is embedded in an HTML document. However, the two languages are very different. Java is a full-fledged programming language for all computing environments, but JavaScript was developed specifically for the Web and is more limited in nature. Microsoft has created an alternative to JavaScript for use in Internet Explorer called *ActiveX*. ActiveX controls are included with some Web pages to provide interactivity or animation.

FIGURE 4.18 • 3D product view

Java applications allow viewers to rotate products on the screen for a complete 360-degree view. This Web page gives you a feel for driving a Mercedes SL55.

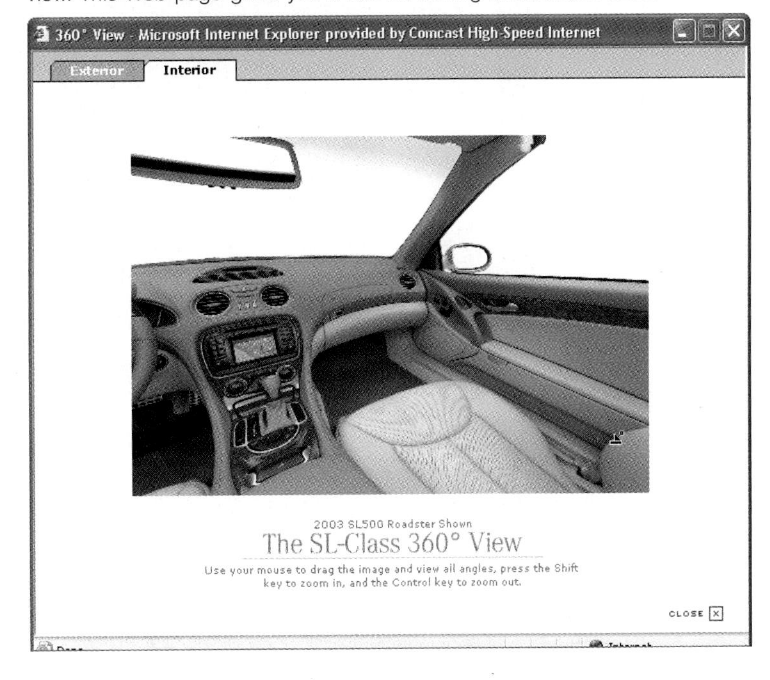

Web Browser Plug-Ins

Some software companies have developed software to extend the capabilities of HTML and Web browsers. HTML files can store only ASCII text. Binary data such as images, sound, and video must be stored in separate files. Web browsers support viewing some image files (.gif and .jpg files) and playing some basic sounds and animation. To employ more advanced Web content, or what some in the industry call *rich content*, such as video and interactive media, a helper application called a plug-in is required. A **plug-in** works with a Web browser to offer extended services such as audio players, video, animation, 3D graphics viewers, and interactive media. When a Web page contains content that requires a plug-in, you are typically provided with the opportunity to download and install the necessary tool at no cost, if you don't already have it. Macromedia's *Flash* is a popular plug-in; Flash enables users to view animations and videos, and interact with games and other multimedia content created with the Flash program (see Figure 4.19).

FIGURE 4.19 • Flash animation

This Flash game provided at the Mini Cooper Web site attracts customers to the site and helps them have fun while they are shopping for a new vehicle.

Many plug-ins available today address the issue of transferring large media files over the Web. The traditional method of viewing Web content is to request a file from a Web server by typing a URL or clicking a link, waiting while the file downloads to your computer, and then viewing the file in the Web browser window. The large size of audio and video files would leave users waiting for minutes, even hours, after they clicked a link before they could view the file. Some systems are able to reduce the wait by compressing the files to smaller sizes and storing the multimedia content in a more efficient manner. Another technique to deliver multimedia without the wait is called content streaming. With **content streaming**, sometimes called streaming media, streaming video, or streaming audio, the media begins playing while the file is being delivered. Problems arise with this technique only when the speed of play outpaces the speed of delivery of your Internet connection.

Television Marketers Turn to Web

With the popularity of high-speed Internet connections and video-capable Web browsers, television programmers are planning to make the Web a lot more like TV. One of the first examples of this trend is *www.living.com*, the all-video Web site published by the E.W. Scripps Company and financed by GMC. It provides programming from its Food Network, Fine Living, HGTV, and DIY Network brands, as well as news clips. GMC's investment in living.com buys them advertising space throughout the site. Alexia S. Quadrani, a senior managing director at Bear Stearns who follows the publishing and advertising industries, predicts that this is the start of a trend. "You are seeing a lot more content go online because there is a demand for it," she said.

More than 34 million homes in the United States, representing 29.9 percent of households, had high-speed Internet connections at the close of 2004, according to eMarketer, an online research provider. By 2008, eMarketer projects, broadband will be in 69.4 million homes, or 56.3 percent of households. This provides fertile soil for a new age of television over the Web.

Web users have already proved that they are willing to watch live sports online for more than an hour at a time, said Bart Feder, president and chief executive at FeedRoom, a provider of broadband video technology to clients such as NBC, Reuters, and Telemundo. Companies such as GMC are ramping up the percentage of advertising dollars that they are investing in the Web. They hope that by providing new and useful technologies and information to Web users, they will increase their visibility and improve their reputation.

Questions

1. How can Web users benefit from sponsored Web sites such as living.com?
2. Consider the nature of television and the nature of the Web. List how these two information delivery systems differ. Which is more effective as an information distributor?
3. What changes will be required of marketers as they make a transition from television to the Web?

Sources

1. *Ives, N. "As TV Moves to the Web, Marketers Follow,"* New York Times, *May 27, 2005, accessed at www.nytimes.com.*
2. *Living.com, accessed June 10, 2005, at www.living.com.*

INTERNET AND WEB APPLICATIONS

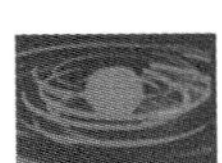

EXPAND YOUR KNOWLEDGE

To learn more about Internet applications, including Web browsers, go to www.course.com/swt2/ch04. Click the link "Expand Your Knowledge", and then complete the lab entitled "Getting the Most Out of the Internet".

The Internet has provided information services to researchers and scholars since 1970. The early Internet introduced a then-new and powerful messaging tool called e-mail that quickly became the most popular Internet application. An Internet tool called FTP (for file transfer protocol) was used to share files. Other tools were developed to catalogue all of the files available on the Internet. For the first 23 years the Internet was a text-based medium accessed using a network command line. Because relatively few people had the ability to access and use a network command line, the Internet remained relatively obscure.

The birth of the World Wide Web in 1993 changed all of that. The World Wide Web and the Web browser provided a graphical user interface to Internet resources that opened the Internet up to the general public. Individuals, groups, organizations, and businesses began publishing information to the Internet for the general public to access using a Web browser. The Web quickly became the main gate for public access to the Internet.

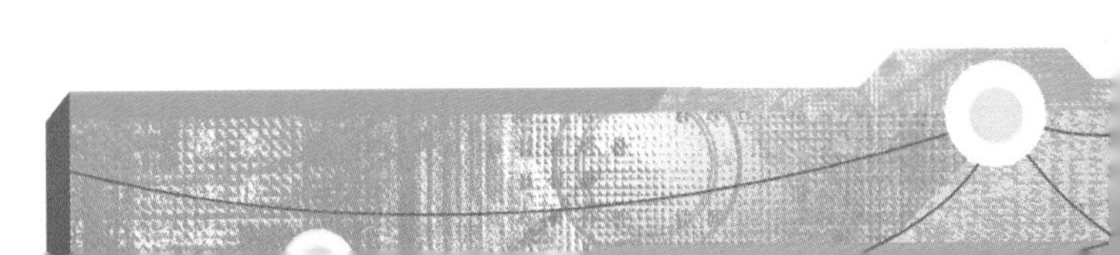

Everything on the Web exists on the Internet; however, the reverse is not true. Many people access e-mail without using a Web browser. Businesses and organizations pass private information over the Internet. Some researchers and scientists still make use of the old command line for using traditional Internet services. Figure 4.20 illustrates the relationship between The Web, the Internet, and Internet applications. The Web exists as an application on the Internet, and services such as communication, e-commerce, and information distribution exist both on the Internet and on the Web.

FIGURE 4.20 • The Web and the Internet

The Web exists as an application on the Internet and supports many Internet services.

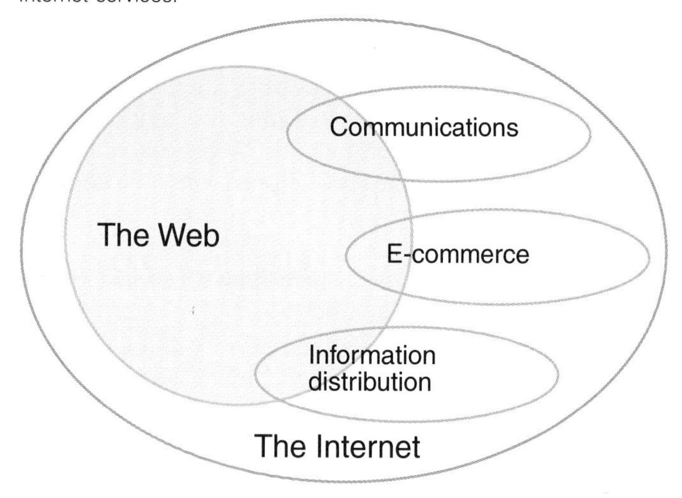

The Web is the ideal tool for sharing and organizing information for the general public. The Web contains millions of Web sites. Each Web site is unique and most can be classified under one of the following categories:

- Search engines, subject directories, and portals
- Communication and collaboration
- News
- Education and training
- E-commerce
- Travel
- Employment and careers
- Multimedia and entertainment
- Information

This section explores each of these Web applications as well as other Internet applications and how they can benefit you personally and professionally.

Search Engines, Subject Directories, and Portals

The fundamental purpose of the Web is to make it easier to find related documents from diverse Internet sources by following hyperlinks. However, the Web has become so large that many complain of *information overload*, or the inability to find the information you need due to the overabundance of unrelated information. In efforts to relieve the strain of information overload, Web developers have provided powerful tools to assist in organizing and cataloging Web content.

A **search engine** is a valuable tool that enables you to find information on the Web by specifying words that are key to a topic of interest—*keywords*. Operators can also be employed for more precise search results. Table 4.4 provides examples of the use of operators in Google searches as listed on Google's help page (*www.google.com/help/cheatsheet.html*).

Search engines scour the Web with *bots* (automated programs) called *spiders* that follow all Web links in an attempt to catalog every Web page by topic. The process is called Web *crawling*, and due to the ever-changing nature of the Web, it is a job that never ends. Google maintains over four billion indexed Web pages on 30 clusters of up to 2000 computers each totaling over 30 petabytes of data.

One of the challenges of Web crawling is determining which of the words on any given Web page describe its topic. Different search engines use different methods. Methods include counting word occurrences within the Web page, evaluating nouns and verbs in the page's title and subtitle, using keywords provided by the page's author in a meta tag, and evaluating the words used in links to the page from other pages. Once the search engine has a reasonable idea of a page's topic, it records the URL, page title, and associated information and keywords in a database.

TABLE 4.4 • Using operators in Web searches

Keywords and operator typed	Search engine interpretation
vacation hawaii	The words *vacation* and *Hawaii*
Maui OR Hawaii	Either the word *Maui* or the word *Hawaii*
"To each his own"	The exact phrase *to each his own*
virus –computer	The word *virus*, but *not* the word *computer*
Star Wars Episode +I	This movie title, including the roman numeral *I*
~auto loan	Loan info for both the word *auto* and its synonyms: *truck*, *car*, etc.
define:computer	Definitions of the word *computer* from around the Web.
red * blue	The words *red* and *blue* separated by one or more words.

After building the search database, the next challenge facing a search engine is to determine which of the hundreds or thousands of Web pages associated with a particular keyword are most useful. The method of ranking Web pages from most relevant to least differs from search engine to search engine. Google uses a popularity contest approach. Web pages that are referenced from other Web pages are ranked higher than those that are not. Each reference is considered a vote for the referenced page. The more votes a Web page gets, the higher its rank. References from higher-ranked pages weigh more heavily than those from lower-ranked pages.

A keyword search at Yahoo!, MSN, or Google isn't a search of the Web but rather a search of a database that stores information about Web pages. The database is continuously checked and refreshed so that it is an accurate reflection of the current status of the Web. With today's rapidly changing and expanding Web, this can be a challenge, and at times you will find that links presented in search results are no longer active.

Today's heated competition in the search engine market is pressing the big players to expand their services. Table 4.5 lists some of the newer search engine services available and being developed.

Industry experts predict that video search will be the next "big thing" and a competitive battleground for search engines. Google is finalizing a service for searching the content of online video clips. This new service will search for relevant video clips online and show you a 10-second preview. If you select a clip, your browser will be sent to the relevant Web site to pay for the whole clip.[5]

A *meta search engine* allows you to run keyword searches on several search engines at once (see Figure 4.21). For example, a search run from *www.dogpile.com* returns results from Ask Jeeves, FAST, FindWhat, Google, LookSmart, Overture, and other search engines.

A *subject directory* is a catalog of sites collected and organized by people rather than automated crawlers. Yahoo.com, a well-known Web site, provides a directory that divides Web topics into 14 general categories with many subcategories and levels of subcategories under each. Subject directories are often called subject trees because they start with relatively few main categories and then branch out into many subcategories, topics, and subtopics. Subject directories contain only a small percentage of all existing Web pages, but because they are created and maintained by people—not bots—they are more likely to contain relevant information. Subject directories are useful for finding general information on popular topics. If you are looking for something more specific, a search engine may be a better tool. For serious scholarly research, online databases of

TABLE 4.5 • Search engine services

Service	What it does
Alerts	Receive news and search results via e-mail
Answers	Ask a question, set a price, get an answer
Catalogs	Search and browse mail-order catalogs
Desktop Search for Enterprise	Search your company's network
Images	Search for images on the Web
Local	Find local businesses and services
Maps	View maps and get directions
Mobile	Search the Web from your cell phone
News	Search thousands of news stories
Personalized Search Page	Customize your search page with current news and weather
Print	Search the full text of books
Ride Finder	Find a taxi, limousine, or shuttle using real-time position of vehicles
Scholar; University Search	Search through journal articles, abstracts, and other scholarly literature; search a specific school's Web site
Search by Location	Filter results by geographic location
Search History	Maintain a history of past searches and the Web sites that paid off
Search Toolbar	Have access to search from the toolbar of your browser or operating system taskbar
Shopping	Find the best deal on consumer products
Video	Search recent TV programs online

FIGURE 4.21 • Meta search engine

Meta search engines such as Dogpile return search results from several popular search engines to take advantage of the best of all search technologies.

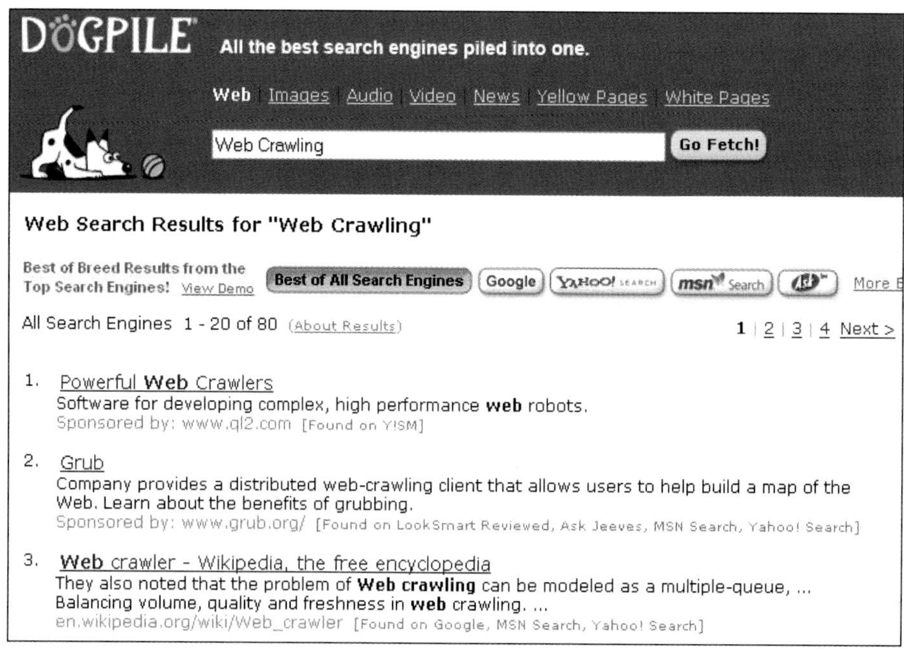

scholarly articles from journals and other publications typically maintained by college libraries offer the best sources of information (more on this later).

Web portals are Web pages that are designed to act as entry points to the Web—the first page you open when you begin browsing the Web. They typically include a search engine, a subject directory, daily headlines, and other items of interest. They can be general or topic specific in nature. Yahoo.com, lycos.com, aol.com, and msn.com are examples of horizontal Web portals; *horizontal* refers to the fact that they cover a wide range of topics. Vertical Web portals focus on special-interest groups. For example, the iVillage.com portal focuses on items of interest to women, and askmen.com is a vertical portal for men (see Figure 4.22).

FIGURE 4.22 • Portals

Portals are entry points to the Web. iVillage.com is a vertical portal that appeals to women's interests and askmen.com for men.

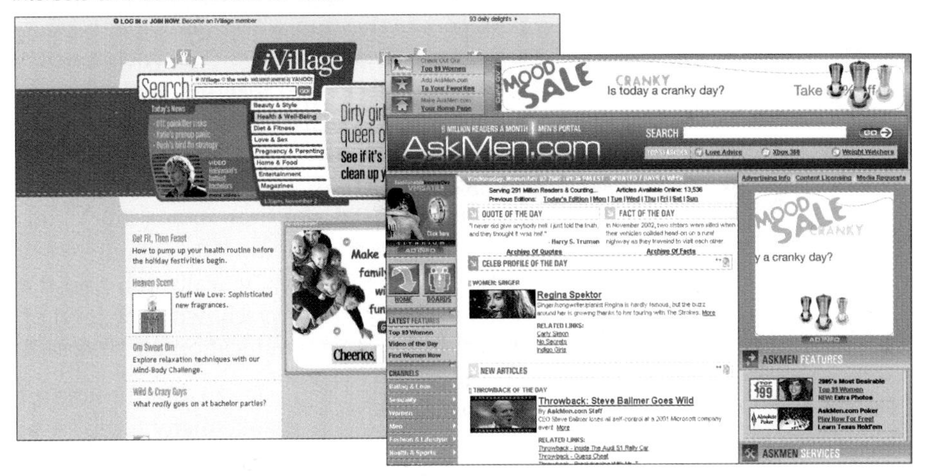

Communication and Collaboration

Earlier, you learned how the digital revolution is bringing in the age of pervasive computing where anyone can have access to information anywhere. Equally important and influential is the birth of *pervasive communications*, the ability to communicate with anyone through a variety of formats from anywhere at anytime. Pervasive communications are the result of advances in wireless communications and Internet communications. This technology is fundamentally altering the ways in which personal and professional relationships are created and nurtured.

Prior to the Internet, forms of two-way communication were limited to face to face, telephone, or printed word—as in mailing a letter. The Internet has broadened our communications options considerably. Today you can phone, e-mail, meet in person, instant message, blog, video chat, podcast, send a letter, post a message, and more. It is useful to consider your communication options and the strengths and weaknesses they exhibit.

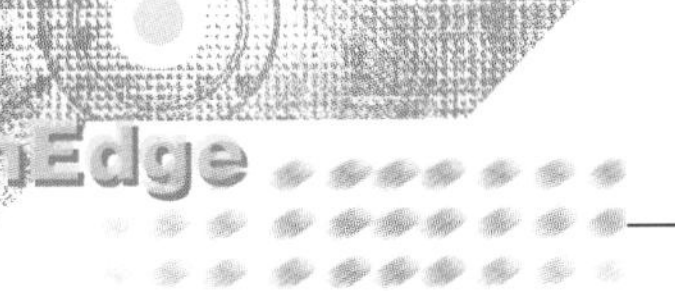

JOURNALISM BY YOUR AVERAGE JOE

A bloom in blogging combined with backlash over corporate media scandals and control has created a new kind of cub reporter: the citizen journalist. As simple as it sounds, average people (i.e., not paid, not professionals) are catching stories the big boys miss and posting their articles on the Net. The concept isn't new, but affordable software for professional-looking Web sites is, and in the open source environment of the Net, any inaccuracies are quickly addressed.

Source:
Citizen journalism takes root online
Just plain folks can create their own news sites
Verne Kopytoff, Chronicle Staff Writer
San Francisco Chronicle/SF Gate
http://www.sfgate.com/cgi-bin/article.cgi?file=/chronicle/archive/2005/06/06/BUGF9D2HAO1.DTL&type=tech
June 6, 2005

There are two forms of communication: synchronous and asynchronous. In **synchronous communication**, people communicate in real time exchanging thoughts in a flowing conversation. Synchronous communication is not always possible as it requires all participants to be engaged in communication at the same time. Face-to-face conversations, telephone conversations, online chat, and instant messaging are examples of synchronous communication. **Asynchronous communication** allows participants to leave messages for each other to be read, heard, or watched, and responded to at the recipient's convenience. Answering machines, voice mail, and e-mail are tools for asynchronous communication.

Although it is sometimes more convenient, asynchronous communication is generally considered a weaker form of communication because the time lapse between thought exchanges can stifle the emergence of new ideas. For example, consider brainstorming sessions. The term *brainstorming session* refers to synchronous communication where participants bounce ideas off each other in order to arrive at optimal solutions. The synchronicity of such communication allows multiple minds to join together and act as one. Such a phenomenon is severely inhibited if conducted asynchronously through e-mail. E-mail is perfect, however, for communications that require time for ponderance, and day-to-day communication such as "I'll pick you up at 8:00," "I agree to the terms of this contract," or "Did you hear about Chuck and Grace?"

Each form of communication should be evaluated in terms of quality, convenience, and time/delay. Typically you find that the quality and speed of communications compares inversely with the level of convenience (see Figure 4.23). Several forms of communication are discussed in the following sections.

FIGURE 4.23 • Evaluating forms of communication

Synchronous forms of communication that require individuals to be present and engaged simultaneously are less convenient but more effective.

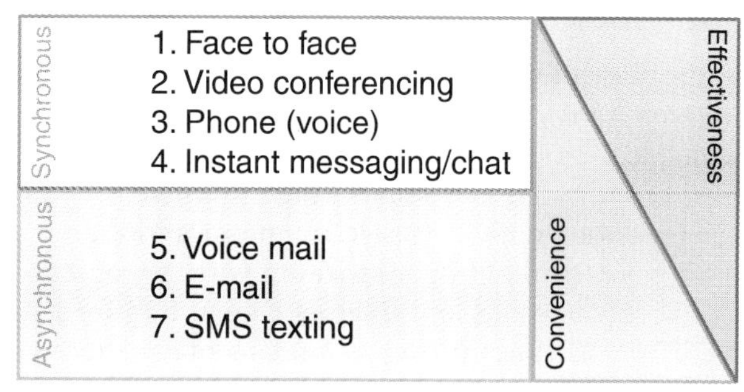

Text Communications. Internet text communications take many forms. Available in both synchronous and asynchronous formats, text communication allows participants to communicate via typed characters. The benefits of communicating electronically by text are that it is cheap and fast and the recipient receives what you type within a fraction of a second. The downside of text communication is that it communicates only words, and lacks the information provided by nonverbal cues—such as voice inflection and facial expression. For this reason, text communication is notorious for creating misunderstandings.

To compensate for the lack of nonverbal cues, a method of conveying underlying sentiment has evolved using emoticons. *Emoticons* (smiley faces) combine keyboard characters to create a sideways facial expression. Table 4.6 shows some examples of commonly used emoticons and what they mean.

TABLE 4.6 • Emoticons

Emoticon	Meaning
:-)	Happy or smiling
:- D	Really happy!
:-}	Embarrassed
:-?	Confused
;-)	Winking

Today, many of the traditionally text-based forms of communication are accessed through graphic interfaces. Many e-mail and instant messaging interfaces provide small animated smiles that can be inserted into messages, like the one next to this paragraph.

The art of creating drawings with typed characters (called ASCII art) was quite popular in the early text-based days of the Internet. Some creative individuals created some amazing effects. Figure 4.24 is one example. There are still hobbyists that tinker with this art form.

There are also programs that can take a photograph and automatically convert it to ASCII art. An Internet search using keywords *ASCII art* yields 1.5 million hits—referenced Web sites—returned.

FIGURE 4.24 • ASCII art

Creating drawings from ASCII characters began as a necessity in the early text-based Internet days, but soon developed into an art form and hobby for some.

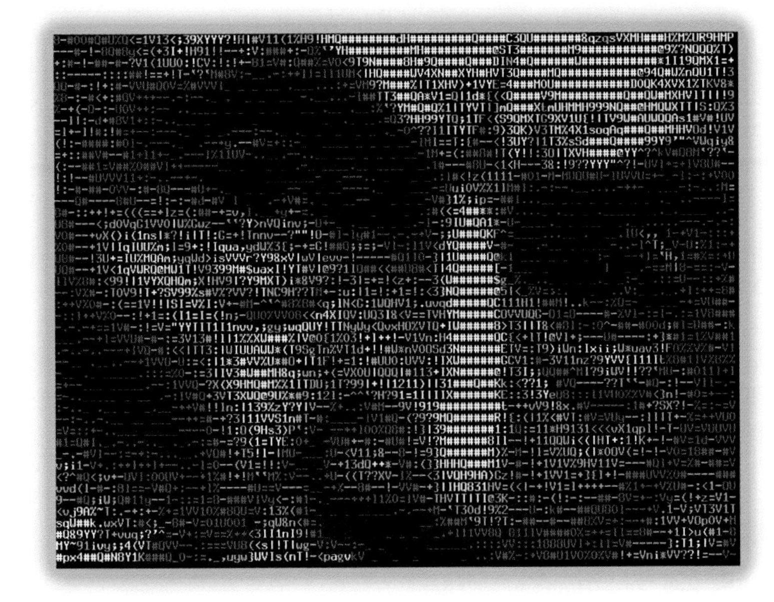

Text communication can also be more time consuming than the spoken word; just how time consuming depends on your typing skills. To minimize the inconvenience of typing, people substitute a number of acronyms for commonly used phrases (see Table 4.7; also search the Web for *Internet acronyms*).

TABLE 4.7 • Internet acronyms

Acronym	Meaning
L8R	Later
FYI	For your information
LOL	Laughing out loud
ROFL	Rolling on floor laughing
TTFN	Ta ta for now

E-mail. **E-mail** (electronic mail) involves the transmission of messages over a computer network to support asynchronous text-based communication. E-mail is still the number one Internet application. Many people are so dependent on e-mail that they must check it on a daily or an hourly basis lest they miss an urgent message. Business travelers access their e-mail using mobile computers, cell phones, and in-flight e-mail services. College students check their e-mail at home, through wireless networks, and on any available public PC on campus. Grandmothers correspond with grandchildren, adults resurrect childhood friendships, and organizations deliver newsletters all using e-mail.

Like most Internet applications, e-mail uses client/server technology. E-mail clients communicate with e-mail servers to send and receive e-mail messages. E-mail messages, like the packets that carry them, include a header and a body. Sometimes e-mail messages can include an attached file (see Figure 4.25). The *e-mail header* contains technical information about the message: destination address, source address, subject, date and time, and other information required by the server. The *e-mail body* is an *ASCII text* message written by the sender to the recipient. The e-mail body may be presented as a text message or as an HTML document—a Web page. An e-mail message body cannot contain *binary data*—data that is encoded for a processor to process. An **e-mail attachment** is typically a binary file, such as an image file, Word document, music file, or spreadsheet, that travels along with an e-mail message but is not part of the e-mail ASCII text message itself. The authors and editors of this book often used e-mail attachments to transfer chapter files to each other.

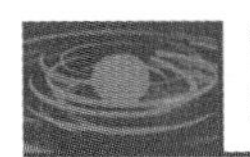

EXPAND YOUR KNOWLEDGE

To learn how to compose and reply to e-mail messages, print e-mail, and more, go to www.course.com/swt2/ch04. Click the link "Expand Your Knowledge", and then complete the lab entitled "E-Mail".

E-mail can be accessed on the Web through free services such as Hotmail, and Yahoo! Mail, or through an e-mail client program such as Microsoft Outlook. Web-based e-mail is convenient for users who like to access their e-mail from any Internet-connected computer. E-mail client software usually provides additional capabilities, such as a calendar, to-do list, and an address book that business users typically find useful. Microsoft Outlook is the most popular e-mail client and is included with Microsoft Office. The e-mail message in Figure 4.25 is viewed in Microsoft Outlook.

FIGURE 4.25 • E-mail components

E-mail messages include a header and body and sometimes an attached file.

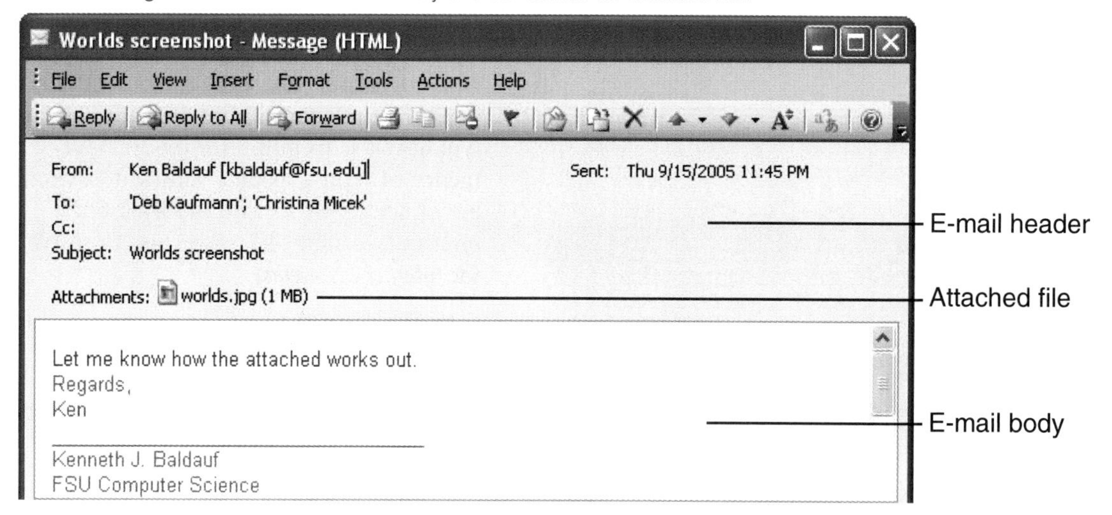

A number of services provide e-mail users with information about their favorite topics. Special-interest groups, called *listservs*, create online communities for discussing topic-related issues via e-mail. E-mail sent to the listserv is forwarded to all members. Another form of broadcast e-mail is subscription-based *newsletters*. For example, you can subscribe to the *New York Times* newsletter (*www.nytimes.com*) and receive the daily news in your e-mail box each morning (see Figure 4.26).

FIGURE 4.26 • The New York Times newsletter

The *New York Times* newsletter subscription service delivers the daily news to your e-mail box.

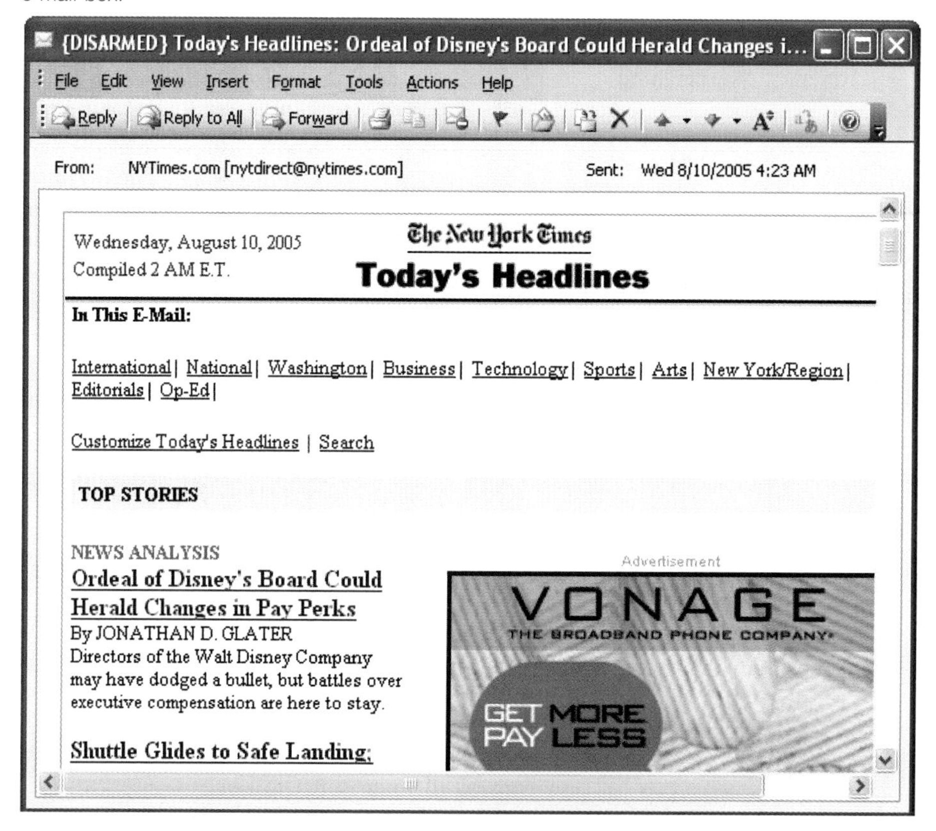

Instant Messaging and Chat. **Instant messaging (IM)** is synchronous one-to-one text-based communication over the Internet. With instant messaging, participants build *buddy lists* or contact lists that allow them to keep track of which people are currently logged on to the Internet. You can send messages to one of your online buddies, which opens up a small dialog box on your buddy's computer and allows the two of you to chat via the keyboard. While chat is typically one to one, and text-based, more advanced forms of chat are emerging, such as group discussions and video imaging (see Figure 4.27). Today's instant messaging software supports the following services:

- Instant messages: Send notes back and forth with a friend who is online
- Chat: Create your own custom chat room to communicate with multiple friends or coworkers
- Nudges, animated winks, and graphics: Helps you get someone's attention or emphasize your point, use a personalized wall paper, and icons
- Web links: Share links to your favorite Web sites
- Images: Look at an image stored on your friend's computer
- Sounds: Play sounds for your friends
- Files: Share files by sending them directly to your friends
- Talk: Use the Internet instead of a phone to verbally speak with friends
- Streaming content: Look at real-time or near-real-time stock quotes and news

FIGURE 4.27 • Instant messaging

ChatBlazer supports IM and much more.

Popular instant messaging services include America Online Instant Messenger (AIM), MSN Messenger, ICQ (another AOL company), and Yahoo! Messenger. All are free services to Internet users and provide client software that you can download for free. The downside is that each of these products is *proprietary*—they do not communicate with each other. So, if you use AOL Instant Messenger, you can chat only with others using AOL Instant Messenger, not people using ICQ,

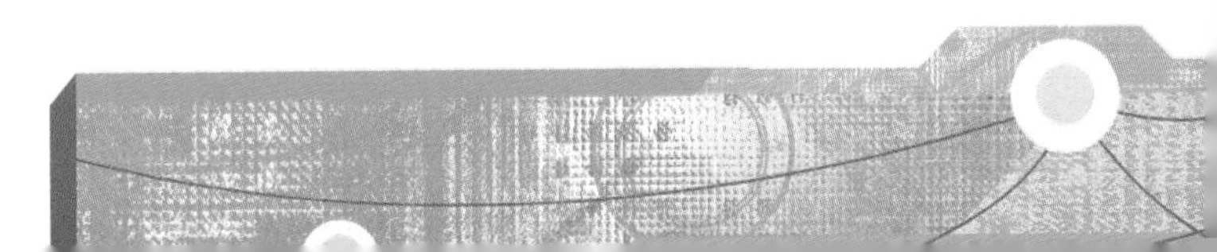

MSN, or Yahoo! Although some software tools have been developed to allow users to combine IM services, they are hampered by the services' unwillingness to use one standard platform.

Instant messaging started out as a fun communication tool for personal use, but it is becoming a serious business tool. Business users are discovering that instant messaging improves productivity and saves money by allowing employees to participate in virtual conferences and collaborate on projects from any location.

Internet **chat** involves synchronous text messaging between two or more participants. Participants log on to a chat server and send each other text messages in real time. As with most forms of group messaging, chat forums are organized by topics. Some services call the various topic-related forums *channels;* others call them *chat rooms*. The most popular public chat utility is called Internet Relay Chat (IRC). IRC has thousands of channels. Any IRC participant can create a channel. At any given time there can be hundreds of thousands of users logged on to IRC. Each channel has a moderator empowered to control the dialog and even kick users out if they get unruly. Other than moderators, chat and many forms of Internet communication are uncensored. If the Internet had a rating system, many chat rooms would be rated for adult use only.

As with most Internet technologies, chat is migrating to the Web. Many public and private chat Web sites are in use today. Two of the most popular are Yahoo! Chat (*http://chat.yahoo.com*) and AOL Chat (*www.aol.com/community/chat/allchats.html*). For a unique chat experience check out *www.worlds.com*. Worlds.com provides a free software download that interacts with the Worlds server to create a virtual world environment (see Figure 4.28). Participants in this chat environment select an *avatar*, a 3D physical representation of themselves, that they navigate through the virtual world, chatting and interacting with other participants and their avatars. There's even a club where avatars can dance the night away.

FIGURE 4.28 • Worlds chat

Chat participants navigate their avatars through a virtual world chatting and interacting with other participants and their avatars.

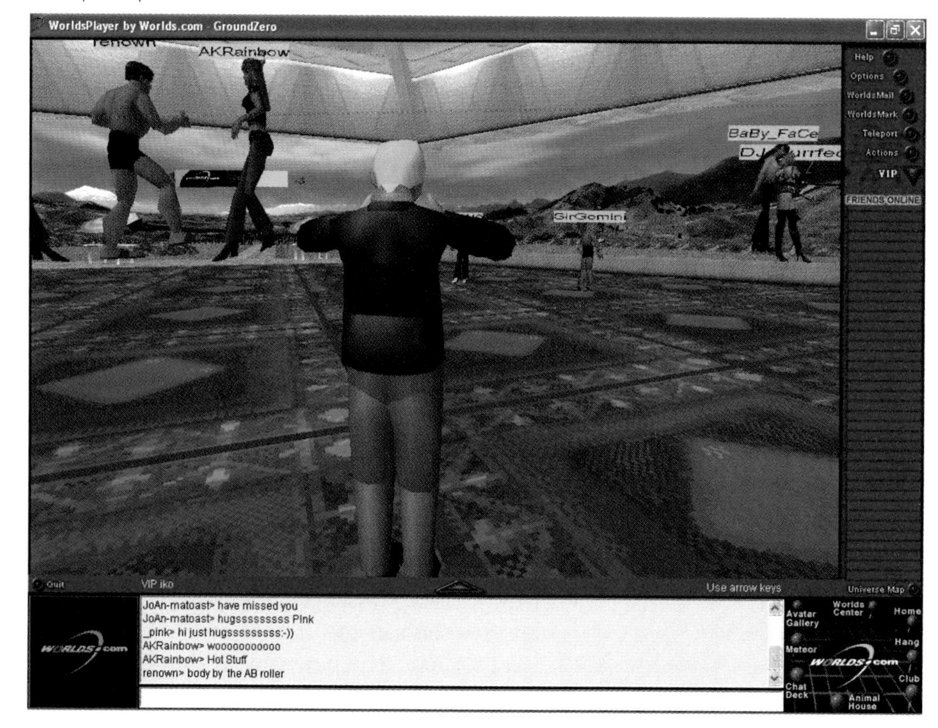

Most chat programs allow you to chat in groups or one on one with a group member. Chat is a popular way of meeting new people on the Internet. However, you should be cautious when making a new acquaintance. The Internet population includes every type of person, including some people who are dangerous. Don't trust that anyone is necessarily who they say they are. Never give your real name, phone number, e-mail address, home address, or other private information to others on the Internet.

SMS Text Messaging. Short Message Service (SMS), more commonly known as **SMS text messaging** or *texting*, is a method of sending short messages, no longer than 160 characters, between cell phones. Using the keypad of the cell phone, users enter a short message and send it to a friend's cell phone. Once a message is sent, it is received by a Short Message Service Center (SMSC), which must then get it to the appropriate mobile device. If the intended recipient is offline or out of range, the SMSC holds the message and delivers it when the recipient returns. Some services allow messages to be sent from the Web to a cell phone and from cell phones to e-mail addresses. Texting is discussed more in the next chapter along with other cell phone services.

Blogs. *Web logs*, more commonly known as **blogs**, are Web sites created to express one or more individual's views on a given topic. Originally blogs took the form of online journals, presenting a person's view on some aspect of life. As privately published blogs became more mainstream, tools were developed to assist anyone in creating a blog. Some blogs allow visitors to post comments. Such blogs can function as discussion boards on a particular topic as interested parties repeatedly check for new comments and leave their own messages in an ongoing dialogue. Bloggers may be famous thinkers and authors such as Noam Chomsky (see Figure 4.29) or everyday folks. Blogs exemplify the power of the Internet to level the playing field and give everyone a voice. Blogs exist on nearly every topic, and express most points of view. You can find many blog listings, such as the one at *www.bigeye.com/blogs.htm* by searching the Web for *blog index*.

FIGURE 4.29 • Blogs

Blogs exist on nearly every topic and present wide-ranging views, such as this one published by a community of people committed to social change.

XML

There are thousands of blogs on the Internet, and it is a challenge to find topics of interest and to keep up with them on a regular basis. A blog distribution system has evolved called *feeds*. Using a technology such as *RSS* (Really Simple Syndication), subscribers can have the daily updates of their favorite blogs delivered to their desktop. RSS uses XML to deliver Web content that changes on a regular basis. It is ideal for keeping up with blogs and news. Programs called aggregators or RSS readers can be downloaded from the Web (see *www.download.com*) and used to subscribe to Web sources. Web sites that support RSS feeds are often marked by a small orange rectangular icon showing XML or SSH, like the one next to this paragraph. Web sites that support SSH feeds include the *New York Times*, Google News, Quotes of the Day, CNET News, *Scientific American*, and thousands of other popular and obscure sources. Table 4.8 lists some popular blogs to which people subscribed when this chapter was written. Check for the latest popular blogs at *www.technorati.com/live/top100.html*.

TABLE 4.8 • Popular blogs

Top 10 blogs
1. BoingBoing (*www.boingboing.com*), A Directory of Wonderful Things: Assorted articles on news, curiosities, technology
2. Instapundit.com (*www.instapundit.com*): A lawyer's blog that focuses on the intersection between advanced technologies and individual liberty
3. Daily Kos (*www.dailykos.com*): Political blog hosted by Markos Moulitsas Zúniga
4. Gizmodo (*www.gizmodo.com*): Cool new technologies are explored.
5. Drew Curtis' Fark.com (*www.fark.com*): A collection of silly news from a variety of sources
6. Engadget (*www.engadget.com*): Cool new technologies are explored.
7. Talking Points Memo by Joshua Micah Marshall: *(http://talkingpointsmemo.com/)* Articles by political columnist Joshua Marshall
8. Davenetics* Pop + Media + Web (*www.davenetics.com*): News and political thoughts by columnist Dave Pell
9. Eschaton (*www.eschaton.com*): Political editorial hosted by Duncan Black, a 32-year-old recovering economist from Philadelphia
10. dooce (*www dooce.com*): An ex-Web designer, now stay-at-home mom discusses her life

FIGURE 4.30 • IP video phone

VoIP provides powerful communications systems within businesses and offers money-saving, feature-packed services for home users.

Voice and Video Communications.

Most telephone companies are implementing Internet technologies in their phone networks. Sprint has announced that it has begun transforming its telephone network so voice calls are transmitted in packets—the same way that data moves over the Internet. The move is designed to lead to a wide range of improved services for consumers, such as online voice mail management. Once voice communications are fully converted to digital, the services offered by telephone systems and the Internet can intermingle. You will be able to access voice mail from your computer and e-mail on your phone. You will no longer think in terms of phones and computers, but rather in terms of communication devices that support voice, text, and video communications in both synchronous and asynchronous modes.

Voice over Internet Protocol (*VoIP*) is the first major step in this direction. *VoIP* is a popular technology that allows phone conversations to travel over the Internet or other data networks. Businesses and residential customers make use of VoIP to merge voice and data networks into one system in order to save money and enjoy additional conveniences made possible by merging Internet and voice communications. Businesses can save large amounts of money by not having to install a phone network. Employees can use special

VoIP handsets that connect to the data network to support standard voice communications and phone conferencing (see Figure 4.30). Some high-end handsets have displays and cameras for video conferencing. VoIP business users can also make use of special software that allows them to check their e-mail and voice mail from any phone or Internet-connected computer.

Video communications are becoming more prevalent as technology advances to support it. Although video phones have been anticipated since the 1950s, the computer industry and high-speed Internet access are finally making the concept a reality. Digital video communications is particularly challenging because of the large quantities of bytes required to produce the video. To work around this limitation, video communications make use of small (3- to 4-inch) video windows, slower rates of frames per second (typically 15 fps or less), and sometimes fewer colors or monochrome images. Such sacrifices in video size and quality have made video communications a reality today. Video communications can take place over a stand-alone video phone, a TV-based video phone, or a PC-based video phone. Many Internet users are investing in inexpensive Web cams to connect to their home PCs for free video chats with friends and family (see Figure 4.31).

FIGURE 4.31 • Apple's iChat

Apple's iChat program provides three-dimensional views for group meetings.

Video conferencing is a technology that combines video and phone call capabilities along with shared data and document access. It is replacing the need for travel in many industries. Through high-speed Internet and private network connections, individuals are able to communicate with associates around the world in "face-to-face" electronic meetings. Several products that support this technology over the Web—called *Web conferencing*—are on the market. Using Web conferencing, groups can see, hear, text chat, present, and share information in a collaborative manner.

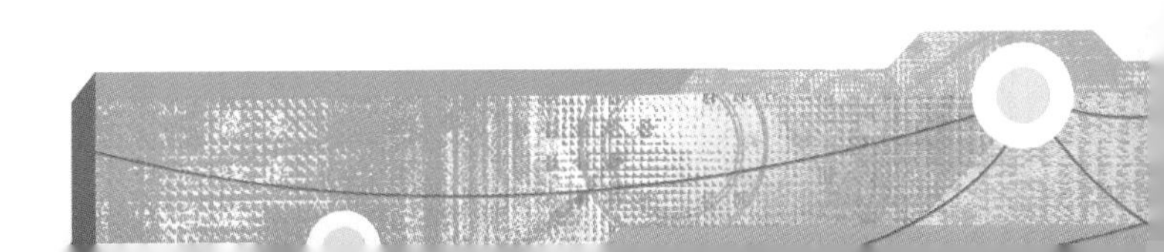

Increasing numbers of professionals are using the power of Internet communications to work from home. **Telecommuting** is the process of working from locations away from the office by using telecommunication technologies. Because most business documents are now digital, and most business communications can be accomplished electronically, there is less need for employees to be physically present in the office. Using technologies that allow employees to access corporate networks from home or anywhere else with Internet access, many of today's employees find it easier to be more productive working away from the office.

There are a number of solutions that provide employees flexibility in choosing work locations. Some employers provide *virtual office telecommuting centers*—office space owned or leased by employers, equipped with computers, and other necessary office equipment, that is shared by telecommuting employees. Some jobs require only partial presence in the office, perhaps only a few hours a week. In such cases, employers can provide shared workspace that employees can use as needed. Companies like FedEx Kinko's provide office facilities and capabilities such as broadband Internet access, conference rooms that can be rented by the hour, video conferencing facilities, phone lines, and PCs for travelers who need access to a connected office environment.

There are good and bad aspects to telecommuting. Working at home offers benefits to employees. Employees that have home obligations may find that telecommuting is the only solution to covering both work and home duties. Telecommuting reduces commute time and expenses, allowing employees to make better use of their time while reducing traffic and pollution. Telecommuters may also save on wardrobe and babysitters. Employers experience benefits as well. One study showed that an average home worker is able to produce seven hours of productive time per day, while in-office workers are productive only six. Telecommuters also require no office space and furniture.

The down side of telecommuting involves the elimination of what is known as "face time." The lack of face-to-face contact with coworkers can lead to a feeling of isolation and being out of the loop. It is difficult to train and keep an eye on new telecommuting employees. It is also more difficult for telecommuters to gain attention and be noticed when not sharing space with coworkers and the boss. For these reasons, telecommuting works better for some careers than others.

News

The Web is a powerful tool for keeping informed about local, state, national, and global news. It allows the public to actively research issues and become more knowledgeable. Traditional news media deliver the news through television, radio, and newspapers. These media provide only the news that they consider of interest to the general public. Items of special or unique interest may be bumped and replaced with more general stories. In contrast, the Web has an abundance of special-interest coverage. It also provides the capacity to drill down into the subject matter. For example, during the war in Iraq, online news services provide news articles in text, audio, and video coverage. Clicking links allows you to *drill down* and find out more about geographic regions by viewing maps; you could link to historical coverage of U.S./Iraqi relations, and you could learn about the battle equipment being deployed. The style of news coverage on the Web is influencing the way news is delivered on television. Major news networks are now dividing the television screen into multiple windows, to more fully present news stories.

Most television news networks are providing video clips of news events from their Web sites through a technology called Webcasting. A *Webcast* takes advantage of streaming video technology and high-speed Internet connections to provide television-style delivery of information over the Web (see Figure 4.32). Webcasting expands the Web's ability to provide detailed news coverage and has the ability to transform Web news services into something resembling interactive TV.

FIGURE 4.32 • Webcasting

Some television news networks are providing video clips of news events from their Web sites through a technology called Webcasting.

Most city newspapers are accessible over the Web, and the major national news agencies, such as Reuters and the Associated Press, also have a Web presence. You can get international news not only from U.S. news sources but also from other countries, providing a wide variety of perspectives on the news.

Some online newspapers only available through subscriptions, such as the *Wall Street Journal Online*, *LexisNexis*, *Consumer Reports*, and *Forrester Research*, are making individual articles available to the general public for small fees.[6] Google and Yahoo! are working with news providers to develop a business model for à la carte news articles. The news publishers provide the search engines access to "deep Web" articles that were previously off limits. An abstract of the article is provided and for a few dollars the user can purchase the entire article.

There are countless special interest news sources that provide industry-specific news and information. Table 4.9 lists a small segment of the wide variety of industry-specific news services available.

Other Web sites provide a wide survey of industry-specific news from the major news sources. For example, Yahoo! provides a categorized menu for industry news at *http://biz.yahoo.com/industry/*.

TABLE 4.9 • News services on the Web

Industry	News Web site
Biotechnology and pharmaceutical	*www.biospace.com/*
Auto	*www.auto.com/index/industry.htm*
Airline, airport, and aviation	*http://news.airwise.com/*
Restaurant	*www.nrn.com*
Audio/visual communications	*www.infocomm.org/*
Hospitality	*www.hotelnewsresource.com/*
Customer relations management	*www.crmdaily.com/*
Oil industry	*www.oilonline.com/news/*
Realty	*www.realtor.org/*
Textile, apparel, footwear	*www.just-style.com/news.asp*

Education and Training

Educational institutions of all types and sizes are using the Web to enhance classroom education or extend it to individuals who are unable to attend.

Primary schools use the Web to inform parents of school schedules and activities. Teachers give elementary school students research exercises in the classroom and at home that utilize Web resources. To make browsing safe for young users, *parental control,* also called *content-filtering* software, such as Net Nanny filter out adult content. By high school, students have integrated the Web into daily study habits. Teachers manage class Web pages that contain information and links for students to use in homework exercises.

Increasing numbers of college-level courses are relying on the Web to enhance learning. Educational support products, such as Blackboard and WebCT, provide an integrated Web environment that includes virtual chat for class members; a discussion group for posting questions and comments; access to the class syllabus and agenda, student grades, and class announcements; and links to class-related material. Some course Web sites go as far as delivering filmed lectures using Webcasting technology. Such environments are used to complement the traditional classroom experience or as the sole method of course delivery. This publisher, for example, has developed content for MyCourse, Blackboard, and WebCT for instructors using this book.

Conducting classes over the Web with no physical class meetings is called *distance education*. Many colleges and universities offer distance education classes, which offer a convenient method for nontraditional students to attend college. Nontraditional students include older students who have job or family obligations that might otherwise prohibit them from attending college. Distance education offers them a way of working through class material on

TEST-DRIVING SURGICAL PROCEDURES

What began as a Web-based educational tool for doctors in dozens of different locations evolved into a marketing tool for major surgery when Webcasting company Slp3d discovered 70 percent of their viewers were consumers. The company broadcasts over the Internet (*www.OR-live.com*) a wide variety of procedures—from organ transplants to cosmetic surgeries—live from the operating room, archiving them online as well. One hospital reported 80 calls the day after a Webcast on spinal disk hernia repair.

Source:
Live surgical Webcasts play to potential patients
By Barnaby J. Feder
The New York Times
http://news.com.com/
Live+surgical+Webcasts+play+to+potential+patients/2100-1038_3-5776140.html?tag=nefd.top
July 6, 2005

a flexible schedule. Some schools are offering entire degree programs online through distance education. At *www.directoryofschools.com* you can find a listing of over 9000 accredited online degree programs.

FIGURE 4.33 • Distance guitar lessons

Using video conferencing technologies, a guitar teacher can teach students around the world.

Kevin Callahan of Seattle (see Figure 4.33) makes use of video conferencing technology to offer guitar lessons online.[7] Seeking to get as intimate an environment as possible, he equipped his Power Mac G5 with Apple's iChat videoconferencing software, iSight camera and microphone package, GarageBand music mixer, and carefully selected lighting equipment. The technology offers his students lessons that are as close as possible to an in-person experience. From his studio in Seattle, Kevin teaches students in Florida, California, Massachusetts, and Spain.

Beyond traditional education, corporations such as NETg and SmartForce offer professional job-skills training over the Web (see Figure 4.34). Job seekers often use these services to acquire specialized business or technical training. Some of the training leads to certification. Certification verifies a person's skill and understanding in a particular area. It has become very important, especially for some technical skill sets, to assure an employer that a job applicant truly has the skills claimed. For instance, if you hire a certified Novell technician, you can be assured that that individual has the knowledge necessary to install and support your Novell computer

FIGURE 4.34 • NETg

Beyond traditional education, corporations such as NETg offer professional skills training over the Web.

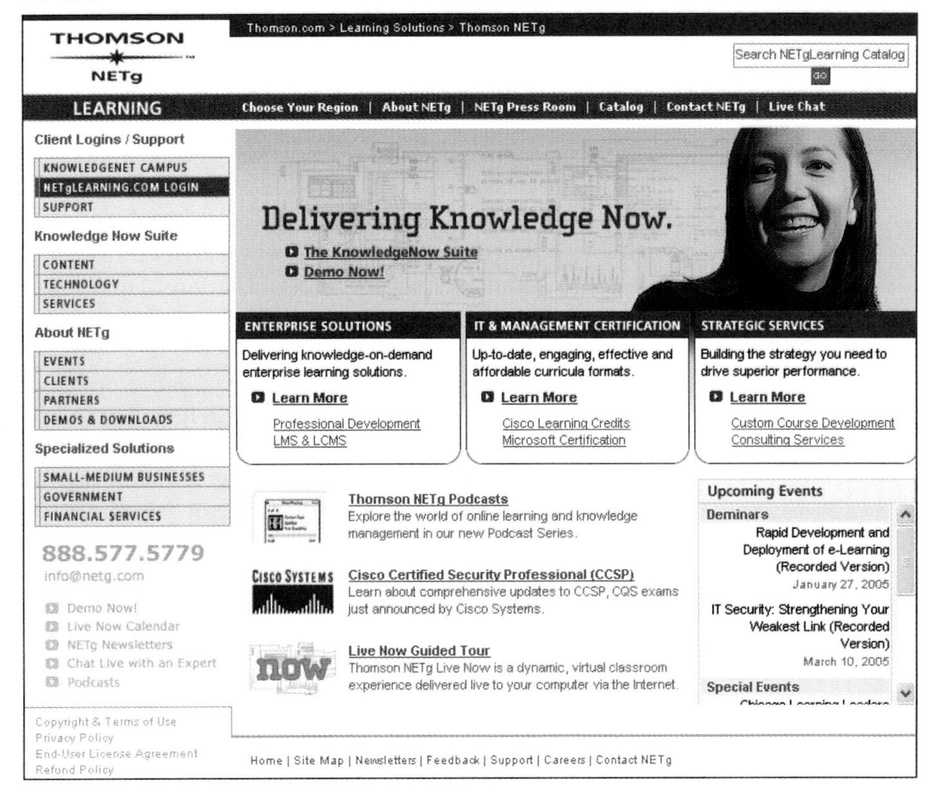

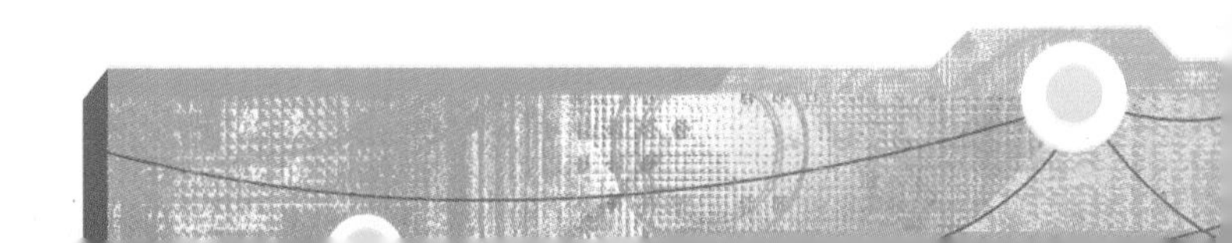

network. Some corporations and organizations contract with NETg or SmartForce to provide on-the-job training for current employees to expand their skills.

E-Commerce

The Web is an integral part of most business information infrastructures. It provides corporate information to employees, shareholders, the media, and the public. It can be and often is used as a primary interface to the corporate database. Its convenience and universal acceptance makes it the ideal platform for communications. It helps support communication among the links in an organization's value chain—the string of companies working together to produce and deliver a product. A value chain typically consists of the suppliers of the raw materials that are used to create a product, the manufacturing unit, transportation and storage providers, marketing, sales, and customer support. The Web acts as the glue that holds all of these units together by providing a central point of access to corporate information. The Web provides a convenient platform for marketing and selling products and is very helpful in collecting marketing data. The Web strengthens communications with customers and improves overall customer satisfaction.

The Web has dramatically changed the manner in which businesses do business and consumers shop for products. Doing business electronically over the Web and private networks is known as *e-commerce*.

Travel

The Web has had a profound effect on the travel industry and the way you plan and prepare for trips. From getting assistance with short trips across town to planning summer long holidays abroad, travelers are turning to the Web to save time and money and overcome much of the risk involved in visiting unknown places.

Many of the success stories of the Web come from the travel industry. Web sites become successful when they uniquely fill a public need. Mapquest.com has certainly performed that function. Offering free street maps for cities around the world, Mapquest assists travelers in finding their way around town and between towns. Provide Mapquest with a departure location and destination, and it shows you the fastest route to take.

Map technologies assist you in all manner of ways. Google Maps has combined with the Boingo wireless Internet service to produce a map service that pinpoints wireless access locations. From *http://wifi-hotspot.wirelessinternetcoverage.com/* you can type a city or zip code to zoom in and pan around the city to find the closest wireless access offered by Boingo.

New mapping software packages from Google and Microsoft overlay road maps onto high-resolution satellite images of earth to produce a zoomable view of the planet (see Figure 4.35). You can view the planet from a mile out in space or zoom in to view a country or state, then zoom in further to view roads and buildings. For example, you can zoom in and actually see swimming pools and chaise lounges along the beach in Miami Beach, Florida. You can zoom in to view natural marvels such as the Grand Canyon and Niagara Falls. Type in an address, and not only view travel directions, but visually fly over the roads to your destination. The satellite photographs used for these services were taken sometime within the last year. Imagine a time in the future when such images are delivered in real time; when you can stand in your back yard and wave to friends on the other side of the country or world and your image is relayed via satellite and Internet!

FIGURE 4.35 • Google Earth

Google Earth allows you to zoom in on satellite imagery to view regions of the planet, natural marvels, city streets, and residential neighborhoods.

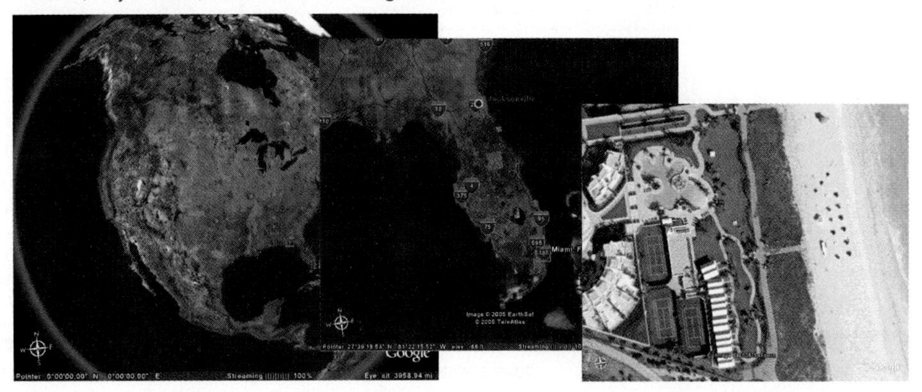

Travel Web sites such as travelocity.com, expedia.com, and priceline.com assist travelers in finding the best deals on flights, hotels, car rentals, vacation packages, and cruises. Provided with dates and locations of travel, most travel Web sites display the available flights and prices from which you can base your choice. Priceline.com takes a slightly different approach. It allows shoppers to name their own price, and then works to find an airline that can meet that price. Once flights have been reserved, travelers can use these Web sites to book hotels and a rental car often at discounted prices.

There are many special-purpose travel Web sites that assist individuals with particular needs. Some categories listed at Yahoo! include backpacking, budget travel, disabilities, ecotourism, hitch-hiking, trail travel, traveling alone, traveling with pets, and vegetarian.

Employment and Careers

Web sites provide useful tools for people seeking employment and for companies seeking employees. Web sites such as careerbuilder.com and monster.com provide resources for choosing a career and finding a job. Most colleges have career and job placement services that make use of the Web to connect graduating students with employers. Consider the Web's role in the job-hunting strategy provided in Table 4.10.

While searching for a way to make a living, don't be taken in by the many get-rich-quick scams that proliferate on the Web and through e-mail. Although there are many "work at home" offers on the Web, only a few are legitimate. You can get information about scams posing as business opportunities in the United States from the Federal Trade Commission (*www.ftc.gov/bcp/conline/pubs/online/netbizop.htm*). In addition, the Internet Fraud Complaint Center, sponsored by the Federal Bureau of Investigation (FBI) and the National White Collar Crime Center (NW3C), provides information on how to spot Internet fraud and an online form for reporting Internet fraud (*www1.ifccfbi.gov*).

TABLE 4.10 • Job hunting strategies: Using the Web to find a career and job

Step	Tips
Select a career	Use online references such as those at *www.jobweb.com/resources* to discover your personal strengths and weaknesses and map them to your ideal job.
Discover who the players are in your chosen career	Search the Web on a given career and industry title and see what companies are represented. Discover online trade journals, and learn as much as you can.
Learn about the companies that interest you	Where better to start than the company's Web site? Many companies list career opportunities and provide you with information on how to apply.
Network with others in the field	There are many online industry-specific discussion groups and forums. Seek them out to make valuable contacts. You can start at *www.hotjobs.com/htdocs/client/splash/communities*.
View job listings at general employment Web sites	Web sites such as *www.hotjobs.com*, *www.joboptions.com*, *www.monster.com*, and *www.careermosaic.com* have large databases of job openings where you can search by profession or keywords. A complete list of the best of these sites can be found at *www.quintcareers.com*.
View job listings at industry-specific employment Web sites	There are hundreds of specialized job Web sites, from employment recruiters of all types to specialized job databank sites that focus on a specific industry.
Create an impressive Web site to represent yourself	Consider purchasing your own domain name—for example, *www.janelle_johnson.com*. Include an attractive welcome page with links to your resume and other details regarding your experience and skills. Employers want hard-copy resumes to be brief. At the bottom of your resume, you can add "Please visit *www.janelle_johnson.com* for more information." Make sure that your Web site is professional in appearance and content.

Multimedia and Entertainment

Faster Internet connections and advances in streaming technology have brought a wide range of media and gaming applications to the Web. The Web has had a dramatic impact on the music and motion picture industries, causing unprecedented changes in marketing and distribution approaches. Many in the software industry are betting that online gaming will increase in its appeal across gender and age markets. As home entertainment equipment begins to merge with home computing equipment, the Web is anticipating the call to deliver entertainment.

Music. *Internet radio* is similar to local AM and FM radio except that it is digitally delivered to your computer over the Internet, and there are a lot more choices of stations. For example, msn.com provides access to hundreds of radio stations in over 35 musical genre categories. All that is required to listen to Internet radio is a media player such as iTunes, the Real media player, or Microsoft Windows Media Player. Some stations charge a subscription fee, but most do not.

Compressed music formats such as MP3 have made music swapping over the Internet a convenient and popular activity. File-sharing software such as KaZaA provides a means by which some music fans copy and distribute music, often without consideration to copyright law. The result is a popular music distribution system that is largely illegal, impossible to control, and cuts deeply into the recording industry's profits. In addition, it is not always safe to

INTERNET SITE EQUALIZES ARTIST OPPORTUNITIES

The subscriber-based Website DeviantArt.com has developed the perfect opportunity for all those whose favorite musician stimulates their creativity. Once the purview of big design firms, now amateurs and professionals compete on equal footing to do cover art for more than 35 headlining musicians from Alanis Morissette to Jurassic 5. If their design is chosen, the artist receives payment, plus bragging rights. The site anticipates expanding the logo competition to include categories from cereal boxes to shoes.

Source:
Artists Ready to Rock the Logos
By David Cohn
Wired News
http://www.wirednews.com/news/business/0,1367,68115,00.html?tw=wn_tophead_5July 11, 2005

swap files with strangers. One study discovered that 6 percent of all the music files downloaded from KaZaA are actually viruses renamed to look like MP3 files. Music industry giants have pulled together to win back customers by offering legal and safe alternatives to electronic music distribution that provide services and perks not offered by file-sharing networks—and at a reasonable price.

Several online music services are available. Apple's iTunes was one of the first online music services to find success. The iTunes Music Store (see Figure 4.36) sold three million songs online to Apple computer users, at 99 cents each, in its first month. Soon after, iTunes extended its service to PC users.

FIGURE 4.36 • Apple's iTunes music service

iTunes was the first online music service to meet with success.

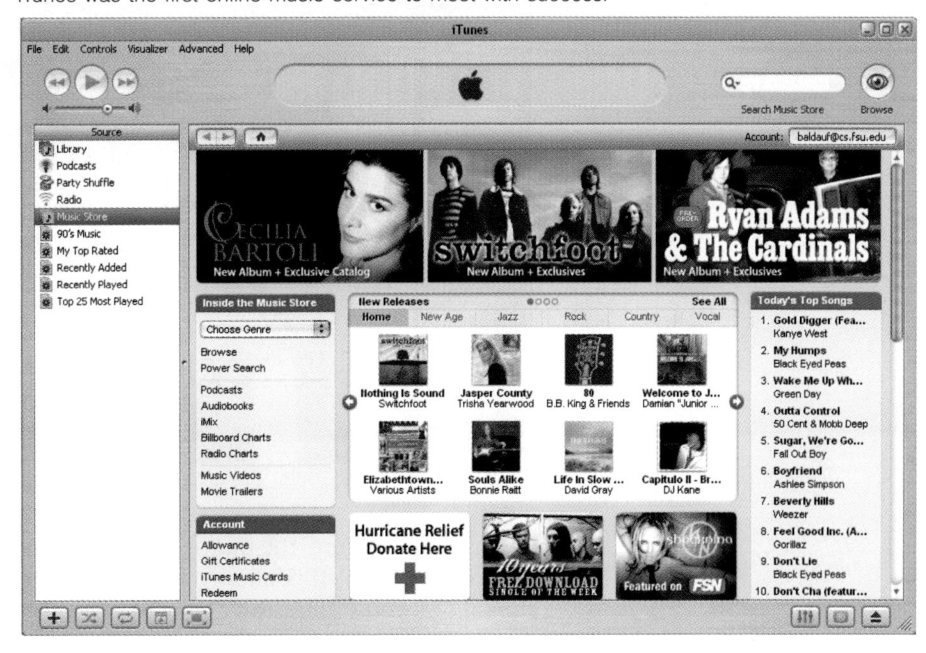

Other popular music services include MusicMatch, Napster, Rhapsody, AOL's MusicNet, and FullAudio's Music Now. These music services offer access to the catalog of the big five labels—Universal Music, Sony Music, Warner Music, BMG, and EMI, which together account for 80 percent of recorded music and offer from 300,000 songs to over a million.

Movies. The movie industry is also making the move to Internet distribution. The large size of video files has so far held video back from being as popular to swap on the Internet as music. However, with the increasing number of broadband connections, movie swapping is becoming increasingly popular. Most big movies are pirated and available on file-sharing networks within weeks or days of release in theaters. They are either stolen by an inside source or someone sneaks a digital camcorder into a theater and records the film off the screen. The Motion Picture Industry of America is so concerned about movie pirating that they have deployed metal detectors and night-vision goggles to some movie theaters.

Like the recording industry, the motion picture industry is undertaking both defensive and offensive tactics to thwart the trend toward illegal file sharing. One tactic is to develop a legitimate Internet distribution system. Movielink.com allows users to rent or purchase movies over the Internet (see Figure 4.37). You browse the movielink.com Web site for movies, just as you would look through a movie rental store. Movies rent or sell at competitive prices and are downloaded to your PC using the Movielink Manager software and are viewed on your PC or a connected television set. Needless to say, unless you have a fast Internet connection and a way to view the movies you download on a large screen, this service probably won't satisfy you. However, as digital convergence leads to Internet-integrated home entertainment centers—where home entertainment equipment merges with home computing and networking equipment—Movielink will be poised to provide what will then be a valuable service.

FIGURE 4.37 • Movielink.com

You can download movies from the Internet at Movielink.com to watch on your PC or connected television.

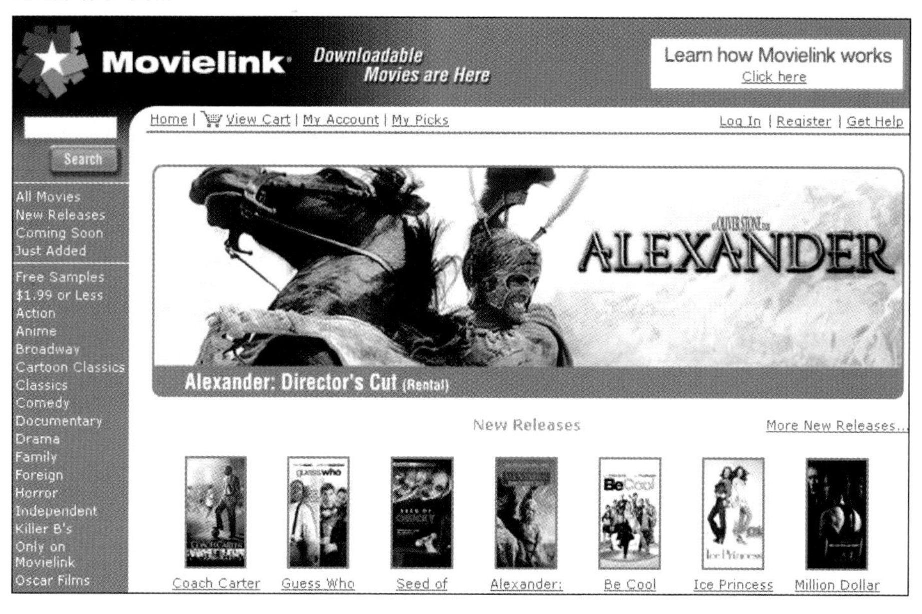

In the meantime, the online DVD rental industry is booming. Subscription services such as Netflix allow members to rent DVDs by mail. A typical subscription allows you to rent three movies at a time. You select the movies from an online catalogue of over 40,000 titles, and the movies are delivered to your door the following day. You can keep the movies as long as you like, and there are no late return fees. Return your movies using the provided prepaid return envelopes and select three more movies. Typical subscription fees range from $10 to $20 per month depending on how many movies you wish to rent at a time.

Games. The Web offers a multitude of games for individuals of all ages. From solitaire to massive multiplayer role-playing games, there is a wide variety of offerings to suit every taste. Of course, the Web provides a medium for downloading single-player games to your desktop, notebook, handheld, or cell phone device (check *www.download.com*), but the power of the Web is most apparent with multiplayer games.

Multiplayer games allow you to interact with other users online (see Figure 4.38). There are a variety of types and platforms. Multiplayer games support from two to thousands of players at a time. The game can be as simple as an online game of checkers. At the time of this writing, of the 123,453 people playing games at *http://zone.msn.com/*, 506 of them are playing checkers—253 simultaneous games. Meanwhile at the Ferion game network (another online gaming site at *www.ferion.com*) 78 players of the 282,000 members are creating empires in space and exploring the virtual galaxy.

Multiplayer online games can be categorized into the following genres:

- Action: Fast-paced games requiring accuracy and quick reflexes
- Board: Games involving play on a virtual game board
- Card: These games involve the use of a virtual deck of cards

FIGURE 4.38 • Ferion

Multiplayer games connect Internet users and allow them to interact in a virtual world

- Flight simulation: Games that involve taking on the role of a pilot in a WWI biplane or a futuristic starship
- Multiuser dimension or multiuser dungeon (MUD): Text-based games that make up for their lack of graphics with diverse and immersive game play
- Role-playing games (RPG): Games in which you take on the persona of a game character
- Sims (simulations): Games in which you create your own character that lives in a simulated environment
- Sports: Games involving sports
- Strategy: Games that require planning, tactics, and diplomacy
- Trivia/puzzle: Games that require a good memory or problem-solving skills

Although most multiplayer games are free, some of the best have fees associated with them. You may need to purchase software, or pay monthly, yearly, or per-play subscription fees.

Game consoles such as Xbox and PlayStation2 provide multiplayer options for online gaming over the Internet. Subscribers can play with or against others that are logged on in 3D virtual environments. They can even talk to each other using a microphone headset. Microsoft's Xbox Live provides features that allow users to keep track of their buddies online, and match up with other players who are of the same skill level.

Although illegal in the United States, online gambling is a billion-dollar global industry.[8] Online casinos, poker rooms, sports and racing bookies, and bingo halls are found in abundance online. Poker is particularly popular since it pits player against player and allows the more skillful to win. Online gaming companies PartyGaming and Empire Online are so large that they sell shares on the London Stock Exchange. Some worry that the convenience of online gambling increases the possibility of addiction and is a risk to the public good.

Information

The Web has become the most popular medium for distributing and accessing information. A recent study by comScore Networks found that "Consumers are increasingly reliant upon online resources to ease and inform major life events."[9] Such events include planning a wedding, buying a home, changing jobs, and having a baby. The Web is quickly becoming the first place you look when faced with a challenge or question.

Besides supporting practical decision making, the Web provides a wealth of information for curious minds. The connected world of the Web has, more than anything else, provided conveniently accessible answers to our questions. Life is full of wonder. Throughout the course of the day, you might wonder about a number of experiences or observations. In days past, you might have gone home and searched for explanations in the family encyclopedia or made a trip to the library. More likely than not, however, you would have found that researching the question wasn't worth the bother, or that the question would simply pass from your mind. But today, you're connected to the world's largest encyclopedia, with many people cracking it open every day. When you go online to check e-mail, you can take a moment to quickly run a search on something you were curious about during the day. As the Web moves to cell phones and other handheld devices, you will be able to look up information at the moment a thought or question strikes. This form of research is known as *curiosity-driven research.* Curiosity-driven research is responsible for most of the world's great inventions.

In another form of research, *assigned research,* you are given a topic to explore for the purpose of education. This type of research is common in college, and is often left open-ended. You are left to choose a topic, typically in a given subject area. The Web supports both forms of research at a number of levels. The following sections provide an overview of research techniques for the Internet.

FIGURE 4.39 • The Yahoo! subject directory

Online subject directories are a great place to explore research topics.

Yahoo! Directory

Arts & Humanities
Photography, History, Literature...

Business & Economy
B2B, Finance, Shopping, Jobs...

Computers & Internet
Software, Web, Blogs, Games...

Education
Colleges, K-12, Distance Learning...

Entertainment
Movies, TV Shows, Music, Humor...

Government
Elections, Military, Law, Taxes...

Health
Diseases, Drugs, Fitness, Nutrition...

News & Media
Newspapers, Radio, Weather...

Recreation & Sports
Sports, Travel, Autos, Outdoors...

Reference
Phone Numbers, Dictionaries, Quotes...

Regional
Countries, Regions, U.S. States...

Science
Animals, Astronomy, Earth Science...

Social Science
Languages, Archaeology, Psychology...

Society & Culture
Sexuality, Religion, Food & Drink...

Selecting and Refining a Research Topic. Selecting a research topic is generally done through many stages of refinement. For example, you can begin by asking yourself the question "What in the universe would I like to explore?" Recall the subject directories found at Web portals such as Yahoo! that were noted earlier in the chapter. These Web portals are good places to begin exploring. For example, the subject directory at *www.yahoo.com* contains the top-level topics shown in Figure 4.39.

If you were to select the Society & Culture listings, you would discover a list of at least 25 subcategories. From this list you might choose Issues and Causes to view yet another list of subcategories in greater detail. With each choice you make in the subcategories, you further refine your topic. At certain points in your research you may choose to back up and take a different path.

Although subject directories may not provide deep information on specific topics, the type required for college-level research, they do provide an overview of a wide breadth of information. Such an overview is useful in exploring and discovering new areas of interest and learning the basics on most topics.

Sources of Information. In addition to the information available through subject directories, there are a number of other sources of information. Web search engines can provide additional Web sites on a chosen topic. As you turn to the Web search engines, you should do so with increased caution. With Web content, things aren't always what they seem. A Web page posted by the Human Cloning Council (HCC) that proclaims the wonderful benefits of human cloning may be a cover for a company that will benefit should human cloning be accepted. In the arena of Web research, you must follow up and check sources for validity.

In judging information provided on the Web, always consider the source. Anonymous postings cannot be trusted. You must identify the provider of the information. If that person or organization is unfamiliar to you, search the Web for information on the provider. Providers of valid and trustworthy information on the Web typically go out of their way to include references to well-known and highly regarded authorities. On the Web, it is better to consider information inaccurate until proven valid rather than the other way around.

Wikipedia: The Communal Encyclopedia

Wikipedia.com has become one of the most popular sites on the Web for accessing information on any topic. Wikipedia is a unique online encyclopedia created and maintained by the general Web community. The English version (*http://en.wikipedia.org*) recently topped 500,000 full-length articles authored and updated by visitors to the site. Every day hundreds of thousands of visitors from around the world make tens of thousands of edits and create thousands of new articles.This collaborative publication is self-policing. It is assumed that experts in any field will correct and improve articles within their domain of knowledge so that each article posted represents the latest and most reliable information on the topic.

The Wiki foundation that owns Wikipedia operates several other free content projects including Wiktionary, Wikibooks, Wikiquote, Wikispecies, and Wikinews. (The name Wiki is from the Hawaiian *wiki wiki*, for *quick*.) The Wiki projects illustrate the potential of the Internet to combine the knowledge of the community into a powerful free resource.

Questions

1. How do you think the reliability of Wikipedia compares to that of a published encyclopedia such as Britannica (*www.britannica.com*)? Which would you trust more?
2. What are the positive and negative aspects of the development processes of both Wikipedia and Britannica?
3. How is it possible to finance a free Web site such as Wikipedia? Would Wikipedia be as popular as it is if it had sponsors and banner ads?

Sources

1. *Lemon, S. "What is:WIKI," ITworld.com, April 1, 2005, www.itworld.com.*
2. *Wikipedia.org Web site, http://en.wikipedia.org, accessed on June 10, 2005.*
3. *Wikipedia stats Web page, http://en.wikipedia.org/wiki/Special:Statistics, accessed on June 10, 2005.*

The most reliable sources of information can be found at your college library's Web site. The books, reference materials, journals, and other periodicals that are housed in your college or local library have undergone quality control evaluation to earn the right to sit on those shelves. Books and periodicals considered fundamental to any given field are typically stocked in the library. Library resources are professionally analyzed and categorized in a logical manner that is easy to navigate. Best of all, the most knowledgeable of researchers, librarians, are available to assist you with your project. Most libraries have online card catalogues that allow you to search for books from the comfort of your home.

Besides online card catalogs, libraries typically provide links to public and sometimes private research databases on the Web. Online research databases allow visitors to search for information in thousands of journal, magazine, and newspaper articles. Information database services are valuable because they offer the best in quality and convenience. They provide full-text articles from reputable sources conveniently over the Web. College and public libraries typically subscribe to many databases to support research. One of the most popular private databases is LexisNexis Academic Universe. LexisNexis provides access to full-text documents from over 5900 news, business, legal, medical, and reference

publications, and you can access the information through a standard keyword search engine. (See Figure 4.40.) The sources from which LexisNexis draws include:

- National and regional newspapers, wire services, broadcast transcripts, international news sources, and non-English language sources
- U.S. federal and state case law, codes, regulations, legal news, law reviews, and international legal information
- *Shepard's Citations* for all U.S. Supreme Court cases back to 1789
- Business news journals, company financial information, SEC filings and reports, and industry and market news

FIGURE 4.40 • LexisNexis

LexisNexis provides thousands of journal articles and other references to support research for those who subscribe.

THE FUTURE INTERNET

The Internet and Web are continuously evolving. Their basis in open standards has provided a fertile environment for innovative development. Anyone is free to design and implement new applications for use on the Internet and Web. This chapter has provided many examples of the way Internet and Web technologies are being used today. This section provides a glimpse at how they might be used in the future.

Internet2 and Beyond

There are significant efforts underway to expand the capabilities of the Internet to support increasing numbers of users and applications at higher speeds. New technologies are being explored by scientists and network engineers to develop the next generation Internet. *Internet2* is a research and development consortium led by over 200 U.S. universities and supported by partnerships with industry and government to develop and deploy advanced network applications and technologies for tomorrow's Internet. The stated goals of the organization are to:

- Create a leading-edge network capability for the national research community
- Enable revolutionary Internet applications
- Ensure the rapid transfer of new network services and applications to the broader Internet community

Internet2 is not a new Internet, but rather a group developing new ways to improve the management and performance of the existing Internet. It includes dozens of research groups developing a variety of network applications in health sciences, arts and humanities, science and engineering, and education.

An offshoot of Internet2 that some call Internet3, but is officially named the *National LambdaRail* (NLR), is a cross-country, high-speed, fiber-optic network dedicated to research in high-speed networking applications (Figure 4.41). The NLR provides a unique national networking infrastructure to foster the advancement of networking research and next generation network-based applications in science, engineering, and medicine. This new high-speed fiber-optic network will support the ever-increasing need of scientists to gather, transfer, and analyze massive amounts of scientific data.

FIGURE 4.41 • National LambdaRail (NLR)

NLR will, for the first time, provide the research community with direct control over a nationwide optical fiber infrastructure.

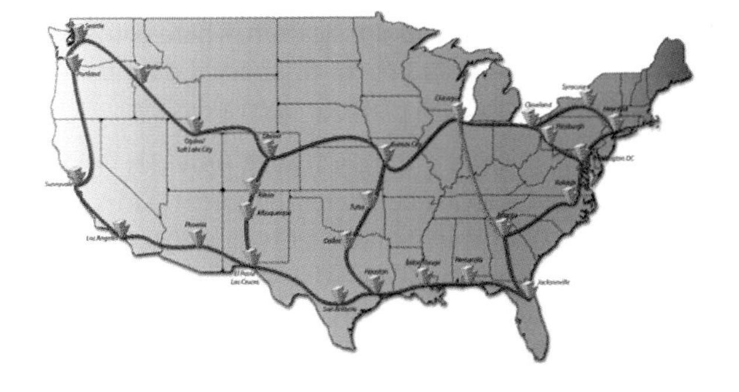

High-Speed Internet Applications

The new and exciting applications being explored on these super-speed networks fall under four categories:

- Interactive collaboration
- Real-time access to remote resources
- Large-scale, multisite computation and data mining
- Shared virtual reality

Some specific examples of these research areas are provided below.

Interactive Collaboration. Through new research in high-speed video and audio networking, individuals are able to collaborate from various locations in a common virtual environment. For example, technology named the Access Grid (*www.accessgrid.org*), uses large-format multimedia displays to create remote visualization environments. The Access Grid provides "designed spaces" that support the high-end audio/video technology needed to provide a compelling virtual collaborative experience that can be used for large-scale distributed meetings, seminars, lectures, tutorials, and training. With Internet2, interactive collaboration is the next best thing to being there.

Internet 2 and interactive collaboration is finding a home in the arts as well. Hillary Herndon, master violist at the New World Symphony, is able to coach viola student, Anna Simeone, in the Conservatory of Music in Pisa, Italy, halfway around the world using video/audio connections over Internet2. The Manhattan School of Music offers group instrumental lessons and custom telementoring sessions with famous musicians on the faculty through the high-fidelity, broadcast-quality streaming audio and video available over Internet2.

Real-time Access to Remote Resources. Internet2 is providing researchers, students, and audiences with access to remote equipment and environments. For example, the Gemini observatories (*www.gemini.edu*) are the result of a multinational project to build twin 8.1 meter astronomical telescopes in Hawaii and Chile (Figure 4.42). The telescopes can be accessed and controlled in real time by astronomers around the world over Internet2. The high-performance connection also allows scientists to collaborate via videoconferencing, and will enable the observatories to share more of their findings with the public through techniques such as virtual observatory tours and live video to museums, planetariums, and classrooms worldwide.

FIGURE 4.42 • Gemini Observatory

Scientists can control the astronomical telescopes at Gemini observatories in Hawaii and Chile from any location on Internet2.

Visitors at Connecticut's Mystic Aquarium immerse themselves in an underwater world 3000 miles away by remotely controlling underwater cameras in California. Using interactive consoles at the Mystic Aquarium's Immersion Institute, visitors control three video cameras on an underwater submersible in Monterey Bay, the largest U.S. marine sanctuary. The live video is encoded and sent at an average rate of 6 megabits per second (Mbps) to the University of California, Santa Cruz, where it travels across Internet2 high-performance networks to the University of Connecticut and on to the Mystic Aquarium.

Large-scale, Multisite Computation and Data Mining. New high-speed network technologies are providing powerful computing solutions for researchers who deal with enormous amounts of data. The National Scalable Cluster Project (NSCP, *www.ncdm.uic.edu*) has pioneered an application which joins computers at various points on Internet2 to analyze data in huge databases, faster than has ever before been possible.

Shared Virtual Reality. Shared virtual reality provides immersive virtual environments over a network to be shared by participants in different locales. For example, Virtual Harlem is a virtual reality environment originally developed in collaboration with University of Missouri-Columbia to supplement African-American literature courses at Central Missouri State University. Students are able to step through a virtual "portal" to the 1925–1935 New York Harlem Renaissance to navigate the city streets, interact with key figures, and listen to music written and popularized during the era.

Shared virtual reality is a natural extension to today's massive multiplayer interactive games. Synthetic worlds may be an inevitable part of our future as people live increasing amounts of their lives online. About 2 percent of the Internet-connected population age 14–28 spend more time in the synthetic Internet world than in the real world. This number is doubling every 18 months. It is only natural that efforts would be made to make our online experience more like the real world.

As these examples illustrate, a faster global Internet will provide opportunities to travel without leaving home. More such examples can be found at *http://apps.internet2.edu*. The future Internet will forge new professional and personal relationships to increase the interaction and collaboration between people from all walks of life with varying world views and experiences. The effects of such a ramp up of communications and information sharing is bound to have profound effects on our civilization.

ACTION PLAN

Remember poor Josh Greene who is disenchanted with the Web and feeling left out among his group of friends?

1. What enriching online activities can Josh use to improve his quality of life?

Josh can communicate with both his local and out-of-town friends online using e-mail, chat, and instant messaging. He can even meet other college students at *www.collegeclub.com* or *www.student.com*. Josh can pick up a new cell phone and service that supports text messaging so that he can text with his friends. He can also involve himself in music and video clubs and multiplayer gaming. Josh can keep up with the latest news from online news services, and use technology news services to find out what's new on the Internet and Web. Josh might decide to broaden his knowledge or skills in a given area by using online training. He might even work towards certification in order to gain an edge when it comes time to look for a job. In fact, in today's competitive job market, it is never too soon to start looking. Josh can start scanning the job sites to help decide on a career and see what is out there.

2. What online activities are legal and safe, and which are not?

When accessing music and movies on the Web, Josh should know that ignoring copyrights can have serious consequences. Josh should also use caution when meeting new people online; not everyone is who they appear to be. Josh shouldn't travel solo to meet someone in person that he has met only online. When using material from the Web in his research paper, Josh should be sure to cite his sources, and never submit someone else's work as his own.

3. How should Josh conduct research on his computer?

Josh could start looking for a topic at an online subject directory. Once he has selected a topic, he can research it using keyword searches at his school library's Web site, in the card catalogue, and in online databases. He can refine his topic and search as he learns more about it. Finally, he can use Web search engines to access public knowledge on the topic, being careful to consider the source.

Summary

LEARNING OBJECTIVE 1

Describe how the Internet developed and how hardware, protocols, and software work together to create the Internet.

The Internet is the largest publicly owned network of networks. It was established in 1969 as the ARPANET under a U.S. government project named ARPA with a goal of connecting the computer networks of four universities. In 1990, it was turned over to the public to be managed by the Internet Society. The birth of the Web in 1993 led to an explosion in the Internet growth rate to the point that it connects millions of computers today.

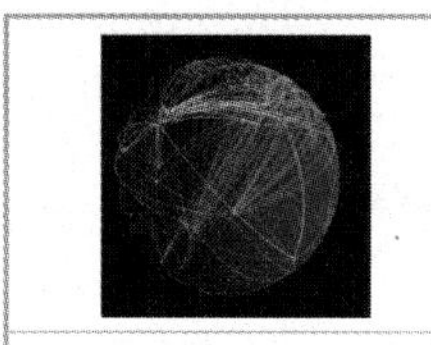

Figure 4.2—p. 177

The Internet combines hardware, protocols, and software to serve its users. Internet hardware consists of network backbones provided by the major telecommunication networks, routers that route packets of data to their destinations, and the computers that request and provide information and services. Internet service providers provide connections to points of presence on the Internet backbone for personal and professional use. The Internet is a packet-switching network that makes use of rules, called protocols, to pass packets of data between computers. Users connect to the Internet either through private networks at work or school that are connected to the Internet, or at home through high-speed cable modem or DSL connections, or slower dial-up connections.

The main protocols of the Internet are TCP/IP. The Internet Protocol (IP) defines the format and addressing scheme used for the packets. Transaction Control Protocol (TCP) enables two hosts to establish a connection and exchange streams of data. The Internet uses IP addresses to identify hosts. Domain names are English representations of IP addresses. The Internet makes use of client/server software to supply users with the information and services they request. Server computers use port numbers to segregate the various types of service requests. Peer-to-peer networks do not use a central server, but rather allow Internet users to create connections directly between their computers.

LEARNING OBJECTIVE 2

Explain the underlying structure of the Web and the technologies that support it.

The Web makes use of the Internet's client/server technology to provide a medium for users to conveniently publish, view, and find information on the Internet. It uses Hypertext Transfer Protocol (HTTP) to allow Web browsers (clients) to access Hypertext Markup Language (HTML) documents, called Web pages, stored on Web servers. HTML documents use hyperlinks to connect to other HTML documents. HTML tags specify the document format and other commands from within HTML documents. Uniform Resource Locators (URLs) are Web page addresses that allow you to access a specific Web page on a server. XML and XHTML are new Web standards that allow for the storage and manipulation of structured data on the Web.

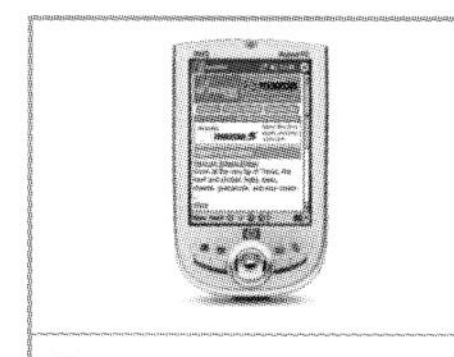

Figure 4.11—p. 188

Web plug-ins work with Web browsers to offer extended services—typically the ability to view audio, animations, or video. Java, JavaScript, and ActiveX are programming languages that allow programs to be included in Web pages, which creates the ability for users to interact with Web content.

Sometimes Web servers store a cookie, or a small text file, on your computer so that it can recognize you upon your next visit to a Web page. Cookies help Web pages present customized information to users. Web pages that are custom created on-the-fly are called *dynamic* Web pages. Others that are the same for every visitor are called *static* Web pages.

Portals are Web pages that provide an entry point to the Web. They typically include headline news and information, along with a search engine and subject directory to help you find information on the Web.

Web authoring software allows you to create HTML documents in a WYSIWYG editor that works just like a word processor. With Web authoring software, you can generate attractive Web pages without having to know HTML.

LEARNING OBJECTIVE 3

Define the categories of information and services that the Internet and Web provide and the forms of communication they support.

The Web provides a wide range of helpful applications. Search engines, subject directories, and portals assist in organizing and finding information on the Web. These applications provide communication and collaboration platforms that allow users to keep in touch, work together, and meet others with similar interests.

Internet and Web communications provide synchronous or asynchronous forms of communication that can be text based, voice based, or video based. Forms of text communications include e-mail, discussion boards, chat, instant messaging, and Short Message Service (SMS), more commonly called texting. Because of the lack of nonverbal cues, e-mail is notorious for creating misunderstanding. Chat involves synchronous text messaging between two or more participants. Instant messaging is the most recent form of chat and allows users to build buddy lists to keep track of and communicate with friends online. SMS text messaging, or texting, is a technology that allows cell phone users to send short text messages to each other. Web logs, more commonly known as blogs, are Web sites created to express one or more individual's views on a given topic. Digital voice communication is creating a merger between Internet and telephone communications. This is particularly apparent in new cell phone technology, and the emergence of video phone services.

The Internet and Web provide local, state, national, and international news on a level equal to if not better than traditional media. Educators and trainers use the Web to support and extend traditional learning environments or for distance education. Businesses rely on the Web to act as a central location for corporate information. The Web is one of the first places people turn to when planning a trip or looking for employment, and it plays an integral role in most job-hunting strategies. The Web supplies all forms of entertainment including music, video, and gaming. It is also used as the primary tool for research, by providing access to online library catalogs, databases, and public opinion.

LEARNING OBJECTIVE 4

Explain what Internet2 is and the types of applications it will provide in the future.

Internet2 is a research and development consortium led by over 200 U.S. universities and supported by partnerships with industry and government to develop and deploy advanced network applications and technologies for tomorrow's Internet. The National LambdaRail (NLR), sometimes called Internet3, is a cross-country, high-speed, fiber-optic network dedicated to research in high-speed networking applications. New high-speed Internet applications emerging from research on Internet2 include interactive communication: video conferencing with individuals around the world, real-time access to remote resources such as scientific equipment or far-away events, large-scale, multisite computation, and shared virtual reality.

Test Yourself

LEARNING OBJECTIVE 1: Describe how the Internet developed and how hardware, protocols, and software work together to create the Internet.

1. The Internet ________________ is made up of a combination of many national and international communication networks.
   a. router
   b. protocol
   c. backbone
   d. server
2. True or False: HTTP is the main protocol of the Internet.
3. Most Internet applications, with the exception of file-sharing software such as KaZaA, make use of ________________ relationships.

LEARNING OBJECTIVE 2: Explain the underlying structure of the Web and the technologies that support it.

4. True or False: Web pages are sometimes called HTTP documents.
5. ________________ are small files of data that are stored and retrieved from your computer by a Web server each time you visit.
   a. cookies
   b. JavaScripts
   c. ActiveX Controls
   d. portals
6. True or False: Content streaming allows you to listen to or view content while it is being downloaded.

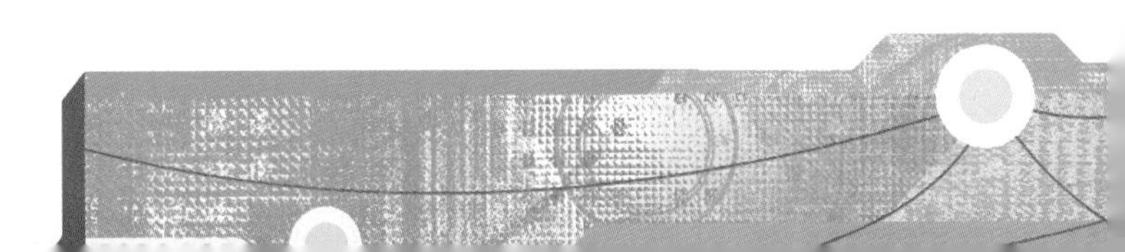

LEARNING OBJECTIVE 3: Define the categories of information and services that the Internet and Web provide and the forms of communication they support.

7. E-mail, voice mail, and discussion boards are all forms of ________________ communication.

8. True or False: Short Message Service is a technology designed for cell phones.

9. Web sites created to express one or more individual's views on a given topic are called ________________ .
   a. portals
   b. search engines
   c. buddy lists
   d. blogs

10. The process of working from locations away from the office by using telecommunication technologies is called ________________ .

LEARNING OBJECTIVE 4: Explain what Internet2 is and the types of applications it will provide in the future.

11. True or False: Internet2 is a separate network from the original Internet.

12. The National LambdaRail (NLR) is a cross-country, high-speed, ________________ network dedicated to research in high-speed networking applications.
   a. fiber-optic
   b. cable
   c. wireless
   d. satellite

Test Yourself Solutions: **1.** c. backbone, **2.** False, **3.** client/server, **4.** False, **5.** a. cookies, **6.** True, **7.** synchronous, **8.** True, **9.** d. blogs, **10.** telecommuting, **11.** False, **12.** a. fiber-optic.

Key Terms

asynchronous communication, p. 200
blog, p. 205
cable modem connection, p. 180
chat, p. 204
client/server , p. 184
content streaming, p. 194
cookies, p. 192
dial-up connection, p. 179
DSL (digital subscriber line) connection, p. 181
e-mail, p. 201
e-mail attachment, p. 201
Extensible Markup Language (XML), p. 189
hyperlink, p. 188
Hypertext Markup Language (HTML), p. 189
Hypertext Transfer Protocol (HTTP), p. 188
instant messaging (IM), p. 203
Internet backbone, p. 178
Internet service providers (ISP), p. 179
IP address, p. 183
peer-to-peer (P2P), p. 186
plug-in, p. 194
port , p. 184
router, p. 179
search engine, p. 196
SMS text messaging, p. 205
synchronous communication, p. 200
TCP/IP, p. 183
telecommuting, p. 208
Uniform Resource Locator (URL), p. 188
video conferencing , p. 207
Web authoring software, p. 191
Web browser, p. 188
Web server, p. 188

Questions

Review Questions

1. What was the motivation behind the creation of the Internet?

2. What three components combined make up today's Internet?

3. What is the Internet backbone, and who provides it?

4. What is the role of an ISP?

5. What are the responsibilities of a router?

6. How does P2P networking differ from client/server?
7. What is the difference between synchronous and asynchronous communication?
8. What is the number one use of the Internet?
9. Why was P2P selected as the network architecture for today's file-sharing networks?
10. What is SMS, and how is it used?
11. What do chat and instant messaging have in common? How do they differ?
12. Why is video conferencing so valuable?
13. What is the primary markup language and protocol of the Web?
14. What is meant by distance education?
15. What information and services can be found on the Web?
16. What are three common sources of information for research?

Discussion Questions

17. Describe how an e-mail message gets from your computer to your friend's computer.
18. How have blogs changed the way that news is delivered?
19. What are the benefits and drawbacks of client/server architecture?
20. What are the benefits and drawbacks of P2P architecture?
21. Why might P2P networks be more hazardous, in terms of viruses and hackers, than client/server architectures?
22. What are the benefits and drawbacks of synchronous communication?
23. What are the benefits and drawbacks of asynchronous communication?
24. What is a common problem with communicating through e-mail? What can be done to help alleviate this problem?
25. What are binary files, and why can't they be sent as e-mail?
26. How will XML and XHTML change the nature of the Web?
27. Why are plug-ins necessary for some Web sites?
28. What are the pros and cons of cookie technology?
29. How do search engines and subject directories differ? What types of scenarios does each support?
30. What is WYSIWYG, and why is it important to Web authors?
31. What is a concern of primary and secondary teachers when it comes to student Web use, and how can this concern be eased?
32. How is the Web used in job hunting?
33. Compare and contrast the pros and cons of library, database, and Web research.

Exercises

Try It Yourself

1. Use your favorite search engine to acquire the URLs of three online music services (you can select from those listed in this chapter) and pay each a visit. Create a spreadsheet that displays the features provided by each service at varying membership levels and their fees. Use formulas and functions to determine which service offers the best deal for your music needs. Write up your results using a word processor.
2. Use your favorite Web search engine to find several online travel services. Evaluate each of the services then, using a word processor, write up a summary of your impressions. Which service do you think is most helpful? Why?
3. Do a search on *plagiarism* at your favorite search engine. Study the search results and find answers to the following questions:
   a. How prevalent is plagiarism on college campuses?
   b. In what ways does the Web support plagiarism?
   c. What are colleges doing to fight plagiarism?

   Write up the results of your study using a word processor.

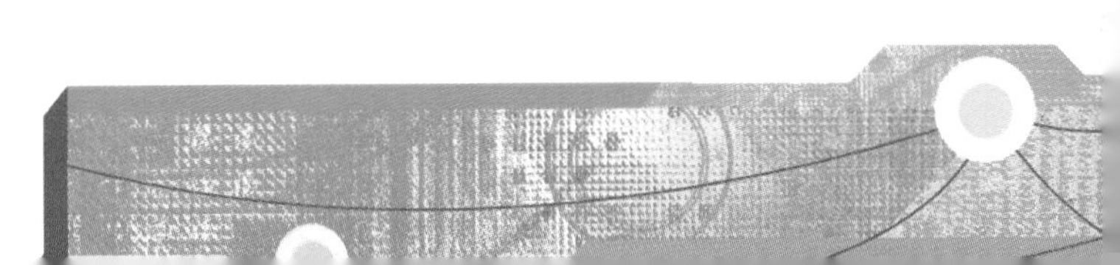

4. Visit *www.opus1.com/www/traceroute.html.* This Web site provides access to a program called Traceroute that allows you to view the path of a packet across the Internet. Traceroute returns the name of each router that the packet encounters on its journey. Use this tool to trace the route of a packet from Tucson, Arizona, (the location of this Web server) to the Louvre in Paris, France. 143.126.211.222 is the IP address of the Louvre's Web site. Type it in the text box and click "Trace that puppy." The resulting list is the routers that your packet encountered on the way to the Louvre. How many routers (hops) did your packet visit? What can you discern from the router information? Can you guess who owns any of the routers and where they are located?
5. Search the Web for a biography on Tim Berners-Lee and use it to write a summary of his professional life.

Virtual Classroom Activities

For the following exercises, do not use face-to-face or telephone communications with your group members. Use only Internet communications.

6. Have each group member use a different search engine with the goal of finding the least expensive roundtrip airfare for a week in Jamaica. The flight can depart from any airport within 100 miles of your present location. Define the departure and return dates of the trip. The person with the lowest fare is the winner and gets to delegate the work of writing up the team results.
7. Choose two instant messaging tools to download and evaluate. If you are unable to install software, you can substitute two Web-based chat utilities. Set up a date and time for all to logon to each tool and try it out. Carry out your evaluation live from within each tool conversing about the good and bad features. Write up reviews of each product.

Teamwork

8. Each team member should use a word processor to write a specific description of the e-mail service(s) that he or she uses to access e-mail and the level of satisfaction with the service. Team members should e-mail their papers to each other as attachments. As a group, determine which members are using appropriate e-mail services and which should switch. Write up the results of your discussion.
9. Have each group member go online in search of employment in his or her chosen field (if you don't have a chosen field, choose something of interest). Find a job posting in your field that lists the pay scale and job description. As a group, decide who found the best paying job and whose job sounds most satisfying. Write up your results.

TechTV

Google Maps Meets Craigslist

Go to www.course.com/swt2/ch04 and click TechTV. Click the "Google Maps Meets Craigslist for Easy Househunting" link to view the TechTV video, and then answer the following questions.

1. What are the advantages and possible risks for Paul in developing independent software linked to established Web sites?
2. What are the advantages and possible risks for Web site owners like Google and Craigslist of having independent software linked to their sites?

Endnotes

[1] Zakon, R. Hobbes' Internet Timeline v8.0 Web site, www.zakon.org/robert/internet/timeline/, accessed on June 10, 2005.

[2] Hamblen, M. "Broadband-over-power-line pilot under way in Houston," *Computerworld*, July 11 2005, www.computerworld.com.

[3] Peters, J. "United Airlines Approved for In-Flight Internet Service," *New York Times*, June 6, 2005, www.nytimes.com.

[4] Darrow, B. "Office 12 To Boost XML Support," Document Security, *Information Week*, June 1, 2005, http://www.informationweek.com/.

[5] Olsen, S. "Google Readying Web-only Video Search," CNET News, June 13 2005, http://news.news.com.

[6] Lacy, S. "Here Comes the iTunes of News", *Business Week, Online*, June 27, 2005, www.businessweek.com.

[7] Wayner, P. "Jerky Pictures and Sound Are History. Videoconferencing Is All Grown Up," *New York Times*, June 16, 2005, www.nytimes.com.

[8] Richardson, B. "The Growing Allure of Online Poker," *BBC News*, June 2, 2005, http://news.bbc.co.uk/1/hi/business/4603351.stm.

[9] Lewin, J. "Consumers Relying on Web During Major Life Changes," *ITworld.com*, April 19, 2005, www.itworld.com.

TELECOMMUNICATIONS, WIRELESS TECHNOLOGIES, AND COMPUTER NETWORKS

Amanda Jackson is a news correspondent recently hired by National Public Radio (NPR). In her previous position with her small town newspaper, she had a desk and PC at which she did most of her work. Her new job entails traveling the world reporting on stories where they occur. Amanda has been given a start-up budget to purchase the reporting equipment she requires. She needs a handheld device for jotting down written notes and digitally recording audio interviews. She needs a cell phone that keeps her connected wherever she travels. She needs a notebook computer to research facts on the Internet and the NPR archives located on the organization's private network.

Amanda is somewhat concerned about the complexities of the technology with which she'll be working. How will she transfer her notes from a handheld PC to her notebook? How will these devices connect to the Internet and the private NPR network? Is it possible for a cell phone to work on international networks?

As you read through the chapter, consider the following questions:

1. What types of handheld device(s) will best suit Amanda's journalistic mobile needs?
2. What type of notebook computer, networking media, devices, and software should Amanda use to connect to the Internet and the NPR network?
3. What networking technologies can Amanda use to transfer files between her devices?

Check out Amanda's *Action Plan* at the conclusion of this chapter.

5

LEARNING OBJECTIVES

1. Understand the fundamentals of data communications and the criteria for choosing a communications medium.
2. Explain how networking media, devices, and software work together to provide data-networking services, and describe the benefits of various types of media.
3. List and describe the most popular forms of wireless telecommunications technologies.
4. List the different classifications of computer networks and their defining characteristics, and understand the basics of wireless home networking.

CHAPTER CONTENT

Fundamentals of Telecommunications

Networking Media, Devices, and Software

Wireless Telecommunications Technologies

Networks and Distributed Computing

Introduction

A network is fundamentally a communication system. As such, it empowers individuals and groups to interact and access resources that would otherwise be more difficult or impossible to access. At home, networks allow household members to share documents, music, photos, and other media among home computers, as well as access a world of resources on the Internet. In an organization, networks act as the circulatory system, providing a flow of information between group members. For organizations, teamwork is synonymous with success. The better the flow of information among group members, the more productive the group becomes. To reach goals more efficiently and effectively, an organization needs to ensure that its network can properly support its information and communication needs.

In selecting the components of a network, you must consider the speed and capacity of the medium that carries the communication signal. The communications medium works in conjunction with communication devices and software to provide a data communication network. There are numerous types of networks, each supporting the unique needs of their environment. Wireless mobile technologies are making it increasingly common to access networks anywhere, anytime. This chapter examines a variety of network types and their components to provide an understanding of how networks help us be connected and more productive in business, at home, and while traveling.

Today's telecommunications networks play a vital role in our daily activities. From the global telecom networks that make up the Internet, to short-range wireless technologies that pass data back and forth between a cell phone and headset, telecommunications technologies keep us connected. In the first part of this chapter, you learn the basics about telecommunications systems, how they work, and how their performance is measured. Next you learn about the components that make up telecommunications systems: media, devices, and software. Section three of this chapter delves into wireless networking technologies including cell phones, Global Positioning Systems, Wi-Fi, WiMAX, and Bluetooth. The chapter then turns its focus to the different types of computer networks and examines some of the benefits of a wireless home network.

FUNDAMENTALS OF TELECOMMUNICATIONS

Communications can be defined as the transmission of a signal from a sender to a receiver by way of a medium such as wires or radio waves. The signal can contain a message composed of data and information. It is important to note two characteristics about communications. First, the message is not communicated directly; rather, it is communicated by way of a *signal*. Second, the signal itself goes through a *communications medium*, which is anything that carries a signal between a sender and receiver.

You can easily recognize these aspects of communications if you consider what happens when humans communicate (see Figure 5.1). When you talk to someone face to face, you send messages to each other. One person may be the sender at one moment in time and the receiver a few seconds later. The same entity, a person in this case, can be a sender, a receiver, or both. This is typical of two-way communication. Some of the signals you use to convey these messages are the sound waves that represent our spoken words. Other signals are nonverbal, such as gestures and expressions. For communication to be effective, both sender and receiver must understand the signals and agree on what they mean. For example, if the sender in Figure 5.1 is speaking in a language the receiver does not understand, or if the sender believes a particular word or gesture has one meaning and the receiver believes it has some other meaning, effective communication cannot occur. Although these facts may seem obvious, they are important building blocks in understanding more complicated forms of electronic communications.

When you talk to someone, the transmission medium is the air. When you read, the transmission medium is the printed page. The traditional telephone converts a signal carried by the medium of air into an electronic signal carried over wires (another medium). A cell phone converts the sound waves created by your voice into a radio signal carried over air, which is received by a base station that transfers the signal to wires, and then perhaps to another cell phone over a radio signal.

As you consider the structure of telecommunications systems, keep these fundamental communications concepts in mind. For example, establishing a communication link between two digital devices requires that they "speak the same language" using protocols, and that software *interprets* the information being transmitted. And, as discussed later in this chapter, the characteristics of the medium—in particular, the speed at which it can carry a signal—is an important consideration.

FIGURE 5.1 • Face-to-face voice communications

In face-to-face voice communications, the transmission medium is the air, and the signal is the sound wave.

Telecommunications and Data Communications

Computing devices communicate with each other via telecommunications and data communications systems. **Telecommunications** refers to the electronic transmission and reception of signals for communications. Some telecommunications devices that you interact with daily include telephones, cell phones, radios, televisions, and networked computers. *Data communications*, a specialized subset of telecommunications, refer to the electronic transmission and reception of digital data, typically between computer systems. A *telecommunications network* connects communications and computing devices. A *computer network* is a specific type of telecommunications network that connects computers and computer systems for data communications. This chapter investigates popular uses of telecommunications networks.

There are three components of telecommunications networks: networking media, networking hardware, and networking software. **Networking media** is anything that carries a signal and creates an interface between a sending device and a receiving device. *Networking hardware devices* (or just *networking devices*) and *networking software* work together to enable and control communications signals between communications and computer devices. As computers, cell phones, televisions, and other devices are connected to a network, they become part of the network infrastructure. Figure 5.2 shows a general model of telecommunications. The model starts with a sending unit, such as a PC, cell phone, or other device that originates the message. The sending unit transmits a signal to a networking device. The networking device can perform a number of functions, including changing or manipulating the signal. The networking device then sends the signal over a medium. The signal is received by another networking device that is connected to the receiving unit, which is a computer system, PC, cell phone, or other device that receives the message. The process can then be reversed, and another message can go back from the receiving unit to the original sending unit.

FIGURE 5.2 • General model of telecommunications

Sending and receiving units use networking devices and media to communicate.

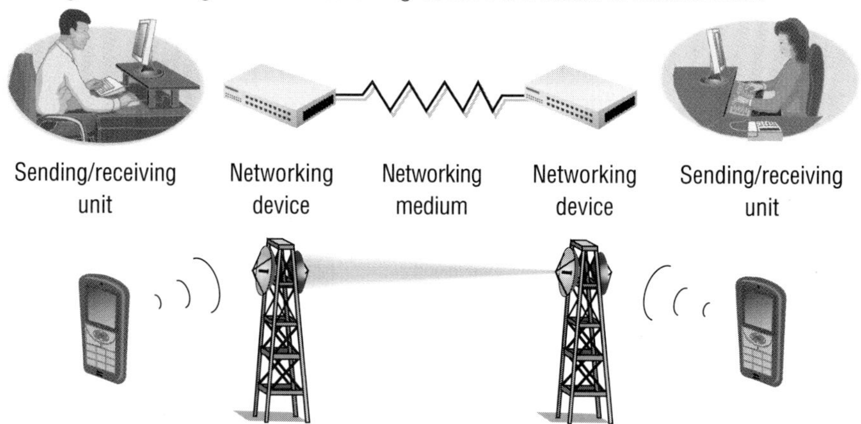

Characteristics of Telecommunications

The characteristics of telecommunications components should be analyzed in terms of speed, quality, and convenience. Telecommunications can allow people to be more productive. For example, being able to access and respond to e-mail during an hour-long daily commute on the bus or train frees up an hour later in the day for going to the gym or some other productive activity. Construction companies have found that investing in a wireless phone system boosts productivity by allowing job superintendents to communicate via phone or notebook computer, sharing problems and solutions, even transferring job site photos to the home office.

Toronto Airport recently decided to provide airlines and personnel with a higher degree of control by installing an integrated wireless network that controls all forms of communication within the airport.[1] Voice, video, and data all travel over one digital network to support the phone system, computer system, check-in kiosks, and security systems. The new system increased passenger throughput by 15 percent. Implementing a wireless network saved the airport hundreds of thousands of dollars over the cost of installing a wired network. In these ways and many others, telecommunications can help solve problems and maximize opportunities.

Types of Signals. If you were to measure the voltage on a telephone wire during a phone conversation, you would see fluctuations in voltages similar to those shown in Figure 5.3a. Such fluctuations occur thousands of times per second in varying intensities that mirror the sound waves of your voice. This type of signal is called an **analog signal**, and it fluctuates continuously.

In contrast, if you measured the voltage on cables used to connect PCs, you would probably see something comparable to Figure 5.3b. The signal in Figure 5.3b at any given time is either high or low. This type of discrete voltage state—either high or low—is called a **digital signal**. The two states are used to represent the state of a bit, high for 1, and low for 0.

Transmission Capacities. Some uses of telecommunications require very fast transmission speeds. For example, when the salesperson swipes your credit card at the checkout counter of a major department store, your purchase information joins with thousands of others that are being routed to the credit card company

FIGURE 5.3 • Analog and digital signals

An analog signal continuously changes over time. A digital signal has a discrete state—either high or low.

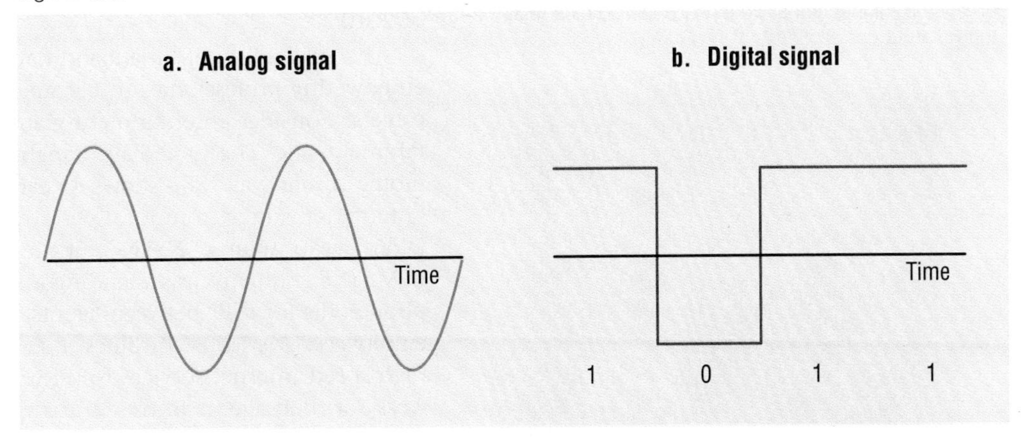

for approval. Such networks must handle high-volume traffic very quickly so as not to keep customers waiting. Other signals, such as those used in a small office or home network, do not require as much transmission capacity and speed.

The speed at which an electronic communications signal can change from high to low is called the signal *frequency* (see Figure 5.4). A faster frequency means a faster data transmission rate. Signal frequency is measured in *hertz (Hz)*, or cycles per second. In computer networks, the data transmission rate is also referred to as the **bandwidth** and is measured in *bits per second (bps)*. Today's bandwidth options fall into one of two categories: narrowband or broadband.

FIGURE 5.4 • Frequency

The speed at which a signal changes from high to low, the signal frequency, dictates the rate of speed at which data is delivered, called the *bandwidth*.

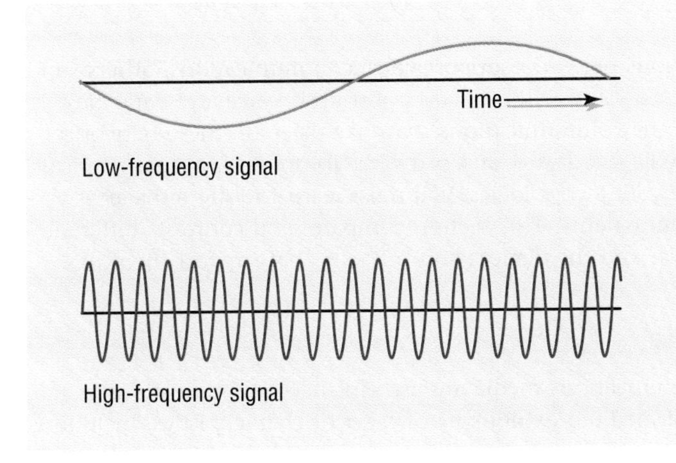

Broadband media are those advertised by Internet service providers as "high-speed." In everyday use, the terms *broadband* or *high-speed Internet* refer to a connection that is always on or active, such as cable or DSL. These connections are significantly faster than dial-up telephone connections (see Figure 5.5). DSL and cable claim maximum download speeds of 3 Mbps and 6 Mbps, respectively. The minimum speed for a media to be considered broadband is hotly debated. Michael Powell, ex-chairman of the U.S. Federal Communications Commission (FCC), stated that a clear, uniformly accepted definition of broadband evades the FCC. "Whatever broadband is, it is fast—the commission has defined it as 200 Kbps. I submit, however, that broadband is not a speed," Powell said. "It is a medium that offers a wide potential set of applications and uses. I think broadband should be viewed holistically as a technical capability that can be matched to consumers' broad communication, entertainment, information, and commercial desires."[2] *Narrowband* would then be considered any medium with a speed less than 200 Kbps, typically the speeds delivered by a 56 Kbps dial-up modem connection.

Not long ago, broadband data transfer speeds were only available at a premium price to businesses and organizations to support high-traffic data communications. Today, inexpensive broadband access is available for homes and small businesses and is rapidly becoming the norm. In 2004, broadband

reached 32 million households in the United States, just below half of residential online accounts. That figure is expected to grow to 88 million, or 78 percent, by the end of 2010.[3]

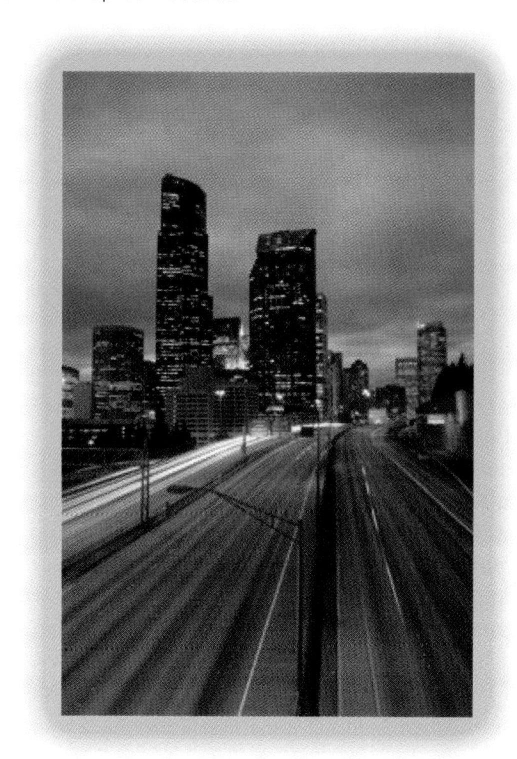

FIGURE 5.5 • Broadband

Like an expressway compared to a neighborhood street, broadband Internet access provides an always-on connection at speeds 25 to 70 times faster than dial-up connections.

The availability of broadband data communications is empowering professionals to become more effective in their careers. Consider entertainment giant, House of Blues Entertainment, Inc. The Canadian branch of House of Blues promotes as many as 1200 shows a year across Canada. Entertainers range from local performers to names like Celine Dion, Bryan Adams, *NSYNC, Britney Spears, The Who, and REM. The company was spending a lot of money annually on phone bills for calls between agents, artists, staff, and employees. The House of Blues took advantage of today's high-speed Internet services to set up a communications network for their agents to use in arranging concerts. Since implementing the high-speed Internet solution, the company has seen monthly long-distance charges fall from $2000 per month to $200 per month, their artists and agents can communicate more conveniently, and everyone in the organization is enjoying the use of an integrated system that allows access to e-mail, voice mail, and faxes over their PCs via broadband Internet.[4]

NETWORKING MEDIA, DEVICES, AND SOFTWARE

Some telecommunications networks support voice communications, others support data communications, and still others support both voice and data. No matter what the type, the communications that take place on these networks require networking media, hardware, and software. To understand the practical benefits of various network technologies, you must learn the strengths and weaknesses of their components. This section compares and contrasts different types of network media, and the devices and software that support them.

Networking Media

Various types of communications media are used for telecommunications networks. Each type of medium exhibits its own set of characteristics, including transmission capacity, speed, convenience, and security. In developing a network, the selection of media depends on the environment and use of the network. Media should support the needs of network users, in the given environment at the least cost, taking into account possible future needs of the network. For example, computer users in a rental apartment would probably not be allowed to run network cables through walls and attic spaces. In this situation, a wireless network setup would be more appropriate. A government security agency may opt for a wired network over a wireless network in order to provide the highest level of security for its data. A business traveler might purchase a wireless cell phone headset so as not to have the inconvenience of tangled wires while rushing through airports. AT&T might decide to send a satellite into orbit, rather than run miles of cable, to provide faster service to a remote location for the lowest cost.

These examples illustrate the many environments and situations in which media considerations come into play. Not only are cost, quality, and speed important considerations, but also one must consider the future. For example, when a business invests in a network, it selects a network that will accommodate a reasonable amount of growth in both workforce and business needs without overestimating these requirements and wasting money. The system selected by the House of Blues discussed earlier was chosen because it was based on standard IP technology that can be easily expanded in the future.

Different communications media connect systems in different ways. Some media send signals along physical connections like cables, but others send signals through the air by light and radio waves.

FIGURE 5.6 • Networking cables

Twisted pair (a), coaxial (b), and fiber-optic (c) networking cables provide support for a variety of bandwidth needs from 1 Mbps to 10 Tbps.

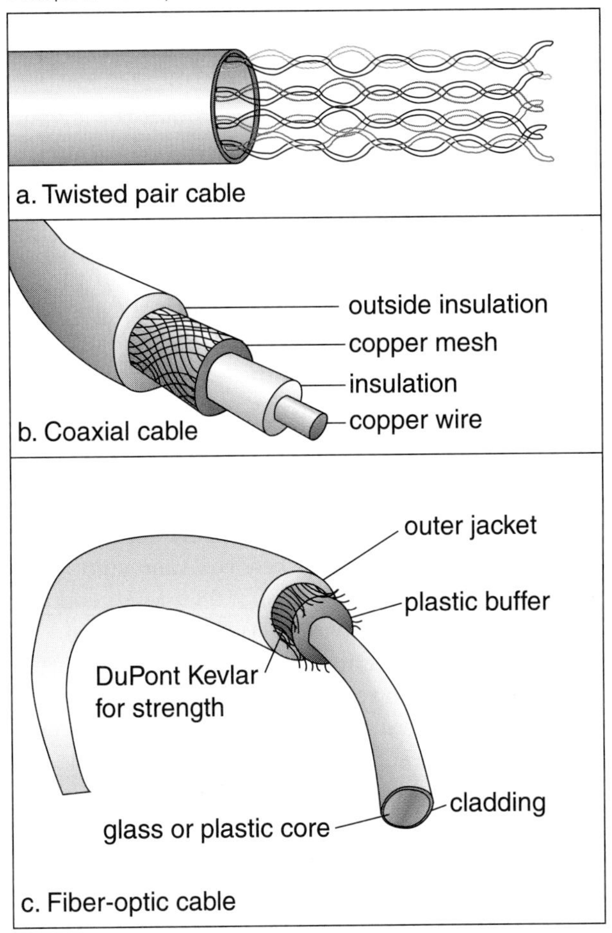

Physical Cables. Different physical cables offer a range in bandwidth from narrow to broadband. Cables have an advantage over wireless options because some cables support much higher data transfer rates than wireless technologies and are considered by some to be more secure. The disadvantage of cables is their physical presence. Cables need to be installed, typically in an inconspicuous manner; they are run underground, undersea, through utility tunnels, strung from pole to pole, through attics and basements, strung above ceiling tiles, pulled down inside walls, through walls, and around the interior of a room. Depending on the environment, laying cable can take a considerable amount of time and effort. Three types of transmission cables are typically used to connect data communications devices: twisted pair cable, coaxial cable, and fiber-optic cable.

Twisted pair cable consists of pairs of twisted wires covered with an insulating layer (see Figure 5.6a). Twisted pair is the type of cable that brings telephone service to your home and is used for dial-up modem connections. It is also used in most wire-based computer networks. Twisted pair comes in varying qualities and categories (Cats) that support a wide range of bandwidths. For example, Cat 1 and Cat 2 is used for home phone systems supporting 1 Mbps for voice communication. Cat 5 is used in many private computer networks to support 100 Mbps for data communications. Cat 7 is used for heavy-traffic networks and supports 10 Gbps. Twisted pair is the least expensive of cable options.

Figure 5.6b shows a typical coaxial cable. You may recognize this as the type of cable provided by cable television services that connects to your cable box. A **coaxial cable** consists of an inner conductor wire surrounded by insulation, a conductive shield (usually a layer of foil or metal braiding), and a cover. Coaxial cable is much faster than Cat 1 and 2 twisted pair cables, and at one time it was the preferred cable for computer networks. However, since the development of faster, less expensive Cat 5 twisted pair cables, coaxial is mostly used just for cable television and radio networks.

Twisted pair and coaxial cables transmit electrical signals over copper or other metal wires. In contrast, **fiber-optic cable**, which consists of thousands of extremely thin strands of glass or plastic bound together in a sheathing (a jacket), transmits signals with light beams (see Figure 5.6c). These high-intensity light beams are generated by lasers and are conducted along the transparent fibers. These fibers have a thin coating, called *cladding*, which effectively works like a mirror, preventing the light from leaking out of the fiber.

Fiber-optic cable has several advantages over traditional copper cable:

- Speed: Fiber-optic cables support data transfer rates of 1.6 terabits per second (Tbps) and have performed as fast as 10 Tbps in lab experiments.
- Size: Fiber-optic strands are much smaller in diameter than many copper wires. More strands can be bundled together in smaller cables than with copper.
- Clarity: Fiber-optic cables do not allow signals to bleed from one strand to another, unlike copper wire that often suffers from such interference.
- Security: Copper wires are easy to tap, but the same is not true for fiber-optic cable.

Fiber optics suffers from one major disadvantage: price. Fiber-optic cable is much more expensive than copper; it is not financially feasible to run it to every household. Once installed, though, fiber-optic cable is less expensive to maintain than copper cable.

Radio Signals and Light. Telecommunications signals can travel through air using radio waves and light. A **radio wave** is an electromagnetic wave transmitted through an antenna at different frequencies. The U.S. Federal Communications Commission assigns different frequencies for different uses in the United States. For example, FM radio, cell phones, baby monitors, and garage door openers all operate at different frequencies. Figure 5.7 illustrates some of the uses of frequencies over the radio spectrum. To show all frequency assignments would take a chart several feet tall (available at *www.fcc.gov/oet/spectrum*).

There are thousands of radio waves passing through the air and your body as you read this. AM and FM radio waves, television, wireless phone conversations, wireless computer networks, Global Positioning Systems, ham radios, CB radios, police and emergency communications, satellite communications, and wireless clock systems all make use of radio waves. All it takes to receive the information being transmitted on a specific frequency is a device with an antenna that allows you to tune into that frequency. For example, when you tune your radio into 98.9 FM, you are selecting the signal being broadcast at 98.9 MHz, and your receiver translates the fluctuations of the radio waves it receives into sound.

Waves sent at the high end of the radio spectrum, between 1 and 300 GHz, are called *microwaves*. These high-frequency waves have numerous uses. As you may have guessed, microwave ovens use microwaves (2.45 GHz) to cook food. Microwave signals are also used for high-speed, high-capacity communication links and satellite communications.

Above microwaves in the radio spectrum comes infrared light and then visible light. Infrared light is also used to carry telecommunications signals, as you'll learn later in the chapter.

Because of the convenience of wireless communications, advances in wireless technologies are occurring at a rapid pace and impacting all of our lives. For this reason an entire section of this chapter is devoted to wireless technologies.

FIGURE 5.7 • The radio spectrum

The FCC assigns different frequencies of the radio spectrum for different uses in the United States. The spectrum is divided into eight bands: Very Low, Low, Medium, High, Very High, Ultra High, Super High, and Extremely High Frequencies.

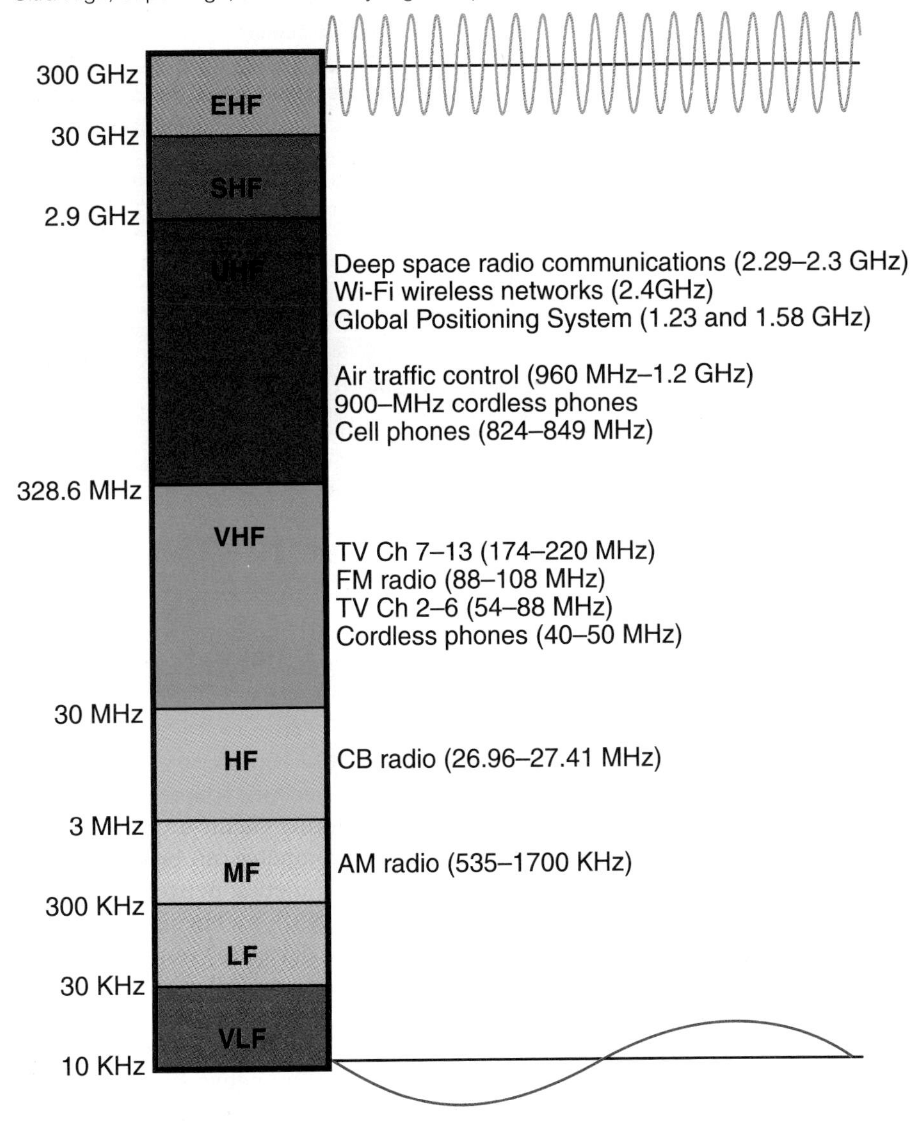

Networking Devices

Not long ago networking devices were of interest only to network technicians. Today, the increase in home networking and wireless networks has provided a reason for everyone to acquire a basic understanding of the most common networking devices such as modems, network adapters, access points, and other network control devices.

Modems. A **modem** modulates and demodulates signals from one form to another, typically for the purpose of connecting to the Internet. A modem can be either an internal or external device. All methods of connecting to the Internet require some type of modem. Dial-up modems are standard on many PCs and support narrowband Internet connections over analog phone lines.

Other special-purpose modems support connections to high-speed networks. A *cable modem* provides Internet access to PCs and computer networks over a cable television network (see Figure 5.8). A cable modem is typically an external

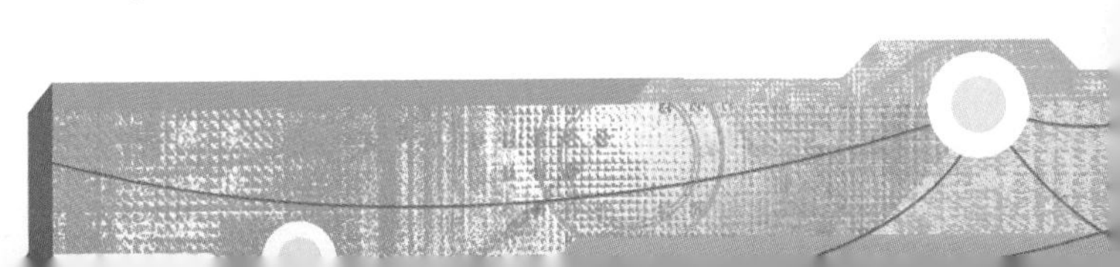

device that has two connections: one to the cable wall outlet and the other to a computer or computer network device. A *DSL modem* is similar to a cable modem, and provides high-speed Internet service over telephone lines.

FIGURE 5.8 • Cable modem

A cable modem shares the cable TV connection to provide Internet access at data transmission rates much faster than a traditional dial-up connection.

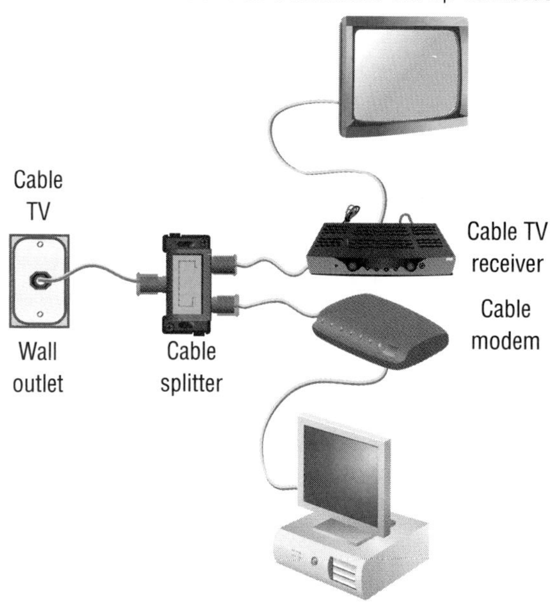

Network Adapters. Computers can connect to networks and the Internet either by modem, network adapter, or a combination of both. A **network adapter** is a computer circuit board, PC card, or USB device installed in a computer so that the computer can be connected to a network. Network adapters come in two basic varieties: network interface cards and wireless adapters. A *network interface card (NIC)* is a circuit board or PC card that, when installed, provides a port for the device to connect to a wired network with traditional network cables. *Wireless adapters* can be a circuit board, PC card, or an external device that connects through a USB port that provides an external antenna to send and receive network radio signals (see Figure 5.9). In addition to connecting devices to computer networks, network adapters are used to connect devices to cable modems and DSL modems.

FIGURE 5.9 • Wireless adapter

This wireless adapter connects to a PC or notebook computer via a USB port.

Network Control Devices. For multiple computers to communicate, special devices are required to control the flow of bits over the network medium and to ensure that information that is sent reaches its destination quickly and securely (see Figure 5.10). A number of different network control devices handle this responsibility. Here are brief descriptions of those that are most commonly used:

- Hubs are small electronic boxes that are used as a central point for connecting a series of computers. A hub sends the signal from each computer to all the other computers on the network.
- Switches are a fundamental part of most networks. They make it possible for several users to send information over a network at the same time without slowing each other down. Hubs and switches are often combined, and the terms are often used interchangeably.

- Repeaters connect multiple network segments, listening to each segment and repeating the signal heard on one segment onto every other segment connected to the repeater. Repeaters are also helpful in situations where a weak signal requires a boost to continue on the medium.
- Bridges connect two or more network segments, as a repeater does, but bridges also help regulate traffic.
- Gateways are network points that act as an entrance to another network.
- Routers are advanced networking components that can divide a single network into two logically separate networks. Although network traffic crosses bridges in its search to find every node on a network, it does not cross routers, because the router forms a logical boundary for the network. Routers are also used to interconnect various types of networks, which has led to their widespread deployment in connecting devices around the world to the Internet.
- Wireless access points connect to a wired network and receive and transmit data to wireless adapters installed in computers. They allow wireless devices to connect to a network.
- A firewall is a device or software that filters the information coming onto a network to protect the network computers from hackers, viruses, and other unwanted network traffic.

FIGURE 5.10 • Network control devices

A variety of special-purpose network control devices have been designed to control the movement of data over networks.

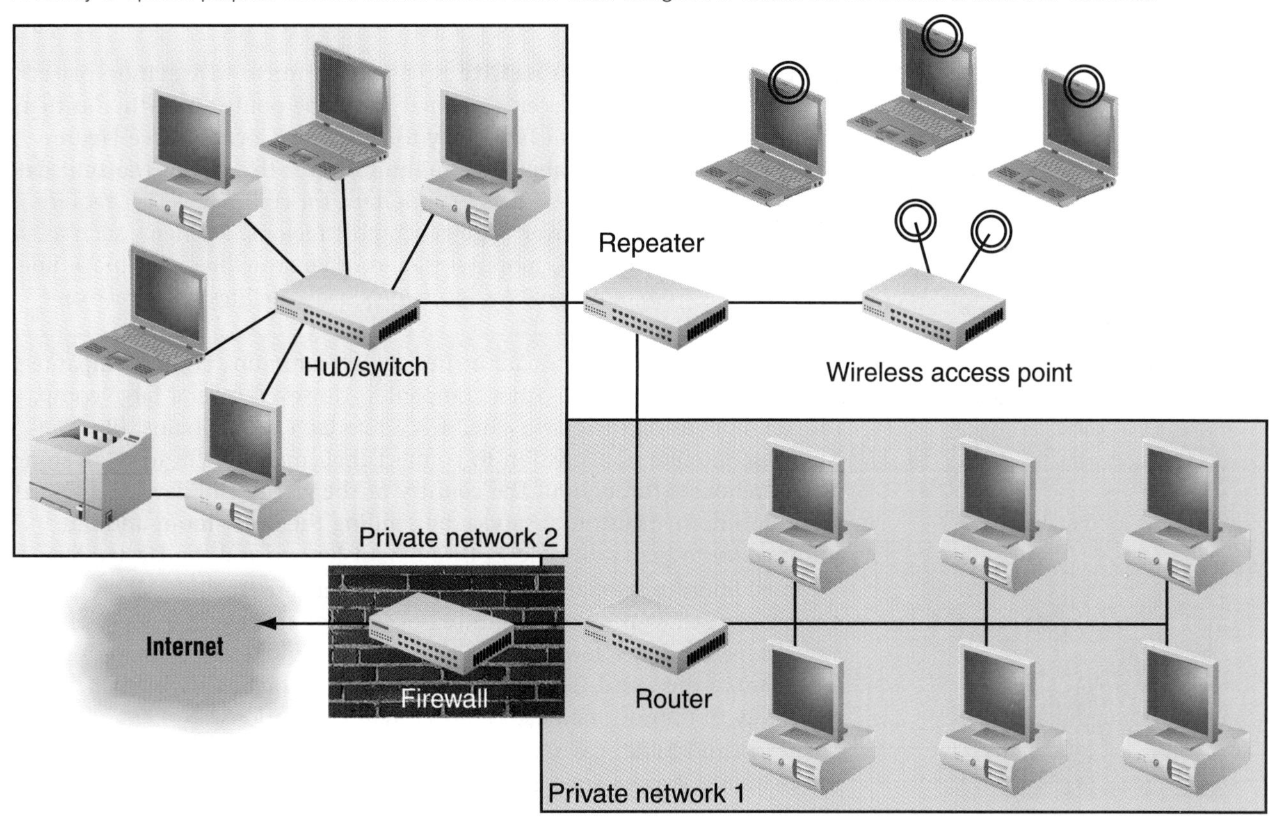

Sometimes network control devices can be combined in a single unit. For example, a home computer network might incorporate a device that includes a router to create a high-speed connection to the Internet, a switch to share Internet access over a wired network and connect computers, and a wireless access point to connect wireless devices (see Figure 5.11).

Industrial Telecommunications Media and Devices

So far, you have learned about devices that most computer users encounter at some point while working with or setting up computer networks. Larger organizations require industrial-strength media and devices to manage large volumes of network traffic.

FIGURE 5.11 • Wireless access point/router/switch

This combination access point/router/switch provides wireless or cable-connected broadband service to network users while controlling network traffic.

Microwave and Satellite Transmission. Microwave and satellite transmissions are sent through the atmosphere and space. Although using these transmission media does not entail the expense of laying cable, the transmission devices needed to utilize them can be quite expensive. **Microwave transmission**, also called terrestrial microwave, is a *line-of-sight* medium, which means that a straight line between the transmitter and receiver must be unobstructed. Microwave transmissions can be sent through the air up to distances of approximately 31 miles or about 50 kilometers. You often see microwave towers alongside interstate roads because the roads provide long, straight unobstructed stretches of land. Weather and atmospheric conditions can impact the quality of a microwave signal. Typically, to achieve longer transmission distances, microwave stations are placed in a series—one station receives a signal, amplifies it, and retransmits it to the next microwave transmission tower. Microwaves can carry literally thousands of channels at the same time. Even at that, a microwave link can carry only a small percentage of the capacity of a single fiber-optic strand. Once the main thoroughfare of the phone networks, today microwave towers support only a fraction of telecommunications traffic and often serve a more useful function as cell phone antenna towers (see Figure 5.12).

FIGURE 5.12 • Microwave towers

Today's communications towers often include large round dish antennas used for microwave communications, as well as tall rods or panels used for cell phone networks—one per service provider.

A **communications satellite** is basically a microwave station placed in outer space. Satellites receive a signal from one point on earth and then rebroadcast it at a different frequency to a different location (see Figure 5.13). The advantage of satellite communication is that it can transmit data quickly over long distances. This is important for companies that require high-speed transmission over large geographic regions. Problems such as the curvature of the earth, mountains, and other structures that block line-of-sight terrestrial microwave transmission make satellites an attractive alternative.

Most of today's satellites are owned by telecommunications companies that rent or lease them to other companies. However, several large companies are now using their own satellites. Some large retail chains, like Wal-Mart, use satellite transmission to connect their main offices to retail stores and warehouses throughout the country or the world. Holiday Inn uses satellites to improve customer service by sending the latest room and rate information to reservation desks throughout Europe and the United States.

In addition to standard satellite stations, there are small mobile satellite systems that allow people and businesses to communicate. These portable systems have a dish a few feet in diameter that can operate on battery power anywhere in the world. This is important for news organizations that require the ability to transmit news stories from remote locations.

Exxon and Mobil gas stations offer a speedy way to pay for gas and merchandise. Wave a Speedpass key fob over the checkout keypad and within two seconds your credit is authorized and you are on your way.[5] Speedpass utilizes two types of wireless networking technologies: (1) radio waves are sent from the key fob to the reader relaying your credit card information and preferences, and (2) credit is quickly checked using high-speed satellite communications.

FIGURE 5.13 • Satellite transmission

Communications satellites are relay stations that receive signals from one earth station and rebroadcast them to another.

Industrial Hardware. Large companies often lease dedicated lines from telecommunications companies to maintain long-distance connections. Unlike a *switched line* that maintains a connection only as long as the receiver is "off the hook," a *dedicated line* leaves the connection open continuously to support a data network connection.

To handle large quantities of data, businesses may lease a T1 line. A *T1 line* supports high data transmission rates by carrying twenty-four 64-Kbps signals on one line. *T3 lines* carry 672 signals on one line and are used by telecommunications companies; some act as the Internet's backbone.

Managing industrial-level network traffic takes a considerable amount of orchestration. Signals sent over these high-speed lines must be processed in a manner that takes full advantage of the medium. Three devices are commonly used to control and protect industrial-level telecommunications:

- Multiplexer: A multiplexer sends multiple signals or streams of information over a medium at the same time in the form of a single, complex signal. A demultiplexer at the receiving end recovers the separate signals. Often a multiplexer and demultiplexer are combined into one device.
- Communications processor: Sometimes called a front-end processor, this device is devoted to handling communications to and from a large computer network. Like a receptionist handling visitors at an office complex, communication processors direct the flow of incoming and outgoing telecommunications.

GALACTIC GOODWILL OR SPACE SPAM?

Those with hopes of attracting the attention of "like-minded" aliens can now send personal messages out into deep space. Using a frequency similar to cell phones and wireless networks, TalkToAliens.com sends e-mails for $20 and phone messages for $4 per minute, and Deep Space Communications Network offers customers five minutes of text and images—or video—for $99. The first transmission in May 2005 lasted 23 minutes and sent 138,000 postings from Craigslist.com.

Source:
Intergalactic Communications:Tele-spamming Our Alien Brethren
By Amanda MacMillan
Popular Science
http://www.popsci.com/popsci/aviation/article/0,20967,1064977,00.html
June 2005

- Encryption devices: An encryption device is installed at the sending computer to alter outgoing communications according to an encoding scheme that makes the communications unintelligible during transport. A decryption device is installed at the receiving computer to decode incoming data and return it to its original state. Encryption protects private data from being accessed by anyone but the intended recipient.

Networking Software

As you have learned in previous chapters, hardware is useless without the software necessary to drive it, and so it is with networks. Networking software performs a number of important functions in a computer network. It monitors the load, or amount of traffic, on the network to ensure that users' needs are being met. It provides error checking and message formatting. In some cases, when there is a problem, the software can indicate what is wrong and suggest possible solutions. Networking software can also provide data security and privacy. Because networking software's main purpose is to support the functioning of the network, it is considered utility software. In its role as a utility, networking software runs mostly unnoticed by network users—with the exception of the network administrator.

A *network administrator*, sometimes called a *system administrator*, is a person responsible for setting up and maintaining the network, implementing network policies, and assigning user access permissions. A large organization might employ dozens of network administrators. With the increasing popularity of home networks, many nontechnical computer users are finding themselves in the role of network administrator. For example, a mother may find herself setting up a home network in a manner that filters out adult Web content from her children's computers. This section discusses a variety of commonly used network utilities.

Network Operating Systems. In Chapter 3, you learned that all computers have operating systems that control many functions. When an application program requires data from a disk drive, it goes through the operating system. Consider a scenario where many computers are accessing resources such as disk drives, printers, and other devices over a network. How does an application program request data from a disk drive on the network? The answer is through the network operating system.

A *network operating system (NOS)* performs the same types of functions for the network as operating system software performs for a computer, such as memory and task management and coordination of hardware. When network equipment such as printers and disk drives are required, the network operating system makes sure that these resources are correctly used. Network operating systems come preinstalled on midrange and mainframe servers. All of today's personal computer operating systems, including Windows, Mac OS, and Linux, can function as NOSs as well.

Network Management Software. In addition to network operating systems, there are a number of useful software tools and utilities for managing networks (see Figure 5.14). With network management software, a manager on a networked desktop can monitor the use of individual computers and shared hardware (such as printers), scan for viruses, and ensure compliance with software licenses. Network management software also simplifies the process of updating files and programs on computers on the network. Changes can be made through a communications server instead of on each individual computer. Some of the many benefits of network management software include fewer hours spent on

routine tasks (such as installing new software), faster response to problems, and greater overall network control.

FIGURE 5.14 • Monitoring network usage

This network traffic monitor displays levels of incoming and outgoing packets over several Internet gateways at FSU over the course of 24 hours.

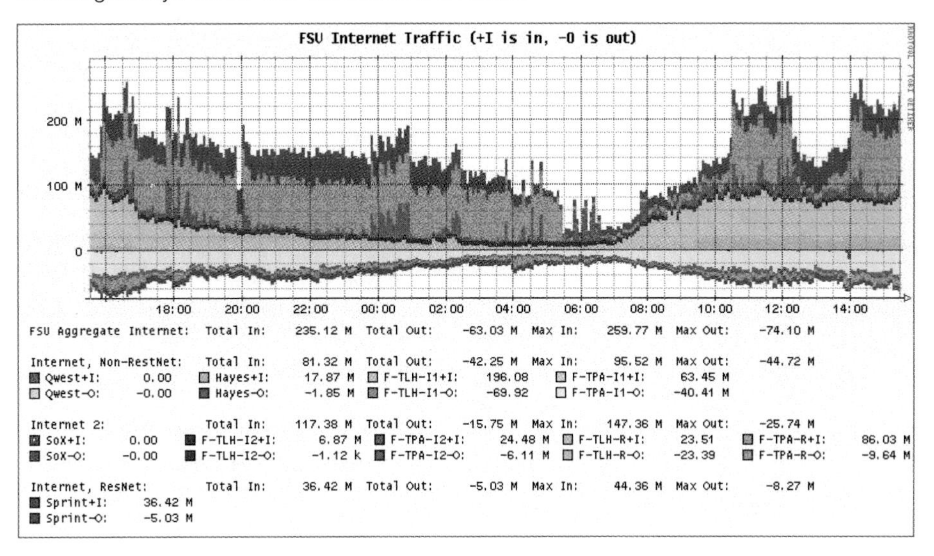

Network Device Software. Routers, switches, firewalls, modems, and other network control devices include software interfaces that allow you to change device settings. Once the device is connected to the network, its software interface can be accessed from a network computer. A wireless access point's software can be accessed to change security settings so that unauthorized users cannot access the network. A router's software can be used to divide up a physical network into multiple virtual networks. A firewall's software can be used to specify what data packets to allow in and which to keep out. Software for a modem might allow you to enable or disable call waiting.

Communications Protocols and Standards. In Chapter 4, you learned that a *protocol* is an agreed-upon format for transmitting data between two devices. In Chapter 2, you learned that a *standard* is an agreed-upon way of doing something within an industry. In the world of networking, protocols and standards are essential for enabling devices to communicate with each other. Protocols define the format of the communications between devices, and standards provide the physical specifications of devices and how they interconnect. In Chapter 4, you also learned that the Internet uses the TCP and IP protocols. TCP/IP has become the default protocols for many private networks as well.

Ethernet is the most widely used network standard for private networks. This standard defines the types of network interface cards, control devices, cables, and software required to create an Ethernet network. Other network standards, such as *token ring*, use unique hardware and software that are not compatible with Ethernet. Special devices, such as routers or gateways, are required to connect networks that use different standards.

WIRELESS TELECOMMUNICATIONS TECHNOLOGIES

Wireless communications and computing are sweeping the globe. It is expected that by 2007 there will be more than 2 billion global mobile subscribers taking advantage of wireless communications and information services.[6] Wireless technologies support today's mobile workforce and are leading us into the era of communications and computing that can occur anywhere at anytime. In the previous chapter you were introduced to the many communications options available over wireless networks such as voice, e-mail, and text messaging. Beyond communications, today's wireless technologies allow you to connect to public and private networks, access television, music, and games, find your current location on a map, and access numerous information services. Wireless technologies put the world at your fingertips on cell phones, handheld and notebook computers, and other special-purpose devices, wherever you may travel.

Cell Phone Technologies

Today's cell phone services are founded on and also limited by cellular network technology. A **cellular network** is a radio network in which a geographic area is divided into cells with a transceiver antenna (tower) and station at the center of each cell, to support mobile communications. If a cell phone user travels from one cell to another, the system judges the cell phone's location based on its signal strength, and passes the phone connection from one cell tower to the next. The signals from the cells are transmitted to a receiver and integrated into the regular phone system (see Figure 5.15).

Each cell tower has a transmission and receiving range of 3–15 miles (about 5–24 kilometers) depending on geography and environment. Because it is not economically feasible to cover the entire planet with towers every few kilometers, cell phone usage is limited to a network's coverage areas, which focus on metropolitan areas and major highways. The inability to provide 100 percent coverage, and the limitations of working with wireless signals that are prone to interference, makes less-than-perfect quality of service for cell phone users. A 2005 *Consumer Reports* survey found that "Only 45 percent of respondents said they were completely satisfied or very satisfied with their cell phone service, a very low showing for any service."[7] However, most cell phone users are willing to accept these limitations in exchange for the power of being able to communicate from most locations at almost any time.

Cell phones provide more than voice communications. Today's cell phone users can choose to subscribe to dozens of communications, information, and entertainment services. Many in the telecommunications industry believe that the cell phone will become an individual's remote control to the world. For example, technicians are researching technology that will allow cell phone users to "point and click" at a cash register to pay a bill, at a billboard to order tickets to the concert advertised, or at a tag on a coat in a store to find out if they can get it for a better price elsewhere. Cell phones are able to plot your location on a map, and given that

NO SPARE CHANGE?

Then feed your meter with your phone. Coral Gables, Florida, is the first city in the United States to adopt a service that allows drivers to do exactly that. Once a parking spot has been secured, enrolled drivers can dial a number, enter their parking space number, and the amount they want to spend. When they depart, they can call back and end the billing cycle. Meter-checking personnel use a wireless handheld device to tell who is paid up and who isn't.

Source:
Paying Meters with Cell Phones Tested
From the Associated Press
New York Times
http://tech.nytimes.com/aponline/technology/AP-High-Tech-Meters.html
June 16, 2005

FIGURE 5.15 • Cellular transmission

If a cell phone user moves from one cell to another, the system judges the cell phone's location based on its signal strength and passes the phone connection from one cell tower to the next.

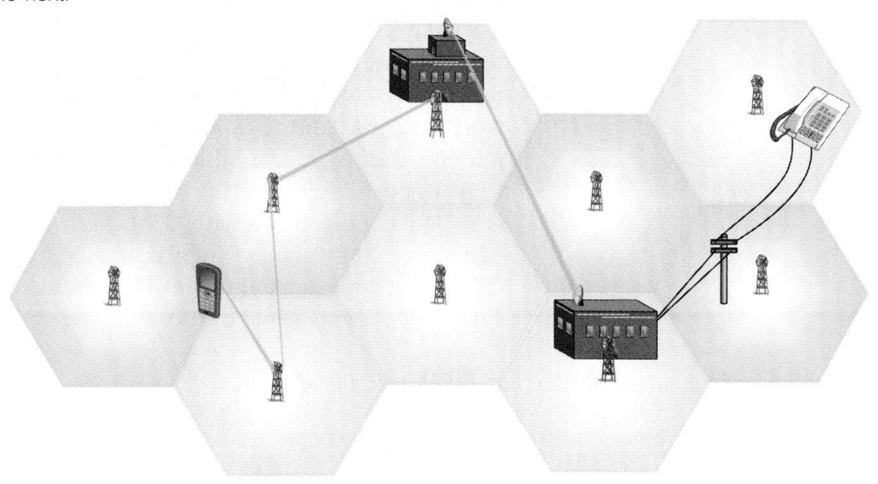

location, direct you to the nearest coffee shop. You can use your cell phone to access files from your home PC or e-mail from your AOL account. Cell phones are becoming a primary information access device.

The new technologies behind third-generation (3G) cell phones are providing the power necessary for many new and useful services. *3G cellular technology* is bringing wireless broadband data services to your mobile phone. Boasting speeds that compete with DSL and cable, 3G networks let you speed through Web pages, enjoy streaming music, watch on-demand video programming, download and play 3D games, and videoconference. Many favorite activities that people enjoy over high-speed Internet at home can now be taken on the road. The only limitation, which is substantial, is the tiny cell phone display.

Deciding on a cell phone service is no small task. The variables involved in the decision are many and complex. When choosing a cell phone service, there are three primary decisions to make: choose a carrier, choose a plan, and choose the phone, features, and services you want. These decisions are interdependent. Each carrier offers different phones, and services are related to phone capabilities. There are two different strategies for deciding on a mobile phone service. You can choose a phone first, and then find out which carriers support it, or you can choose a carrier first, then choose one of the phones that the carrier supports. Because the carrier is perhaps the most important piece of this puzzle, the discussion begins there.

REAL-LIFE FACSIMILE

After pondering the paradox of everyone having mobile phones, but no one engaging with people around them, a design student came up with The Social Fabric software. The program displays on a mobile phone screen a group of avatars representing friends and family. Their body language shows how recently they've been contacted. Frequent contact results in an avatar that looks alert and meets your gaze. A turned back means you might want to send flowers.

Source:
See If You're a Good Friend
By Daniel Terdiman
Wired News
http://www.wirednews.com/news/culture/0,1284,68020,00.html?tw=wn_tophead_7
July 5, 2005

Cellular Networks and Carriers. A **cellular carrier** is a company that builds and maintains a cellular network and provides cell phone service to the public. There are many carriers providing cell phone service worldwide. The most popular U.S. carriers are Verizon, T-Mobile, Cingular/AT&T, and Sprint/Nextel.

Different carriers offer different cell phones, features, coverage areas, and services. It is important to take time in selecting a carrier as your carrier will define your cell phone experience. It is typically costly and inconvenient to change your carrier once you've signed on.

Today's cell phone networks are nearly all digital. The areas of coverage that still use the old analog network technology are being rapidly updated to digital technology. The predominant digital networking standards for cell phone networks are GSM and CDMA. *GSM* is the most popular global standard for mobile phones, used by over a billion people across more than 200 countries. The *CDMA* networking standard is predominantly used in the United States where it is in equal competition with GSM. Verizon and Sprint cell phone networks are based on CDMA, while Cingular and T-Mobile use GSM. Phones designed for CDMA networks have different characteristics than those using GSM networks. Table 5.1 presents some of those differences.

TABLE 5.1 • Comparison of GSM and CDMA cell phones

GSM phones	CDMA phones
• Provide better international coverage • Provide more talk time on a battery charge (5 hours) • Do not support analog networks, which may mean spotty coverage if you travel to rural areas in the United States • Provide more features across more handsets and more innovative designs • Are more portable due to a subscriber identity module (SIM) that contains subscriber information, address book, and other personal information (see Figure 5.16) that can be easily transferred to other phones	• Provide better coverage in the United States • Provide less talk time on a charge (3 hours) • Provide analog backup for more reliable coverage in rural areas • Tend to have folding-case designs, which, in general, have an edge in voice quality • First to offer third-generation (3G) services that can trim the time it takes to upload and download files

FIGURE 5.16 • GSM phone SIM card

Many GSM phones have a subscriber identity module (SIM) card that stores subscriber information and personal data and can be transferred to other GSM phones.

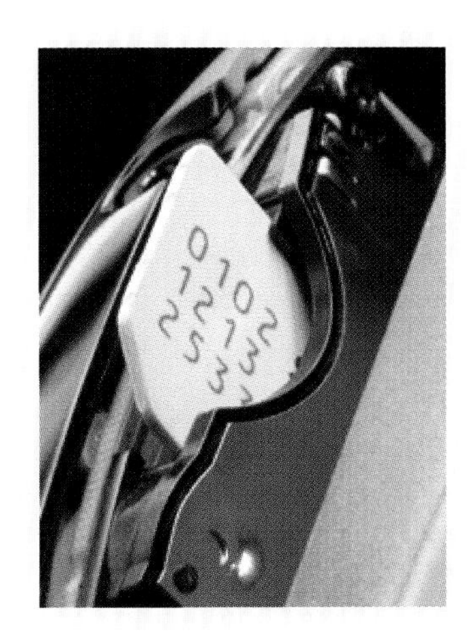

When choosing a carrier, it is useful to consult user survey data, such as that provided by *www.consumerreports.com*, as well as the opinions of friends. Check the coverage maps provided by the carrier's Web site to make sure that there is coverage in the areas you will be using your phone. Finally compare the carriers' networks (GSM or CDMA), phones, rates, and plans to find which best suits your needs.

Cellular Service Plans. Cell phone carriers offer service plans for every type of phone user. The trick is in determining your usage habits and needs. Do you plan to use your cell phone locally, regionally, nationally, or internationally? What time of day and days of the week will you use your phone? How many minutes during working days each week will you use your cell phone? These questions need to be addressed before selecting a plan.

Most carriers provide a choice of a "pay as you go" plan without a contract, or 1- or 2-year subscription plans with a contract. Pay as you go is a prepay system in which you buy minutes up front that must be used in a given time frame, such as 130 minutes for $25,which must be used in 90 days. Users that opt to sign a subscription contract must choose a subscription plan that best matches their usage habits for the least amount of money. The longer you sign on for (2 years instead of 1 year, for example), the more money you save.

Calling plans are defined by usage in three time frames:

- Whenever, or anytime, minutes with no time restrictions
- Weeknight minutes, typically Monday through Friday, 9:00 p.m. to 5:59 a.m. or 6:59 a.m.
- Weekend minutes, typically Saturday 12:00 a.m. to Sunday 11:59 p.m.

Dealing with Wireless on Campus

Colleges are racing to provide wireless Internet access on campus to allow students, faculty, administrators, and staff to be more productive and as an incentive in recruiting new students. Many colleges are providing wireless Internet access in classrooms to students who bring their laptop computers. Some teachers are concerned that having Internet access may be distracting to students during lectures. While having access to the Internet can be a valuable teaching tool, some feel there should also be a method of cutting off access in order to refocus the student's attention on the business at hand.

Vanderbilt University in Nashville developed a solution for this issue that allows teachers to cut off access to the Internet for students in their classroom at any time through a Web-based interface. The new system has received mixed reviews from students and faculty. Most students were upset that their Internet access was cut off during class time. One student complained "I'm an adult. I pay my tuition. I should be able to do what I want in class." Faculty, for the most part, were in favor of the change. Those who didn't wish to deprive students of Internet access could simply leave their students connected. The system cost roughly $15,000 to implement.

Questions

1. Do you agree with the student quoted in this article? Should college students have the ability to do whatever they want on their computers during lectures? Why or why not?
2. Would the effectiveness of a lecture be diminished if half the students in attendance were shopping on eBay and checking their e-mail? In what way?
3. If you were the teacher, would you allow your students Internet connectivity during your lectures?

Sources

1. *Graychase, N. "Vanderbilt Wi-Fi: Blocked When Needed,"* Wi-Fi Planet, *www.wi-fiplanet.com, June 1, 2005.*
2. *Vanderbilt University Law School's Information Technology Web page, http://law.vanderbilt.edu/it/, accessed June 23, 2005.*
3. *Bluesocket Web site, www.bluesocket.com, accessed June 23, 2005.*

Figure 5.17 illustrates some average plans from the major vendors at the time of this writing. There will undoubtedly be changes in these features by the time this book is in print, but the chart provides a typical sample of plans and features. As you can see, different carriers have differing ways of packaging their services. Cingular advertises "w/Rollover." *Rollover* is a feature that allows you to apply unused minutes from one month to the next. Because carriers charge you by the minute for any time used over your monthly allotment, rollover makes it possible to avoid those charges.

Notice also that Cingular has a specification for "mobile to mobile." This refers to "in network" phoning to other Cingular subscribers. Verizon calls this "IN calling." In other words, according to these ads, both Verizon and Cingular allow you to phone anyone else that uses the same provider anytime for free. Finally, notice that Verizon includes a specification that tells you how much each additional minute over your allotment costs. It is important to find out about hidden usage fees such as additional minutes and roaming. All carriers charge for additional minutes, but it is not always apparent how much that charge is. *Roaming fees* are incurred when you use your phone outside your carrier's network.

FIGURE 5.17 • Cell phone plans

Different carriers provide different plan options and incentives to cover every type of usage and win over customers.

Carrier	Plan name	Monthly cost	Whenever minutes	Weeknight/weekend minutes & In network (IN)
T-Mobile Get more from life	Basic	$19.99	60	0/500
T-Mobile Get more from life	Basic Plus	$29.99	300	0/Unlimited
T-Mobile Get more from life	Get More	$39.99	600	Unlimited/Unlimited
cingular raising the bar	Nation 450 w/ Rollover	$39.99	450	5,000/Unlimited Unlimited mobile to mobile
cingular raising the bar	Nation 900 w/ Rollover	$59.99	900	Unlimited/Unlimited Unlimited mobile to mobile
verizonwireless We never stop working for you.	America's Choice	$39.99 ($0.45 each additional minute)	450	Unlimited IN calling *AND* night & weekend home airtime minutes
verizonwireless We never stop working for you.	America's Choice	$59.99 ($0.40 each additional minute)	900	Unlimited IN calling *AND* night & weekend home airtime minutes
verizonwireless We never stop working for you.	America's Choice	$79.99 ($0.35 each additional minute)	1350	Unlimited IN calling *AND* night & weekend home airtime minutes

If two plans from different carriers seem roughly comparable in terms of features, you should view the coverage map of both carriers (see Figure 5.18) and examine their phones and services. You should also find out which carrier your friends and family use, as phoning within the network could save you money.

All of the plans discussed so far are national plans that allow you to use your cell phone anywhere there is coverage in the continental United States. While national plans are the most popular, there are also local plans, that allow use only in one metropolitan area and regional plans that cover a few adjoining states. The more limited your coverage area the more money you save. Carriers also provide family plans that offer discounts for each additional handset that is added to the plan. Each additional handset has its own phone number and set of features and services associated with it. There are also specialty plans for international travelers who require Internet, corporate network, and e-mail access.

Cellular Handsets, Features, and Services. Advances in cell phone handsets have been fast and furious over the past five years with manufacturers such as Nokia, Samsung, Motorola, Audiovox, Kyocera, LG, and Sony competing intensely to gain an edge in the largest technology market. Cell phone handsets range in price from free (with service subscription), to over $500 for the latest and greatest smart phones—phones with handheld computing capabilities. There are two basic styles of cell phone: *Flip phones* open and close and offer a compact design that protects the keypad and display when closed. *Candy bar* style phones can be smaller than flip phones, and because they don't flip open and closed may be easier to handle.

FIGURE 5.18 • Cellular coverage

Shoppers should carefully view coverage maps of services they are comparing, which include national coverage as well as enlarged maps of local coverage.

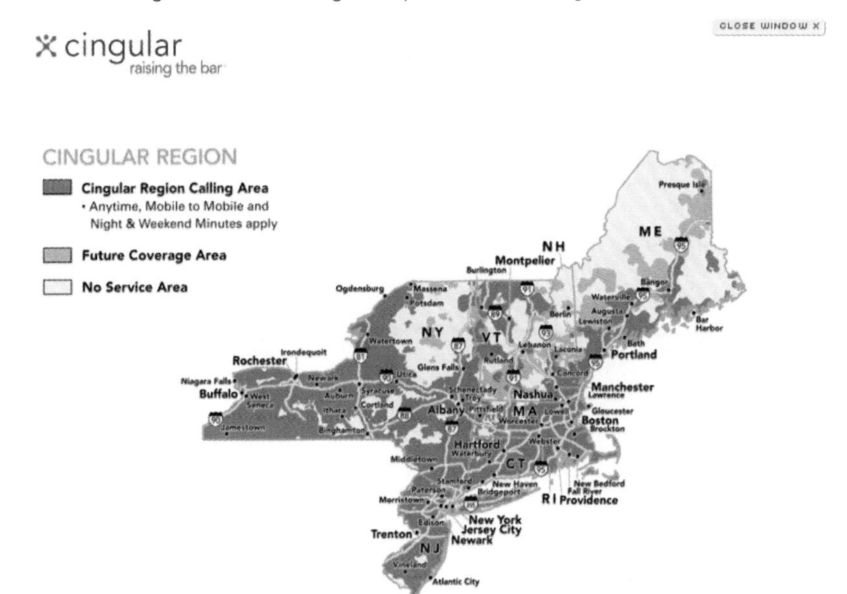

Cingular Wireless GSM handset required on Cingular Region plans.

Your phone's display does not indicate the rate you will be charged. Please review your coverage map for areas included in and out of plan. Map depicts an approximation of outdoor coverage. Map may include areas served by unaffiliated carriers and may depict their licensed area rather than an approximation of the coverage there. Actual coverage area may differ substantially from map graphics, and coverage may be affected by such things as terrain, weather, foliage, buildings and other construction, signal strength, customer equipment and other factors. Cingular does not guarantee coverage. Charges will be based on the location of the site receiving and transmitting the call, not the location of the subscriber. Future Coverage, if depicted above, is based on current planning assumptions but is subject to change and may not be relied upon.

FIGURE 5.19 • Cell phone features

This Samsung cell phone on the left includes a full keyboard, and the Motorola ROKR supports iTunes downloads.

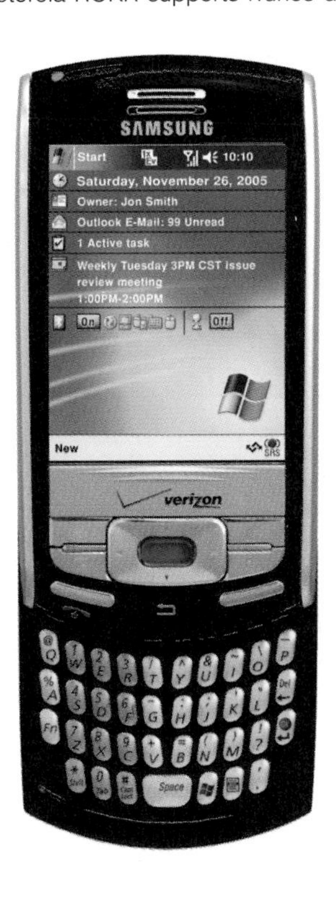

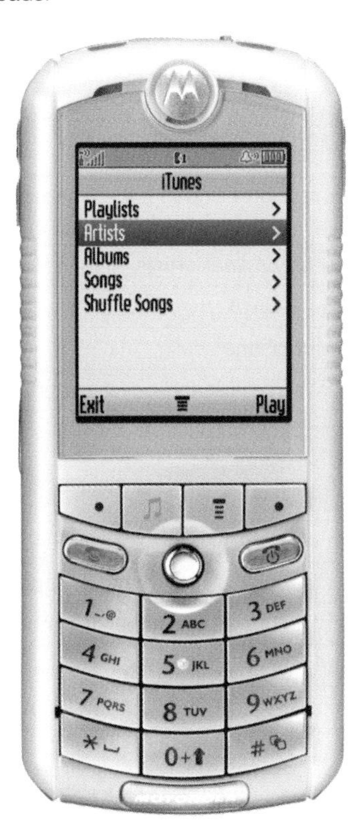

Cell phone manufacturers work hand-in-hand with carriers to provide the functionality for the services they offer, while the carriers work to provide services for the new technologies that manufacturers develop (see Figure 5.19). Because phones are matched to services, users typically select a phone and subscribe to services at the same time in order to make sure that the two match. For example, you wouldn't want to purchase a plan that includes photo messaging but then buy a phone that doesn't have a camera. Note also that you are not typically able to take a phone with you if you move to a different carrier; another reason to choose your carrier carefully. Table 5.2 lists features that are available on some of today's cell phones, and the list is growing.

TABLE 5.2 • Features available on some cell phones

Cell phone features
• Bluetooth wireless technology
• Camera
• Changeable face plate
• Color display
• External display (flip phone)
• External caller ID
• Infrared and USB
• Internet ready, PDA
• MP3 player
• Quad band world phone (850/900/1800/1900 MHz)
• QWERTY keyboard
• Speakerphone
• Stereo headset
• Text to speech
• Touch screen
• Video camera
• Video player
• Voice dialing
• Synchronize with PC

Cell phone carriers provide features that can be added on to monthly service plans at additional cost. The most popular add-on service is text messaging. Messages can be sent to and from cell phones or between cell phones and Internet e-mail addresses. Table 5.3 lists cell phone add-on services that provide additional communication power, entertaining media, access to information, and even safety.

After selecting a carrier, plan, phone, and add-on services, you can also select cell phone accessories. Cell phones typically come with a case of some kind, but you may wish to select a nicer, leather case, or one with a belt clip. A car charger is useful for charging your phone while in transit. A headset or earbud (Figure 5.20) is useful for hands-free phoning. Some U.S. states are making these a requirement for using cell phones while driving. Phones and headsets that contain Bluetooth chips can connect wirelessly to other Bluetooth devices (see the section on Bluetooth later in the chapter). If your cell phone plays digital music or other media, you may wish to purchase stereo headphones and an additional memory module for storing more music.

TABLE 5.3 • Cell phone add-on services

Feature	Description
Text messaging	Between phones or e-mail addresses
Picture and video messaging	Send images and videos phone to phone, or phone to e-mail
Instant messaging	Send messages to friends on AIM, ICQ, MSN Messenger, or Yahoo Messenger
Push to talk	Connect using a walkie-talkie style interface
E-mail	Use carrier's e-mail service or your own corporate server
Airfone	Transfer your calls to the Airfone located in aircraft
Personalized ring tones	Hear music or sound clips when your phone rings
Caller tunes, ringback tones	Have your friends hear music or sound clips when they call you rather than the usual ring tone
Hi-fi ringer	Hear tunes as recorded by original artists when your phone rings
Wallpaper	Customize the look of your display
Games	Play video games on your phone
Broadband entertainment	Enjoy streaming video over broadband on your phone
Basic text info	Access news, weather, sports, horoscope, movie listings, and so on from leading providers such as CNN and ESPN
411 directory information	Live operators answer your questions
Alerts	Schedule content to be delivered such as news, sports scores, stock alerts, and so on
Narrowband Internet	Access Web content at 56 Kbps from phone or connected PC
Broadband Internet	Access Web content at 2 Mbps from phone or connected PC
Roadside assistance	Free jump start, tire changes, refuels, lockout assistance, and towing from wherever you stall
Phone insurance	Have your phone replaced for free if it is lost, stolen, or damaged

FIGURE 5.20 • Hands-free phoning

Cell phone headsets provide for hands-free phoning, which are especially useful when driving.

Pagers

Pagers are small, lightweight devices that receive signals from transmitters. Different pagers have varying levels of functionality. The most basic pager simply beeps, flashes, or vibrates to get the attention of the user. More sophisticated pagers accept numeric or text messages. Some pagers allow the owner to listen to voice messages, or send messages as well as receive them.

There are different types of paging systems. National and regional systems set up transmission towers, much like cell phone networks, to cover large geographic areas. On-site paging systems use small desktop transmitters to send pages over a small wireless network that covers a range of up to 2 miles (about 3 kilometers).

On-site paging systems are finding a variety of uses in businesses and organizations. Pagers are being used in businesses to assist in customer service. For instance, in restaurants, emergency rooms, and golf courses, pagers can be given to customers and used to indicate when a table, doctor, or tee time is available. Pagers can also be used to call servers when food is ready to be served, or to call store managers to a checkout counter for a price check or check validation. Pagers can eliminate the need for invasive and annoying public address announcements, while supporting finely tuned orchestration between staff members and the provision of high-quality service to the customer.

Global Positioning Systems

A **Global Positioning System (GPS)** uses satellites to pinpoint the location of objects on earth. Using a GPS receiver and a network of 24 satellites, the GPS can tell you the exact location of the receiver on the earth's surface. The GPS satellites orbit the earth in such a way that at any given time and location on earth, four satellites are visible to a GPS receiver. By measuring the distance from the receiver to each satellite, and calculating those distances with the known position of each satellite, the receiver is able to determine its location on Earth. Like the Internet, GPS was originally developed for national security and later extended for public use.

Early GPS receivers were expensive and used only in environments where determining your location was a matter of life and death, such as in far-traveling ships. Now, you can purchase a GPS receiver for under $130, and they are becoming increasingly popular in a variety of applications. Available as small handheld devices, add-ons for handheld computers and cell phones, and as in-dashboard devices for automobiles, GPS receivers are primarily used to assist travelers in getting from one place to another. GPS software can display a traveler's location on a city map and give suggestions for the shortest routes to destinations (see Figure 5.21).

FIGURE 5.21 • Handheld GPS receiver

A handheld GPS receiver shows your exact location on a map.

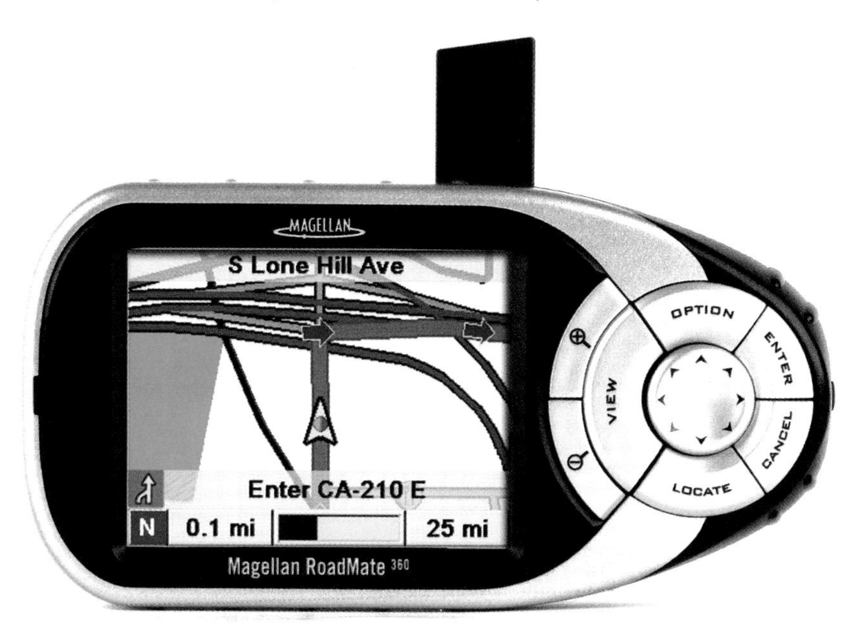

KITTY GONE A GALLIVANTING?

No problem. Now available is a collar that uses GPS to report an animal's whereabouts back to the owner via cell phone network. Owners can not only check on their best friend's location by text messaging a special number, they can check the temperature to make sure their cats are cool and their dogs aren't hot. A special multidirectional antennae means the collar picks up the satellite signals no matter where the animal is.

Source:
Creature Beacon Reunites Lost Pets and Owners
By Will Knight
NewScientist.com news service
http://www.newscientist.com/article.ns?id=dn7578
June 24, 2005

Many car rental companies provide GPS units in order to assist customers from out of town with navigating in an unfamiliar city. The Hertz company developed a GPS service they call Never Lost (*http://hertzneverlost.com*). The GPS receiver in the vehicle is programmed with the desired destination, and the system acts as a navigator providing specific instructions along the way.

GPS has also been used to map the planet's surface. For example, a GPS receiver was taken to the peak of Mt. Everest in 1999 to determine the exact height of the mountain. It was measured to be 29,035 feet above sea level, 7 feet taller than the previously accepted height calculated in the 1954 Survey of India. The GPS also revealed that Mt. Everest is moving northeast at approximately 2.4 inches a year.

A newer application of GPS technology assists others in locating a person in danger. An amazing 359,000 children are kidnapped in the United States each year, with many more attempted kidnappings reported. A GPS receiver for children made by Wherify (*www.wherifywireless.com*) looks like a child's toy wristwatch, but is actually a marvel of wireless technology. The watch includes buttons that initiate a 911 emergency response and a GPS receiver that communicates with the Wherify wireless network. Parents can go to the Wherify Web site to view their child's location on a city map or satellite image. The watch also acts as a pager and tells perfect time since it is wirelessly synchronized with an atomic clock (see Figure 5.22).

FIGURE 5.22 • GPS child safety

The Whereify GPS Locator for Children assists children in getting emergency help and parents in finding lost or kidnapped children.

To keep herself safe during late-night walks, Kursty Groves, a design student at the Royal College of Art in London, developed the Techno Bra, an undergarment device that is able to detect rapid jumps in heart rate caused by dangerous, distressing situations, and automatically alerts the authorities with the location of the victim using GPS and cell phone technologies embedded in the undergarment. Other similar devices are being developed that monitor the health of individuals and notify rescue services if a heart attack or seizure occurs.

The GPS provides convenience and safety; however, as with all technology that accesses personal data (such as a person's location at a given time), the use of GPS technology may infringe on an individual's right to privacy. Consider, for instance, the possibility of someone stashing a device like the Wherify locator in your car or backpack in order to track your movements.

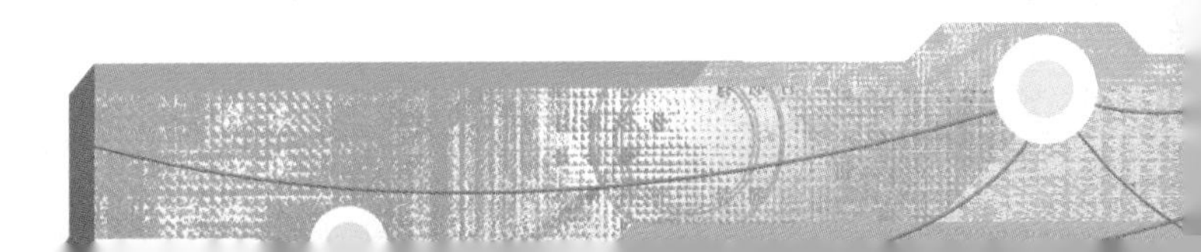

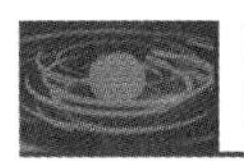

EXPAND YOUR KNOWLEDGE

To learn more about wireless technology, go to www.course.com/swt2/ch05. Click the link "Expand Your Knowledge," and then complete the lab entitled "Wireless Networking."

Wireless Fidelity and WiMax

Wireless fidelity (Wi-Fi) is wireless networking technology that makes use of access points to wirelessly connect users to networks within a range of 250–1000 feet (75–300 meters). The Wi-Fi standards, also known as the 802.11 family of standards, were developed by the Institute of Electrical and Electronics Engineers (IEEE) to support wireless computer networking within a limited range at broadband speeds. Table 5.4 shows the data rates and frequency bands of the most popular 802.11 standards. Note that 802.11g devices can interoperate with 802.11b devices, and in many cases they have replaced them, but neither 802.11b or 802.11g can interoperate with 802.11a devices. Most wireless *hotspots* use the 802.11b/g standard.

TABLE 5.4 • Popular IEEE Wi-Fi standards

Wi-Fi standard	Maximum speed	Frequency band	Notes
802.11a	54 Mbps	5 GHz	Less potential for RF interference than 802.11b and g; relatively shorter range
802.11b	11 Mbps	2.4 GHz	Not interoperable with 802.11a; relatively larger range (fewer access points required) than 802.11a; slower than 802.11g
802.11g	54 Mbps	2.4 GHz	May replace 802.11b; better security features and faster data rate; not interoperable with 802.11a

Wi-Fi technology uses wireless *access points* that broadcast network traffic using radio frequencies to computers equipped with Wi-Fi cards or adapters (see Figure 5.23). Wi-Fi has a maximum range of about 1000 feet (300 meters) in open areas and 250 to 400 feet (75 to 125 meters) in closed areas. Areas covered by Wi-Fi are often called *hotspots*. By positioning wireless access points at strategic locations throughout a building, campus, or city, Wi-Fi users can be continuously connected to the network and Internet, no matter where they roam on the premisis. Wi-Fi has become increasingly popular for home networks. It provides an affordable and simple way to network home computers without the need to run cables throughout the house.

Many new notebooks come equipped with Wi-Fi capability thanks to Intel's Centrino technology. Centrino combines three technologies for convenient mobile computing: the Intel PRO/Wireless Network Connection, the Intel Pentium M Processor, and the Intel 855 chipset family to offer fast processing, longer battery life, and wireless networking capabilities. Apple's PowerBook G4 notebook comes with Wi-Fi capability built in as well.

Having a Wi-Fi-equipped notebook is useless however, unless you have an access point to connect to. Access points are popping up in many locations where people tend to congregate. Internet cafes and even some McDonald's restaurants provide hotspots for their customers. Your school may offer hotspots on campus for wireless network and Internet access (check *www.intel.com/personal/products/mobiletechnology/unwiredcolleges.htm* for Intel's list of most unwired campuses). If you do a lot of traveling you might want to look into Boingo. Boingo is a subscription service that for around $21.95 per month provides access to 18,000 hotspots in airports, hotels, cafes, coffee shops, and other public locations in over 300 U.S. cities and 43 states.

Some cities, states, and even countries are providing public hotspots. A 2005 study found Seattle and San Francisco to be the most "wireless" cities in the United States, providing the most Wi-Fi access points. Also in the top 10 were

FIGURE 5.23 • Wi-Fi access points

Wi-Fi access points are distributed around a geographic area to provide a network connection wherever you roam.

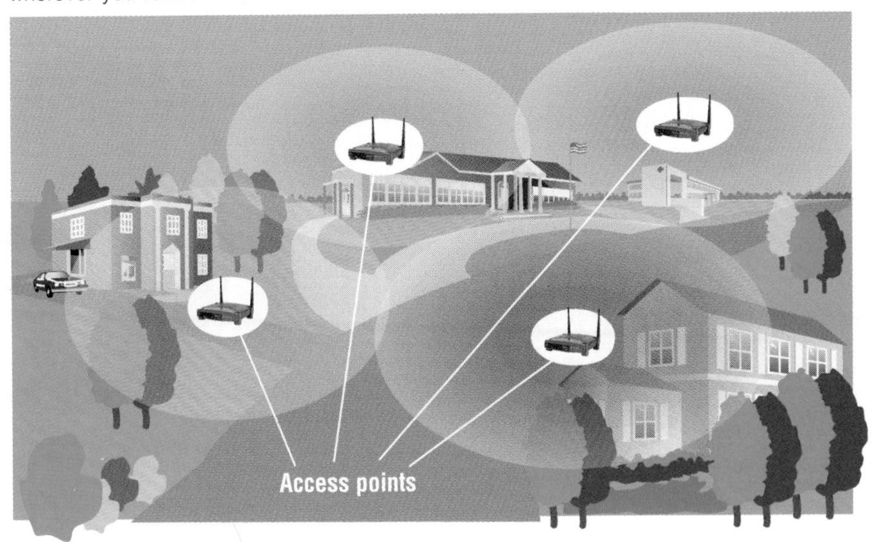

Austin, Portland, Toledo, Atlanta, Denver, and Minneapolis. Many small towns are providing wireless access as well, hoping to develop a high-tech reputation and attract businesses and new residents. The Polynesian island of Niue in the South Pacific has claimed be the first country to offer free nationwide Wi-Fi Internet access to its citizens and visitors.

A new type of voice communications technology that takes advantage of Wi-Fi network and VoIP technologies is offered by Vocera Communications. The Vocera Communications system uses lightweight communications devices similar to the original Star Trek communicator, clipped to a lapel or pocket, or worn on a lanyard around the neck (see Figure 5.24). The Vocera device (which they call a badge) uses speech recognition to support hands-free communications. It is activated when you speak a command word. For instance, to initiate a conversation with Jim and Mary, the user says "Vocera, get me Jim and Mary." Jim and Mary's badges alert them to your page and create a three-way connection for hands-free, wireless voice communications. The system allows users to be tracked geographically around a site and supports text messaging. The device is being marketed in several industries, including hospitals where immediate communication among the staff can mean the difference between life and death.

WiMAX, also known as *IEEE 802.16,* is the next generation wireless broadband technology that is both faster and has a longer range than Wi-Fi. WiMAX is built on Wi-Fi standards and is able to interoperate with Wi-Fi networks. A WiMAX antenna (Figure 5-25) has a 31-mile (50-kilometer) range and is a perfect technology to provide citywide high-speed Internet access. WiMAX is a good fit for large cities such as New York, Chicago, and Las Vegas that want to provide citywide Internet access to pedestrians. In fact, Seattle's famous Space Needle has become home to a WiMAX antenna that is providing connection speeds double and triple those provided over cable modem, at premium prices to business users citywide.[8] Cell phone manufacturer Nokia has suggested that cell phone towers might include WiMAX transceivers to complement traditional cell phone voice lines and support future increases in media downloads. WiMAX has potential to be the next "big thing" in wireless technologies. Both Intel and Nokia are working to push WiMAX into the mainstream.[9]

FIGURE 5.24 • The Vocera Communications badge

Using VoIP, this hands-free device allows users to communicate with and locate others on the premises.

FIGURE 5.25 • WiMAX antenna

A WiMAX antenna can send a high-speed Internet signal up to 31 miles (50 kilometers).

Bluetooth

Bluetooth technology is becoming quite well known. Go shopping for a new cell phone, handheld, or notebook computer, and there is a good chance that you will find Bluetooth listed among the specifications. The Bluetooth specification was developed by the Bluetooth Special Interest Group (BSIG), a trade organization composed of leaders of the telecommunications, computing, and network industries, such as 3Com, Agere, Ericsson, IBM, Intel, Microsoft, Motorola, Nokia, and Toshiba. **Bluetooth** (named after a 10th-century Danish king) enables a wide assortment of digital devices to communicate directly with each other wirelessly over short distances. Table 5.5 lists some Bluetooth-enabled devices.

Bluetooth-enabled devices communicate directly with each other in pairs. Up to seven devices can be paired simultaneously. The pairings may be created automatically or manually. For example, you might use a wireless headset to chat on a cell phone stored in your backpack. Your handheld computer can be set to automatically synchronize with your personal computer when within range. The Lexus is among several cars that can pair with your cell phone to provide hands-free phone use. When you start the car, the car automatically connects with your cell phone and displays your call info on an LCD display in the dash. You can scroll through your address book, make calls, and check messages

TABLE 5.5 • Bluetooth-enabled devices

Some Bluetooth-enabled devices		
Personal computers	Mobile phones	Automobiles
Printers	Handheld computers (PDAs)	Microwave ovens
Keyboard and mouse	Digital cameras	Refrigerators
	Portable MP3 players	Washers and dryers
	Headphones and headsets	
	Speakers	

all through a touch screen on the dash (see Figure 5.26), using controls on the steering wheel, or by speaking voice commands. The car utilizes a microphone and speaker on the driver's side for phone conversations. BMW, DaimlerChrysler and Ford offer some form of Bluetooth connectivity in all or most of their cars in the United States. GM, Honda, Toyota, Lexus, and Volkswagen/Audi each offer at least one car with integrated Bluetooth.[10]

FIGURE 5.26 • Lexus and Bluetooth

Once the Lexus pairs with your cell phone, you can make calls using the touch screen, controls on the steering wheel, or voice commands.

Bluetooth communicates at speeds of up to 1 Mbps within a range of up to 33 feet (10 meters). Bluetooth can also be used to connect devices to a computer network using access points like Wi-Fi. Bluetooth and Wi-Fi compete in some areas, but have unique qualities. Manufacturers are installing Bluetooth chips in a wide variety of communications and computer appliances to allow device-to-device connections. For example, six participants sitting around a conference table could exchange notes or business cards among their notebook, tablet, or handheld computers. In contrast, Wi-Fi is generally used to connect devices to a network and the Internet.

Infrared Transmission

Another type of wireless transmission, called *infrared transmission*, involves sending signals through the air via light waves. As mentioned earlier, these light waves are longer than the visible spectrum but shorter than radio waves. Infrared transmission requires a direct line-of-sight connection and operates at short distances. For example, your television remote control uses infrared to send signals to your TV, but it must be relatively close and have an unobstructed line-of-sight to the TV. Most notebook and handheld computers support the infrared data communications standard from the IrDA (Infrared Data Association) and include infrared ports called *IrDA ports*. External IrDA devices can be purchased and connected to desktop computers and printers. To transfer data between devices, you simply line up the infrared ports of the two devices within a couple feet of each other to create the connection. Once a connection is established, the operating system provides instructions or a wizard to allow you to share files.

Although infrared is slower than both Bluetooth and Wi-Fi, it has an advantage in that it uses light rather than broadcast technology. This makes it ideal for secure data transmissions that you do not wish to have intercepted by spying devices. For instance, transferring credit card information to a cash register could be performed with infrared because the signal is directed to a specific device and is less likely to be hacked. ZOOP (*www.mzoop.com*), a South Korean company, provides a universal mobile payment system for cell phone users in South Korea and Japan. The system stores credit card information on the user's cell phone that can be transferred to cash registers, vending machines, and mass transit vehicles to pay for merchandise and services with a push of a button and an infrared signal.

Radio Frequency Identification (RFID)

Radio frequency identification (RFID) uses tiny transponders in tags that can be attached to merchandise or other objects and read using an RFID transceiver or reader for the purpose of identification. Primarily used to track merchandise from supplier to retailer to customer, it is anticipated that RFID will eventually replace the bar code as the primary identification system for merchandise. The most common type of RFID tag consists of a transponder as small as a grain of sand, with an antenna embedded on a paper tag (see Figure 5-27). Most commonly used tags are passive; that is, the tag itself is not self-powered, but gets its power from the magnetic field used by the reader, and uses that power to transmit data to the reader. Wal-Mart has required its suppliers to include RFID tags on all shipping crates and pallets. The tags save time when taking inventory and allow retailers to maintain up-to-the-minute inventory control to keep shelves full and hot items in stock.

Besides its use in retail inventory control, RFID is being used in a number of other areas where automatic identification is useful. Hospitals like the Jacobi Medical Center in New York City are employing RFID patient wristbands to not only enhance patient care and staff working conditions, but save on money and mistakes.[11] The Lucile Packard Children's Hospital at Stanford University uses RFID to track the location of its newborn patients with a system that ensures the IDs won't be removed without permission. Delta and other airlines are integrating RFID into their baggage-handling systems to more efficiently track the location of passenger baggage.[12] Here are some other examples:

- RFID devices are being injected under the skin of pets, where they remain for the lifetime of the pet to assist in tracking the pet should it become lost.

FIGURE 5.27 • Radio frequency ID

RFID tags save time for Wal-Mart's distribution system. Automatic RFID scanning makes it unnecessary to open a box to find out what is inside.

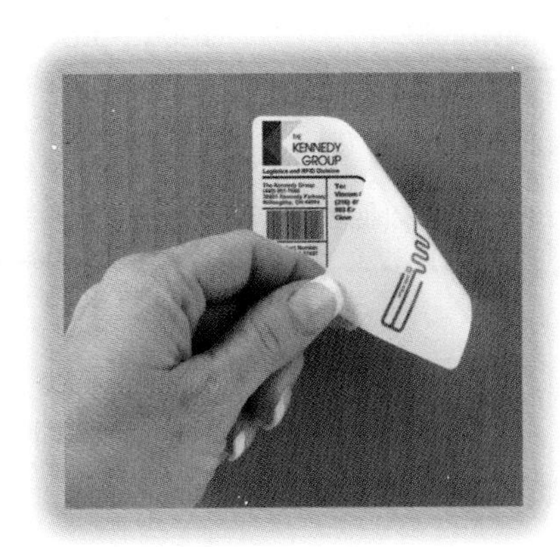

- RFID is used in tagging and tracking wildlife for scientific research.
- RFID is being used to identify people in controlling access to secure locations and information.
- RFID is used in cars to automatically pay tolls as the vehicle cruises through the toll gate.
- The ScripTalk talking prescription reader uses RFID labels on prescription medicine bottles and a reader that speaks the name of the medicine to assist the blind, elderly, and visually impaired.
- In parking lot control, RFID tags are being affixed to authorized vehicles to automatically raise the gate upon approach.
- Championchip USA[13] provides RFID tags for marathons, such as the Boston Marathon, that are attached to runner's shoes to track runners throughout the race and help determine who crosses the finish line first.

With the backing of huge retailers like Wal-Mart, it is clear that RFID technology is poised to take off. There are two issues that might slow the development of this technology: the inability of manufacturers to produce tags fast enough to cover individual merchandise items and concerns over privacy rights. Privacy advocates are concerned about the ramifications of having RFID tags embedded in merchandise. Might it be possible for someone to point an RFID reader at your backpack to see what is inside? Compounding these concerns is talk by some of embedding RFID tags in all automobile license plates, drivers licenses, and money. RFID tags injected under the skin are being used to track ex-convicts and employees in high-security facilities.

LEAVE THE DRIVING TO...THE CAR

Using video to look for stop lights, distance-sensing lasers to keep track of driver blind spots, and radar sensors and microprocessors to reduce the effects of impact in a collision, cars may soon outperform most drivers. DaimlerChrysler's radio-networked cars exchange information with one another at almost unlimited range by relaying through other vehicles. Already on the streets of Europe is their SmartCar, which uses a wireless local area network system and GPS to warn of problems ahead.

Source:
Intelligence: Behold the All-Seeing, Self-Parking, Safety-Enforcing, Networked Automobile
By Paul Horrell
Popular Science
http://popsci.com/popsci/futurecar/article/0,20967,679165,00.html
June 2005

NETWORKS AND DISTRIBUTED COMPUTING

Most of you are familiar with the benefits of sharing information between computers from your experience with the Internet and Web. Information sharing is just one of the benefits of networking.

Within a private network, computing resources are shared in order to maximize computing power. You have learned that a computer includes devices for input, processing, storage, and output. These components can be distributed throughout a computer network. For instance, you may do your work on a terminal that connects you to a server shared by all employees in the organization. In such a scenario the input and output devices are on your desk, but the processing and storage may be handled at some other location. You may store your files on a network drive that exists on a file server and send your print jobs to a printer in a shared area of your office (see Figure 5.28).

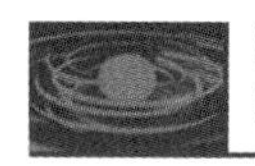

EXPAND YOUR KNOWLEDGE

To learn more about networks, go to www.course.com/swt2/ch05. Click the link "Expand Your Knowledge," and then complete the lab entitled "Networking Basics."

FIGURE 5.28 • Distributed network

On a distributed network you may store your files on a network drive that exists on a file server and send your print jobs to a printer in a shared area of your office.

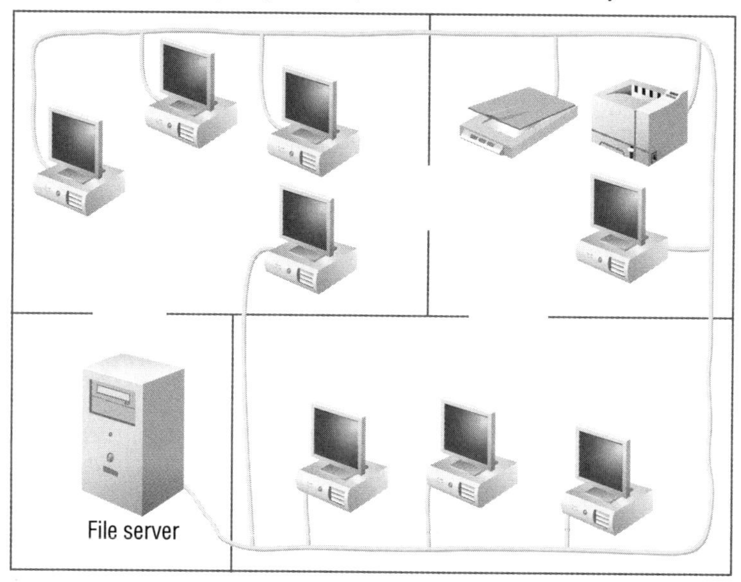

Computer Networking Concepts

Devices attached to a network are called *nodes*. Personal computers attached to a network are often called *workstations*. Workstations typically have access to two types of resources: local and network resources. *Local* resources are the files, drives, and perhaps a printer or other peripheral device that are connected directly to the workstation and accessible on or off the network. *Network* resources, also called *remote* resources, are resources that the workstation accesses over the network.

Network resources are often installed on a workstation in a manner that makes it difficult to tell which resources are local and which are on the network. For example, you might save a document on the F drive. This drive could exist on your own workstation or any other workstation on the network. When you click the Adobe Illustrator icon to start the program, that program might be loaded from the local hard drive or from some other computer on the network. Hiding the underlying network structure from the user makes it invisible, or transparent, and uncomplicated to use. *Transparency* provides a more intuitive and user-friendly computing environment by hiding the complexities of the underlying system from the user. Rather than feeling like you are working on a 100-node network, you get the sense of working on one very powerful computer.

Distributed Computing. *Distributed computing* refers to computing that involves multiple remote computers that work together to solve a computation problem or perform information processing. Large businesses and organizations, sometimes called *enterprises*, make extensive use of distributed computing. In enterprises, distributed computing generally has meant putting various steps of business processes at the most efficient places in a network of computers. The user interface processing is done on the PC at the user's location, business processing is done on a remote computer, and database access and processing is done on another computer that provides centralized access for many business processes. Typically, this kind of distributed computing uses the client/server network model.

In Chapter 4 you learned that the Internet uses *servers* to serve up Internet resources, such as Web pages and e-mail, and it uses clients, such as Internet Explorer, to access those resources. Private networks also make use of client/server systems. *Database servers* store organizational databases and respond to user queries with requested information. *File servers* store organizational and user files, delivering them to workstations on request. *Application servers* store programs such as word processors and spreadsheets and deliver them to workstations to run when users click the program icon.

Hewlett-Packard (HP), IBM, Sun Microsystems, and other enterprises provide a form of distributed computing through a technology called *blade computing*. Blade computing takes advantage of the fact that of the many PCs installed in an enterprise, typically around 30 percent are not being used at any given time. In an organization with 1000 PCs, 300 of those PCs are not really needed. Rather than removing 300 PCs from employees' desks and asking employees to share, PCs are replaced with stripped-down network PCs called *thin clients* that cost less than half a full-blown PC. A *thin client* includes a keyboard, a mouse, a display, and a small system unit that supplies only enough computing power to connect the device to a server over the network. Thin clients connect to clusters of blade servers. *Blade servers* are like PC motherboards that are rack-mounted together in groups of up to 20 to a case (see Figure 5.29). When an employee sits down to work at a workstation, the thin client connects with one of the available blade servers to provide a typical PC work environment. The underlying technology is transparent, so unless they know they are using a blade server, employees think they are using a standard PC. In this type of system, rather than purchasing 1000 PCs, an organization can purchase 1000 inexpensive thin clients to connect to a blade server that supports up to 700 users simultaneously. This system, paired with a file server that stores user files, is significantly less costly than 1000 regular PCs, but offers identical services to users. System administrators find blade systems much easier to maintain than traditional PCs.

In some cases, each blade can act as a multiuser system. For example, clothing retail giant Arcadia, based in the United Kingdom, supports the information system needs of 2000 outlet stores, with 1800 users operating the systems in its London and Leeds headquarters. Arcadia implemented six IBM BladeCenter systems, with a total of 80 dual-processor HS20 blade servers. Each blade supports around 40 users. They chose a blade system for its modular design. In the unlikely event of a hardware failure, the failed blade unit can simply be swapped out; a process that takes less than a minute. Arcadia also likes the fact that as they grow, all they need to do is add additional blades to support their network needs.[14]

FIGURE 5.29 • Blade server and thin client

HP's blade server includes many separate circuit boards that handle the processing for each thin client workstation.

Network Types

Networks are classified by size in terms of the number of users they serve and the geographic area they cover. From a network that links two personal devices that serve an individual user, to international enterprise networks that serve large corporations, to the Internet, which serves the entire world, different types of networks are uniquely designed to accommodate the specific needs of their environment. Types of networks include personal area networks, local area networks, virtual private networks, metropolitan area networks, wide area networks, and global networks. This section also discusses home networking technologies and electronic data interchange (EDI), a network technology that provides links between businesses.

Personal Area Network (PAN). A **personal area network (PAN)** is the interconnection of personal information technology devices, typically wirelessly, within the range of an individual (typically around 33 feet or 10 meters). The Bluetooth networking standard discussed earlier brought attention to the conveniences offered by PAN technology. Using PAN technology you might be able to use one set of wireless headphones for both your digital music player and your cell phone. A PAN allows your notebook computer to automatically communicate with your cell phone to connect to the Internet. Your PAN could interact with other PANs to transfer meeting notes to others at a committee meeting or to allow your friends to listen to music from your digital music player on their headphones.

PAN Gets More Personal

Prior to Bluetooth, the expression *PAN* was coined by Thomas Zimmerman, a student at the MIT Media Lab back in 1996, for a unique new networking device he developed and embedded in a pair of gym shoes. Two individuals wearing the customized gym shoes could pass information from one device to the other simply by touching each other. Zimmerman's PAN uses a small transceiver, about the size of a thick credit card, and the natural salinity of the human body to conduct an electrical current through a person's body to carry data. The current is so tiny that the user cannot feel it, but can carry up to 400,000 bits per second.

While Bluetooth has become the default technology for today's PANs, some of Zimmerman's ideas are surfacing once again. In June 2004 Microsoft patented a technology they call Personal Area Network that uses human skin for transmitting power and data to devices worn on the body and for communicating data between those devices. Around the same time News.com.au published a story entitled "Exchange e-mails through handshake" that described technology from Japanese telecom giant DoCoMo that turns the human body into a broadband-paced link that allows e-mail addresses to be exchanged through a simple handshake. With DoCoMo's technology a cell phone in a backpack could send data through the material of the backpack and a person's clothing to the person's skin, to be transmitted through touch to another person's device in his or her backpack.

Although it may be a while before this technology is on the streets, it is clear that some big companies are taking it seriously.

Questions

1. How might you use technology like Microsoft's PAN to support device-to-device data communications?
2. List two situations where you might find it useful to pass digital information to someone through touch.
3. What benefit might this technology have over Bluetooth technology?

Sources

1. *IBM User System Ergnomics Research Web site, www.almaden.ibm.com/cs/user/pan/pan.html, accessed June 23, 2005.*
2. *Loney, M. "Microsoft Patents Bodybus,"* CNet News, *http://news.com.com, June 23, 2004.*

Local Area Network (LAN). A network that connects computer systems and devices within the same building or local geographical area is a **local area network (LAN)**. A local area network can use various designs, called *topologies*, such as ring, star, bus, or a combination of these, as illustrated in Figure 5.30. There are more local area networks than any other network type. LANs are used in homes, businesses, and other institutions and organizations.

Local area networks can include personal computers, servers, printers, and other network capable devices. Devices connect to LANs through network interface cards or wireless network adapters. Larger networks make use of servers to store databases, files, and programs. When a person on the network uses a program or data stored on the server, the server transfers the necessary programs or data to the user's computer. While servers are typically large multiuser computers, a server can be a computer of any size, even a personal computer—any computer that serves up information or services to others on the network can be a server.

Since the rise in popularity of the Internet and Web, many LANs are incorporating familiar Internet technologies to create intranets. An **intranet** uses the protocols of the Internet and the Web—TCP/IP and HTTP, along with Internet

FIGURE 5.30 • Local area network

A LAN can include various topologies, such as the star and bus configurations shown here, as well as servers and assorted network hardware.

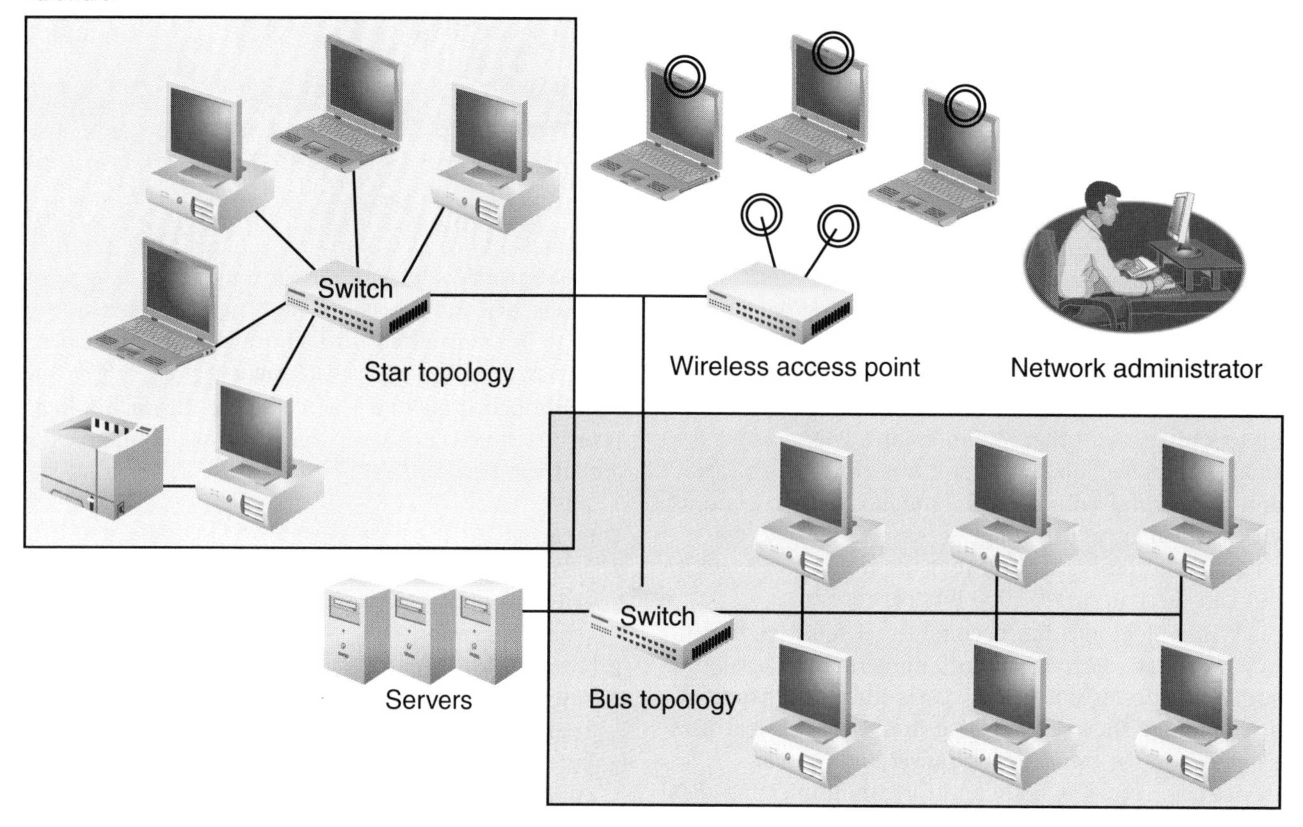

services such as Web browsers—within the confines of a private network (see Figure 5.31). In an intranet, employees might access confidential documents using a Web browser, while those same documents remain secure from the outside world.

Enterprises typically allow users within their intranet to access the public Internet through firewalls that screen messages to maintain the security of the private network. An intranet may be extended beyond the confines of the LAN to connect with other networks to create a virtual private network. A **virtual private network (VPN)** uses a technique called *tunneling* to securely send private network data over the Internet. A VPN may be used to connect an organization's networks dispersed around the world into one large intranet.

Intranet content can be extended to specific individuals outside the network, such as customers, partners, or suppliers, in an arrangement called an *extranet.* Extranets are sometimes implemented through a simple login procedure on a Web server. For example, Wal-Mart provides key suppliers with access to its intranet so that they can see what products are selling fastest and ramp up production to meet the demand.

VPNs can also be used to allow employees access to the corporate intranet from home and while on the road. A large business or organization might hire an *enterprise service provider (ESP)* to set up a *network access server (NAS).* Users are provided with software that connects to the NAS, VPN, and ultimately the corporate intranet. Some services are set up so that if you have access to the Internet, then you also have secure access to your private intranet.

FIGURE 5.31 • Intranet

In an intranet, a Web server provides confidential data to LAN users, while keeping the data safe from those outside the organization through the use of a firewall.

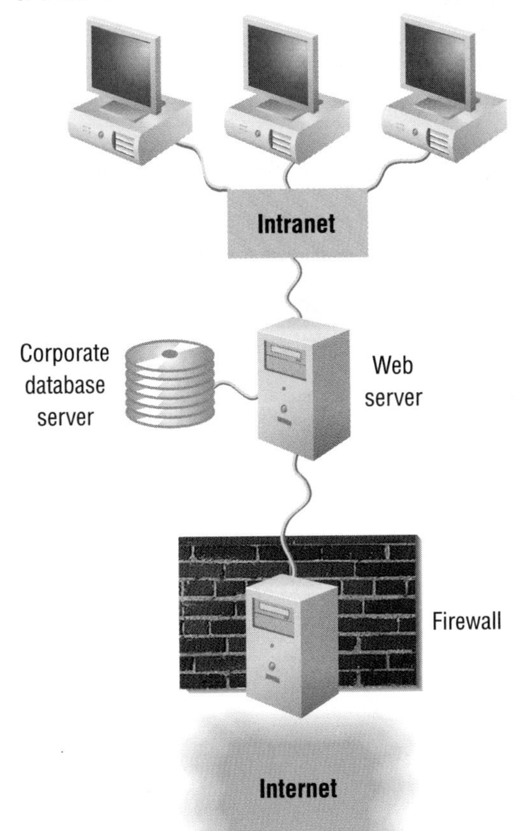

Metropolitan Area Network (MAN). A **metropolitan area network (MAN)** connects networks within a city or metropolitan-size area into a larger high-speed network. Many cities supply local businesses with access to a MAN to improve local commerce and communications. Often a MAN acts as a stepping stone to larger networks, such as the Internet. The WiMAX wireless high-speed technology introduced earlier is perfect for setting up a MAN as it has a range of up to 30 miles (50 kilometers). The city of Minneapolis is investing $15 to $20 million to create a wireless and fiber-optic MAN. The purpose is to improve government communications by linking every city building, police car, and housing inspector. The MAN will also be available to individuals in the city for $18 to $24 a month.[15]

Wide Area Network (WAN). A **wide area network (WAN)** connects LANs and MANs between cities, across country, and around the world using microwave and satellite transmission or telephone lines. A LAN becomes a WAN when it extends beyond one geographic location to another geographic location (see Figure 5.32). When you make a long-distance phone call, you are using a wide area network. Cingular, MCI, Sprint, and other telecommunications companies are examples of companies that offer WAN services to the public. Companies, organizations, and government agencies also design and implement WANs for private use. These WANs usually consist of privately owned LANs connected over a dedicated line provided by a telecommunications company. For example, your college may maintain a LAN that covers the campus. The college network engineers may have laid fiber-optic cable across campus and connected it to networks in each building to provide high-speed networking to students, faculty, and staff. This network is owned and controlled by the college. If your college should decide to open a branch campus across the state, administrators may decide to join the LAN of the main campus with the LAN of the branch campus. Using a dedicated line leased from the phone company, your college creates a WAN by joining the two LANs.

Global Networks. A WAN that crosses an international border is considered a global or international network. The Internet is the most obvious global network, but as an increasing number of businesses are entering global markets, private global networks are becoming more prevalent.

However, creating and maintaining a global network has its challenges. In addition to requiring sophisticated equipment and software, global networks must meet specific national and international laws regulating the electronic flow of data across international boundaries, often called *transborder data flow.* Some countries have strict laws restricting the use of telecommunications and databases, making normal business transactions such as payroll costly, slow, or even impossible. Other countries have few laws restricting the use of telecommunications or databases. Other governments and companies can avoid their own country's laws by processing data within the boundaries of other countries, sometimes called *data havens,* that have few restrictions on telecommunications or databases. For example, the popular file-sharing service, KaZaA, has been able to escape prosecution because it maintains its servers in Denmark, has its domain registered in Australia, and runs its software from the South Pacific island nation of Vanuatu, a well-known tax haven.

FIGURE 5.32 • Wide area network (WAN)

A WAN connects LANs between cities, across countries, and around the world, typically using lines leased from telecom companies.

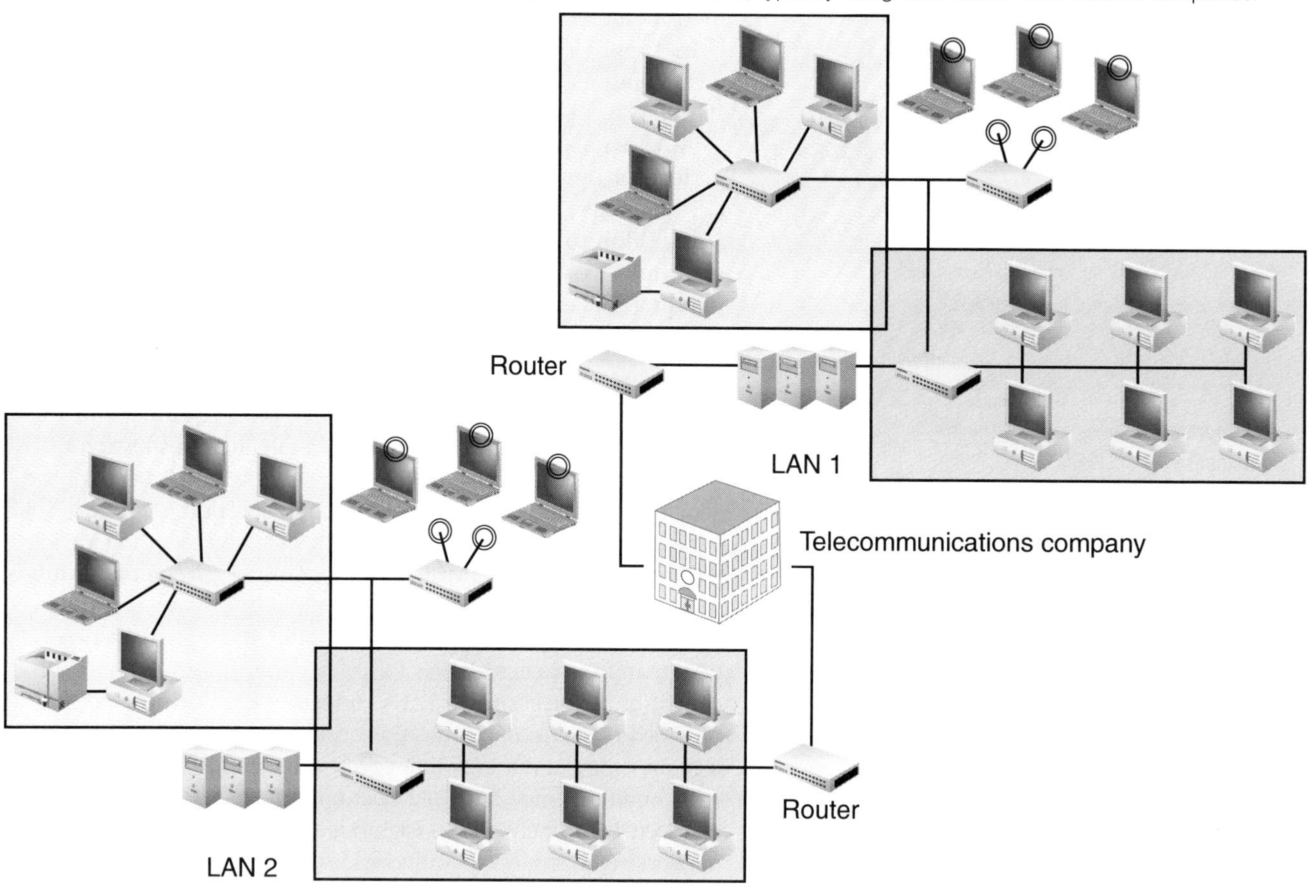

Despite the obstacles, there are numerous private and public international networks. United Parcel Service (UPS), for example, covers over 200 countries with its international UPSnet network. UPS drivers use handheld computers called *DIADs* (delivery information acquisition devices) to send real-time information about pickups and deliveries to central data centers via the global network (see Figure 5.33). The GPS integrated into the system is not yet activated, but will notify drivers if they are about to deliver a package to the wrong location. In addition to the 90,000 handheld computers for drivers, UPSnet uses 15 mainframes, 2,202 mid-range computers, more than 115,00 personal computers, and 8,700 servers. The global network allows data to be retrieved by customers to track packages or to be used by the company for faster billing, better fleet planning, and improved customer service.

Electronic Data Interchange (EDI). Connecting corporate computer systems among organizations is the idea behind electronic data interchange (EDI). *EDI* uses network systems and follows standards and procedures that allow output from one system to be processed directly as input to other systems, without human participation. With EDI, the computers of customers, retailers, manufacturers, and suppliers can be linked. For example, as the cashier scans the barcode of the new jeans that you are purchasing, the inventory count of that item is decreased by one. The effect of the decrease may bring the total inventory amount for that item below a set threshold, indicating that more jeans need to be ordered. Because the retail store's computer system is connected to the manufacturer's computer system, the order for more jeans can be made automatically.

FIGURE 5.33 • UPS DIAD

A poster child for wireless networking, the latest version of the UPS DIAD is equipped with a barcode scanner, GPS receiver, a GPRS or CDMA radio, an infrared port, Bluetooth, and Wi-Fi to assist drivers in organizing and delivering packages.

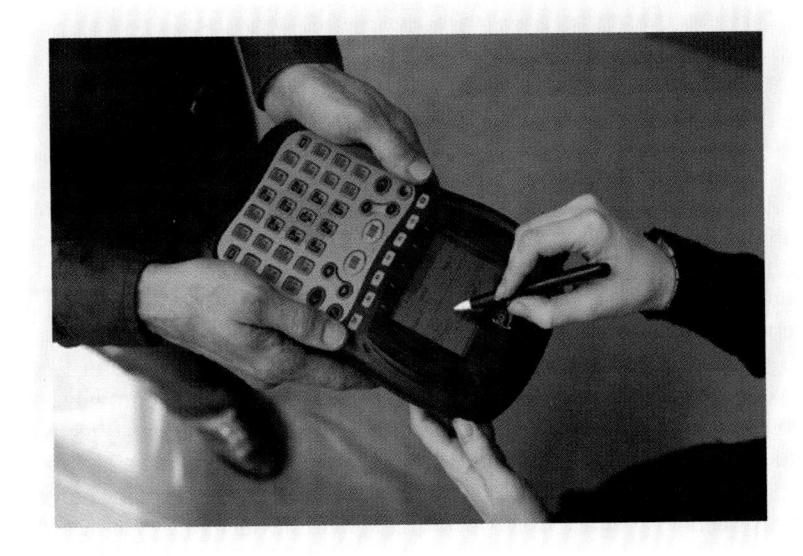

The manufacturer's computer system receives the order and places the order in the queue to be processed. The manufacturer's computer system may be connected to their suppliers' computer systems. So as they begin running low on indigo dye, for example, an order for more is automatically generated and delivered to the dye producer.

For some industries, EDI is becoming a necessity. For many large companies, including General Motors and Dow Chemical, computer input often originates as output from another computer system. Some companies only do business with suppliers and vendors using compatible EDI systems, regardless of the expense or the effort involved. As more industries demand that businesses have this ability to stay competitive, EDI will cause massive changes in the work activities of companies. Companies will have to change the way they deal with processes as simple as billing and ordering, while new industries will emerge to help build the networks needed to support EDI.

Home Networks

Like televisions, telephones, and automobiles, personal computers have become such an integral part of many people's lives that many households have more than one. As PC prices continue to decrease, it is increasingly common to find one PC per individual in a household. Whether the individuals are family members or roommates, they can benefit from connecting their computers in a home network. Home networks allow residents to:

- Share a single Internet connection.
- Share a single printer between computers.
- Share files such as images, music, and programs.
- Back up copies of important files to another PC for safekeeping.
- Participate in multiplayer games.
- Share output from devices such as a DVD player or Webcam.

In the past, setting up a home network was an intimidating challenge. The computer industry has recognized home networking as one of the most important and lucrative markets of this century and has made available many new technologies that vastly improve the ease with which a home network can be installed.

PERIPATETIC PRODUCTIVITY

Wireless and networking technologies have created a new brand of employee: the teleworker. Companies like Sun Microsystems, IBM, and Cisco have developed programs where employees don't just work from home, they work from wherever they are, sometimes at a branch, sometimes on the commuter train. Sun reported saving $69 million in 2005 just by cutting back on office space and administrative costs, and proponents say people are more productive at home where interruptions are fewer.

Source:
Work is where you hang your coat: Sun leads way in telework—working not just from home but anywhere
By Carolyn Said
San Francisco Chronicle, July 18, 2005

Setting up a Wireless Home Network

To set up a wireless home network you need a high-speed Internet connection, a wireless access point/router, and wireless network adapters for all devices that will connect to the network (see Figure 5.34).

Access points and network adapters come with software that is easy to install, and it works with your operating system to automatically configure your network. By following simple instructions, you can have your home network set up within an hour.

Because the range on access points is limited, it is best to find a location for the access point that is central to the area in which you will use the network. If your cable modem or DSL modem is not located in a central location, you need to move it by either switching it to another connector, or running a length of cable to a central location. Make sure that the location you select for your access point is not near any electronic equipment that may interfere with the signal, such as stereo speakers or power transformers.

Connect the access point to the modem, and power it up. Read the instructions for installing the wireless network adapter to your PC. Once installed, you can access the setup software on the access point and configure your network settings. Wireless network adapters can be installed on all PCs in the residence. You can also use wireless network adapters for printers, televisions, and stereos.

Once your computers are communicating with each other over the network, network users must determine what resources to make available to others on the network. Today's operating systems provide methods for specifying which drives, folders, directories, files, and printers you wish to share over the network. How you share and access network resources varies from operating system to operating system.

A primary concern about wireless networks, and perhaps their only drawback, is security. By default, Wi-Fi networks are set with no security in place. Once set up, they broadcast their existence to anyone within range, inviting them to join the network. With a click of the mouse you, your neighbors or someone passing on the street can all join the network. This makes it convenient to set up the network and get working, but additional steps are required to secure it so only intended individuals may use it. Chapter 11 provides instructions for locking down your home wireless network to keep your information secure.

Home Networking Technologies. Many of the technologies discussed in this chapter are applicable to home networks. Home networks are typically based on Ethernet standards and can be wired or wireless. Some technically inclined people may opt to run twisted pair cable through the walls and attic space of their homes to set up a business-quality network. However, most home users are taking advantage of more convenient setups, such as:

- Phone-line networking: Also called *HomePNA* (for phone-line networking alliance), phone-line networking takes advantage of existing phone wiring in a residence. Computers share the phone line with telephones, utilizing different frequencies so that both can be used simultaneously. "If you can plug in a phone, you can network your home" is the slogan of the Home Phoneline Networking Alliance.[16]
- Power-line networking: Also called *HomePLC* (for power-line communication), Power-line networking takes advantage of the home's existing power lines and electrical outlets to connect computers. As with Home PNA, HomePLC divides up the frequencies of the household electrical system and uses some for network communications and the others for electrical current.

FIGURE 5.34 • Wireless home network

A typical wireless home network uses a wireless access point/router to connect the network to an ISP. Wireless adapters connected to each computer communicate with the access point.

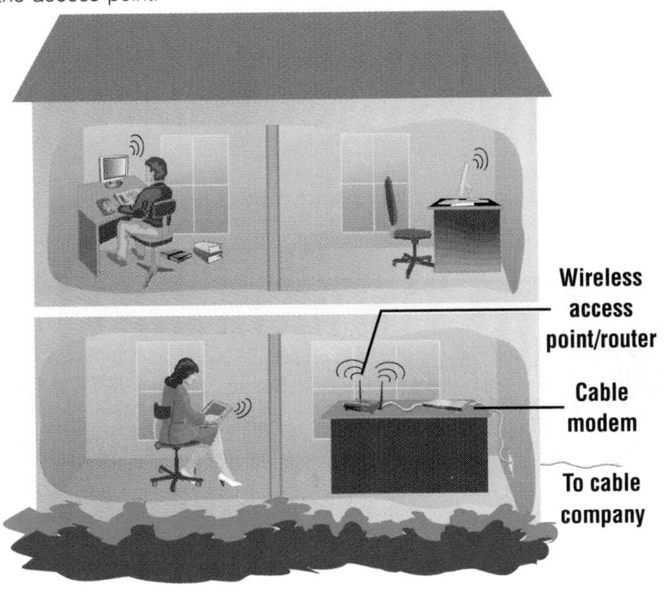

- Wireless networking: The popularity, convenience, and freedom of Wi-Fi has vastly overshadowed phone-line and power-line technologies. With speeds up to 54 Mbps, home wireless networks easily support everything a typical home user might want to do on a network.

Home Network Applications. The rapid rise in wireless home networks is spurring the development of new technologies that use Wi-Fi networks to provide useful residential services. Two of the most prevalent are technologies that assist in (1) data storage and access, and (2) wireless media distribution throughout the home. Here is a brief description of products that target these needs.

Like businesses, families are beginning to find a need for centralized file servers. For example, family members may take hundreds of digital photos that they would like to catalog and share. Rather than trying to remember whose computer holds which photos, it makes sense to store them all in one central location. Over a home network this can be accomplished by setting up a file server, connecting it to the router/access point, and naming it a common drive letter on all network PCs. So, for instance, if you save your photos to the P drive on any computer in the house, it is stored on one central location on the network.

In addition to storing and backing up data and files, home networking technologies can be used to distribute media files throughout your home. The SoundBlaster wireless music system allows you to transmit your MP3 music from your computer to any stereo or powered speaker system in your home over a Wi-Fi network (Figure 5.35). Connect the system to your stereo using standard stereo cables, install the software on your PC, and control the music using a remote control. The remote control lets you browse through your song lists to select the music you want to hear. For Apple users, the AirPort Express can be used to share music from your Apple computer with other computers or stereo systems in the house. Other companies such as Linksys, Gateway, and Prismiq also have Wi-Fi devices that play either audio files or both audio and video files, and sometimes streaming media from the Internet. Other technology allows you to send television signals from TV to TV in order to share a cable or satellite connection without wires.

FIGURE 5.35 • Wireless home stereo

With this SoundBlaster equipment, you can broadcast MP3 music files from your computer to your home stereo system.

ACTION PLAN

Remember Amanda Jackson from the beginning of this chapter? She was the NPR news correspondent shopping for field equipment. Here are answers to the questions about Amanda's situation.

1. What types of handheld devices best suit Amanda's journalistic needs?

Some of today's high-end smart phones include voice recording, note taking with a stylus or QWERTY keyboard, and cell phone capabilities. She will want to choose a Quad band world phone that works on a variety of international networks. Because GSM networks are an international standard, she should probably look to GSM carriers such as Cingular and T-Mobile. When choosing a cellular service plan, she should invest in one that supports global travel. Because her cell phone will be her lifeline, she probably wants to include several add-on features and services. Certainly a GPS locator will be of great use when visiting foreign cities. A phone with a camera may be of use when gathering facts for her stories. She may also want to use her phone for relaxation by storing MP3s and accessing streaming video—and don't forget headphones!

2. What type of notebook computer, networking media, devices, and software should Amanda use to connect to the Internet and the NPR private network?

Amanda should consider a notebook with Centrino technology. Centrino includes built in Wi-Fi networking capabilities and extended battery life. If she prefers an Apple notebook, the PowerBook G4 also has built in Wi-Fi network connectivity. Wi-Fi networks are popping up all over, and there is a good chance Amanda can find an access point wherever she travels. As a backup she should make sure that her notebook includes an Ethernet cable port. Amanda might consider subscribing to the Boingo service that provides access to 18,000 hotspots worldwide. Amanda could also add broadband Internet access to her cell phone service and connect to the Internet through her phone. Amanda may need to install VPN software on her notebook in order to connect to the NPR private network over her Internet connection.

3. What networking technologies can Amanda use to transfer files between her devices?

Amanda might consider getting a Bluetooth-capable cell phone and a Bluetooth adapter for her notebook. This allows for a wireless connection between the devices to transfer data and to connect to the Internet. Smart phones typically come with a desktop cradle that can be used to transfer data to a notebook via the USB port.

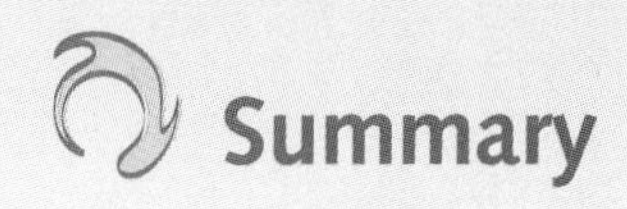

Summary

LEARNING OBJECTIVE 1

Understand the fundamentals of data communications and the criteria for choosing a communications medium.

Communication takes place between sender and receiver by way of a signal that travels through a communications medium. Telecommunications refers to the electronic transmission of signals for communication. Data communications is a type of telecommunications that involves sending and receiving bits and bytes that represent data. A computer network connects computers for data communications.

Figure 5.1—p. 233

Telecommunications involves three components: networking media, devices, and software. A data-bearing signal travels over the media between devices that act as relay points. Network software controls the devices to manage telecommunications signals in an economic and efficient manner.

Telecommunications networks manipulate both analog and digital signals. The transmission speed of a given medium is dictated by the signal frequency, measured in hertz (Hz), and described in terms of the number of bits per second (bps) that the medium can deliver. The range of frequencies that can be sent over a given medium is known as its bandwidth.

LEARNING OBJECTIVE 2

Explain how networking media, devices, and software work together to provide data networking services, and describe the benefits of various types of media.

Network media include cables and wireless signals. The most common types of cables used in telecommunications are twisted pair cables, coaxial cable, and fiber-optic cable. Fiber-optic cable is the fastest cable medium. Wireless communications media includes radio waves and infrared light. The U.S. Federal Communications Commission assigns different frequencies of radio waves for different uses. High-frequency signals are called *microwaves.*

Figure 5.6—p. 237

Networking devices include modems, network adapters, network control devices, RFID devices, and pagers. Modems connect computers to various types of communication media. Network adapters are circuit boards, PC cards, or external devices that allow a computer to connect to a computer network. Hubs, switches, repeaters, bridges, gateways, and routers are used to control computer network traffic. A wireless access point connects wireless devices to a Wi-Fi network. A firewall can be either a device or software that filters the information coming onto a network to protect network computers from hackers, viruses, and other unwanted network traffic.

Microwave transmission sends signals through the air from tower to tower across the land or up to satellites that retransmit the signal to another location on earth. Several other communications devices and media are designed for industrial use. T1 and T3 carrier lines are used to support high-demand network traffic. Multiplexers and communications processors assist in managing the flow of information in networks with large quantities of network traffic. Encryption devices secure network traffic by encrypting data on the network so that it is unintelligible to all but intended receivers.

A network administrator is a person responsible for setting up and maintaining the network, implementing network policies, and assigning user access permissions. A network operating system (NOS) is installed on network servers and workstations, and controls the computer systems and devices on a network, enabling them to communicate with each other. Ethernet is the most widely used network standard for private networks.

LEARNING OBJECTIVE 3

List and describe the most popular forms of wireless telecommunications technologies.

A cellular network is a radio network in which a geographic area is divided into cells with a transceiver antenna (tower) and station at the center of each cell to support mobile communications. When choosing a cell phone service, there are three primary decisions to make: choose a carrier, choose a plan, and then choose a phone with particular features and services. A cellular carrier is a company that builds and maintains a cellular network and provides cell phone service to the public. The predominant digital networking standards used for cell phones networks are GSM and CDMA. GSM is the most popular global standard for mobile phones and is used by over a billion people across more than 200 countries. The CDMA networking standard is predominantly used in the United States where it is in equal competition with GSM. Different carriers offer different cell phones, features, coverage areas, and services.

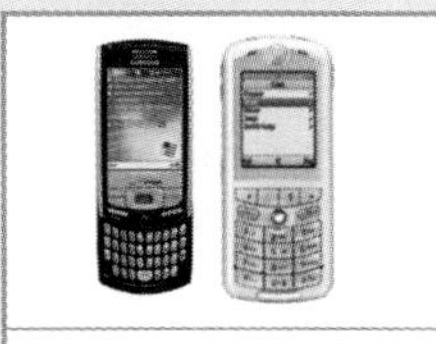

Figure 5.19—p. 251

Pagers are small lightweight devices that receive signals from transmitters. Global Positioning Systems (GPS) use a constellation of satellites to pinpoint the location of a GPS receiver on earth. Wireless fidelity (Wi-Fi) uses radio signals to connect computers to a network, which is typically connected to the Internet. WiMAX is the next generation wireless broadband technology that is both faster and has a longer range than Wi-Fi. Bluetooth-enabled devices use radio signals to communicate between personal and mobile devices. Infrared transmission uses infrared light to transfer data between devices at close range without wires. A radio frequency identification (RFID) device is a tiny microprocessor combined with an antenna that is able to broadcast identifying information to an RFID reader.

LEARNING OBJECTIVE 4

List the different classifications of computer networks and their defining characteristics, and understand the basics of wireless home networking.

Many large businesses and organizations use the client/server network architecture. Server computers are used to distribute data, files, and programs to users, or clients, on the network. Computers connected to a network are called *workstations*, or nodes. A workstation has access to the local resources and network, or remote resources.

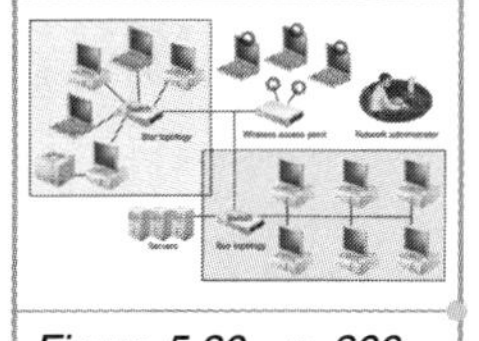

Figure 5.30—p. 266

Networks are classified based on size. From smallest to largest they are PAN, LAN, MAN, WAN, GAN. A personal area network (PAN) is the interconnection of information technology devices within the range of an individual. A network that connects computer systems and devices within the same geographical area is a local area network (LAN). A metropolitan area network (MAN) connects networks within a city or metropolitan-size area into a larger high-speed network. Wide area networks (WANs) tie together geographically dispersed LANs. WANs that cross international borders are called *global networks*. An intranet is a private network, set up in an organization, based on Internet protocols. When intranets include specific outside parties, it becomes an extranet. Intranets sometimes use the Internet to connect geographically dispersed networks in a virtual private network (VPN) using tunneling technology.

Through electronic data interchange, or EDI, networks owned by different organizations can be joined and programmed to communicate so that the output of one system is processed as input by the other. EDI allows organizations to automate many time-consuming tasks.

Home networks are used to share hardware, files, and a common Internet connection. A modem, which provides the link between the ISP and the home computer, can be connected to a single computer or all computers within the home network. Home network technologies include phone-line networks (HomePNA), power-line networks (Home PLC), and wireless networks. Wireless home networks typically require a wireless access point and a wireless adapter for each computer on the network. Wireless networks require additional setup to ensure that the signals sent and received are secure and not accessible to others outside the network.

Test Yourself

LEARNING OBJECTIVE 1: Understand the fundamentals of data communications and the criteria for choosing a communications medium.

1. Networking ________________ is anything that carries an electronic signal and interfaces between a sending device and a receiving device.

2. In computer networks, the data transmission rate is also referred to as the ________________ and is measured in bits per second (bps).
   a. hertz
   b. frequency
   c. bandwidth
   d. broadband

3. True or False: A 56 Kbps modem delivers broadband performance.

LEARNING OBJECTIVE 2: Explain how networking media, devices, and software work together to provide data-networking services, and describe the benefits of various types of media.

4. True or False: For exceptionally high data transmission rates one should look to fiber-optic cabling.

5. A network ________________ is a computer circuit board, PC card, or USB device installed in a computing device so that the computing device can be connected to a network.

6. A network ________________ is a person responsible for setting up and maintaining the network, implementing network policies, and assigning user access permissions.
   a. technician
   b. supervisor
   c. engineer
   d. administrator

LEARNING OBJECTIVE 3: List and describe the most popular forms of wireless telecommunications technologies.

7. True or False: To subscribe to a cell phone service you must sign at least a 1-year contract.

8. Many GSM phones have a ________________ card that stores subscriber information and personal data and can be transferred to other GSM phones.
   a. USB
   b. SMS
   c. SIM
   d. GPS

9. A(n) ________________ tells you the exact location of the receiver on the earth's surface.

10. ________________ enables a wide assortment of digital devices to communicate wirelessly over short distances.
    a. Bluetooth
    b. Wi-Fi
    c. Fiber optics
    d. Microwave transmission

LEARNING OBJECTIVE 4: List the different classifications of computer networks and their defining characteristics, and understand the basics of wireless home networking.

11. True or False: A private network made accessible to select outsiders is called an *intranet*.

12. A(n) ________________ is often used by enterprises to allow employees to access the corporate intranet from home and while on the road.
    a. virtual private network (VPN)
    b. extranet
    c. metropolitan area network (MAN)
    d. router

13. True or False: EDI uses network systems and follows standards and procedures that allow output from one system to be processed directly as input to other systems, without human intervention.

14. The ease of installation of a(n) ________________ network is making it the obvious choice for most home networks.

Test Yourself Solutions: **1.** media, **2.** c. bandwidth, **3.** False, **4.** True, **5.** adapter, **6.** d. administrator, **7.** False, **8.** c. SIM, **9.** global positioning system (GPS), **10.** a. Bluetooth, **11.** False, **12.** a. virtual private network (VPN), **13.** True, **14.** Wi-Fi wireless

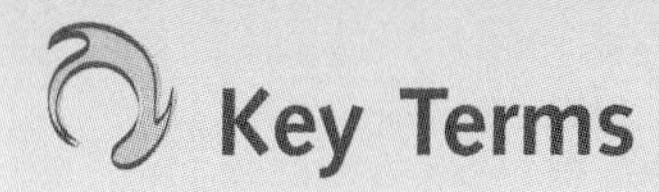

Key Terms

analog signal, p. 234
bandwidth, p. 235
Bluetooth, p. 258
broadband, p. 235
cellular carrier, p. 247
cellular network, p. 246
coaxial cable, p. 237
communications satellite, p. 242
digital signal, p. 234
Ethernet, p. 245
fiber-optic cable, p. 238
Global Positioning System (GPS), p. 253
intranet, p. 265
local area network (LAN), p. 265
metropolitan area network (MAN), p. 267
microwave transmission, p. 242
modem, p. 239
network adapter, p. 240
networking media, p. 233
personal area network (PAN), p. 264
radio frequency identification (RFID), p. 260
radio wave, p. 238
telecommunications, p. 233
twisted pair cable, p. 237
virtual private network (VPN), p. 266
wide area network (WAN), p. 267
WiMAX, p. 257
wireless fidelity (Wi-Fi), p. 256

Questions

Review Questions

1. Besides a sender and receiver, what other components are required for communication to take place?
2. How do analog and digital signals differ?
3. What do you need to keep in mind when deciding on a networking medium?
4. List the three types of cables discussed in this chapter in order of lowest to highest bandwidth.
5. Name three advantages and one disadvantage of using fiber-optic cable as compared to coaxial and twisted pair.
6. Under what conditions might a company consider using a telecommunications satellite over microwave towers?
7. What are the fundamental differences between Wi-Fi and Bluetooth technologies?
8. Name two personal computer devices that are used to connect to networks.
9. What concerns do privacy advocates have about RFID?
10. What is the purpose of a firewall?
11. What is distributed computing?
12. What is the difference between local resources and remote resources on a network workstation?
13. What are three decisions required when purchasing cell phone service?
14. List network types in order of size.
15. What is the most popular type of network?
16. How does an intranet differ from the Internet?
17. What unique concerns are associated with global networks?
18. What are the advantages and disadvantages of GSM network cell phone systems?
19. What equipment is required to set up a wireless home network?
20. List six add-on cell phone services.

Discussion Questions

21. What are some negative effects of insufficient bandwidth for residential networks and professional business networks?
22. What personal and professional benefits are afforded by Bluetooth technology? What effect can Bluetooth have on the life of a person in your future profession?
23. How does a client/server network system assist in managing information in a large organization?

24. Envision yourself in your future career. What role might telecommuting play in your weekly activities? Will you be able to do some or all of your work from a home office? How?

25. What role does teamwork play in your career area? How can a computer network assist team members in their work?

Exercises

Try It Yourself

1. Sit down at a network computer on campus. Use a file management utility (such as My Computer on Microsoft Windows) to determine which disk drives are local resources and which are network resources. What other local and network resources are available on the computer? Printers? Scanners? If you are unsure, ask a computer lab assistant or network administrator. Use a word processor to create a document that lists the location of the computer lab that you used and who manages the network (either a group or person). Include a two-column table listing local resources in the left column and network resources in the right. Use proper column headings and make your document look like an official report.

2. Find the Web page of the organization that provides your campus network. Find the network usage policy or agreement for your campus network. Use a word processor to list the five most interesting activities *not* allowed on your campus network and your rationale for why these rules might exist. Why is a network usage policy necessary?

3. Compare and contrast plans, phones, features and services of two major cell phone carriers. Create a report using a word processor to tell which carrier best meets your needs and explain why.

4. Visit the Web site of the department of your major (or intended major). Create a document that lists the computing environment provided to you by the department. Does your department have its own computer lab, or are students expected to use campus labs? Why do you think this is? Does your field require special computing or networking hardware or software?

5. Use a spreadsheet and the Web to compare the costs of setting up a wired Ethernet network compared to setting up a wireless network. The Ethernet network should include three 10/100 Fast Ethernet adapters, a 10/100 Fast Ethernet hub (with four ports), and three 50-foot Category 5 (Cat5) RJ-45 cables. Your wireless network should include an 802.11g wireless access point and three 802.11g wireless network adapters (USB). You might start your research at *www.cdw.com*. Which network is cheapest? Which network is fastest? Which network would you prefer? Why?

6. Conduct a Web search on *Bluetooth*. Research several informative pages on the topic with at least one positive and one negative perspective. Write a 2-page paper on how Bluetooth is being used, along with a summary of positive and negative comments about it.

7. Conduct a Web search on *network security*, following the links to network security Web sites in an effort to determine what issues are of greatest concern in network security. What appear to be the top five concerns in the field of network security? Rank them in order of importance, and include a brief description of each.

Virtual Classroom Activities

For the following exercises, do not use face-to-face or telephone communications with your group members. Use only Internet communications.

8. Select a U.S. college, and visit the college's Web site. Find information about computer access provided by the college. Does the college provide public computer labs? Does it provide wireless Internet access? If so where? Are the dorms networked? How can notebook computers be connected to the campus network? Use a word processor to list interesting statistics, such as how many computer labs are available, how many computers are in each lab, what types of computers, and so on. Include anything that you find interesting and unique about the college's network setup. Each group member should distribute his or her findings to the group. Then, hold a group discussion and vote to determine which college has the best setup.

9. If you have set up or maintain a home computer network, create a document that lists the networking difficulties that you have experienced along with the benefits that the network has provided. Include in your document a detailed description of the type of network you have and the equipment you use. If you have no experience with home networking, interview someone who does, and write up his or her comments. Swap stories with other group members, and write a summary of shared experiences.

Teamwork

10. Scour the Web in search of the cheapest 802.11g wireless network setup. The wireless setup should include the components discussed in this chapter (see the "Setting up a Wireless Home Network" section), it must accommodate two desktop PCs and two notebook computers, and it should connect all network users to a cable modem Internet connection. The team member who comes up with the cheapest network that meets the requirements wins! Make sure to include shipping costs.
11. The team should place itself in the role of system administrators of a corporate network. Each team member should work independently to design a network-usage policy that restricts employees from wasting time on the corporate network for personal needs, while not overly restricting them. You should address issues such as personal e-mail, personal Web browsing, access to the network from home, and so on. After listing individual ideas for important issues and policies, the team should get together and share their ideas. Work to merge everyone's policies into one cohesive corporate network usage policy. Do not include policies that do not have full team support, but rather list those separately for further discussion in class.

TechTV

Free Wi-Fi

Go to www.course.com/swt2/ch05 and click TechTV. Click the Free Wi-Fi link to view the TechTV video, and then answer the following questions.

1. Summarize the problem that the "last mile" represents. How does Tim Cozar's system solve the problem for San Bruno, California?
2. How do you think Comcast and other professional ISPs react to home-brewed Neighborhood Area Networks? What strategies might they use to discourage this trend?

Endnotes

[1] Cisco Industry Solutions Web page, http://www.cisco.com/en/US/strategy/government/homeland_security.html, accessed June 22, 2005.

[2] Fusco, P. "FCC National Broadband Policy," *ISP-Planet*, November 7, 2001, http://www.isp-planet.com/politics/2001/national_broadband_policy.html, accessed June, 15, 2003.

[3] Burns, E. "Broadband Population Growth Continues," *ClickZ Network*, http://www.clickz.com/stats/sectors/broadband/article.php/3509516, June 2, 2005.

[4] Nortel Small and Medium Business Success Stories Web Site, http://www.nortel.com/solutions/smb/success.html, accessed June 22, 2005.

[5] Speedpass Web site, http://www.speedpass.com/how/index.jsp, accessed September 23, 2005.

[6] Greenspan, R."Global Mobile Population Growing," *ClickZ*, http://www.clickz.com/stats/sectors/wireless/article.php/3377511, July 6, 2004.

[7] "Cellular service: Best carriers," *Consumer Reports*, http://www.consumerreports.org, February 2005.

[8] Wexler, J. "Space Needle Becomes WiMAX Tower," *NetworkWorld*, http://www.networkworld.com/newsletters/wireless/2005/0509wireless2.html, May 11, 2005.

[9] van Grinsven, L. "Nokia and Intel Push to Get WiMAX Out This Year," Reuters, http://www.reuters.com, June 10, 2005.

[10] Rothman, W. "Your Phone Is Calling Your Car," *New York Times*, http://www.nytimes.com, May 4, 2005.

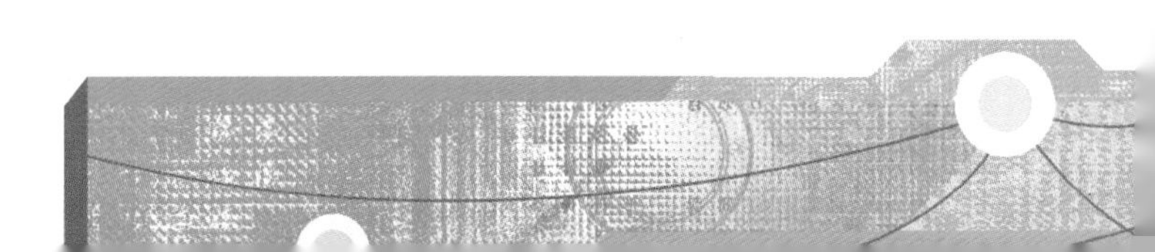

[11] "Hospital Gains Efficiency with Innovative RFID Pilot," Siemens Business Services Case Study, http://www.sbs-usa.siemens.com/press/docs/jacobimedical-casestudy.pdf, accessed June 29 2005.

[12] Collins, J. "Delta Plans U.S.-Wide RFID System," *RFID Journal*, http://www.rfidjournal.com/article/articleview/1013/1/1/, July 2 2005.

[13] Championchip USA Website, http://www.championchip.com, accessed June 23, 2005.

[14] "Arcadia Stocks up on Blades for Reliability and Operational Flexibility," IBM Success Stories Web site, http://www-306.ibm.com/software/success/cssdb.nsf/CS/DNSD-645E2R?OpenDocument&Site=eserverxseries, accessed June 23, 2005.

[15] "Wi-Fi Minneapolis," *Minneapolis Star Tribune*, http://www.startribune.com, April 12, 2005.

[16] HomePNAWeb site, http://www.homepna.org/.

Digital Media for Work and Leisure

Ana Arguello was in her first semester of college. Away from home, on her own for the first time, she had a real sense of freedom—along with a bit of apprehension. Most of her friends had chosen a major and knew what they wanted to do in life. Ana enjoyed so many different things that she had trouble choosing just one area to focus on. She had always been an expressive, artistic person. She had a good sense of style and real artistic talent. Besides sketching and painting, she had developed skills in using Photoshop and enjoyed creating interesting digital images. She also played the piano; she had taken lessons since she was in elementary school, and over the years she had developed an enthusiasm for music. She even received an award in the talent show during her senior year in high school. Much of Ana's music collection was on her notebook computer.

Computers had been very much a part of Ana's life. Her mom and dad worked as computer consultants, and her older twin brothers seemed to always be playing video games. By the time Ana went to college, her computer skills surpassed almost all of her friends'. She enjoyed the challenge of computer games and had mastered most of the major titles. She had a personal digital photo collection that easily numbered in the thousands; photos that she had taken on her cell phone and digital camera. She had taken and edited digital video and loved going to the movies or watching movies on her notebook. Ana thought about majoring in a computer-related field, but wondered if she wouldn't be happier majoring in art or music.

As you read through the chapter, consider the following questions:

1. What careers and majors might Ana consider in digital media?
2. How can Ana use digital media software and services to manage her media collections?
3. How might Ana enjoy digital media as a hobby and as entertainment?

Check out Ana's *Action Plan* at the conclusion of this chapter.

LEARNING OBJECTIVES

1. Understand the uses of digital audio and today's digital music technologies.
2. Describe the many uses of 2D and 3D digital graphics and the technologies behind them.
3. Explain the technologies available to acquire, edit, distribute, and print digital photos, and list new advances in video technologies and distribution.
4. Discuss how interactive media is used to educate and entertain.

CHAPTER CONTENT

Digital Music and Audio

Digital Graphics

Digital Photography and Video

Interactive Media

Introduction

Digital technology and information systems are tremendously useful in many practical ways. Digital media brings these systems to life with stunning and vivid imagery, powerful sound and music, and realistic, interactive animated 3D environments. If technology were alive, multimedia might be considered its heart and soul. Digital media provides a technical venue for people to express themselves through audio and visual output. This chapter provides an overview of all areas of digital media, including digital music and audio, 2D and 3D digital graphics and animation, digital photography and video, and interactive media such as video games and interactive TV. This chapter examines state-of-the-art media technologies and how they affect us in our personal and professional lives.

You don't need to be artistically or musically inclined to use and appreciate digital media. **Digital media** encompass digital technologies of all kinds that serve and support digital music, video, and graphics (see Figure 6.1). Digital music players, digital cameras, video game consoles, DVD and CD players, and cell phones are all devices that serve up digital media. MP3 music, DVD movies, digital photos and artwork, cell phone ring tones and wallpaper, screen savers, motion picture special effects, animated television, and movies are examples of digital media. Media player software such as iTunes, Windows Media Player, and RealPlayer; paint and drawing software; photo- and video-editing software; and voice and music recording software are examples of digital media software. When different digital media types are combined, such as animation or video and audio, it is called *multimedia* or sometimes *rich media*. Digital media that can be controlled, manipulated, or in some way interacted with is called *interactive media*.

Digital media is transforming the manner in which we access entertainment. A study by Arbitron and Edison Media Research shows that consumers are choosing to spend more time on the Internet, and less time with television and radio.[1] The study indicates that consumers have a preference for on-demand media. *On-demand media* refers to the ability to view or listen to programming or music at any time rather than at a time dictated by television and radio schedules. On-demand media is made possible by broadband Internet access at home and by cell phone, digital video recorders, portable digital music and video players, satellite radio, CDs and DVDs, and of course, the Web.

FIGURE 6.1 • Digital media

Digital media impacts our daily lives in many ways.

Digital media is fundamental to today's entertainment industry and key to many professions. Designers, engineers, and architects make use of digital graphics software to design 3D products and projects. Desktop publishers and Web designers use digital graphics software to develop attractive 2D print and Web pages. Manufacturers and retailers use digital media to sell products and support their customers. Scientists use digital media to simulate and interact with inaccessible objects and environments. Almost everyone uses or is affected by digital media in one manner or another.

Thousands of digital media software applications are available for both professional and personal use. This chapter discusses four general categories of digital media: digital music and audio, digital graphics, digital photography and video, and interactive media.

DIGITAL MUSIC AND AUDIO

EXPAND YOUR KNOWLEDGE

To learn more about working with audio, go to www.course.com/swt2/ch06. Click the link "Expand Your Knowledge," and then complete the lab entitled, "Working with Audio."

Digital audio is any type of sound, including voice, music, and sound effects, recorded and stored digitally as a series of 1s and 0s. **Digital music** is a subcategory of digital audio that involves recording and storing music.

The ability to digitize sound has dramatically altered our phone networks, radio, television, and the entertainment and music industries. Digital phone networks digitize speech and send it as bits over cables or through the air. The radio industry is being transformed by satellite and Internet-delivered digital radio services. Digital audio in the form of voice, sound effects, and music is embedded in television programs, motion pictures, animated media, and computer games to provide high-quality and sometimes dramatic realism. The digitization of music has fundamentally altered the production and distribution mechanisms within the music industry, providing musicians with powerfully creative tools, improving the quality of recorded music, and providing listeners with more convenient access. The digitization of music and audio has also created new challenges to the creative and intellectual property rights of artists and production companies.

This section examines the impact of digital music and audio on various professions. Special attention is given to digital music production and distribution as well as thorough coverage of popular digital music devices, software, and services. The section begins by explaining how sound and music is encoded to travel over wired and wireless networks and the Internet, and how it is stored on CDs, DVDs, hard drives, and flash drives.

Digitizing Music and Audio

In the natural world, sound is the displacement of air particles caused by vibration and sensed by the eardrum. One way to quantify sound is by measuring the amount of air particle displacement and charting it over time to create a graph, called an analog sound wave (see Figure 6.2). The term *analog* refers to signals that vary continuously. An analog sound wave can be transmitted electrically using varying voltages of electricity as is done over traditional telephone networks, or varying a radio signal as is done for AM/FM radio. Another more recent way of quantifying sound is to represent sound waves with numbers, digitally, through a process called *analog to digital conversion (ADC)*.

FIGURE 6.2 • Analog to digital conversion

A sound wave is sampled by measuring its amplitude at consistent time intervals, and storing the amplitude values as a list of binary numbers.

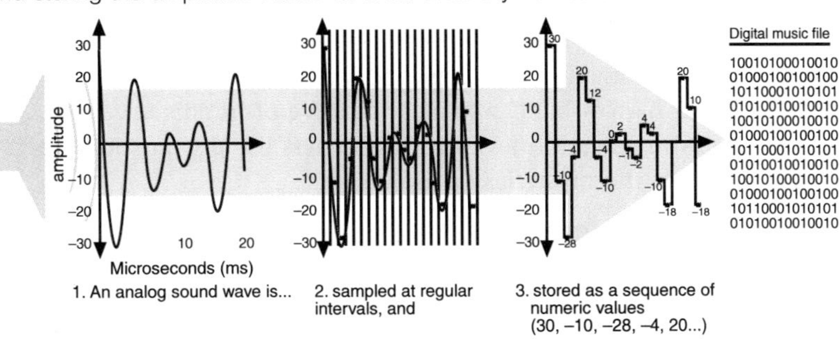

ADC uses a technology called *sampling* to encode a sound wave as binary numbers. When you digitize, or sample, a sound wave, you measure and record its amplitude (height) at regular time intervals called the sampling rate; the shorter the time interval, the higher the sampling rate and more accurate the reproduction of the sound. For example, the sampling rate for audio CDs is 44,100 samples per second, whereas the sampling rate of your voice on a digital cell phone is 8000 times per second.

Digitized sound is transformed back into its analog form in a process called *digital to analog conversion*. Although the recreated sound wave is not an exact duplicate of the original live sound, a sampling rate of 44,100 times per second is close enough to the original sound to satisfy our less-than-perfect ears.

Digitized sound has tremendous advantages over analog sound in a number of ways. It can be easily duplicated and transmitted without any degeneration. It has a relatively limitless life span. It is easy to manipulate and process and can be encrypted for secure communications. Digital phones and media recorders/players include *analog-to-digital converters* and *digital-to-analog converters* to translate sound and music back and forth between analog and digital representation.

Digital Sound for Professionals

A number of nonentertainment professionals make use of digital sound technology in their work. Digital sound devices can help professionals who must rely on their ears to do their work more thoroughly and precisely.

Digital Voice Recorders. Professionals in many fields use portable *digital voice recorders* (see Figure 6.3) to capture dialog for future reference. Journalists, lawyers, investigators, and others whose work involves interviewing others, rely on digital voice recorders to keep their facts straight. Doctors and other professionals use digital voice recorders to record notes for future transcription.

Digital voice recorders store recordings in standard digital sound formats that can be transferred to a computer for transcription or editing. Sound files can be edited for broadcast, played back and transcribed to text documents for print or Web publishing, or filed away for future reference. Transcription is typically handled manually by typing the text as the digital voice recording is played back. Footswitches are available that allow professional transcribers to pause, rewind, and fast forward digital recordings on a PC while typing. While software is available for automating the process of transcribing recorded speech to typed characters, it does not yet provide the high quality of human transcription.

FIGURE 6.3 • Digital voice recorder

Portable digital voice recorders can store hours of interviews or verbal notes for future transcription.

Digital Sound in Scientific Research. Scientists have used digital audio to study various natural phenomena. The Australian Marine Mammal Research Centre used digital recordings to study whale songs off the east coast of Australia. Using three underwater microphones called *hydrophones* placed at different locations, the researchers were able to track the songs of individual whales as they migrated. Using sophisticated digital processing equipment, the researchers discovered that male whale song "is highly structured, and, at any one time, all the males in the population sing the same song using the same sounds arranged in the same pattern. Over time, however, this pattern changes, but all the singers make the same changes to their songs. After a few years the song may be quite different, but all the singers are still singing the same new song."[2]

Researchers at the SETI (Search for Extra Terrestrial Intelligence) institute predict that humans will detect an extraterrestrial transmission within the next 20 years.[3] Similar to listening for a station as you twist your car radio's tuning knob, SETI researchers use digital technologies to scan millions of radio channels gathered from space for narrow-band signals, indicative of intelligent origin. In almost all scientific and medical professions, researchers use digital sound technologies to help decipher the mysteries of the universe.

Digital Sound in Law Enforcement. Digital sound plays an important role in law enforcement. *Forensic audio* uses digital processing to denoise (remove nonessential sounds and audio interference), enhance, edit, and detect sounds to assist in criminal investigations. Among the many tools available to forensic audio specialists, the spectrographic sonogram is perhaps the most valuable. A spectrographic sonogram provides a visual fingerprint for various sounds in a recording. For instance, when a tape head engages the tape, it leaves a distinct impression (fingerprint) that can be used to determine if a tape was tampered with. Gunshots, car engines, and voices all have a unique pattern when viewed as a spectrographic sonogram. Figure 6.4 shows the spectrographic sonogram of a human voice. The process of identifying a recorded human voice is known as voice-print identification. The technique is considered highly reliable and has been used as evidence in more than 7000 criminal cases. Forensic audio has been helpful in the hunt for Al-Qaeda terrorist leader Osama bin Laden. Forensic audio specialists studied recorded speeches claiming to have been made by bin Laden to determine the authenticity of the recording to confirm that he is still alive.

FIGURE 6.4 • Spectrographic sonogram of a human voice

Gunshots, car engines, and voices all have a unique pattern when viewed as a spectrographic sonogram.

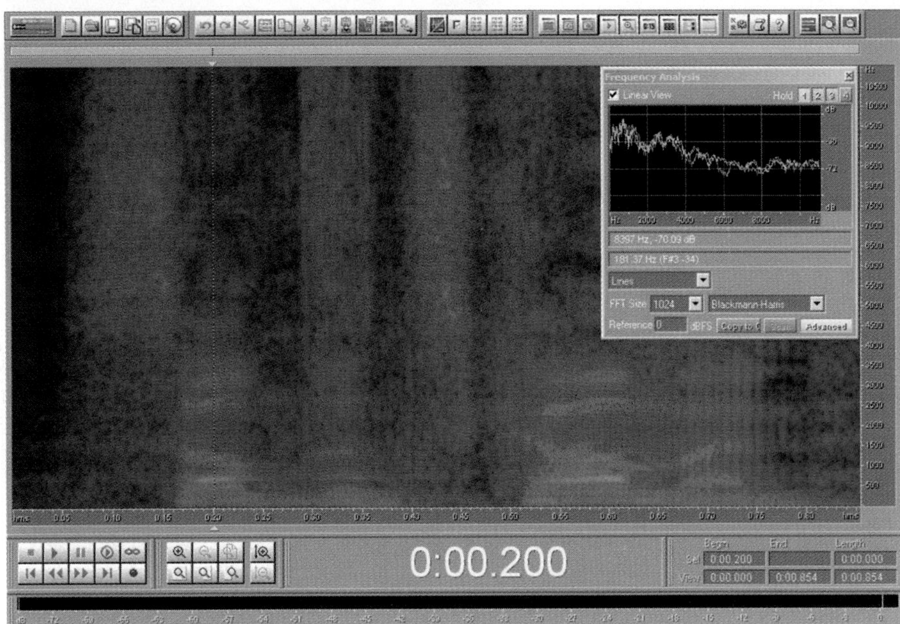

Digital Sound in Entertainment and Communication. The professional production and editing of digital audio takes place in sound production studios. A *sound production studio* uses a wide variety of audio hardware and software to record and manipulate music and sound recordings. Today's sound production studios typically record sound to digital media such as tapes or disks and then use digital sound-editing equipment to perfect what has been recorded. Figure 6.5 shows a topnotch Hollywood professional sound studio. Many professionals make their living in such studios working as digital sound engineers.

Table 6.1 lists the wide array of services provided by sound production studios. Some large studios provide all of these services, but smaller studios may specialize in a specific area. Each service requires specific hardware and software tools. For example, recording the soundtrack for a motion picture requires video production tools that allow the sound engineer to synchronize the soundtrack to the action in the film. Studios that specialize in recording music are known simply as *recording studios*. Sound studios that work with motion pictures are typically referred to as *production studios*. As you can see from the list of services in

FIGURE 6.5 • Sound production studio

Sound production studios use a wide variety of audio hardware and software to record and manipulate music and sound recordings.

Table 6.1, sound production plays an important role in many media and entertainment industries. Sound production is an important part of music, movies, radio, television, video games, and the Internet.

TABLE 6.1 • Sound production studio services

Sound production services	Description
Music recording	Recording of music performance using multitrack recording equipment
Movie soundtrack production	Recording of musical performance for motion picture accompaniment
Movie sound effects	Addition of special sounds, such as footsteps, breaking glass, and explosions, to motion pictures to lend additional realism and impact
Voice-over recording and dubbing	Voice recordings that are provided behind video or other media productions that may be narrative or synchronized to the lip movements of the actors
Postproduction sound engineering	Includes movie sound track production, movie sound effects, voice-over recording and dubbing, and also the overall editing of motion picture sound tracks to create the final product
Radio commercials	Recording of voice, music, and sound effects for radio commercials
Music and sound effects for computer video games	Same services as offered for motion pictures, but applied to video games
Audio for distance learning and training	Recording narration, lectures, and other sound needs for educational software and distance learning
Multimedia and Internet audio	Audio recordings for delivery over the Internet, such as Internet radio and audio accompaniment to Flash and video presentations
Audio restoration and enhancement	Improving the recorded quality of old or damaged recordings

Digital Music and Audio Production

Although digital audio processing has many industrial and professional applications, it is most strongly connected with the music industry. Today's recording studios are high-tech digital processing centers. Even the most "unplugged"-sounding acoustic music recordings utilize digital sound-processing techniques to enrich and purify the sound so that it sounds as though you are right there sitting with the musicians.

Professional Music Production. Today's recording studios use analog-to-digital converters to transform the recorded sound of voices, violins, horns, and other acoustic instruments to digital signals that can then be manipulated. Studios record music using multitrack recording devices. *Multitrack recording* devices treat each instrument or microphone as a separate input, or track. The engineer uses a *mixing board* (the large panel with many dials, buttons, and sliders in Figure 6.5) to adjust the sound quality of each instrument separately. Multitrack recording allows the instruments to be recorded either all at once or separately. For example, a jazz quintet might decide to record the rhythm section—bass, guitar, and drums—first. After the rhythm section tracks are recorded ("laid down"), the solo instruments and vocals can be added one at a time. Using multitrack recorders, studio engineers are able to mix many separate instrument tracks together to create the finished product. Using digital signal processing, the engineer can mold the sound of each instrument, adjusting the tone quality and adding effects. In what is called the "final mix," after all tracks have been recorded, the engineer plays the recording and applies changes to the volume levels of each track to balance the sound of the instruments and bring listeners' attention to specific instruments at specific times. As the engineer "mixes" the song, the computer stores the settings. The final product is then transferred to CD, or some other storage medium.

FIGURE 6.6 • Synthesizer keyboard

This Yamaha synthesizer keyboard includes hundreds of digitally sampled instrument sounds and synthesized sounds.

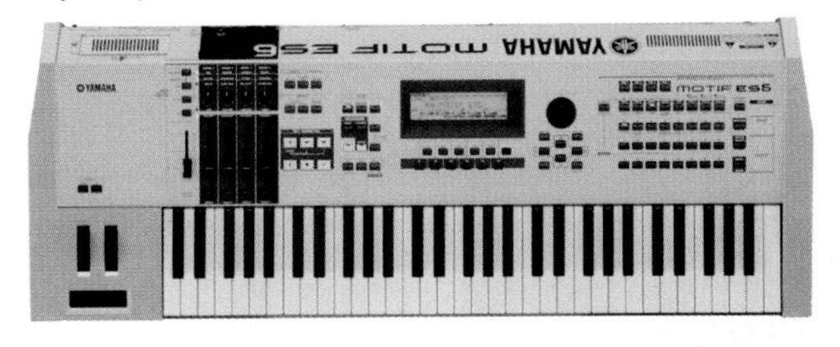

Digital music instruments, such as synthesizers and samplers, produce musical sounds electronically. A **synthesizer** electronically produces sounds designed to be similar to the sounds of real instruments; they can also produce new sounds unlike any that a traditional instrument could produce. A **sampler** digitally records real musical instrument sounds and allows them to be played back at various pitches using an electronic keyboard. Synthesized instrument sounds have an electronic quality to them that rarely fools anyone into thinking that they are the actual instruments. In contrast, it is often difficult for average listeners to tell a sampled sound from the real instrument. Professional-grade synthesizer keyboards, like the one in Figure 6.6, use sampled sounds for the instruments that they emulate. If you play a middle C on a synthesizer keyboard with the trumpet sound selected, an actual digital recording of a trumpet playing a middle C is produced.

Musicians and sound engineers sometimes use *drum machines* to record drum beat patterns by tapping on pressure-sensitive buttons or pads to produce sampled drum sounds that can be played back in a looping pattern. With a little training and practice even a novice can lay down a basic repeating drum pattern; it won't necessarily rival a real drummer's performance, but it can be useful for practice and composition. A **sequencer** allows musicians to create multitrack recordings with a minimal investment in equipment. Using a sequencer, a musician can first record a drum track, then record a bass track over the drum track, and continuing to layer instrument track upon instrument track to produce a recording that sounds like a full band of musicians. Many solo performers make use of drum machines and sequencers to provide accompaniment for their performance. In this way, one person singing (and perhaps strumming a guitar) can sound like an entire band. A small band can use sequencers and synthesizers to sound like an entire orchestra.

Synthesizers, samplers, drum machines, and sequencers are only a few examples of the many devices available for creating digital music. Most music production studios have large racks of interconnected digital audio devices, called *outboard devices*, to process digital music and audio signals. The **musical instrument digital interface (MIDI)** protocol was implemented in 1983 to provide a standard language for digital music devices to use in communicating with each other. MIDI commands include basic control commands such as "Note on," "Note off," "Program change" (to change instrument sounds), and others. Using MIDI a musician can connect and control many devices all from a single synthesizer keyboard or computer.

Home Recording Studios. *Integrated digital studios* package many digital recording devices in one unit for convenient home recording. For example, the Yamaha AW2400 includes a 16-channel mixing board, a 40-GB hard drive on which to record, an analog-to-digital converter, sequencing and sampling software, a CD burner, and many other more-technical features to take music from home performance to a distribution CD. A musician could purchase this device for about the same price as the cost of one day in a professional studio.

MIDI cards allow computers to be connected to digital music devices and are available for Windows PCs; they are included as standard equipment in most Apple computers. All of the digital music devices described in this section—drum machines, sequencers, samplers, and synthesizers—are available as personal computer software. Most personal computers come equipped with *on-board synthesizers* on the computer's sound card that include all the standard synthesizer keyboard sounds. Using sequencing software, such as Apple's Logic Platinum software, shown in Figure 6.7, a personal computer can become a self-contained recording studio.

FIGURE 6.7 • Sequencing software

Sequencing software, such as Apple's Logic Platinum software, allows a personal computer to become a self-contained recording studio.

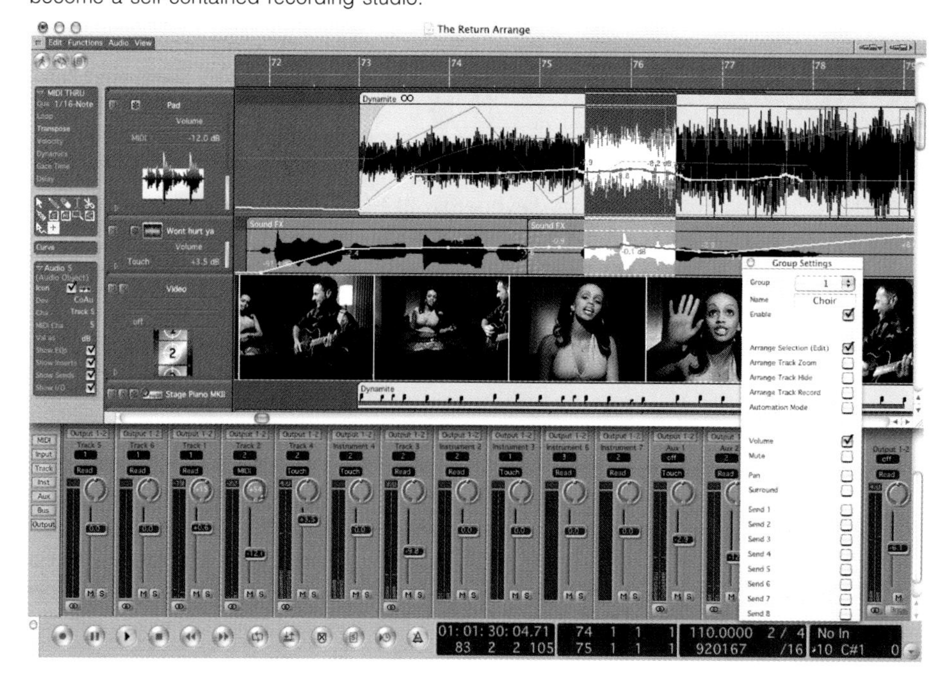

Sgt. William Thompson IV, a U.S. Army reservist who was deployed in Iraq, found comfort and solace after his daily patrols by creating music on his Apple G4 notebook computer.[4] The soldier, a third-generation jazz musician from New Orleans, created aural landscapes of his daily experiences in Iraq and posted them on the Web (*www.wativ.com*). Using his iPod, he sampled the street sounds of Iraq to incorporate into his recordings. The result is a haunting, "kind of freaky," and often intense "soundscape." Creating these digital recordings served as therapy for Thompson and provided the world with insight into a musician's sense of life in a war zone.

Many musicians are connecting their home computers to low-priced home recording equipment to create their own professional-grade home recording studios. Home recording studios save musicians from having to pay premium rates for services that may not be required. Home studios allow many musicians to perform, record, produce, and distribute their music independently. As with most of today's new technology, advances in digital music technology are empowering musicians with tools that were previously available only to professionals. Table 6.2 provides an idea of the equipment used to turn a home PC into a high-quality home recording studio. Web sites like GarageBand.com provide a distribution channel for independent artists to get their home-recorded, or professionally recorded independent music to the public.

TABLE 6.2 • Home recording studio equipment

If you already have a PC with lots of hard drive space and a CD burner, as well as your own musical instruments, this equipment will transform your PC into a recording studio.

Equipment	Approx. cost (2006)	Notes
Microphone with cord and stand	$120	Can cost as little as $12 or as much as $1000. Decent mics, like the brand MXL (www.mxlmics.com), can be found for around $100. Multiple mics may be needed if you wish to record several musicians at once
Headphones	$30	Headphones, like AKG's studio line (www.akg.com) are needed while recording to prevent feedback that would occur with standard speakers
Powered speakers	$200	Good-quality speakers, such as those from Alesis (www.alesis.com) and JBL (www.jblpro.com), are needed for listening to and mixing recorded music
Powered mixing board	$120	Mixing boards, such as those from Mackie (www.mackie.com) and Yamaha (www.yamaha.com), range in price from $50 to $10,000 depending on the amount of channels and features
PC audio interface	$150	An audio interface, such as those offered by m-audio (www.m-audio.com), connects the mixing board to your PC and converts the analog signal to a digital one; it can connect through Firewire or USB ports, or a PCI sound card
Software	$150	Home studio software such as Cakewalk Sonar for Windows (www.cakewalk.com), and GarageBand for Mac, allow you to mix and manipulate multitrack recordings on your PC
Total	$770	

One example of the empowering nature of digital music can be found in New York City's East Village. At the Openair Bar, musicians and musician "wannabes" gather Sunday nights for a musical jam session. But these musicians don't play musical instruments: they perform on notebook computers equipped with software sequencers, synthesizers, and other digital sound tools, connected to the sound system to create layers of spontaneous sound that combine into an "electronica" symphony.

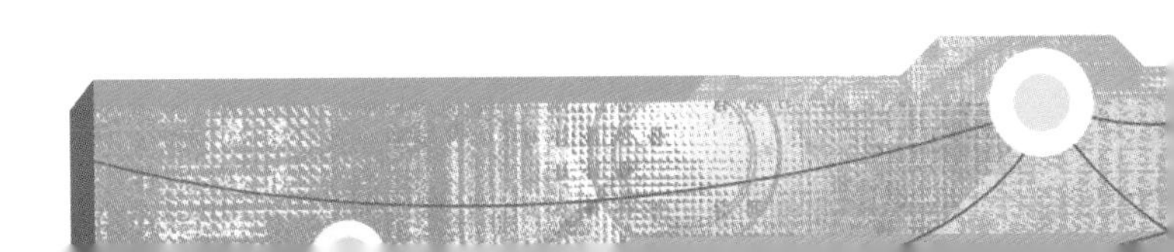

Podcasting. Even nonmusicians are turning to PC-based home recording in a relatively new digital broadcast phenomenon called *podcasting*. A **podcast** is an MP3 audio file that contains a recorded broadcast distributed over the Internet. Podcasting gets its name from the Apple iPod, but the technology can be used on any media player that supports the MP3 format. Podcasts can contain professional voice interviews with celebrities or experts, recordings made available by the producers of regular radio programs or talk shows, music demos, instructional training, self-guided walking tours, personal discussions and commentaries, and more. Increasingly, novices who want to share their views, humor, talents, or musical taste with the world are using podcasts. Table 6.3 lists some popular podcasts at this writing.

TABLE 6.3 • Some popular podcasts

Title/URL	Desciption
The Peanut Gallery *www.thepeanutgallery.info*	Features a new radio story every Sunday evening
5 Minutes with Wichita *www.5minuteswithwichita.com*	Humorous interviews with bluegrass and country music stars
WeFunk Radio *www.wefunkradio.com*	Hip-hop and Funk radio show from Montreal, featuring live DJs and MCs, and classic, rare, and underground selections
All Things Considered *www.npr.org*	National Public Radio's flagship program
The Zeph Report *www.zephnet.com*	Minister Zeph speaks on the inspirational word for today
The Dawn and Drew Show *www.dawnanddrew.com*	Husband and wife podcasting duo coming to you from the town of Wayne, Wisconsin
Science@NASA Feature Stories *http://science.nasa.gov*	Stories of exciting NASA research
Radio Motion *http://radiomotion.com*	An eclectic mix of tech-house, techno, acid, electro, and disco house

There are thousands of free podcasts available on the Web. Most are regular publications with new releases available daily, weekly, or monthly. There are a number of ways to find and access podcasts. Podcast directories such as Podcast Alley (*www.podcastalley.com*), iPodder.org, and the Podcast Directory (*www.podcast.net*) allow you to browse through podcasts by topic. Figure 6.8 shows categories listed at the Podcast Directory. The numbers next to each category indicate how many podcasts are available in that area.

A podcast can be downloaded from a Web site as an MP3 file and listened to using a PC media player, or transferred to a portable MP3 player. Podcasts are also distributed using RSS technology. RSS, described in Chapter 5, is the technology used to subscribe to blogs. Software called a *podcast aggregator*, such as iPodder (*http://ipodder.sourceforge.net*), Doppler (*www.dopplerradio.net*), and PlayPod for Mac (*www.iggsoftware.com/playpod*), uses RSS to allow you to subscribe to your favorite podcasts. Apple iTunes also supports podcasts and advertises access to thousands of podcast radio shows. Because podcast files can be large, podcast aggregators can be set up to download your favorite podcasts while you sleep so that they are ready to quickly transfer to your player when you awaken. Podcasts are so popular that many of today's popular media players plan to support podcast aggregation services. Microsoft's latest version of Windows supports RSS technology as does Mac's OS X, which opens the door to all kinds of blog and podcast applications and indicates a bright future for these technologies.

Creating a podcast can be accomplished by anyone with the desire to do so. When you visit a podcast directory, you find hundreds of podcasts created by amateurs who wish to share their unique perspectives with the world. To create professional-quality podcasts you need a headset with a noise-canceling

FIGURE 6.8 • The Podcast Directory

There are thousands of free podcasts in many categories available on the Web.

microphone. Most PCs have ports that accommodate these devices. Digital recording software is required to record a podcast. Software such as Audacity (*http://audacity.sourceforge.net*) for Windows and GarageBand for Apple meet the basic needs for podcasting. Serious podcasters should invest in software specifically designed for podcasting such as Propaganda (*www.makepropaganda.com*) or iPodcast Producer (*www.industrialaudiosoftware.com*). These packages not only allow you to record your voice but also add music, sound effects, and other audio in real time. They also provide the tools to take your podcast online for others to access.

Once a podcast has been recorded, the MP3 file must be made available for users to access over podcast aggregators (see Figure 6.9). Online services such as FeedBurner (*www.feedburner.com*), Ourmedia.org, and TD Scripts.com provide wizards that allow you to put your podcast online for free. Some fee-based sites, such as Audioblog.com and Liberated Syndication (*www.libsyn.com*) charge monthly fees starting at $5.

Podcasting is another liberating force in the growing on-demand culture. Rather than having to wait for a given time to listen to your favorite programming, podcasts allow you to listen at your convenience.

FIGURE 6.9 • Free podcasts on iTunes

iTunes provides access to thousands of podcasts and manages podcast subscriptions.

Digital Music and Audio Formats, Storage Media, Players, and Software

Producing, distributing, and enjoying digital music and audio depends on four technology components: standardized media file formats, storage media on which to store the files, player devices, and software that can read and manipulate the files. This section examines the digital music components that support music and podcasts.

Digital Music and Audio File Formats. Today, music is most often distributed on CDs, with an increasing trend towards distribution over the Internet. Digital music is burned to CD in a special format designed for audio CDs. Compact disk audio (.cda) files are designed to be read and manipulated by CD players and computer media players. Other music file formats are specifically designed for use on a computer. For example, the beeps and bells that are a standard part of the Windows interface are stored in .wav files; Apple computers use the .aiff music file format.

Larger audio files, such as a typical 3-minute song, a voice recording, or a half-hour radio program, require too much computer storage if stored in their native format. Three typical audio CDs worth of music by your favorite artist take up more than 1 GB of space on your hard drive. Compression technologies are applied to audio files to greatly reduce their size with little or no loss to sound quality. *Sound compression* removes those frequencies that are beyond the range of human hearing and in so doing reduce the size of the digital music file. The most recognized compressed digital audio file format is MP3. The **MP3** file format compresses CDA music files to less than 10 percent of their original size. A 32-MB music file on a music CD can be compressed down to a 3-MB MP3 file. A CD that holds up to 20 standard CDA songs could hold as many as 200 average size MP3 songs.

The new era of legal Internet music distribution has developed into something of a war of audio formats. The traditionally popular MP3 format is still preferred by many people who have most of their digital music in this format. Microsoft primarily supports the WMA format in the services provided by Windows Media Player.

Apple iTunes, the most popular music download service only delivers M4P files. WMA and M4P formats provide better sound quality than the MP3 format because they use more advanced compression technology. Users who collect music from varying sources find themselves with a collection of music files in a variety of formats. Unfortunately, there are very few media players that support all three formats. Until the big digital music providers agree on one standard format, users will not be able to transfer their music across platforms and devices.

Digital Music and Audio Storage Media. The days of vinyl records and cassette tapes—both analog media—are fading. These analog media are being displaced by digital music on digital media. The clarity of digital sound, combined with the low price and longevity of digital media, make it an obvious choice over traditional analog media. There are three primary forms of audio storage media: CDs, hard drives, and flash memory.

Because most music is available on CD, the CD player remains the most popular digital music player. People use CD players to listen to music at computers, in living rooms, in cars, and while walking about. However, today's new generation of CD players have additional functionality that allows them to play not only traditional music CDs but also homemade CDs containing MP3, WMA, or M4P files. People can create their own mix of artists and songs on a CD-R (recordable CD) or CD-RW (rewritable CD), and are no longer bound by song sequences on store-bought CDs. Acquiring digital music from various sources, arranging the songs into personally appealing playlists, and recording the playlists to a CD-R has become a common activity known as "rip-mix-burn" and is yet another method of enjoying on-demand media.

FIGURE 6.10 • The Apple iPod

The Apple iPod is designed to hold music, podcasts, photos, and sometimes videos.

Some music lovers are doing away with the need for CDs and are storing their favorite music directly on their computer hard drive. Once on the computer, music tracks can be played on the PC or sent over a network to a home entertainment center, transferred to a portable MP3 player, or burned to a CD to listen to while away from the computer. Today's high-capacity portable digital music players, like the iPod, make use of small hard drives and microdrives to store gigabytes of music, room enough for hundreds, even thousands of songs. Less expensive digital music players make use of flash memory. Flash memory MP3 players come in various sizes to accommodate up to several gigabytes of music.

The size of the storage medium is a major factor in deciding which MP3 player to purchase. Smart shoppers should first determine how they will be using the player, and how much music they wish to store. An iPod with a 60 GB hard drive could store your entire music (and photo) collection with ease (see Figure 6.10). To put this in perspective, if you completely filled up the iPod's 60 GB hard drive with digital music and listened to every song in order, it would take more than 41 days before the last song finished. Perhaps you don't need to have your entire collection with you everywhere you go. If you do, you will certainly want to maintain a copy on another drive in case you lose your iPod or it becomes damaged. Many people prefer to keep their entire collection on a PC, and transfer only a portion of their music to their portable player for access while away. As wireless connection speeds continue to increase, some experts predict that you will soon be able to store and manage your personal digital music collection on the Internet and access it from anywhere wirelessly. In such an environment, storage capacity of portable players would not be an issue.

Digital Music, Audio Players, and Software. Our on-demand culture is becoming increasingly comfortable with the freedom that digital music provides. Digital music players and systems for all environments allow us to exercise full control over our digital music collections. **Media player software** such as iTunes, Windows Media Player, and RealPlayer, allows users to organize and play digital music, audio, and video files on PCs and media devices. Media player software can search for all music files on a PC and sort and arrange them by artist, genre, or album title. It also allows users to create custom playlists that can be transferred to a CD-R or portable player. Most popular media players provide a music download service so that users can conveniently build their digital music collections.

Media player software has been designed for desktop and notebook PCs, handheld PCs, and even cell phones (see Figure 6.11).

FIGURE 6.11 • MP3 cell phone

This Samsung phone includes media player software and a media card to store audio files.

Portable MP3 music players come in a variety of shapes and sizes, from lightweight players that strap to your upper arm while you're exercising, to heavyweights that can store 60 GB or more of music. Portable MP3 players work hand-in-hand with PC media player software. Songs and other audio files are transferred from PC to player through USB or FireWire ports. Most portable players use a rechargeable battery.

MP3 music players are showing up in all kinds of forms. Oakley's THUMP sunglasses (Figure 6.12) provide shade from the sun and MP3 and WMA music through attached flip-up earphones.

There are numerous software utilities available for working with digital music files, many of which are built into popular media players. *Jukebox software* allows computer users to categorize and organize their digital music files for easy access. The process of transferring music from CD to MP3 or other digital audio format is called *ripping* a CD. *Ripper software* can be used to translate your favorite music CDs to MP3 files on your hard drive. Digital music files can be transferred from one format to another using *encoder software*.

Many of today's MP3 players can also be used as storage devices. When a player is connected to a PC, the operating system recognizes it as an additional storage device and assigns it a drive letter. This makes it possible to transfer files of any type to your MP3 player for temporary storage, and use it like a USB drive.

FIGURE 6.12 • Sunglasses that rock

Oakley's THUMP sunglasses have flip-up lenses and earphones and deliver MP3 music.

Some digital music enthusiasts use their home PC as their primary home stereo, utilizing a wireless home network. Systems like Sonos (*www.sonos.com*) and Apple AirTunes allow you to send music from your PC to powered speakers anywhere in your house. Sonos provides a handheld remote that is used to program music selections on your PC from any location on your wireless network. Powered speakers are connected to Sonos zone players placed around the home. You could play relaxing music in the living room for your roommate, while you listen to Salsa music in the kitchen as you prepare dinner—all from the same PC over your home wireless network.

Digital Music and Audio Distribution

Today music distribution and acquisition are at a crossroad. The application of digital compression to digital music brought us MP3 music files and the rise of file-sharing services like Napster and KaZaA. Because the vast majority of commercial music in the United States is protected as intellectual property under copyright, our society was placed in an untenable situation in which over 61 million otherwise law-abiding citizens were accessing music through illegal file distribution networks and illegally copied CDs. Illegal downloading and copying is said to have cost the music industry $6 billion between 1998 and 2003, and a loss of 22 percent of the entire music market. Such a contradiction between popular habit, law, and industry certainly could not exist for long.

To resolve the contradiction between law and social practice, and to protect their legal rights, the Record Industry Association of America (RIAA) has sued thousands of individuals involved in illegal MP3 file sharing while simultaneously developing new distribution models for digital music.

In 2004 the music industry finally caught up with the Internet generation and developed means for people to legally download music. Apple's iTunes was the first service to capture the attention of music downloaders, soon followed by several similar services. The number of people legally downloading music grew by 75 percent within the year following the introduction of iTunes. The number of legal music downloads from iTunes surpassed 200 million in 2004,[5] and 300 million in 2005 (a rate of 1.39 million songs per day)[6] proving that increasing numbers of people are willing to pay for music once again. In 2005, iTunes was the second most popular music download Web site, outperforming all but one of the illegal file-sharing networks.[7]

Online Music Services. Online music services work with the recording industry to distribute music legally over the Internet. There are two types of online music services: download and subscription. The big four online music services at the time of this writing are iTunes, MSN Music, Napster, and Rhapsody. All four services offer 99 cent per song downloads. Napster and Rhapsody are subscription services that provide additional features (see Figure 6.13). To determine which service would best suite your needs, you must first understand two important technologies: streaming audio and digital rights management.

FIGURE 6.13 • Rhapsody music service

Rhapsody provides a subscription service that allows subscribers to access more than a million full-length songs from the most popular labels.

Streaming audio is an Internet technology that plays audio files as they are in the process of being delivered. In many cases the audio file is not permanently stored on your computer, but rather plays from an Internet source. Streaming audio protects copyrights because users who listen to streaming audio do not actually possess the file they listen to and cannot copy it or burn it to CD. Many music services provide free or inexpensive features that allow users to listen to streamed Internet radio or even personalized playlists. However, all legal online services must treat downloads of complete music files very differently as these files may be copied and burned to CD and the possibility of copyright infringement exists.

For online services to allow users to download and own music files, they first needed to assure the recording industry that the songs they sell cannot be illegally copied. The technology invented to protect intellectual property in digital files is called *digital rights management (DRM)*. Microsoft's WMA and Apple's M4P file formats include DRM technology. For the big online services this implies that songs you download in these formats can only be played on software and portable devices that support the DRM technology used by the provider. When deciding on a music service, you should check to see that the DRM used by the service is compatible with your portable digital music player. DRM also limits music files to a maximum of seven copies to CD.

Streaming audio and DRM provide the technological foundation for a new legitimate system of music distribution that provides convenient access to vast libraries of digital music while protecting music from illegal copying and distribution. Monthly subscription services such as Napster and Rhapsody provide access to more than one million songs for under $10 a month. Although the monthly fee does not entitle you to own the music, you have access to it anytime you are connected to the Internet, and sometimes even when not, as long as you remain a subscribing member. These services allow you to create

Selecting a Digital Music Service

When selecting a legal digital music service, you need to consider many options and decide which are valuable to you. Subscription services often provide a free month for users to evaluate the service before having to pay. Below is a list of some of the options currently made available by today's most popular services.

- **Downloads**: All services provide the ability to purchase and download songs. $0.99 is the going rate. Visitors can listen to 30-second snippets of tunes to decide whether they wish to purchase them. Purchased music is constrained from being illegally copied with digital rights management (DRM) technology.
- **Subscriptions**: Monthly subscription services allow members to listen to full versions of more than a million songs provided by the most popular recording labels. Songs are only available to you while you remain a subscriber. These services also offer the ability to purchase, download, and own music.
- **Music-to-go service**: Some subscription services also offer higher-priced monthly subscription service that allows users to download and listen to music on select portable devices—as long as they continue membership.
- **Proprietary software**: Most services require the use of their software to manage and play music. Purchased music can be listened to on software and devices that support the DRM used by the service.
- **Selection**: The top services advertise access to more than a million songs.
- **Portable digital music player support**: Purchased music can only be listened to on portable players that support the DRM of the service. The iPod supports songs downloaded from iTunes. Creative's ZEN player is one of the players that supports downloaded music in WMA format.
- **Internet radio**: The most popular services provide free access to streamed Internet radio while connected.
- **Podcasts**: Some services provides free access to thousands of downloadable podcasts.
- **Tools**: All services provide tools that assist in organizing music and finding new music based on your tastes and preferences. Subscription services may also allow users to share play lists with other users.

playlists, search for artists and songs, and they provide suggestions for new music you may like based on your listening patterns. For music you find and wish to own, you can pay (typically $0.99 per song) to own it. You maintain ownership of songs you purchase even if you quit the service.

All of the popular legitimate digital music services provide the option of paying for and owning music. The major difference between service types is that some have monthly subscription fees that allow you to listen to millions of songs without owning them. Those without monthly subscription fees, such as iTunes and MSN Music, only allow you to listen to 30-second snippets of songs in their music store in order to decide whether or not to purchase them. Those people accustomed to sharing files freely and illegally have to get used to the fact that when you purchase music legitimately, you still do not have the freedom to copy it and share it with friends. Copyrights do not support such sharing, and neither do the DRM systems. DRM restricts the ways in which purchased music can be copied.

Satellite Radio. **Satellite radio** is a form of digital radio that receives broadcast signals via a communications satellite (see Figure 6.14). Unlike traditional AM and FM radio, so long as there is no major obstruction between the receiver and satellite, the user can listen to the same channels from any location in range—which could cover entire hemispheres of the planet.

FIGURE 6.14 • Satellite radio

A satellite radio receiver can be stationary, mobile, or a combination of both, such as this Delphi XM receiver that can be connected to powered speakers at home, in the car, or in the boat.

Satellite radio services charge a monthly subscription fee and offer stations featuring commercial-free music, comedy, news, talk, and sports programming. Channels playing Top 40 hits are augmented with specialty channels that provide music not typically available on commercial radio, such as reggae, hip-hop, bluegrass, jazz, classical, soul, and classic country. Satellite radio offers news from such mainstream providers as CNN, Fox News, and BBC World Service, as well as a wide variety of talk, comedy, sports, and old-time radio shows, and exclusive live performances. Two popular satellite radio services in the United States are XM and Sirius.

There are several options for satellite radio receivers, a number likely to greatly expand as this medium becomes more popular. The most popular receivers are in-home, in-car, and mobile. In-home systems connect to your standard stereo, or can be purchased as stand-alone systems that include powered speakers. In-car systems connect to your car audio system, typically wirelessly by tuning your car radio into a specific station to connect to the satellite receiver then using the satellite receiver to tune in a satellite station. Mobile systems look similar to any portable music player except for the fact that they contain an antenna that must be positioned to receive the satellite signal. Some systems have been designed to plug into a car unit and a home unit, so that you can take the receiver with you and use it in multiple environments.

Satellite radio providers are looking for ways to make satellite music ubiquitous, and it is likely that you will see satellite receivers embedded in MP3 players, DVD players, cell phones, and other consumer electronics in the near future.[8, 9]

Other Legal Digital Music Distribution Channels. Pressure from the music industry, government, and society, along with convenient and useful legal online music services, account for much of the retreat from illegal music pirating. Other factors are at work as well. A number of enterprising companies are offering convenient and attractive legal music services. For example, Starbucks coffee shops are providing music download stations for their customers. For $8.99, while sipping coffee, a Starbucks customer can peruse a library of music, and burn seven songs to CD to go ($0.99 for each additional song).[10]

However, many of the incentives for moving to legal download services affect primarily professionals with extra cash and do little to assist teens and students with limited financial resources. Some colleges, like the University of Washington, are offering free music download services to their students. Dell and Napster have teamed up to create special music delivery systems for colleges.[11] Using Dell blade servers that hold digital music supplied by Napster, students can access their favorite music from the local college network. While the college must pay for the music and hardware, the cost is offset by higher levels of student satisfaction, fewer instances of lawsuits from the recording industry, and far less data traffic because fewer students are downloading music from the Internet.

Recording companies are also working to make CDs more attractive to consumers. Consider the dual-disk release of Miles Davis' "Kind of Blue." Besides including the original music recordings, the package includes a film documentary on the making of the album. Apple's "iTunes Originals" offers song collections recorded for iTunes users by famous artists such as Sarah McLachlan and Dave Matthews; some include video interviews with the artists.

The combination of useful and attractive features, low price, and a guilt-free conscience seems to be the right recipe for attracting consumers to legal forms of music acquisition.

DIGITAL GRAPHICS

EXPAND YOUR KNOWLEDGE

To learn more about digital graphics, go to www.course.com/swt2/ch06. Click the link "Expand Your Knowledge," and then complete the lab entitled, "Working with Graphics."

Digital graphics refers to computer-based media applications that support creating, editing, and viewing 2D and 3D images and animation. At first glance, digital graphics might appear to be the exclusive domain of artists. But, in reality, many people, artistic and nonartistic alike, are finding themselves called upon or inspired to create digital artwork for personal and professional use. This section begins by explaining the technology behind digital graphics and then explores several ways in which digital graphics may be useful to you. A section is devoted to each of the primary types of digital graphics: vector graphics, 3D modeling, and animation. Digital photography and video are addressed in the next section.

Digitizing Graphics

The simple fact that graphic images can be represented digitally with binary numbers has brought us flatbed scanners, digital cameras, digital camcorders, digital cable TV, the DVD movie, and digital video recorders, not to mention phenomenal advances in movie production techniques. The fundamental technology behind all of these technologies rests on a small point of light or dot of ink called a pixel.

Digital images are made up of a grid of small points called **pixels** (short for picture element). *Pixel* is a good term to remember as it has practical value when used to determine the quality of displays, printers, scanners, and digital cameras. Representing an image using bytes is simply a matter of storing the color of each pixel used in the image. Images stored in this manner are called **bit-mapped graphics** or *raster graphics*. Colors are expressed using numbers that represent combinations of intensities of red, blue, and green, with one byte (256 levels of intensity) for each. Using three bytes provides 16,777,216 (2^{24}) possible colors. Whenever all three bytes are set to the same values, the resulting color is white or a shade of gray (see Figure 6.15).

FIGURE 6.15 • RGB

Digital graphics uses three bytes to express the color used in each pixel of an image. Each of the three bytes represents the intensity of red, green, and blue (RGB), respectively.

Intensity level			
Red	*Green*	*Blue*	**Resulting color**
255	255	255	
255	0	0	
0	255	0	
0	0	255	
255	0	255	
255	255	0	
255	123	0	
192	192	192	
128	128	128	
0	0	0	

Bit-mapped graphics are ideal for representing photo-realistic images as they are able to capture minute details in an image. They do have some drawbacks however, in that they are difficult to edit and enlarge. *Pixilation*, or fuzziness, occurs when bit-mapped images are made larger than the size at which they are captured.

Simpler images, such as those used in clip art, can be represented using vector graphics. **Vector graphics** use bytes to store geometric descriptions that define all the shapes in the image. Although vector graphics are impractical for representing photo-quality images, they are preferred for creating and storing drawings (see Figure 6.16). Vector graphics are easier to edit and manipulate than bit-mapped graphics and use far fewer bytes to store an image.

FIGURE 6.16 • Bit-mapped vs. vector graphics

Bit-mapped graphics (top) store images by specifying the color of each pixel and lose their clarity as they are enlarged; vector graphics (bottom) store pictures as defined shapes and retain clarity when enlarged.

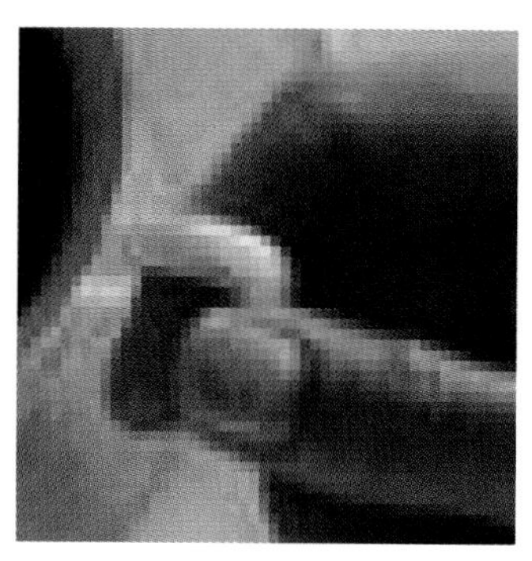

Animation and video (moving pictures) are stored in the computer as a series of images called *frames*. When shown in quick succession, the frames create the illusion of movement. Television uses a rate of 30 frames per second (fps).

Graphics File Formats

There are many graphics file formats, most of which are proprietary, that is, they depend on the software with which they were created. For example, if you work on a photo in Adobe Photoshop, the file is saved by default as a .psd file which can be opened and manipulated only with Adobe products.

There are a few graphics file formats that have become standards across the industry:

- Windows bitmap (.bmp) format is used by the Windows operating system and recognized by most graphics software. It is not, however, recommended for use in generating images for the Web because not all browsers recognize it as a valid Web format.
- Graphics Interchange Format (.gif) is the standard for the Web. It is limited in that it supports only 255 colors.
- Joint Photographic Experts Group (.jpg or .jpeg) is the other standard for the Web and is also the default format on most digital cameras.
- Portable Network Graphics (.png) was developed to replace GIF on the Web, but has so far been unable to do so. This file format supports more colors than GIF and has other benefits as well.
- Tagged Image File Format (.tiff or .tif) is an industrial-strength format used by many professional photographers and graphic artists.

Independent Film Producer Robert Rodriguez

Independent film producer Robert Rodriguez is known as a troublemaker and rebel in the industry. Armed with an arsenal of digital filmmaking tools and talent, he produces, directs, scores, and embellishes his movies with lavish special effects. You probably recognize many of his movie titles: *El Mariachi, Desperado, The Faculty, Spy Kids* (1, 2, and 3D), *Sin City*, and *The Adventures of Sharkboy and Lavagirl*. He is known for bucking the system by quitting the Directors Guild of America when they attempted to rein him in, and for producing high-quality movies on a shoestring budget through digital technology.

Rodriguez had humble beginnings, financing his first picture by taking work as a human lab rat testing a new experimental cholesterol drug. With the $7000 he earned he wrote, produced, directed, edited, photographed, and scored the movie *El Mariachi*, which went on to win the Audience Award at the 1993 Sundance Film Festival. At the time, it was the lowest-budget film ever released in the mainstream market.

With the money made from *El Mariachi*, Rodriguez invested in an arsenal of digital filmmaking tools including Sony HD cameras, a Discreet visual effects system, four Avid digital editing machines, and XSI animation modeling software. Using his new digital studio, he created the box office smash *Spy Kids* in 2001, which grossed more than $110 million. He used the money to set up Troublemaker Digital Studios in Austin, Texas. He rented two soundstages and converted his garage into a postproduction suite with 10 monitors, editing equipment, and a storyboard machine.

"Having finished the *Spy Kids* series," Rodriguez says, "I was looking for a good effects challenge." In 2004 he took on *Sin City* and created an eye-popping film of an underworld packed with tough cops, femme fatales, and seedy lowlifes. The movie implemented effects that could never have been accomplished using a traditional camera. Rodriguez shot the actors against a green screen and added most of the backgrounds digitally in postproduction.

Even after Rodriguez has demonstrated the power of digital filmmaking, many Hollywood purists still dismiss the technology as detracting from the story. To dispel this notion, Rodriguez invited his pal Quentin Tarantino, known for his storytelling talent, to direct a digital scene in Sin City. Rodriguez explains: "Quentin did a scene where the actors are in a car and it's raining. Instead of worrying about all that stuff, the car and the rain were added later, and he could just get the performance." Tarantino conceded, telling Rodriguez, "Mission accomplished. I'm glad you brought me down here." Tarantino now says he'll soon be shooting his own digital feature.

Questions:

1. In what ways do you think digital technologies allow Robert Rodriguez to produce films for less money?
2. What are your impressions of digital effects in movies? Do they detract or add to the impact of the story? Why and how?
3. Why do you think the Director's Guild would object to Rodriquez' and his method of producing movies?

Sources:

1. *Ashcraft, B. "The Man Who Shot Sin City,"* Wired, *April 2005, http://www.wired.com/wired/archive/13.04/sincity.html?pg=1.*
2. *IMDb Web site, http://www.imdb.com/name/nm0001675/, accessed July 12, 2005.*
3. *Troublemaker Studios Web site, www.loshooligans.com, accessed July 12, 2005.*

Some graphics file formats support different types of file compression to reduce the file size. There are two types of compression: *lossless compression* allows the original data to be reconstructed without loss, while *lossy compression* accepts some loss of data to achieve higher rates of compression. Savings in file size can be considerable and is essential for fast-loading Web pages. For example, a 7 MB .tif file can be reduced to a 605 KB compressed .jpg file with little noticeable loss in quality.

File sizes and image quality can also be controlled by adjusting the color depth. Figure 6.17 provides an example of varying levels of color depth.

FIGURE 6.17 • Color depth

Three versions of the same image: (a) millions of colors (324 KB), (b) 256 colors (47 KB), (c) 16 colors (14 KB)

Uses of Digital Graphics

Digital graphics provides personal and professional benefits in several ways. They are used as a form of creative expression, as a means of visually presenting information, as a means to communicate and explore ideas, as a form of entertainment, as a tool to assist in the design of real-world objects, and as a method of documenting life.

Creative Expression. *Digital art* is a relatively new form of art that uses computer software as the brush and the computer display as the canvas. Digital art can employ real or abstract images. The image shown in Figure 6.18 illustrates digital art depicting a surrealistic world in a superrealistic style. Other artists are taking advantage of aspects of computers that are not available to traditional visual artists. For example, artwork displayed at the HotWired RGB Gallery (*http://hotwired.wired.com/rgb/gallery/*) is described as an "interactive experiment" that allows users to interact with artwork that may include stills, animation, and sound. Taken a step further, using 3D media, artists can move beyond the two-dimensional plane to allow the viewer to virtually enter the artwork to further interact with it.

FIGURE 6.18 • Digital art

Digital art can employ real or abstract images.

Photographers and videographers use digital graphics software to edit digital images for the purpose of creative expression. Artists of all genres are using computers in some aspect of their work. Sculptors use computer models to assist in planning their projects. Artists use the Internet as a platform for collaboration, publishing, and selling artwork of all kinds.

Amateurs as well as professionals enjoy using their computers to create artwork for the home and for friends. It is increasingly common to receive greeting cards made on a PC that incorporate personalized messages and photos. Digital photography makes it easy and inexpensive for nonprofessionals to experiment with and have fun taking pictures and enhancing them with special effects.

Presenting Information. Commercial graphics designers depend on graphics software to create visually appealing designs. Web designers are specialists in creating attractive Web page designs. They use graphics software to design buttons, backgrounds, and other stylistic elements that combine to make an appealing Web page. Desktop publishers use *desktop publishing* software to design page layouts for magazines, newspapers, books, and other publications (see Figure 6.19). Other graphics designers use computers to design company logos, product packaging, television and printed advertisements, billboards, and artwork for other commercial needs.

FIGURE 6.19 • Desktop publishing

Desktop publishing software is used to design page layouts for magazines, newspapers, books, and other publications.

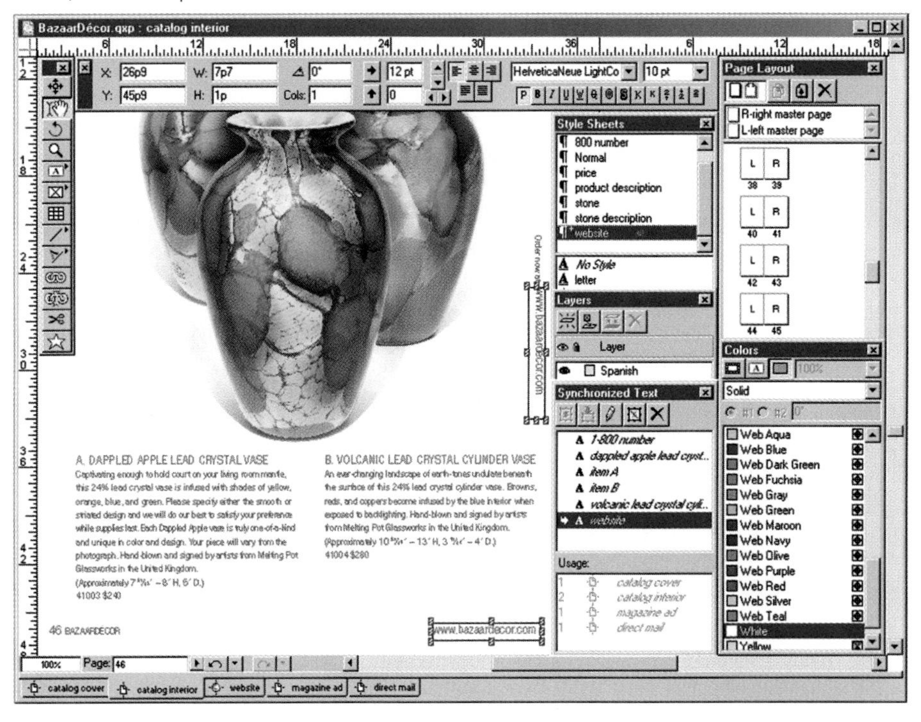

Digital video is an increasingly popular way of presenting news and information. Broadband Internet users can view digital video clips that present news and sports coverage. RealNetworks (*www.real.com*) offers free and subscription services that provide news and entertainment video clips. Microsoft offers a competing service from its *www.msn.com* Web site.

Communicating Ideas. Pictures, photographs, illustrations, graphs, animation, and video can communicate ideas in a more powerful manner than the printed or spoken word alone. Political cartoons might be the best example of this point. They quickly and concisely illustrate biting wit in a manner that would be difficult to accomplish with words alone. Professionals from nearly every career area use graphics to make a point. Teachers and others who provide educational presentations may use graphic presentation software to provide visual accompaniment to their presentations. Educators may incorporate video or animated simulations during presentations.

Illustrators use graphics software to illustrate children's stories in order to assist in developing a child's imagination. Technical artists illustrate instruction manuals and draw product specifications so customers can more easily assemble and understand the workings of products. All of these professionals use computer-generated graphics to communicate ideas.

Some ideas can be described only in graphic form. Maps are created to graphically represent our world and are available on the Internet for nearly every location on Earth. GPS (Global Positioning System) technology is used to accurately represent distances and elevations to the inch. Digital topography takes measurements and elevations as input and produces maps, even 3D maps, from the data.

Exploring New Ideas. **Scientific visualization** uses computer graphics to provide visual representations that improve our understanding of some phenomenon. There is a wide range of applications of scientific visualization, everything from the presentation of football team statistics to predict the winner of the Super Bowl to studying the interaction of subatomic particles. Scientific visualization can be used to represent quantities of raw data as pictures. Meteorologists use visualization to study weather patterns such as the movement of air and pressure around and within tornadoes and hurricanes. Cardiologists use it to study irregular patterns in heartbeats. Scientific visualization can at times turn into beautiful imagery such as shown in Figure 6.20. This picture, from a large gallery of visualization graphics available at *www.ericjhellergallery.com* illustrates the flow of electrons through a nanowire.

Entertainment. Artists in a variety of entertainment industries make use of digital graphics to produce products for the enjoyment of their audiences. Cartoonists and comic book artists can produce their products much more quickly using graphics software than they can by sketching and coloring. Animators no longer need to work with markers on transparent sheets, drawing hundreds of individual pictures that take only a matter of seconds to play on the screen. Now animators draw cartoon characters on a computer and program them to move across the screen.

The motion picture industry is capitalizing on the power of computing to create effects that are not possible in the real world. Gradually, the motion picture industry is moving towards digital cinema, which will do away with outdated motion picture projection methods in favor of digital projection.

OFF THE MAP

Things change, but maps are static—something that has plagued many an emergency worker or military general. Now researchers have developed a system that brings maps to "life" with hardware that provides up-to-the minute info, including photographs or video feed. The system uses an overhead camera, image recognition software, and an overhead projector to "project" current images onto an ordinary table-top map. From flood status to traffic accidents, operators can quickly assess any emergency situation in real time.

Source:
Augmented reality brings maps to life
By Will Knight
NewScientist.com
http://www.newscientist.com/article.ns?id=dn7695
July 19, 2005

FIGURE 6.20 • Electrons in a nanowire

This scientific visualization shows electrons flowing in a microscopic nanowire by assigning varying colors to the crests and troughs of the electron waves injected into the wire at the bright spot.

Designing Real-World Objects. **Computer-assisted design (CAD) software** assists designers, engineers, and architects in designing three-dimensional objects, from the gear mechanism in a watch to suspension bridges. CAD software provides tools to construct 3D objects on the computer screen, examine them from all angles, and test their properties. CAD is able to turn designs on the computer into blueprint specifications for manufacturing. Urban and regional planners use CAD to design neighborhoods, shopping malls, and other large-scale construction (see Figure 6.21).

CAD is not used just for 3D manufacturing, however. Professionals in the textile industry rely on computers to help design the graphic patterns for material. Computer systems are also used to apply those designs to the fabric.

FIGURE 6.21 • Computer-assisted design (CAD)

CAD is able to turn designs on the computer into blueprint specifications for manufacturing.

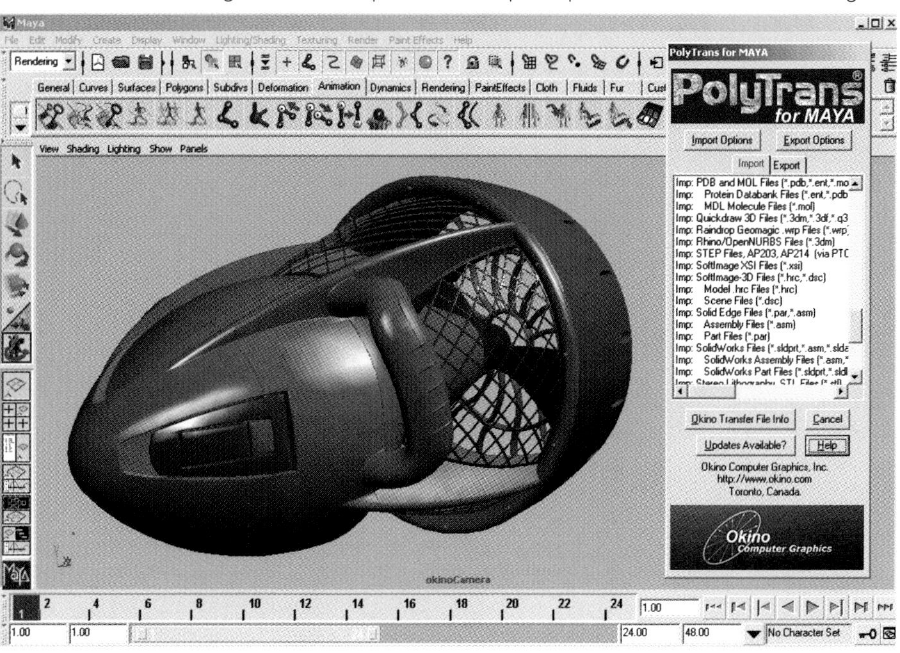

FIGURE 6.22 • Vector graphics

Vector graphics use an object-oriented approach that recognize pictures as being made up of layers of multiple objects. This image uses repeated copies of 10 objects.

Documenting Life. Photos and home movies preserve visual memories and act as a witness to special moments in our lives. Photojournalists as well as amateurs are moving to digital cameras and image processing for collecting, managing, and manipulating photographs and films for historical and sentimental value. Once digitized, these images are easily copied and shared with others over the Internet.

Vector Graphics Software

Vector graphics software, sometimes called *drawing software*, provides tools to create, arrange, and layer graphical objects on the screen. Vector graphics uses an object-oriented approach that recognizes pictures as being made up of layers of multiple objects, some in the foreground and some in the background. Figure 6.22 illustrates the object-oriented approach of vector graphics. If you examine this image closely, you can see that there are only five primary objects in the picture: a butterfly, a blossom, a rainbow fish, a starfish, and a seahorse. There are also five secondary images: a tree leaf, seaweed, a bubble, a palm tree, and a pine tree. The primary and secondary images are copied repeatedly at different sizes, locations, orientations, and levels of transparency and are interwoven together over a background to make a complex picture.

The primary benefit of the object-oriented approach of vector graphics is that objects in the picture can be manipulated independently. Consider the difference between creating a picture using colored markers on paper and creating a picture using paper cutouts. With markers, once something is drawn, it cannot be moved, and making a mistake means starting over. With paper cutouts, you have the freedom to arrange the objects on the page anyway you like. You can place the objects one atop the other to create layers and the illusion of depth, and if you don't like the way one arrangement looks, you can simply rearrange the items. Working with vector graphics is like working with paper cutouts, with the additional benefit of being able to manipulate the properties of each object. For instance, the repeated images of the butterfly in Figure 6.22 are rotated, resized, and presented at varying levels of transparency.

Vector graphics artists not only view pictures as combinations of objects, but also view the objects in the picture as combinations of objects. Figure 6.23 illustrates the individual vector objects that combine to create a realistic image of a gardenia blossom. When an artist combines objects to create larger objects, vector graphics provides a method to group the objects so that, for example, the artist can treat the completed flower in the figure as one object. The artist can copy, rotate, and manipulate the flower in dozens of ways. Objects can be grouped together, arranged one atop another, and transformed by being moved, stretched, sized, or rotated.

FIGURE 6.23 • Vector objects

Vector objects can be composed of many subobjects grouped together.

Vector graphics software also provides filtering and effects tools that an artist can use to further manipulate objects in a picture. Filter tools allow you to adjust the color of an object by altering the levels of brightness, contrast, hue, and saturation. Effects tools range from subtle effects, such as changing the sharpness or blurriness of edges within a drawing, to dramatic effects, such as changing a picture so that it looks as though you are viewing it through a glass block.

Popular professional-level vector graphics software packages include Adobe Illustrator and Corel Draw. Both are full-feature, professional-level graphics applications (and sell at professional-level prices). Less expensive vector graphics packages are available from Macromedia, Ulead, and other vendors.

Three-Dimensional Modeling Software

The latest **3D modeling software** provides graphics tools that allow artists to create pictures of realistic 3D models. Three-dimensional modeling takes the object-oriented approach of vector graphics to the next level. The process of changing two-dimensional objects such as those created with vector graphics applications into 3D models involves adding shadows and light. Three-dimensional modeling is often referred to as *ray tracing* because the software must trace beams of light as they would interact with the models in the real world. Three-dimensional modeling also requires that surface textures be defined for models. Surface textures are an important element in the interaction between light and a model. Consider the way light interacts with the model in Figure 6.24. Although this figure looks a lot like a

FIGURE 6.24 • Ray tracing

Ray tracing uses the software's ability to calculate how light interacts with models in the real world. Notice the detail, shadows, and reflections in this computer-generated image from digital artist Liam Kemp.

photograph it is actually a computer drawing by digital artist Liam Kemp (*www.this-wonderful-life.com*). Notice the effect of light on the skin and the lens of the eye in this picture. The model is shaded to produce a 3D effect, and the surfaces of the model provide textures and reflections of the space around her. Also notice, as you view this picture, that the source of the light is apparently coming from somewhere off to your right, and looks like natural light coming perhaps through a window. All of these attributes are defined in the 3D modeling software.

Creating 3D digital art takes place in a scene on a 3D stage (see Figure 6.25). The artist starts by defining a light source: the position of the light in the scene, the style of lighting—natural, spotlight, fluorescent, candle, table lamp, and so on—and the intensity of the light. Models are inserted into the scene either by selecting from a library of predesigned 3D models, by creating the model from scratch through a process of manipulating virtual wire frames, or by using 3D scanners to import images of real models into the computer. The artist selects surface textures for each model and positions the models on the stage. Finally, the artist selects a background for the scene, and the software renders the scene. *Rendering* is the process of calculating the light interaction with the virtual 3D models in the scene and presenting the final drawing in two dimensions to be viewed on the screen or printed.

FIGURE 6.25 • Three-dimensional graphics production

Producing 3D digital art takes place in a scene on a 3D stage. The artist defines a light source and places objects on the stage.

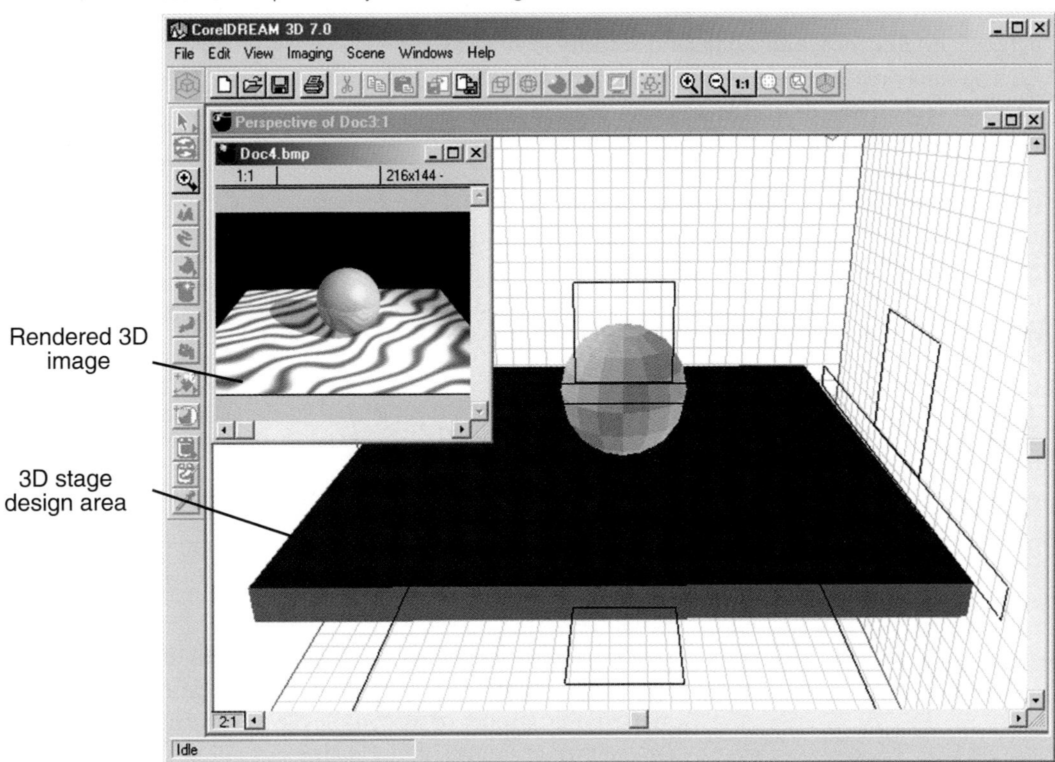

Three-dimensional graphics act as a foundation for other technologies. For example, CAD software uses 3D graphics to design and view manufactured products. Virtual reality applications use 3D graphics to build virtual worlds in which users can interact with the 3D models. The fundamental component in computer-generated animation for video games and motion pictures are 3D graphics. A popular 3D graphics development application is Strata 3-D.

Google Digitizing the World

Google Earth is an application that combines satellite imagery of the earth with map overlays (see Figure 6.26). The software allows you to type any address, city, state, or country and view high-resolution satellite photos of the location, with roads, landmarks, and other information from maps superimposed on the images, which can be zoomed and panned. The software can be used to get travel directions, find local businesses, measure distances, and a host of other useful applications. Google Earth Pro is used by professionals in commercial and residential real estate, architects, engineers, construction managers, insurance adjusters and investigators, the media, state and local government officials, as well as professionals in defense, intelligence, and homeland security.

The next step for Google Earth is to provide 3D views of city streets. With the addition of such a feature, using Google Earth, you could zoom in on a particular location to street level and virtually drive through the 3D rendering of the actual city.

The first city that Google is digitizing into virtual 3D is San Francisco. Google is using trucks equipped with lasers and digital cameras to create a realistic 3D online version of the city. The trucks drive along every San Francisco street using the lasers to measure the dimensions of buildings, and digital cameras to photograph them at every angle. The measurements are used to create a 3D framework of the buildings in the city onto which the digital photos are overlaid.

So far, Google is having minor difficulties with the process due to people and traffic getting in the way and throwing off the measurements. They feel they can get accurate measures with a second drive-through but want to perfect the technology so that a single drive through is all that is required. Researchers at Stanford are involved in similar research to create 3D interactive versions of actual cities.

Questions

1. What are some benefits of being able to virtually stroll through a city?
2. What will be the major challenge of a service that provides a virtual walk-through of a city? Why do you think Google is so interested in making it possible to map a city with one drive-through?
3. What threats to privacy do these services present? Should they be allowed? Why or why not?

Sources:

1. *Foremski, T. "Scoop! Smile for the Google 3D mapping truck,"* Silicon Valley Watcher, *June 8, 2005, www.siliconvalleywatcher.com.*
2. *Google Earth Web site, http://earth.google.com, accessed July 12, 2005.*
3. *Fast 3D City Model Generation Web site, www-video.eecs.berkeley.edu/~frueh/3d/, accessed July 12, 2005.*
4. *The Stanford CityBlock Project Web site, http://graphics.stanford.edu/projects/cityblock/,accessed July 12, 2005.*

Computer Animation

Digital graphics animation involves displaying digital images in rapid succession to provide the illusion of motion. Graphic animations can be as simple as a stick man jumping rope, or as complex as major motion pictures such as *Toy Story, Finding Nemo, Ice Age,* and *Madagascar*—all of which were completely computer generated (Figure 6.27). Animated graphics employ either 2D or 3D objects, with 3D animations requiring the most advanced graphics software and processing power to create.

2D Computer Animation. You may have seen simple animations on the Web, simple drawings that repeat the same motion over and over, endlessly. This is the most basic form of animation, called an *animated GIF* (pronounced *jiff*). Animated GIFs are created with simple tools that allow the artist to draw several images that when played in succession create the illusion of motion. The images are stored in

FIGURE 6.26 • Google Earth

Google Earth is working to provide realistic 3D imagery of major cities.

one GIF file, and when viewed in a Web page, they are played sequentially to present the animation. The artist can set the animated GIF to loop endlessly, and he or she can control the speed of the sequence. The animated GIF format allows for as many images as are required for the animation; however, the more images loaded into the GIF file, the larger the file becomes and the longer it takes to load.

FIGURE 6.27 • Pixar's *Ice Age*

The digitally animated motion picture *Ice Age*, from Pixar, was created completely within the confines of a computer system.

For more complex Web animations, artists must turn to programming languages such as Java, or Web animation development platforms such as Macromedia Flash. Because most animators are not computer programmers, Flash has become the most popular tool for creating animated Web content. Flash provides a timeline tool that is used to cue movement in the animation (see Figure 6.28). Unlike animated GIF tools, Flash automates the frame production process. For example, a picture of a rocket-bug can be placed in the upper-left corner of the Flash workspace at time unit 1; at time unit 10, the rocket-bug could be placed at the right side of the workspace. Flash will "fill in" the movement from left to right evenly over the time between 1 and 10 according to your instructions; for example you might instruct Flash to move the rocket-bug in an *S* pattern.

FIGURE 6.28 • Macromedia Flash

Macromedia Flash provides a timeline tool that is used to cue movement in the animation. In this image, the rocket-bug zooms across the screen over the duration of the timeline.

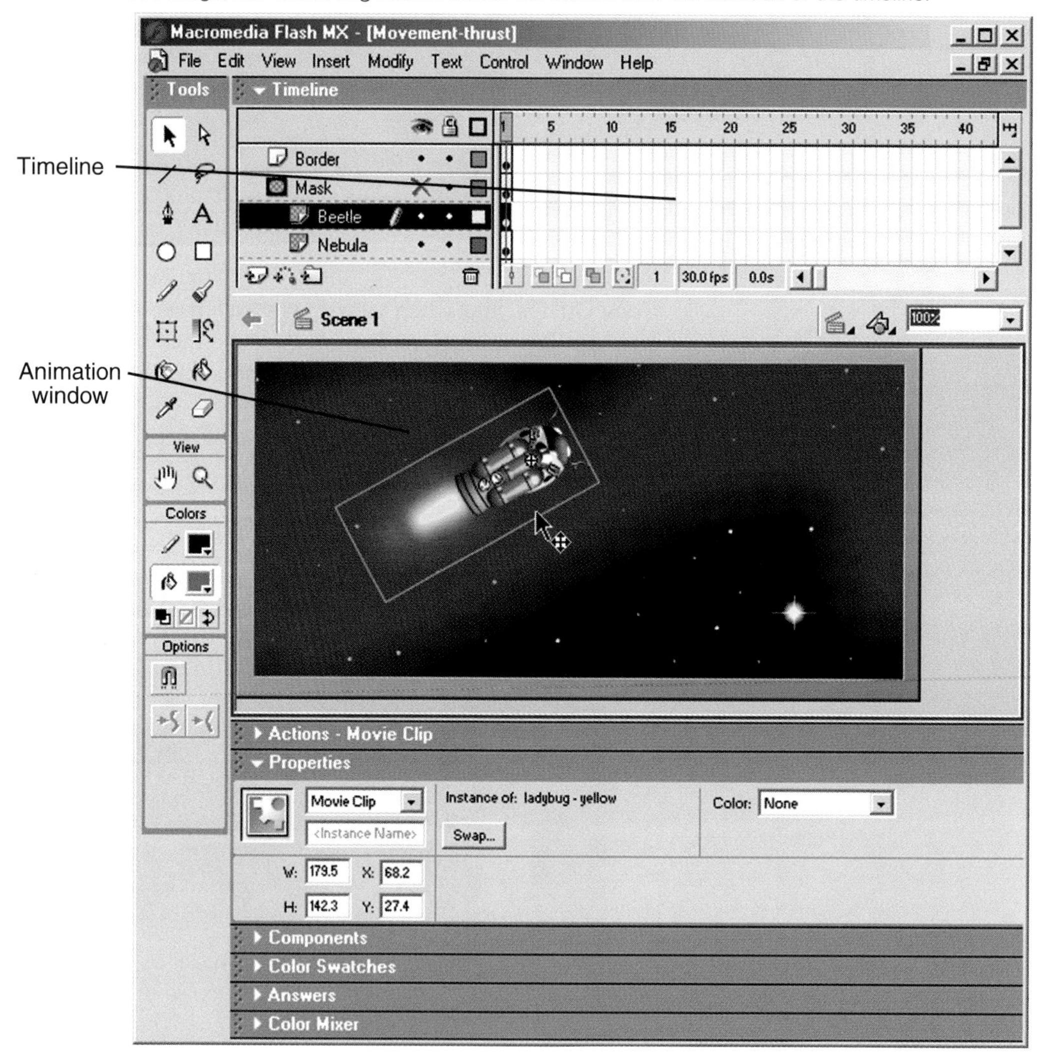

Flash can be used for simple or advanced Web animations and applications. Full-blown animated films can be produced, as well as interactive games. Other tools used to create animations on the Web include Macromedia Shockwave and Apple Quicktime.

Three-Dimensional Computer Animation. Three-dimensional computer animation is much more complex than 2D computer animation. *Three-dimensional computer animation* includes all of the complexity of 3D graphic rendering, multiplied by the necessity to render 24 3D images per second to create the illusion of movement. Three-dimensional animation software is typically packaged with 3D modeling software. These animation programs, such as LightWave from NewTek and Mental Ray from SoftImage, are used by professionals to create popular animated television shows, commercials, and movies, and range in price from $1,595 to $13,000.

Three-dimensional models of people, animals, and other moving objects, created with the modeling software, are provided with the ability to move using avars. *Avars* are points on the object that are designed to bend or pivot at specific angles. Avars are used at joints in the model's skeleton to provide articulation—movement at the joints. Avars are also used for muscular movement in the skin. The character Woody, from Disney's *Toy Story*, used 100 avars in his face to depict facial expressions. Avars are controlled either by special input devices manipulated by the artists or through software. Figure 6.29 shows an avar placed on the eyelid of a virtual frog.

FIGURE 6.29 • Animated 3D objects

An avar placed on the eyelid of this frog allows for a blinking motion.

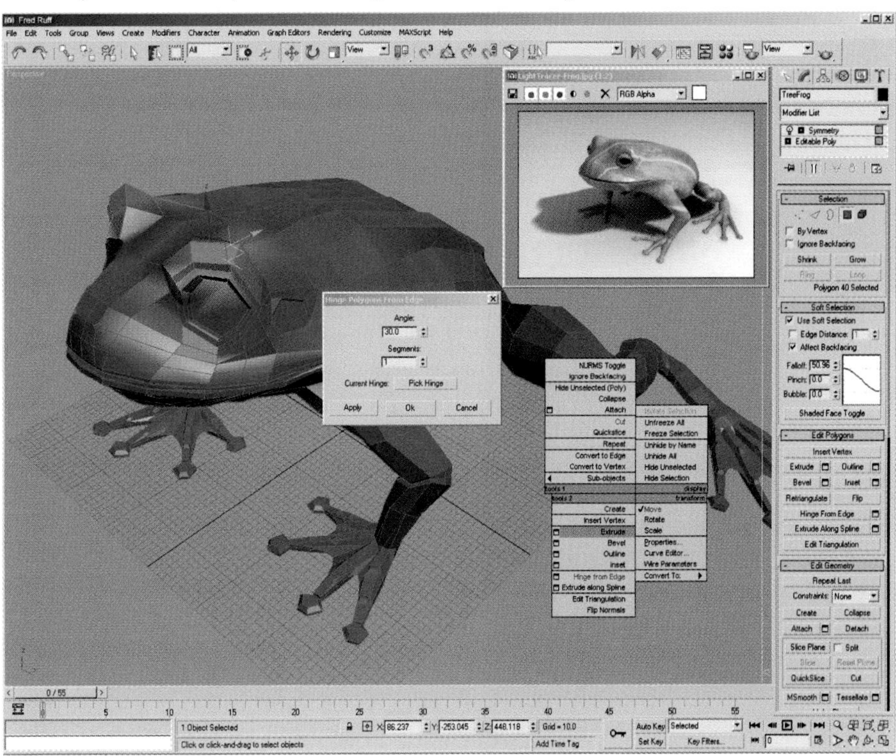

Three-dimensional computer animation software is able to move animated characters according to the direction of the artist and director. Rather than drawing and painting individual frames, animators are more like puppeteers who

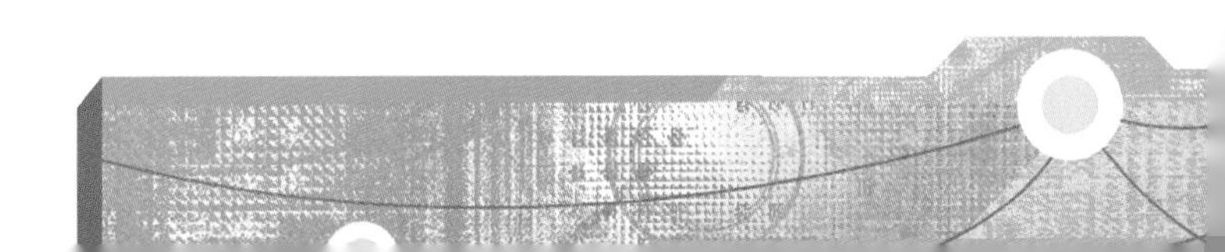

direct the animated characters around the set. After the action is recorded, rendering computers apply the 3D effects and lighting to each frame. Even with powerful computers, high-quality rendering, such as that used for major animated motion pictures, takes hours per frame.

Pixar, the company that brought us *Toy Story, A Bug's Life, Monsters Inc, Finding Nemo, The Incredibles,* and *Cars,* uses a powerful computer system called the RenderFarm. Pixar's RenderFarm has supercomputing power consisting of 1024 Intel Xeon processors housed in eight BladeRack supercomputing clusters running Pixar's own RenderMan software. The RenderFarm features two terabytes of memory and 60 terabytes of disk space.

DIGITAL PHOTOGRAPHY AND VIDEO

So far this chapter has covered only vector graphics and the method of creating and editing artwork, graphics, and animation by manipulating objects. Another method of working with photographic images and video is *digital imaging.* Instead of using vector graphics, digital imaging uses bit-mapped, or raster, images. Recall from earlier in the chapter that bit-mapped images store the color code for each individual pixel in the image. When you work with bit-mapped images, you lose the capacity to work with objects. Editing bit-mapped images is more like coloring with markers than arranging cutouts.

Digital Photography

Digital photography has become a very popular form of digital media primarily because inexpensive digital cameras and camera phones are widely available. Nearly everyone is taking digital photos and enjoying the instant gratification of viewing the image as soon as it is taken—something previously not possible. Gradually, photo albums are transforming from dusty bound volumes on bookshelves to volumes stored on hard drives and online services. The number of photos being generated is at its highest peak ever, and hundreds of new devices, software, and services are rapidly being introduced to the market in support of the digital photo fever. This section surveys the latest technologies for creating, editing, sharing, and printing digital photos.

Creating Digital Photos. Digital photos are created, or acquired, using a digital camera (see Figure 6.30) or scanner and are saved as bit-mapped images. Digital cameras are ranked by the amount of *megapixels* they can capture (1 megapixel = 1 million pixels). An inexpensive digital camera can capture around 2 megapixels, and the best can capture over 8 megapixels.

Cameras are available from dozens of manufactures in a wide range of styles and qualities. Inexpensive disposable digital cameras are even available for people who don't want to invest a lot of money or who forget their own camera while out and about.[12] Table 6.4 lists digital camera specifications to consider.

FIGURE 6.30 • Coolpix

About the size of a wallet and weighing less than most cell phones, this Nikon Coolpix takes up to 6-megapixel photos and includes a 2.5" display, 3X optical zoom, 640 x 480 movies at 15 frames per second, 17 scene modes, and a voice annotation feature, all for less than $300.

TABLE 6.4 • Digital camera features and considerations

Feature	Specifications and comments
Resolution	Three megapixels or more is best for recreational use, higher than four megapixels for professional use
Price	$150 to $500 for recreational-grade cameras, thousands of dollars for professional-grade cameras
Lens type	Zoom with a range encompassing at least 38 mm to 114 mm
Storage media	Flash memory card such as CompactFlash, Memory Stick, xD-Picture Card, or Secure Digital/ MultiMediaCard; each megabyte holds approximately two to three high-resolution photos
Photo file format	JPEG for recreational use, TIFF for professionals
Interfaces	USB, NTSC/PAL television connection
Exposure controls	Automatic, programmed scene modes, exposure compensation (for tweaking the automatic exposure)
Focus controls	Automatic for recreational use; pros will want to include manual control as well
Flash modes	Automatic, fill, red-eye reduction
Software	Photo editor, photo stitching (combining photos into panoramas), photo album creator, slide show
Multimedia	VGA (640 x 480), 30-fps video-clip recording
Other useful features	Ultracompact design, Webcam capabilities, voice recording, wireless transfer, large LCD picture display
Battery	Varying size rechargeable batteries are available for extended use

Memory cards that come with digital cameras often have limited capacity, so it is usually a good idea to purchase a higher-capacity card. A 64 MB card can store over 100 photos depending on the selected resolution. Even with a higher-capacity card, travelers often take along a laptop computer to download photos over the course of long trips, so they can clear and reuse the camera's memory card. New high-capacity photo storage devices are another option (see Figure 6.31). Photos can be transferred directly from camera to the device. Some versions of the iPod can be used for photo storage and offer the added benefit of providing music.

FIGURE 6.31 • Multimedia storage viewer

Handheld devices like the Epson P-2000 can store hundreds of photos transferred from a PC or directly from your digital camera.

Editing Digital Photos. **Photo-editing software**, such as Adobe Photoshop and Photoshop Elements, Microsoft Picture It!, and Ulead PhotoImpact, includes special tools and effects that you can use to improve or manipulate bit-mapped photograph images. Photo-editing software allows you to:

- Alter the hue and saturation of the colors in the photograph to give skin tones a healthy glow, or enrich colors in a landscape.
- Smooth surfaces or remove flaws in surfaces.
- Remove "red eye"—the effect of the camera flash on the eyes of subjects, which causes eyes to appear red in a photograph.
- Smooth edges and sharpen focus.
- Crop and realign photos.

Photo-editing tools also include special effects, such as the ability to transform photos into watercolor paintings. Or you can use edge detection to remove an object from one photo and insert it in another. The ability to combine photo images to create fictitious photos has made it difficult to trust any photograph at face value (see Figure 6.32). Photos no longer stand up as evidence in the courtroom as they once did.

Dartmouth College professor Haney Farid believes that he may be able to put trust back in digital images. Although it is often impossible to detect tampering with the naked eye, Dr. Farid has developed a method of analyzing the values of the bits that compose an image to determine if it has been altered. Dr. Farid and his students developed a statistical model that recognizes the mathematical regularities inherent in natural images. Because these statistics fundamentally change when images are altered, the model can be used to detect digital tampering.[13]

FIGURE 6.32 • Faked photos

Has SeaWorld really added a giant goldfish to their killer whale performance?

Most digital cameras come with photo-editing software designed for personal use, providing easy-to-use tools for cleaning up photos and integrating them into frames, greeting cards, calendars, and other popular printed forms. This software often includes wizards that walk the user through the photo-editing process. Figure 6.33 illustrates such a program. Professional-grade photo-editing software, such as Adobe Photoshop, provides many tools for retouching and editing photographs to professional standards.

FIGURE 6.33 • Personal photo editing

Photo-editing software designed for personal use provides easy-to-use tools for touching up photos and adding effects.

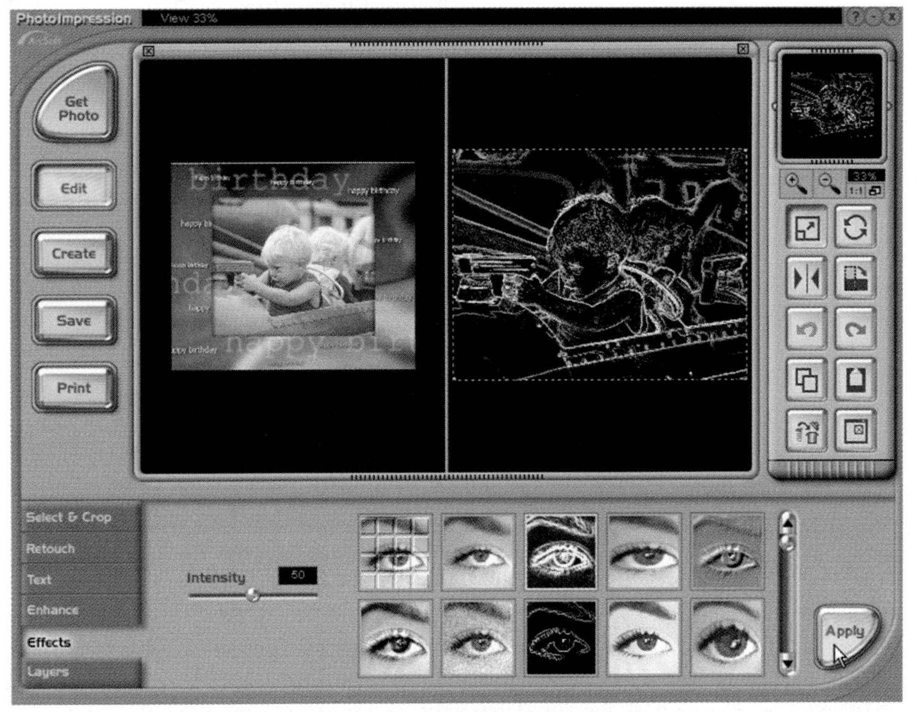

Viewing and Sharing Digital Photos. There are several ways to view digital images and share them with friends. It is not always convenient to have everyone gather around the computer to view photos. However there are systems designed to display photographs to groups. Many of today's cameras provide ways to display the images on a television as a slide show. As PCs become more integrated with entertainment systems, it will become increasingly common to gather around the television, or media display, to view family photos. Windows XP Media Center Edition is set up to support just such a service. There are even digital picture frames. The Ceiva Digital Photo Receiver (*www.ceiva.com*), looks like a regular desktop picture frame but is in reality an Internet-connected digital photo display. The picture frame connects to the Internet via a phone line. Each night it automatically dials up the Ceiva system to download new photos to the picture frame that have been sent to you by friends and family. Photos are sent to the picture frame using Web-based software or directly from camera phones. When you awaken, a new montage of photos is progressively displayed for your enjoyment throughout the day. Other digital picture frames read photos directly from media cards or CD.

Perhaps the most convenient way to share photos is over the Internet. E-mailing photos is problematic because photo files are typically large and can dramatically slow down e-mail delivery. It is much more practical and considerate to post photos on the Web for others to view as time permits. Many companies offer such services, such as Yahoo! Photos (*http://photos.yahoo.com*), Google's Picasa (*http://picasa/google.com*), and Kodakgallery (*www.kodakgallery.com*). Over 1.3 billion pictures were uploaded to online photo sites from personal computers in 2004.[14] The services also provide relatively inexpensive photo prints.

Some services also offer the ability to upload and view photos from cell phones. For example, Yahoo! Mobile Photos (*http://mobile.yahoo.com/photos*) allows users to upload photos from cell phones directly to their Yahoo! Photos accounts. This is quite valuable as it is often difficult to transfer photos from a cell phone to a PC. There is no charge for the service, and no limit to the number of photos that can be transferred.

Printing Digital Photos. As with most aspects of the rapidly expanding digital photo industry, options for printing photos are bountiful. In 2004, consumers spent an estimated $8.2 billion on printing photos.[15] In the early years of digital photography most photographers simply printed their own photos using high-quality paper and a standard inkjet printer. However, most standard inkjet printers do not provide the quality of professionally printed photos. Today many printers are designed specifically for printing photos, but still many people prefer to have their digital photos printed by a professional service due to the low price, convenience, and high quality. Their low price rivals the cost of purchasing paper and toner to print your own photos. Also, many users find it difficult and time consuming to match the quality provided by professional printing services.

Yahoo! has recently teamed with Target stores to offer photo pickup service. Prints of digital photos uploaded to *http://target.photos.yahoo.com* can be picked up at your nearby Target store (Figure 6.34). Costco has a service that allows its members to upload images from their computers to Costco's Web site and then pick up the prints at the store within the hour.

Photo kiosks are another popular option for printing digital photos. In 2005 there were 75,000 photo kiosks in the United States. That number is expected to increase to 121,000 by 2008.[16] Photo kiosks read photos from CDs or any of the popular media cards, they allow you to crop and edit each image, and then print out your photos—perhaps while you do a little shopping. Kiosks are popular for their ease of use, print quality, speed, and the amount of control they

FIGURE 6.34 • Target Yahoo! Prints

Target and Yahoo! have teamed up to provide digital photo uploading, storing, printing, and pickup services.

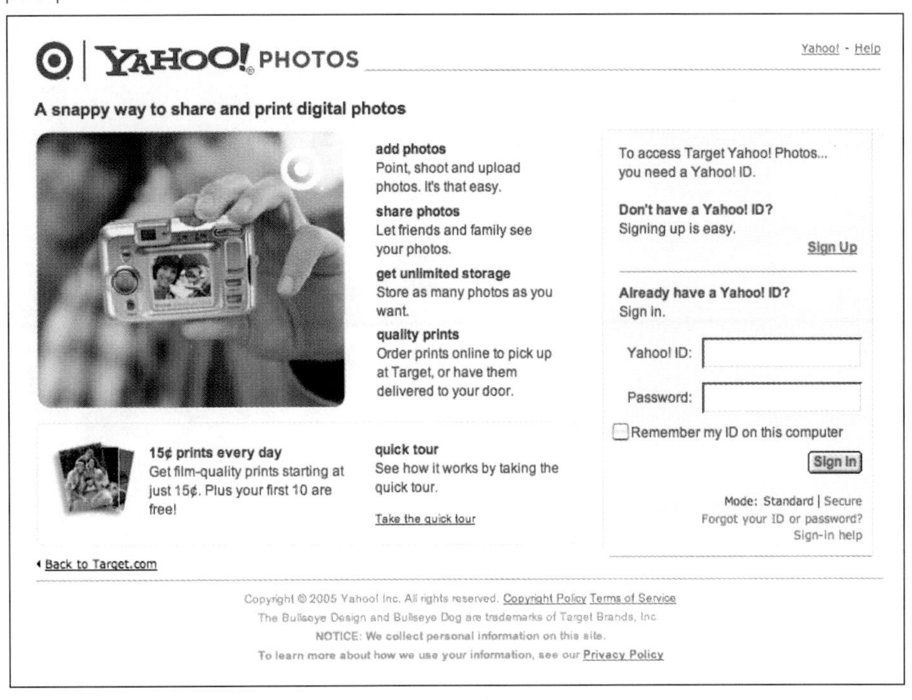

offer the user. Home photo printer manufacturers have designed several home printers that function in much the same manner as photo kiosks in an attempt to provide consumers with the conveniences they desire.

Digital Video

To learn more about digital video, go to www.course.com/swt2/ch06. Click the link "Expand Your Knowledge," and then complete the lab entitled, "Working with Video."

Advances in digital technologies and broadband communications and information systems are having a profound effect on the creation, distribution, and enjoyment of video. Digital video is becoming increasingly accessible for personal enjoyment as well as professional use.

Both digital photography and digital video are used in a variety of research and other professional areas. *Forensic graphics* is used to create animations and exhibits to use in courts of law to explain theories and present evidence. Figure 6.35 shows a forensic graphics storyboard used to reconstruct the spread of an industrial fire. Forensic graphics experts also study photos and videotaped evidence in an effort to solve crimes. For example, a forensic graphics expert might study video of a convenience store robbery to ascertain the identity of the thieves. Video cameras mounted at intersections are increasingly being used to catch drivers who run red lights by recording the license number of the vehicle in the video image.

By studying digital video, athletes and trainers can review the movement of athletes and determine how to perfect their abilities. Digital video is used to study pedestrian and traffic patterns. Scientists also use it to study the activity of microbes and other organisms.

Digital video is also finding a home in the heart of consumers as a primary form of entertainment. Taking in a movie has become a favorite pastime for all ages. Digital cable networks, on-demand television, and DVDs have brought popular motion pictures to home television. Large high-definition displays and surround sound systems are providing theater-quality experiences at home.

FIGURE 6.35 • Forensic graphics

Forensic graphics storyboards are used to reconstruct the spread of an industrial fire.

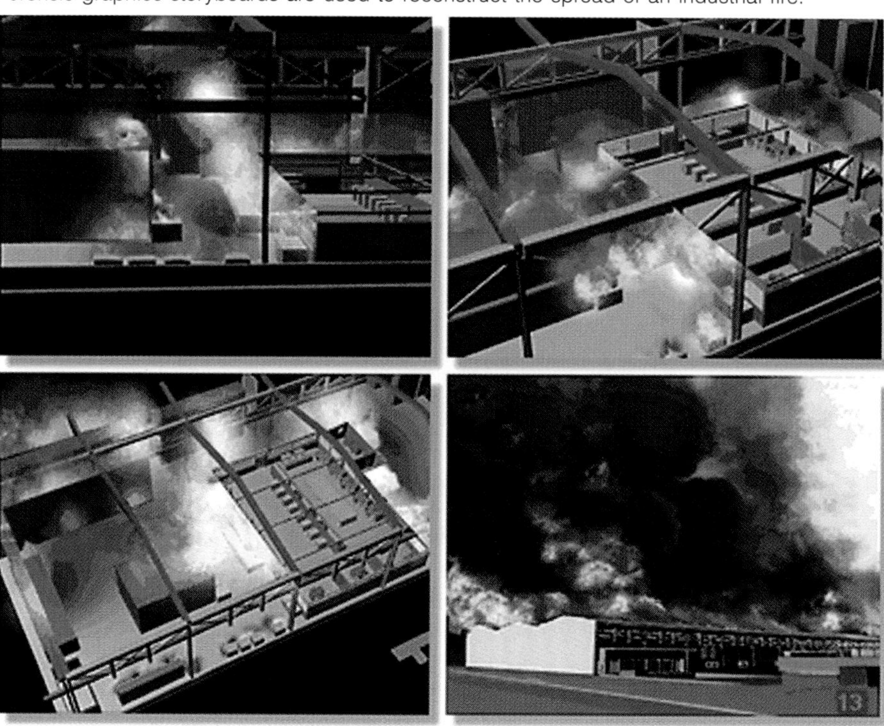

Broadband Internet and service providers are bringing motion pictures, television programming, and other video services to PCs and cell phones. Digital camcorders allow us to capture the important moments of our lives in motion.

Creating Digital Video. Digital video cameras, called digital camcorders, are available in a wide range of prices with wide-ranging capabilities. At the low end of the spectrum are disposable camcorders that sell for under \$30.[17] At the opposite end of the spectrum are high-definition camcorders selling for over \$3000.

Because most popular camcorders cost between \$300 and \$500, many consumers are making do with the video capture capabilities of digital cameras and cell phones at half the price. Most digital cameras and some camera phones can capture short video clips at relatively low resolution (320 × 240 or 160 × 120). However, when it comes to recording soccer games, ballet recitals, and weddings, camera phones and digital cameras don't have the memory capacity to capture a full-screen lengthy video that is typically desired; a camcorder is the tool for the task. For this reason, many consumers purchase their first camcorder when they have their first child.

LET THERE BE LIGHT

On a film set, where time equals dollar signs and several digits, lighting is the hardest thing to control. A new computer graphics tool, however, allows filmmakers to simulate lighting conditions for scenes their actors weren't in, or even to add effects to film already shot. The software program uses algorithms to choose and layer different frames—each shot under different, rapidly changing, lighting conditions—to come up with the right light for the right mood.

Source:
From the Lab: Information Technology
By Monya Baker (editor)
Technology Review
http://www.technologyreview.com/articles/05/08/issue/ftl_info.asp
August 2005

Editing Digital Video. With the increase in popularity of digital camcorders, increasing numbers of home PC users are finding it useful to edit their digital videos on their home PC. Much of the footage captured on video tape is often not worth saving. **Video-editing software** allows professional and amateur videographers to edit bad footage out and rearrange the good footage to produce a professional-style video production. Popular video-editing software packages include Studio Moviebox for Windows from Pinnacle Systems; Adobe products such as Premier, Video Collection, and After Effects available for both Windows and Apple computers; Windows Movie Maker; and Final Cut Pro for Apple. Many of today's camcorders include the ability to edit video directly on the camera. This provides the considerable convenience of not needing to download and edit on a PC. The tools provided are simple to use but not as powerful as video-editing software.

Digital video can be transferred to a computer directly from a camcorder or read from a DVD or VCR tape. Video takes up a significant amount of space on the hard drive—roughly a GB for one hour of video, for this reason most home users do not store many videos on their PC, but rather copy them to CD or DVD after they finish editing.

Video-editing software uses a *storyline* on which to build a video production. A storyline allows the videographer to arrange video scenes sequentially and specify the transition effects between each scene. Figure 6.36 illustrates the video-editing process. The videographer can add still images, background music, and text to the storyline as well as video scenes. When the storyline is complete, it is saved to one of many possible video formats on a hard drive and usually burned to DVD to be viewed on TV.

FIGURE 6.36 • Video-editing storyline

Using video-editing software, a videographer arranges scenes and transitions between scenes on a storyline to create a video production.

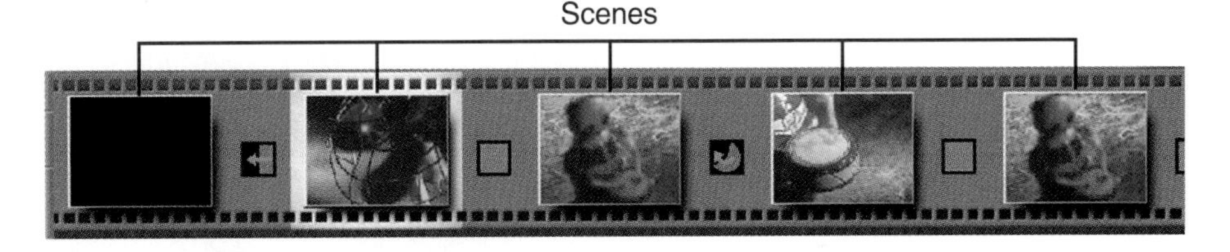

The increase in home video production has led to a new form of expression on the Web called vlogs. *Vlog* is short for *video log*, and is the video version of a blog. Just as podcasts might be considered a form of audio blog, a vlog presents a person's ideas or tells a story through video. A search at your favorite search engine on *vlog* is sure to turn up interesting sites such as the one at *http://x.nnon.tv/*.

Digital movies, which might otherwise require hundreds of gigabytes or even terabytes of storage space, are compressed down to less than 10 GB in order to fit onto a DVD. Digital video compression involves analyzing each frame of the video and storing only the pixels that change from frame to frame. Consider a nightly news show. Such video typically has little or no change to the background behind the news reporter. So, when compressed, the bits that make up the background only need to be stored once for the frames in which they remain unchanged. A video frame might display a million pixels but

BLOGOSHPERE TO VLOGOSPHERE–NOT JUST A MISPRONUNCIATION

Although it hasn't yet reached the pervasiveness of blogging, vlogging is fast creating its own sphere on the Internet. From politicos to proud parents, all manner of folks are bypassing distribution through cable or TV and self-publishing video blogs online. Aficionados think it can change the world by allowing access to subjects not touched my mainstream media. They've started their own convention—Vloggercon—and gathered together for summer vloggercues to talk technology and frames per second.

Source:
Blogging + Video = Vlogging
By Katie Dean
Wired News
http://www.wirednews.com/news/digiwood/0,1412,68171,00.html?tw=wn_tophead_1
July 13, 2005

only have 1000 pixels that change from the previous frame reducing the storage requirements from a million to a thousand.

Digital Video, Television, and Movie Services. Besides the traditional forms of taking in movies—theater and television, many new forms of digital video delivery are becoming available. In earlier chapters you learned about Internet companies that provide motion picture download and DVD rental services, and cell phone services that provide video clip downloads such as music videos, and sports clips (see Figure 6.37). It is clear that just as digital technologies are transforming the music industry, they are transforming the movie and video industries as well.

FIGURE 6.37 • TV on the phone

Television programming is increasingly available anywhere, anytime: at home, in the car, on your PC, or on your phone.

The rapid adoption of high-speed Internet connections is making trading pirated copies of movies online easier and more widespread. Still, sharing movies is not as widespread as sharing music because of the size of movie files: it can take an hour or more to download one movie. Major motion picture studios are scrambling to develop an online distribution model and invent strong incentives to offer consumers convenient legal options for accessing movies online, and many have digitized hundreds of movie titles in preparation for online sales.[18]

The Akimbo service (*www.akimbo.com*) makes use of a set-top box about the size of a VCR to download movies and television programming from the Internet and play them on your TV. The Akimbo box contains an 80-GB hard drive and requires a high-speed Internet connection. At this writing, the box costs $199, and the service is $9.99 per month.

Mobile television services are also on the rise. Microsoft's video download service provides daily television programming, entertainment clips, and other digital content for viewing on mobile Windows-based devices.[19] The service can also be used to view shows recorded on a TiVo digital video recorder using a service called TiVoToGo.[20] The new iPod can store 150 hours of video supplied by the iTunes service in cooperation with Pixar and Disney (which owns ABC-TV).[21]

Not to be left out of the race to digital convergence, cell phone companies are offering video and television programming over 3G networks. For example Verizon's Vcast service provides video entertainment, news, weather, and sports from major media outlets for $15 per month.[22]

Sirius Satellite Radio is also working to provide video programming over its satellite network. Delphi is working with Sirius to develop an in-car video system for the service.[23] Several other companies are working on developing mobile media for automobiles. Comcast is planning a service that allows customers to fill up a removable hard drive with television programming from the Comcast TV network to be watched on in-car systems. Other companies are installing Wi-Fi and media storage systems in cars to allow users to wirelessly transfer music and video from home media centers to auto media centers.

The Sony Location Free TV uses a base station that connects to a digital television service to broadcast the television signal wirelessly to PCs or the wireless television display (see Figure 6.38). Other companies, like Orb Networks, offer software solutions that establish one PC, connected to the television network, as the receiver and broadcasts television programming to other PCs around the house over a Wi-Fi network.

FIGURE 6.38 • Location-free TV

Connect the Sony location-free TV base station to your digital television network and watch TV on a wireless display that you can take with you anywhere around the house—or view TV on any PC around the house.

Boeing is supporting service to airlines that offers live television broadcasts to notebook PC users on its international flights. Travelers can access four channels of live television on their laptop PCs over its Connexion service. The service will expand into more flights and eventually offer live sporting events.[24]

Even Google is getting into the television business. The Internet search company is working to develop technologies that will do for television programming what it has done for the Internet. With so much video of every kind being digitized and networked, a tool that assists in sorting through and finding video clips will be invaluable.[25]

All of these examples point to the rapid integration and convergence taking place between computing, television, and communications networks. In the near future, you will be able to watch television at home, in the car, in an airplane, or in the park on a notebook or a cell phone display. It will be delivered by television providers, phone providers, Internet providers, or more likely a combination of all. Comcast, a company that started out as a television provider, now also provides Internet and telephone services. Sprint and other telecoms provide phone, Internet, and now television programming. It is easy to foresee that through partnerships, takeovers, and mergers, it is very possible that in the near future there may be only a handful of companies competing to provide all of our digital electronics connections.

INTERACTIVE MEDIA

Interactive media refers to digital media presentations that involve user interaction for education, training, or entertainment. Interactive media is unique in that the audience is not passively observing the media; it is created specifically for the audience to take part in the creative or educational process. Interactive media typically combines digital audio and digital video for a full digital media experience.

When interactive media incorporates 3D graphic animation, the result is *virtual reality*. Virtual reality, discussed further in later chapters, produces a simulated environment in which the human participant can move and manipulate objects. Adding surround sound makes the interactive experience more realistic.

Video games are a large portion of the interactive media market (Figure 6.39). Other forms of interactive media provide computer-based tutorials and training, and still others are commercial applications that support the sales of

FIGURE 6.39 • Three-dimensional interactive game environments

Three-dimensional interactive games put you inside the virtual world where you exercise control over the action.

products and provide customer support. Interactive media is becoming increasingly important as technology advances are able to support its large demands for processing speed, storage, and bandwidth. New development tools also better support interactive media over the Internet and Web. This section provides an overview of interactive media and its value in our lives.

Education and Training

Research shows that most individuals are able to comprehend complex ideas more thoroughly and quickly when able to interact with them using digital media. For example, engineering professors at the University of Missouri turned to interactive media to assist students who were having difficulty understanding the theories behind stress transformation—the internal stresses and forces that loads place on building materials. They developed an interactive media tool that allowed students to witness the effects of stress on different materials and the associated stress transformation equations. By visually associating the stress placed on virtual objects with the equations, students could more easily understand how abstract concepts relate to the real world. Results from the study showed that when lecture, interactive media, and reading textbooks are compared, students felt that they learned best in lectures and preferred interactive media over textbook reading. Educational research also shows that everyone learns differently. For this reason many curricula are including digital media and interactive multimedia components.

Many teachers are finding the simulated worlds of virtual reality games a fertile environment for learning. The Muzzy Lane Software Company develops gaming software for the classroom to help high school and college students learn about history and develop thinking skills. Making History is a multiplayer simulation that puts players in control of European governments before, during, and after World War II. The game integrates learning into the software through player experience rather than traditional methods that preach to the player.[26]

The hand-eye coordination developed through the use of video games can be translated to skills in various professions. Dr. James Rosser Jr., a top surgeon and director of the Advanced Medical Technologies Institute at Beth Israel Medical Center in New York, uses video games to help train laparoscopic surgeons. Surgeons who play video games three hours a week have 37 percent fewer errors and accomplish tasks 27 percent faster, he says, basing his observation on results of tests using the video game Super Monkey Ball.[27]

Video games and simulations can help train professionals before they go into dangerous situations. STATCare lets medics bandage wounds, apply tourniquets, administer intravenous fluids, inject medications, and make all of the other assessments they would be required to do in an actual battlefield. HazMat:Hotzone teaches firefighters how to respond to a chemical-weapons attack. Forex Trader shows financiers the ins and outs of currency trading, and college administrators use Virtual U to "wrestle with angry professors and meddlesome state legislators."[28]

REALLY REMOTE CONTROL

Help is handy for those who get anxious if they have to miss their favorite TV show while traveling. Whether in Biloxi or Bangkok, a person can now redirect their regular programming via the Internet with a product called Slingbox. Slingbox digitally redirects the transmission signal from any channel received on the home TV to anywhere in the world using a computer and a broadband or satellite connection. Slingbox's software automatically adjusts for varying line speeds.

Source:
Box slings TV to on-the-road computer:
Device allows user to catch shows from remote location
By Benny Evangelista
San Francisco Chronicle
http://www.sfgate.com/cgi-bin/article.cgi?file=/chronicle/archive/2005/07/18/BUG10DNHGV1.DTL&type=tech
July 18, 2005

Interactive media is being used in many education environments:

- In the traditional classroom setting, it helps students learn difficult concepts.
- In distance learning, it sometimes takes the place of lecture demonstrations.
- In museums, it allows the public to interact with virtual objects and environments that are not possible to interact with in real life.
- In skills training, interactive multimedia simulations are used to train pilots and others in a host of skills-based occupations.

Interactive lessons can be presented on PCs or public kiosks. They allow learners to work at their own pace and often in their own environment. Often multimedia permits interaction that is not feasible or convenient in real life.

Commercial Applications of Interactive Media

Companies have jumped on interactive media to offer customers additional services and benefits. Interactive media plays a large role in Web-based e-commerce. It provides the fundamental technology for 3D product viewing, which allows online customers to thoroughly examine products. Travelers can take virtual tours of resort destinations prior to booking a room; furniture buyers can "try out" different fabrics on a sofa, and hair stylists can show you "the new you" before snipping. Web marketers even use interactive games to get users to click banner ads.

Interactive media is also used to provide product support and customer training. For example, Mercedes provides its customers with interactive media on CD that allows them to explore the parts and features of their new automobile. In the construction industry, electronic blueprints provide a 3D view of a home. The owner can click an option to view the home's electrical system and rotate the image to determine how and where to add an additional wall outlet. Interactive 3D is used in interior design to place objects within a virtual home and determine what looks best.

Interactive multimedia is playing a large role in attracting customers to Web sites, and once there, it plays a role in keeping them there. The Web site for the Toyota Prius (see Figure 6.40) uses interactive multimedia in nearly every conceivable way, short of taking it for an actual test drive. It includes 360-degree interior and exterior views of the car, educational animations that explain the unique gas-electric hybrid engine design, and a wizard that allows you to custom design your own Prius.

Interactive Video Games

In the area of interactive multimedia entertainment, the computer video game reigns supreme. Video games employ nearly every aspect of digital media discussed in this chapter. Gaming takes place on computers or special purpose gaming devices. **Video game consoles**, such as such as the Nintendo GameCube, the Sony Playstation 2 (PS2), and Microsoft's Xbox, are high-powered multiprocessor computers designed to support 3D interactive multimedia. They come equipped with a fast microprocessor that works in conjunction with a graphics coprocessor to support fast-paced gaming action. The microprocessor is specially designed to handle the high demands of live action 3D rendering. These units also include memory, storage similar to a PC, and an optional Internet connection to connect with gamers in other locations.

FIGURE 6.40 • Interactive media sells cars

The Toyota Prius Web site uses interactive multimedia to enable visitors to rotate and manipulate the vehicle on the screen and even take it on a test drive.

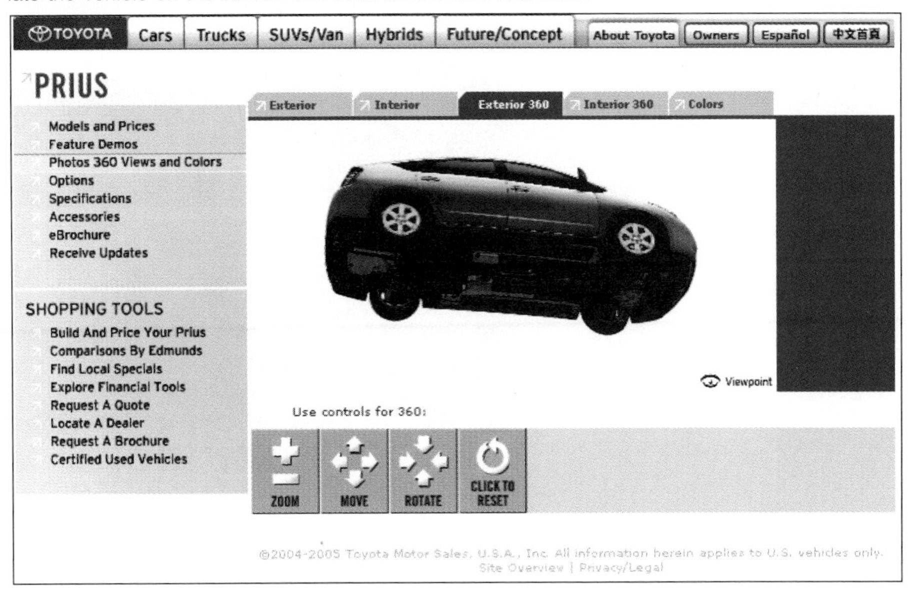

The quality of the gaming experience has dramatically increased over the past few years. "Epic worlds are alive with detail, from thunderous skies rumbling over a mountain range to tiny blades of grass rustling together in the breeze. Vibrant characters display depth of emotion to evoke more dramatic responses, immersing you in the experience like never before."[29]

The gaming market has ridden a wave of success right through the downturn in the economy in 2001, and it doesn't look as though it will slow down anytime soon. Since 1995 and the release of the Sony PlayStation, over 3 billion video game consoles have been sold. Global sales for video games are expected to increase from $20.7 billion in 2002 to as much as $30 billion by 2007.

Gaming is becoming an increasingly popular mobile activity due to new next generation handheld game devices such as the PlayStation Portable and Nintendo DS. The Sony PlayStation Portable (PSP) brings PlayStation-quality 3D graphics to a handheld device that doubles as a music and movie player (Figure 6.41). Movies are available for the PSP from Sony, Disney, and Buena Vista Home Entertainment.[30]

In Japan, Nintendo has established 1000 Wi-Fi hotspots where owners of its DS portable game machines can play games with others online for free.[31] They are planning to set up similar hotspots in other countries.

Cell phones are bringing the gaming world to increasing amounts of adults. Short, uncomplicated games like Tetris, Solitaire, and PacMan allow users to pass the time between appointments without needing to invest a lot of time and effort. In reaction to the popularity of cell phone gaming, Nokia is equipping all of its SmartPhones with gaming capabilities.[32]

Video game development requires a team effort from specialists in a variety of areas: game designers, artists, sound designers, programmers, and testers. Putting a new game on the market requires a large financial investment and is a big risk—a risk that many developers are willing to take in pursuit of the rewards that come with having a hit. A successful game must engage the user by allowing progress through the game at just the right pace, with action sequences timed at just the right intervals. Successful video games are easy to learn but difficult to master.

FIGURE 6.41 • Playstation Portable

The Playstation Portable brings high-quality 3D graphics to a slick handheld design.

In efforts to engage teen and older gamers, games have progressively moved to more violent themes (Figure 6.42). The most recent edition of the popular but controversial Grand Theft Auto, San Andreas, became all the more popular for the stir it created over its use of extreme violence and experienced record sales maintaining the games number one position in units sold. Some people have become concerned about the effects of prolonged exposure to violence on game users.

FIGURE 6.42 • Video game violence

With the popularity of games such as Grand Theft Auto, many people wonder how prolonged exposure to violence might be affecting the young people who play it.

Interactive TV

Interactive TV has been touted as "the next big thing" in interactive multimedia entertainment. Various features of interactive TV are available to cable subscribers in select areas. **Interactive TV** is a digital television service that includes one or more of the following: video on demand, personal video recorder, local information on TV, purchase over TV, Internet access over TV, and video games over TV. *Video on demand (VoD)* allows digital cable customers to select from hundreds of movies and programs to watch anytime they choose. The movie or program is stored using a set-top box and can be paused, rewound, and treated as a DVD. *Personal video recorders* (PVR), such as Tivo and Replay TV, provide large hard drive storage to record dozens of movies and programs to be watched at your leisure. *Local information on TV* provides local community news and information. *Purchase over TV*, sometimes called *t-commerce*, allows viewers to make purchases over their cable TV connection much as computer users make purchases on the Web. *Internet access over TV*, through services like WebTV, allows viewers to navigate the Web on their television sets. Cable TV services provide access to video games through *video games over TV*.

As with the Web, interactive TV has the potential to provide individual targeted marketing. Cable providers are considering ways they can collect customer information in order to provide customers with television commercials that are targeted to individuals' interests. Using *behavioral profiling*, cable TV providers can observe customers' viewing patterns in order to develop an understanding of their interests. Imagine watching television and being shown only commercials for products and services that interest you. Some cable providers consider targeted marketing to be a core component of interactive television.

Microsoft has paved the way for interactive TV with Windows XP Media Center Edition. This version of Windows runs on media center PCs, such as the one shown in Figure 6.43, that use a television, or preferably a plasma display. The media center PC connects to your television service and acts as an electronic program guide, a Web browser, television receiver, personal video recorder and player, DVD player, digital music player, and digital photograph displayer. Each type of media is supported by programs that sort, catalog, and schedule your media content.

FIGURE 6.43 • HP Media Center PC

This media center PC includes removable hard drives and numerous media ports that allow convenient connections to media devices, stereos, and televisions.

ACTION PLAN

Remember Ana Arguello from the beginning of this chapter? She was the college freshman considering how to merge her artistic talent and computer skills into a college degree. Here are answers to the questions about Ana's situation.

1. What careers and majors might Ana consider in digital media?

Enjoying computers does not necessarily imply that you enjoy programming computers. Ana might try a beginning computer programming class to see if she likes it, but it sounds as though her creative talents and interests in the arts may overshadow her interest in the workings of a computer. Artists, musicians, photographers, videographers, and designers all rely heavily on computers. No matter what Ana chooses, she will continue working closely with computers, in fact her considerable computer skills gives her an advantage over others in whatever major she chooses. Ana's experience and interests lend themselves to working in graphic design, music production, music composition, motion pictures, art, and photography. There are many professions that make use of the combination of Ana's skills to develop Web sites, video games, movies, and all sorts of popular media products.

2. How can Ana use digital media software and services to manage her media collections?

Ana can make use of online music services to access her favorite music and develop playlists to match her every mood. Ana might also find other musicians online with whom she could collaborate. She can download free software to manage her digital photo collection and make photo albums to share with her friends and family online. Ana can make use of online movie services to download and enjoy her favorite movies and to discover new favorites.

3. How might Ana enjoy digital media as a hobby and as entertainment?

Ana can make use of powerful tools to create and edit digital music, graphics, photos, and video for professional-grade results. Ana can enjoy music, pictures, and movies produced by others through online services that provide media to her notebook computer, television, and cell phone.

Summary

LEARNING OBJECTIVE 1

Understand the uses of digital audio and today's digital music technologies.

Digital audio includes both digital music and digital sound. A sound wave can also be represented with numbers, digitally, through a process called analog-to-digital conversion (ADC). ADC uses a technology called *sampling* to encode a sound wave as binary numbers. Digital sound is useful in professions including scientific research, law enforcement, entertainment, and communication. The professional production and editing of digital audio takes place in sound production studios. Sound production is an important part of music, movies, radio, television, video games, and the Internet. Postproduction dubbing and voice-over recording are used to apply music, sound effects, and voice to films. Audio restoration and enhancement is used to improve the quality of old or damaged recordings. Forensic audio uses digital processing to denoise, enhance, edit, and detect sounds to assist in criminal investigations.

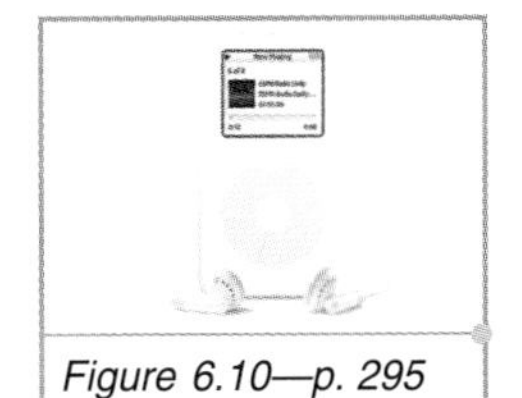

Figure 6.10—p. 295

Digital technology is applied to all aspects of the music industry: creation, production, and distribution. Digital instruments such as synthesizer keyboards, samplers, and drum machines are used in the creation of digital music. A recording studio employs many digital sound production tools, such as sequencers and outboard devices, to record and manipulate digital music. The MIDI (Musical Instrument Digital Interface) protocol provides a standard language for digital music devices to use in communicating with each other. Many people are taking advantage of digital audio production at home to create home recordings and podcasts.

A wide array of digital music software and devices are available today to provide users with more control over their music-listening experience. The production, distribution, and enjoyment of digital music and audio depends on four technology components: standardized media file formats, storage media on which to store the files, players, and software that can read and manipulate the files from the media. Online music services work with the recording industry to distribute music legally over the Internet. Satellite radio and other digital music distribution systems are providing the public with many options for enjoying digital music.

LEARNING OBJECTIVE 2

Describe the many uses of 2D and 3D digital graphics and the technologies behind them.

Digital graphics refers to computer-based media applications that support creating, editing, and viewing 2D and 3D images and animation. Graphics can be digitized by storing the color of each pixel used in the image as a combination of intensities of red, blue, and green, as is done in bit-mapped graphics, or storing a description of the shapes that compose the image, as in vector graphics. Digital graphics is used as a form of creative expression, as a means of presenting information in a visually pleasing fashion, as a means to communicate and explore ideas, as a form of entertainment, as a tool to assist in the design of real-world objects, and as a method of documenting life. Vector graphics software, sometimes called drawing software, provides tools to create, arrange, and layer graphical objects on the screen to create pictures. Vector graphics software also provides filtering and effects tools to further manipulate objects in the picture. Three-dimensional modeling software provides graphics tools that allow artists to create pictures of realistic 3D models. Ray tracing involves calculating how light interacts with surfaces in a picture to create shadows and reflections. Digital graphics animation involves displaying digital images in rapid succession to provide the illusion of motion. Three-dimensional computer animation includes all of the complexity of 3D graphic rendering, multiplied by the necessity to render 24 3D images per second to create the illusion of movement.

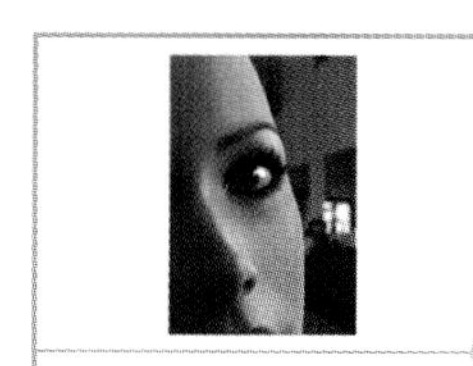

Figure 6.24—p. 310

LEARNING OBJECTIVE 3

Explain the technologies available to acquire, edit, distribute, and print digital photos, and list new advances in video technologies and distribution.

Digital photos are acquired from a digital camera or scanner. There are many factors to consider when choosing a digital camera. Photo-editing software provides editing tools for manipulating, enhancing, and repairing digital photographs. Photo editing also includes special effects, such as the ability to remove an object from one photo and insert it in another. Personal photo-editing software provides easy-to-use tools for cleaning up photos and integrating them into frames, greeting cards, calendars, and other popular printed forms. Digital photos can be stored and shared with others using free Web services. Digital photos can be printed at home on a standard printer or a specially designed photo printer. Many Web-based services and photo kiosks are available for convenient photo printing.

Figure 6.30—p. 317

Digital video is acquired using a digital camcorder. Video-editing software uses a storyline on which to build a video production. The videographer cuts scenes out of the video footage and drags the scenes to a position on the storyline. After cutting and pasting scenes onto the storyline, the videographer defines the transitions between scenes to create a professional-looking video production. There are many new services that provide access to motion pictures, video clips, and television programming on the Web that enable people to view video almost anywhere.

LEARNING OBJECTIVE 4

Discuss how interactive media is used to educate and entertain.

Interactive media refers to multimedia presentations that involve user interaction for education, training, or entertainment. Interactive media is unique because it empowers the audience to take part in the creative or educational process. Companies use interactive media to offer customers additional services and benefits.

Figure 6.42—p. 330

Computer video games make up a large portion of the interactive media market. Most gaming takes place on game consoles wired to television sets, rather than on computers. The three big gaming consoles on the market today are Nintendo GameCube, Sony Playstation 2 (PS2), and Microsoft Xbox. Today's gaming consoles are high- powered multiprocessor computers designed to support 3D interactive multimedia. Video game development requires a team effort from specialists in game design, art, sound design, programming, and testing. Interactive TV is likely to be the next big thing in interactive multimedia. Interactive TV includes digital services such as video on demand, personal video recorder (PVR), purchase over TV, Web TV, and video games over TV.

Test Yourself

LEARNING OBJECTIVE 1: Understand the uses of digital audio and today's digital music technologies.

1. Professional production and editing of digital audio takes place in ________________ .
2. The ________________ protocol was implemented in 1983 to provide a standard language for digital music devices to use in communicating with each other.
   a. Integrated Digital Studio (IDS)
   b. Digital Music Media (DMM)
   c. Musical Instrument Digital Interface (MIDI)
   d. Compact Disc Audio (CDA)
3. True or False: Sharing copies of copyright protected music is legal.
4. The ________________ music format compresses CD music files to less than 10 percent of their original size.
5. A ________________ is a device that allows musicians to create multitrack recordings with a minimal investment in equipment.
   a. drum machine
   b. synthesizer
   c. sequencer
   d. MIDI
6. A ________________ is an MP3 audio file that contains a recorded broadcast distributed over the Internet.

LEARNING OBJECTIVE 2: Describe the many uses of 2D and 3D digital graphics and the technologies behind them.

7. ________________ software, sometimes called *drawing software*, provides tools to create, arrange, and layer, graphical objects on the screen to create pictures.

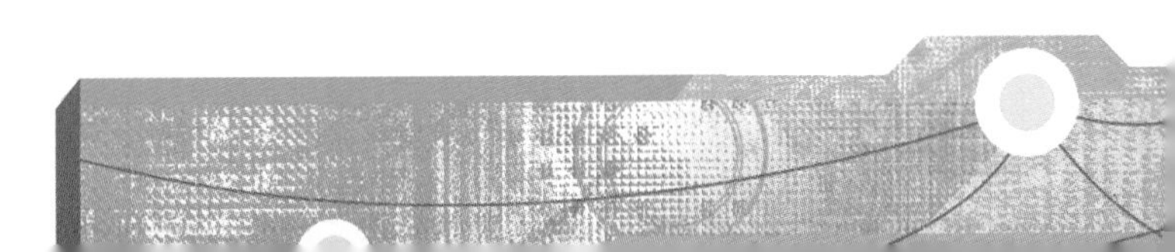

8. _______________ is the process of calculating the light interaction with the virtual 3D models in a scene and presenting the final drawing in two dimensions to be viewed on the screen or printed.
   a. rendering
   b. animating
   c. compiling
   d. light tracing
9. True or False: Even with powerful computers, high-quality rendering such as that used for major animated motion pictures takes hours per frame.
10. _______________ software is able to turn designs on the computer into blueprint specifications for manufacturing.
11. Using _______________ graphics you can layer objects in an image and arrange them like paper cutouts on a page.
   a. bit-mapped
   b. 3D
   c. animated
   d. vector

LEARNING OBJECTIVE 3: Explain the technologies available to acquire, edit, distribute, and print digital photos, and list new advances in video technologies and distribution.

12. Digital cameras are ranked by the amount of _______________ they can capture.
   a. photos
   b. light
   c. pixels
   d. color
13. True or False: Most online photo-sharing services charge a modest fee per month for storing digital photos.
14. _______________ includes special tools and effects that are designed for improving or manipulating bit-mapped photograph images.
15. True or False: People can download television shows to their PCs.

LEARNING OBJECTIVE 4: Discuss how interactive media is used to educate and entertain.

16. Microsoft Xbox is a _______________ .
   a. video game console
   b. home entertainment system
   c. digital cell phone
   d. handheld computer
17. True or False: A media center PC acts as a centralized entertainment control system for the home.

Test Yourself Solutions: 1. sound production studio, 2. c. Musical Instrument Digital Interface (MIDI), 3. False, 4. MP3, 5. c. sequencer, 6. podcast, 7. Vector graphics, 8. a. rendering, 9. True, 10. CAD, 11. d. vector, 12. c. pixels, 13. False, 14. Photo-editing software, 15. True, 16. a. video game console, 17. True.

Key Terms

3D modeling software, p. 310
bit-mapped graphics, p. 301
computer-assisted design (CAD) software, p. 307
digital audio, p. 285
digital graphics, p. 301
digital media, p. 284
digital music, p. 285
interactive media, p. 326
interactive TV, p. 331
media player software, p. 296
MP3, p. 294
musical instrument digital interface (MIDI) , p. 290
photo-editing software, p. 318
pixels, p. 301
podcast, p. 292
sampler, p. 289
satellite radio, p. 300
scientific visualization, p. 306
sequencer, p. 289
synthesizer, p. 289
vector graphics, p. 301
video game consoles, p. 328
video-editing software, p. 322

Questions

Review Questions

1. Name and define the two subcategories of digital audio.
2. Define digital media.
3. What services are provided by a sound production studio?
4. What audio services cater to the motion picture industry?
5. What digital tools and instruments do musicians use to create music?
6. Who produces podcasts?
7. Name three methods of transferring music to your computer.
8. How does digital graphics differ from digital imaging?
9. What types of professionals make use of CAD software?
10. How do vector graphics differ from bit-mapped graphics?
11. What does it mean to render 3D models?
12. What are animated GIFs, and where do you usually find them?
13. What benefits does Macromedia Flash offer for Web content?
14. What are avars and how do they give life to 3D animations?
15. Name four forms of interactive media.
16. What are some commercial uses of interactive media?
17. What are the three most popular models and manufacturers of video game consoles?
18. Provide three examples of digital convergence in digital media.
19. What types of specialists are typically involved in video game production?
20. Name five services associated with interactive TV.
21. What is video on demand?

Discussion Questions

22. How is forensic audio used to catch criminals?
23. What is the purpose of MIDI?
24. How has digital music empowered musicians?
25. How has digital music empowered music fans?
26. What is MP3, and why is it so popular?
27. How has digital music affected music distribution?
28. What is scientific visualization, and how is it used?
29. What is digital cinema, and how will it affect the motion picture industry?
30. Why are vector graphics considered to be object oriented?
31. Why is the process of 3D modeling sometimes called ray tracing?
32. What benefits does photo-editing software provide photographers?
33. How is video editing used to create professional-quality video productions?
34. What benefits does Windows XP Media Center offer?

Exercises

Try It Yourself

1. Visit *http://www.yamaha.com/yamahavgn/CDA/Home/YamahaHome/*, click Keyboards and Digital Instruments. Explore the products, then view MOTIF. On the MOTIF page click MOTIF Synthesizers, then click the link for Media Clips. Listen to the sampled instrument recordings presented there. List your five favorite sounds. Listen to the Acoustic instrument sounds. Do they sound real? Write your impressions in a few paragraphs and submit them to your teacher.

2. Visit *www.irtc.org* and *www.hotwired.com/rgb/gallery* and view the digital artwork that you find there. Write a few paragraphs on which piece most affects you or interests you from each site and why. If none interest you, explain why not.
3. Visit *www.mindavenue.com*, download and install the Axel player, then go to the Gallery. Try out some of the interactive productions and games. List the five that you feel are the most interesting. Write a paragraph summary of your impression of the technology and its value on the Web.
4. Visit *www.pixar.com* and view the link that explains Pixar's animation and movie making process. Summarize the steps of the animation process in a word-processing document, and submit it to your instructor.

Virtual Classroom Activities

For the following exercises, do not use face-to-face or telephone communications with your group members. Use only Internet communications.

5. Have the group members answer the questions following this exercise. Compile the data to derive statistical information about your group. Add group member song amounts to get a total amount of songs downloaded for your group. Divide that number by 10 (the average number of songs on a CD) to determine how many CDs worth of music has been downloaded by your group. Then, multiply that number by $15 to determine how much money your group has saved and the record industry has lost. Report your findings, without attaching individual names to specific responses, to your instructor. Your instructor can compile the statistics for the entire class.

   Questions:
   a. Do you use file-sharing networks to download music? If yes, approximately how many music files are currently in your collection?
   b. Do you use file-sharing networks to supply music to others?
   c. Do you believe that sharing music in this way should be legal?
6. Using a virtual classroom application or other chat utility, as a group, visit *www.worth1000.com*—the Web site of faked and creative photos. Explore the images you find there, and decide which image everyone in the group enjoys most. Share it with your instructor.

Teamwork

7. Try or view a demonstration of the video game Grand Theft Auto by visiting a video game retail store or accessing it in some other way. Learn what the goals are in the game and the details of the challenges. Have a group debate to decide whether or not the game desensitizes users to violence. After everyone has a chance to express their opinions, take a vote and submit the results to the instructor.
8. Use a group member's digital camera (or borrow one from a friend) to take a group photo. Distribute the photo electronically to all group members, and using photo-editing software, have a contest to see who can edit the photo in the most interesting way. Submit all entries to the instructor.
9. Assign various legal music download services to group members: Napster, Rhapsody, iTunes, MSN Music. Have group members write critiques of their assigned service and share your findings and summarize the group results in charts and slides.
10. Assign various online photo-sharing services (those mentioned in the chapter) to group members. Have group members write critiques of their assigned service and share your findings.

TechTV

Army's Virtual World

Go to www.course.com/swt2/ch06 and click TechTV. Click the "Army's Virtual World" link to view the TechTV video, and then answer the following questions.

1. The "war on terror" has ushered the United States into a new form of battle where individual soldiers, rather than generals and platoon leaders, are often responsible for making important decisions in the field. Describe the new multimedia system presented in this video and explain how the U.S. military is using it to train soldiers to make decisions quickly.
2. How might this method of using virtual simulations to train soldiers be applied to other training scenarios?

Endnotes

[1] Lewin, J. "Broadband users choosing Internet over TV and radio,", *ITworld.com*, March 29 2005, www.itworld.com.

[2] Whale Dreams Web Site, http://www.abc.net.au/oceans/whale/song.htm. accessed July 3, 2005.

[3] "We're All Ears: Listening For ET," *Science Daily* Web site, January 1 2005, http://www.sciencedaily.com/releases/2005/01/050104104855.htm.

[4] Mandel, H. "U.S. Soldier Composes Aural Landscapes of Iraq," National Public Radio, September 20, 2005, http://www.npr.org/templates/story/story.php?storyId=4856428.

[5] "Download Pirates on Retreat," *Haymarket Publishing Services Ltd*, July 6, 2005, www.lexis-nexis.com.

[6] Smith, T. "Apple music store downloads top 300m," *The Register*, March 2 2005, www.theregister.co.uk.

[7] "Study: iTunes more popular than many P2P sites," *CNET News*, June 7 2005, http://news.news.com.

[8] "XM Sees Satellite Radio Built Into Myriad Gadgets," *Reuters*, March 30 2005, www.retures.com.

[9] "iPod could get satellite radio," *Macworld Daily News*, May 26, 2005, http://www.macworld.co.uk/.

[10] Groom, N. "Starbucks Expanding Custom CD Service to 45 Stores," *Reuters*, October 14, 2004, www.reuters.com.

[11] DiCarlo, L. "Dell, Napster Target College Downloads," *Forbes*, July 6, 2005, www.forbes.com.

[12] Graham, J. "A disposable digital camera enters the market at $19.99," *USA Today*, August 19, 2005, www.usatoday.com/tech/news/2004-08-18-puredigital_x.htm.

[13] "Investigating digital images," *Dartmouth News*, July 1, 2004, http://www.dartmouth.edu/~news/releases/2004/07/01.html.

[14] "Target, Yahoo in Online Picture Developing Pact," Reuters, April 21, 2005, www.reuters.com.

[15] Selingo, J. "Photo Prints? Everyone Wants Your Business," *New York Times*, June 8 2005, www.nytimes.com

[16] See note 15 above.

[17] Marshal, M. "Pure Digital delivers disposable camcorder," *SiliconBeat*, June 8, 2005, www.siliconbeat.com.

[18] Hansell, S. "Forget the Bootleg, Just Download the Movie Legally," *New York Times*, July 4 2005, www.nytimes.com.

[19] Pruitt, S."Microsoft Launches Mobile Video Downloads," *PCWorld*, March 30, 2005, www.pcworld.com.

[20] "TiVo Goes Windows Mobile," *Pocket PC City*, June 13, 2005, www.pocketpccity.com.

[21] Markoff, J, and Holson, L, "With New IPod, Apple Aims to Be a Video Star", *New York Times*, October 13, 2005, www.nytimes.com

[22] Haley, C. C. "Verizon Wireless Flicks Content Switch," *Wi-Fi Planet*, January 7, 2005, www.wi-fiplanet.com.

[23] Lawson, S. "CES: Satellite video, Internet coming to vehicles," *ITWorld*, January 6 2005, www.itworld.com.

[24] Williams, M. "Boeing Eyes In-Flight Live TV," *PCWorld*, December 17, 2005, www.pcworld.com.

[25] Olson, S. "Coming soon: Google TV?", *ZDNet*, November 30, 2004, www.zdnet.com.au.

[26] Larson, C. "To Study History, Pupils Can Rewrite It," *New York Times*, May 27, 2004, www.nytimes.com.

[27] Berkowitz, B. "Doctors Use Video Games to Hone Skills," *Reuters*, December 24, 2004, www.reuters.com.

[28] Andy Sullivan, "Video Games Teach More Than Hand-Eye Coordination", *Reuters*, November 28 2004, www.reuters.com

[29] Xbox 360 Web site, accessed July 15, 2005, at http://www.xbox.com/en-US/xbox360/factsheet.htm

[30] "Disney to Release Films for Sony PSP Game Device", *Reuters*, March 16 2005, www.reuters.com

[31] "Nintendo to create WiFi hotspots for DS handheld", Reuters, June 7 2005, www.reuters.com

[32] "Nokia branches out on gaming platform", Reuters, May 18, 2005, www.reuters.com

DATABASE SYSTEMS

When Mary Bolger found her grandfather's recipes in a small box in the attic, she never thought those few papers would turn into money. She made up a batch of his special butter-cheese spread, her friends loved it, and before she knew it she was selling it to local grocery stores in Waukesha, Wisconsin. After college, Mary got a job working for an upscale retail store, but she still sold her Wisconsin's-Best brand of cheese. Now she is thinking of quitting her job to go into her cheese business full time. But keeping track of everything is a problem. Mary currently has a personal computer with word processing, spreadsheet, and database programs.

As you read through the chapter, consider the following questions:

1. How can Mary keep track of the different cheeses, butters, spices, crocks, and other supplies she needs?
2. If she starts a small business, how will she be able to pay her employees, pay her bills, and keep track of other business transactions?

Check out Mary's *Action Plan* at the conclusion of this chapter.

LEARNING OBJECTIVES

1. Understand basic data management concepts.
2. Describe database models and characteristics.
3. Discuss the different types of database management systems and their design and use by individuals and organizations.
4. Describe how organizations use database systems to perform routine processing, provide information and decision support, and how they use data warehouses, marts, and mining.
5. Discuss additional database systems, including distributed systems and Web-based systems.
6. Describe the role of the database administrator (DBA) and database policies and security practices.

CHAPTER CONTENT

Basic Data Management Concepts
Organizing Data in a Database
Database Management Systems
Using Database Systems in Organizations
Database Trends
Managing Databases

Introduction

People and organizations need a way to store data and to convert that data into important information. Databases serve this function. Without databases, today's businesses could not survive. Databases are also used in medicine, science, engineering, the military, and most other fields. Databases are also useful to individuals for keeping track of items in an apartment, music in a CD collection, and expenses used to prepare a budget and complete a tax return. To use a database, you must first understand basic data management concepts.

BASIC DATA MANAGEMENT CONCEPTS

Recall that data consists of raw facts, like sales or weather statistics. For data to be transformed into useful information, such as quarterly profits or hurricane predictions, it must first be organized in some meaningful way. As you learned in Chapter 1, a *database* is a collection of data organized to meet users' needs. Throughout your career, you will directly or indirectly access a variety of databases ranging from a simple list of music in your collection to a fully integrated database at work.

A database can help individuals and organizations maximize data as a valuable resource. Databases also help individuals and organizations achieve their goals. The Panzano restaurant in downtown Denver uses a database to store customer information, such as birthdays, anniversaries, and preferences for certain types of food and beverage.[1] When a customer arrives for a meal, Panzano's can suggest meals that the customer may want, or offer a special congratulatory tidbit for a birthday or anniversary. UPS, a large shipping company, uses a database to process almost 60 million transactions every day (see Figure 7.1).[2] When you shop online, you access the merchants' inventory databases that store thousands of products. Database software has also been used to help save endangered species and track terrorists.

Partially as a response to the threat of terrorism, some governments are increasingly using databases to track and prevent unwanted people from entering their countries. The U.S. government, for example, uses a database containing visual images of fingerprints to track tens of thousands of suspected terrorists or visitors of "national security concern."[3] The National Security Entry-Exit Registration System attempts to close the borders to suspected terrorists by comparing the fingerprints of entering visitors against a comprehensive database. Sharing attributes and data items can also be a critical factor in coordinating responses across diverse functional areas of an organization. Problems, however, can arise when data privacy concerns and laws are different in different countries. Countries with more strict privacy laws may not want to share data with countries where the data may be available to the public.

You usually access databases using software called a **database management system (DBMS)**. A DBMS consists of a group of programs that manipulate the database and provide an interface between the database and the user or the database and application programs. DBMSs are available for many sizes and types of computers. They are covered in more detail later in the chapter.

A database, a DBMS, and the application programs that utilize the data in the database make up a *database system* or database environment. Some of the functions of a DBMS include:

- Storing important data for individuals, groups, and organizations, including numbers, text, visual images, audio signals, and so on
- Performing routine tasks, such as helping an individual prepare income tax forms or helping an organization produce paychecks for its employees
- Providing information to help people, groups, and organizations make better decisions by asking questions of the database and creating reports
- Ensuring that the data is protected and safe from attacks and unauthorized access

FIGURE 7.1 • Door to door

UPS uses state-of-the-art databases to help keep track of millions of packages a day. Customers can use the Web and software that links to these databases to see exactly where their packages are.

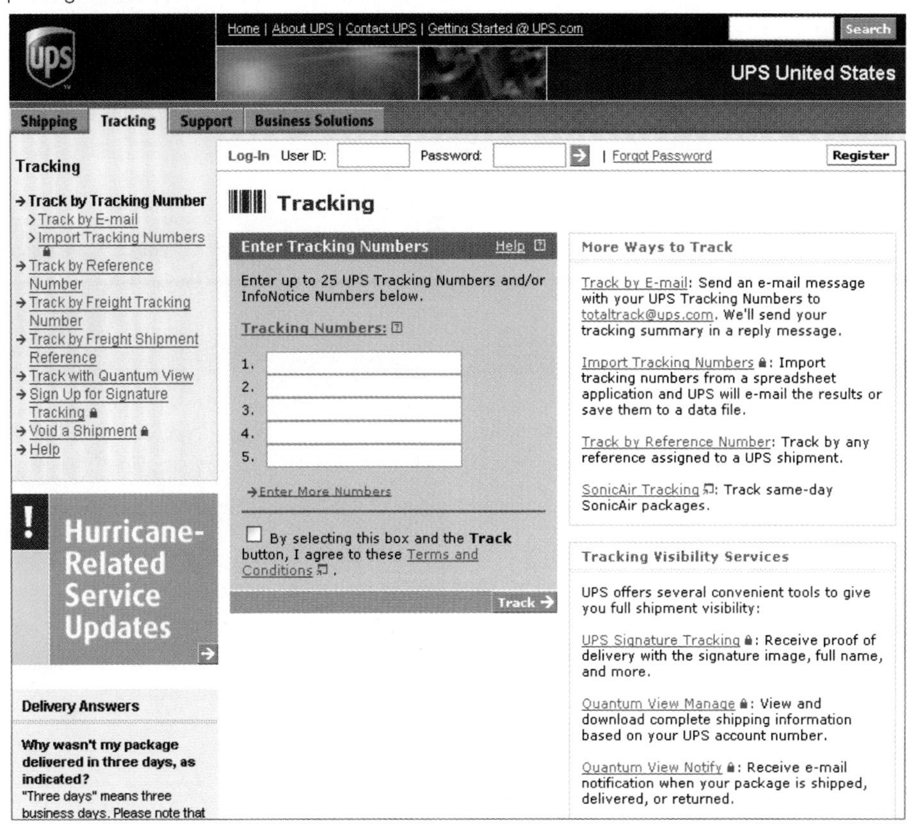

Data Management for Individuals and Organizations

Individuals use databases to develop monthly budgets, store phone numbers and addresses, keep track of important dates, keep track of valuables for possible insurance claims, and get information about organizations such as hospitals. Medicare, for example, is developing a database that can be accessed through the Internet to provide information about the quality of hospital care.[4] The database site will give statistics about 17 accepted hospital quality measures on treating heart attacks, pneumonia, and other diseases. According to Mark McClellan, head of the Centers for Medicare and Medicaid, "This is another big step toward supporting and rewarding better quality, rather than just paying more and supporting more services." Others have questioned whether the database will improve health care quality for all patients.

Without data and the ability to process it, an organization would not be able to successfully complete most activities. It would not be able to generate reports to support decision makers to help achieve organizational goals. Businesses would find it difficult to pay employees, send out

TechEdge

HOW NOW MAD COW?

The cattle industry has developed a database of over a million livestock producers and "tens of millions of animals," including hogs and poultry, to help identify cattle suspected of carrying mad cow disease. The trace-back system (as opposed to the proposed federal animal identification system) tracks down an infected animal's herdmates within 48 hours of an outbreak of the brain-wasting disease. It is anticipated it will cost about $100 million a year to maintain.

Source:
Riding Herd with a Database
Wired News
http://www.wirednews.com/news/technology/0,1282,68140,00.html?tw=wn_tophead_10
July 08, 2005

bills, order new inventory, or produce information to assist managers in the decision-making process. Databases have made it possible to map the structure of DNA and for scientists to share their research (see Figure 7.2); without databases, the human genome project would not be possible.

FIGURE 7.2 • Databases and research

Through the use of extensive research databases, scientists all over the world have access to the most up-to-date data available relating to the human genome and mouse genome projects.

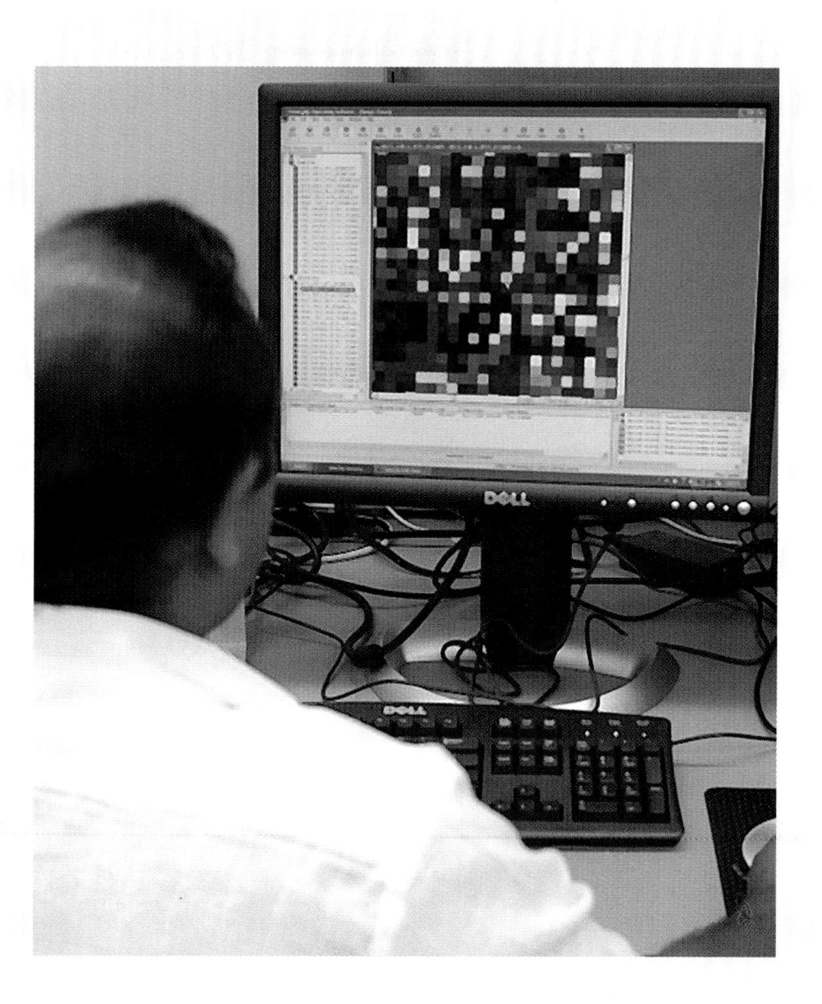

The Hierarchy of Data

Recall that in a computer, a byte is used to represent a character, which is the basic building block of information. Characters can be letters, uppercase or lowercase (A, a, B, b, C, c,...Z), numeric digits (0, 1, 2,...9), or special symbols (.![+][-]/€¥...).

In a database, characters are put together to form a field, the smallest practical unit in most databases. A **field** is typically a name, number, or combination of characters that in some way describes an aspect of an object (an individual, a music CD, employee, a shelf, a truck) or activity (a sale). Every field has a *field name* and can have either a fixed or variable length. For example, EmployeeNumber is a field name for an employee number that is a fixed eight characters long. PartDescription is the field name for a part description, where the length of the description can vary, depending on the part.

A collection of related fields that describe some object or activity is a **record**. You can create a more complete description of an object or activity by combining fields that represent various characteristics of objects or activities into records. For instance, an employee record is a collection of fields about one employee. One field would be the employee's name, another her address, and still others her phone number, pay rate, earnings to date, and so forth.

A collection of related records is a **file**, also called a *table* in some databases. For example, an employee file is a collection of all of an organization's employee records. Likewise, an inventory file is a collection of all inventory records for a particular organization. Employee and inventory files are relatively permanent files. These are examples of *master files*, permanent files that are updated over time. Organizations also have temporary files that hold data that needs to be processed, such as the transactions of paying employees or taking sales orders. These are called *transaction files* and are temporary files that contain data representing transactions or actions that must be taken. A file containing the number of hours employees worked last week or the sales orders from yesterday are examples of transaction files. They must be processed to pay employees or fill orders. Transaction files often cause changes to master files. An inventory master file containing the current amount of an inventory item has to be adjusted (reduced in this case) to reflect new sales of the inventory item contained in the order transaction file. If a customer orders five cookbooks from an online bookstore, the online bookstore must subtract five cookbooks from its inventory master file to keep the master file current and accurate.

At the highest level of this hierarchy is a *database*, a collection of integrated and related files or tables. Together, characters, fields, records, files, and databases form the *hierarchy of data* (see Figure 7.3). Characters are combined to make a field, fields are combined to make a record, records are combined to make a file, and files are combined to make a database. It is important to remember that a database houses not only all of these levels of data but the relationships among them.

FIGURE 7.3 • The hierarchy of data

The hierarchy of data represents the idea that characters are combined to make a field, fields are combined to make a record, records are combined to make a file, and files are combined to make a database.

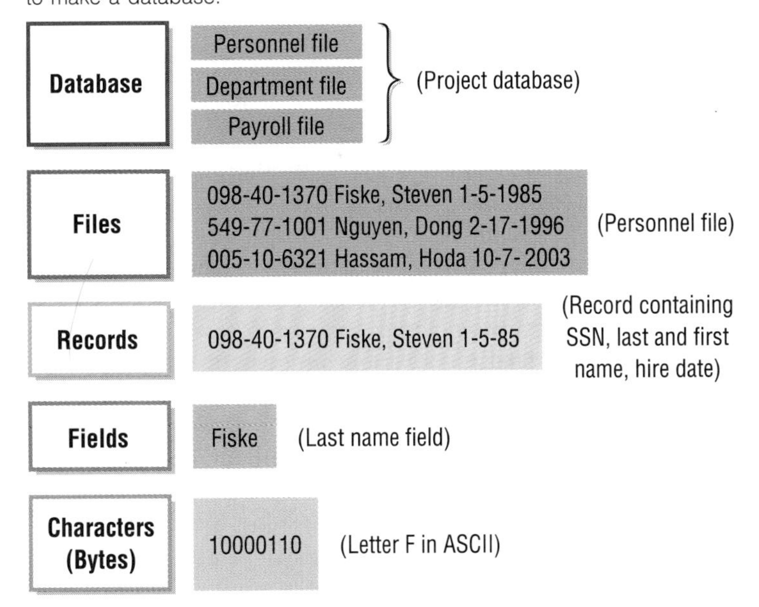

Data Entities, Attributes, and Keys

Databases use entities, attributes, and keys to store data and information. An *entity* is a generalized class of people, places, or things (objects) for which data is collected, stored, and maintained. Examples of entities include music CDs, the contents of an apartment, employees, inventory, and customers. Most organizations store data about entities.

An *attribute* is a characteristic of an entity. For example, employee number, last name, first name, hire date, and department number are attributes for an employee (see Figure 7.4). Inventory number, description, number of units on hand, and location of the inventory item in the warehouse are examples of attributes for items in inventory; customer number, name, address, phone number, credit rating, and contact person are examples of attributes for customers. Attributes are usually selected to capture the relevant characteristics of entities such as employees or customers. The specific value of an attribute, called a *data item*, can be found in the fields of the record describing an entity. Federal databases, for example, often include the results of DNA tests as an attribute in databases of convicted criminals. The 3rd U.S. Circuit Court of Appeals has ruled that DNA samples taken from criminals on release are legal because of the government's desire to develop a national DNA database.[5]

FIGURE 7.4 • Attributes and keys

The attributes in this entity include employee number, last name, first name, hire date, and department number. Employee number is the primary key in this entity because it uniquely identifies each employee.

Employee number	Last name	First name	Hire date	Dept. number
005-10-6321	Hassam	Hoda	10-7-2003	257
549-77-1001	Nguyen	Dong	2-17-1996	650
098-40-1370	Fiske	Steven	1-5-1985	598

Entities (records)

Primary key field

Attributes (fields)

As discussed, a collection of fields about a specific object is a record. A *key* is a field in a record that is used to identify the record. A **primary key** uniquely identifies the record. No other record can have the same primary key. The primary key is the main or principal key used to distinguish records so that they can be accessed, organized, and manipulated. For example, a credit card company would use the customer's credit card number as the primary key in its database because it uniquely identifies each customer. There may be many Smiths, but only one 1331 2322 2344 2313. For privacy reasons, most schools have switched from using student Social Security numbers as primary keys to a unique assigned student ID number. In an employee record such as the ones shown in Figure 7.4, the employee number is an example of a primary key.

Once you have identified a unique primary key for each record, it can be used to relate data in a database. For example, you could have an employee file containing a record for each employee that contains the employee number, pay rate, deductions to gross pay, years used by the organization, and similar permanent employment data. You could also have an hours-worked file that contains the

Hyundai Shoots for #1 with Executive DBMS

Hyundai Motor Company has set a lofty goal of becoming the top auto producer for the 21st century. It is counting on its data-driven computer systems to take it there. The 30-year-old company has rapidly become one of the world's largest auto producers. Its business strategy revolves around using advanced technology to produce top quality, reliable vehicles that satisfy customers. With over 47,000 employees and capital exceeding $350 million, Hyundai Motor Company is the largest independent manufacturing plant in Korea.

Hyundai employs an enterprise-wide DBMS, called an enterprise information and management system (EIMS), that supplies valuable information to its manufacturing plants in Korea, Tokyo, Peking, Detroit, and Frankfurt, and subsidiaries such as Hyundai Motor America, Hyundai Auto Technical, Hyundai Motor Finance Company, and Hyundai Motor India. Hyundai counts on the decision support information stored in its data warehouse to direct decisions made at all levels of management.

The system employs several different reporting tools to keep top managers informed of the state of the company. They receive periodic reports on the company's progress toward long-term goals. Exception reports provide a warning system, allowing executives to view all levels of sales and production volumes and to locate performance problems by comparing figures with an established warning point. Many other reports support decision making and provide timely access to management-level information.

To assist decision makers, the EIMS data warehouse stores information from all organizational divisions, including:

- Organizational charts, personnel records, and staff counts from the Human Resource Department
- Daily and monthly sales, market share, and market analysis from domestic sales
- Foreign exports, daily exports, local sales, inventory, and competitive analysis from foreign sales
- Production per factory and per model, target achievement, and factory operation from production

Hyundai has found that the best way to protect capital investments is to invest in systems to manage valuable data and information. Rather than reacting to market movements, Hyundai can be proactive and innovative, in other words, to lead rather than follow.

Questions

1. How can Hyundai's EIMS assist managers in setting car sticker prices?
2. In a far-flung organization such as Hyundai, how can database reports be distributed uniformly and consistently across the organization?

Sources

1. *Mohamed, A. "Hyundai Boosts Network Access and Control,"* Computer Weekly, *February 22, 2005, p. 44.*
2. *"Technology Drives Decisions at Hyundai," SAS Institute Web site, http://www.sas.com/success/hyundai.html, accessed on March 6, 2004.*
3. *Hyundai Web site, www.hyundai.com, accessed March 6, 2004; "Hyundai Helps Chinese Auto Industry Mature,"* The Korea Herald, *October 28, 2003, www.lexis-nexis.com.*

employee number and hours worked for employees who worked this week. Using the employee number as the primary key, both files can be related and used to compute and print the pay checks for each employee working this week. Multiplying pay rate in the employee file times the hours worked in the hours-worked file and subtracting any deductions gives the employee's pay for this week.

In some cases, multiple fields may be combined as a primary key. Consider a personal CD inventory database. Fields may include artist, CD title, music style, year released, and rating. Which one of these fields would you select for a primary key? The artist field may tempt you, but what happens if you have several

CDs by the same artist? The field would no longer be unique for each record. The CD title is another option. But, consider the number of CDs that have been released with the title *Greatest Hits*. The solution is to use the combination of artist and title to create the primary key. This way the Red Hot Chili Peppers *Greatest Hits* can be uniquely identified from Miles Davis *Greatest Hits* and Miles Davis *Greatest Hits* from Miles Davis *Kind of Blue*.

Simple Approaches to Data Management

There are a number of simple ways to manage data. For example, simple data management software packages are easy to use and update. You can use personal information managers (PIMs) to keep track of phone numbers, addresses, Web sites, and e-mail addresses (see Figure 7.5). PIMS are often stored on personal computers and small handheld devices such as personal digital assistants (PDAs). A PIM also contains a calendar, in which you can enter appointments. A home budget software package is another example of a simple data management system. You can enter expense items, and the software computes totals and compares your actual expenses to your budget. Although these data management systems are simple to use, their structure can be very difficult or impossible to change. For example, the home budget program might limit you to a certain number of income and expense categories.

FIGURE 7.5 • A relatively simple data manager

A personal information manager (PIM) helps people keep track of phone numbers, addresses, Web sites, and e-mail addresses. It also has a calendar application.

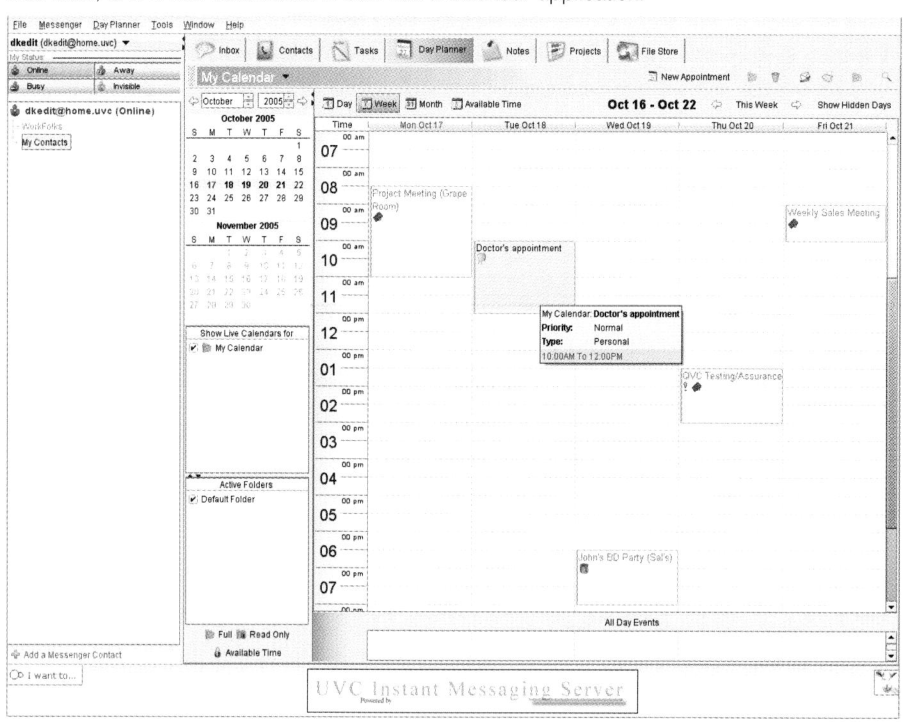

The Database Approach to Data Management

In a *database approach* to data management, multiple application programs share a pool of related data. Rather than each application having its own separate data files, each shares a collection of data files that are maintained in a database stored in a central location or on a data network. A grocery store chain, for example, can use the database approach to allow multiple applications to run from a common database system. The approach allows the grocer to obtain and analyze customer receipts from hundreds of stores in a digital format. Storing data in one centralized database is more efficient and less prone to errors or mistakes.

Using the database approach to data management requires a database management system (DBMS). Recall that a DBMS consists of a group of programs that can be used as an interface between a database and the user or the database and application programs. Typically, this software acts as a buffer between the application programs and the database itself. Figure 7.6 illustrates the database approach. With the database approach, a centralized database system contains all the data for the organization or individual. The DBMS interacts directly with the database and passes data to various applications. A DBMS also reduces *data redundancy*. Data redundancy occurs when data is copied, stored and used from different locations. Redundant data is difficult to manage because any required change must be applied to all copies of the data, and often times, a copy is overlooked and becomes inaccurate. Inaccurate data can be more dangerous to an organization or individual than no data at all. Because a centralized database managed by a DBMS stores data only once, the problem of data redundancy is solved. Reducing copies of data also means that less storage is required.

Many modern databases are organization-wide, encompassing much of the data of the entire organization. Table 7.1 lists some of the primary advantages of the database approach. The primary disadvantage of implementing a centralized database system is the initial cost of installation and ongoing maintenance. Most companies and organizations find that the advantages more than justify the initial and ongoing costs.

FIGURE 7.6 • The database approach to data management

In the database approach, a centralized database system contains all the data for the organization or individual. The DBMS interacts directly with the database and passes data to various applications.

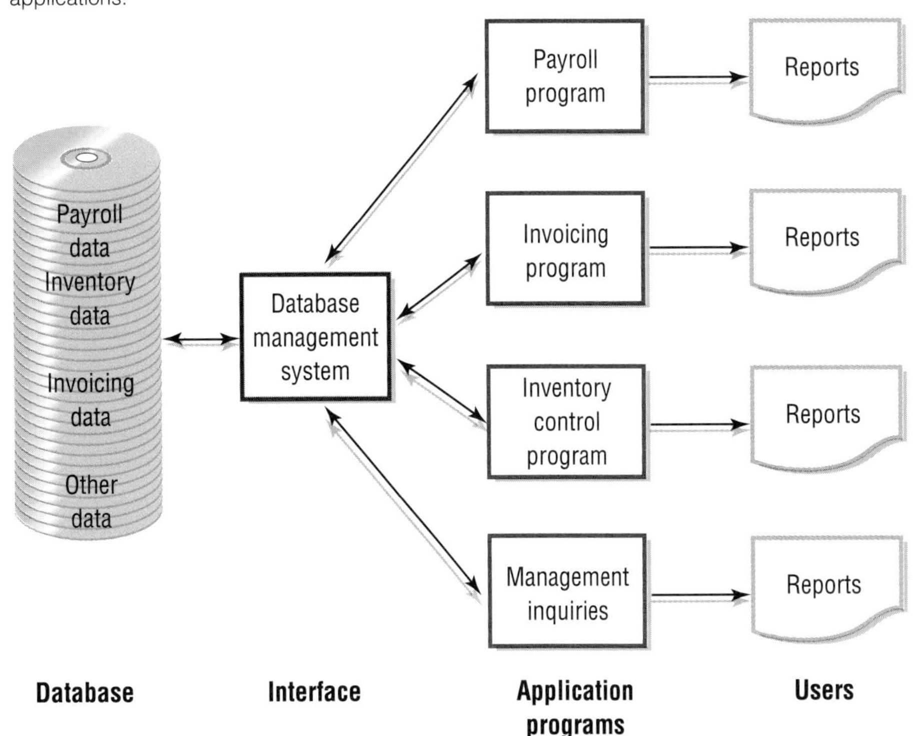

TABLE 7.1 • Advantages of the database approach

Advantages	Explanation
Reduced data redundancy	The database approach can reduce or eliminate data redundancy. Data is organized by the DBMS and stored in only one location. This results in more efficient utilization of system storage space.
Improved data integrity	Before DBMSs, some changes to data were not reflected in all copies of the data kept in separate files. This is prevented with the database approach because there are not separate files that contain copies of the same piece of data.
Easier modification and updating	With the database approach, the DBMS coordinates updates and data modifications. Programmers and users do not have to know where the data is physically stored. Data is stored and modified once. Modification and updating is also easier because the data is stored at only one location in most cases.
Data and program independence	The DBMS organizes the data independently of the application program. With the database approach, the application program is not affected by the location or type of data. Introduction of new types of data not relevant to a particular application does not require the rewriting of that application to maintain compatibility with the data file.
Better access to data and information	Most DBMSs have software that makes it easy to store and retrieve data from a database. In most cases, simple information commands can be given to get important information. Relationships between records can be more easily investigated and exploited, and applications can be more easily combined.
Standardization of data access	A primary feature of the database approach is a standardized, uniform approach to database access. This means that all application programs use the same overall procedures to retrieve data and information.
Better overall protection of the data	The use of and access to centrally located data is easier to monitor and control. Security codes and passwords can ensure privacy by allowing only authorized people to have access to particular data and information in the database.
Shared data and information resources	The cost of hardware, software, and personnel can be spread over a large number of applications and users.

ORGANIZING DATA IN A DATABASE

Because many of today's organizations are large and complex, it is critical to keep data organized so that it can be effectively utilized. A museum, for example, needs to organize its database to make sure it can catalogue and retrieve all of its items or pieces. Without a good organization of the database, it would be difficult or impossible for the museum to know exactly what items were part of its collection and where each item was located. This type of information could be critical if the museum has a break-in and an insurance company needs to determine what may have been stolen. It is also critical during audits of the museum's assets.

A database should be designed to store all data relevant to the organization and provide quick access and easy modification. When building a database, careful consideration must be given to these questions:

- Content: What data is to be collected and at what cost?
- Access: What data is to be provided to which users when appropriate?
- Logical structure: How is the data to be arranged so that it makes sense to users?
- Physical organization: Where is the data to be physically located?
- Management and coordination: Who is responsible for maintaining an accurate database system, including the development of data modeling?

EXPAND YOUR KNOWLEDGE

To learn how to create a database and use it to answer queries, go to www.course.com/swt2/ch07. Click the link "Expand Your Knowledge" and then complete the lab entitled "Databases."

The Relational Database Model

The structure of the relationships in most databases follows a logical database model. Of several models developed over the years for designing database structures, relational models are the most popular.

The overall purpose of the relational model is to describe data using a standard tabular format. In a database structured according to the **relational model**, all data elements are placed in two-dimensional tables called *relations* that are the logical equivalent of files. The tables in relational databases organize data in rows and columns, simplifying data access and manipulation (see Figure 7.7). Notice in the figure that the tables are "related" by a common element, Dept. number. Having tables related by common elements allows them to be linked to produce useful information.

FIGURE 7.7 • The relational database model

In the relational model, all data elements are placed in two-dimensional tables. As long as they share at least one common element, these tables can be linked to produce useful information.

Data Table 1: Project Table

Project number	Description	Dept. number
155	Payroll	257
498	Widgets	632
226	Sales Manual	598

Data Table 2: Department Table

Dept. number	Dept. name	Manager SSN
257	Accounting	421-55-9993
632	Manufacturing	765-00-3192
598	Marketing	098-40-1370

Data Table 3: Manager Table

SSN	Last name	First name	Hire date	Dept. number
005-10-6321	Hassam	Hoda	10-7-2003	257
549-77-1001	Nguyen	Dong	2-17-1996	650
098-40-1370	Fiske	Steven	1-5-1985	598

Once data has been placed into a relational database, data inquiries and manipulations can be made. Two common data manipulations are selecting data and joining tables. *Selecting* involves choosing data based on certain criteria. Suppose a Project table contains the project number, description, and department number for all projects being performed by an organization as seen in Figure 7.7. A president of the company might want to find the department number for Project 226, which is a sales manual project. Using selection, the president can see only the data for number 226 and determine that the department number for the department completing the sales manual project is 598.

Joining involves combining two or more tables. For example, you could combine the Project table and the Department table to get a new table with the project number, project description, department number, department name, and the Social Security number for the manager in charge of the project.

Being able to relate tables to each other through common data elements is one of the keys to the flexibility and power of relational databases. Suppose that the president of a company wants to find out the name of the manager of the sales manual project and how long the manager has been with the company. (See Figure 7.8.) The president would make the inquiry to the database, perhaps via a desktop personal computer. The DBMS starts with the project description and searches the Project table to find out the project's department number. It then uses the department number to search the Department table for the department manager's Social Security number. The department number is also in the Department table and is the common element that allows the Project table and the Department table to be related. The DBMS then uses the manager's Social Security number to search the Manager table for the manager's hire date. The final result: the manager's name and hire date are presented to the president as a response to the inquiry. Creating relationships between tables is especially useful when information is needed from multiple tables, as in this example. The manager's Social Security number, for example, is only maintained in the Manager table. If the Social Security number is needed, it can be obtained by following the relationships to the Manager table.

FIGURE 7.8 • Relating data tables to answer an inquiry

In finding the name and hiring date of the manager working on the sales manual project, the president needs three tables: Project, Department, and Manager. The project description (Sales Manual) leads to the department number (598) in the Project table, which leads to the manager's SSN (098-40-1370) in the Department table, which leads to the manager's name (Fiske) and hiring date (1-5-1985) in the Manager table.

Project Table

Project number	Description	Dept. number
155	Payroll	257
498	Widgets	632
226	Sales Manual	598

Department Table

Dept. number	Dept. name	Manager SSN
257	Accounting	421-55-9993
632	Manufacturing	765-00-3192
598	Marketing	098-40-1370

Manager Table

SSN	Last name	First name	Hire date	Dept. number
005-10-6321	Hassam	Hoda	10-7-2003	257
549-77-1001	Nguyen	Dong	2-17-1996	650
098-40-1370	Fiske	Steven	1-5-1985	598

Data Analysis. Having the ability to create relationships between tables becomes very important when trying to ensure that the content of a database is "good." Good data should be nonredundant, flexible, simple, and adaptable to a number of different applications. The purpose of data analysis is to develop data with these characteristics. **Data analysis** is a process that involves evaluating data to identify problems with the content of a database. Consider a database for a fitness center that contains the customer's name, phone number, gender, dues paid, and date paid (see Figure 7.9). As the records in Figure 7.9 show, Brown and Thomas paid their dues in September. Thomas has paid his dues in two installments. Note that no primary key uniquely identifies each record. As you will see next, this is a problem that must be corrected

FIGURE 7.9 • Fitness Center dues

These database entries for a fitness center contain the name, phone number, gender, dues paid, and date paid.

Name	Phone	Gender	Dues Paid	Date Paid
Brown, A.	468-3342	Female	$50	September 15th
Thomas, S.	468-8788	Male	$25	September 15th
Thomas, S.	468-5238	Male	$25	September 15th

This database was designed to keep track of the dues that fitness center members paid in September. Because Thomas has paid dues twice, the data in the database is now redundant. The name, phone number, and gender for Thomas are repeated in two records. Notice that the data in the database is also inconsistent: Thomas has changed his phone number, but only one of the records reflects this change. Further reducing this database's reliability is the fact that no primary key exists to uniquely identify Thomas' record. The first Thomas could be Sam Thomas, but the second might be Steve Thomas. These problems and irregularities in data are called *anomalies*. Data anomalies often result in incorrect information, causing database users to be misinformed about actual conditions. These anomalies and others must be corrected.

To solve these problems, you can add a primary key, called *member number*, and put the data into two tables: a Fitness Center Members table with gender, phone number, and related information, and a Dues Paid table with dues paid and date paid (see Figure 7.10). As you can see in Figure 7.10, both tables include the member number field that can be used to create a relationship between the two.

With the relations in Figure 7.10, the redundancy has been reduced and the potential problem of having two different phone numbers for the member has been eliminated. Also note that the member number gives each record in the Fitness Center Members table a primary key. Because there are two dues paid ($25 each) with the same member number (SN656), you know this is the same person, not two different people.

STREET-SMART DATA CENTER

New York's finest have a new tool for finding the bad guys. The first of its kind in the country, the Real Time Crime Center is a digital database of millions of crime records and billions of public records that can spit out lists of possible suspects and similar crimes even before officers arrive at the scene. Its repertoire also includes satellite imaging and computerized mapping systems to determine geographic patterns of crime.

NYPD's Digital Crime-Fighter
Associated Press
Los Angeles Times
http://www.latimes.com/technology/la-na-nypd15jul15,1,246030.story?coll=la-headlines-technology&ctrack=1&cset=true
July 15, 2005

FIGURE 7.10 • Data analysis at a fitness center

To solve data problems, add a primary key, called member number, and put the data into two tables: a Fitness Center Members table with gender, phone number, and related information; and a Dues Paid table with dues paid and date paid.

Fitness Center Members Table

Member no.	Name	Phone	Gender
SN123	Brown, A.	468-3342	Female
SN656	Thomas, S.	468-5238	Male

Fitness Center Dues Paid Table

Member no.	Dues paid	Date paid
SN123	$50	September 15th
SN656	$25	September 15th
SN656	$25	September 15th

The process of correcting data problems or anomalies is called *normalization*. It ensures that the database contains "good data." Normalization normally involves breaking one table into two or more tables in order to correct a data problem or anomaly, as in Figure 7.10. Normalization is a very important technique in database management and is yet another advantage to using a relational database. The details of normalization, however, are beyond the scope of this text.

Object-Oriented Databases

Another database approach is the object-oriented model. An **object-oriented database** uses the same overall approach as objected-oriented programming, first discussed in Chapter 3. With this approach, the data and the processing instructions are stored in the database. For example, an object-oriented database could store both monthly expenses and the instructions needed to compute a monthly budget from the monthly expenses. A traditional DBMS might only store the monthly expenses. An object-oriented database uses an *object-oriented database management system (OODBMS)*. See Figure 7.11.

Object-oriented databases also offer the ability to reuse and modify existing objects to develop new database applications. Using existing and pretested objects for new database applications can result in fewer errors. J.D. Edwards is using an object-oriented database to help its customers make fast and efficient forecasts of future sales and to determine if they have enough materials and supplies to meet future demand for products and services. With the object-oriented database, customers can quickly get a variety of reports on available inventory

FIGURE 7.11 • Object-oriented database

An object-oriented database integrates a database with object-oriented programming capabilities.

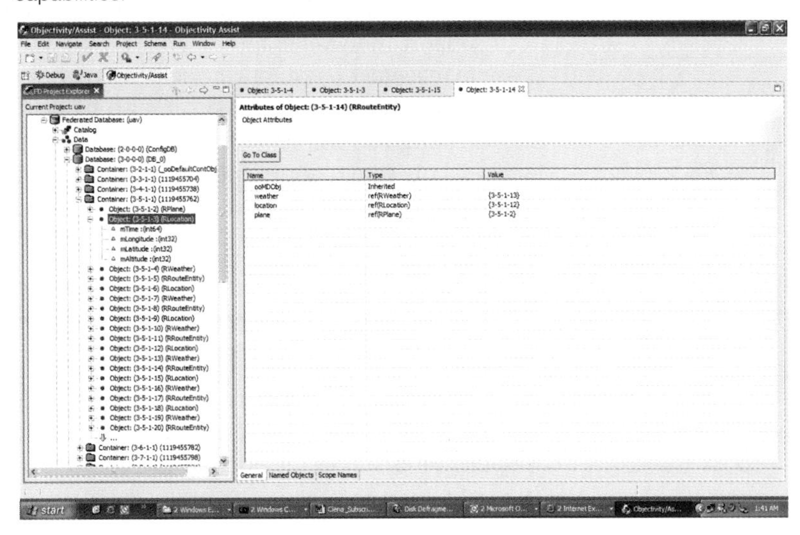

and supplies. CERN, the European Organization for Nuclear Research, is developing an object-oriented database from Objectivity (www.objectivity.com). The object-oriented database will hold a huge 5 petabytes of raw data obtained from a particle accelerator to help scientists gain insights into the structure of matter. The data will be collected and stored over the next 10 to 15 years.

Some modern DBMSs, such as Oracle and DB2, incorporate object-oriented capabilities in a relational database, and are known as *object-relational databases*.

Database Characteristics

The information needs of individuals or organizations have an impact on what type of data is collected and what type of database model is used. Important characteristics of databases include the amount of data, the volatility of the data, and how immediately the data needs to be updated.

- The database size or *amount* depends on the number of records or files in the database. The size determines the overall storage requirement for the database.
- *Volatility* of data is a measure of the changes, such as additions, deletions, or modifications, typically required in a given period of time.
- *Immediacy* is a measure of how rapidly changes must be made to data. Some applications, such as providing concert ticket reservations, require immediate updating and processing so that two customers are not booked for the same seat. Other applications, such as payroll, can be done once a week or less frequently and do not require immediate processing. If an application demands immediacy, it also demands rapid recovery facilities in the event the computer system shuts down temporarily.

The above characteristics are important for any individual or organization. They determine the requirements of the database and the database system that is needed for the amount of data, the volatility of the data, and the possible need for rapid changes. These characteristics are important in selecting and designing a database management system, discussed next.

DATABASE MANAGEMENT SYSTEMS

EXPAND YOUR KNOWLEDGE

To learn more about database systems, go to www.course.com/swt2/ch07. Click the link "Expand Your Knowledge" and then complete the lab entitled "Advanced Databases."

Creating and implementing the right database system ensures that the database can support individual and organizational goals. For example, an effective database management system can help doctors provide better patient care—a key goal of any hospital. A DBMS can also help streamline paperwork.

Creating and implementing the right database system involves determining how data is stored and retrieved, how people will see and use the database, how the database will be created and maintained, and how reports and documents will be generated. But how do you actually create, implement, use, and update a database? What type of database is needed?

Overview of Database Types

Database management systems can range from small, inexpensive software packages to sophisticated systems costing hundreds of thousands of dollars. A few popular alternatives include flat file, single user, multiuser, general-purpose, and special-purpose systems. Open-source databases are also available.

Flat File and File Organizers. A *flat file* has no relationship between its records and is often used to store and manipulate a single table or file. Flat files don't use any of the database models discussed earlier. Flat files often use a standard text file format, delineating records by line breaks—one record per line, and fields by some predefined character. The most common type of flat file format is called *CSV* for *comma-separated values*. The contents of a file named addressbook.csv might look something like this when viewed with a text editor:

```
Smith,Jon,1881 Mango Ct,Atlanta,GA,31123
Peters,Linda,2991 Montrose,Chicago,IL,60645
```

Databases, and even spreadsheet software can read a .csv file, using the line breaks and commas to interpret the data as a table. Some Web-based software use flat files to store data in an organized manner without the need of setting up a complicated database. Flat files can also delineate fields by using fixed-width columns rather than commas. Flat files are standardized and understood by many different database and spreadsheet applications (see Figure 7.12), so they are handy for sharing data between different types of software.

FIGURE 7.12 • Flat file capabilities

Many spreadsheet programs have flat file capabilities.

Order ID	Product	Unit Price	Quantity	Discount
10248	Queso Cabrales	$14.00	12	0.00%
10248	Singaporean Hokkien Fried Mee	$9.80	10	0.00%
10248	Mozzarella di Giovanni	$34.80	5	0.00%
10249	Tofu	$18.60	9	0.00%
10249	Manjimup Dried Apples	$42.40	40	0.00%
10250	Jack's New England Clam Chowder	$7.70	10	0.00%
10250	Manjimup Dried Apples	$42.40	35	15.00%
10250	Louisiana Fiery Hot Pepper Sauce	$16.80	15	15.00%
10251	Gustaf's Knäckebröd	$16.80	6	5.00%
10251	Ravioli Angelo	$15.60	15	5.00%
10251	Louisiana Fiery Hot Pepper Sauce	$16.80	20	0.00%
10252	Sir Rodney's Marmalade	$64.80	40	5.00%
10252	Geitost	$2.00	25	5.00%
10252	Camembert Pierrot	$27.20	40	0.00%
10253	Gorgonzola Telino	$10.00	20	0.00%
10253	Chartreuse verte	$14.40	42	0.00%
10253	Maxilaku	$16.00	40	0.00%
10254	Guaraná Fantástica	$3.60	15	15.00%

A *file organizer* is a database that goes beyond the capabilities of a flat file to efficiently store and/or retrieve data. File organizers are typically used with personal computers. TaskTracker is a file organizer that gives rapid access to recent documents. OneNote by Microsoft is another example of a file organizer used to store, organize, and retrieve normal text, handwritten notes, photos, and drawings (Figure 7.13). These capabilities are important for some people who use Tablet PCs.

FIGURE 7.13 • File organizer

A file organizer allows you to efficiently organize and access data on a personal computer.

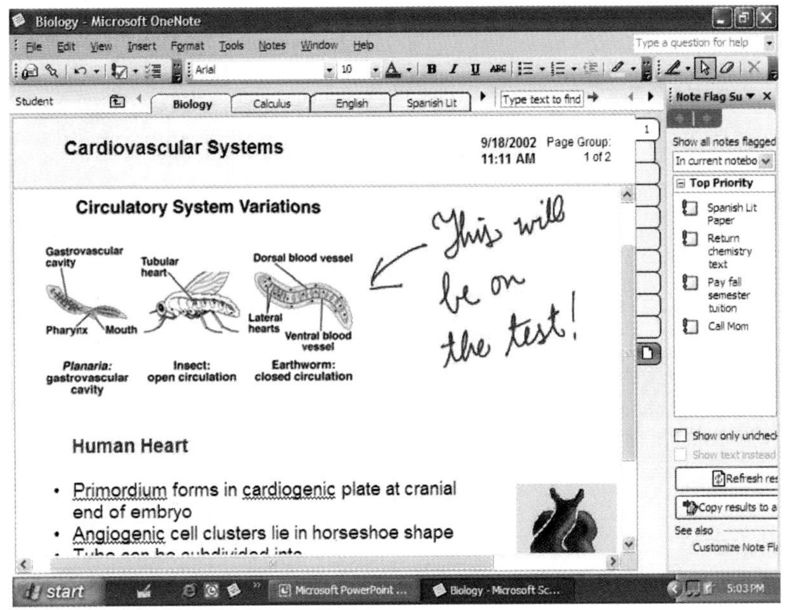

Single User. Databases for personal computers are most often for a single user. Only one person can use the database at any time. Access and Quicken are examples of popular single-user DBMSs through which users store and manipulate personal data. Access is used for a wide variety of personal data, and Quicken is used for financial data (see Figure 7.14). Microsoft InfoPath is another example of a single-user database.[6] The database is part of the Microsoft Office suite and helps people collect and organize information from a variety of sources. InfoPath has built-in forms that can be used to enter expense information, time-sheet data, and a variety of other information. Oracle has developed Database Lite 10*g* for laptop computers and other mobile or wireless devices.[7] "Companies are constantly looking for new ways of gaining competitive advantage and reducing costs by streamlining their business processes. Making information available directly and in real time to employees on the front line—serving or selling to customers—is an increasingly critical step to gaining such efficiencies," says Oracle's chief technology officer (CTO).

FIGURE 7.14 • Quicken is a single-user financial DBMS

Quicken is an example of a single-user DBMS used by many people to store and manipulate financial data.

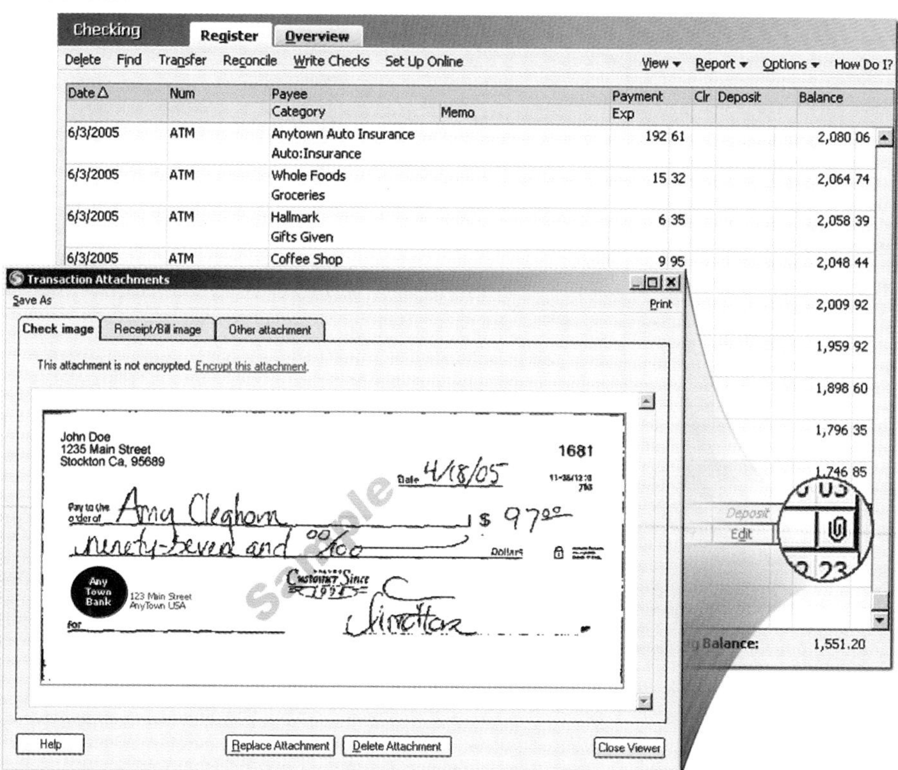

Multiuser. Networked computer systems need multiuser DBMSs. Hundreds or thousands of people on a network or the Internet, for example, cannot use a Microsoft Access database at the same time because it is not a multiuser database. Multiuser databases are more powerful, expensive, and allow dozens, hundreds, even thousands of people to access the same database system at the same time. Popular vendors for multiuser database systems include Oracle, Sybase, Microsoft, and IBM (see Figure 7.15). AutoTradeCenter, Inc., a forum for auto dealers to buy and sell cars directly with other dealers, lease companies, or rental companies, has linked its Web site to an Oracle database.[8] The database supports about 24,000 franchise dealerships and 80,000 independent dealerships.

FIGURE 7.15 • Multiuser database

A multiuser database like Sybase can accommodate thousands of users and data distributed among many locations.

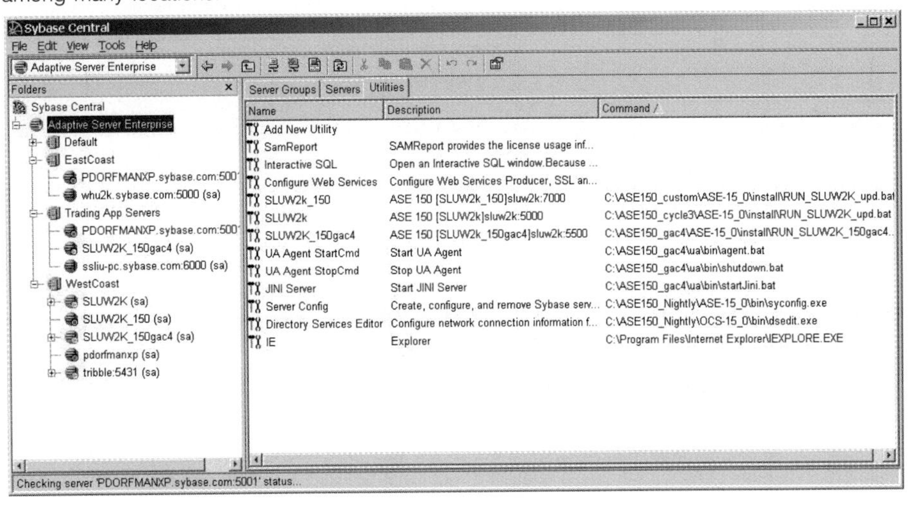

General-Purpose and Special-Purpose Databases. A *general-purpose database* can be used for a large number of applications. Oracle, Sybase, and IBM databases are examples of general-purpose databases that can be used by businesses, the military, charitable organizations, scientific researchers, and most other organizations for many different types of applications. Oracle is currently the market leader in general-purpose databases, with about 43 percent of the $15 billion database market.[9]

In contrast, a *special-purpose database* is designed for one purpose or a limited number of applications. The Israeli Holocaust Database (*www.yadvashem.org*) is a special-purpose database available through the Internet and contains information on about 3 million people in 14 languages.[10] The Hazmat database is another special-purpose database developed by The National Occupational Health and Safety Commission in Australia and contains information on about 3500 hazardous materials and various national exposure standards to hazardous materials.[11] CaseMap organizes information about a case, and LiveNote is used to display and analyze transcripts. These databases help law firms develop and execute good litigation strategies. J.P. Morgan Chase & Co. uses an "in-memory" database to speed trade orders from pension funds, hedge funds, and various institutional investors.[12] In-memory databases use a computer's memory instead of a hard disk to store and manipulate important data. The database can process more than 100,000 queries per second. Art and Antique Organizer Deluxe is a specialized-database for cataloguing art works and antiques.[13] Another special-purpose database by Tableau can be used to store and process visual images. According to Christian Chabot, the company's CEO, "Now you can query and walk through a database visually."[14]

Another example of a special-purpose database is QuickBooks, an accounting program for small businesses. In many cases, the overall structure of the database and the reports that can be generated are already developed for easy use. MindManager is a special-purpose database used to organize projects and people's thoughts and ideas (see Figure 7.16).

Open-Source Database Systems. Like other software products, there are a number of open-source database systems, including PostgreSQL and MySQL.[15] MySQL is the most popular open-source DBMS and is used by travel agencies, manufacturing companies, and other companies. MySQL use has increased by more than 30 percent in some years, compared with a 6 percent growth for

FIGURE 7.16 • MindManager

MindManager is a special-purpose database used to organize projects and people's thoughts and ideas.

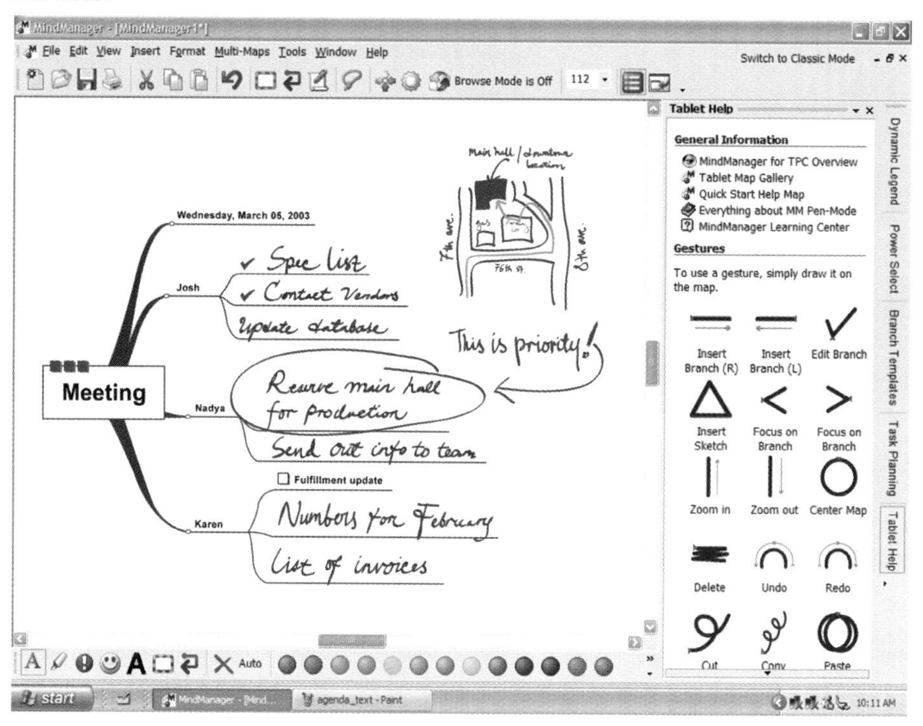

other popular database packages.[16] According to Charles Gary of the MetaGroup, "I get calls now from customers, not [asking,] 'Should I use an open-source database?' but instead it's, 'Should I use PostgreSQL or MySQL?'"[17]

Table 7.2 lists some popular DBMSs from various vendors.

TABLE 7.2 • Popular database management systems

DBMS	Vendor	Type	Computer Type
Access	Microsoft Corp.	Relational	PC, server, PDA
Approach	Lotus Development Corp.	Relational	PC, server
DB2	IBM Corp.	Object-relational	PC, midrange server, mainframe
FileMaker	FileMaker, Inc.	Relational	PC, server, PDA
Informix	IBM Corp.	Relational	PC, midrange server, mainframe
Ingres	Computer Associates International, Inc.	Relational	PC, midrange server, mainframe
MySQL	MySQL AB	Relational; open source	PC, midrange server, mainframe
Oracle	Oracle Corporation	Object-relational	PC, midrange server, mainframe, PDA
SQL Server	Microsoft Corp.	Relational	Server
Sybase	Sybase Inc.	Relational	PC, midrange server, PDA
Versant	Versant Corp.	Object-oriented	PC, midrange server
Visual FoxPro	Microsoft Corp.	Relational	PC, server

Open-Source Database for Continental

Still recovering from the threat of a terrorist attack and dealing with high fuel costs, many airlines are looking to cut costs and increase efficiency to survive. Continental Airlines is no exception. One solution was to use open-source software.

Continental picked Linux for its operating system. It also selected the newer 64-bit open-source database management system by MySQL. According to an analyst for Gartner, Inc., a research and consulting firm, "They're leading-edge. You're not even talking about hundreds of companies that are using 64-bit MySQL." MySQL has allowed Continental to streamline its operations and save time and money. For example, it used to take about 20 minutes for a Continental employee to reissue an airline ticket. With a MySQL database application, customers can now get tickets reissued through the Internet. The new database application also gives customers a consistent price, whether they complete the transaction through a Continental employee or the Internet. Switching to a total open-source solution went smoothly and wasn't difficult.

Continental is now working on developing a database application to streamline the refund process, and other applications are expected to follow. According to Gary Hein, an analyst for the Burton Group, "Success breeds success. They're hesitant to do the first one. But after they do, they say, 'Wow, that's hard not to do.'"

Questions

1. How was MySQL able to help Continental Airlines?
2. What are possible disadvantages of using an open-source database management system?

Sources

1. *Silwa, C. "Open Ticket for Continental,"* Computerworld, *April 4, 2005, p. 24*
2. *Staff, "Continental Airlines Is Pushing,"* Aviation Week & Space Technology, *April 11, 2005, p. 16*
3. *Babcock, C. "MySQL Tries Subscriptions,"* Information Week, *February 14, 2005, p. 15.*

Database Design

Before data can be stored, manipulated, and retrieved, the database must be logically designed. At a minimum, this requires field, record, and table design. All database management systems have the ability to perform these important design functions. In addition, other aspects of the database can be designed, including the input and output interfaces.

MOBILE MEDICAL RECORDS

Cost and the concern that the vendor might not stay in business have kept U.S. doctors reluctant to adopt electronic filing systems. Medicare feels so strongly this kind of database is crucial to good care, that it is offering its own database software free to all comers. The system tracks routine tests, issues warnings if a wrong prescription is about to be filled, and allows for record access from anywhere in the country.

U.S. Will Offer Doctors Free Electronic Records System
By Gina Kolata
The New York Times
http://www.nytimes.com/2005/07/21/health/21records.html?
July 21, 2005

Field Design. As mentioned earlier, a field is typically a name, number, or a combination of characters. The purpose of *field design* is to specify the type, size, format, and other aspects of each field. Some popular field types include:

- Numeric: A *numeric field* contains numbers that can be used in making calculations. The number of CD-RW disks you purchase at a computer store and the price are examples of numeric fields. The hours you work at a local restaurant and your age are examples of numeric data. Numeric data can include real numbers, which may contain a decimal point, such as the price of a new bicycle, $349.95, and integers, which are whole numbers without decimal points, such as your age or the number of CD-RW disks you purchase.

- Alphanumeric: *Alphanumeric* or character type data includes characters or numbers that cannot be manipulated or used in calculations. Examples include your name, your street address, your Social Security number, and the color of your hair. Numbers that are not used in calculations, such as student or employee ID numbers, are usually entered as alphanumeric data.
- Date: Databases allow you to enter dates, such as 06/12/06, into the database. Once entered, *dates* can be sorted or even used in computations, such as how many days occur between two dates.
- Logical: A *logical* piece of data contains items, such as *yes* or *no*. Only these logical operators can be included in the field. For example, a university database might include a field that tracks whether students have met their English requirements for graduation. The field can contain a *yes* or a *no*.
- Computed: A *computed* field, also called a calculated field, is determined from other fields, instead of being entered into the database. Gross pay from a job, for example, is a computed field. It is determined by multiplying your hourly rate times the number of hours you worked. Other computed fields include grade point average, your total bill at a restaurant, and the total energy expended in a scientific experiment.

Record and Table Design. Recall that a record is a collection of related fields, and a table is a collection of related records. In any database, you must identify the exact fields that are contained in each record, and the types of records that may be included in each table. For example, you might want to develop a database for your music CD collection. Each record in the database could contain an item number that you specify, the artist, the name of the CD, when the CD was produced, when you purchased the CD, the general condition of the CD, and where you store the CD. Similar record designs can be specified for a database of the members in a volunteer group that builds homes, a list of items in your apartment for insurance purposes in case of a fire or theft, the results of a scientific experiment in your nutrition class, and the required courses you must complete to graduate.

Database tables can contain a few or millions of records and fields. Tables can be organized in a variety of ways, and different tables can be related as discussed earlier in the section on relational databases. You can sort on one or more fields to help organize a table. For a table containing the first and last names of members of a student group, you could sort on last name as the primary sorting key and the first name as the secondary sorting key. Adams would be sorted before Brill, and Crystal Adams would appear before Jake Adams. As mentioned before, tables can also be selected and joined. See Figure 7.17.

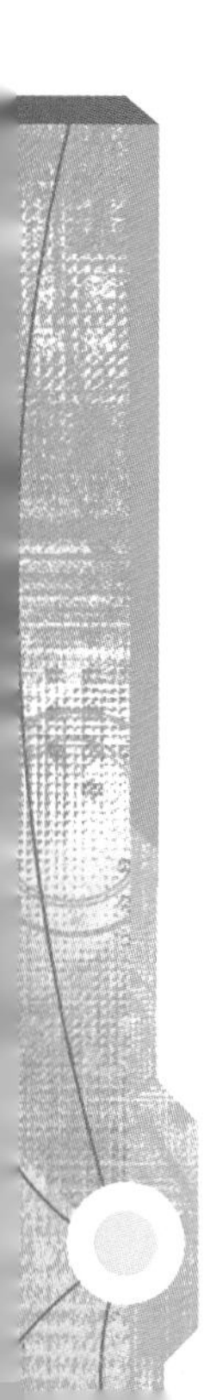

Input and Output Interface Design. Designing effective interfaces is a convenient and powerful database design feature in most database management systems. Designers can create easy-to-understand forms that users can fill out for each record (see Figure 7.18), and after users completely enter data into a field, the database automatically goes to the next field, without requiring the user to press Enter for each field.

Reports and other outputs from the database can also be designed to be powerful and packed with useful information. A university can have a report that lists only those students who haven't paid their fees; a civil engineering firm could design a report that lists stress points in an old bridge; and a medical blood test report could be designed to include a column with all tests that are

FIGURE 7.17 • Database tables

The information in the tables of a relational database can be queried and manipulated in many different ways.

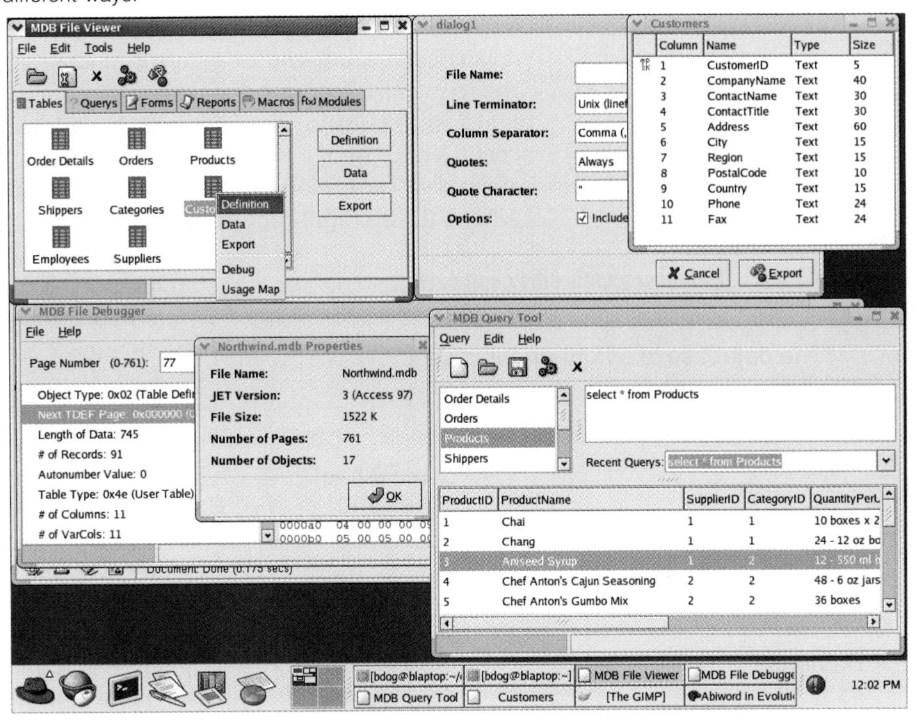

outside of normal ranges. These types of reports are an important aspect of providing information support to administrators, military generals, engineers, doctors, and corporate executives.

FIGURE 7.18 • Input and output design

Database designers can create forms that are easy for users to fill out.

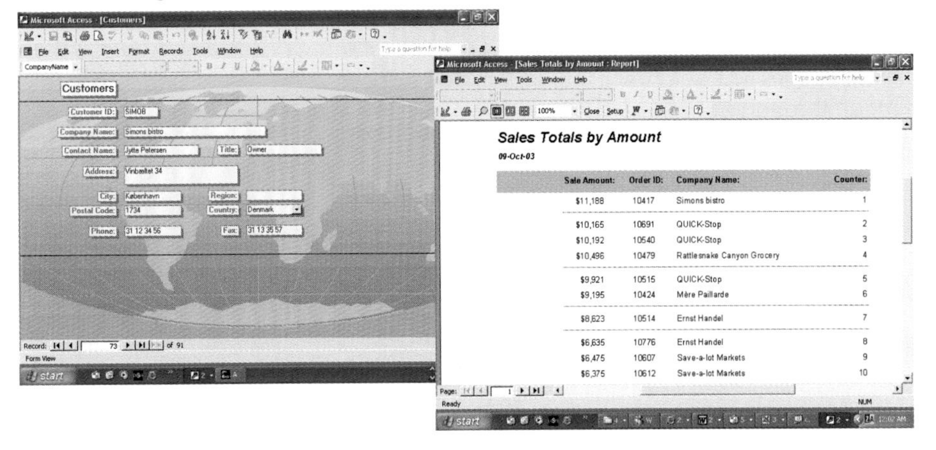

Using Databases with Other Software

Database management systems are often used in conjunction with other software packages or the Internet. A database management system can act as a front-end application or a back-end application. A *front-end application* is one that directly interacts with people or users. A *back-end application* interacts with other programs or applications; it only indirectly interacts with people or users.

Some instructors, for example, use a database as a front-end to a spreadsheet program that contains student grades. The instructor and several teaching assistants can enter student grades into the database. The data is then transferred to a spreadsheet program that computes a student's grade. Researchers often use a database as a front-end to a statistical analysis program. The researchers enter the results of experiments or surveys into a database. The data is then transferred to a statistical analysis program to test hypotheses or compute correlation coefficients. Using a database as a front end can increase data accuracy and avoid data corruption in spreadsheet programs, statistical analysis programs, or other back-end applications (see Figure 7.19).

FIGURE 7.19 • Databases with other software

A database can be used as a front-end or a back-end application.

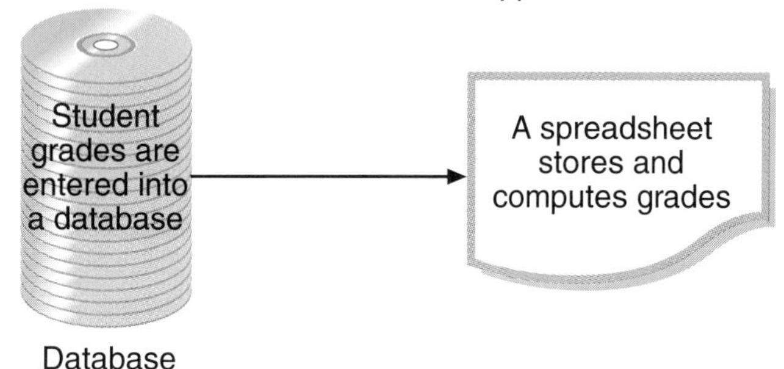

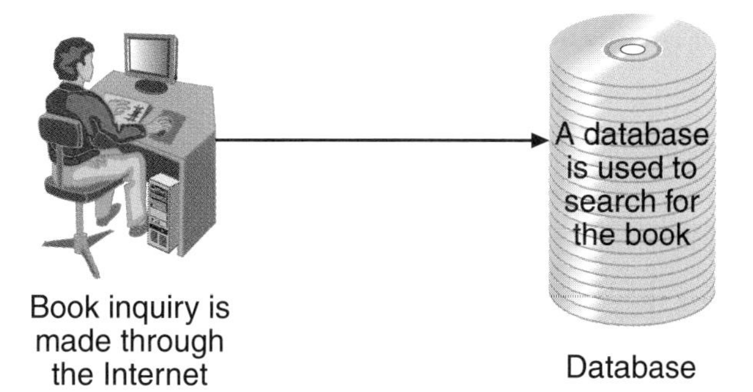

Databases can also be used as the back-end application. A call center operator may use a software package to enter questions from callers needing help or information. The software then interacts with one or more databases (the back-end) to get the requested information. When people request information from a Web site on the Internet, the Web site can interact with a database (again the back-end) that supplies the desired information. For example, you can connect to a library Web site to find if the library has a book you want to read. The library Web site then interacts with a database that contains a catalogue of library books and articles to determine if the book you want is available.

System designers are increasingly using the Web as the front end to database systems. The Web page *www.google.com* acts as the front end to the huge Google database that resides at the back end of the system. On corporate intranets, a Web browser serves as the perfect front end to private database systems because employees are already comfortable with using Web browsers and require little training to use the system.

Data Accuracy and Integrity

As first discussed in Chapter 1, to be of value, data must be accurate. **Data integrity** means that data stored in the database is accurate and up to date. There are many possibilities for inaccuracies in today's information-rich society. Many people have been unable to get credit cards, car loans, or home mortgages because data stored about them by credit bureaus was wrong. In a manufacturing company, a clerk might enter the hours an employee worked as 4 hours instead of 40 hours. In some cases, a retail store may report that someone didn't pay his or her bills, when the bills were paid in full. These types of errors were caused by inaccurate data being entered into a database. The results are inaccurate output. This is called **garbage in, garbage out (GIGO)**.

In other cases, data is simply old and no longer valid. Managers have made multimillion-dollar decisions based on out-of-date data. A manager might decide to build a new manufacturing plant based on old sales data that was much higher than today's numbers. A nurse may give the wrong drugs to a patient because the prescribed treatment was old and reflected the patient's past situation, or a surgeon might amputate the wrong leg because of a mistake in the patient's records.

In other cases, people intentionally enter wrong data for their own gain. In the corporate scandals of the early 2000s, a number of executives and accountants falsified records to inflate profits or hide expenses to get higher bonuses based on the company's stock performance (see Figure 7.20). Some medical researchers have falsified research results to get more articles published, which often results in getting tenure, a promotion, or salary increases.

FIGURE 7.20 • Damaging data integrity

In the corporate scandals of the early 2000s, a number of executives and accountants falsified records to inflate profits or hide expenses to get higher bonuses based on the company's stock performance. Many of the people involved had to testify in court and some were imprisoned.

Database management systems must be programmed to detect and eliminate data inaccuracies whenever possible. Databases can be programmed to check for inconsistent data. For example, if the combined expenses of various departments or branches are greater than the total expenses reported on a company's income statement, something is probably wrong. If everyone in a department was reported to have worked only 4 hours last week, a clerical error has probably been made. It is likely that 4 hours was entered into the database instead of 40 hours. Although it is not possible to eliminate all data inaccuracies, good database design and development requires that checks and balances be set up to detect and eliminate errors whenever possible.

Creating and Modifying a Database

One of the first steps in creating a database is to outline the logical and physical structure of the data and relationships among the data in the database. This description is called a **schema** (as in schematic diagram). A schema can be part of the database or a separate schema file. The DBMS can reference a schema to find where to access the requested data in relation to another piece of data. Schemas are entered into the DBMS (usually by database personnel) via a data definition language. A *data definition language (DDL)* is a collection of instructions and commands used to define and describe data and data relationships in a specific database.

Another important step in creating a database is to establish a **data dictionary**, a detailed description of all data used in the database. The data dictionary includes information such as the name of the data item, who prepared the data, who approved the data, the date, a description, other names, the range

of values for the data, the data type (numeric or alphanumeric), and the number of positions or space needed for the data. Figure 7.21 shows a typical data dictionary entry.

FIGURE 7.21 • A typical data dictionary entry

The data dictionary includes information, such as the name of the data item, who prepared the data, who approved the data, the date, a description, other names, the range of values for the data, the data type (numeric or alphanumeric), and the number of positions or space needed for the data.

```
          NORTHWESTERN
          MANUFACTURING
PREPARED BY:          D. BORDWELL
DATE:                 04 AUGUST
APPORVED BY:          J. EDWARDS          DATE: 13 OCTOBER
DATE:                 13 OCTOBER
VERSION:              3.1
PAGE:                 1 OF 1

DATA ELEMENT NAME:    PARTNO
DESCRIPTION:          INVENTORY PART NUMBER
OTHER NAMES:          PTNO
VALUE RANGE:          100 TO 5000
DATA TYPE:            NUMERIC
POSITONS:             4 POSITIONS OF COLUMNS
```

Some of the typical uses of a data dictionary are to:

- Provide a standard definition of terms and data elements: This can help in the programming process by providing consistent terms and variables to be used for all programs. Programmers know what data elements are already captured in the database and how they relate to other data elements.
- Assist programmers in designing and writing programs: Programmers do not need to know which storage devices are used to store needed data. Using the data dictionary, programmers specify the required data elements. The DBMS locates the necessary data.
- Simplify database modification: If for any reason a data element needs to be changed or deleted, the data dictionary points to those specific programs that utilize the data element that may need modification.

A data dictionary helps achieve the advantages of the database approach in the following ways:

- Reduced data redundancy: By providing standard definitions of all data, it is less likely that the same data item is stored in different places under different names.
- Increased data security: It is more difficult for unauthorized people to gain access to sensitive data and information.
- Faster program development: Programmers don't have to develop names, descriptions, value ranges, and other data attributes for data items because the data dictionary does that for them. Programmers don't have to check to make sure that the same name is not being used for another purpose or that the same date item doesn't have two or more names. With some programs requiring hundreds or thousands of data names, this can save a significant amount of time.

- Easier modification of data and information: Modifications to data are easier because users don't need to know where the data is stored. The person making the change indicates the new value of the variable or item, such as part number, that is to be changed. The database system locates the data and makes the necessary change.

Updating a Database

Databases are updated by adding, modifying, and deleting records. A paleontologist looking for dinosaur bones can add records to her database at each new dig, or an ornithologist can search for and update information on particular birds in an online database (see Figure 7.22). A university can modify your student records to include courses you just completed last semester along with your new GPA. A company can delete customers who paid their bills completely from their accounts receivables table, which lists all customers who still owe the company money from past sales. Continual database updating is absolutely essential to maintain a high degree of data accuracy and integrity. As mentioned earlier, a front-end application can be used to enter the changes, which are then transferred to the database.

FIGURE 7.22 • Ornithological database

Amateurs or professionals looking for information on birds can search this online database of over 2 million entries.

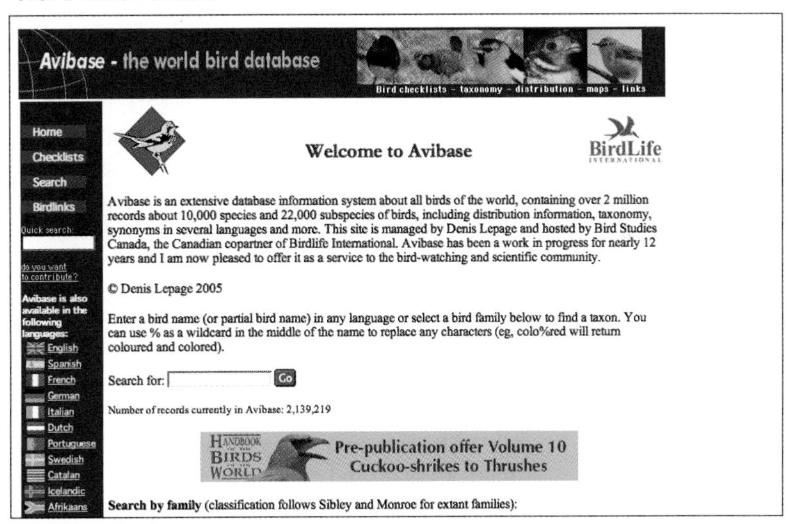

Manipulating Data and Generating Reports

Once a DBMS has been installed and a database and table(s) created, the system can be used via specific commands in various programming languages or queried using a data manipulation language. In general, a *data manipulation language (DML)* is a specific language provided with the DBMS that allows people and other database users to access, modify, and make queries about data contained in the database and to generate reports. Many databases use *query by example (QBE)* to give you ideas and examples of how queries can be made. QBE is a very visual approach to making queries or getting answers to questions by entering names, values, and other items into a window. For example, you could enter CustomerName and CustomerPhoneNumber into the field row of a QBE window of a large database containing customer name, customer identification number, customer address, customer phone number, and additional information. QBE then returns a smaller table containing customer names and customer phone numbers. QBE can also be used to quickly get a list of all customers who live in a particular city or a list of all customers that generated more than $10,000 in sales last month. As you can see, QBE makes manipulating databases much easier and faster than learning formal DMLs such as SQL, which is discussed next (see Figure 7.23).

FIGURE 7.23 • Query by example

Query by example allows users to fill in criteria or click items to return information and generate reports.

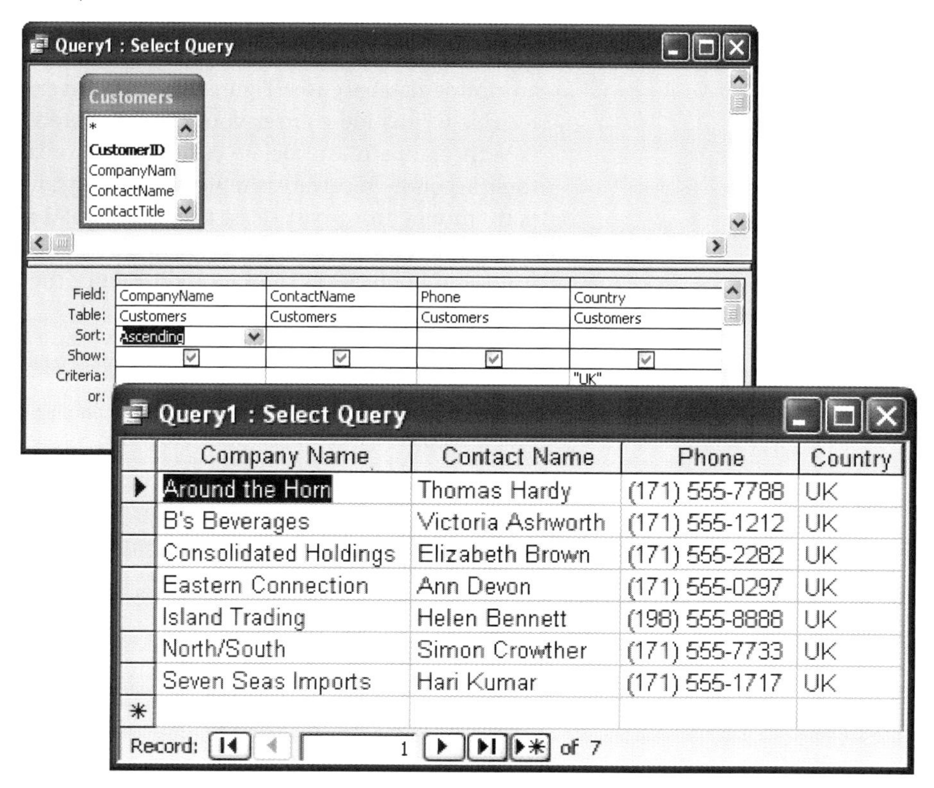

In the 1970s, D. D. Chamberlain and others at the IBM Research Laboratory in San Jose, California, developed a standardized data manipulation language, called **Structured Query Language (SQL)**, pronounced *sequel* In 1986, the American National Standards Institute (ANSI) adopted SQL as the standard query language for relational databases. Today, SQL is an integral part of popular databases on both mainframe and personal computers. The following query is written in SQL:

```
SELECT * FROM EMPLOYEE WHERE JOB_CLASSIFICATION = 'C2'
```

This query tells the DBMS to select all (*) columns from the EMPLOYEE table for which the JOB_CLASSIFICATION value is equal to C2.

SQL is actually a programming language, but it is easy for nonprogrammers to understand and use; note the English-like commands. Programmers can use SQL on systems ranging from PCs to the largest mainframe computers. SQL statements also can be embedded in many programming languages, such as the widely used language COBOL. Programs that include embedded SQL also make it easier for end users to create their own reports. Table 7.3 contains examples of SQL commands.

TABLE 7.3 • Examples of SQL commands

SQL Command	Description
`SELECT ClientName, Debt FROM Client WHERE Debt>1000`	This query displays all clients (ClientName) and the amount they owe the company (Debt) from a database table called Client for clients that owe the company more than $1000 (WHERE Debt>1000).
`SELECT ClientName, ClientNum, OrderNum FROM Client, Order WHERE Client.ClientNum = Order.ClientNum`	This command is an example of a join command that combines data from two tables—the client table and the order table (FROM Client, Order). The command creates a new table with the client name, client number, and order number (SELECT ClientName, ClientNum, OrderNum). Both tables include the customer number, which allows them to be joined. This is indicated in the WHERE clause that states that the client number in the client table is the same (equal to) the client number in the order table (WHERE Client.ClientNum = Order. ClientNum).
`GRANT INSERT ON Client to Guthrie`	This command is an example of a security command. It allows "Bob Guthrie" to insert new values or rows into the Client table.

Database Backup and Recovery

Database backup and recovery are important functions of any DBMS. Some database experts believe that organizations have an obligation to provide secure and reliable databases by using adequate database backup procedures. A *database backup* is a copy of all or part of the database. For example, if you have a database containing the items in your apartment, you can create a database backup by making a copy of the entire database on another disk drive, CD, DVD, or other storage device. It is also possible to make a partial backup of only the data that has changed since the last backup. Database backup software can automate the database backup process. Some backup software makes copies of a database every day or more frequently. Database backup hardware can also be used. Some individuals and organizations purchase special hard disks for data backup. Maxtor, for example, is a popular external hard disk and software combination that people can use to backup a database and recover from database malfunctions. The more often you make backup copies the safer you are from potential database problems.

Database recovery is the process of returning the database to its original, correct condition if the database has crashed or been corrupted. If the database containing the items in your apartment crashes or becomes unusable, you can recover by using the backup copy. Most database backup software and hardware have excellent recovery capabilities to restore the database if something goes wrong. In some cases, it is possible to give one command or push a button to restore a damaged database. Some organizations use *redundant array of independent disks (RAID)* to store duplicate data on multiple disks or a *storage area network (SAN)* to connect multiple storage devices on high-speed networks to make recovering from a database failure faster and more efficient.

USING DATABASE SYSTEMS IN ORGANIZATIONS

You have explored a number of database applications in this chapter, from an individual entering a list of music CDs to large corporations keeping track of business operations. This section takes a closer look at how databases are used in organizations, including transaction processing, information and decision support, and a variety of other areas.

Routine Processing

All organizations need to process routine transactions. Consider a small precious gemstone business. Companies have developed routine processing applications to help these types of businesses. Polygon, for example, has developed the ColorNet database to help traders of rare gemstones perform routine processing activities, including buying and selling precious gemstones.[18] According to the CEO of Polygon, "If I'm looking for a 2-carat sapphire, [that request] goes out to other users, and I'll also find matches on our own database. It will be that much easier to search for colored stones right on the Internet."

Organizations of all sizes, including small precious gemstone companies, have to pay their employees. A small business needs to send out bills quickly to maintain a healthy cash flow, and a religious organization might want to send out a monthly newsletter. A manufacturing company needs to pay their suppliers for parts and raw materials. These are all routine processing activities that can be implemented with a database system.

Information and Decision Support

Database systems are a valuable tool for producing information that supports decision making for people and organizations to further their goals. A new database by Intellifit can be used to help shoppers make better decisions and get clothes that fit when shopping online.[19] The database contains true sizes of apparel from various clothing companies that do business on the Web. The process starts when a customer's body is scanned into a database at one of the company's locations, typically in a shopping mall (Figure 7-24). About 200,000 measurements are taken to construct a 3-D image of the person's body shape. The database then compares the actual body dimensions with sizes given by Web-based clothing stores to get an excellent fit. According to one company executive, "We're 90 percent (accurate) about the sizes and the styles and the brands that will fit you best."

FIGURE 7.24 • Intellifit

A whole-body scanner inputs your precise measurements, compares them to a database of clothing, sizes, and styles, and makes suggestions for apparel that will fit just right.

Manipulating the data in a database into valuable information has helped many people achieve their goals. A hotel chain, achieves better customer satisfaction using a database system to customize service. This type of database system can provide detailed information about customers. A hotel receptionist in New York, for example, might apologize to a customer for not having her room cleaned up as desired during a recent stay at one of the chain's hotels in Orlando. The receptionist at the New York hotel might offer the customer a special rate or provide additional service as a result of the data about the customer's Orlando stay in the hotel's database (see Figure 7.25). The Food Allergen and Consumer Protection Act that takes effect in 2006 requires that food manufacturing companies keep an accurate database of ingredients, formulas, and food preparation techniques to be made available to the public.[20] The new law requires that food companies clearly label their products for major allergens.

FIGURE 7.25 • Hilton's OnQ

Hotels in the Hilton Family of Brands use proprietary OnQ™ technology to understand and deliver on guest preferences. Customers opt in to provide profile information, which is combined with history from past stays. The information is encoded on tags that can be read at check-in and used by hotel team members to enhance the guest's stay.

Data Warehouses, Data Marts, and Data Mining

To realize the potential of a database to provide information and decision support, a number of technologies have been developed. A **data warehouse** is a typically large database that holds important information from a variety of sources. It is usually a subset of multiple databases maintained by an organization or individual. The first data warehouses were developed at PacTel Celluar, Aetna Casualty, and Blue Cross/Blue Shield in the 1980s. Today, data warehouses are used by many companies and organizations. A hardware store, for example, can use a data warehouse to analyze pricing trends. This can help the store determine what inventory to carry and what price to charge. As a result, the store might lower the price of wheelbarrows and sell twice as many each year. With a data warehouse, all you have to do is ask where a certain product is selling well and a colorful table showing sales performance by region, product type, and time frame automatically pops up on the screen.[21]

Data warehouses can also get data from unique sources. Oracle's Warehouse Management software, for example, can accept information from radio-frequency identification (RFID) technology, which is being used to tag products as they are shipped or moved from one location to another.[22] Some believe that RFID could help companies discover important information.[23] Instead of recalling hundreds of thousands of cars because of a possible defective part, automotive companies could determine exactly which cars had the defective parts and recall only the cars with the bad parts. The savings would be huge. The storage requirements for RFID could also be huge. If a company like Wal-Mart stored all of its inventory using RFID, one day of data could require a very large 7 million terabytes of storage.

A *data mart* is a small data warehouse, often developed for a specific person or purpose. It can be generated from a data warehouse using a database management system (see Figure 7.26). **Data mining** is the process of extracting information from a data warehouse or a data mart. The DBMS can be used to generate a variety of reports that assist people and organizations in making decisions and achieving their goals. The FBI, for example, is using the ClearForest database package to support data warehousing and data mining of the Terrorism Intelligence Database. Data mining has also been used in the airline-passenger profiling system used to block suspected terrorists from flying. Data mining is also used by the Terrorism Information Awareness Program, which attempts to detect patterns of terrorist activity.

The Hartford Life Insurance Co. uses data mining to get detailed information about its customers to increase sales and profits.[24] The data-mining tool helped the company generate record sales, up over 40% from the previous year. According to Victoria Severino, chief information officer for Hartford, "We model against these trends and come up with an unexpected risk scenario of the guarantees we offer. We also model risk based on a policyholder's behavior."

FIGURE 7.26 • Generating business intelligence from data warehouses and data marts

Data warehouses are generated from a database and other sources; a data mart is a small data warehouse, often developed for a specific person or purpose. Data mining can be used to generate business intelligence.

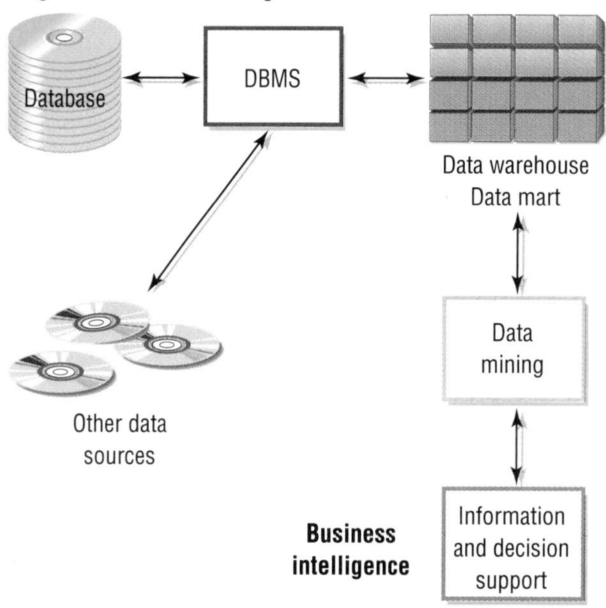

In a business setting, data mining can yield outstanding results. Often called *business intelligence,* a term first coined by a consultant at Gartner Group, the business use of data mining can help increase efficiency, reduce costs, or increase profits. The business-intelligence approach was first used by Procter & Gamble in 1985 to analyze data from checkout scanners. Today, a number of companies use the business intelligence approach. Hundreds of hotels use business intelligence software to get valuable customer information used to develop marketing programs. Companies like Ben and Jerry's store and process huge amounts of data. The company collects data on all 190,000 pints it produces in its factories each day, with all the data being shipped to the company's headquarters in Burlington, Vermont. In the marketing department, the massive amount of data is analyzed. Using business intelligence software, the company is able to cut costs and improve customer satisfaction (see Figure 7.27). The software allows Ben and Jerry's to match the over 200 calls and e-mails received each week with ice cream products and supplies. Today, the company can quickly determine if there was a bad batch of milk or eggs. The company can also determine if sales of Chocolate Chip Cookie Dough is gaining on the No. 1 selling Cherry Garcia.

FIGURE 7.27 • Business intelligence software

BI "dashboards" help executives quickly receive statistics and trends in a mostly graphic format.

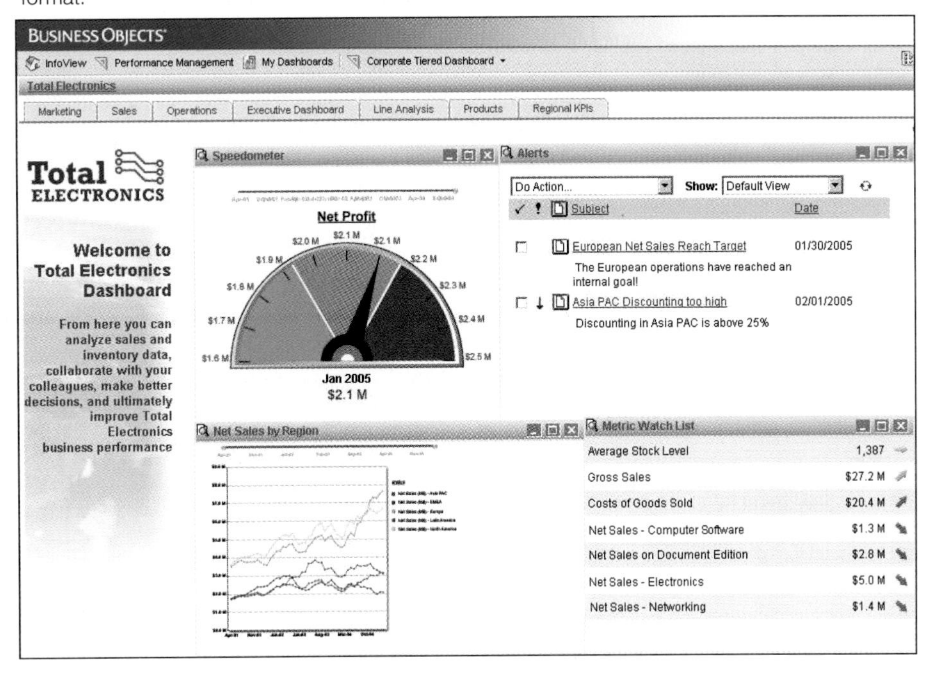

DATABASE TRENDS

The types of data and information that people and organizations need can change. A number of trends in the development and use of databases and database management systems will meet these changing needs.[25] For example, more and more organizations are finding they need to coordinate databases at different locations and to access databases through the Internet. Some people and organizations need to store audio and video files in an organized database. These trends are explored in this section.

Distributed Databases

With a **distributed database**, also called a *virtualized database,* the actual data may be spread across several databases at different locations. To users, the distributed databases appear to be a single, unified database. Distributed databases connect data at different locations via telecommunications. A user in the Milwaukee branch of a shoe manufacturer, for example, might make a request for data that is physically located at corporate headquarters in Milan, Italy. The user does not have to know where the data is physically stored. He or she makes a request for data, and the DBMS determines where the data is physically located and retrieves it (see Figure 7.28). Novell, the network company, and Red Hat, a company that distributes the Linux operating system, are joining together to develop open-source technology for virtualized systems and databases.[26] The technology will use servers that run the Linux operating system.

A distributed database creates additional challenges in maintaining data security, accuracy, timeliness, and conformance to standards. Distributed databases allow more users direct access at different user sites; thus, controlling who accesses and changes data is sometimes difficult. Also, because distributed databases rely

FIGURE 7.28 • A distributed database

For a shoe manufacturer, computers may be located at corporate headquarters, in the research and development center, the warehouse, and in a company-owned retail store. Telecommunications systems link the computers so that users at all locations can access the same distributed database no matter where the data is actually stored.

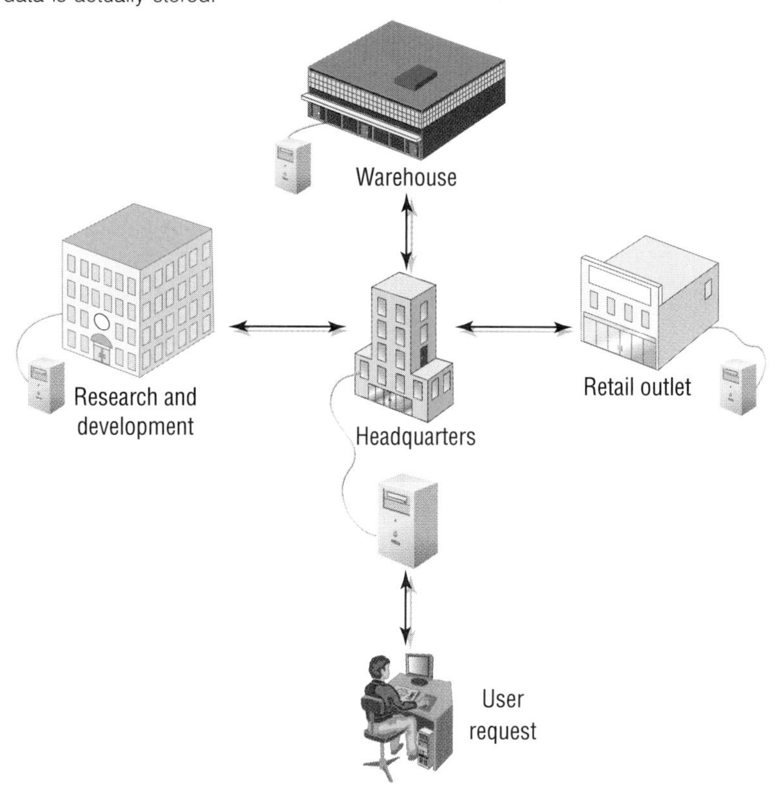

on telecommunications to transport data, access to data can be slower. To reduce the demand on telecommunications media, some organizations build a replicated database. A *replicated database* is a database that holds a duplicate set of frequently used data. At the beginning of the day, an organization sends a copy of important data to each distributed processing location. At the end of the day, the different sites send the changed data back to be stored in the main database.

Database Systems, the Internet, and Networks

Anyone with even limited experience with the Internet knows that there is a vast amount of raw data and important information available on the Web. In most cases, a traditional database, such as a relational database, is used. As mentioned before, the database is often the back-end application. From a user's perspective, the traditional database is invisible and behind the scenes. The Web is then used as the front end. All requests made to the database are done through the Internet (see Figure 7.29).

A number of Internet development tools can be used to interact with a traditional database. Most of these tools were introduced in Chapter 4. HTML, XML, and other Web development tools can all be used to develop a Web site to be a front-end interface to a traditional database system.

Access to databases through the Internet and private networks is allowing many companies and organizations to collaborate in ways that would not have been possible just a few years ago. General Electric and Intermountain Health Care are developing a comprehensive database on medical treatments and clinical protocols for doctors over the Internet.[27] The database is expected to cost about $100 million to develop and should help physicians more accurately diagnose patient illnesses. The Health Record Network Foundation, a joint partnership between the medical and business schools of Duke University, is developing a pilot program to make electronic health records (EHR) available over networks and the Internet.[28] The database would allow physicians and other authorized people to access patient records from remote locations.

Some people, however, are concerned about the accuracy and privacy of the information in databases that are accessible from the Web or linked to other private networks. One database expert believes that up to 40 percent of Web sites that connect to corporate databases are susceptible to hackers' taking complete control of the database.[29] By typing certain characters in a form on some Web sites, a hacker is able to give SQL commands to control the corporate database.

FIGURE 7.29 • Database access through the Internet

This online shopping site uses the Web as the front end to a database of thousands of products and reviews. From a user's perspective, the back-end database is invisible and behind the scenes.

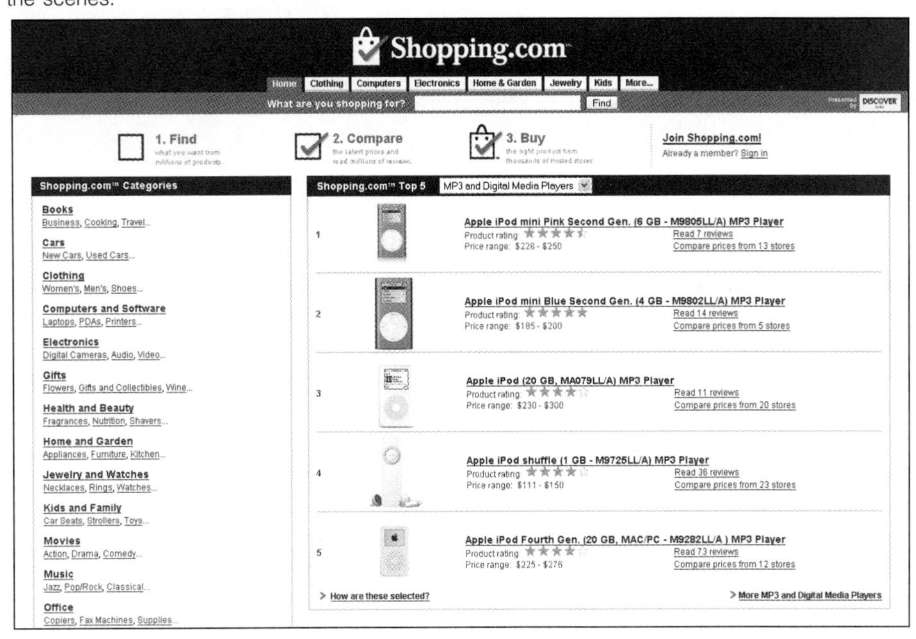

FIGURE 7.30 • Visual database

Visual databases of fingerprints have become an essential tool for law enforcement and crime investigations.

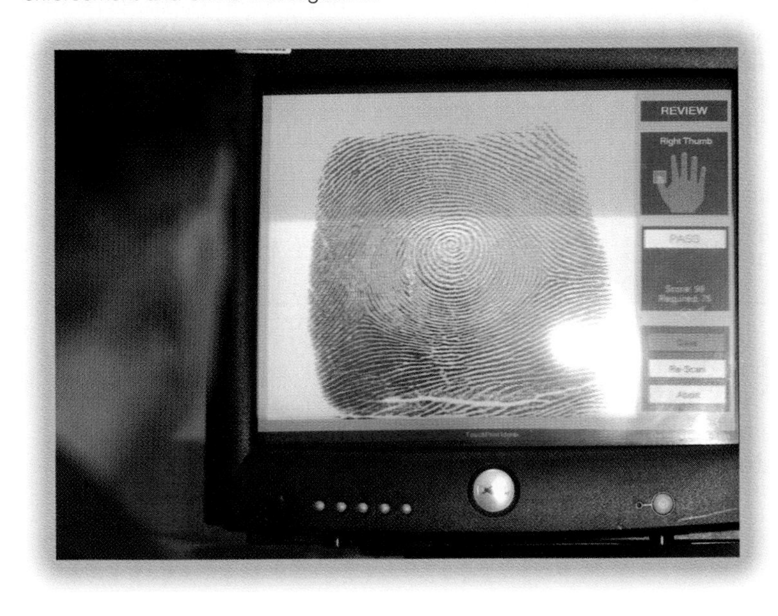

Visual, Audio, Unstructured, and Other Database Systems

In addition to text and numbers, organizations are increasingly finding a need to store large amounts of visual and audio data in an organized fashion. Music companies, for example, need to store and manipulate sound from recording studios. One university has developed an audio database and processing software to give singers a voice makeover. The database software can correct pitch errors and modify voice patterns to introduce vibrato and other vocal characteristics. Drug companies often need to analyze a large number of visual images from laboratories. Other visual databases allow petroleum engineers to analyze geographic information to help them determine where to drill for oil and gas. A visual-fingerprint database can be used to solve a cold-case murder that has gone unsolved for decades (see Figure 7.30).

An *unstructured database* contains data that is difficult to place in a traditional database system. The data can include notes, drawings, fingerprints, medical abstracts, sound recordings, and other data that is difficult or impossible to store in a traditional database, such as Access or

JOB TECHNOLOGY

Web-Based DBMS Empowers Cruise Line Personnel

Recently, Holland America Cruise Lines traded its old, complex mainframe database system—one that only a computer programmer could understand—for a new system that ordinary employees could interact with directly. The goal of the upgrade was to increase revenues by $1 million annually by speeding information to employees for more efficient sales, marketing, and revenue management. Prior to the upgrade, database programmers and tech staff would prepare and distribute weekly reports to address other employees' ad hoc inquiries and would generate scheduled reports to assess the company's revenue and inventory, says Paul Grigsby, senior revenue manager at Holland America. The new reporting system, called WebFocus, connects to the same mainframe database as the old DBMS but provides more powerful analysis, querying, and reporting tools that are accessed through an intuitive Web-based user interface.

The company is limited, however, because of the size of the existing mainframe DBMS. It plans to load the information into a streamlined data warehouse, where it can be accessed more quickly. Speed and accessibility, after all, are the attributes that will increase productivity and revenues. "The real value of the new system at Holland America," Bill Hostmann, an analyst at Gartner Inc., says, "is how it makes it easier to develop and distribute business management information to more users in a timely fashion around the world than ever before and thereby more fully leverage existing IT investments."

Questions

1. How does the new system allow the computer staff to work more efficiently?
2. The change in information access at Holland America Line is indicative of a general trend in many industries: noncomputer employees are assuming traditional computer staff responsibilities. Do you think this trend evolved purely out of efforts to save money by reducing computer staff, or are there substantial benefits to bringing computer power to the people? What might those benefits be?
3. Is it realistic to expect nontechnical employees to acquire higher-level technical skills? Will the nontechnical staff be willing and able to assume the task?

Sources

1. *Staff, "Holland America Promotes 108-day Grand Voyage,"* Travel Trade Gazette UK & Ireland, *January 7, 2005, p. 10*
2. *Sangini, M. "Cruise Line Changes BI Tack,"* Computerworld, *October 6, 2003, www.computerworld.com; Information Builder's Web site, www.informationbuilders.com, accessed March 5, 2004.*
3. *Holland America Line Web site, http://www.hollandamerica.com/, accessed March 5, 2004.*

DB2. As mentioned previously, OneNote by Microsoft is an example of a database that can store and retrieve unstructured data. EverNote is a free database that can store notes and other pieces of information.[30] In the future, you will see more databases that can handle unstructured data.

MANAGING DATABASES

EXPAND YOUR KNOWLEDGE

To learn more about backup, go to www.course.com/swt2/ch07. Click the link "Expand Your Knowledge" and then complete the lab entitled "Backing Up Your Computer."

Managing a database is complex and requires great skill. Hiring a good database administrator (DBA), concentrating on important and strategic aspects of databases, training database users, and developing good security procedures are all important.

Database Administration

Databases and database management systems are typically installed and coordinated by an individual or group responsible for managing one of the most valuable resources of any organization: its data. These **database administrators (DBAs)** are skilled and trained computer professionals who direct all activities related to an organization's database, including providing security from intruders.[31] DBAs must work well with both programmers and nonprogramming users of the database. Most database administrators are responsible for the following areas:

- Overall design and coordination of the database
- Development and maintenance of schemas
- Development and maintenance of the data dictionary
- Implementation of the DBMS
- System and user documentation
- User support and training
- Overall operation of the DBMS
- Testing and maintaining the DBMS
- Establishing emergency or failure-recovery procedures

Today, DBAs are fine-tuning databases to increase performance and reduce costs. For example, some organizations are trimming the size of their databases to maintain good performance and reduce costs.[32] According to a project manager at Kennametal, "Our overweight database was months away from crashing due to exceeding our production disk-space capacity." The company was able to save about $700,000 in additional hardware and storage costs by trimming its database. Other performance considerations include the number of concurrent users that can be supported and the amount of memory that is required to execute the database management program. One database was able to process 1,184,893 transactions per minute on average, setting a new world record for speed.[33]

Newer database management systems can help database administrators monitor and control the database. Some DBMSs, like IBM's DB2 for example, have the ability to monitor a database's performance and to make changes or adjustments to increase productivity and efficiency.

TechEdge

WHEN GOOD DATABASES GO BAD

The good news: our Social Security number can identify us. The bad news: our Social Security number can identify us. Recent—and massive—data leaks show the number's increasing vulnerability as a "dangerous master key" to our bank and/or credit accounts for those with grand larceny on their mind. One possible solution is a system that creates an encrypted numeric token to represent a person's identity in a transaction, and then discards it when the transaction is completed.

Social Security number access being eyed
by Brian Bergstein
AP Technology Writer
Seattle Times
http://seattletimes.nwsource.com/APWires/business/D8BL6RO8K.html
July 29, 2005

Government Regulations and Database Costs

Increasing government regulations are forcing many businesses to scrutinize their data, database systems, and storage technologies. U.S. government agencies such as the Internal Revenue Service have been concerned about the accuracy and security of records storage since the invention of the computer. Recently the mounting concerns of varying government agencies have been transformed into laws that are costing businesses big bucks to implement.

In the wake of corporate scandals and in an effort to make investment brokers and dealers accountable for their practices, the Securities and Exchange Commission (SEC) supported passage of the Sarbanes-Oxley Act. Among the act's many provisions is a requirement for brokers and dealers to log and record all electronic communications, such as e-mail, using "write-once, read many," or WORM, technology. WORM technology assures that stored data cannot be tampered with later.

The SEC regulatory requirements compelled Jay Cohen, corporate compliance officer at The Mony Group, to install an EMC Centera storage server system and database application called AXS-One Email and Instant Messaging Management Solution software suite to track e-mails. Cohen explains. "All the external e-mail from the sales force, either incoming or outgoing, goes into the AXS-One Email archival system." The system permanently stores each communication as a record with a unique identifier in a database. Administrators can then run queries on the e-mail data for content surveillance, security, and audit trails.

Another law, the USA Patriot Act, has placed a significant burden on financial service firms. They are now responsible for monitoring new customers to ensure that they are not laundering money for terrorists. Under scrutiny by regulators, such firms are employing data-mining applications that analyze risks and use complex algorithms to identify unusual customer trends within transactions.

More recently, the Food and Drug Administration issued a ruling that requires pharmaceutical companies to apply bar codes to thousands of prescription and over-the-counter drugs dispensed in hospitals. The FDA believes the move will save lives by reducing medical errors but estimated that it would hit the nation's 6000-plus hospitals with a $7 billion bill for bar code readers, databases, and management tools.

Questions

1. Why has the government stepped up efforts to regulate record keeping in financial and pharmaceutical industries?
2. With many pharmaceutical organizations already strapped for cash is it fair for the government to force them to comply with expensive new regulations? Who should bear the financial burden of secure and private record keeping?

Sources

1. *Mannion, P. "Cost of Sarbanes-Oxley Compliance Decried," April 11, 2005, p. 6*
2. *Mearian, L. "Sidebar: Regulations, Volume and Capacity Add Archiving Pressure,"* Computerworld, *February 16, 2004, www.computerworld.com .*
3. *Hoffman, T. "IBM Tailors Bundle for Preserving Corporate Data: Integrated System to Aid in Regulatory Compliance Efforts,"* Computerworld, *February 23, 2004, www.computerworld.com.*
4. *Brewin, B. "FDA Mandates Bar Codes on Drugs Used in Hospitals,"* Computerworld, *February 26, 2004, www.computerworld.com.*

Database Use, Policies, and Security

With the proliferation of low-cost hardware and off-the-shelf database and other software packages, traditional end users are now developing computer systems to solve their own problems. *End-user computing* may be broadly defined as the development and use of application programs and computer systems by non-computer systems professionals. Concerns with end-user computing are generally related to issues of training and control. As you've seen, data contained within an organization's databases is usually critical to the basic functioning of the organization. It is often proprietary in nature, confidential, and of strategic importance. Therefore, the following end-user computing issues must be addressed in terms of database policies and use:

- What data can users read, update, or write in a database?
- Under what circumstances can data be transferred from a personal computer or small computer system to the large mainframe system? (This data transfer is called *uploading*.)
- Under what circumstances can data be transferred from the large mainframe or server system to personal computers or small computer systems? (This type of data transfer is called *downloading*.)
- What procedures are needed to guarantee proper database use and security?

Because there are so many users of any one database, potential data security and invasion of privacy problems have become increasingly important. About 150,000 customers had their personal information stolen by criminals from the ChoicePoint database.[34] ChoicePoint has about 19 billion records on people with data on Social Security numbers, military records, real estate deeds, motor vehicle registrations, addresses, and more. Politicians have called for ChoicePoint to tell impacted people, as well as hearings on how to protect people from this type of database theft in the future.

ACTION PLAN

Remember Mary from the beginning of the chapter? Here are answers to her questions.

1. **How can Mary keep track of the different cheeses, butters, spices, crocks, and other supplies she needs?**

   Mary could use the database or flat file capabilities of a spreadsheet or word-processing program. She could also use a general database system designed for a personal computer or a specialized database system for small businesses to keep track of this information. Because Mary is thinking about starting a small business, a specialized database system would likely be the best option for her.

2. **If she starts a small business, how will she be able to pay her employees, pay her bills, and keep track of other business transactions?**

   Starting a small business requires more than the database capabilities of a spreadsheet or word-processing program. Mary could develop everything she needs for a small business using a general database program, like Microsoft Access or FileMaker Pro, but she should also consider a specialized database system for small businesses, such as QuickBooks.

Summary

LEARNING OBJECTIVE 1

Understand basic data management concepts.

Data is one of the most valuable resources an organization possesses. Data is organized into a hierarchy that builds from the smallest element to the largest.

The smallest element is the byte, which represents a character. A group of characters, such as a name or number, is called a field. A collection of related fields is a record; a collection of related records is called a file. The database,
at the top of the hierarchy, is an integrated collection of records and files.

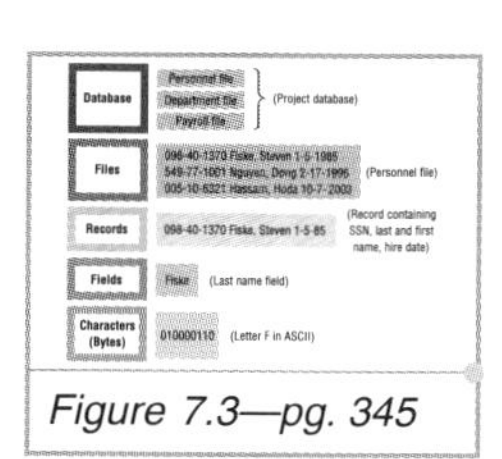

Figure 7.3—pg. 345

An entity is a generalized class of objects for which data is collected, stored, and maintained. An attribute is a characteristic of an entity. Specific values of attributes, called data items, can be found in the fields of the record describing an entity. A primary key uniquely identifies a record.

The database approach has a number of benefits, including reduced data redundancy, improved data consistency and integrity, easier modification and updating, data and program independence, standardization of data access, and more efficient program development. A DBMS consists of a group of programs that manipulate the database and provide an interface between the database and the user of the database and application programs.

LEARNING OBJECTIVE 2

Describe database models and characteristics.

Database designers can use a data model to show the relationships among data. One of the most flexible database models is the relational model. Data is set up in two-dimensional tables. Tables can be linked by common data elements, which are used to access data when the database is queried. Each row represents a record. Columns of the tables are called attributes, and allowable values for these attributes are called the domain. Basic data manipulations include selecting and joining.

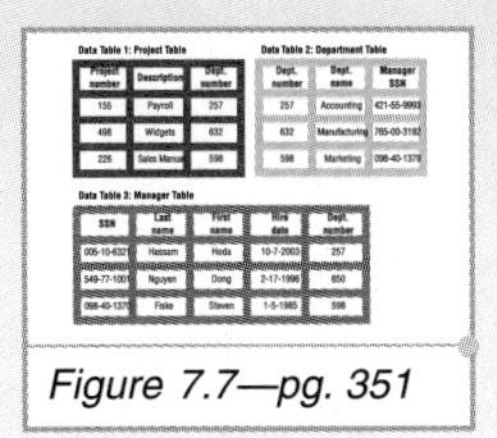

Figure 7.7—pg. 351

The object-oriented model stores data as objects, which contain both the data and the processing instructions needed to complete the database transaction. The objects can be retrieved and related by an object-oriented database management system (OODBMS). Object-oriented databases offer the capability to reuse and modify existing objects to develop new database applications.

Data analysis is used to uncover problems with the content of the database. Problems and irregularities in data are called anomalies. The process of correcting anomalies is called normalization. Normalization involves breaking one file into two or more tables in order to reduce redundancy and inconsistency in the data.

LEARNING OBJECTIVE 3

Discuss the different types of database management systems and their design and use by individuals and organizations.

A DBMS is a group of programs used as an interface between a database and application programs. Database types include flat file, single-user, multiuser, and special-purpose databases. Schemas are used to describe the entire database, its record types, and their relationships to the DBMS.

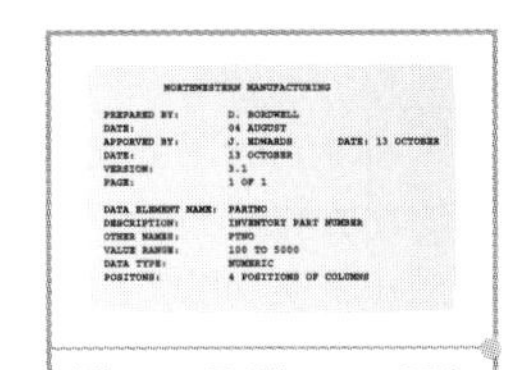

Figure 7.21—pg. 366

Schemas are entered into the computer via a data definition language (DDL), which describes the data and relationships in a specific database. Another tool used in database management is the data dictionary, which contains detailed descriptions of all data in the database.

Once a DBMS has been installed, the database may be accessed, modified, and queried via a data manipulation language (DML). A more specialized DML is the query language; the most common are Query by Example (QBE) and Structured Query Language (SQL). QBE and SQL are used in several popular database packages.

LEARNING OBJECTIVE 4

Describe how organizations use database systems to perform routine processing, provide information and decision support, and how they use data warehouses, marts, and mining.

Most organizations use a database system to send out bills, pay suppliers, print paychecks, and perform other routine transaction processing activities.

Perhaps the biggest potential of a database system is to provide information and decision support. The data contained in a database can be filtered and manipulated to provide critical information to a wide range of organizations.

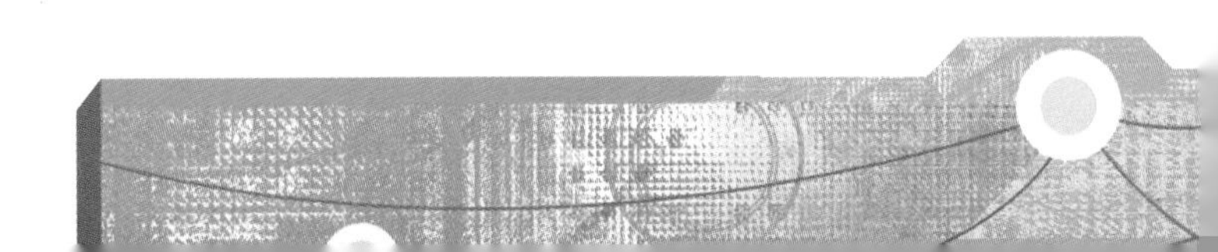

Information is usually obtained and decision support is usually provided using data warehouses, data marts, and data mining. A data warehouse is a database that holds important information from a variety of sources. A data warehouse normally contains a subset of the data stored in the database system. A data mart is a small data warehouse. Data mining is the process of extracting information from data warehouses and data marts.

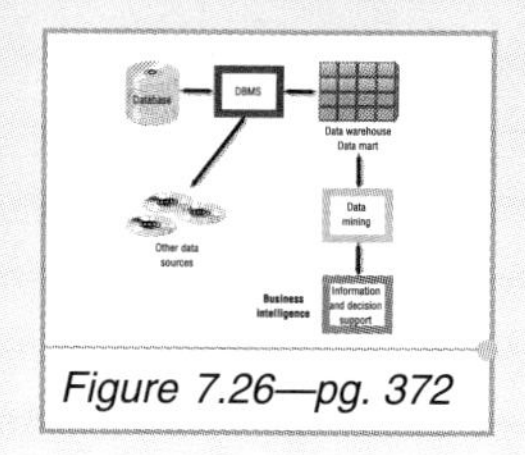

Figure 7.26—pg. 372

LEARNING OBJECTIVE 5

Discuss additional database systems, including distributed systems and Web-based systems.

A distributed database allows data to be spread across several databases at different locations.

Database systems are often used in conjunction with the Internet and networks. In many cases, the Internet or network is used as the front end, where requests for information and data are made. The database management system is the back end, providing the needed information and data.

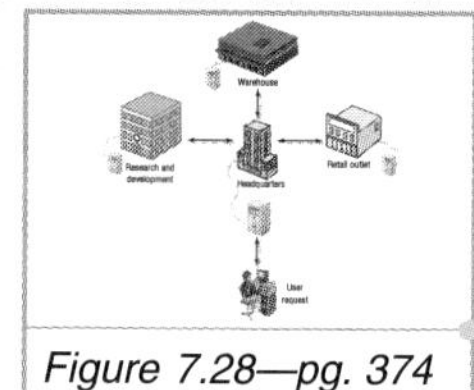

Figure 7.28—pg. 374

An increasing amount of the data used by organizations is in the form of visual images, which can be stored in image databases. Audio databases are used to store audio data, including voice and music. Some organizations are using virtual databases that can integrate separate databases into a unified system that acts like a single database.

LEARNING OBJECTIVE 6

Describe the role of the database administrator (DBA) and database policies and security practices.

Management of the database is part of database administration. Database administrators (DBAs) are responsible for database use, policies, and security. They help control DBMS design, implementation, and maintenance. They also establish security and control measures, monitor and tune the database, and perform many other aspects of database use and control.

Test Yourself

LEARNING OBJECTIVE 1: Understand basic data management concepts.

1. A(n) ________________ consists of a group of programs that perform the actual manipulation of the database.
2. In the hierarchy of data, characters that are put together form a:
   a. file
   b. field
   c. record
   d. database
3. True or False: Employee and inventory files are examples of transaction files.
4. True or False: A primary key uniquely identifies a record.

LEARNING OBJECTIVE 2: Describe database models and characteristics.

5. A(n) ________________ database contains both the data and the processing instructions for that data.

6. What type of database model places data in two-dimensional tables?
   a. hierarchical
   b. network
   c. entity
   d. relational

7. A(n) ______________ uniquely identifies a database record.

8. ______________ is a process that involves evaluating data to uncover problems with the content of a database.

LEARNING OBJECTIVE 3: Discuss the different types of database management systems and their design and use by individuals and organizations.

9. True or False: A database that can be used to store and manipulate a single table in a spreadsheet or word-processing program is called a single-user database.

10. A situation where inaccurate input causes inaccurate output in a database is often called ______________.

11. True or False: Oracle is an example of a general-purpose database.

12. What consists of commands used to change a database?
    a. the data definition language
    b. the physical access language
    c. the data manipulation language
    d. the logical access language

LEARNING OBJECTIVE 4: Describe how organizations use database systems to perform routine processing, provide information and decision support, and how they use data warehouses, marts, and mining.

13. A(n) ______________ is typically a subset of an organization's database.

14. True or False: A data mart is a small data warehouse.

15. What is the process of extracting information from a data warehouse or data mart?
    a. logical access
    b. physical access
    c. decision support
    d. data mining

LEARNING OBJECTIVE 5: Discuss additional database systems, including distributed systems and Web-based systems.

16. With a(n) ______________ the actual data may be spread across several databases at different locations.

17. With many databases, the Internet is used as a ______________ to receive requests made to the database system.

18. An ______________ contains data that is difficult to place in a traditional database system, such as drawings, fingerprints, and sound recordings.

LEARNING OBJECTIVE 6: Describe the role of the database administrator (DBA) and database policies and security practices.

19. True or False: Database systems are typically installed and coordinated by the chief information officer.

20. The ______________ must work well with both programmers and nonprogramming users of the database.

Test Yourself Solutions **1.** database management system, **2.** b. field, **3.** False, **4.** True, **5.** object-oriented database, **6.** d. relational, **7.** primary key, **8.** Data analysis, **9.** False, **10.** garbage in, garbage out (GIGO), **11.** True, **12.** c. the data manipulation language, **13.** data warehouse, **14.** True, **15.** d. data mining, **16.** distributed database, **17.** front end, **18.** unstructured database, **19.** False, **20.** database administrator (DBA)

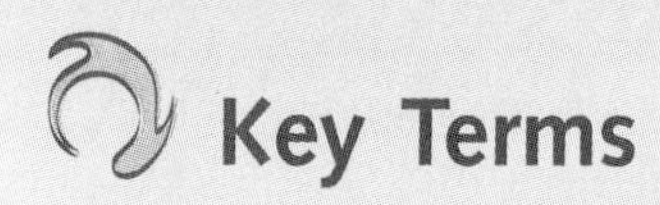

Key Terms

data analysis, p. 352
database administrator (DBA), p. 377
database management system (DBMS) , p. 342
data dictionary , p. 365
data integrity , p. 364
data mining, p. 371
data warehouse, p. 371
distributed database , p. 373
field, p. 344
file, p. 345
garbage in, garbage out (GIGO), p. 364
object-oriented database, p. 354
primary key , p. 346
record, p. 345
relational model, p. 351
schema, p. 365
Structured Query Language (SQL) , p. 368

Questions

Review Questions

1. What is a database management system?
2. What is a field? What is a record?
3. What are entities and attributes? What is a key?
4. What is a primary key?
5. Describe simple approaches to data management.
6. What are the advantages of the database approach?
7. Describe the relational model.
8. Describe the characteristics of a relational database model.
9. What is an object-oriented database system?
10. What are the important characteristics of databases?
11. What is a flat file? What is a single-user database?
12. What is the purpose of a data definition language (DDL)? A data dictionary?
13. What is query by example? What is SQL?
14. Describe at least three widely used open-source database systems.
15. How are databases used in organizations?
16. What is a data warehouse? What is a data mart?
17. How is data mining used by organizations?
18. What is a distributed database system?
19. How are the Internet and networks used with database systems?
20. List and describe the newer types of database systems. What types of data might they house?
21. Explain the responsibilities of a database administrator.

Discussion Questions

22. Why is a database a necessary component of a computer system? Why is the selection of DBMS software so important to organizations?
23. In what way do database systems apply to your personal life?
24. What databases on your campus contain your name? Off campus? What is your primary key on campus?
25. What is a data model?
26. How could you use a database management system in a career of your choice?

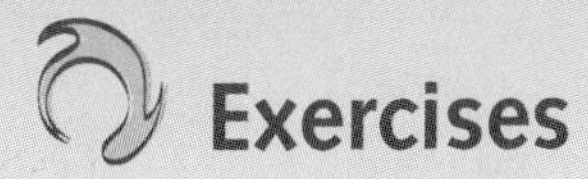

Exercises

Try It Yourself

1. You are a database administrator in the Information Systems Department of a large university. Your organization is currently using a relational database system. Use your word-processing program to write a letter to your supervisor(s) outlining the need to migrate toward object-oriented database management systems (OODBMSs). Discuss the advantages of OODBMSs over relational database systems, and discuss the disadvantages of migrating to an OODBMS.
2. Using the database of your choice, create a database of the top 10 jobs you would like to have. Your database should include the company or organization, location of the company or organization, possible salary, location, and similar attributes. Sort your database according to salary.
3. A university needs to design and implement a relational database to maintain records for students and courses. Some of the essential data fields are student identification number, student name, major, credits completed, and student GPA (grade point average). Using the chapter material on designing a database, show the logical structure of the relational table or tables for this proposed database. In your design, include any additional fields that you feel are necessary for this database, and show the primary key for every table. Fill in the database tables with sample data for demonstration purposes (10 records). Once your design is complete, insert the data into a database, such as Microsoft Access.
4. A video movie rental store is using a relational database to store the following information on movie rentals to answer customer questions: movie ID number, movie title, year made, movie type (comedy, drama, horror, science fiction, or western), rating (G, PG, PG-13, R, or X), and quantity on hand. Develop a database to store this information for 30 movies. Using the database, create another database with only movies rated PG-13. Develop a report of all horror movies.
5. Based upon the database design from the above exercise, design a data-entry screen that could be used to enter information into this database. Also include some examples of typical queries the salespeople would use to respond to customers' requests.

Virtual Classroom Activities

For the following exercises, do not use face-to-face or telephone communications with your group members. Use only Internet communications.

6. With distance learning, students from around the world can take the same course. It is also possible to have several different instructors located around the world. With your group, develop a brief report that describes how several instructors at different locations could integrate their databases of assignments, tests, and student grades.
7. On the Internet, research one database system that you could use, such as Access or Quicken. Write a paper describing the features of each database system. Using a spreadsheet program, summarize the costs of each database system. All group members should participate in developing the paper and spreadsheet without any face-to-face meetings.
8. With your team, develop a report that compares two traditional databases, like Oracle, with two open-source databases. Include descriptions, advantages, and disadvantages of each database.

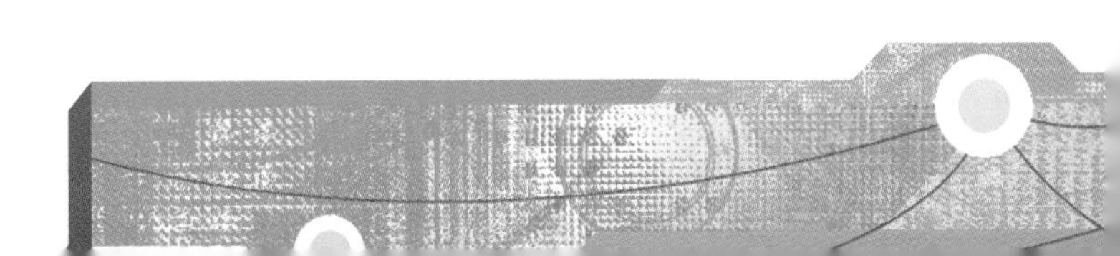

Teamwork

9. In a team of three or four classmates, interview three users, programmers, or system analysts from different organizations that use DBMSs. Determine what DBMSs they are using. How did these people choose the DBMSs that they are using? What do they like about their DBMSs and what could be improved? Considering the information obtained from these people, select one DBMS and briefly present your selection and the rationale for your selection to the class.
10. Using the Internet or your school library, research object-oriented databases. Use a database system to summarize your findings. At a minimum, your database should have columns on the database name, the cost to purchase or lease, and a brief description.

TechTV

Video: Predicting Huge Surf

Go to www.course.com/swt2/ch07 and click TechTV. Click the "Predicting Huge Surf" link to view the TechTV video, and then answer the following questions.

1. The Mavericks Surf Contest has been hailed by *Sports Illustrated* as "the Super Bowl of big wave surfing." What role do databases play in Jeff Clark's ability to predict the best day and time for the contest, when the surf will be at its peak?
2. Clark uses data provided on the Web from weather services and ocean buoys, and then "processes" the data into useful information that guides his decisions. What human elements does he say are used in the processing, and how reliable is the end result?

Endnotes

[1] Alsever, J. "Restaurants Keep Tabs on Diners," *Rocky Mountain News,* February 22, 2004, p. K1.

[2] Benjamin, M. "UPS Eyes a Future Going Far Beyond Package Delivery," *U.S. News & World Report,* January 26, 2004, p. ee2.

[3] Fisher, D. "New DHS Border Plan Scrutinized," *eWeek,* January 12, 2004, p. 1.

[4] Rundle, R. "Government Puts Data Comparing Hospitals Onto Public Web Site," *The Wall Street Journal,* April 1, 2005, p. B1.

[5] Duffy, S. "Taking DNA From Felon Upheld; Court Cites Need for Data," *Delaware Law Weekly,* March 22, 2005, p. D3.

[6] McAmis, D. "Introducing InfoPath," *Intelligent Enterprise,* February 7, 2004, p. 36.

[7] Savvas, A. "Oracle Database Lite Offers Mobile Access," *Computer Weekly,* January 18, 2005, p. 14.

[8] Fonseca, B. "ATC Database Upgrade Supports Growth, Improves Reliability," *eWeek,* January 26, 2004, p. 33.

[9] Staff. "Oracle Tops Relational Database Market," *ComputerWire,* March 16, 2005.

[10] Staff. "Israeli Holocaust Database Online," *Library Journal,* January 15, 2005, p. 38.

[11] Staff. "Hazmat Database Released," *Factory Equipment News,* January 2005.

[12] Whiting, R. "Stock Trades Get a Boost," *InformationWeek,* January 12, 2004, p. 47.

[13] Staff. "Art and Antiques Organizer Deluxe," *PC Magazine,* March 22, 2005, p. 12.

[14] Zaino, J. "Analysis That's Easy on the Eyes," *InformationWeek,* March 21, 2005, p. 52.

[15] Perez, J. "Open-Source DBs Go Big Time," *Intelligent Enterprise,"* February 7, 2004, p. 8.

[16] Whiting, R. "Open-Source Database Gaining," *InformationWeek,* January 12, 2004, p. 10.

[17] Fonseca, B. "Database Opening Up," *eWeek,* January 12, 2004, p. 12.

[18] Novellino, T., "Polygon Launches Colored Gemstone Database," *Business Media,* February 25, 2005.

[19] Staff. "Company Offers High Tech Way to Get Clothes to Fit," *CIO Insight,* March 18, 2005.

[20] Vijauyuan, J. "New Law Prods Food Makers to Focus on Data Management," *Computerworld,* January 24, 2005, p. 16.

[21] Staff. "The State of Business Intelligence," *Computer Weekly,* February 3, 2004, p. 24.

[22] Sullivan, L. "Oracle Embraces RFID," *Information Week,* February 2, 2004, p. 8.

[23] Mitchell, K. "Databases Can't Handle RFID," *Computerworld,* February 7, 2005, p. 8.

[24] Staff. "Hartford Life's Condor System Mines Databases for the Gold," *Insurance Networking News,* April 4, 2005, p. 29.

[25] Gray, J. & Compton, M. "Long Anticipated, the Arrival of Radically Restructured Database Architectures Is Now Finally at Hand," *CM Queue,* 2005, 3 (3).

[26] Thibodeau, P. "Novel, Ret Hat Eye Virtualization for Linux," *Computerworld,* January 10, 2005, p. 14.

[27] Kranhold, K. "High-Tech Tool Planned for Physicians," *The Wall Street Journal,* February 17, 2005, p. D3.

[28] Havenstein, H. "Push for Web-based Health Records Launched," *Computerworld,* February 7, 2005, p. 5.

[29] Saran, Cliff, "Code Issue Affects 40% of Websites," *Computer Weekly,* January 13, 2004, p. 5.

[30] Poor, Alfred, "Three Ways to Get (and stay) Organized," *PC Magazine,* March 8, 2005, p. 50.

[31] Fonseca, Brian, "DBA Boundaries Blurring," *eWeek,* January 26, 2004, p. 995.

[32] Robb, Drew, "The Database Diet," *Computerworld,* March 8, 2004, p. 32.

[33] Staff, "Oracle and HP Set World Record," *VAR-Busines,* January 26, 2004, p. 64.

[34] Weber, Harry, "Breach of data affects 145,000," *Rocky Mountain News,* February 22, 2005, p. 2B.

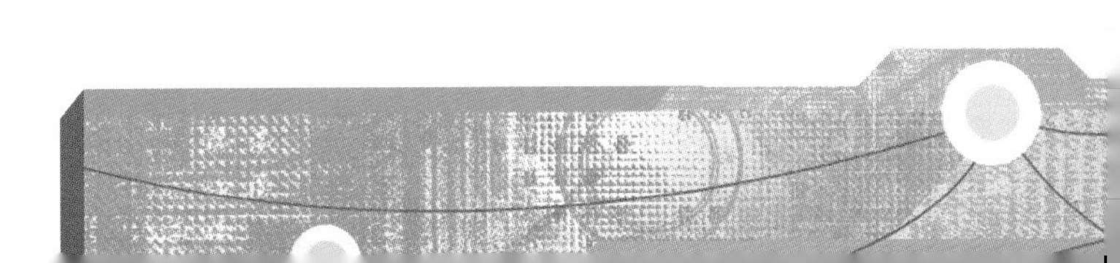

E-COMMERCE

The Forrero name is well known as a maker of fine handcrafted leather products. You can find Forrero products in many gift shops around the country. Alejandro Forrero is the youngest of many generations of leather artisans and is currently completing his college education, which he plans on using to the benefit of the family business. The computer course that Alejandro is taking has got him thinking about new possibilities for the family business. Currently, Forrero products are marketed at wholesale prices only to retail clothing and gift shops, which then sell them to consumers at twice the wholesale cost. Alejandro is considering the possibility of selling Forrero leather products direct to consumers on the Web. A professionally designed Web site would provide great publicity for the company, the family would be able to make more money per sale, and they would be able to offer better prices to their customers. Selling direct to consumers through the Web could dramatically change the family business! He is excited to share the idea with his parents, but wants to do it right and make them proud. He needs to do some research and write a proposal that provides details on the costs and benefits of taking the family business online.

As you read through the chapter, consider the following questions:

1. What will it take to put the Forrero family business online?
2. What benefits might the Forrero family enjoy from an online presence? Do these benefits outweigh the costs?
3. Besides taking the business online, what other ways might e-commerce assist the Forrero family business?

Check out Alejandro's *Action Plan* at the conclusion of this chapter.

8

LEARNING OBJECTIVES

1. Define e-commerce, and understand its role as a transaction processing system.
2. List the three types of e-commerce, and explain how e-commerce supports the stages of the buying process and methods of marketing and selling.
3. Discuss several examples of e-commerce applications and services.
4. Define m-commerce, and describe several m-commerce services.
5. List the components of an e-commerce system, and explain how they function together to provide e-commerce services.

CHAPTER CONTENT

The Roots of E-Commerce

Overview of Electronic Commerce

E-Commerce Applications

Mobile Commerce

E-Commerce Implementation

Introduction

E-commerce has provided a fresh platform for business that has changed the way businesses and consumers think about buying and selling. Increasingly, buyers and sellers are turning to their computers to buy and sell products and are enjoying the benefits. Conducting business online offers convenience and savings to both buyers and sellers. This chapter explores the impact of e-commerce and m-commerce on consumers and businesses, what it takes to set up a successful e-commerce Web site, and the challenges and issues faced by e-commerce participants.

Electronic commerce, or **e-commerce**, refers to systems that support electronically executed business transactions. E-commerce has changed the world and continues to change the world in very dramatic ways. Philadelphia University Human Geography professor Steven Dinero and his research team have been studying the people in the remote Alaskan viilages of Arctic Village and Nulato as they transition from a nomadic, hunter-gatherer way of life to a settled existence.[1] So impressed was he by the villager's artistic skills, that Dr. Dinero secured a $600,000 grant from the National Science Foundation to connect the craftswomen of Arctic Village with the global marketplace. ArcticWays.com is an e-commerce site where shoppers can acquire rarely seen, handmade items from these small villages in bush Alaska (see Figure 8.1). While the team introduces Alaskan grandmothers to e-commerce, it is also teaching their grandchildren computer and Internet technology skills. The NSF grant included funds to run computer camps focusing on e-commerce and Web design for local high school students. Consider the impact that e-commerce has had on this village as well as the impact this unique culture may now have on the rest of the world.

FIGURE 8.1 • ArcticWays e-commerce

The ArcticWays Web site supports the sale of native crafts from villagers in bush Alaska.

The 1990s was a decade of amazing growth for the Internet. During this decade, due largely to a decision by Congress, the Internet expanded from primarily supporting research for scholars to providing a platform on which to conduct business. The release of the first Web browser in 1991 made the Internet more accessible to the general public. The Internet boom really began in 1994 with the birth of Amazon.com and the first online shopping malls, radio stations, cyberbanks, and a host of other commercial ventures. About that time, the term *e-commerce* became a popular buzzword. E-commerce is often used in reference to Web-based shopping, but in reality it encompasses much more than that. Table 8.1 lists some e-commerce milestones.

TABLE 8.1 • E-commerce milestones[2]

1991
- Commercial Internet Exchange (CIX) Association established to support business on the Internet
- First commercial Web browser, Mosaic, made available

1994
- First Web-based e-commerce transaction—Netmarket sells a copy of Sting's "Ten Summoner's Tales" for $12.48
- Pizza Hut launches Pizza Net in Santa Cruz, CA, enabling customers to order pizza deliveries online
- Amazon.com opens for business as an online book store

1995
- Microsoft adds secure connection capability (SSL) to Internet Explorer
- eBay conducts its first online auction on Labor day

1996
- Microsoft and IBM release merchant system software that supports Internet transactions

1998
- Cable modems become affordable for residential broadband access
- AOL generates $1.2 billion for retail partners

1999
- U.S. Dept. of Commerce begins tracking e-commerce sales ($5.3 billion in the 4th quarter)

2003
- Amazon.com ships more than one million copies of *Harry Potter and the Order of the Phoenix*—the largest distribution of a single item in e-commerce history
- Apple launches iTunes, the first attempt to bring digital music to e-commerce

2004
- 57 million U.S. taxpayers file returns on the Web

2005
- Mobile commerce takes off with $22.2 million in global revenues earned through mobile devices

Since the 1990s, e-commerce has done nothing but gain momentum. It has been projected that e-commerce will continue its climb at an annual growth rate of 15 percent with increasing numbers of consumers turning to the Web rather than the mall. In 2004 e-commerce accounted for 7 percent ($136.6 billion) of total U.S. retail sales; by 2010 it is expected that e-commerce will account for 13 percent ($331 billion) of total U.S. retail sales. Figure 8.2 illustrates the forecasted growth of e-commerce in relation to the forecasted growth in the economy.

FIGURE 8.2 • U.S. e-commerce projections

In 2004 e-commerce accounted for 7 percent ($136.6 billion) of total U.S. retail sales; by 2010, it is expected that e-commerce will account for 13 percent ($331 billion).

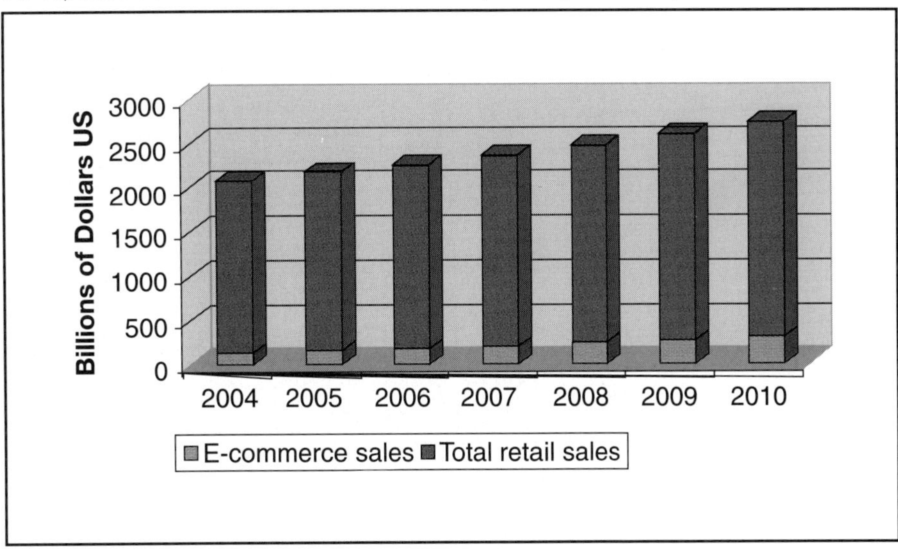

The United States leads the world in e-commerce, with over twice as much sold over the Web as all other countries combined.[3] E-commerce plays an important role in Western Europe as well as in Asia. Considerable growth in e-commerce is expected in Asia and Latin America over the next several years. E-commerce is contributing significantly to the growth of a global economy by leveling the international economic "playing field."

THE ROOTS OF E-COMMERCE

Even though the Internet and the Web were responsible for the boom in electronic commerce, note that the words *Internet* and *Web* are not part of the definition of e-commerce. Although most consumers with Internet access purchase products over the Internet,[4] these purchases are only a small percentage of the electronic business transactions that take place. Many electronically executed transactions take place off the Internet, on private networks. In fact, the activity of electronically executing business transactions predates the Internet.

E-Commerce History

In the 1960s, banks began to use computers and magnetic ink recognition to automate check processing, a step that significantly reduced staffing needs and increased efficiency and accuracy. Soon a variety of industries began to use computers to keep accounting ledgers, administer payroll, create management reports, and schedule production. Computer-based information systems became an accepted part of streamlining business processes.

In the 1970s and 1980s, businesses extended their computer-based information systems beyond their corporate walls to connect with other companies' systems using electronic data interchange (EDI). EDI uses private communications networks called value-added networks (VANs) to transmit standardized transaction data between business partners and suppliers (see Figure 8.3). Automating transactions using EDI drastically reduced the amount of paperwork and the need for human intervention. This was the true beginning of e-commerce, even if it would take another 20 years for the term to be coined.

FIGURE 8.3 • Electronic data interchange (EDI)

EDI uses private communications networks called value-added networks (VANs) to transmit standardized transaction data between business partners and suppliers.

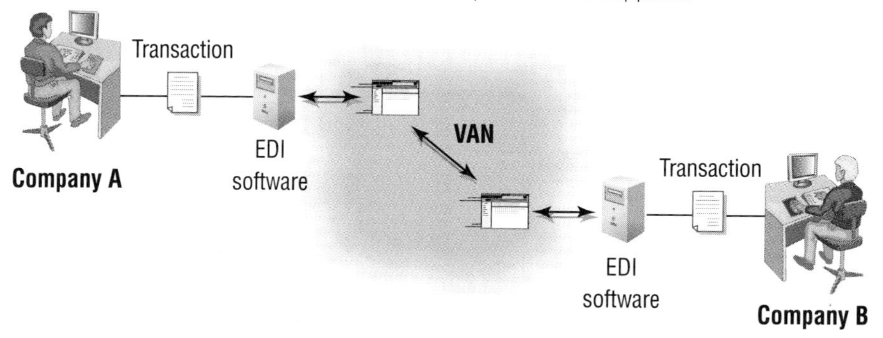

Even though it improved business-to-business transactions, EDI had some problems and was costly. Only businesses that paid for a VAN connection could participate. For e-commerce to really take off, businesses and people needed an inexpensive universal network to which everyone could connect. The Internet provided the ideal platform for conducting EDI transactions as well as other forms of transaction processing between businesses. The invention of the Web provided the first opportunity for businesses to conduct transactions with consumers over a computer network. Businesses and consumers alike embraced doing business online. The far-reaching implications of the Web as a tool for executing business transactions soon became clear.

To fully understand the benefits of e-commerce, you need to examine its fundamental purpose: to execute transactions.

Transaction Processing

A *transaction* is an exchange involving goods or services such as buying medical supplies for a hospital or purchasing and downloading music on the Internet. E-commerce is a form of a transaction processing system. A *transaction processing system (TPS)* is an information system used to support and record transactions such as paying for products, or paying an employee. The transaction information collected by the TPS is fundamental to the operation of other information systems that support important decision making. For example, the company contracted to provide food services for your campus uses a TPS to collect sales information from the cashier's point-of-sale terminal in the school cafeteria (see Figure 8.4). That sales information is then processed by another information system, such as a management information system (MIS), to determine which food items are selling best. Items that don't sell may be discontinued and replaced with new items. Through this approach, the food service provider can use the transaction information to continuously improve its service to customers.

FIGURE 8.4 • The value of transaction processing

The transaction data collected through point-of-sale terminals can be used to assess which products are selling well and which are not.

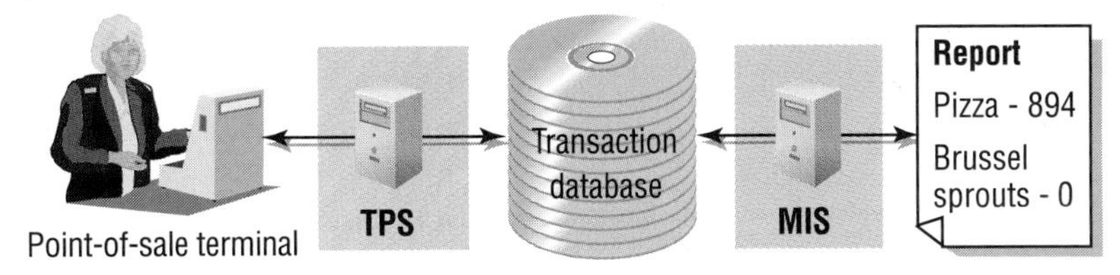

Transaction processing takes place in many different environments, and the systems that support them must be created to suit the environment. The electronic checkout system at Amazon.com is a TPS, as is the checkout system at your local bookstore. The payroll system that calculates an employee's pay and cuts a check is also a TPS. The ATM at your bank is a TPS, as is the keypad on a gas station filling pump. Transaction processing includes capturing input data, making calculations, storing information in a database, and producing various forms of output such as receipts and purchase orders.

There are two methods of processing transactions: batch and online. In *batch processing*, transactions are collected over time and processed together in batches. Batch processing is useful in situations where transactions take place away from the computer system, or when processing would slow down the collection of transaction data. For example, a sales representative operating from a booth at a trade show might record orders on a laptop computer and enter the orders into the main system in a batch upon returning to the home office and connecting to the corporate network.

With *online transaction processing* the processing takes place at the point of sale. For example, as you pay for your concert ticket, the seat you choose is marked as reserved in the concert hall database and your payment is recorded in

the day's earnings. Online transaction processing is critical to time-sensitive transactions such as selling concert tickets and making flight reservations, as well as for college class registration systems. If the transaction isn't processed immediately, seats could become double-booked or classes filled beyond capacity.

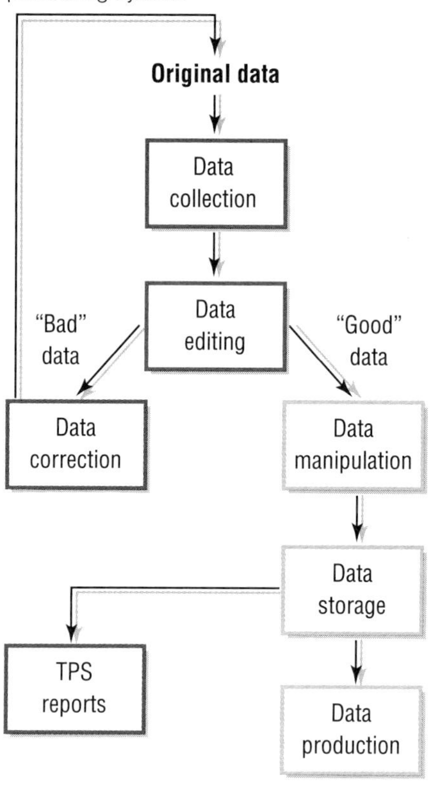

FIGURE 8.5 • The transaction processing cycle

Data processing activities of a transaction processing system.

The Transaction Processing Cycle

E-commerce and all other forms of transaction processing systems share a common set of activities called the *transaction processing cycle*. The stages of the transaction processing cycle are shown in Figure 8.5. They include the following:

- Data collection: This first step is the process of capturing transaction-related data. For example, an order processing system would need to know the item number, quantity, and payment method.
- Data editing: This second step is the process of checking the validity of the data entered. For example, an invalid credit card number would be caught in the data-editing stage of the transaction processing cycle.
- Data correction: The third step is implemented if an error is found in the entered data. Typically, a descriptive error message is displayed along with a request for a correction: "The credit card number you entered is invalid; please check the number and try again."
- Data manipulation: This step involves processing the transaction data. A TPS is typically only required to do simple processing such as adding up the value of the items being purchased, calculating taxes, and determining if the items requested are in inventory, along with performing any other calculations necessary to allow the transaction to proceed.
- Data storage: In this step, databases are altered to reflect the transaction.
- Data output: Data output comes in the form of receipts, picking lists for the warehouse, and other documents.

To conclude the transaction, documents may be produced or displayed. For example, after making an online purchase, verification is displayed on the screen, an electronic receipt is sent to your e-mail address, and a document called a picking list is produced in the warehouse that tells the workers what to pack and where to ship it.

Different Transaction Processing for Different Needs

There are a variety of transaction processing systems and subsystems that serve many functions within an organization. The two primary categories of TPS are order processing systems and purchasing systems. An *order processing system* supports the sales of goods or services to customers and arranges for shipment of products. A *purchasing system* supports the purchase of goods and raw materials from suppliers for the manufacturing of products. Each of these systems is composed of subsystems that interact to address the needs of an organization. Figure 8.6 illustrates how the TPSs interact within an organization to address the needs of the organization. Notice how the inventory control system acts to connect the order processing system and purchasing system.

FIGURE 8.6 • Transaction processing system interaction

Transaction processing typically makes use of many interconnected systems and subsystems.

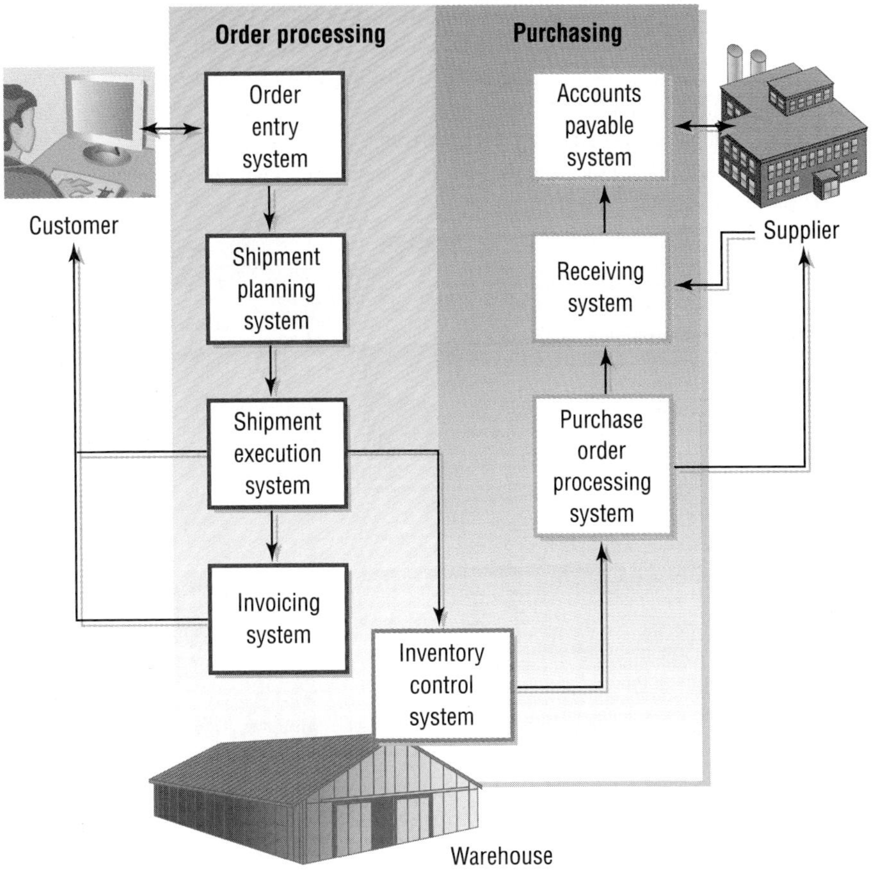

In each step of the process and in each subsystem, vendors strive to carry out the action in a streamlined manner with the least amount of effort and cost and the highest amount of speed and quality. In this way, a business can gain an edge over the competition.

OVERVIEW OF ELECTRONIC COMMERCE

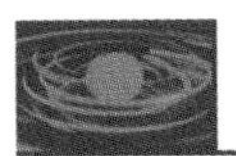

EXPAND YOUR KNOWLEDGE

To learn more about e-commerce transactions, e-commerce security, and other aspects of e-commerce, go to www.course.com/swt2/ch08. Click the link "Expand Your Knowledge," and then complete the lab entitled, "Electronic Commerce."

Just as there are different types of transaction processing systems, there are also different types of e-commerce: business-to-consumer, business-to-business, and consumer-to-consumer. In this section, you will look at e-commerce from the buyer's and seller's perspective, including the benefits and challenges of effectively conducting e-commerce.

Types of E-Commerce

When most of us think of e-commerce, companies like Amazon.com come to mind. Founded on a great idea—selling books online—and a deep understanding of both technology and business, Amazon.com has succeeded where thousands of others have failed. With annual net sales that surpass traditional retailers such as Sears, Amazon.com has expanded into dozens of retail arenas and is proof that success can be found in doing business on the Web.

Amazon.com is an example of business-to-consumer e-commerce. **Business-to-consumer e-commerce**, or **B2C**, makes use of the Web to connect individual consumers directly with sellers to purchase products (see Figure 8.7). B2C e-commerce is sometimes called *e-tailing*, a takeoff on the term *retailing*, as it is the electronic equivalent of a *brick-and-mortar* retail store.

FIGURE 8.7 • Business-to-consumer e-commerce

Peapod is a B2C service that allows customers in select cities to do their grocery shopping online and have their groceries on their doorstep within hours.

Although B2C is the most visible form of e-commerce, it does not generate as much transaction traffic as business-to-business e-commerce. **Business-to-business e-commerce**, or **B2B**, supports transactions between businesses across private networks, the Internet, and the Web. Because businesses conduct frequent and high-volume transactions, B2B e-commerce is especially valuable. Figure 8.8 illustrates the dramatic difference between B2C and B2B sales between 1999 and 2003. With most businesses scrambling to automate their transactions through e-commerce, B2B e-commerce sales have continued to increase at dramatic rates. Businesses make use of B2B e-commerce to purchase:

- Raw materials for production of products
- Tools, parts, and machinery for the production line
- Office furnishings, equipment, and products
- Transportation and shipping services

Many B2B transactions take place over EDI networks. As mentioned earlier, there is a growing trend for EDI transactions to take place over the Internet, rather than over private networks. Wal-Mart has requested that its more than 10,000 suppliers implement Internet-based EDI—changing from the previously used private network. The suppliers were given one year to comply. EDI allows Wal-Mart to place product orders directly to supplier's information systems without human intervention. In some cases, these orders can be automatically placed by software that recognizes when inventory becomes low. The convenience of automating B2B transactions saves buyers and sellers significant time and money.

The third form of e-commerce is epitomized by the popular trend of consumers selling their own belongings on eBay. **Consumer-to-consumer e-commerce**, or **C2C**, uses the Web to connect individuals who wish to sell their personal belongings with people shopping for used items. Although eBay supports all forms of e-commerce, many credit it with the increasing popularity of C2C e-commerce.

FIGURE 8.8 • B2C vs. B2B

Historically, B2B has far outpaced B2C in sales.

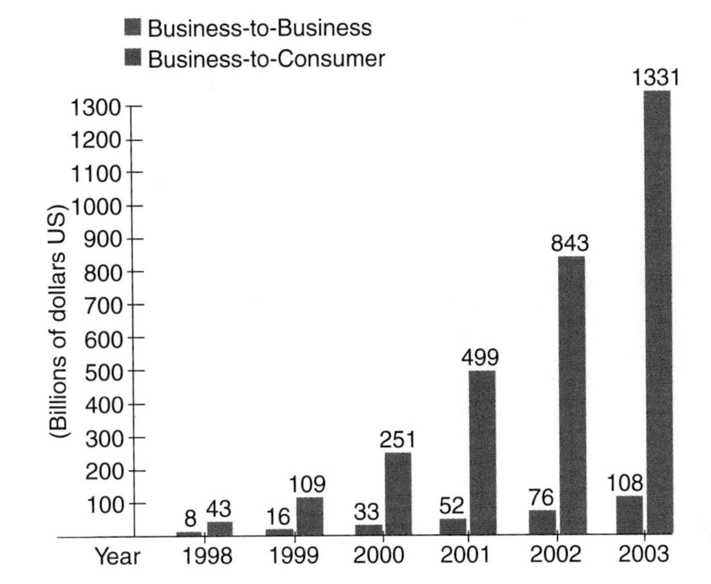

E-Commerce from the Buyer's Perspective

The process of buying or acquiring goods or services takes place in six distinct stages (see Figure 8.9). E-commerce can assist buyers with each of these stages of buying. Consider the following scenario.

1. Realizing a need: You spy an ad on the Web for a satellite radio system that offers over 100 channels of music, sports, and information. That is something you simply must have.
2. Researching a product: Running a Web search on *satellite radio*, you learn that acquiring satellite radio requires the purchase of a receiver, antenna, and a monthly service subscription. You read several user reviews and consult a friend who has just purchased a satellite radio system.
3. Selecting a vendor: Upon further Web research you discover that because this is new technology, there are not a lot of choices for service and hardware. Between the two digital radio broadcasting networks, you choose the service that is least expensive. Out of several receivers, you select an all-inclusive package that includes a portable receiver that can be connected to your car stereo system and home stereo system. With the manufacturer's name and part number in hand you search the Web for the vendor offering the best price (including shipping), best reputation, and fastest and cheapest delivery. Fortunately, the vendor offering the lowest price also has a good reputation and is highly recommended by the 574 customers who have ranked its service.
4. Providing payment: You proceed to the vendor's Web site and, finding that there are units in stock that can be shipped immediately; you place the item in your electronic shopping cart and proceed to check out. You provide the vendor with your shipping information and credit card number using the secure electronic checkout form, and the transaction is approved and completed.

5. Accepting delivery: Three days later a package arrives at your doorstep. After installation and setup, you're enjoying CD-quality music piped to your automobile from a satellite while cruising down Main Street.
6. Using product support: Should your new satellite radio receiver break down while under warranty, or if you have questions about setting it up, you can visit the manufacturer's Web site or phone the company to gather information or arrange an exchange.

FIGURE 8.9 • The six stages of buying goods

E-commerce can assist consumers with each of the six stages of the buying process.

Consider the benefits that e-commerce provided you over traditional forms of shopping. Without the Web, you might or might not have heard about satellite radio. Without the Web, you would have had to rely on trade magazines or store salespeople as resources for information about the product. Perhaps the biggest advantage offered by the Web is the ability to comparison shop (see Figure 8.10). Without the Web, you would have been limited to local merchants, who may or may not take advantage of monopoly power with unreasonable price markups.

On the other hand, shopping locally provides the advantage of taking possession of the product at the point of purchase. Local retailers may also offer benefits that are difficult to duplicate over the Web such as demonstrations, installation, and the ability to exchange a faulty product for a new one without shipping delays. One of the challenges of e-commerce is to offer online shoppers all the benefits found with local merchants with the added convenience of shopping from home.

FIGURE 8.10 • Comparison shopping on the Web

E-commerce empowers buyers with strong support of comparison shopping to find the best deal.

E-Commerce from the Seller's Perspective

Sellers strive to influence and support the stages of the buying process using the following business practices. Note how e-commerce assists the seller in meeting objectives.

1. Market research to identify customer needs: Sellers may monitor the flow of Web traffic, and solicit customer opinions using Web-based forms in order to conduct market research.
2. Manufacturing products or supplying services that meet customer needs: B2B e-commerce is used to acquire raw materials for the manufacturing of products. Often the products sold or services provided are complemented by or even dependent on Web technology. For example, software, e-books, digital music, and movies can all be delivered via the Internet. Airlines, shipping companies, and banks provide free and valuable services to their customers on the Web.
3. Marketing and advertising to make customers aware of available products and services: Sellers actively use the Web for advertising, as you saw in the satellite radio example, in order to make customers aware of products and services they may desire.
4. Providing a method for acquiring payments: Banks provide merchant accounts to e-tailers for safe and secure credit card transactions over the Web (more on this later in the chapter).
5. Making arrangements for delivery of the product: E-tailers work with shipping companies like UPS to provide several shipping options to customers at varying price levels. As noted earlier, some services and products can be delivered via download over the Internet.
6. Providing after-sales support: E-tailers and manufacturers may provide product support on their Web site, or through telephone, e-mail, or online chat. Manufacturers' warranties are typically the same for items purchased online as in a local store.

From the seller's perspective, the process of producing and selling goods is sometimes referred to as supply chain management. *Supply chain management* involves three areas of focus: demand planning, supply planning, and demand fulfillment (see Figure 8.11). *Demand planning* involves analyzing buying patterns and forecasting customer demand. *Supply planning* involves producing and making logistical arrangements to ensure that you are able to meet the forecasted demand. *Demand fulfillment* is the process of getting the product or service to the customer. E-commerce is ideally suited for streamlining these processes, saving sellers time and money, while providing more accurate information.

FIGURE 8.11 • Supply chain management

Supply chain management involves three areas of focus: demand planning, supply planning, and demand fulfillment.

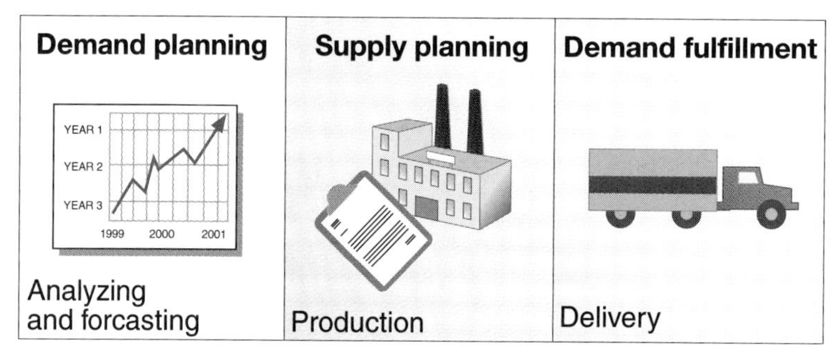

Benefits and Challenges of E-Commerce

E-commerce offers advantages to both buyers and sellers. Buyers enjoy the convenience of shopping from their desktop, or in the case of B2B, fully automated order placement. Sellers value e-commerce because it dramatically extends their markets. Farmyard Nurseries provides a dramatic example. Prior to e-commerce, this small nursery that grows a wide variety of specialty plants in Llandysul, West Wales, did the majority of its business (90 percent) with local residents. A year after expanding to the Web, Farmyard Nurseries found that the majority of its business now came from the rest of Wales (36 percent) and the United Kingdom (37 percent), with distribution beginning to extend around the world (see Figure 8.12).

B2C e-commerce levels the playing field between large and small businesses, making it much easier for new companies to enter a market and for small businesses to gain market share from large businesses. The Web is an equalizer in that it allows businesses to win over customers with high-quality services and low prices, rather than through size and monopoly power. Consider, for instance, a young entrepreneur who decides to open a small hardware store. What chance for success would you give such a business considering that local competition consists of the two largest superwarehouse, home-improvement chains in the world? Could the small private business compete with the superstores' discount prices and wide selection? What if the young entrepreneur specialized

SURF'S NOT UP

According to Jupiter Research, only 5 percent of Web site visitors translate into a sale. While such traffic has earned the not-so-flattering moniker of *dumb surf*, it has inspired customer service as the new frontier in e-commerce. Overstock.com has found when a live rep is involved, the average purchase doubles in value. It is tricky, however, because studies also show the majority of those shopping online do so from home or at work and like the privacy of it.

Source:
E-Tailing: It's All About Service. Special Report: Retailing's New Tech
By Sarah Lacy
Business Week
http://www.businessweek.com/technology/content/jul2005/tc2005076_1187.htm
July 6, 2005

FIGURE 8.12 • E-commerce can dramatically extend a business's market

Farmyard Nurseries' customers consisted of mostly local residents before e-commerce, but expanded to include customers from all over the world after implementing e-commerce.

Before e-commerce sales

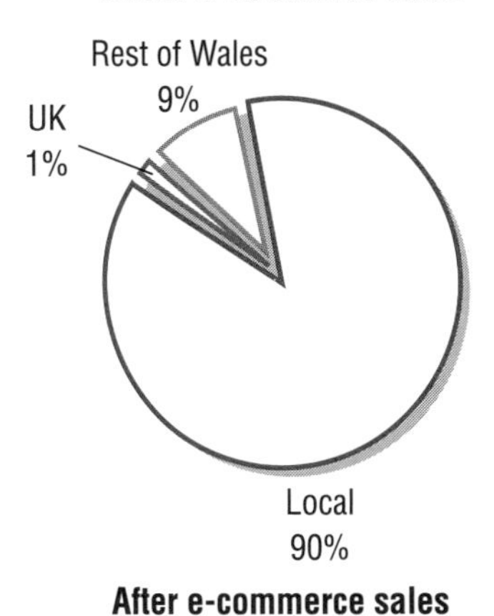

After e-commerce sales

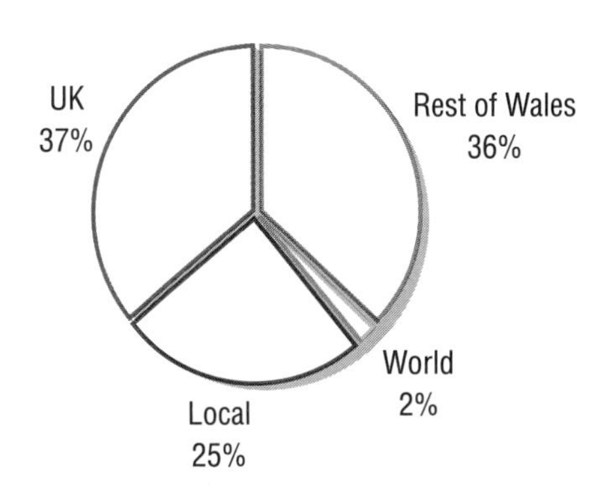

in unique and hard-to-find items on a professionally managed e-commerce site? This would expand the potential customer base for this business from people in the local community to the entire Internet population. Certainly, the chances for this small business would improve in the online environment.

In some cases, e-commerce has extended a lifeline to businesses selling products or services that were made obsolete by technology. For example, realizing that photographic film and regular cameras are gradually being replaced by memory sticks and digital cameras, Kodak is working to change its image from that of a film company to that of a picture company. It put its efforts and investments into digital camera technology and an online presence, where it provides products and tools for digital photographers (see Figure 8.13). The Kodak Web site includes tools and space for visitors to display their digital photos. Kodak discovered that by providing methods for people to share photographs on the Web, they create a greater demand for professional prints of

ORDER FULFILLMENT MORE THAN WITCHCRAFT

Amazon.com presold online more than 1.4 million advance copies of the sixth book in the Harry Potter series, ranking it as the largest new book release in Amazon's history. While such demand has the accountants chortling, it also means retailers of such mass appeal items must have the infrastructure to handle thousands of simultaneous orders. Amazon's own sales and inventory system received U.S. patent office approval the same month Potter's latest adventure topped their sales charts.

Source:
'Harry Potter' Brings Magic to E-Commerce Muggles
By Sean Michael Kerner
Technology Review
http://www.internetnews.com/ec-news/article.php/3520711
July 15, 2005

those photos, enlargements, and other specialized services such as photo T-shirts and coffee mugs. The additional revenue generated by these digital photo services has more than made up for the lost revenue from film.

FIGURE 8.13 • The Kodak Web site

Kodak changed its image from that of a film company to that of a picture company by investing in digital camera technology and an online presence, where they provide products and tools for digital photographers.

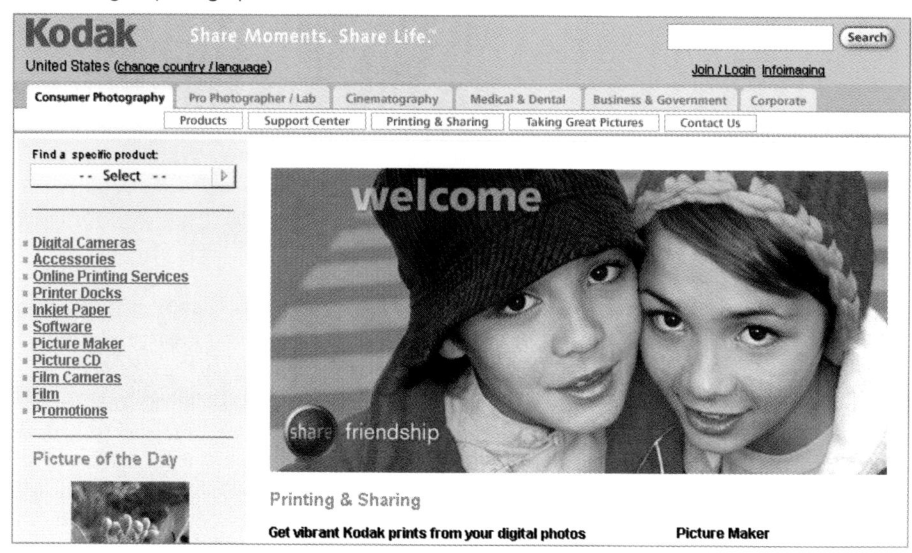

Although there are numerous advantages to e-commerce, there are also some challenges. Established businesses that wish to expand to the Internet need to alter systems and business practices to accommodate the new method of transaction processing that e-commerce requires. The larger and more established the business, the more costly this change can be. Consider the changes that must have been required at Farmyard Nurseries (Figure 8.12) when the business shifted from serving local customers to serving the world. Now consider an equal upheaval in a business with 15,000 employees.

There are also issues of security, privacy, and reliability. E-commerce can only survive if those involved can trust the system. Scams, identity theft, and fraud are a prevalent and real concern for all involved in e-commerce.

Finally, there are social concerns. Not everyone has equal access to the technology of e-commerce. People with low incomes and people living in less developed countries often don't have computers, mobile telephones, and Internet access. Many people in the United States and European Union do not have high-speed, broadband Internet access. The differences between those who have access to technology and those who do not are deepened with the increased use of e-commerce—those without access to the Internet are denied the benefits of e-commerce technology. In addition, accessibility advocates warn that as society uses increasingly smaller devices for day-to-day business, accommodations must be made for individuals who are unable to manipulate such small devices due to old age, poor vision, or other physical limitations.

E-COMMERCE APPLICATIONS

E-commerce is playing an increasingly important role in our personal and professional lives. It allows us to discover new and interesting products that may not be available in our own community. For items that are available locally, it allows us to find better deals. We use e-commerce to monitor our bank accounts and transfer electronic funds. Businesses use e-commerce to streamline transaction processes and reach new customers. This section provides a categorical and comprehensive view of e-commerce and applications from both the buyer's and seller's perspective.

Retail E-Commerce: Shopping Online

As previously discussed, e-tailing has dramatically influenced the way people shop by providing customers with product information and the ability to comparison shop for most products. Price battles continually rage on the Web to the benefit of consumers. Web sites such as mySimon.com, DealTime.com, PriceSCAN.com, PriceGrabber.com, and NexTag.com provide product price quotations from numerous e-tailers to help you to find the best deal (see Figure 8.14). It may still be that the best deal is found at your local warehouse store. In such a case, shopping online provides the assurance that you are getting the best deal. In some cases, consumers can use prices found online to negotiate a better price with a local dealer. E-commerce can empower consumers.

FIGURE 8.14 • mySimon

mySimon provides product price quotations from numerous e-tailers to help you find the best deal on many different products.

Some items are easier to sell online than others. Some purchases are determined not by holding a product in your hands, but by the description or demonstration of the product. Such is the case with books, digital music, computer software, and games—items that sell easily on the Web. Tangibles, such as DVD players, blue jeans, and automobiles, are more difficult to sell online. However,

E-Commerce and Customized Products

One way to get customers to shop online is to provide additional services. Several e-tailors are providing their customers with the ability to customize their online purchase. At Nike.com, shoppers can design their own shoes, choosing everything from the color of six or more shoe components, including the famous Nike swoosh, to personalizing the tongue with a word or phrase. The service appeals to individuals who wish to have unique shoes that express their individualism.

At the Lands' End Web site, customers can order custom clothing constructed to their exact measurements. Customers complete a brief online profile by answering a few simple fit questions and questions about lifestyle, such as exercise habits. Using the information provided, Lands' End creates a mathematical model of the customer's body based on information in their database containing more than 5 million detailed sets of body measurements. The garment is then cut and sewn to a pattern based on the unique fit profile.

Many customers enjoy the special attention provided by these customizing services, and don't mind paying a little extra for them. Both Lands' End and Nike have had to design new manufacturing systems to support piece-by-piece manufacturing rather than using the traditional mass production systems.

Questions

1. List three products that you would like to be able to customize along with the features that could be changed.
2. Would you be interested in designing your own pair of Nike's? How much more would you be willing to spend to do so?
3. Henry Ford brought us the assembly line and mass production—a key propellant of the industrial revolution. What characteristics of today's society are causing us to regress to piecemeal manufacturing?

Sources

1. *Kahn, M. "Nike Says Just Do It Yourself,"* Reuters, *May 30, 2005, http://www.reuters.com.*
2. *Nike Web site, www.nike.com, accessed July 28, 2005.*
3. *Lands' End Web site, www.landsend.com, accessed July 28, 2005.*

examining tangible items online has become significantly easier with the development of 3D virtual imagery. Three-dimensional virtual imagery allows a Web user to rotate objects on the screen and view them from every angle. This technology has been a tremendous help to businesses selling tangible products online.

There are several approaches to e-tailing. A business can set up its own electronic storefront, such as *www.landsend.com* or *www.circuitcity.com*, and provide visitors access to an electronic catalogue of products, an electronic shopping cart for items they wish to purchase, and an electronic checkout procedure. Another e-tailing option is to lease space in a cybermall. A *cybermall* is a Web site that allows visitors to browse through a wide variety of products from varying e-tailers. Cybermalls are typically aligned with popular Web portals and include Yahoo!'s *http://shopping.yahoo.com*, AOL's *www.instore.com*, and MSN's *http://shopping.msn.com*.

Online Clearinghouses, Web Auctions, and Marketplaces

Online clearinghouses, *Web auctions*, and *marketplaces* provide a platform for businesses and individuals to sell their products and belongings. Online clearinghouses such as *www.ubid.com* provide a method for manufacturers to liquidate stock and consumers to find a good deal. Outdated or overstocked items are put on the virtual auction block for customers to bid on. User's place bids on the objects. The highest bidder(s) when the auction closes gets the merchandise—often for less than 50 percent of the advertised retail price. Credit card numbers are collected at the time that bids are placed. A good rule to keep in mind is not to place a bid on an item unless you are prepared to buy it.

The most popular auction/marketplace is eBay.com. eBay provides a public platform for global trading where practically anyone can buy, sell, or trade practically anything. eBay offers a wide variety of features and services that enable members to buy and sell on the site quickly and conveniently. Buyers have the option to purchase items in auction-style format, or they can purchase items at a fixed price. On any given day, more than 12 million items are listed on eBay across over 18,000 categories. Figure 8.15 shows information on an item up for bid on eBay.

FIGURE 8.15 • eBay auction item

Information about auction items on eBay include how much time is left in the auction, the current highest bid, as well as information about the item and seller.

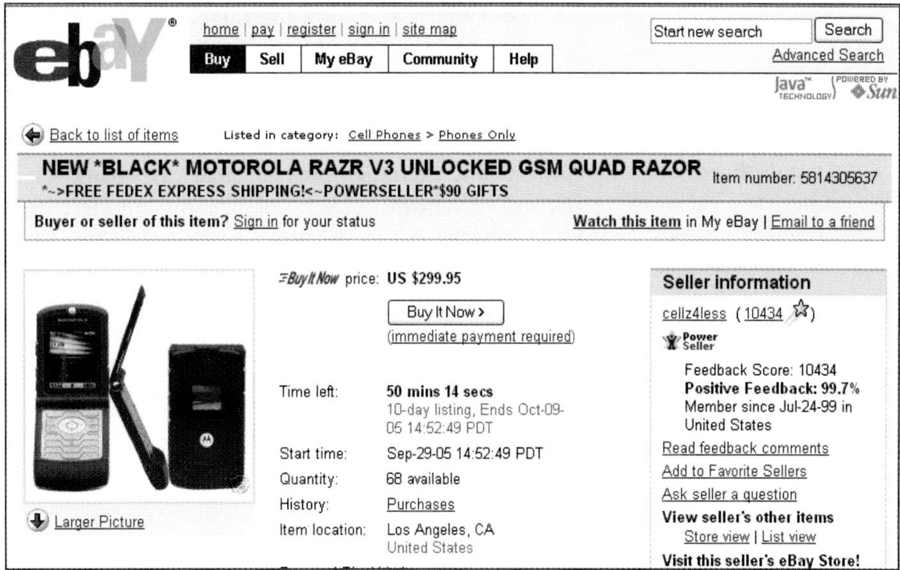

Auction houses such as eBay accept limited liability for problems that buyers or sellers may experience in their transactions. Transactions that make use of eBay's PayPal service are protected. Others may be risky. Participants should be aware that the possibility of fraud is very real in any such Internet dealings.

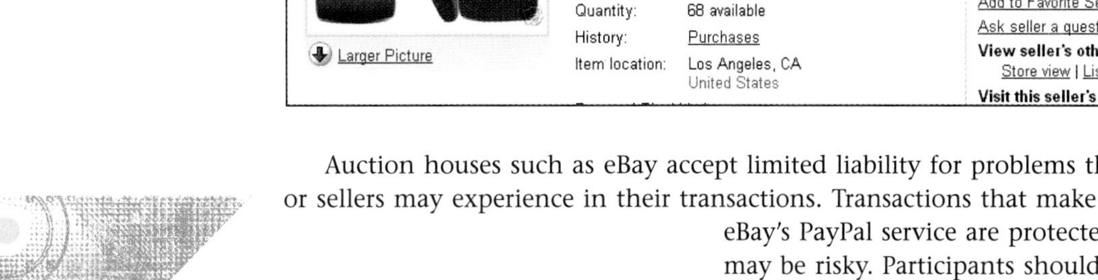

SINGER SELLS NOT SOUL, BUT SELF ONLINE

In a bold venture, a young East Indian singer raised money for his debut album by auctioning off shares of himself on e-Bay. For £3000, investors receive a share in one-quarter of the singer's lifetime earnings in music. The fine print says this includes all CD and DVD sales, concert income, or anything he is paid for: radio, TV, press, or other media. It is a gamble, but the investment includes copyright, which extends for 60 years postmortem in the United States.

Source:
Singer launches career on eBay
BBC News
http://news.bbc.co.uk/2/hi/technology/4682315.stm
July 18, 2005

B2B Global Supply Management and Electronic Exchanges

By now you know that the Internet serves as an ideal way for businesses to connect. The real challenge lies in organizing relationships between businesses. For example, pretend you have a great idea for a new product, such as a new lightweight, motorized, campus scooter. You have test marketed the product, which students on campus greeted with great enthusiasm; you've done all the necessary background work and preparation; you've found a wealthy investor; and you are ready to go into production. You have a list of parts required for the manufacturing of your scooter. But how do you go about finding suppliers that are both reputable and inexpensive? The solution for your scooter manufacturing business is the same as for any business: global supply management.

Global supply management (GSM) provides methods for businesses to find the best deals on the global market for raw materials and supplies needed to manufacture their products. There are many GSM products and services available on the Web that promise to lower a businesses costs by providing connections to a wide variety of reputable suppliers along with negotiation tools that allow a business to be assured that it is getting the best deal. Ariba (*www.ariba.com*), a GSM company, advertises that its services and software can cut total supply costs by 45 percent.

Some businesses join together with others in their industry to pool resources in Web-based electronic exchanges. An *electronic exchange* is an industry-specific Web resource created to provide a convenient centralized platform for B2B e-commerce among manufacturers, suppliers, and customers. Electronic exchanges promote cooperation between competing companies for greater industry-wide efficiency and effectiveness. Through an electronic exchange, a manufacturer has access to a wide variety of industry-specific suppliers and services. Once business relationships are established between members, the electronic exchange provides the framework for fast and efficient transactions.

Covisint (*www.covisint.com*) is an electronic exchange for automotive manufacturers. Founded by DaimlerChrysler AG, Ford Motor Company, General Motors, Nissan, and Renault, Covisint has created alliances between automotive manufactures and suppliers and contracted several of the largest technology providers to create "the most successful business-to-business electronic exchange the world has ever seen."[5] Covisint members have access to online catalogues and auctions, tools that assist in quality management and problem solving, and an industry portal (see Figure 8.16). The industry portal is software installed on a member's computer system that provides secure access to the electronic exchange's services.

Marketing

Internet users are well aware of marketing on the Internet. Banner ads threaten to overshadow Web content, pop-up ads accumulate as users visit Web pages, and e-mail boxes brim with spam. Although some Internet marketing is intrusive, there are positive aspects of e-commerce. For example, advertisements make it possible for services such as Google and Yahoo! to be available free of charge.

E-commerce has affected the marketing process perhaps more than any other area of business. The Web is used for:

- Unsolicited advertising to make buyers aware of products
- Access to product information (call it solicited advertising) through business Web sites, which allow buyers to find information about products they are actively pursuing
- Market research, to find out what consumers want

The focus of marketing in many organizations is gradually shifting from television and print media to the Web. In some markets, the Web is overtaking television in providing the public with information and entertainment. CBS News, which has struggled for market share on television, has reinvented itself as a multiplatform digital network with the Web as its primary delivery point. CBS said it was "bypassing cable television in favor of the nation's fastest-growing distribution system—broadband."[6]

FIGURE 8.16 • Covisint

The Covisint portal provides convenient access to valuable automotive industry resources.

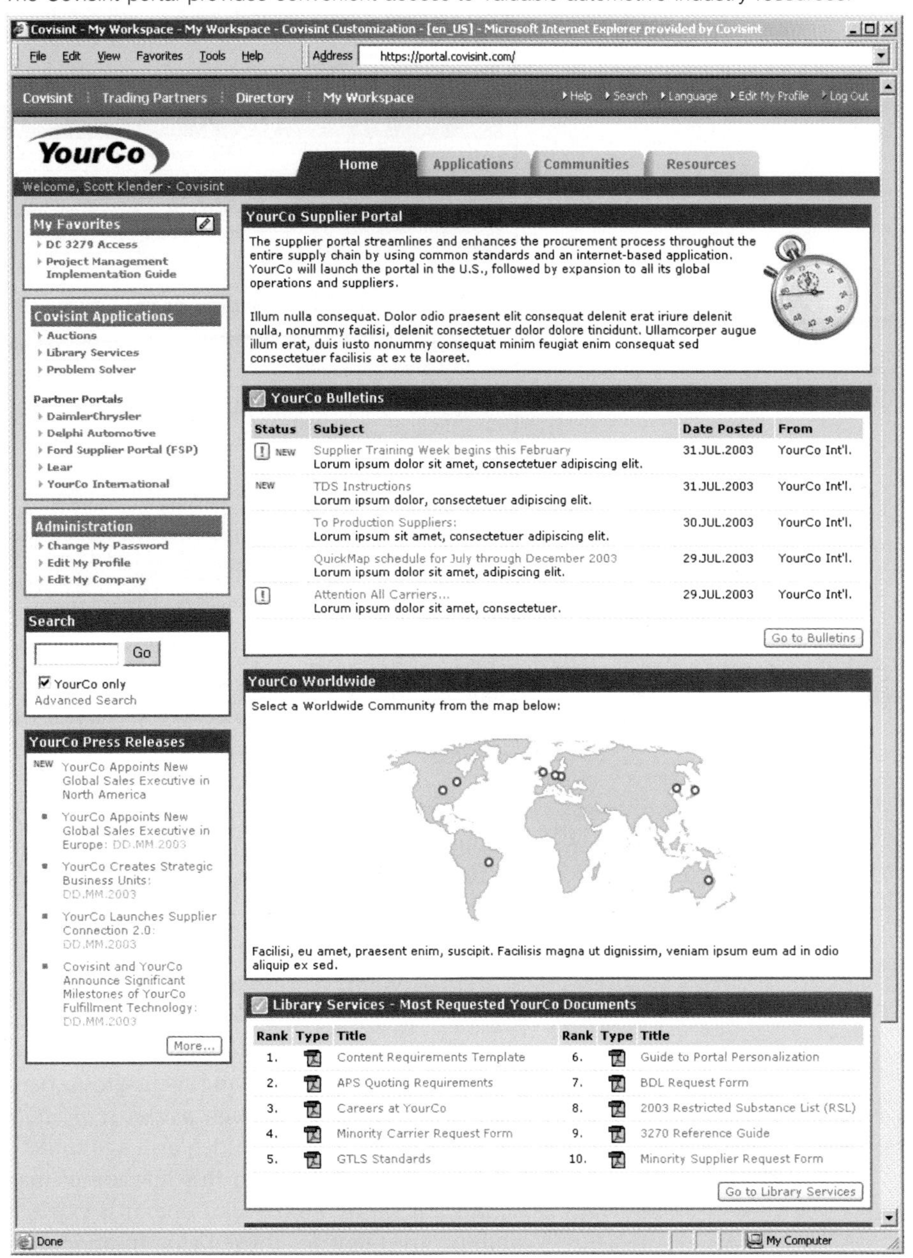

Access to Product Information. The ability to get better deals and make informed decisions is turning many people into Web researchers. In some cases Web-based product research leads to e-commerce purchases; in others, it leads to more traditional-style purchases. For example, when shopping for a car, many people compare and contrast a variety of makes and models on Web sites such as Edmunds.com, CarSmart.com, and FightingChance.com; on auto manufacturers' Web sites; or on ConsumerReports.org (see Figure 8.17). Upon deciding on a make and model, the purchaser might get some price quotes on the Web (*www.pricequotes.com*). Ultimately, however, buyers will visit a local auto dealer to test drive and purchase a vehicle.

FIGURE 8.17 • Consumer Reports

Consumer Reports advertises unbiased product testing and reviews. Many consumers find it worth the subscription fee to access its vehicle ratings.

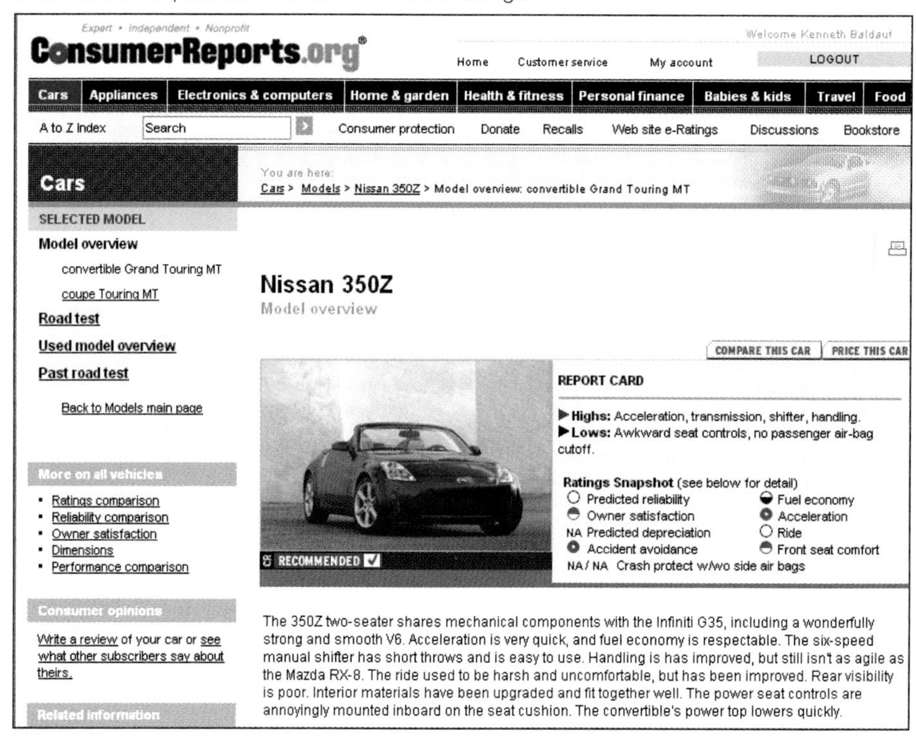

Whether dealing with an auto salesperson, a mechanic, a loan officer, a financial advisor, or any number of professionals, the Web can empower consumers with knowledge in areas previously reserved for professionals. The Web can better equip people to ask intelligent questions and confront those who might otherwise take advantage of uninformed consumers.

Market Research. Traditional marketing research takes place by observing what products customers purchase or by interviewing customers to find out how they feel about specific products. Interviews may take place on the street, by phone, by mail, or by paying focus-group participants to try a product and provide their opinions. Through *market segmentation,* customer opinions are divided into categories of race, gender, and age to determine which segment a product appeals to most. Although this method of market research is useful, it has some significant shortcomings. For one, it is expensive for those doing the research and inconvenient for those being interviewed. These restrictions make it difficult for market researchers to obtain the views of a representative cross-section of the population. Another problem is that market segmentation caters to a majority and may exclude individuals who think differently from their peers.

E-commerce allows market segmentation to take place at the level of each individual consumer. E-commerce tools make it possible to follow a visitor around a Web site, monitor which areas and products draw the visitor's attention, and monitor how much time the visitor spends in each area. This data is often collected without the customer's knowledge—so it is in no way an inconvenience and requires very little investment on the e-tailer's part. With the use of cookies (data files placed on the user's computer by a Web server), an e-tailer can maintain a history of customer preferences and highlight products that have proven historically to best hold the customer's interest. For example, a person who purchases a Harry Potter book at Amazon.com will be greeted by advertisements for other Harry Potter products on future visits as well as ads for other items that Harry Potter fans tend to purchase.

It is also possible to develop a customer profile based on broader Internet browsing patterns. Marketing companies such as DoubleClick provide advertising servers that display banner ads and pop-up ads on client Web sites (see Figure 8.18). For example, *The New York Times* (*www.nytimes.com*) and the *Washington Post* (*www.washingtonpost.com*), along with most other free online services, have hired DoubleClick to handle their sponsors' Web advertisements. Each time you visit these sites, DoubleClick inserts an advertisement from a sponsor into space provided on the Web site. Often, a different ad will be in the banner each time you visit.

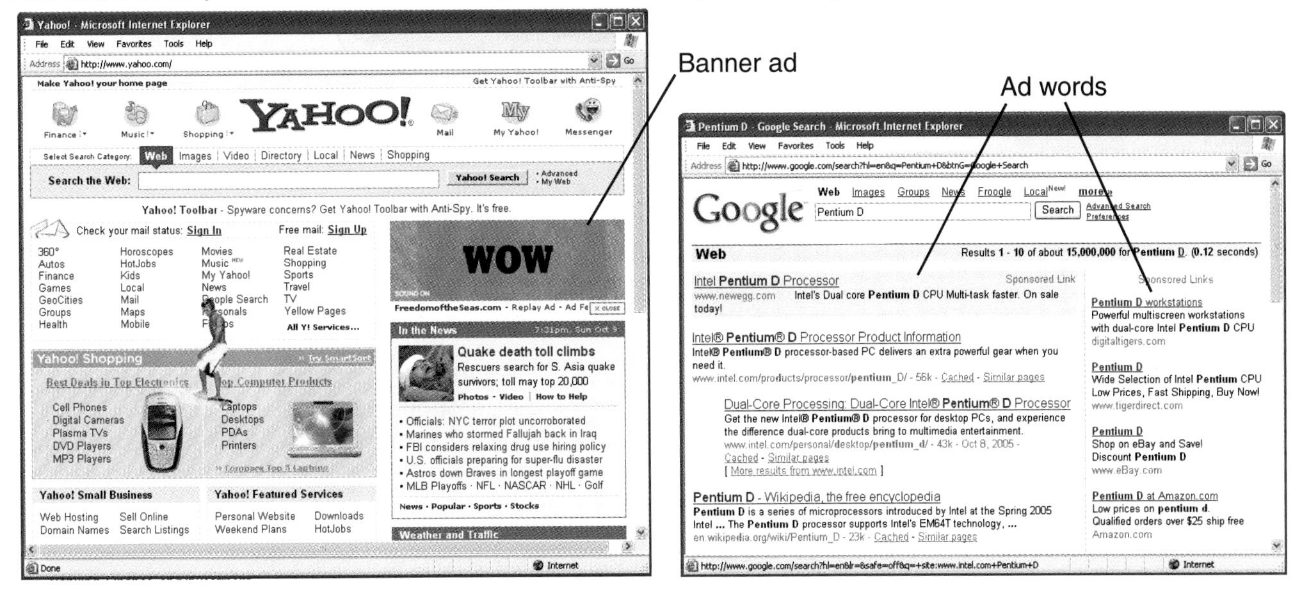

FIGURE 8.18 • Web ads

DoubleClick provides banner ads for most Web sites that use them. Google's approach to sponsorship and advertising, called AdWords, does away with the bandwidth-consuming banner ads used by most other sponsored Web sites.

Web advertising and marketing companies collect large amounts of data that can be mined to determine buying patterns and trends. This provides significantly more detailed market research than traditional methods.

Banking, Finance, and Investment

Since banks have moved online, managing money has never been easier. Online banking provides convenient access to bank balance information, the ability to transfer funds, pay bills, and obtain account histories. Most bank Web sites provide a way to link your online financial records with financial software on your home PC, such as Quicken, in order to make use of advanced tools that assist with financial management.

Electronic funds transfer has become popular for paying bills and receiving paychecks. Electronic bill payments save companies so much money that many are considering requiring it of their customers. British Telecommunications (BT) is encouraging its customers to pay bills online after learning that it would save 45 cents per bill. If BT could get 90 percent of its more than 21 million customers signed up and paying bills online, it could save close to $110 million

annually. Most banks include an automatic bill payment service that can be configured from the bank's Web site. You can also set up automatic payments at the Web sites of companies to whom you are making payments or through third-party services that perform the transfers for you for a small monthly fee.

Investing activities have also moved to the Web in a big way. Online brokerages offer low-cost stock trades and tools to assist you in making investment decisions (see Figure 8.19).

FIGURE 8.19 • Investing online

Most online trading services provide a simple form to buy and sell stocks.

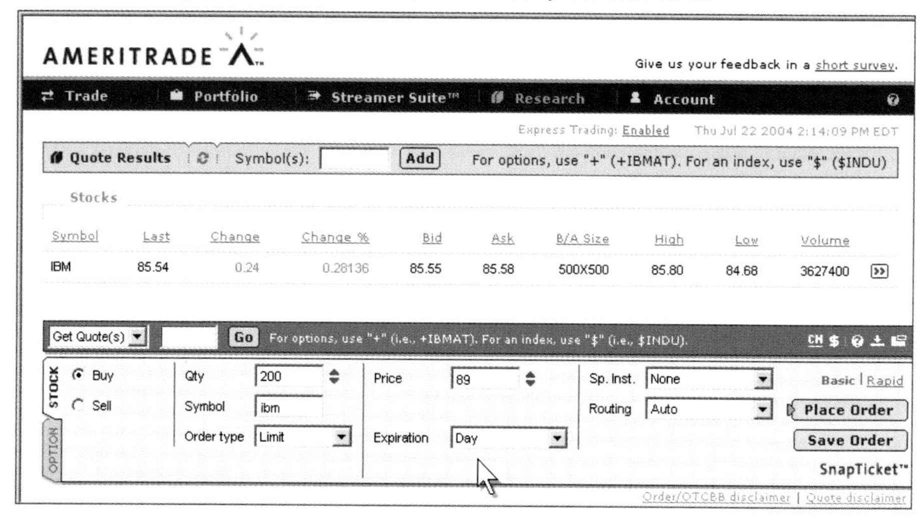

Online brokerages are able to execute trades fast, within seconds, allowing customers to buy or sell at the moment of opportunity. Most services offer helpful software that provides free quotes and streaming news for a first look at the stories that shape the market. Market research tools are available to help in their decision-making process. For example, Ameritrade provides the following tools for its customers:

- QuoteScope: Helps you visualize market trends
- Ameritrade Streamer: Provides up-to-the-second market information
- Advanced Analyzer: Provides charting, screening, and analyzing of options
- Trade Triggers: Set orders in advance to be sent automatically

Popular online brokerages include Ameritrade (*www.ameritrade.com*), ShareBuilder (*www.sharebuilder.com*), Brown & Co. (*www.brownco.com*), Harris Direct (*www.harrisdirect.com*), and E*Trade (*http://us.etrade.com*). For more information about online trading, visit *www.investingonline.org*. You can also find information about various online brokers at The Motley Fool (*www.fool.com*).

MOBILE COMMERCE

Mobile commerce, or **m-commerce**, is a form of e-commerce that takes place over wireless mobile devices such as handheld computers and cell phones. Although most of the principles and practices of e-commerce extend to m-commerce, m-commerce presents unique opportunities and challenges. This section examines the technologies on which m-commerce is built and provides examples of how m-commerce is used.

M-Commerce Technology

M-commerce depends on the proliferation of mobile communications and computing devices. Through m-commerce, you can purchase and download songs to your cell phone, which you can then listen to through a wireless headset and transfer to your home stereo. Other m-commerce examples include using your cell phone or handheld computer to trade stocks, purchase concert tickets, have flowers delivered, call a cab, or book a flight.

Although m-commerce may be accessed through Internet-connected handheld computers, the growing popularity of cell phones and advances in cell phone technology are driving m-commerce research and development. The next generation of cell phones, sometimes referred to as *smart phones*, include the computational power, bandwidth, and functionality required for m-commerce. In addition to taking advantage of next-generation cell phone technology, m-commerce makes use of other technologies and standards:

- Wireless Application Protocol (WAP): A communication standard used by developers to create m-commerce applications
- Wireless Markup Language (WML): A part of WAP that is similar to HTML and is used to create Web pages designed to fit on the small displays of mobile devices (see Figure 8.20)
- Infrared or Bluetooth wireless networking technology: Enables wireless, private, close range, device-to-device communications

FIGURE 8.20 • WAP and WML

WAP is used to deliver games and other applications to your cell phone, and WML is used to design graphics for a cell phone's small display.

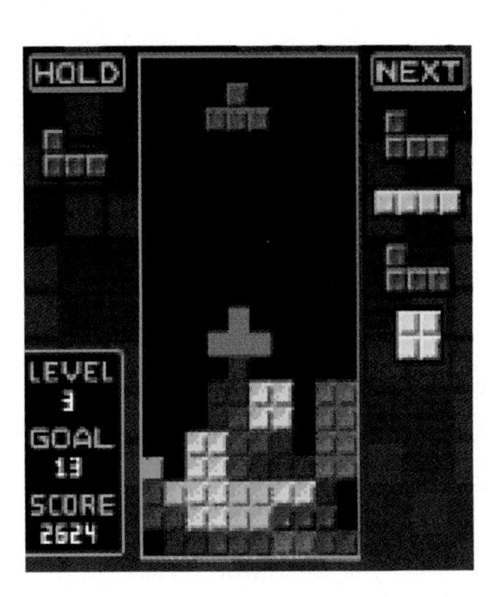

M-commerce technology received a big boost when industry leaders set aside competitive attitudes to join together in support of the open standards on which m-commerce is built. The *Open Mobile Alliance (OMA)* comprises hundreds of the world's leading mobile operators, device and network suppliers, information technology companies, and content providers, which have joined together to create standards and ensure interoperability between mobile devices.[7]

Types of M-Commerce and Applications

There are four methods at present for delivering m-commerce services to cell phones and other portable devices:

- Directly from cell phone service providers
- Via mobile Internet or Web applications
- Using Short Message Service (SMS) text messaging or Multimedia Messaging Service (MMS)
- Using short-range wireless technology, such as infrared

Through Cell Phone Service Providers. Any cell phone user is aware that cell phone service providers sell a lot more than just the ability to communicate with others. Cell phone service providers make a lot of money from text, picture, and video messaging. Ring tones, wallpaper, games, and news alerts provide additional revenues for the phone company as well as the companies providing the service or product. These inexpensive services represent the beginnings of m-commerce. The introduction of handsets with broadband capabilities is bringing m-commerce into its next phase—media! Almost all cell phone service providers now offer services that allow customers to access streamed music and video clips from such media companies as Sony, Fox, Disney, ESPN, and CNN.

Analysts predict that music and video downloads will become the most successful m-commerce products. Of these two, music is expected to become more popular. Video requires a user's full attention; something of a rarity when a person is mobile. IDC research expects the U.S. wireless music market to surge to $1.2 billion with over 50 million customers and subscribers by 2009.[8]

Over the Mobile Web. Web-based and application-based m-commerce allow the mobile user to interact with a seller's e-commerce system over the Internet or Web (see Figure 8.21). Because smart phones can run Java applications, users can download m-commerce applications to their cell phones and use them to purchase products or services.

Many m-commerce applications can be accessed through your mobile Web browser, in much the same way that you access e-commerce applications through regular Web browsers. Web developers use WML to create small Web pages designed to fit the mobile display and provide easy navigation through the mobile device interface. M-commerce Web sites typically focus on delivering services that are useful to users on the go such as news, stock-tickers, local weather, traffic information, and flight schedules and delays.

Location-based m-commerce applications make use of global positioning system (GPS) or the cell network to track your current location in order to provide location-related services. Services include weather reports, road maps, lists of nearby merchants, lists of nearby friends, and traffic reports. For example, for a small monthly fee the MyTraffic service from Traffic.com delivers updated traffic information by text to a cell phone or other wireless device.[9]

Through SMS Text Messaging. Cell phone text messaging or texting can be used to order merchandise or services, or it can be used for advertisements. Spam over SMS texting is raising concerns in the United Kingdom, where texting has been in use longer than in the United States. It is estimated that more than half of British firms are using text messaging as a marketing tool. U.S. firms are likely to follow suit. In the United Kingdom, text messaging service charges are picked up by the sender. Here in the United States, however, many services require the receiver to pick up the bill, making mobile spam all the more disconcerting. Some believe that an "opt-in" approach to mobile spam is the solution. "Text messaging is a double-edged sword. It can ruin your brand if

FIGURE 8.21 • Web-based m-commerce

Web-based m-commerce makes use of mobile Web browsers in cell phones and PDAs to support traditional e-commerce applications.

you misuse it or be deadly effective in creating customer loyalty if used correctly," says Pamir Gelenbe, Director of Development at FlyTXT, a U.K. text marketing firm.

Through Short-Range Wireless Data Communications. Short-range wireless technology enables some interesting m-commerce applications. The concept of a cashless society has been a dream for quite some time. In such a society, no one would need to carry wallets or money, and transactions would take place through automatic debits and credits to consumer and merchant accounts. Today such transactions can indeed take place using a small device such as a cell phone and short-range wireless data transmission, such as provided by infrared. Smart phones with Bluetooth technology are becoming a practical approach to a cashless society. Referred to as a *proximity payment system,* devices such as Vivo, shown in Figure 8.22, allow customers to transfer funds wirelessly between their mobile device and a point-of-sale terminal.

E-commerce and m-commerce applications are limited only by what we can imagine. Although each application is suited to a unique service or product, all applications provide extended service to buyers and increased revenues to sellers.

FIGURE 8.22 • Proximity payment system

Proximity payment systems such as Vivo store credit card information on your cell phone and transfer funds wirelessly to point-of-sale terminals.

E-COMMERCE IMPLEMENTATION

Implementing e-commerce requires a significant amount of investment and expertise. When a company decides to invest in e-commerce, it places itself at considerable risk. Because e-commerce is highly visible to the public, if it is executed poorly, it could tarnish the reputation of an otherwise well-respected company. E-commerce not only requires significant hardware and networking capabilities to accommodate heavy traffic, but also expertise in system administration, software development, Web design, and graphics design. Anything less than a professional approach in any of these areas could mean embarrassment for the company. For this reason, companies typically either hire specialists in these areas or contract the work out to professional e-commerce hosting companies.

An **e-commerce host** is a company that takes on some or all of the responsibility of setting up and maintaining an e-commerce system for a business or organization. Hosting services range in price from $7.95 per month to thousands of dollars per month, depending on the size of the business and the services offered by the host (see Figure 8.23). Companies like ValueWeb provide everything needed for a simple e-commerce online business, online catalogue, shopping cart, and transaction processing for about $49.95 per month. Large companies with more complicated systems might contract IBM or Sprint to work with their business to develop Web e-commerce solutions at a considerably higher cost. For example, Office Depot contracted IBM to design and implement a Web-based system that would accommodate transactions with customers and businesses (see Figure 8.24).

FIGURE 8.23 • E-commerce host

A typical e-commerce host will set up and maintain an e-commerce system for around $49.95 per month.

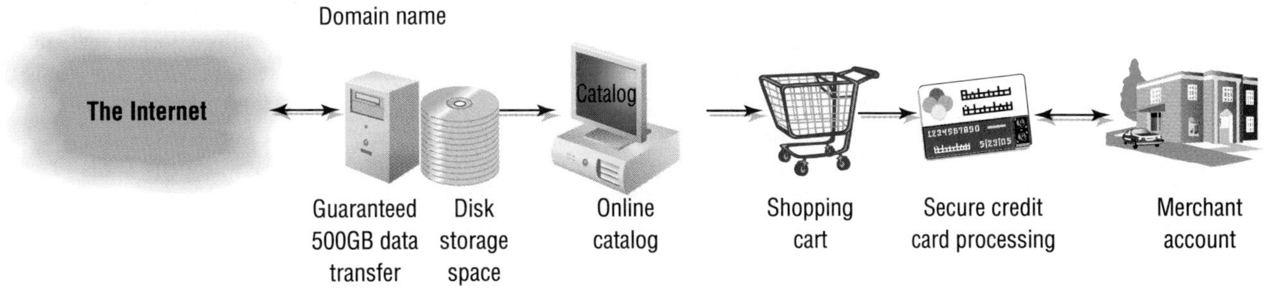

Regardless of whether you are creating and implementing your own e-commerce Web site or contracting the work out to a professional e-commerce hosting company, you need to understand the basics of e-commerce infrastructure, hardware and networking, and software issues. You also need to have a grasp of how to build Web site traffic, electronic payment systems, international markets, as well as understand the security and privacy issues involved.

FIGURE 8.24 • Office Depot: A large e-commerce system

Office Depot contracted IBM to design this e-commerce system to support both B2C and B2B.

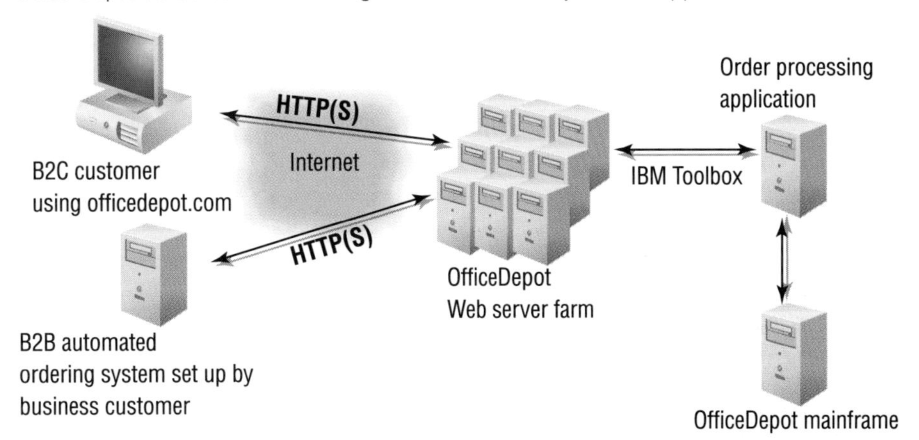

Infrastructure

E-commerce typically requires significant changes in an organization's infrastructure. Organizations expanding to the Web find that all areas of the business are affected to some degree: manufacturing, financial departments, sales, and customer service all need to adapt procedures to support doing business online. Wayne Cross found this to be the case when he began using e-commerce to support his resort, The Springs Retreat, outside Melbourne, Australia. The Springs Retreat offers accommodations, a restaurant, a café, conference facilities, a spa, and a golf course, and is staffed by 12 full-time employees. Cross, the managing director, took advantage of state-provided financial incentives and invested in a high profile Web site. The Web site (*www.thesprings.com.au*), which provides information and reservation services, is marketed in print and television media. The Web site has generated an additional $85,000 in revenue for the resort and saved $33,760 in its first year. The extra business is keeping the staff hopping, but fortunately other areas of the business are now less demanding. For example, the success of the Web site has eliminated the need for printed marketing material, producing additional savings in materials and time. Online reservations made at the Web site and outsourced to a reservations service save the staff time in reservation processing.

A business that uses e-commerce needs to employ people who are technically savvy and able to understand how technology can assist in meeting the goals of the organization. Knowledge of the technology is necessary even for a small business using an e-commerce hosting company to maintain a Web site. An e-commerce hosting company is only responsible for providing the tools of e-commerce, not for the success of the business. Large business owners and managers benefit as well, but usually enhance their knowledge through a team of experts assigned to support e-commerce operations.

B2C e-commerce often connects manufacturers directly with consumers, cutting out the middleman. This usually means drastic changes in manufacturing, storage, and shipping. Rather than shipping bulk products to retailers, B2C e-commerce requires shipping individual products directly to consumers. This may result in minor changes to manufacturing processes and major changes in shipping and storage practices. An extreme example can be found in Izumiya Co. Ltd. of Japan. Izumiya is a retail chain offering food, clothing, books, furniture, and housewares, that provides an amazing service (see Figure 8.25). If you're in Hakone, Japan, and you realize that you don't have enough sushi to accommodate your dinner guests, you can log on to *www.izumiya.co.jp*, and order sushi or any other item in Izumiya's inventory to be on your doorstep within an hour. Such a service requires significant support from inventory management systems, storage facilities, distribution networks, and support staff. To handle online orders, an entirely new system is often needed, one that works in harmony with the existing system that services in-store customers.

FIGURE 8.25 • B2C e-commerce

Izumiya provides 1-hour grocery delivery service to Japan's residents' doorsteps.

Hardware and Networking

Web-based e-commerce requires enough computing power and network bandwidth to support the Web traffic your site generates. Underestimating the amount of Web traffic leads to network stalls and long wait times that leave visitors frustrated. A typical e-commerce Web site employs one or more server computers and a high-speed Internet connection. Businesses that choose to outsource to a Web hosting company are typically guaranteed operation 24 hours a day, 7 days week, accommodating a specific number or volume of users. Such hosting companies typically use *load-balancing* among servers so that if one goes down, the others pick up the slack. They also have backup power sources, such as an uninterruptible power supply, backed up by a diesel generator to keep the system up and running in case of power failure.

Software

Several categories of software are associated with e-commerce, from the low-level software that controls the functioning of the Web server to high-level software used to design Web pages and graphics. To succeed at e-commerce, you need to understand Web server software and utility programs, e-commerce software, Web site design tools, graphics applications, Web site development tools, and Web services.

Web Server software. The primary purpose of *Web server software* is to fulfill requests for Web pages from browsers. For e-commerce applications, Web servers also provide security by encrypting sensitive transaction data such as credit card information. Web servers work with custom-designed programs and databases to provide e-commerce functionality. The two most popular Web server applications are Apache and Microsoft Internet Information Services (IIS).

Web Server Utility Programs. *Web server utility programs* provide statistical information about server usage and Web site traffic patterns. This information can be used to gauge the success of a Web site, its products, and its services.

E-Commerce Software. *E-commerce software* is designed specifically to support e-commerce activities. It includes:

- Catalog management software: Used for organizing a product line into a convenient format for Web navigation
- Electronic shopping cart software: Allows visitors to collect items to purchase
- Payment software: Facilitates payment for the selected merchandise and arranges shipping

Web Site Design Tools. Web site design tools are typically what-you-see-is-what-you-get (WYSIWYG) applications or wizards that make it simple to graphically lay out a Web page design. It is highly desirable for a Web site to have a consistent look and feel. Typically, design tools allow you to develop a standard template to be used in creating all Web site pages within a site.

Graphics Applications. Graphics applications are particularly important in Web site design. Graphics applications allow the Web developer to design and create the graphic elements that give a Web site its style and overall appearance. Menus, menu buttons, corporate logos, backgrounds, and other graphic elements combine to give a Web site a professional look. Businesses are wise to hire professional Web graphic designers when coming up with the initial design of a Web site.

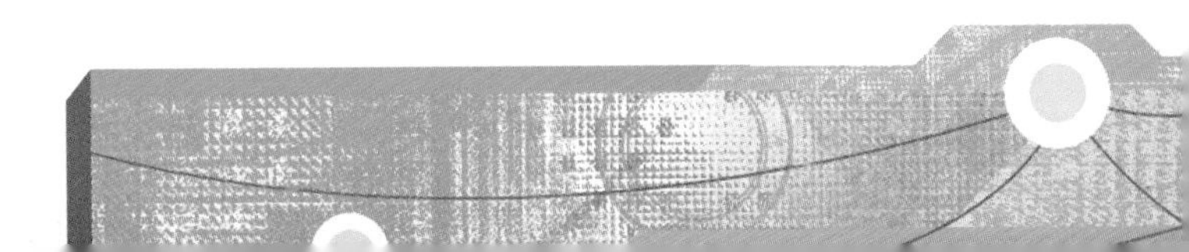

Web Site Development Tools. Web site development tools include *application programming interfaces (APIs)* that allow software engineers to develop Web-driven programs. Using programming languages such as Java, C++, or Perl, software engineers develop applications that allow Web pages to be custom created and delivered as users call up a URL. *Web-driven* programs allow users to interact with Web sites to access useful information and services. Often, Web-driven programs access data from a secure corporate database.

Amazon.com uses Web-driven programs to create personalized Web pages on the fly (see Figure 8.26). For example, if you frequent Amazon.com, it customizes its home page each time you visit, highlighting those items of interest to you. When you search for products using the keyword search engine, the results pages is created by a Web application accessing the Amazon inventory database.

FIGURE 8.26 • Personalized Web pages

Amazon.com uses Web-driven programs to create personalized Web pages on the fly.

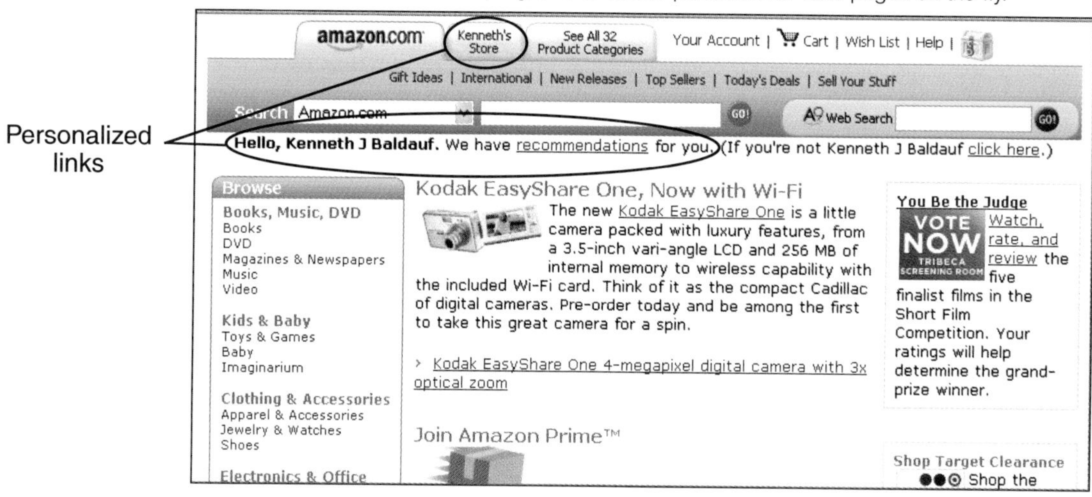

Web Services. **Web services** are programs that automate tasks by communicating with each other over the Web. Recall that when computers combine resources over a network to solve problems it is called *distributed computing*. Although the concept of distributed computing has been a part of networking for some time, until now there has been no framework for easily developing distributed computing applications for the Web. Web services provide that framework. Web service development is based on Extensible Markup Language (XML) and other languages and protocols used to organize and deliver structured data and govern the communication between applications.

Through Web services, systems developers are able to provide tools for automating trivial or repetitive tasks that traditionally require human interaction. For example, Microsoft has developed a calendar service that allows users to share their appointment books with others on the Web. Using this service, you could easily make appointments with your dentist, hair stylist, or mechanic through your Web browser without the need to speak with a receptionist (see Figure 8.27).

Web services are becoming increasingly important in transaction processing, because they are ideal for automating the exchange of information between computer systems. For example, a manufacturer could use a Web service to order materials from a supplier. The Web service on the supplier's system could then notify the manufacturer of whether or not the item is in stock and when to expect delivery.

FIGURE 8.27 • Web services

A calendar Web service running on your home PC could interact with the calendar Web service installed on a computer in the dentist's office to allow you to make an appointment without the need to speak to a receptionist.

Microsoft, IBM, and Sun have invested heavily in Web services development. Microsoft offers a Web services development platform called .NET (pronounced *dot net*). IBM offers a development platform called WebSphere, and Sun offers the Java Web Services Developer Pack (Java WSDP). These development platforms allow program developers to more easily develop Web services using a variety of programming languages. The Web Services Interoperability Organization (WS-I) has been established to promote Web services interoperability across platforms, operating systems, and programming languages. It includes dozens of big technology companies, including Hewlett-Packard and Oracle. Web services technology has become a standard part of application integration, allowing systems of varying types to communicate in the support of many of today's e-commerce applications.

Web services have, in no small manner, contributed to the distribution of work subprocesses across different locations, which has in turn contributed to outsourcing and the rise of the global work force. Thanks to the Internet and Web services, companies are able to outsource portions of their work processes to the lowest bidder, which may be located anywhere in the world. Using Web services, a project can be managed from the home location while contributors around the world work together as though they were in the same room. Everyone involved in the production process has access to common resources through Internet connections and Web services.

For example, Wild Brain (*www.wildbrain.com*) is a cutting-edge animation studio in San Francisco that produces animated films for Disney and others. Using the Internet and Web services, Wild Brain connects talented individuals in locations around the world to produce films. The design and direction takes place in the home studio in San Francisco; writers submit their work from their homes located around the world; sound recording of the character actors is done in New York or Los Angeles; and the character animation is done in Bangalore, India. Such collaboration would not be possible without the seamless integration of the work processes that Web services make possible (Figure 8.28). Such distributed work arrangements are becoming commonplace in many industries. In some industries a company wouldn't be able to compete without outsourcing portions of their work processes to companies that can provide the service for much less money.

Building Traffic

The expression "If you build it, they will come," does not hold true in e-commerce. A new e-commerce Web site is just one among billions of Web sites, and will remain undiscovered unless brought to the attention of the online community. Once a Web site gains attention, it must provide content that keeps users coming back. A stagnant Web site will quickly lose visitors. E-commerce companies use several approaches to build traffic to their Web sites. These include the 3Cs approach, keywords and search engines, partnerships, and marketing to build Web site traffic.

FIGURE 8.28 • Web services and outsourcing

Web services provide individuals with the ability to interact through connected systems that in turn makes it possible to hire talented professionals who reside in any location.

The 3Cs Approach. Many B2C e-commerce businesses use the *3Cs approach* for capturing the interest of the online community: *content*, *community*, and *commerce*. The underlying assumption of this approach is that people prefer Web sites that offer free and useful information and services over those that offer only a sales pitch. A 3Cs Web site provides useful content to the public. For example, a Web site sponsored by a health food company might offer valuable suggestions for good health, references on medical research, and a calculator that allows visitors to calculate daily caloric needs. The Web site might be updated weekly to encourage return visits. As the Web site builds traffic, an online, health-minded community is formed. This community can be nurtured by providing a free membership option. Members provide you with information to receive e-mail news bulletins, and have access to other special members-only privileges. A bulletin board or chat utility visitors can use to communicate could help to foster a feeling of community. Finally, in addition to the content and community being provided, the sponsor can offer its visitors a catalog of products to assist them in meeting their health objectives. The General Nutrition Center Web site (*www.gnc.com*) in Figure 8.29 provides a good example of the 3Cs approach and other traffic-building tactics.

Keywords and Search Engines. People discover many Web sites through the use of search engines. There are steps that business owners can take to better ensure that their company's Web site appears in search results. To begin with, a business should choose its name and its product names in a manner that best describes its purpose and features. A hair salon with the name Quality Hair Stylings would be recognized by a search engine for what it is much more easily than one with the name Haute Headz.

Second, a business should select a descriptive domain name; if possible, it should be a name that is the same or similar to the business name—*www.qualityhairstylings.com*. Domain names can be acquired from one of the accredited registrars listed at *www.icann.org/registrars/accredited-list.html*.

FIGURE 8.29 • Building Web site traffic

General Nutrition Center (GNC) uses smart strategies to build traffic to their Web site. It follows the 3Cs approach by offering content, community, and commerce to its visitors; it has a descriptive company name, has partnered with other companies to assist in marketing (drugstore.com), and advertises in print and on television.

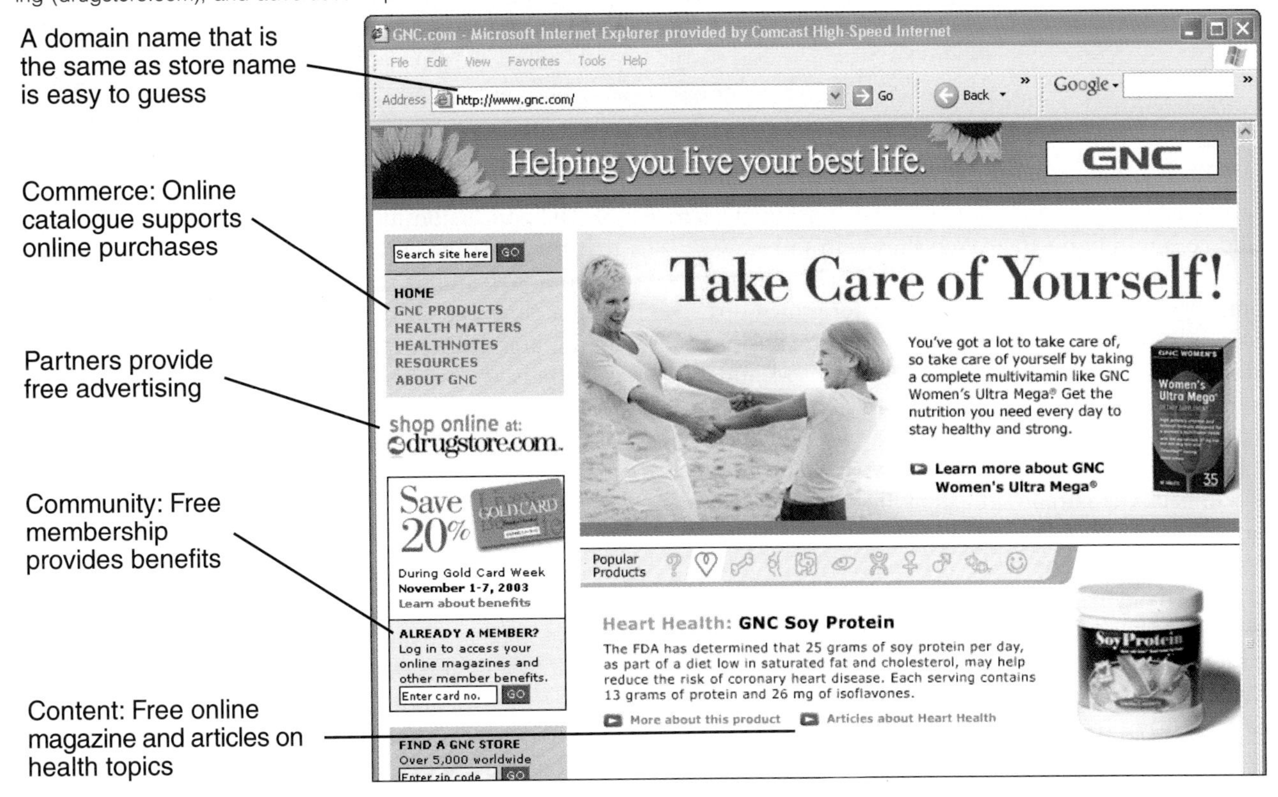

Third, business-related keywords can be listed in the HTML code of a Web page in a meta tag. *Meta tags* are read by search engines and Web servers, but are not displayed on the page by a Web browser. Here is the meta tag from *www.hersheys.com*:

```
<META NAME="keywords" content="hershey chocolate kisses,
american candies samples, reese's peanut butter cup,
chocolate products, visit hershey, baking recipes, dessert
recipes, chocolate chip cookies, hershey foods corporation
jobs, hershey's web sites, fundraising, hershey park,
hershey's new products, hershey coupons, hershey's sweet
treats, chocolate food snacks, milton snavely hershey,
chocolate factory, cocoa bean, gum &#38; mints,
confectionery, fun kids games, halloween, trick or treats,
candy stores, chocolate gifts, hershey syrup, chocolate
world, investments, hershey souvenirs">
```

Some search engines rely on meta tags to define the important keywords on the Web page. Web page URLs are stored in the search engine database according to the terms listed in the meta tag. Searches run on keywords such as *Halloween* or any of the other keywords listed in the hersheys.com meta tag would return the Hershey URL.

JOB TECHNOLOGY

Blogs: Benefit or Backfire?

Many businesses are experimenting with using blogs to build community on their Web sites with mixed results. The hobby store eHobbies that sells remote-controlled helicopters and other toys for grown-ups recently added a blog to their Web site. The business uses the blog to post photos from trade shows and shots of employees, mainly with the intent of projecting a homespun image of the company. Seth Greenberg, chief executive of the company, explains "It lets us pull back the curtain and show how we're a company of hobbyists who love participating in the things they're buyers for," he said. "It humanizes us."

One challenge for businesses that use blogs is in keeping the blog from competing with the commerce portion of the site. Business managers are afraid that when Web site visitors follow a link to the blog, they may never return to the original Web site to shop. Such was the dilemma faced by executives of Ice.com, an online jeweler based in Montreal. Ice.com had a very successful blog line with thousands of visitors a day, but it was not linked directly to or from their Web site. People stumbled onto the blog through search engines and links from other blogs. Ice.com wanted to keep the dialog on the blog uncensored, but were afraid that connecting the blog too closely to their business site would risk insulting some of their clientele.

The conclusion most businesses have reached is that blogs can be very good at attracting new customers to a Web site. It is wise, though, for a business to keep a close eye on the blog in order to keep it clean, useful, and inoffensive to all current and potential customers.

Questions

1. In what ways might a blog benefit an e-commerce business?
2. In what ways might a blog backfire on an e-commerce business?
3. Consider your favorite brand of clothing or favorite Web retailer. Would you be interested in participating in a blog if the company were to provide one? What information might you be able to share with others or learn from others about the company's products?

Sources

1. *Tedeschi, B."Blogging While Browsing, but Not Buying",* New York Times, *July 4, 2005.*
2. *eHobbie's blog,http://ehobbies.blogs.com/rc, accessed July 28, 2005.*
3. *Ice.com's blog,http://blog.ice.com, accessed July 28, 2005.*

Most search engines use a combination of techniques and information, which include the title tag and meta tags, to classify Web sites, as discussed in Chapter 4. To ensure that your site is listed in a search engine or directory, you can use a link at the site that allows you to submit your URL. For example, at Yahoo.com, there is a link at the bottom of the page for "How to Suggest a Site."

Marketing. If a business is unable to generate traffic to its Web site using free methods, such as search engines and partnerships, it often invests in advertising. Marketing companies can be hired to advertise a Web site online using banner ads, pop-up ads, and e-mail. Search engines such as Google and Yahoo! offer paid advertising services (described earlier) that make your Web site more likely to be noticed than the hundreds of other links that are displayed as the result of a search. Web sites are also advertised offline using traditional advertising media such as magazines, newspapers, radio, and television.

Electronic Payment Systems

In most e-commerce transactions, the customer typically pays with either a credit card number (B2C) or a merchant account number (B2B). Using a secure encrypted connection, the buyer enters the account information in a Web form and submits it. The merchant's Web server passes the information along to the payment processing system, which checks the validity of the account number and balance in the account. If the credit card checks out, the transaction is processed (see Figure 8.30). The only difference between online credit card processing and in-store processing is that laws require online retailers to wait until shipping the product to deduct funds from the credit card. Because of this, e-tailers put a hold on the funds until the product is shipped, at which times the funds are transferred.

FIGURE 8.30 • Online credit card transaction

Online vendors use payment processing systems to check credit and arrange for the transfer of funds.

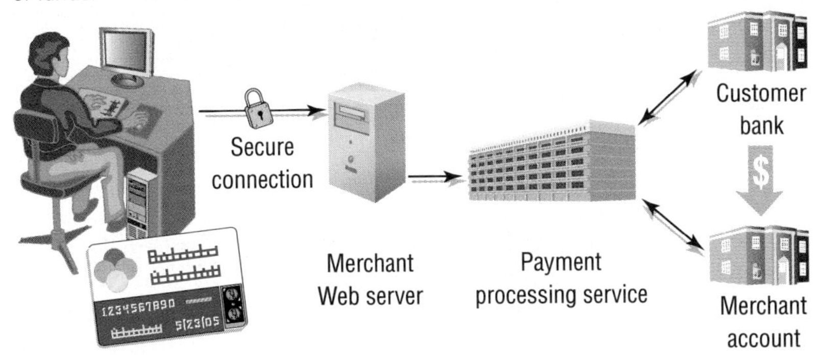

For individuals who do not have a credit card or do not wish to provide their credit card information to merchants online, or in the case of C2C, where the seller does not have the ability to process credit card transactions, there is electronic cash. **Electronic cash** or **e-cash** is a Web service that provides a private and secure method of transferring funds from a bank account or credit card to online vendors or individuals for e-commerce transactions. PayPal is the best-known e-cash provider. Owned by eBay, PayPal allows members (membership is free) to transfer funds into their PayPal account from a bank account or credit card (see Figure 8.31). The funds in your PayPal account are e-cash that you can use for purchases from online vendors or purchases made on eBay. You can also e-mail PayPal e-cash to individuals. For instance, say you're short of cash at a lunch outing with a friend. You can get your friend to pick up your tab and reimburse him through PayPal when you get back to your computer. Your friend receives an e-mail from PayPal in your name with instructions for how to transfer the cash into his bank account.

E-cash has two fundamental benefits. It provides:

- Privacy by hiding your account information from vendors; the e-cash provider is the only one who knows this information
- A method for e-commerce transactions in circumstances where the seller cannot process a credit card or the buyer does not own a credit card

Although most large online vendors do not support PayPal, there are indications that the service is gaining more widespread acceptance. Recently, Bango, a mobile content provider, contracted with PayPal to support purchases from mobile phones.[10] Users will be able to pay for games, ringtones, videos, and wallpaper using their PayPal accounts.

FIGURE 8.31 • E-cash

E-cash systems such as PayPal allow users to purchase products online without providing the vendor with a credit card number. You can also use e-cash to transfer funds between individuals.

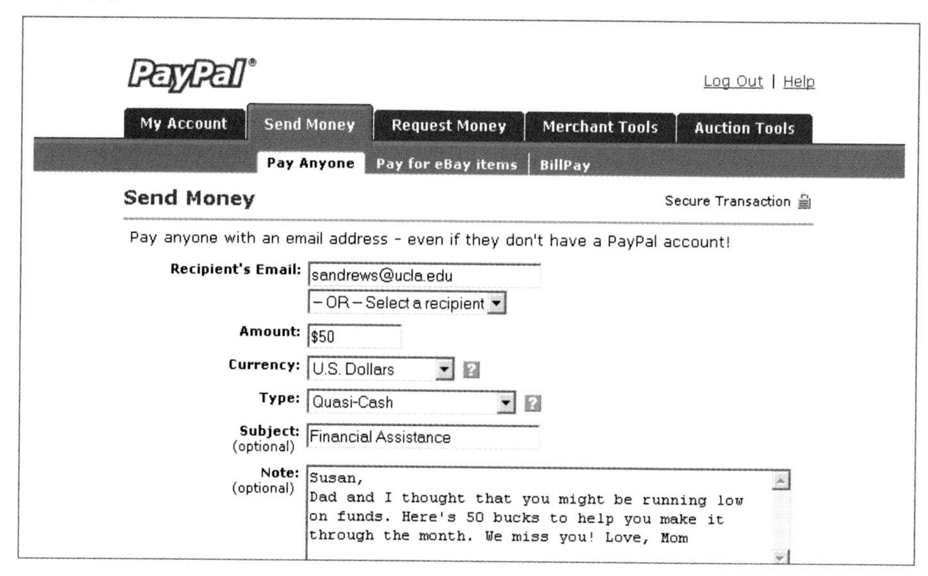

Several software vendors have created electronic wallet applications to make online transactions more convenient. An *electronic wallet*, or *e-wallet*, is an application that encrypts and stores your credit card information, e-cash information, bank account information, name, address—essentially, all the personal information required for e-commerce transactions—securely on your computer. The convenience provided by e-wallets is significant; rather than having to fill out forms with each online purchase, you simply click a button that withdraws the funds from your e-wallet. However, in the several years that e-wallet applications have been available, there has been no agreed-upon standard for the technology, so online merchants have been unwilling to support it.

Smart cards, credit cards with embedded microchips, play a small but increasing role in e-commerce payment. The "smart" Visa card (*http://usa.visa.com/personal/cards/credit/smart_visa.html*) is used with a card reader connected to your PC to make secure e-commerce purchases. When making a purchase at a supporting merchant's Web site, a swipe of the card sends a temporary transaction number, rather than your credit card number, to the vendor for processing. The chip on the smart Visa card also works in conjunction with software on Visa's Web site to access detailed account information. Smart cards like Visa's provide higher levels of security and privacy than traditional credit cards. Account information is only available to the cardholder, and transactions can only be made with a card swipe. Customers can use the free USB reader provided by Visa. Many companies feel that there is a future in smart cards, and some have even gone so far as to incorporate card readers into computer keyboards (see

COIN OF THE NOT-SO-REAL REALM

Without a central bank to control flow, gamers using real money to buy currency issued by various cyberworlds are encountering high inflation and fluctuating exchange rates. The estimated $880 million spent annually on such currencies also makes this a matter of keen interest to tax authorities. Economic experts anticipate this will drive regulation of the exchange, and cyberspace nations issuing currency will eventually be required to report all sales—just as if the money were real.

Source:
Virtual gaming's elusive exchange rates
By Daniel Terdiman
CNET News
http://news.com.com/Virtual+gamings+elusive+exchange+rates/2100-1043_3-5820137.html
August 5, 2005

Figure 8.32). This method of making online purchases is much more secure than typing in a number, because users without physical possession of the card are unable to use it. With traditional credit cards, anyone with knowledge of the card number and expiration date can use it to make purchases online.

FIGURE 8.32 • The Compaq keyboard and smart card reader

This keyboard and smart card reader allows Web users to make online purchases with the swipe of a card.

MasterCard has teamed with Nokia to place smart card technology in smart phones in a service they call PayPass.[11] A trial of the technology in Dallas allows participants to pay for products and services at several Dallas businesses by simply waving their cell phone over a reader at the checkout counter. The technology is currently being tested in the local bus network in the city of Hanau, near Frankfurt, Germany. Customers are able to store digital bus tokens in their phones and wave the phone past a reader for bus entry. It is expected to greatly reduce the congestion and lines at bus stations.

Companies such as Seimens, Mobileway, and Simpay are designing ways to pay for m-commerce services over your cell phone. Simpay's system debits a cell phone user's account to pay for digital goods such as ringtones, MP3 music files, games, or parking meter payments.

International Markets

Taking a business online automatically turns your business into a global enterprise. Internet users of all nationalities will have access to your products and may wish to purchase them. Business owners need to determine if and how to market their products in the global market. Recent trends have indicated that business outside the United States can be lucrative. For example, Amazon.com's international sales will soon equal its U.S. sales.

The first consideration of a global e-commerce strategy is to make sure that visitors of all nationalities and cultures feel at home and comfortable while viewing your Web content. Although English is widely spoken and understood around the world, local references and colloquialisms may confuse and alienate international visitors.

A more costly approach is to create multiple versions of your Web site, each in a different language and catering to a different cultural bias (see Figure 8.33). Lands' End, for example, has launched separate Web sites in the United Kingdom, Japan, Germany, France, Ireland, and Italy. This process, called *localization,* requires hiring international Web developers to assist with translation and cultural issues. Each time the Web site is updated, this may require editing all of the associated international Web sites.

Localization can be applied to more detailed geographies such as areas of a country, city, or zip code. Consider the various versions of The Weather Channel and its associated Web site that are created to focus on different areas of the United States as well as different countries in the world.

Once an international market is courted and won over, the e-tailer must be able to carry out transactions in foreign currency, including pricing items accordingly, applying the correct national taxes, and accommodating the possibly complex issues of international shipping. E-commerce hosting companies can assist with some of the transaction details. Shipping companies also assist e-businesses with the complications of international shipping. FedEx, for

FIGURE 8.33 • Localization

Some e-businesses create multiple versions of their Web site, each in a different language and catering to a different cultural bias.

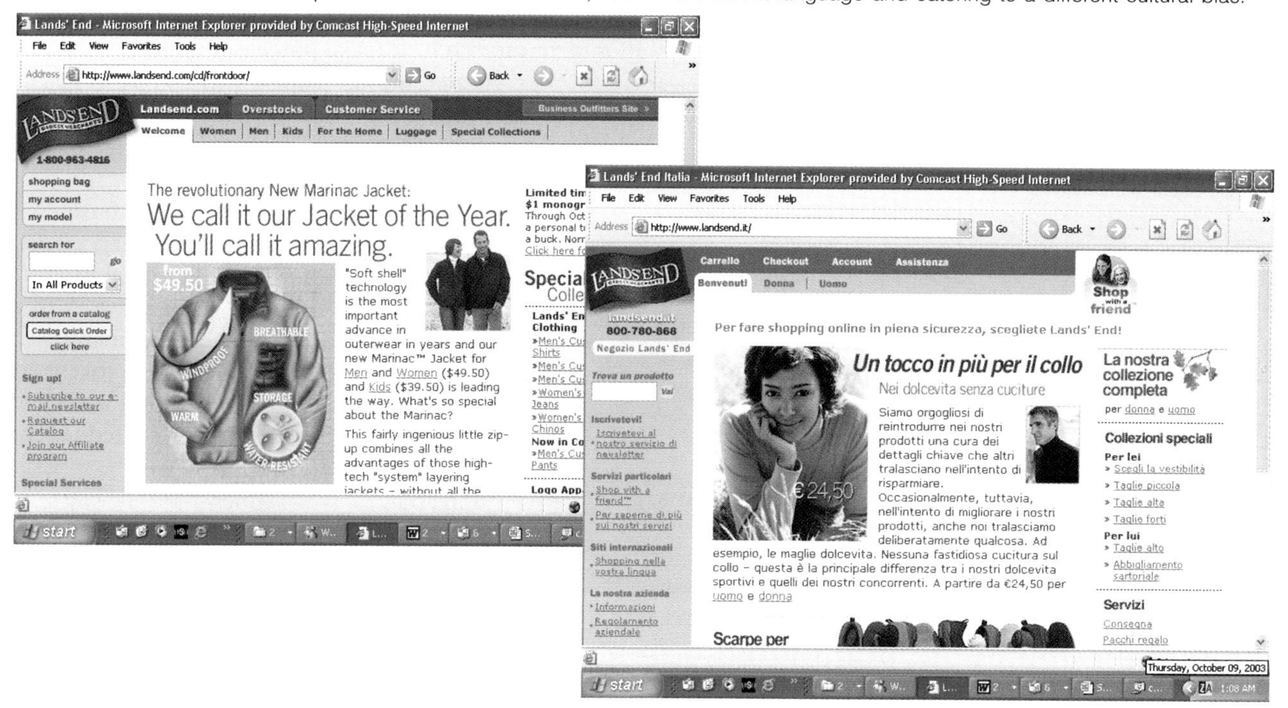

example, provides a free service called FedEx Global Trade Manager. The application helps shippers understand global trade regulations and prepare the appropriate import or export forms based on the commodity being shipped and the countries of origin and destination.

E-Commerce Security Issues

Figure 8.34 displays a typical Web form used to submit bank account information to PayPal. Would you feel comfortable filling out and submitting this form? If you were to submit this form, would it be possible for a hacker to intercept your account information on its way to PayPal? Are you certain that the Web server is really owned and protected by PayPal? How can PayPal tell that the account information you are providing is really yours? According to research firm Gartner, one out of three Internet users is buying less online because of security concerns.[12] There are significant and legitimate concerns regarding e-commerce privacy and security. Security concerns arise from the dangers of carrying out electronic funds transfers over a public network without buyer and seller identity verification. In addition, e-businesses are vulnerable to hacker attacks that can put them temporarily or permanently out of business. Identity verification, securing data, attacks, vulnerabilities, and business resumption planning are discussed next. Specific threats to information security, as well as how to protect against them, are discussed in Chapter 11.

Identity Verification. Because e-commerce transactions occur electronically, and at times automatically, it is important that the identities of the two or more participants in a transaction are positively verified. Consumers need to make sure that they are giving their credit card number to a legitimate and trustworthy business, and businesses need to confirm that the customer is the owner of the credit card being used. Transaction data must be accessed only by intended

FIGURE 8.34 • Web form for submitting bank account information

Is it safe to provide your bank information using this form?

PayPal®
Log Out | Help
My Account | Send Money | Request Money | Merchant Tools | Auction Tools
Overview | Add Funds | Withdraw | History | Profile

Add Bank Account

The safety and security of your bank account information is protected by PayPal. We protect against unauthorized withdrawals and will notify you by email whenever you deposit or withdraw funds from this bank account.

Bank Name:
Account Type: Checking / Savings
Routing Number: (Is usually located between the ⑆ symbols on your check.)
Account Number: (Typically comes before the ⑈ symbol. Its exact location and number of digits varies from bank to bank.)
Retype Account Number:

U.S. Check Sample
Routing Number | Check # | Account Number
⑆ 211554485 ⑆ | 0012 | 1456874801⑈

Add Bank Account | Cancel

parties, and not be intercepted by outsiders. A variety of technologies are available to assist individuals and businesses in meeting these goals, including digital certificates and encryption.

A **digital certificate** is a type of electronic business card that is attached to Internet transaction data to verify the sender of the data. Digital certificates are provided by *certification authorities* such as VeriSign (*www.verisign.com*) and Thawte (*www.thawte.com*). They can be used to verify the sender of e-mail and other forms of Internet communication. Digital certificates for use in encrypting Web communications and credit card transactions cost the provider between \$350 and \$1400, depending on the level of security and type of communications. Certificates cost more for e-commerce than for nonbusiness use. Digital certificates for personal e-mail are provided by Thawte for free.

A digital certificate contains the owner's name, a serial number, an expiration date, and a public key. The public key is used in encrypting messages and digital certificates. **Encryption** uses high-level mathematical functions and computer algorithms to encode data so that it is unintelligible to all but the intended recipient. Through the use of a public key, a large number, and a private key, kept by the certification authority, an encrypted message can be decrypted back into its original state.

Securing Data in Transit. Digital certificates combined with *Secure Sockets Layer (SSL)* and a more recent version of SSL called *Transport Layer Security (TLS)* technologies allow for encrypted communications to occur between Web browser and Web server (see Figure 8.35). This combination of technology is what is used to secure usernames, passwords, and credit card information when they are typed into a Web form and sent to a Web server. The presence of an SSL connection is usually indicated by a URL that uses *https://* rather than *http://*. Also, a closed lock icon appears at the bottom of the browser window when the connection is secure. The many threats to e-commerce are further discussed in Chapter 11.

FIGURE 8.35 • Secure Sockets Layer (SSL)

SSL encrypts data sent over the Web and verifies the identity of the Web server.

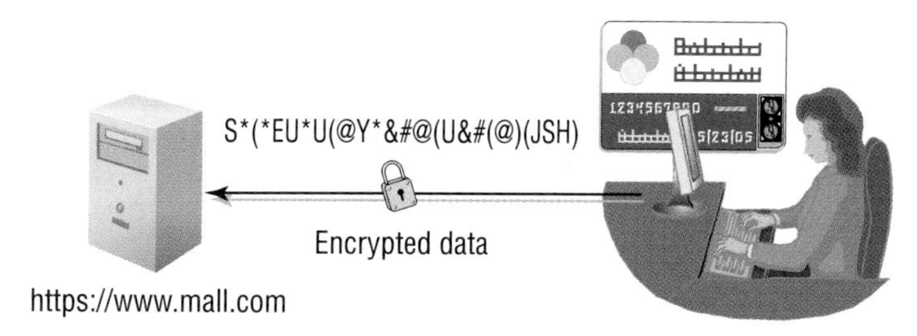

Business Resumption Planning. Transaction processing systems, especially order processing systems, are so important to a business that great pains are taken to ensure that they stay up and running. Web businesses are particularly vulnerable because every minute that the Web site is out of commission could mean losses of thousands of dollars. *Business resumption planning (BRP)* takes into account every conceivable disaster that could negatively impact the system and provides courses of action to minimize their effects. Through the use of backup power systems, backup computer systems, and security software, BRP takes into account natural disasters such as flood, fire, and earthquake, and man-made disasters such as employee strikes and accidental or intentional sabotage. The goal of business resumption planning is to protect data and keep key systems operational until order is resumed.

ACTION PLAN

1. What will it take to put the Forrero family business online?

The Forrero's need to make changes in infrastructure to support sales directly to consumers in addition to bulk sales to retailers. This will most likely require additional personnel. Alejandro can act as the director of e-commerce operations once he graduates and can implement needed changes in the manufacturing, storage, and shipment procedures.

The Forrero's can outsource the set up and maintenance of the Web site to an e-commerce hosting company at a cost of under $100 per month. They may need to contract with a graphic artist and Web designer to design a Web site that has a professional appearance. The Forreros may also wish to invest in advertisements for their new Web site.

2. What benefits might the Forrero family enjoy from an online presence? Do these benefits outweigh the costs?

By taking their business online, the Forreros' business will acquire greatly increased international exposure. Online sales will represent a new source of revenue for the business. These sales can be used to fund the Forreros' overall e-commerce investment. The Forreros' Web site will increase the Forrero brand-name recognition, improving sales in retail gift shops on which the business depends. Although these benefits seem significant, it will take time to build traffic to the Web site and reap the associated rewards. If the Forreros take it slowly, investing minimally at first, and reinvesting the revenues back into the system, an e-commerce venture should pay off.

3. Besides taking the business online, what other ways might e-commerce assist the Forrero family business?

The Forreros can streamline transactions and save time and money by using the e-commerce systems of their suppliers, and implementing an e-commerce system with their retail buyers. Retailers will appreciate efforts to streamline the ordering process, giving Forrero a competitive advantage over the competition.

Summary

LEARNING OBJECTIVE 1

Define e-commerce, and understand its role as a transaction processing system.

E-commerce refers to systems that support electronically executed transactions over the Internet, Web, or a private network. Prior to the Internet, e-commerce took place using electronic data interchange (EDI) over private value-added networks (VANs). Today e-commerce takes place mostly over the Internet and Web. E-commerce is a form of transaction processing system. A transaction processing system (TPS) supports and records transactions and is at the heart of most businesses. All transaction processing systems share a common set of activities called the *transaction processing cycle*. This includes data collection, data editing, data correction, data manipulation, data storage, and document and report

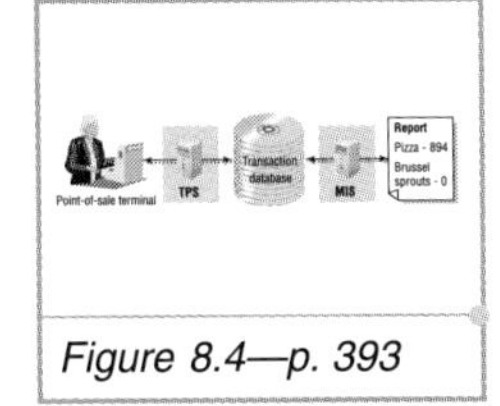

Figure 8.4—p. 393

production. There are different TPSs for different needs. An order processing system supports the sales of goods or services to customers and arranges for shipment of products. A purchasing system supports the purchasing of goods and raw materials from suppliers for the manufacturing of products.

LEARNING OBJECTIVE 2

List the three types of e-commerce, and explain how e-commerce supports the stages of the buying process and methods of marketing and selling.

The three main types of e-commerce are business-to-consumer (B2C), business-to-business (B2B), and consumer-to-consumer (C2C). B2C, sometimes called e-tailing, involves retailers selling products to consumers. B2B services often take place privately between manufacturers and suppliers. C2C Web sites allow consumers to sell items and trade items with each other. Of these three types of e-commerce, B2B supports the greatest number of business transactions.

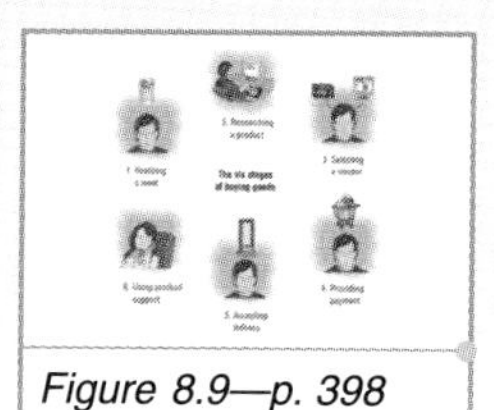
Figure 8.9—p. 398

E-commerce supports the six stages of buying a product: (1) realizing a need, (2) researching a product, (3) selecting a vendor (4) providing payment, (5) accepting delivery of the product, and (6) taking advantage of product support. It also supports the seller's efforts to support these stages. The seller's considerations are sometimes referred to as supply chain management. Supply chain management involves three areas of focus: demand planning, supply planning, and demand fulfillment.

LEARNING OBJECTIVE 3

Discuss several examples of e-commerce applications and services.

Retail Web sites allow consumers to comparison shop to find the best deals. Wholesale Web sites provide a way for suppliers to do business with manufacturers and retailers. Clearinghouses provide good deals to consumers on overstocked or outdated items and assist businesses in clearing their inventory. Web auctions sell merchandise to the highest bidders, and online marketplaces provide a method for consumers to sell their own possessions.

Figure 8.13—p. 402

Manufacturers join together in global supply management services and industry-specific electronic exchanges to combine resources to connect with suppliers. E-commerce has made market research much easier and less intrusive, but somewhat worrisome to privacy advocates. E-commerce brings convenience to banking and investing by allowing people to monitor their finances and make investments online.

LEARNING OBJECTIVE 4

Define m-commerce, and describe several types of m-commerce services.

Mobile commerce, or m-commerce, is a form of e-commerce that takes place over wireless mobile devices such as smart phones. M-commerce uses a smart phone's Internet capabilities for commerce. Through text messaging, Web applications, and short-range wireless networking, smart phones can access m-commerce services. M-commerce developers use special protocols and languages such as the Wireless Application Protocol (WAP) and Wireless Markup Language (WML) to create m-commerce applications that can run on small handheld displays.

Figure 8.22—p. 414

Four methods are currently being pursued for m-commerce service delivery. The first takes advantage of the connection between cell phone and cell phone service provider. The second utilizes the cell phone's ability to run applications and access the Internet and Web services. The third uses a cell phone's Short Message Service (SMS) text messaging and Multimedia Messaging Service (MMS). The last uses short-range wireless technology, such as infrared, and a portable device's ability to use it to communicate with other devices such as cash registers.

LEARNING OBJECTIVE 5

List the components of an e-commerce system, and explain how they function together to provide e-commerce services.

E-commerce requires investment in networking, hardware, and a wide variety of software. It requires changes in infrastructure and can include hiring of specialists such as system administrators, Web designers, and graphic artists. Many firms decide to outsource their e-commerce operations to e-commerce hosting businesses.

Hardware and networking services for e-commerce must be robust and trustworthy so that service is never interrupted. Software required for e-commerce includes Web server software to deliver Web pages and services; Web server utilities to provide statistical information about Web traffic; e-commerce software to provide a merchandise catalogue, shopping cart, and checkout services; and the software used to design Web pages, graphics, and Web applications. Web services use new technology, such as XML, to support distributed computing that allows applications to talk to each other over the Internet, automating many processes.

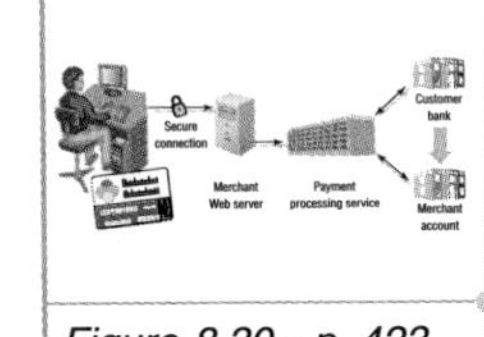

Figure 8.30—p. 423

Businesses use a variety of techniques to build traffic to a Web site. Providing interesting content, building community, and providing commerce (the 3Cs) is one technique. Choosing appropriate names for a business, products, and domain helps search engines recognize a Web site for what it is. Keywords can be included in a Web page's meta tags, which are visible to search engines but not to people viewing the Web page. E-tailers can also profit from forming alliances with others on the Web and promoting each other's Web sites.

E-commerce electronic payment systems typically handle credit card transactions. E-cash supports financial transactions over the Web without the need for credit cards; an electronic wallet stores transaction information in an encrypted file on your PC. Smart cards store financial information in a microprocessor embedded in a credit card and provide a more secure and convenient method of carrying out e-commerce transactions.

E-commerce sites that participate in the global market must provide Web sites that cater to the needs of various cultures. Consumers and e-tailers should be aware of security issues associated with e-commerce, including identity verification and methods of securing transaction data transmitted over the Internet. Business resumption planning takes into account every conceivable disaster that could have a negative impact on the system and provides courses of action to minimize their effects.

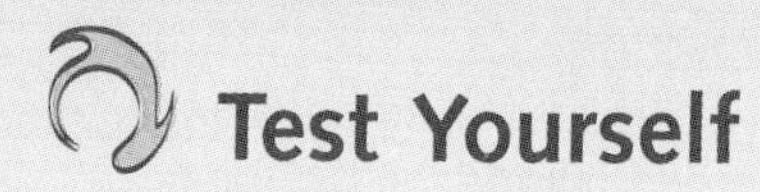

Test Yourself

LEARNING OBJECTIVE 1: Define e-commerce, and understand its role as a transaction processing system.

1. ________________ refers to systems that support electronically executed transactions over the Internet, Web, or a private network.
2. Prior to the Internet, electronic transactions between businesses were often carried out using ________________ over a value-added network.
   a. order processing system
   b. electronic data interchange (EDI)
   c. business resumption planning
   d. source data automation
3. True or False: An airline's flight reservation system is likely to use the online method of transaction processing.

LEARNING OBJECTIVE 2: List the three types of e-commerce, and explain how e-commerce supports the stages of the buying process and methods of marketing and selling.

4. Many B2B transactions take place over ________________ networks that allow the output of one computer system to act as input into another.
   a. TPS
   b. EDI
   c. e-tail
   d. B2B
5. True or False: B2C comprises a larger segment of e-commerce transactions than B2B.
6. If you were to sell your textbook to someone on eBay you would be utilizing a type of e-commerce called ________________ .

LEARNING OBJECTIVE 3: Discuss several examples of e-commerce applications and services.

7. A(n) ________________ is a Web site that allows visitors to browse through a wide variety of products from varying e-tailers.
8. An industry-specific Web resource created to provide a convenient centralized platform for B2B e-commerce between manufacturers, suppliers, and customers is called a(n):
   a. Web portal
   b. Web auction
   c. electronic exchange
   d. online marketplace
9. True or False: E-commerce supports a more detailed level of market segmentation than traditional market research.

LEARNING OBJECTIVE 4: Define m-commerce, and describe several types of m-commerce services.

10. ________________ is a form of e-commerce that takes place over wireless mobile devices such as smart phones.
11. A ________________ payment system uses infrared to beam credit card information from a cell phone.
    a. e-wallet
    b. e-cash
    c. proximity
    d. smart card
12. True or False: Music is expected to be the next big product for m-commerce.

LEARNING OBJECTIVE 5: List the components of an e-commerce system, and explain how they function together to provide e-commerce services.

13. A(n) ________________ is a company that takes on some or all of the responsibility of setting up and maintaining an e-commerce system for a business or organization.
14. ________________ are programs that automate tasks by communicating with each other over the Web.
    a. Web services
    b. Electronic exchanges
    c. Web-driven programs
    d. 3C programs
15. True or False: E-cash is most useful for C2C e-commerce.

Test Yourself Solutions: 1. E-commerce, 2. b. EDI, 3. True, 4. b. EDI, 5. False, 6. C2C, 7. cybermall, 8. c. electronic exchange, 9. True, 10. m-commerce, 11. c. proximity, 12. True, 13. e-commerce host, 14. a. Web services, 15. True

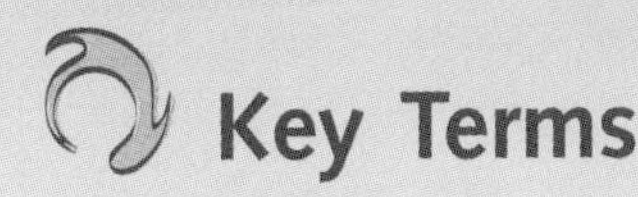

Key Terms

business-to-business e-commerce (B2B), p. 396
business-to-consumer e-commerce (B2C), p. 396
consumer-to-consumer e-commerce (C2C), p. 396
digital certificate, p. 427
e-commerce host, p. 414
electronic cash (e-cash, digital cash), p. 423
electronic commerce (e-commerce), p. 390
encryption, p. 427
mobile commerce (m-commerce), p. 410
smart cards, p. 424
Web services, p. 418

Questions

Review Questions

1. What types of transactions benefit from batch processing?
2. What types of transactions benefit from online processing?
3. List the steps of the transaction processing cycle.
4. What takes place in the data editing stage of the transaction processing cycle?
5. What types of documents or displays might a TPS produce?
6. What is the difference between e-commerce and m-commerce?
7. Name the three types of e-commerce, and provide examples of each.
8. What are the six stages of buying?
9. What are the three areas of supply chain management?
10. What device is closely associated with m-commerce? What next-generation features of this device are used in m-commerce?
11. List five methods of selling items on the Web.
12. What are the three methods of providing m-commerce services?
13. What is a problematic application of m-commerce?
14. What is the primary benefit of Web services?
15. List the types of software associated with implementing e-commerce.
16. What is a primary hardware and network concern of an e-commerce vendor?
17. List four methods of building traffic to a Web site.
18. List two benefits of e-cash.
19. How can smart cards make shopping on the Web more secure?
20. What are the goals of business resumption planning?

Discussion Questions

21. Describe benefits that e-commerce provides for buyers and sellers.
22. Describe challenges that e-commerce faces.
23. List five types of e-commerce applications, and provide examples of each.
24. What are the benefits for both suppliers and manufacturers of joining an electronic exchange?
25. What are the pros and cons of online marketing?
26. How can e-commerce affect the infrastructure of an organization?
27. What are the security risks for buyers who participate in e-commerce?
28. What are the security risks for sellers who participate in e-commerce?

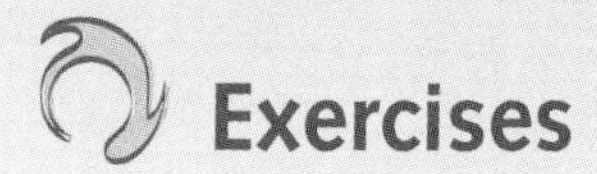

Exercises

Try It Yourself

1. Write a proposal for a new e-commerce Web site to sell a product or service of your choosing.
   a. Select a business name that is creative and descriptive.
   b. Use *www.enameco.com* (or any of the accredited registrars at *www.icann.org/registrars/accredited-list.html*) to find an available domain name for your Web site that is logical and descriptive.
   c. Describe the product(s) you intend to sell and why you think that it will sell well on the Web.
   d. Describe the type of content (one of the 3Cs) that you will provide to visitors to keep them coming back to your Web site.
   e. Describe what tools you will make available to build community (another of the 3Cs).
   f. List other types of online businesses that sell complementary products and would make good partners for your business.
   g. List 12 keywords that you would want search engines to associate with your product Web site.

2. Surf the Web to find two Web sites (other than *www.gnc.com*) that follows the 3Cs approach to Web development. Use a word-processing program to describe what elements of each Web site contribute to content, community, and commerce. Provide your opinion on which Web site does a better job of making you want to visit again, and why.

3. Use a paint program to design a logo and Web page template for an imaginary e-commerce Web site. Include a Web site menu for navigation around the Web site.

4. Search the Web on the keyword *m-commerce*. Write a description of an interesting m-commerce application you discover.

5. Consider an electronics device that you might like to purchase over the next year (for example, a smart phone, digital camera, MP3 player, and so on). Use *www.mysimon.com* to learn more about the product and determine your preferred vendor. Write a couple paragraphs explaining your logic for choosing the device and vendor. Include the features of the particular product that you found attractive, and why, out of all the vendors selling this product, you decided on the one you did.

6. Consider an electronics device that you might like to purchase over the next year (for example, a smart phone, digital camera, MP3 player, and so on). Use *www.ubid.com* and *www.ebay.com* to shop for the item. After finding a good deal, search for the same item (using manufacturer and product name along with model number) on the Web using a standard search engine or *www.mysimon.com*. Were you able to find the same item? If so, was the deal you found at eBay or uBid as good as you originally thought? Write a summary of your findings.

Virtual Classroom Activities

For the following exercises, do not use face-to-face or telephone communications with your group members. Use only Internet communications.

7. Each group member should select an object that he or she is interested in shopping for online. Be very specific in regard to the item's specifications (for example, a 3.2 megapixel digital camera, a pair of black leather moccasins with beads, and so on). Compile everyone's choices into a list and conduct a Web scavenger hunt. The person who finds the most items at the lowest total cost is the winner.

8. List the group members alphabetically and assign member numbers according to the ordering. Each group member should use *www.paypal.com* to e-mail the next group member $1. The last group member should e-mail the first group member. You need to set up a PayPal account and add funds to it from your checking account or credit card prior to sending the e-cash. After collecting each other's dollar, each group member should write a review of the PayPal service. You can delete your account information on PayPal after this experiment if you so desire.

9. Each group member should visit the Web site of a cell phone service provider: Cingular, Sprint, T-Mobile, or Verizon. Create a document that lists those features and services from each provider that support m-commerce. Compare your findings and decide if any one carrier seems to offer better services.

Teamwork

10. Team members should each set out on the Web to find the best monthly price on e-commerce hosting. You can start by looking at *www.cnet.com* and looking under Internet Services and then E-commerce hosting. Follow links and view advertised prices and specifications. The service should include e-commerce software (for example, Miva), a merchant account, a payment gateway service, 500 MB of space, a 30-GB data transfer rate, 24-hour technical support, and data backup services.

11. Each team member should share a favorite online place to shop with the group. Compile a list of all Web sites and have all team members explore and rank the list of Web sites in order from best to worst. Each team member should create a document listing the Web sites in order of preference with a paragraph explaining what aspects of the favored Web site sold him or her on the site. Include comments on the 3Cs.

TechTV

Finding the Best Deals Online

Go to www.course.com/swt2/ch08 and click TechTV. Click the "Finding the Best Deals Online" link to view the TechTV video, and then answer the following questions.

1. What services do Web sites such as Epinions.com offer consumers that are unavailable through traditional brick-and-mortar shopping?
2. Describe the scam the reporter warned about regarding e-commerce vendor reviews and ratings.

Endnotes

[1] Kuchinskas, S. "E-Commerce on Ice," InternetNews.com, June 13, 2005, http://www.internetnews.com/ec-news/article.php/3512096.

[2] Gilbert, A. "E-Commerce Turns 10," Cnet News.com, August 11, 2004, http://news.com.com.

[3] E-Commerce Statistics Web site, http://www.10xmarketing.com/Learning-Center/Internet-Statistics/E-Commerce-Statistics.html, accessed July 25, 2005.

[4] Johnson, C. "2004 US eCommerce: Wrapping Up the Year," Forrester Research, January 27, 2005, http://www.forrester.com.

[5] Covisint Web site, http://www.covisint.com/about/alliances/, accessed July 28, 2005.

[6] Kuchinskas, S. "TV Moves to the Internet," *InternetNews.com*, July 12, 2005, http://www.internetnews.com/ec-news/article.php/3519611.

[7] Open Mobile Alliance Web site, http://www.openmobilealliance.org.

[8] "Grooves On The Move: IDC Forecasts U.S. Wireless Music Market To Surge By 2009", IDC press release, July 18, 2005, http://www.idc.com/getdoc.jsp?containerId=prUS00195505.

[9] Sharkey, J. "All Traffic, All the Time and Just a Click Away," *New York Times*, July 19, 2005, http://www.nytimes.com.

[10] Oates, J. "Bango Calls on PayPal," *The Register*, September 12, 2005, www.theregister.co.uk.

[11] "Nokia Announces Contactless Payment And Ticketing," *Mobiledia.com*, February 9, 2005, http://www.mobiledia.com/news/25560.html.

[12] Gray, T. "Sunny Outlook Holds For E-Commerce," July 14, 2005, www.internetnews.com/ec-news/article.php/3520001.

INFORMATION, DECISION SUPPORT, ARTIFICIAL INTELLIGENCE, AND SPECIAL-PURPOSE SYSTEMS

Although only a sophomore in college majoring in history, Maurice Duchane had already decided what he wanted to do after graduation. He would follow his father into the commercial real estate business as a sales agent. He would always have a passion for European history and France, where his family was originally from, but he wanted the freedom and excitement of real estate.

Maurice's father increasingly had clients who lived hundreds or even thousands of miles from the Duchane's hometown of New Orleans, Louisiana. This became a problem, because it was hard to get clients excited about office buildings and rental apartments when they weren't there to see them first hand. Clients wanted to see the inside and outside of buildings from a variety of perspectives. In addition, many real estate investors need to work closely with sellers, the seller's real estate sales agent, attorneys, architects, and even builders. A real estate investor, for example, might want to purchase an existing apartment complex from a seller—the person who owns the building. This can involve working with the real estate agent representing the seller and the seller's lawyer. In other cases, a real estate investor may want to build a new apartment complex, which usually involves working with an architect to design the apartment complex and a builder or contractor that constructs the apartment complex. Maurice's father needed a convenient way to bring everyone together to close a deal. Maurice would also like to be able to show potential clients the sights of New Orleans.

As you read this chapter, consider the following:

1. What type of system would help show potential clients a realistic view of commercial real estate?
2. How can Maurice's father get investors, attorneys, sellers, their agents, architects, and builders together at the same time to close real estate deals or respond to natural disasters such as hurricanes?
3. What can Maurice do to give clients a tour of New Orleans over the Internet?

Check out Maurice's *Action Plan* at the conclusion of this chapter.

LEARNING OBJECTIVES

1. Define the stages of decision making and problem solving.
2. Discuss the use of management information systems in providing reports to help solve structured problems.
3. Describe how decision support systems are used to solve nonprogrammed and unstructured problems.
4. Explain how a group decision support system can help people and organizations collaborate on team projects.
5. Discuss the uses of artificial intelligence and special-purpose systems.

CHAPTER CONTENT

Decision Making and Problem Solving

Management Information Systems

Decision Support Systems

The Group Decision Support System

Artificial Intelligence and Special-Purpose Systems

Introduction

Decisions lie at the heart of just about everything you do. You can't get through a day without making many decisions—some large, some small. In our careers, some decisions can have great significance and affect many other people. A doctor who makes the right diagnosis and prescribes the right treatment can improve lives and sometimes save them. The decisions made by scientists, engineers, and technicians at NASA have life and death consequences for astronauts. Management information and decision support systems can provide a wealth of information to help people make better decisions. Business executives, engineers, environmental specialists, military personnel, music producers, scientists, librarians, and people in most careers can benefit from getting better reports and decision support from computer systems.

Increasingly, people work in small teams. Meetings can be a big waste of time or a productive tool to accomplish organizational objectives. Group decision support systems can help people work effectively in groups, avoiding the pitfalls of negative group behavior, while taking advantage of working in teams. In addition, artificial intelligence and special-purpose systems can help individuals and organizations achieve their goals. Expert systems, for example, can be used to diagnose complex medical problems, saving lives in some cases. Knowing a little about what these systems can do and how they function can help you use them more effectively.

Although people can use computer systems to do tasks more efficiently and less expensively, the true potential of computer systems is in providing information to help people and organizations make better decisions. In criminology, police and investigators use computer systems to help solve crimes, and in engineering, engineers use computer systems to design better electric circuits. In business, computer systems help sales representatives target customers with the greatest potential. In this chapter, you explore how computer systems can provide a wealth of information and decision support to individuals and organizations, regardless of their field or career. To understand how computer systems can provide decision support, you first need a basic understanding of decision making and problem solving.

DECISION MAKING AND PROBLEM SOLVING

One of the highest compliments is to be recognized by your friends and coworkers as a "real problem solver." In general, problem solving is the most critical activity an individual or organization undertakes (see Figure 9.1). It makes the difference between success and failure, profit and loss. Increasingly, computers are used to analyze decision-making styles and approaches.[1] Dr. Laibson of Harvard University, for example, is using computer-based magnetic resonance imaging (MRI) techniques to explore how people can make better decisions. One day, this use of computers could unlock the mysteries of human decision making.

FIGURE 9.1 • Decision making and strategy

Decision making and problem solving are critical activities for individuals and organizations.

The process of problem solving begins with decision making. **Decision making** is a process that takes place in three stages: intelligence, design, and choice. **Problem solving** includes and goes beyond decision making. In Figure 9.2, you can see that the problem-solving process includes implementation and monitoring, in addition to all three phases of decision making. Here is a brief description of the problem solving process.

- Intelligence stage: During this stage, a person or group of people identify and define potential problems or opportunities. Information is gathered that relates to the cause and scope of the problem or opportunity. The problem or opportunity environment is investigated, and things that might constrain the solution are identified.
- Design stage*:* During this stage alternative solutions to the problem are developed. In addition, the feasibility and implications of these alternatives are evaluated.
- Choice stage: This stage requires selecting a course of action.
- Implementation stage: In this stage of the problem-solving process, action is taken to put the solution into effect.
- Monitoring stage: In this stage, the decision makers evaluate the implementation of the solution to determine whether the anticipated results were achieved and to modify the process in light of new information learned during the implementation stage. The monitoring stage involves a feedback and adjustment process.

FIGURE 9.2 • The problem-solving process

The three phases of decision making—intelligence, design, and choice—combine with implementation and monitoring to result in problem solving.

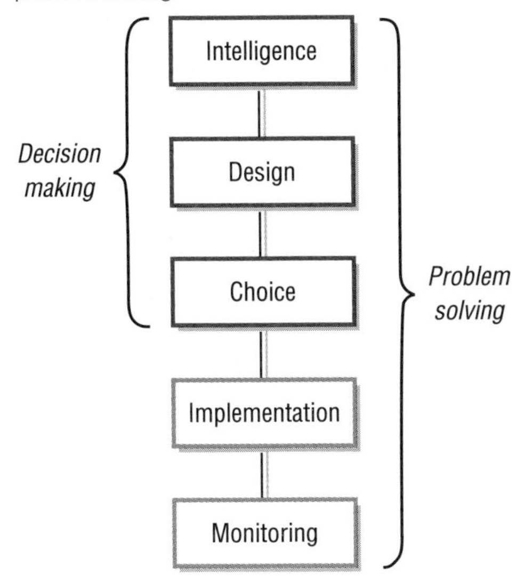

Individuals and organizations can use a reactive or proactive approach to problem solving. With a *reactive problem-solving approach,* the problem solver waits until a problem surfaces or becomes apparent before any action is taken. With a *proactive problem-solving approach,* the problem solver seeks out potential problems before they become serious. For example, an FBI agent could predict a future terrorist attack using credible intelligence information, which could save thousands of lives and millions or even billions of dollars in damage (Figure 9.3). Both reactive and proactive problem solvers can turn problems into opportunities, but taking a proactive approach means you can identify opportunities earlier and exploit them faster. In reality, most organizations and individuals use a combination of reactive and proactive problem-solving approaches.

Programmed Versus Nonprogrammed Decisions

In the choice stage, a number of factors influence the decision maker's selection of a solution. One factor is whether or not the decision can be programmed. **Programmed decisions** are ones that are made using a rule, procedure, or quantitative method. Programmed decisions are easy to automate using traditional computer systems (see Figure 9.4). It is simple, for example, for a music store to program a computer to order more CDs of a popular artist when inventory levels fall to 10 units or less. *Management information systems (MISs)* are often used to support programmed decisions by providing reports on problems that are routine and where the relationships are well defined; these types of problems are called *structured problems.*

FIGURE 9.3 • Proactive approach

The FBI collects and analyzes many different types of data to help them predict, prepare for, and avoid possible security threats.

Nonprogrammed decisions, however, deal with unusual or exceptional situations. In many cases, these decisions are difficult to represent as a rule, procedure, or quantitative method (see Figure 9.4). Determining what research projects a genetics company should fund and pursue is an example of a nonprogrammed decision. Determining the appropriate training program for a new employee, deciding whether to buy a townhouse or rent an apartment, and weighing the benefits and disadvantages of various diets for cancer patients are additional examples of nonprogrammed decisions. Each of these decisions contains many unique characteristics for which the application of rules or procedures is not so obvious. Today, *decision support systems* and *expert systems* are being used to solve a variety of nonprogrammed decisions in which the problem is not routine and rules and relationships are not well defined. These are called *unstructured problems*.

FIGURE 9.4 • Programmed and nonprogrammed decisions

Inventory control systems, illustrated in the screen on the left, are typically programmed decisions. In contrast, some medical decisions are nonprogrammed decisions.

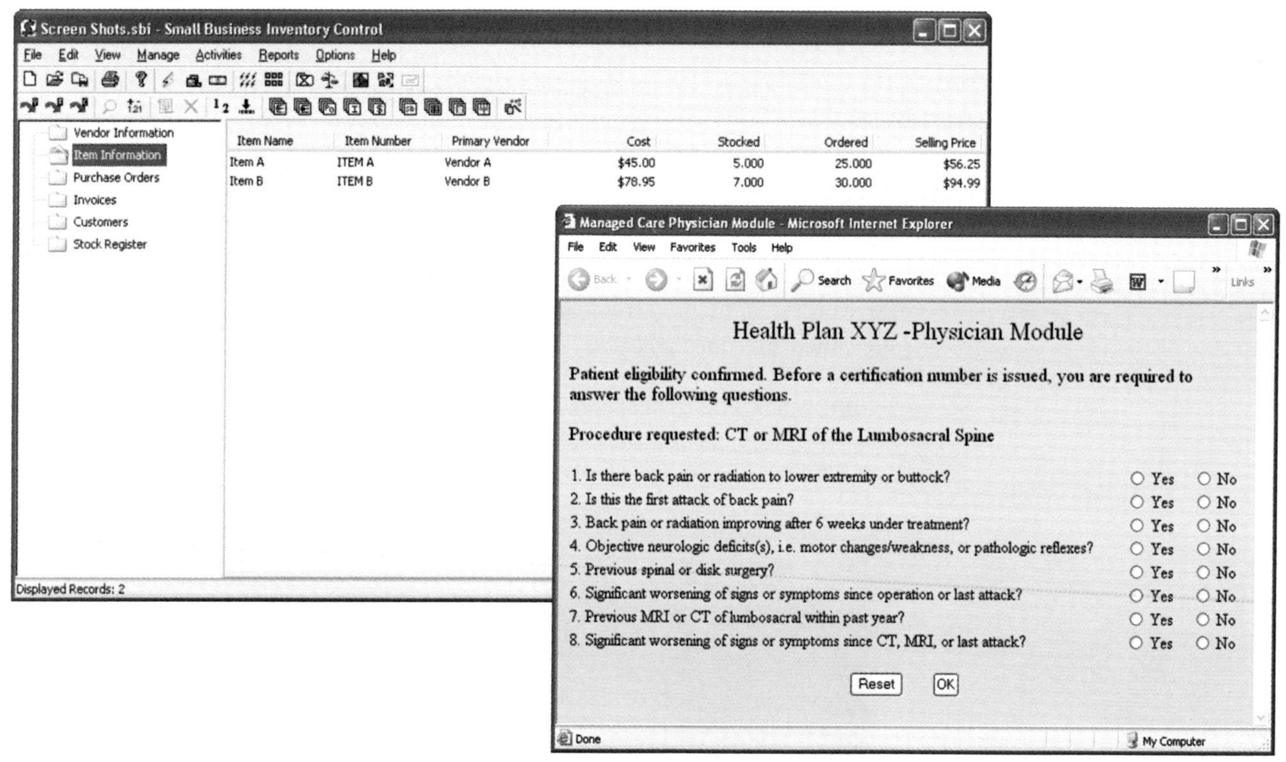

Optimization and Heuristic Approaches

Optimization and heuristic approaches are popular problem-solving methods used in decision support systems. An **optimization model** finds the best solution, usually the one that will best help individuals or organizations meet their goals. For example, an optimization model can find the best way to schedule nurses at a hospital or the appropriate number of products an organization should produce to meet a profit goal, given certain conditions and assumptions. Optimization models utilize problem constraints. A limit on the number of available work hours in a manufacturing facility is an example of a problem constraint. Optimization can be used to achieve huge savings. Here are just a few examples. Bombardier Flexjet, a company that sells fractional ownership of jets, used an optimization program to save almost $30 million annually to better

A "TaylorMade" Information and Decision Support System

For some people, being an ambassador for the best-selling golf club manufacturer would be a dream job. Traveling to all the best country clubs, playing one round of golf after another with the pros while they try out your clubs, and retiring to the clubhouse for dining and more golf chat might sound like heaven on earth. The reality, however, is not as glamorous. The sales associates for TaylorMade, makers of the number-one driver on the PGA tour, spend 60 percent of their time taking inventory in golf shops and nearly all the rest of their time traveling to visit the next customer.

TaylorMade's management team was struggling to keep up with the data. It had a group of "order management system programmers" who produced business reports for managers. The group was constantly buried under backlogged report requests. Managers were afraid to request reports because of the group's get-in-line attitude. Clearly, it was time for an overhaul of their information system.

TaylorMade hired a system development team to create an information management system to provide the sales associates and managers with the information they require to make informed decisions. Scheduled key-indicator reports provide important information for guiding sales strategies. Managers in the manufacturing division rely on exception reports to let them know which types of golf clubs are running low in the warehouse. The sales force uses demand reports to find out who their best customers are and on what areas of the market they should focus. All TaylorMade employees are able to access reports from the new system through a convenient Web-based self-service interface. They no longer need to wait in line for the "order management system programmers" to fulfill report requests.

Today, TaylorMade sales associates travel equipped with much more than golf clubs. Their handheld PCs provide them with the latest sales and inventory information. They can call up charts that summarize the customer's purchases for the year and can view open orders, recent shipments, and orders in transit. They are fully equipped to answer any question that may come up. The handhelds are equipped with bar-code scanners as well. The scanners dramatically reduce the amount of time the reps spend taking inventory and entering inventory data. Once scanned, the information is automatically uploaded to the home office's database.

Questions

1. How has TaylorMade increased the efficiency of its sales associates?
2. How have computer systems increased TaylorMade's ability to gain a competitive advantage in the golf equipment market?

Sources

1. *Staff. "TaylorMade Tour Pros Tee Off In Global Effort," BrandWeek, March 25, 2005, TaylorMade Web site, www.taylormadegolf.com.*
2. *"KANA Honored with CRM Magazine 2004 Service Leader Award for Web Self-Service, Business Wire, March 2, 2004; Business Objects Customers in the Spotlight Web site, www.businessobjects.com/customers/spotlight/taylormade.asp.*
3. *Kana ROI Success Stories Web site, accessed April 26, 2004, www.kana.com.*

schedule its aircraft and crews.[2] Hutchinson Port Holdings, the world's largest container terminal, saved even more—over $50 million annually.[3] The company used optimization to maximize the use of its trucks. The company processes a staggering 10,000 trucks and 15 ships every day. Deere, a manufacturer of commercial vehicles and equipment, increased shareholder value by over $100 million annually by enhancing customer satisfaction and using optimization to minimize inventory levels.[4] Using Solver in Microsoft's Excel spreadsheet can be used to solve optimization models (see Figure 9.5).

FIGURE 9.5 • Optimization as an approach to problem solving

Solver in Microsoft's Excel spreadsheet can be used to solve optimization models.

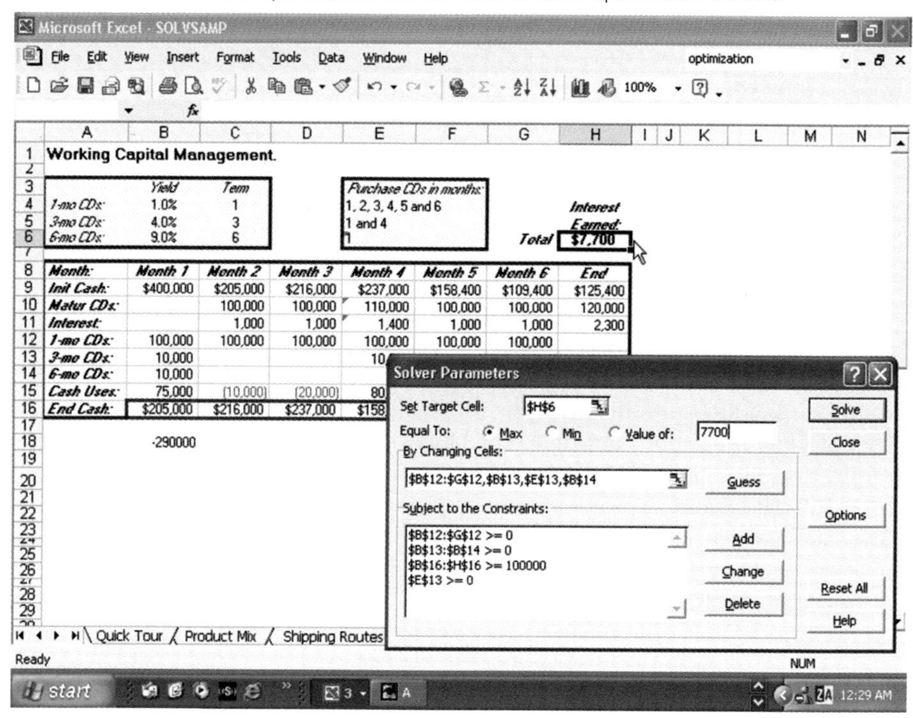

Heuristics, often referred to as "rules of thumb"—commonly accepted guidelines or procedures that usually find a good solution, but not necessarily the optimal solution—are often used in the decision-making process. An example of a heuristic rule in the military is to "get there first with the most firepower." A heuristic procedure used by baseball team managers is to place those batters most likely to get on base at the top of the lineup, followed by the "power hitters" who'll drive them in to score. An example of a heuristic used in business is to order four months' supply of inventory for a particular item when the inventory level drops to 20 units or less. See Figure 9.6. Even though this heuristic may not minimize total inventory costs, it may be a very good rule of thumb that produces advantageous results by avoiding stockouts without too much excess inventory. Trend Micro, a provider of antivirus software, has developed an antispam product that is based on heuristics that claims to have a 90 to 95 percent success rate in blocking spam.

FIGURE 9.6 • Heuristics

Heuristics (rules of thumb) are useful in almost every enterprise, from determining when to automatically restock inventory to deciding on the batting order of baseball players.

MANAGEMENT INFORMATION SYSTEMS

A **management information system (MIS)** is often used to support programmed decisions made in response to structured problems. The primary purpose of an MIS is to help individuals and organizations achieve their goals by providing reports and information to make better decisions. Individuals, for example, might want to get reports from their doctor on the results of a routine physical exam or a report comparing features and prices on a particular make and model car at auto dealers within 100 miles of their home. A swimming coach at a university might want a report showing all swimmers who have grades below a certain GPA. Filtering and analyzing highly detailed data can produce these reports.

Inputs to a Management Information System

Data entering the MIS can originate from either internal or external sources. The most significant internal source of data for the MIS is the transaction processing system (TPS). External sources include other organizations, governmental agencies, foreign countries, and outside individuals. New federal regulations from the IRS on how public and private organizations report their activities is an example of an external source of information.

One of the major activities of the TPS is to capture and store the data resulting from ongoing transactions. With every transaction, various TPS applications make changes to and update the organization's files and databases. Databases also supply the data needed for the MIS. A database containing research results on new drugs, for example, can be used by a pharmaceutical company to produce a variety of reports about which drugs seem to be working as expected. Outputs of a MIS—various reports—are discussed next. See Figure 9.7.

FIGURE 9.7 • Inputs and outputs of an MIS

An MIS uses internal and external data to produce reports that support decision making.

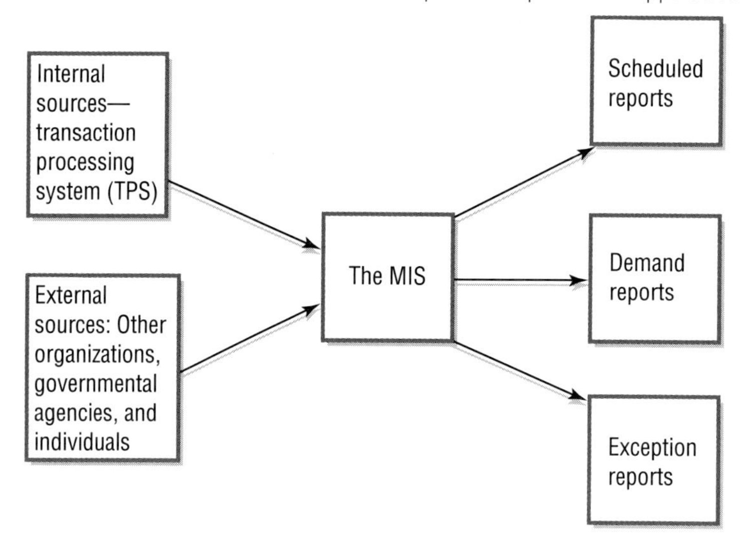

Outputs of a Management Information System

The output of most management information systems is a collection of reports, as shown in Figure 9.7.[5] These reports can be produced on paper, or, increasingly, distributed and viewed electronically. Providence Washington Insurance Company, for example, is using ReportNet from Cognos to reduce the number of paper reports and their associated costs.[6] According to Ed Levelle, CIO of Providence Washington, "Our intention was to replace paper reports with an Internet-based reporting process." The new reporting system creates an "executive dashboard" that shows current data, graphs, and tables to help managers make better real-time decisions. See Figure 9.8.

FIGURE 9.8 • An executive "dashboard"

This MIS reporting system puts many kinds of real-time information at managers' fingertips to aid decision making.

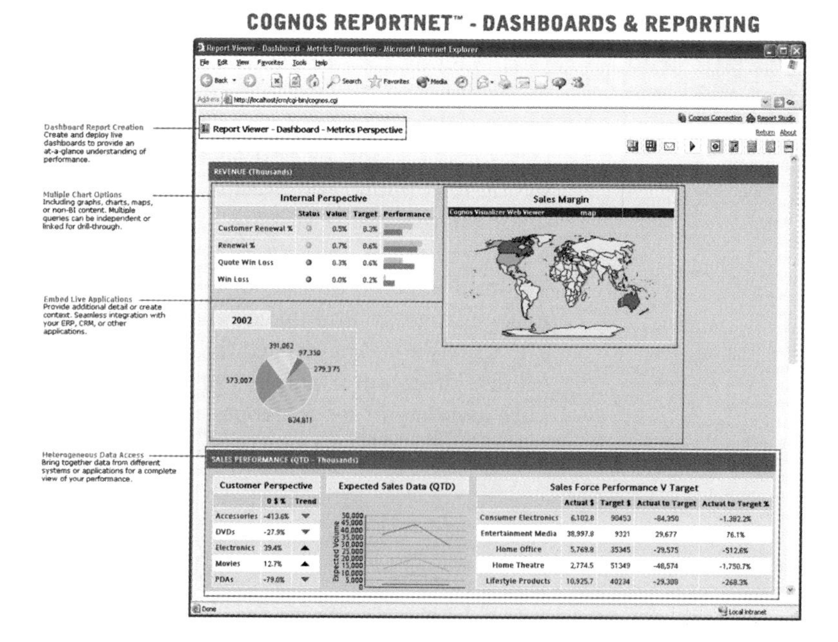

The main types of reports produced by an MIS are scheduled reports, demand reports, and exception reports.

Scheduled reports. **Scheduled reports** are produced periodically or on a schedule, such as daily, weekly, or monthly. An investor can receive monthly statements summarizing the performance of his or her stock and bond holdings (see Figure 9.9). A university student receives a report at the end of each semester or quarter summarizing his or her grades. The 200-year-old New York Stock Exchange is investigating the use of scheduled reports of trading.[7] The new system should cut costs and be more efficient.[8] Other financial MISs attempt to detect stock market fraud and abuse.[9]

FIGURE 9.9 • Scheduled reports

A scheduled report can summarize stock and bond positions every month.

MONTHLY PORTFOLIO PERFORMANCE

Name	Symbol	Quote	Shares	Current market value	Previous market value
Fiserve Inc.	FISV	$28.86	8	$230.88	$245.21
The Sports Authority	TSA	$25.23	11	$277.53	$251.55
Oracle	ORCL	$12.19	15	$182.85	$176.20
Total				$691.26	$685.55

A *key-indicator report*, a special type of scheduled report, summarizes the previous day's critical activities, and is typically available at the beginning of each workday. The president of the United States, for example, receives daily reports in the morning on national security, terrorism, the economy, and many other areas. Key-indicator reports can summarize inventory levels, production activity, sales volume, and the like. Key-indicator reports allow people to take quick, corrective action when it is most needed.

Demand reports. **Demand reports** are developed to give certain information at a person's request. In other words, these reports are produced on demand. An executive, for example, may want to know the inventory level for a particular item, such as the number of Antelope 800 Trek bicycles with a 20-inch frame that are in stock (see Figure 9.10). A demand report can be generated to give the requested information. You can get a statement of your credit from one of the credit bureaus on demand. The Laurel Pub Company, a bar and pub chain in England with over 630 outlets, uses demand reports to generate important sales data when requested.[10] The company expects to save about £500,000 over a 5-year period. Finding the right classes and professors can be critical for college students. Today, students can get demand reports on both. Other examples of demand reports include reports requested by executives to show the hours worked by a particular employee, total sales for a product for the year, and so on. The Atlanta Veterans Administration is putting some medical records on the

Internet to make them available on demand.[11] Dr. David Bower, chief of staff of the hospital, uses the new system to look at a patient's X-ray, read the medical files online, make a diagnosis, and prescribe treatment from his home, saving valuable time.

FIGURE 9.10 • Inventory demand report

An executive might look at this inventory demand report to learn the inventory level for Green Trek Antelope 800 bicycles with a 20-inch frame.

INVENTORY DEMAND REPORT

Model	Size	Color	Quantity in stock
Antelope 800	20 inch frame	Green	4

Exception reports. **Exception reports** are reports that are automatically produced when a situation is unusual or requires action. For example, a patient will get an exception report from a clinic only if a blood test shows a possible problem (Figure 9.11). A manager might set a parameter that generates a report of all inventory items with fewer than 50 units on hand. The exception report generated by this parameter would contain only those items with fewer than 50 units in inventory. A bank can use exception reports to get a list of customer inquiries that have been open for a period of time without some progress or closure. As with key-indicator reports, exception reports are most often used to monitor aspects critical to an organization's success. In general, when an exception report is produced, an appropriate individual or executive takes action.

Exception reports are also used to help fight terrorism.[12] The Matchmaker System scans airline passenger lists and displays an exception report of passengers that could be a threat, so authorities can remove the suspected passengers before the plane takes off. The system was developed in England as part of its Defence Evaluation Research Agency.

FIGURE 9.11 • Exception report

This exception report from a medical lab shows only blood test results that are out of normal range.

EXCEPTION REPORT

Facility: Dallas Primary Care
Physician: Welch
Patient: Ben Bechtold

Test Name	Units	Results	Range	Flag (H–high; L–low)
Sodium	mEq/L	133	(135–146)	L
Cholesterol	mg/dL	265	(140–200)	H

Ancient Company Profits from Better Decisions and Reports

The Kano has been brewing sake in the small town of Mikage, Kobe, Japan, since 1659. Their sake wine, named *Kiku-Masamune* (chrysanthemum sake) became a staple in Japan. Over the centuries it has played an important role in Japanese culture and religion. In 1877, Kiku-Masamune became a global commodity when the company began exporting to England. Today, Kiku-Masamune sake is exported to many countries, and until recently it enjoyed a monopoly position in the sake market.

Kiku-Masamune contracted with a U.S. systems developer to design an information and decision support system customized for the company's needs. "Before implementing the system, it was like we were only seeing pieces of the puzzle but never the big picture. Now we're able to identify where changes are needed in our contracts and drill down to the information we need to use our sales resources most effectively," says Yoshihiko Handa, Sales and Marketing Director.

The new system produces scheduled reports annually, quarterly, monthly, weekly, and daily that provide information on key indicators such as performance by retailer and profits by product and customer. Kiku-Masamune uses this information to determine which of its customers require attention and to expand its customer base. The new system also provides the company with information to better manage its manufacturing process. Kiku-Masamune can provide customers with "just in time" delivery of sake, which saves both Kiku-Masamune and its customers the expense of storing large inventories.

Kiku-Masamune uses its new information and decision support system to price its premium sake astutely, to guide decision making, and help optimize sales opportunities. Effective information gathering and information sharing are clearly at the center of the company's new-found calm amid the storm of competition.

Questions

1. How can computer management information systems guide a company to achieve a competitive advantage within an industry?
2. What characteristics of an information and decision support system make it valuable?

Sources:

1. Kiku-Musamune Web site, accessed April 29, 2005, *www.kikumasamune.com.*
2. "Sake Producer Provides Old-World Taste and New-World Competitive Edge," *Business Objects Success Stories*, accessed April 29, 2005, *www.businessobjects.com/customers.*
3. Business Objects Business Intelligence Web page, accessed April 29, 2005, http://www.businessobjects.com/products/bistandardization/ default.asp.

DECISION SUPPORT SYSTEMS

As discussed in Chapter 1, a *decision support system (DSS)* is an information system used to support problem-specific decision making. The focus of a DSS is on decision-making effectiveness when faced with unstructured or semistructured problems. For a TV producer, it could be better ratings through a more comprehensive analysis of viewer desires. For a business, it could mean higher profits, lower costs, and better products and services. In the sports world, AC Milan, a football (soccer) league champion, uses a DSS to reduce injuries by 90 percent.[13] The software models help players determine their diets, training schedule, and mental attitude. According to the coach, skeptical athletes changed their minds about the program once it was in place: "When they [the players] realized that it would make them more healthy and prolong their playing careers, they became much more accepting."

Overall, a decision support system should assist people and organizations with all aspects of decision making. Moreover, the DSS approach realizes that people, not machines, make decisions. DSS technology is used primarily to support making decisions that can solve problems and help achieve individual and organizational goals. For example, in addition to being trained to diagnose disease, today's health-care providers also have to be trained in how to handle a terrorist attack. The U.S. Department of Human and Health Services (HHS) has developed a decision support system, called ePocrates, that not only offers doctors instant access to patients' medical information, but can also send alerts to doctors and other medical providers in the event of a bioterrorist event (see Figure 9.12). Cornell University uses a DSS to monitor the journals and newspapers it purchases for its library.[14]

FIGURE 9.12 • ePocrates

Products like ePocrates offer doctors instant access to medical information and aid in the treatment and diagnosis of patients and to help in the event of a bio-terrorist event.

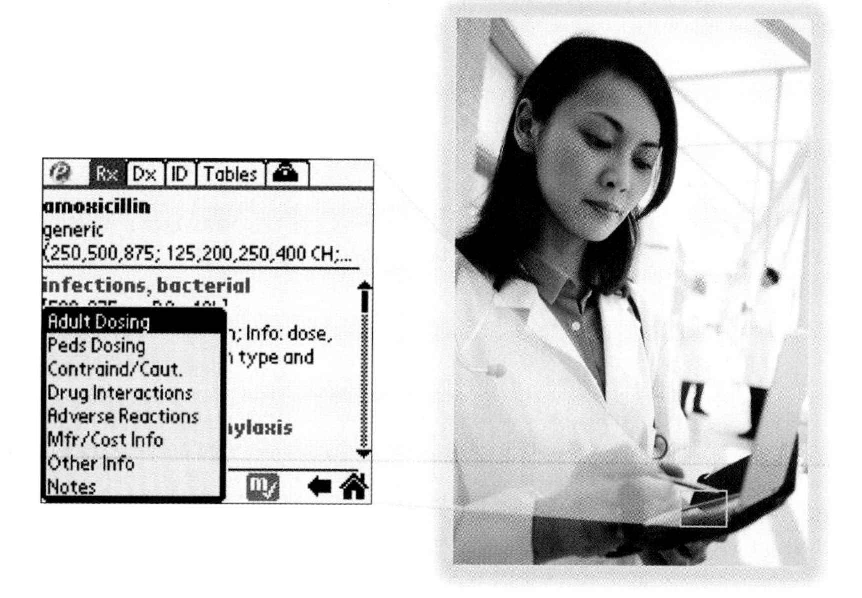

Characteristics of a Decision Support System

Decision support systems have a number of characteristics. In general, a decision support system can:

Handle a range of data from small amounts to large amounts. For instance, advanced database management systems have allowed people to search databases of any size for information when using a DSS. TUI, a travel company in Europe, uses a DSS with a large amount of data to make better decisions to help reduce costs and increase efficiency.[15] According to the company's financial director, "We will make significant cost savings. It will allow us to become more efficient and provide us with one version of the truth." A DSS is also flexible enough to solve problems where only a small amount of data is required.

Obtain and process data from different sources. Some data sources may reside in databases on personal computers; others could be located on different mainframe systems or networks. DSSs have the capability to access data external to the organization and integrate this data with internal data.

Provide report and presentation flexibility. One of the reasons that DSSs were developed was that TPS and MIS were not flexible enough to meet the full variety of decision makers' problem or information needs. Whereas other information systems produce primarily fixed-format reports, DSSs have more widely varied formats. People can get the information they want, presented in a format that suits their needs. Furthermore, output can be presented on computer screens or produced on printers, depending on the needs and desires of the problem solvers. One diabetes Web site, for example, developed a DSS to provide customized treatment plans and reports. ING Direct, a financial services company, uses a flexible DSS to summarize the bank's financial performance (see Figure 9.13).[16] According to one bank executive, "What we really needed was measurement and tracking so that we could determine how successful we were and make modifications to our plans on a real-time basis."

FIGURE 9.13 • ING Direct

ING Direct, a financial services company that operates primarily online, uses a DSS to help it respond quickly to enhance performance and provide excellent service and competitive rates to customers.

A decision support system can provide whatever orientation a person prefers, be it textual or graphical. Some prefer a straight text interface, but others may want a decision support system that helps them make attractive, informative graphical presentations on computer screens and in printed documents. Today's decision support systems can produce text, tables, line drawings, pie charts, trend lines, and more. By using their preferred orientation, people can both use a DSS to get a better understanding of a situation, if required, and to convey this understanding to others.

WHO'S NAUGHTY AND WHO'S NICE

Special data-gathering systems help retailers define their customer base—who buys what, who returns what—and form more precise marketing strategies. Best Buy identified four "typical" shoppers: "Buzz," young and high tech; "Barry," wealthy professional; "Ray," family man; and "Jill," soccer-mom. They are redesigning each of their 660 stores to fit whichever one is predominant in that area. When the Santa Rosa, California, store added lower, softer music and pink umbrellas for Jills, sales climbed 30%.

Source:
Retailer's mantra: Profile thy shopper
By Ariana Eunjung Cha
The Washington Post
http://seattletimes.nwsource.com/html/businesstechnology/2002443975_retailprofile18.html
August 18, 2005

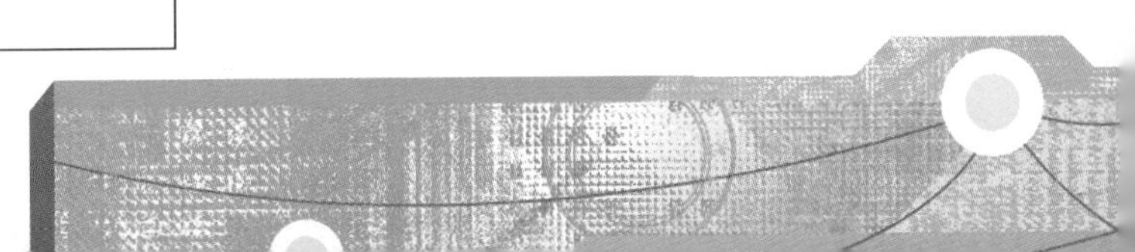

FIGURE 9.14 • Trading DSS

Brokers and traders use a DSS to decide when to buy and sell.

Perform complex, sophisticated analysis and comparisons using advanced software packages. Marketing research surveys, for example, can be analyzed in a variety of ways using analysis programs that are part of a DSS. Many of the analytical programs associated with a DSS are actually stand-alone programs. The DSS provides a means of bringing these together. Toyota uses a sophisticated 3D DSS to design plants around the world.[17] The DSS was used to improve robotic work processes and avoid having them collide on the factory floor. As futures and options trading is becoming totally electronic, many trading firms are using DSS software to perform sophisticated analysis and make substantial profits for traders and investors. Some DSS trading software is programmed to place buy and sell orders automatically without a trader manually entering a trade, based on parameters set by the trader. See Figure 9.14.

Support optimization and heuristic approaches. For smaller problems, decision support systems have the capability to find the best (optimal) solution. For more complex problems, heuristics are used. With heuristics, the computer system can determine a very good—but not necessarily the optimal—solution. A state, for example, can order drug companies to distribute hundreds of millions of dollars of scarce drugs to hospitals and clinics as a result of the settlement of a drug case using heuristics. By supporting all types of decision-making approaches, a DSS gives the decision maker a great deal of flexibility in getting computer support for decision-making activities.

Perform "what-if" and goal-seeking analysis. **What-if analysis** is the process of making hypothetical changes to problem data and observing the impact on the results. Consider an emergency disaster plan. What-if analysis could determine the consequences of a hurricane slamming into New Orleans with heavy flooding or the consequences of the hurricane hitting the Florida coast instead. With what-if analysis, a person can make changes to problem data (where the hurricane hits) and immediately see the impact on the results (the damage done and the lives lost.) See Figure 9.15. **Goal-seeking analysis** is the process of determining what problem data is required for a given result. For example, suppose a financial manager has a goal to earn a return of 9 percent on any investment, and she is considering an investment with a certain monthly net income. Goal seeking allows the manager to determine what monthly net income (problem data) is needed to have a return of 9 percent (problem result).

FIGURE 9.15 • The value of what-if analysis

What-if analysis could determine the consequences of a hurricane slamming into New Orleans with heavy flooding or the consequences of the hurricane hitting the Florida coast instead.

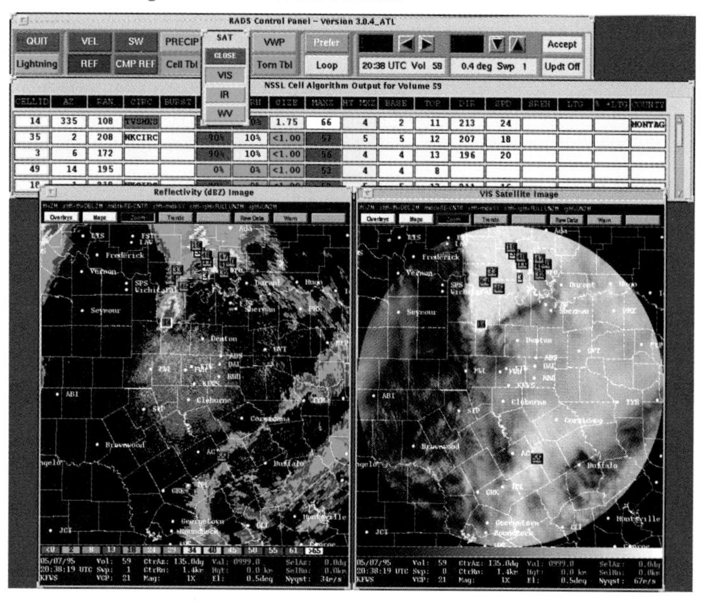

Simulation. With *simulation*, the DSS attempts to mimic an event that could happen in the future. Simulation uses chance or probability. For example, there is a certain chance, or probability, that it will rain tomorrow or during the next several months. Sometimes, a problem is so complex that a normal decision support system is just too difficult to develop. One popular alternative is to develop a computer simulation that allows people to analyze various possibilities and scenarios. Corporate executives and military commanders often use computer simulations to allow them to try different strategies in different situations. Corporate executives, for example, can try different marketing decisions in various market conditions. Military commanders often use computer war games to fine-tune their military strategies in different war conditions. Air traffic controllers can hone their skills and emergency responses in various scenarios. See Figure 9.16.

FIGURE 9.16 • Simulation

Air traffic controllers use simulation to learn and practice skills and emergency responses without endangering anyone.

TABLE 9.1 • Characteristics of decision support systems

DSS characteristics
Large amounts of data
Different data sources
Report and presentation flexibility
Geared toward individual decision-making styles
Modular format
Perform sophisticated analysis
Graphical orientation
Optimization and heuristic approach
What-if and goal-seeking analysis
Simulation

Of course, not all DSSs work the same or have all of the characteristics summarized in Table 9.1. Some are small in scope and only take advantage of some of the characteristics in Table 9.1. An agricultural department for a state university, for example, can develop a DSS based on an Excel spreadsheet to help farmers decide what to plant to maximize their revenues. Other small-scale DSSs can provide patients and their families with important medical records and reports, which can be critical to patient involvement and recovery from disease.

THE GROUP DECISION SUPPORT SYSTEM

The DSS approach has resulted in better decision making for all kinds of individual users. However, many DSS approaches and techniques are not suitable for a group decision-making environment. Although not all people are involved in committee meetings and group decision-making sessions, some people can spend more than half of their decision-making time in a group setting. Such people need effective approaches to assist with group decision making, such as a formal decision room (see Figure 9.17). A **group decision support system (GDSS)**, also called a *computerized collaborative work system*, consists of the hardware, software, people, databases, and procedures needed to provide effective support in group decision-making settings. A group of petroleum engineers located around the world, for example, might be collaborating on a new refinery. A GDSS can be used to help the engineers design the new refinery by providing a means to communicate with each other and share design ideas.

FIGURE 9.17 • Group decision support system

High-tech rooms like the one shown here create a group dynamic that allows people to come together to explore solutions to complex problems.

Characteristics of a GDSS

A GDSS has a number of characteristics that go beyond the traditional DSS. These systems try to build on the advantages of individual support systems while responding to the fact that new and additional approaches are needed in a group decision-making environment. Some GDSSs allow the exchange of information and expertise among people without meetings or direct face-to-face interaction. The characteristics of a typical GDSS include:

- Flexibility
- Anonymous input
- Reduction of negative group behavior
- Support of positive group behavior

GDSS Software or Groupware

GDSS software, often called **groupware**, helps with joint work group scheduling, communication, and management. One popular groupware package, Lotus Notes, can capture, store, manipulate, and distribute memos and communications that are developed during group projects. By using groupware, all group members can share information to accomplish joint work, even when group members are located around the globe. Increasingly, groupware is being used on the Internet. WebEx (*www.webex.com*), Genesys Meeting Center (*www.genesys.com*), and GoToMeeting Corporate (*www.gotomeetingonline.com*) are examples of groupware products available on the Web.[18] Using groupware gives every employee rapid access to a vast source of information. Wyndham International, for example, uses collaboration and e-meeting software from Centra Software to save time and money.[19] The company estimates that it has saved more than $1 million so far in travel and communications costs. According to a company manager, "We use it for weekly and monthly conferences, and because it's voice over IP, we save about $10,000 to $15,000 a month on our telephone bill." See Figure 9.18. Artificial intelligence and special-purpose systems, discussed next, can also give people access to vast sources of information.

FIGURE 9.18 • Groupware

Groupware enables team members to meet and collaborate across the office, across town, or across the globe.

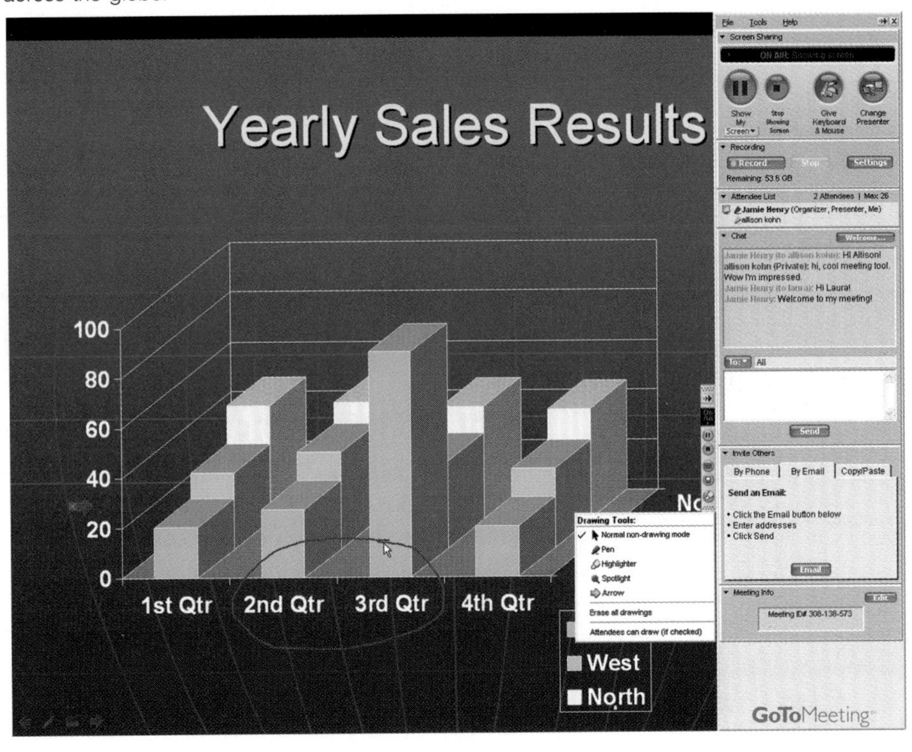

ARTIFICIAL INTELLIGENCE AND SPECIAL-PURPOSE SYSTEMS

At a Dartmouth College conference in 1956, John McCarthy proposed the use of the term *artificial intelligence (AI)* to describe computers with the ability to mimic or duplicate the functions of the human brain. Many AI pioneers attended this first AI conference; a few predicted that computers would be as "smart" as people by the 1960s. The prediction never materialized, but the benefits of artificial intelligence can be seen today. Advances in AI have led to systems that work like the human brain to recognize complex patterns.[20]

An Overview of Artificial Intelligence

Science fiction novels and popular movies have featured scenarios of computer systems and intelligent machines taking over the world (Figure 9.19). The movie *A.I.* explored the conflicted emotions of a boy with a robotic body but human emotions, a futuristic glimpse of the issues you may someday face. Fictional accounts of humanlike computers are entertaining, but humans are far from creating computer systems that can completely replace human decision makers. Even so, today you are seeing the direct application of many computer systems that use the notion of AI. These systems are helping to make medical diagnoses, explore for natural resources, determine what is wrong with mechanical devices, and assist in designing and developing other computer systems. In this section you explore the exciting applications of artificial intelligence and look at what the future might hold.

FIGURE 9.19 • Timeline of AI in the movies

Many popular movies have featured scenarios of computer systems and intelligent machines taking over the world. People seem to be both fascinated and scared by the idea of artificial intelligence.

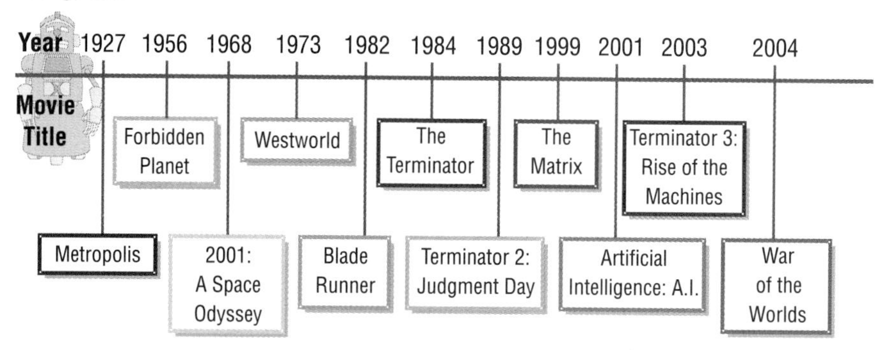

Artificial Intelligence in Perspective

Artificial intelligence systems are computer systems that simulate human thought and behavior. They include the people, procedures, hardware, software, data, and knowledge needed to develop computer systems and machines that demonstrate characteristics of intelligence and behavior. Researchers, scientists, and experts on how humans think are often involved in developing these systems. As with other computer systems, the overall purpose of artificial intelligence applications is to help individuals and organizations achieve their goals. According to Eric Lander, director of the Broad Institute at MIT and Harvard, "Computers looked at DNA sequences and decided the next experiment that needed to be done was to close gaps in the genome. The computer knew what experiments to order up, without human input."[21]

The Difference Between Natural and Artificial Intelligence

Today, there are profound differences between natural and artificial intelligence, but the differences are declining in number, as shown in Table 9.2. One of the driving forces behind AI research is an attempt to understand how human beings actually reason and think. It is believed that the ability to create machines that can reason will only be possible once the human processes for doing so are fully understood. According to Steve Grand, who developed a robot called Lucy and was given an award for his work in artificial intelligence, "True machine intelligence, let alone consciousness, is a very long way off."[22] Asimo, a robot from Honda, is making progress (*http://world.honda.com/asimo*). According to Honda's head of the European Research Institute, "Asimo is a marvelous walking machine, a masterpiece of engineering, but the next stage is to enable it to develop the ability to think for itself."[23]

The *Turing Test* attempts to determine if the responses from a computer with intelligent behavior are indistinguishable from responses from a human. No computer has passed the Turing Test, developed by Alan Turing, a British mathematician. The Loebner Prize offers money and a gold medal for anyone developing a computer that can pass the Turing Test (see *www.loebner.net*).

TABLE 9.2 • A comparison of natural and artificial intelligence

Attributes	Natural intelligence (human)	Artificial intelligence (machine)
Acquire a large amount of external information	High	Low
Use sensors (eyes, ears, touch, smell)	High	Low
Be creative and imaginative	High	Low
Learn from experience	High	Low
Be forgetful	High	Low
Make complex calculations	Low	High
Be adaptive	High	Low
Use a variety of information sources	High	Low
Transfer information	Low	High

Components of Artificial Intelligence

AI is a broad field that includes several key branches, such as robotics, vision systems, natural language processing, learning systems, neural networks, genetic algorithms, intelligent agents, and expert systems (see Figure 9.20). Many of these areas are related; advances in one can occur simultaneously with advances in another, or result in advances in others.

FIGURE 9.20 • A conceptual model of artificial intelligence

AI is a broad field that includes several key branches. Many of these areas are related; advances in one can occur simultaneously with advances in another, or result in advances in others.

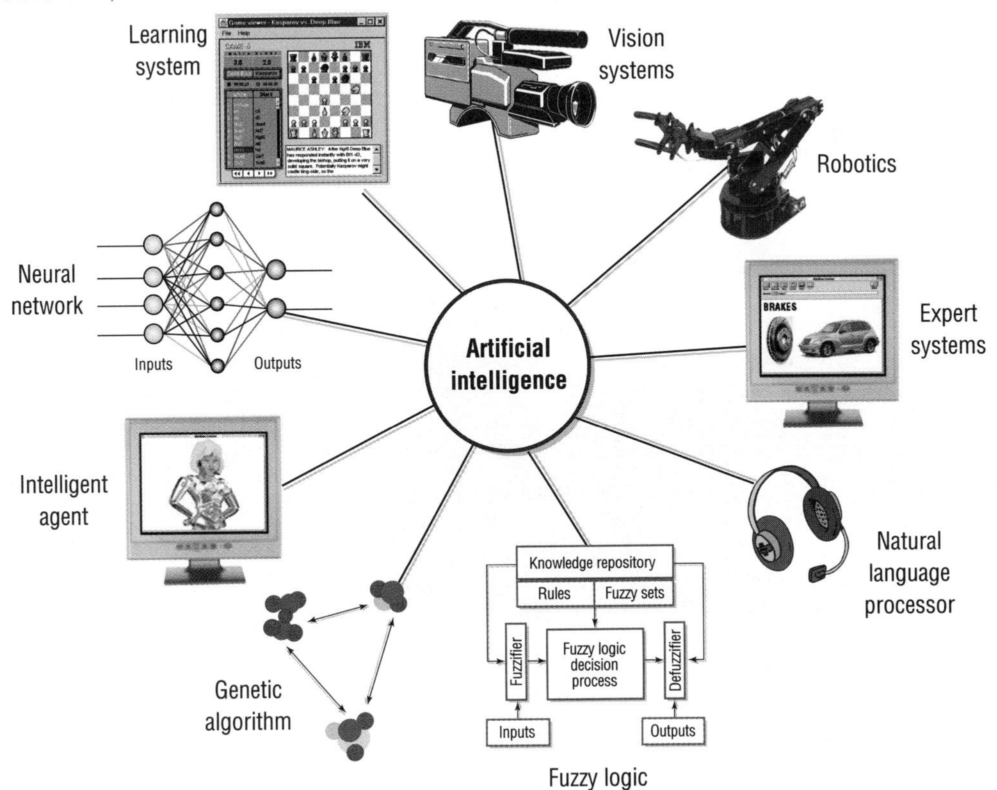

Robotics

Robotics involves developing mechanical or computer devices to perform tasks that require a high degree of precision or are tedious or hazardous for humans. Contemporary robotics combines both high-precision machine capabilities and sophisticated controlling software. The controlling software in robots is what is most important in terms of A.I.

There are many applications of robots, and research into these unique devices continues. For many businesses, robots are used to do the 3 Ds—dull, dirty, and dangerous jobs.[24] Manufacturers use robots to assemble and paint products. The U.S. Army is involved in developing medical robotics to allow doctors to perform surgery in combat areas via remote control. Not only does this technique make it safer for doctors in combat situations, it allows them to "be" in several places at once. A surgical system, for example, can allow doctors to operate using a robotic arm (see Figure 9.21). Sitting at a console, the surgeon can replace a heart valve or remove a tumor. The robotic arm can be accurately controlled and requires only a small incision in the patient, making surgery more precise and the recovery easier. Some surgical robots cost more than $1 million and have multiple surgical arms and sophisticated vision systems.[25] Researchers at the University of West England have designed an experimental robot that gets its power from digesting houseflies.[26] The scientists hope the self-sustaining robot, called EcoBot, will allow future robots to function for long periods of time without the need for electricity. MIT's Kismet is an example of a robot that displays human characteristics. (*www.ai.mit.edu/projects/humanoid-robotics-group/site-index.html*).

FIGURE 9.21 • Robotic surgery

Some surgeons are using robotic technology to help them operate more precisely.

Vision Systems

Another area of AI involves vision systems. **Vision systems** include hardware and software that permit computers to capture, store, and manipulate visual images and pictures. A California wine bottle manufacturer, for example, can use a computerized vision system to inspect wine bottles for flaws. The vision system can save the bottle producer both time and money. The U.S. Justice Department makes use of vision systems to perform fingerprint analysis, with almost the same level of precision as human experts. The speed with which the system can search through the huge database of fingerprints has brought a quick resolution to many unsolved mysteries.

Vision systems can be used to give robots "sight" (see Figure 9.22). A sophisticated robot and vision software can be used to attach the rear windscreen on cars. Generally, robots with vision systems can recognize black and white and some shades of gray, but do not have good color or three-dimensional vision. Other systems concentrate on only a few key features in an image, ignoring the rest. It may take many years before a robot or other computer system can "see" in full color and draw conclusions from what it sees, the way humans do.

FIGURE 9.22 • ASIMO (advanced step in innovative mobility)

Honda's ASIMO robot has eyes and a vision system sophisticated enough that it can use light switches, turn door knobs, and work at tables.

Natural Language Processing

Natural language processing, often referred to as *speech recognition*, allows a computer to understand and react to statements and commands made in a "natural" language, such as English. With natural language processing, it is possible to speak into a microphone connected to a computer and have the computer convert the electrical impulses generated from the voice into text files or program commands. NaturallySpeaking from ScanSoft, for example, can be used to dictate commands and text into a variety of applications, including Word, Word Perfect, PowerPoint, Internet Explorer, AOL, Outlook, and other programs. Recent versions of Microsoft Office include a speech recognition tool. Accordant Health Services uses natural language processing to allow insurance customers to ask questions using the company's Web site.[27] Computers convert questions, such as "Why did I get a bill from my doctor?" into answers. Help desks that use natural language processing can be frustrating when a person calling with a problem needs to speak to a human.

Some brokerage firms have search engines that use natural language processing to allow customers to get their questions answered through the brokerage firm's call center. Perhaps you've encountered natural language processing systems when using directory assistance, or making collect or calling card calls. Restoration Hardware has developed a Web site that uses natural language processing to allow its customers to quickly find what they want on its site.[28] The natural language processing system corrects spelling mistakes, converts abbreviations into words and commands, and allows people to ask questions in English. In addition to making it easier for customers, Restoration Hardware has seen an increase in revenues as a result of the use of natural language processing.

Three major challenges of natural language processing require artificial intelligence solutions. The first is interpreting ambiguous words, words that sound alike but are spelled differently, such as "here" and "hear." The second is sentence parsing, which involves determining where one word ends and the next begins. The third challenge is being able to interpret the unique ways in which people pronounce words. Although researchers have not completely overcome these challenges, speech recognition technology has advanced enough that many computer users are considering it as an alternative to typing (see Figure 9.23).

NOT QUITE ROSIE THE ROBOT—YET

The home robot called Nuovo represents a breakthrough in size and affordability of electronic helpers. Developed by a Japanese firm, Nuovo is 15 inches high and sells for about $6000. Besides winking, blinking, and nodding, it plays music, tells the time, takes pictures, shakes hands, walks, talks, and is *very* polite. While the next model still won't be able to walk Astro the dog, it will read appointments, e-mail messages, traffic reports, and headlines off the Internet.

Source:
I, Roommate: The Robot Housekeeper Arrives
By Mark Allen
The New York Times
http://www.nytimes.com/2005/07/14/garden/14robot.html?
July 14, 2005

Learning Systems

Another part of AI deals with **learning systems**, a combination of software and hardware that allows the computer to change how it functions or reacts to situations based on feedback it receives. For example, a number of computerized games have learning abilities, such as Twenty Questions (*www.20q.net*). If the computer does not win a particular game, it remembers not to make the same moves under the same conditions. Learning software requires feedback on the results of its actions or decisions. At a minimum, the feedback needs to indicate whether the results are desirable (winning a game) or undesirable (losing a game). The feedback is then used to alter what the system does in the future.

FIGURE 9.23 • Speech recognition technology

Speech recognition technology has advanced enough that many computer users are considering it as an alternative to typing.

Neural Networks

An increasingly important aspect of AI involves neural networks. A **neural network** is a computer system that can act like or simulate the functioning of a human brain. The systems use massively parallel processors in an architecture that is based on the human brain's own meshlike structure. Neural networks can process many pieces of data at once and can learn to recognize patterns. The systems then program themselves to solve related problems on their own. Some of the specific features of neural networks include:

- The ability to retrieve information even if some of the neural nodes fail
- Fast modification of stored data as a result of new information
- The ability to discover relationships and trends in large databases
- The ability to solve complex problems for which all of the information is not present

Neural networks excel at pattern recognition, and this ability can be used in a wide array of situations. Pattern recognition can be used to help prevent terrorism by analyzing and matching images from multiple cameras focusing on people or locations.[29] Sandia Laboratories has developed a neural network system to give soldiers in a military conflict real-time advice on strategy and tactics.[30] Neural network software designed for investors looks for stock market patterns and advises brokers when to buy or sell. Some hospitals use neural networks to determine a patient's likelihood of contracting cancer or other diseases. Companies making robots can use neural networks to improve motor coordination and movement in robots. The neural network software can give the robot a smooth walk and allows it to get up if it falls.

Fuzzy Logic

Computers typically work with numerical certainty; certain input values always result in the same output. However, in the real world, as you know from experience, certainty is not always the case. To handle this dilemma, a specialty research area in computer science, called *fuzzy sets* or **fuzzy logic**, has been developed. Research into fuzzy sets has been going on for decades. A simple example of fuzzy logic might be one in which cumulative probabilities do not add up to 100 percent, a state that occurs frequently in medical diagnosis. Another example of fuzzy logic involves unclear terms, such as tall or many. Fuzzy logic deals with ambiguous criteria or probabilities and events that are not mutually exclusive. Fuzzy sets and fuzzy logic theory allow people to incorporate interpretations and relationships that are not completely precise or known.

Genetic Algorithms

A **genetic algorithm** is an approach to solving large, complex problems where a number of algorithms or models change and evolve until the best one emerges. The approach is based on the theory of evolution, which requires variation and natural selection. The first step in developing a genetic algorithm solution is to change or vary a number of competing solutions to a problem. This can be done by changing the parts of a program or combining different program

JOB TECHNOLOGY

Amazon Leverages Artificial Intelligence against Fraud

For large retailers like Amazon.com, losses to credit-card fraud are substantial. Amazon.com has 35 million customers who access its products through five Web sites adapted for different countries and languages: www.amazon.com, www.amazon.fr, www.amazon.co.uk, www.amazon.de, and www.amazon.co.jp. Amazon turned to SAS Institute to develop the foundation for its state-of-the-art fraud-detection system.

Fraud-detection techniques are not typically publicized—the less people know about them the more effective they are. Even Amazon won't disclose the details of its AI system. The company is happy to disclose, however, the effectiveness of its new AI fraud-detection system. According to Jaya Kolhatkar, Amazon.com's director of fraud detection, the new system greatly reduced the cases of fraud at Amazon. In the first six months of the system's use, fraud rates were halved.

The system developed for Amazon uses classic AI techniques, such as neural networks, to analyze patterns in the data to "learn" which patterns represent fraud, much the same way humans learn. The system works on multiple levels. One portion of the system analyzes transactions as they occur, while the credit-card number is being approved. Another subsystem crunches data in Amazon's huge transaction database, looking for fraudulent activities in past transactions.

"Fraudsters generally follow similar patterns of behavior," says Kolhatkar. "That makes it easier to detect fraud because you can look for corresponding patterns in transaction and customer data." For example, Amazon knows that fraudsters tend to purchase goods—such as electronics—that they can dispose of easily. Also, they do not ship the goods to the same address that is used for billing, so an order not shipped to the billing address might be an indication of a fraudulent transaction. They also tend to use the fastest possible shipping method. Of course, use of these features does not mean that fraud has occurred, but combined with other indicators, they would be pointers to follow up on.

Amazon's fraud-detection system analyzes the behavior patterns of fraudsters and builds predictive scores that indicate the likelihood of fraudulent behavior. "We run these scores against the customer database," says Kolhatkar. "We then use SAS to prioritize the results. Obviously, we have to investigate a case of potential fraud very thoroughly before beginning legal action, so we prioritize the results of running the fraud scores and begin with the highest priority cases."

Questions

1. How are AI systems empowering businesses to attain goals previously unattainable?
2. What are the strengths and limitations of AI systems, and how do they relate to the strengths and weaknesses of humans?

Sources:
1. *Cohen, A. "Amazon Outlet,"* PC Magazine, *May 10, 2005, p. 85.*
2. *"Amazon.com Calls on SAS for Fraud Detection,"* Citigate ICT PR, *May 20, 2003, www.lexis-nexis.com.*
3. *"Growth of Internet Fraud Is Driving New Technologies to Safeguard Online Payments,"* PR Newswire, *December 10, 2003.*
4. *Amazon.com Web site, accessed April 29, 2005, www.amazon.com.*

segments into a new program. If you think of program segments as building blocks similar to genetic material, this process is similar to the evolution of species, where the genetic makeup of a plant or animal mutates or changes over time (see Figure 9.24).

The second step is to select only the best models or algorithms, which then continue to evolve. Programs or program segments that are not as good as others are discarded. This part of the process is similar to natural selection, where only the best or fittest members of a species survive and continue to evolve.

FIGURE 9.24 • Genetic algorithms

If you think of program segments as building blocks similar to genetic material, the process of a genetic algorithm is similar to the evolution of species, where the genetic makeup of a plant or animal mutates or changes over time.

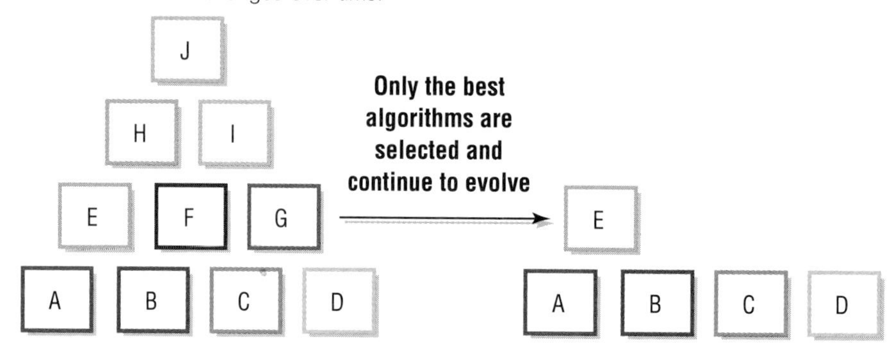

This process of variation and natural selection continues until the genetic algorithm yields the best possible solution to the original problem. For example, some investment firms use genetic algorithms to help select the best stocks or bonds. Genetic algorithms are also used in computer science and mathematics. Genetic algorithms can also help companies determine which orders to accept for maximum profit.[31] The approach helps companies select the orders that will increase profits and take full advantage of the company's production facilities.

Intelligent Agents

An **intelligent agent** (also called an intelligent robot or *bot*, an abbreviation for robot) consists of programs and a knowledge base used to perform a specific task for a person, a process, or another program. Like an agent who searches for the best endorsement deals for a top athlete, an intelligent agent often searches to find the best price, the best schedule, or the best solution to a problem. Often used to search the vast resources of the Internet, intelligent agents can help people find information on an important topic or the best price for a new digital camera. Intelligent agents, like Cybelle and others, can be found at *www.agentland.com* (see Figure 9.25). Intelligent agents have been used by the U.S. Army to route security clearance information for soldiers to the correct departments and individuals. What used to take days when done manually now takes hours. Intelligent agents can also be used to make travel arrangements, monitor incoming e-mail for viruses or junk mail, and coordinate the meetings and schedules of busy executives. Some companies use intelligent agents to find job candidates by searching millions of resumes in a database by title, company, gender, and other factors to find the best job candidates.

REST ON THE SEVENTH DAY?

A European research consortium is virtually playing games to garner insight into the evolution of human society. Hosted by a network of 50 computers, their system unites experts in linguistics, artificial intelligence, computer science, and sociology to see if approximately 1000 virtual agents in a simulated world will develop language and culture. Rules include the need to eat to live, the ability to learn through trial and error, and being fruitful and multiplying.

Source:
Simulated society may generate virtual culture
by Will Knight
NewScientist.com news service
http://www.newscientist.com/article.ns?id=dn7674
July 14, 2005

Expert Systems

An **expert system (ES)** acts or behaves like a human expert in a field or area. Charles Bailey, one of the original members of the Library and Information Technology Association, developed one of the first expert systems in the mid-1980s to search the University of Houston's library

FIGURE 9.25 • Internet-based intelligent agent

Intelligent agents on the Internet can help people find information on an important topic or the best price for a new digital camera.

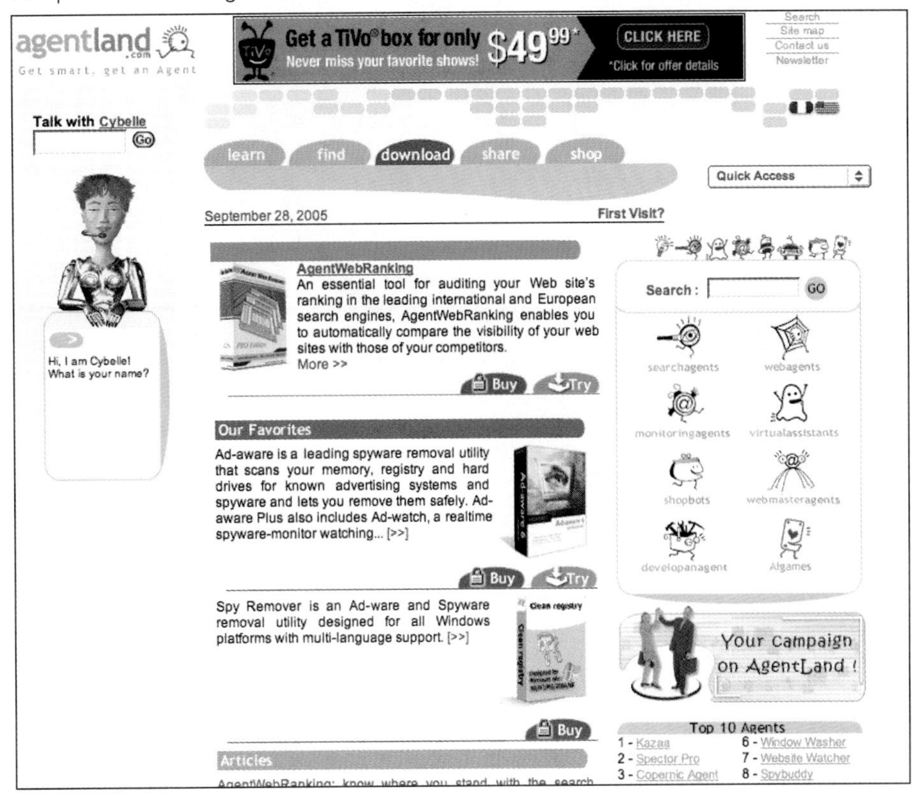

to retrieve requested resources and citations. Computerized expert systems have been developed to diagnose diseases given a patient's symptoms (see Figure 9.26), suggest the cause of a mechanical failure of an engine, predict future weather events, and assist in designing new products and systems. FocalPoint, by TriPath Imaging (*www.tripathimaging.com*), is an expert system that examines Pap smears for signs of cervical cancer. The FDA has approved FocalPoint for primary cancer screening. Expert systems are also used in the automotive industry to solve hard to diagnose problems.

Like human experts, computerized expert systems use heuristics, or "rules of thumb," to arrive at conclusions or make suggestions. However, one challenge for expert system developers is capturing knowledge and relationships that are not precise or exact. For example, expert systems have been used to help power plants reduce pollutants while maintaining profits.[32] They have also been used to determine the best way to distribute weight in a ferryboat to reduce the risk of capsizing or sinking. Expert systems can be used to spot defective welds during the manufacturing process.[33] The expert system analyzes radiographic images and suggests which welds could be flawed. The CIA is testing the software to see whether it can be used to detect possible terrorists when they make hotel or airline reservations.

FIGURE 9.26 • Expert systems

Medical expert systems help diagnose patients and suggest treatments.

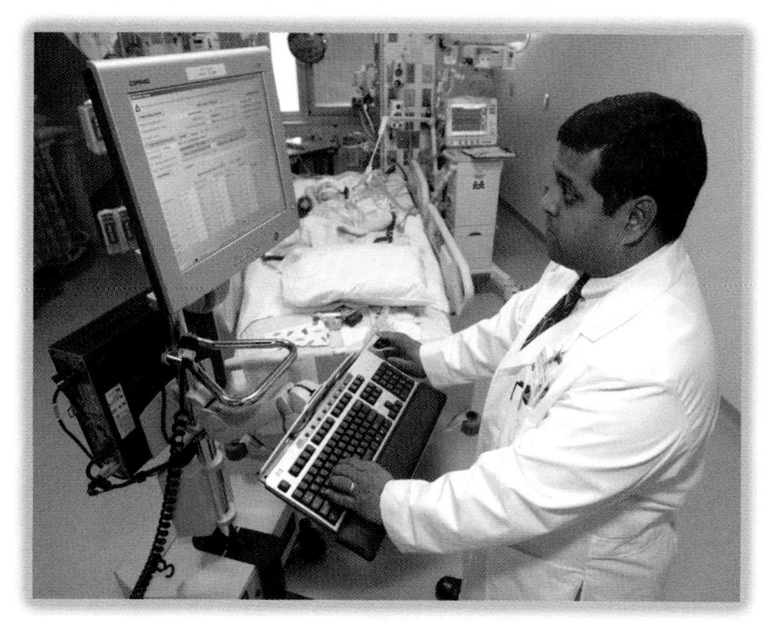

Specialized Systems

In addition to the management information, decision support, and artificial intelligence systems discussed in the preceding sections, there are a number of specialized computer systems that benefit individuals and organizations. These specialized systems include virtual reality systems, geographic information systems, game theory systems, and a variety of other specialized systems.

Virtual reality systems. **Virtual reality** is a computer-simulated environment or event. The idea was to have a three-dimensional world totally created by a computer system (see Figure 9.27). Originally, the term referred to *immersive virtual reality* in which the user becomes fully immersed in an artificial, three-dimensional world that is completely generated by a computer. By using a special headset with a computer simulated view and stereo sound, you could become immersed in the computer simulation as if it were the real world. Other devices, such as special gloves and motion detectors, could change the world you were seeing and hearing through the headset. These computer simulations were used to totally immerse you in another (virtual) world to play realistic games or take a tour of a building. The computer simulation made you feel like you were actually fighting an enemy or walking around inside a building. The military and major airlines uses virtual reality to simulate combat or airline flight emergencies.

FIGURE 9.27 • Virtual reality systems

Virtual reality systems use computers to simulate an environment or event.

Today, virtual reality is used to describe a computer simulation that is generated on the Internet or using a computer system, such as a personal computer. These forms of virtual reality allow you to take a tour of a home or condo anywhere around the world, see inside the human body, or be on stage with a rock band. Virtual reality, for example, was used to design a $90 million addition to the Denver Art Museum.[34] According to the contractor, "We're effectively building this with 3D tools. It has improved quality. It has definitely improved production time." The virtual reality software can be used to view every beam and duct in the large building. The software can also show the picture, length, and diameter of the 50,000 bolts that are being used. DaimlerChrysler uses virtual reality to help it simulate and design factories.[35] According to the CIO of DaimlerChrysler, "We are piloting a digital plant. As the engineers and designers are developing new products, the digital factory is simulating production." The company believes that this use of virtual reality can reduce the time it takes to move from an idea to production by about 30 percent. An architect can use virtual reality to show a real estate investor the outside and inside of a building before it is constructed. Using virtual reality, the investor can view the grounds, landscaping, and rooms of the proposed building.

Geographic Information Systems. A **geographic information system (GIS)** is capable of storing, manipulating, and displaying geographic or spatial information, including maps of locations or regions around the world (see Figure 9.28). A 911 operator can use a GIS to quickly determine the specific location of a caller with an emergency. The GIS converts the caller's phone number to the specific address of the caller and the directions for the police or an ambulance to get to the caller in a minimum amount of time. A business can use a GIS to display sales information for a specific region of the country. Higher sales can be displayed in red, and lower sales can be displayed in green or blue. This can give sales representatives a visual image of where to find sales opportunities and target their efforts. The military can use a GIS to target enemy forces on a battlefield. Using a GIS, along with other GPSs (global positioning systems), a tank group can quickly identify enemy positions and equipment in a constantly and rapidly changing battle situation. This helps a military force devastate an enemy position while avoiding friendly-fire casualties. GPSs were first used by the military. The technology uses satellites to determine one's location. GISs are also used in urban planning, social work, criminology and law enforcement, and a variety of other fields.

FIGURE 9.28 • Geographic information system

Geographic information systems store, manipulate, and display geographic information. This GeoSim application shows a 3-D model of the city of Philadelphia, based on GIS data.

Biologically Inspired Algorithms Fight Terrorists and Guide Businesses

Soon people may be collaborating with intelligent machines to anticipate future human actions. While scientists have been using computers to forecast predictable natural events such as weather and earthquakes for years, the technology has recently been turned to predicting human behavior. Scientists could use such technology to explore the future of humanity, businesses can use it to anticipate the moves of their competitors, and government and law enforcement can use it to catch the bad guys.

As with many revolutionary technologies, funding for this research is coming from the U.S. government, through the Defense Advanced Research Projects Agency (DARPA), the same organization that produced the Internet. Research in this new intelligence technology is taking place as part of a $54 million program known as Genoa II. The goal of the project is to employ machine intelligence to anticipate future terrorist threats. Researchers hope to make it possible for humans and computers to "think together" in real time to "anticipate and preempt terrorist threats," according to official program documents.

"In Genoa II, we are interested in collaboration between people and machines," said Tom Armour, Genoa II program manager at DARPA, "We imagine software agents working with humans. . .and having different sorts of software agents also collaborating among themselves."

The challenge lies in getting computers to mimic the human brain's ability to reduce complexity. While computers are good at carrying out rules-based algorithms such as those required for playing chess, they aren't so good at more complex deciphering, such as finding a word hidden in a picture. Researchers hope to change that through the use of biologically inspired algorithms. "One way to make computers more intelligent and lifelike is to look at living systems and imitate them," says Melanie Mitchell, an associate professor at Oregon Health & Science University's School of Science & Engineering in Portland.

Questions

1. Why may computers be able to do a better job of predicting human behavior than humans?
2. What precautions would be wise when developing systems that study human behavior to predict future events?

Sources:

1. *Morris, J. "DARPA Technologies Saving Lives in Iraq,"* Aerospace Daily & Defense Report, *March 14, 2005, p. 5.*
2. *Verton, D. "Using Computers to Outthink Terrorists,"* Computerworld, *September 1, 2003, www.computerworld.com.*
3. *A4Vision Web site, accessed May 31, 2004, www.a4vision.com.*

Game Theory Systems. **Game theory** involves developing strategies for people, organizations, or even countries that are competing against each other. Two competing businesses in the same market can use game theory to determine the best strategy to achieve their goals. The military can use game theory to determine the best military strategy to win a conflict, and individual investors can use game theory to determine the best investment strategies when competing against other investors in a government auction of bonds. Ground-breaking mathematical work was done on game theory by John Nash, and made popular in the book and film *A Beautiful Mind.*

Other Specialized Systems. In addition to the preceding applications, there are a number of other exciting special-purpose systems. **Informatics** combines traditional disciplines, like science and medicine, with computer systems and

FIGURE 9.29 • A special-purpose personal transportation system

Using sophisticated software, sensors, and gyro motors, Segway can transport standing people through warehouses, offices, and downtown sidewalks.

technology. *Bioinformatics,* for example, combines biology and computer science. Also called computational biology, bioinformatics has been used to help map the human genome and conduct research on biological organisms. Using sophisticated databases and artificial intelligence, bioinformatics is being used to help unlock the secrets of the human genome, which could eventually prevent diseases and save lives. Stanford University has a course on bioinformatics and offers a bioinformatics certification. *Medical informatics* combines traditional medical research with computer science. Journals, such as *Healthcare Informatics*, reports research on ways to reduce medical errors and improve health care delivery by using computer systems and technology in medicine. The University of Edinburgh in Scotland has a School of Informatics. The school has courses in the structure, behavior, and interactions of natural and artificial computational systems. The program combines artificial intelligence, computer science, engineering, and science.

The Segway personal transporter is another example of a special-purpose system. Using sophisticated software, sensors, and gyro motors, the device can transport standing people through warehouses, offices, and downtown sidewalks (see Figure 9.29).[36] Segway doesn't have formal controls for going faster or making other kinds of motion-related changes. The inventor, Dean Kamen, has other inventions to his credit—he developed the first wearable kidney dialysis pump for home use and the iBOT, a computer-controlled wheelchair.

Cyberkinetics is experimenting with a small microchip that could be embedded in the brain of patients with spinal cord injuries and wired to a computer.[37] Patients are asked to think about moving the cursor on a computer screen while the system tries to record the physical responses to their thoughts. If successful, the chip might be able to move computer cursors or perform other tasks to help quadriplegics perform tasks they couldn't perform otherwise.[38]

Another technology is being used to create "smart containers" for ships, railroads, and trucks.[39] NaviTag and other companies are developing communications systems that allow containers to broadcast the contents, location, and condition of shipments to shipping and cargo managers. Burlington Northern and Santa Fe Railway Company use standard radio messages to generate shipment and tracking data for customers and managers.[40] The railroad company used its radio system to link 14,000 miles of track in 27 states. The system's messages are input to a voice-response and speech-recognition system that is tied to the company's back-end processing systems, which tracks shipments.

COLD BEDSIDE MANNER, BUT RED HOT DIAGNOSTICS

Informatics for Integrating Biology and the Bedside (I2B2) is the long way of referring to the 5-year and $20 million AI system Harvard is developing to look for the genetic roots of disease. I2B2 will scan over 2.5 million private medical files from Boston-area hospitals to look for links between patient DNA and disease. Early tests showed the computer was as accurate as the admitting doctor in diagnosing the reasons for hospital admission.

Source:
Harvard project to scan millions of medical files
By Gareth Cook, Globe Staff
Boston Globe
July 3, 2005
http://www.boston.com/news/education/higher/articles/2005/07/03/harvard_project_to_scan_millions_of_medical_files/

Many other specialized devices are used by companies for a variety of purposes. "Smart dust" was developed at the University of California at Berkeley with Pentagon funding.[41] Smart dust involves small networks powered by batteries and an operating system, called TinyOS. The technology can be used to monitor temperature, light, vibration, or even toxic chemicals. Small radio transceivers can be placed in other products, like cell phones. The radio transceivers allow cell phones and other devices to connect to the Internet, cellular phone service, and other devices that use the technology. The radio transceivers could save companies hundreds of thousands of dollars annually. Microsoft's Smart Personal Objects Technology (SPOT) allows small devices to transmit data and messages over the air.[42] SPOT is being used in wristwatches to transmit data and messages over FM radio bands. The new technology, however, requires a subscription to Microsoft's MSN Direct information service. The software can track a driver's speed and location, allow gas stations to remotely charge for fuel and related services, and more. Special-purpose bar codes are also being introduced in a variety of settings. For example, to manage office space efficiently, a company gives each employee and office a bar code. Instead of having permanent offices, the employees are assigned offices and supplies as needed, and the bar codes help to make sure that an employee's work, mail, and other materials are routed to the right place. Companies can save millions of dollars by reducing office space and supplies. Manufacturing experiments are also being done with ink-jet printers to allow them to "print" 3D parts.[43] The technology is being used in Iowa City to print new circuit boards using a specialized ink-jet printer. The printer sprays layers of polymers onto circuit boards to form transistors and other electronic components.

ACTION PLAN

Remember Maurice from the beginning of the chapter? Here are answers to his questions.

1. **What type of system would help show potential clients a realistic view of commercial real estate?**

   Although there are a variety of tools that can be used, virtual reality on an Internet site is an excellent choice to show clients the inside and outside of buildings. Virtual reality technology allows potential clients or investors to take a "virtual tour" of the building and landscape.

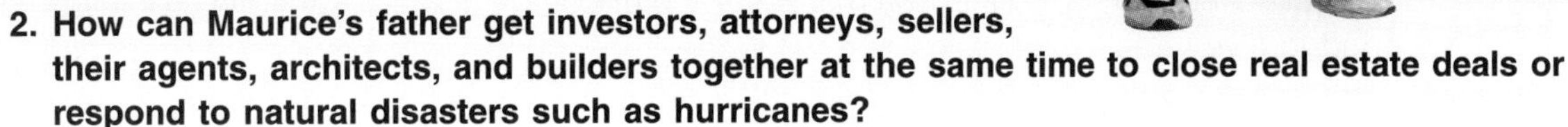

2. **How can Maurice's father get investors, attorneys, sellers, their agents, architects, and builders together at the same time to close real estate deals or respond to natural disasters such as hurricanes?**

   A group decision support system could help get everyone together. Groupware, such as Lotus Notes, can be used to facilitate group meetings and work. Spreadsheets, legal documents, and real estate analysis can be shared using this approach. In addition, a GIS can be used by investors, attorneys, sellers, and their agents to analyze different real estate possibilities or deal with disasters such as hurricanes.

3. **What can Maurice do to give clients a tour of New Orleans over the Internet?**

   Using virtual reality, Maurice can build an Internet site or have one built that shows clients various points of interest in New Orleans. This Internet site can also be linked to other Internet sites developed by the city and other companies and organizations.

Summary

LEARNING OBJECTIVE 1

Define the stages of decision making and problem solving.

The true potential of computer systems is in providing information to help people and organizations make better decisions. Decision making is divided into three phases: intelligence, design, and choice. Problem solving takes decision making a step further and involves taking action by implementing the choice made by the decision maker and monitoring the effects of the decision.

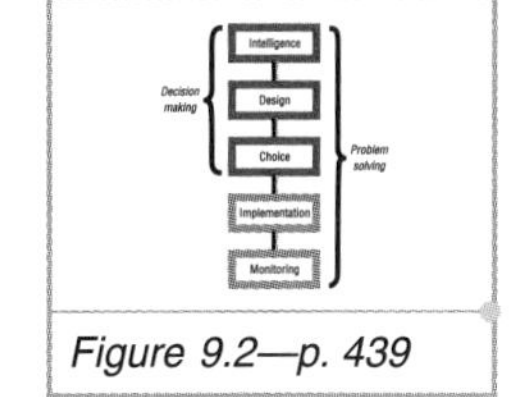

Figure 9.2—p. 439

The types of decisions made by an organization can range from structured, programmed decisions to unstructured, nonprogrammed decisions. In some instances, when the optimal solution to a problem must be determined, optimization models assist in finding the best solution. In other cases, a solution that meets a basic set of criteria may be acceptable, although not optimal. This heuristic approach is often used because it is more cost effective than the optimization approach. In addition, individuals and organizations can use a reactive or proactive approach to problem solving. With a reactive problem-solving approach, the problem solver waits until a problem surfaces or becomes apparent before taking any action. With a proactive problem-solving approach, the problem solver seeks out potential problems before they become serious.

LEARNING OBJECTIVE 2

Discuss the use of management information systems in providing reports to help solve structured problems.

A management information system is an organized collection of people, procedures, databases, and devices that provide managers and decision makers with information to help achieve organizational goals. An MIS can help an organization achieve its goals by providing managers with insight into the regular operations of the organization so that they can control, organize, and plan more effectively and efficiently. The output of most management information systems is a collection of reports that are distributed to managers. These reports include scheduled reports, demand reports, and exception reports. Scheduled reports are produced periodically or on a schedule, such as daily, weekly, or monthly. A key-indicator report is a special type of scheduled report that summarizes the previous day's critical activities. Demand reports are developed to give certain information at a manager's request. Exception reports are reports that are automatically produced when a situation is unusual or requires management action.

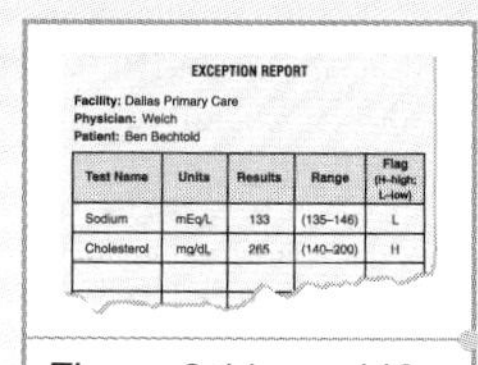

EXCEPTION REPORT

Facility: Dallas Primary Care
Physician: Welch
Patient: Ben Bechtold

Test Name	Units	Results	Range	Flag (H–high; L–low)
Sodium	mEq/L	133	(135–146)	L
Cholesterol	mg/dL	265	(140–200)	H

Figure 9.11—p. 446

LEARNING OBJECTIVE 3

Describe how decision support systems are used to solve nonprogrammed and unstructured problems.

A decision support system (DSS) is an organized collection of people, procedures, software, databases, and devices working to support managerial decision making. DSSs provide assistance through all phases of the decision-making process. Decision support systems can handle a large amount of data, obtain and process data from different sources, provide report and presentation flexibility, have both textual and graphical orientation, perform complex and sophisticated analysis, support optimization and heuristic approaches, and perform what-if and goal-seeking analysis.

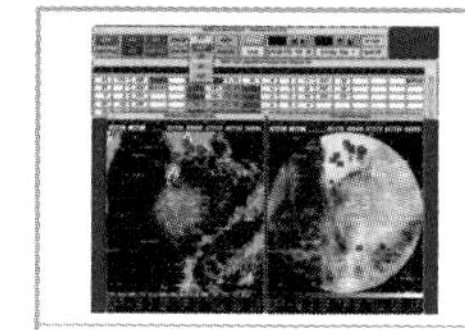

Figure 9.15—p. 450

LEARNING OBJECTIVE 4

Explain how a group decision support system can help people and organizations collaborate on team projects.

A group decision support system (GDSS), also called a *computerized collaborative work system*, consists of hardware, software, people, databases, and procedures needed to provide effective support in group decision-making settings. GDSSs are typically easy to learn and use and can offer specific or general decision-making support.

A GDSS also has some unique components, such as compound documents, groupware, and telecommunications links. Groupware is specially designed software that helps generate lists of decision alternatives and performs data analysis. These packages let people work on joint documents and files over a network. The characteristics of a GDSS include special design, ease of use, flexibility, anonymous input, reduced negative behavior, and support of positive group behavior.

Figure 9.17—p. 452

LEARNING OBJECTIVE 5

Discuss the uses of artificial intelligence and special-purpose systems.

Artificial intelligence (AI) is a broad field that includes several key components, including robotics, vision systems, natural language processing, learning systems, neural networks, fuzzy logic, genetic algorithms, intelligent agents, and expert systems. Robotics involves developing mechanical or computer devices to perform tasks that require a high degree of precision or are tedious or hazardous for humans.

Vision systems include hardware and software that permit computers to capture, store, and manipulate images and pictures. Natural language processing allows the computer to understand and react to statements and commands made in a "natural" language, such as English. Learning systems use a combination of software and hardware that allows the computer to change how it functions or reacts to situations based on feedback it receives. A neural network is a computer system that can act like or simulate the functioning of a human brain. A genetic algorithm is an approach to solving large, complex problems, where the algorithms or models change and evolve until the best one emerges. Expert systems act or behave like a human expert in a field or area. Fuzzy logic entails dealing with ambiguous criteria or probabilities and events that are not mutually exclusive. An intelligent agent consists of programs and a knowledge base used to perform a specific task for a person, a process, or another program. Like an agent who searches for the best endorsement deals for a top athlete, an intelligent agent often searches to find the best price, the best schedule, or the best solution to a problem.

Figure 9.27—p. 462

There are a number of special-purpose systems, including virtual reality, geographic information systems, game theory, and other special-purpose systems. Virtual reality is a computer-simulated environment or event. It can be used to design buildings, help the military, or play exciting games. A geographic information system is capable of storing manipulating, and displaying geographic information, including maps of locations or regions around the world. Game theory involves developing strategies for people, organizations, or even countries competing against each other.

Test Yourself

LEARNING OBJECTIVE 1: Define the stages of decision making and problem solving.

1. With the ________________ approach, the problem solver waits until a problem surfaces before any action is taken.
2. True or False: Problem solving is a component of the decision-making approach.
3. The second stage in the decision-making process is:
   a. design
   b. choice
   c. intelligence
   d. goal seeking
4. The final stage of the problem solving process is ________________ .
5. True or False: A programmed decision is easy to computerize using traditional computer systems.
6. What is commonly referred to as a "rule of thumb"?
   a. heuristic
   b. optimization
   c. goal seek
   d. nonprogrammed decision

LEARNING OBJECTIVE 2: Discuss the use of management information systems in providing reports to help solve structured problems.

7. True or False: Data entering the MIS can be considered to originate from either internal or external sources.

8. What types of reports are produced periodically?
   a. demand
   b. scheduled
   c. heuristic
   d. optimization
9. A(n) ________________ is a special type of scheduled report that summarizes the previous day's critical activities.
10. True or False: A heuristic report is automatically produced when a situation is unusual or requires management action.

LEARNING OBJECTIVE 3: Describe how decision support systems are used to solve non-programmed and unstructured problems.

11. A(n) ________________ is used when people or organizations face unstructured or semistructured problems.
12. What is the process of determining the problem data required for a given result?
    a. heuristic
    b. optimizing
    c. goal-seeking
    d. simulation
13. True or False: With heuristics, the DSS attempts to mimic an event that could happen in the future.
14. What type of system has both a textual and graphical orientation?
    a. transaction processing system
    b. management information system
    c. neural network system
    d. decision support system
15. ________________ is the process of making hypothetical changes to problem data and observing the impact on the results.

LEARNING OBJECTIVE 4: Explain how a group decision support system can help people and organizations collaborate on team projects.

16. What type of software is used to allow two or more individuals to work together effectively in a group?
    a. groupware
    b. decisionware
    c. cooperative software
    d. teamware
17. True or False: Anonymous input in a GDSS is used to make sure that the person giving input is not known to other group members.

LEARNING OBJECTIVE 5: Discuss the uses of artificial intelligence and special-purpose systems.

18. ________________ allows the computer to understand and react to statements and commands made in English or a similar language.
19. What type of artificial intelligence is used to simulate or act like the functioning of the human brain?
    a. fuzzy logic
    b. expert systems
    c. neural networks
    d. genetic algorithms
20. True or False: A genetic algorithm is an approach to solving problems where a number of models change and evolve until the best one emerges.
21. A ________________ acts like a human expert in a field or area.
22. True or False: Game theory involves developing strategies for people, organizations, or even countries competing against each other.

Test Yourself Answers: 1. reactive approach, 2. False, 3. a. design, 4. monitoring, 5. True, 6. a. heuristic, 7. true, 8. b. scheduled, 9. key-indicator, 10. False, 11. decision support system, 12. c. goal-seeking, 13. False, 14. d. decision support system, 15. What-if analysis, 16. a. groupware, 17. True, 18. Natural language processing, 19. c. neural networks, 20. True, 21. expert system, 22. True.

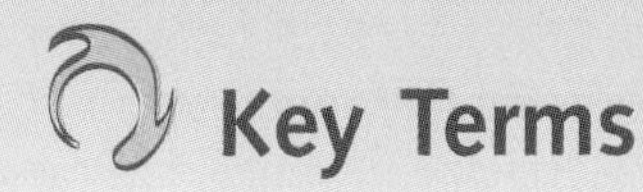

Key Terms

decision making, p. 438
demand report, p. 445
exception report, p. 446
expert system (ES), p. 460
fuzzy logic, p. 458
game theory, p. 464
geographic information system (GIS), p. 463
genetic algorithm, p. 458
goal-seeking analysis , p. 450
group decision support system (GDSS), p. 451
groupware, p. 452
heuristic, p. 442
informatics, p. 464
intelligent agent, p. 460
learning system, p. 457
management information system (MIS), p. 443
natural language processing , p. 457
neural network , p. 458
nonprogrammed decision, p. 440
optimization model, p. 440
problem solving , p. 438
programmed decision, p. 439
robotics , p. 456
scheduled report, p. 445
virtual reality , p. 462
vision systems, p. 456
what-if analysis, p. 450

Questions

Review Questions

1. Describe the stages of decision making.
2. What is the overall purpose of the design stage of decision making?
3. Describe the stages of problem solving.
4. What is a programmed decision?
5. What is the heuristic approach?
6. What is a management information system (MIS)?
7. What are the inputs to an MIS?
8. Describe the outputs of management information systems.
9. What is the difference between a scheduled report and a demand report?
10. What is a key-indicator report?
11. Define a decision support system. What are its characteristics?
12. Describe what-if analysis.
13. How can goal-seeking analysis be used?
14. Describe the overall approach of simulation.
15. What is a group decision support system (GDSS)?
16. List five characteristics of a GDSS.
17. What is the difference between a DSS and a GDSS?
18. Describe the use of groupware.
19. What are the components of artificial intelligence?
20. What is the difference between artificial and natural intelligence?
21. Describe three new developments in artificial intelligence and explain why they are important.
22. What is robotics? What are neural networks?
23. What are the capabilities of an expert system?
24. What is virtual reality? What is a geographic information system?

Discussion Questions

25. Think of an important decision you made in the last few years. Describe the results of decision making and problem solving steps you used.
26. Discuss the difference between scheduled, demand, and exception reports.
27. How can MISs be used to help individuals and organizations make better decisions?
28. What functions do decision support systems (DSS) support in organizations?

29. List one or two career areas that interest you. Describe how a DSS might be used to help you achieve your career goals.
30. How is decision making in a group environment different from individual decision making, and why are information systems that assist in the group environment different? What are the advantages and disadvantages of making decisions as a group?
31. What are the characteristics of intelligent behavior?
32. How could robots be used in the military or law enforcement?
33. Give an example of how expert systems can be used in a field of interest to you.
34. What type of college or university courses might benefit from virtual reality?
35. Describe two special-purpose devices that would make your life more enjoyable or easier.

Exercises

Try It Yourself

1. You have been asked to set up some reports to help your school better plan course offerings. Your school's registration system collects data not only on what classes students have enrolled in, but also on what classes students tried to enroll in but were denied enrollment because the class was full. Using the data in the following table, create two reports: (1) an exception report listing all courses that more than three students couldn't get into because the class was full, and (2) a scheduled report listing the students who are enrolled in each class (a class roster). This scheduled report will be printed out every day during registration and can also be used on demand after registration is over.

Course#	StudentID	Status	Date	Reason
370	5987	Enroll	10/13/06	
370	9237	Enroll	10/15/06	
567	1629	Enroll	10/14/06	
567	2863	Denied	10/15/06	Full
567	4631	Enroll	10/14/06	
567	4731	Denied	10/15/06	Full
567	5987	Enroll	10/13/06	
567	9237	Denied	10/15/06	Full
567	9832	Enroll	10/13/06	
963	3958	Denied	10/13/06	Full
963	5678	Denied	10/13/06	Full
963	9832	Denied	10/13/06	Full

2. A university has designed and implemented a database to maintain records for students and courses. Some of the essential fields are student identification number, student name, student address, majors, credits completed, course number, course name, faculty member teaching the course, student GPA, and grades. Using this database, design and implement a decision support system query for department heads and administrators to determine the number of credit hours that are generated by each instructor. Course credit hours are calculated by multiplying course credits times the number of students. Instructor credit hours are calculated by adding together the credit hours for each course. Administrators can use this information as a method of determining the number of course sections and the workload of instructors. You can use a database management system or spreadsheet to design the report.
3. Use the Internet to explore the decisions that are required for your chosen career area or field. Also use the Internet to explore expert systems. Using your word-processing program, describe how you could develop an expert system for your career area or field. Develop 10 heuristics, or "rules of thumb," that show how your expert system could make decisions given certain conditions or situations.
4. Use the Internet or your online university library to search for new artificial intelligence applications. Write a 2-page report describing two of your most interesting findings.
5. Investigate the use of game theory on the Internet or an academic journal. Use your word-processing program to describe how you could use game theory in your career area or field.

Virtual Classroom Activities

For the following exercises, do not use face-to-face or telephone communications with your group members. Use only Internet communications.

6. As discussed in the chapter, a nonprogrammed decision can involve unusual or exceptional situations. Your virtual classroom group should first list three nonprogrammed decisions that a freshman in college might encounter. For each nonprogrammed decision, develop a brief description of the factors to consider during the intelligence and design phases of decision making to help the freshman make the best decision.
7. Groupware is often used to help groups make decisions. Using the Internet, investigate the use of groupware in two or more areas or organizations. The members of your virtual classroom group should summarize the features, advantages, and disadvantages of each groupware product.

Teamwork

8. Your team is to design an information system for a local bookstore chain. Currently there is a system installed to process sales transactions, but an integrated information system does not exist. Andy Masters, president of the local bookstore, is planning on expanding from four stores to 10 within the next year. To manage this growth, and to keep track of the regular operations of the business, he wants to have a management information system that links together all stores. Prepare a brief memo to Andy explaining five different reports that the information system should produce. Include at least one demand and exception report. Your group should create a layout of the bookstore's new management information system using presentation graphics software.
9. Your team should develop a small expert system with five or six rules for a career or job that is of interest to your team.
10. Your team should develop a spreadsheet to be used to estimate the expenses to move off campus next year. The model must allow input of a variety of rent amounts, food costs, transportation costs, and so on, as well as tuition and books. Each team member should put in his or her current estimated costs. Assume that tuition and books are fixed at this year's amounts. After finding an average of the group members' current costs, perform two what-if scenarios. For the first, assume no change in costs but 2 percent inflation. For the second, assume 5 percent inflation. Do the projection for a year. Create a word-processed document to explain your model to a novice student who might use it to guide decision making for next year.
11. Your team should brainstorm at least five ideas for expert systems that would be useful on campus. Try to develop rules you might use to make decisions (say, to choose classes for a semester). This exercise will show the difficulty of using multiple experts to build a knowledge base and the difficulty in defining rules for a knowledge base.

Ben Casnoscha

Go to www.course.com/swt2/ch09 and click TechTV. Click the Ben Casnoscha link to view the TechTV video, and then answer the following questions.

1. How does Ben Casnoscha's company, Comcate, epitomize today's information economy—the trend towards profiting from nontangible, information-based services?
2. Comcate provides an e-mail and tracking system for individuals and organizations to use for communicating with government agencies. Why has this business prospered? What demand does it fulfill?

Endnotes

[1] Coy, P. "Why Logic Often Takes a Back Seat," *Business Week,* March 28, 2005, p. 94.

[2] Lacroix, Y. et al. "Bombarier Flexjet Significantly Improves Its Fractional Aircraft Ownership Opereations," *Interfaces,* January-February, 2005, p. 49.

[3] Murty, K. et al. "Hongkong International Terminals Gains Elastic Capacity, *Interfaces,* January-February, 2005, p. 61.

[4] Troyer, L. et al, "Improving Asset Management and Order Fulfillment at Deere," *Interfaces,* January-February, 2005, p. 76.

[5] Havenstein, H. "Business Objects Adds Crystal Reports Tools," *Computerworld,* January 17, 2005, p. 4.

[6] MacSweeney, G. "ProvWash Removes Paper," *Insurance and Technology,* February 1, 2004, p. 12.

[7] Kelly, K. "Big Board Chief Mulls Changing Trading System," *The Wall Street Journal,* January 30, 2004, p. C1.

[8] Kelly, K. "A Little Scary," *The Wall Street Journal,* February 2, 2004, p. C1.

[9] Huber, N. "FSA to Target Stock Market Abuse," *Computer Weekly,* January 20, 2004, p. 14.

[10] Thomas, D. "Pub Chain Cuts a Better Deal," *Computer Weekly,* January 27, 2004, p. 10.

[11] Cropper, C. "Between You, The Doctor, and the PC," *Business Week,* January 31, 2005, p. 90.

[12] Warren, P. "Software Could Aid Anti-Terrorism Fight," *Computing,* January 15, 2004, p. 1.

[13] Thomas, D. "Software Improves AC Milan's Game," *Computing,* January 29, 2004, p. 6.

[14] Goldsmith, C. "Reed Elsevier Feels Resistance to Web Pricing," *The Wall Street Journal,* January 10, 2004, p. B1.

[15] Nash, E. "Travel Firm Uses BI Tools to Compete," *Computing,* January 29, 2004, p. 11.

[16] Ramsaran, Cynthia, "ING Implements Cognos for Intelligence," *Bank Systems & Technology,* February 1, 2004, p. 42.

[17] Brown, S. "Toyota's Global Body Shop," *Fortune,* February 9, 2004, p. 120B.

[18] Lipschutz, R. "Instant," *PC Magazine,* March 22, 2005, p. 120.

[19] Rosencrance, L. "Meet Me in Cyber Space," *Computerworld,* February 21, 2005, p. 23.

[20] Quain, J. "Thinking Machines, Take Two," *PC Magazine,* May 24, 2005, p. 23.

[21] Begley, S. "This Robot Can Design, Perform and Interpret A Genetic," *The Wall Street Journal (Eastern Edition),* January 16, 2004, p. A7.

[22] Sourbut, E. "An Agreeable Android," *New Scientist,* February 28, 2004, p. 49.

[23] Excell, J. "Interview – Robo Doc," *The Engineer,* March 5, 2004, p. 30.

[24] Staff, "Send In The Robots," *Fortune,* January 24, 2005, p. 140.

[25] Wysocki, B. "Robots in the OR," *The Wall Street Journal,* February 26, 2004, p. B1.

[26] Staff, "Fly-Eating Robot Powers Itself," *CNN Online,* December 29, 2004.

[27] Staff, "Industry Watch," *Health Management Technology,* February 2004, p. 10.

[28] Lunt, P. "Online Retailer Restores Web Sales," *Transform Magazine,* January 1, 2004, p. 32.

[29] Shihav, A. I., et al, "Distributed Intelligence for Multiple-camera Visual Surveillance," *Pattern Recognition,* April 2004, p. 675.

[30] Johnson, C. "Neural Software Could Become Soldier's Best Friend," *Electronic Engineering Times,* February 2, 2004, p. 51.

[31] Hyung, R. et al, An Agent for Selecting Optimal Order Set in EC Marketplace," *Decision Support Systems,* March 2004, p. 371.

[32] Huang, Z. et al, "Development of an Intelligent Decision Support System for Air Pollution Control," *Expert Systems With Applications,* April 2004, p. 335.

[33] Liao, T. W. "Fuzzy Reasoning Based Automatic Inspection of Radiographic Welds," *Journal of Intelligent Manufacturing,* February 2004, p. 69.

[34] Fillion, R. "Easy as 1, 2, 3-D," *Rocky Mountain News,* March 7, 2005, p. 9A.

[35] Saran, C. "Digital Factories Use Virtual Reality to Track Car Production," *Computer Weekly,* March 2, 2004, p. 18.

[36] Armstrong, D. "The Seqway," *The Wall Street Journal,* February 12, 2004, p. B1.

[37] Moukheiber, Z. "Mind Over Matter," *Forbes,* March 15, 2004, p. 186.

[38] Pollack, A. "With Tiny Brain Implants, Just Thinking May Make It So," *The New York Times,* April 13, 2004, [need page numbers].

[39] Machalaba, D. et al, "Thinking Inside the Box," *The Wall Street Journal,* January 15, 2004, p. B1.

[40] Brewin, B, "A Railroad Finds Its Voice," *Computerworld,* January 26, 2004, p. 37.

[41] Boyle, M. "Smart Dust Kicks Up a Storm," *Fortune,* February 23, 2004, p. 76.

[42] Manes, S. "New Twist for the Wrist," *Forbes,* February 2, 2004, p. 94.

[43] Weiss, T."Printer Magic," *Computerworld,* January 26, 2004, p. 31.

SYSTEMS DEVELOPMENT

Life for Linda Perez has been good. Two years after getting her undergraduate degree in business, she has already been promoted to account manager at Mutual Insurance in Tampa, Florida. In addition to her base salary, Mutual is offering Linda a generous commission for each new insurance customer. Linda would like the company to develop an insurance and risk analysis program that would help her and other account managers generate even more business. The new computer system would contain a database of existing and potential customers containing their information and insurance needs. Linda recently met James Jones, a systems analyst with the company, at a party. She discussed her ideas for the program with him, and he was very excited about it and gave her even more ideas for how she could also use the program to improve customer service. Now she wants to present her ideas to the company.

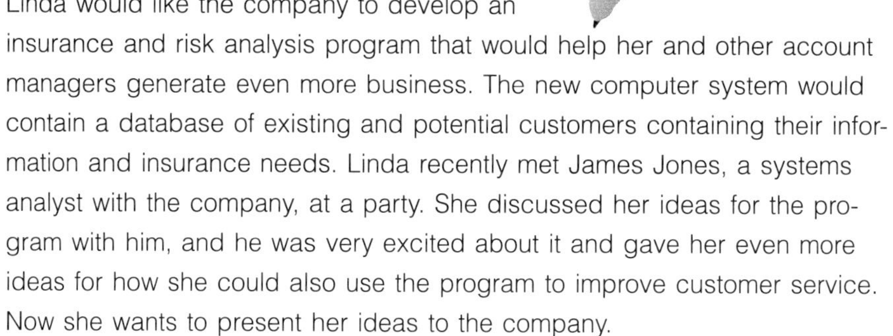

As you read through this chapter, consider the following questions:

1. How should Linda approach her company to get the new program developed?
2. If the company decides to develop the new program, how should Linda be involved in the process?
3. What role would James and people like him play in developing the new program?

Check out Linda's *Action Plan* at the end of this chapter.

LEARNING OBJECTIVES

1. Describe the systems development life cycle, who participates in it, and why it is important.
2. Discuss systems development tools.
3. Understand how systems development projects are investigated.
4. Describe how an existing system can be evaluated.
5. Discuss what is involved in planning a new system.
6. List the steps to implement a new or modified system.
7. Describe the importance of updating and monitoring a system.

CHAPTER CONTENT

An Overview of Systems Development

Tools and Techniques for Systems Development

Systems Investigation

Systems Analysis

Systems Design

Systems Implementation

Systems Maintenance and Review

Introduction

Throughout this book, you have seen how computer systems have helped people and organizations in a variety of settings: engineering, science, the arts, business, library studies, sociology, criminology, architecture, music, the military, and many other fields. But how are these computer systems acquired or developed? And how can the people and organizations that use these systems make sure those systems truly help them achieve their goals? The answer is a process called *systems development.*

This chapter introduces the systems development process, including systems investigation, analysis, design, implementation, maintenance, and review. In this chapter, you will see how you can be involved in systems development to advance your career and help your company or organization. You will also see how computer systems professionals, including systems analysts and computer programmers, work together to develop effective computer systems. In addition, you will see how you can use the systems development process to obtain the systems and software you need. In the pages that follow, you will see how systems development can be used to realize the true potential of computer systems in almost every field or discipline.

Systems development is the activity of creating new or modifying existing systems. It refers to all aspects of the process—from identifying problems to be solved or opportunities to be exploited to the evaluation and possible refinement of the chosen solution. Throughout this book, you have seen the results of systems development in numerous examples and boxes. All of these uses of computer systems are a direct result of the systems development process discussed in this chapter.

AN OVERVIEW OF SYSTEMS DEVELOPMENT

Systems development efforts can range from a small project, such as purchasing an inexpensive computer program, to a major undertaking, such as installing a huge system that includes hardware, software, communications systems, and databases, and requires new computer systems personnel. Some government and military systems development projects can involve millions of dollars.[1] Hilton Hotels invested about $13 million to improve the performance of its Web site.[2] Its Web site systems development effort allowed the company to develop specific content for its hotels' different locations and cultures. According to a senior vice president of Hilton International, "We're now much more able to take our worldwide network of hotels and market them to where our customers live." Emery, a worldwide shipping company, was able to develop a new transportation system that increased profits by $80 million in North America. The New York Stock Exchange (NYSE) has invested about $2 billion to speed trading and reduce costs.[3] Stock trades are currently completed with human brokers on a trading floor. In the future, electronic trading could take over, slashing trading costs to investors. Table 10.1 summarizes some additional systems development projects.[4]

TABLE 10.1 • A few successful systems development projects

Organization	Description
U.S. Air Force	The systems development effort created a new e-mail system that halved costs and allowed 2000 U.S. Air Force personnel who were once needed for the old e-mail system to move to other jobs.
Reliant Pharmaceuticals LLC	The large drug company developed a sales force automation system that used voice recognition. The new system reduced the time to respond to sales calls from 8 weeks to less than a day.
Northrop Grumman Corp	The large defense and space company used systems development to integrate diverse applications and allow managers to collaborate on important projects. The systems development effort saved the company almost $60 million in the first year. Ten-year savings are projected to be $600 million.
Visa	The credit-card company used systems development to create a system to help resolve credit-card charge disputes with customers and clients. The Web-based application streamlines the processing of charge disputes, saving time and money. Visa, which manages over 300 million credit and debit cards, estimates that it will save more than $1 billion over the next five years.
State of Ohio	A new worker's compensation system was developed, and it reduced worker's compensation transaction costs by more than 40 percent.

Source: Data from "Best in Class," *Computerworld,* March 15, 2004, p. 3.

Organizations have used different approaches to developing computer systems. In some cases, these approaches are formalized and captured in volumes of documents that describe what is to be done. In other cases, less formal

A Technology Makeover with Systems Development

Boehringer Ingelheim is among the world's 20 largest pharmaceutical companies. With $7.6 billion in revenue and 32,000 employees in 60 nations, Boehringer has diversified into segments that include manufacturing and marketing pharmaceuticals (such as prescription medicines and consumer health care products), products for industrial customers (such as chemicals and biopharmaceuticals), and animal health products.

Top managers decided to totally revamp the company's computer systems. It took 14 months to roll out the new system, and many employees needed intensive training. In the end, the results were well worth the investment of time and money. The software provided a standard system used across all of Boehringer's business segments and offered convenient Web access to current information. Boehringer is now able to complete monthly reports just two hours after the close of business at the end of each month.

Boehringer is committed to providing employees at all levels of the company with access to the applications and information they need to meet their objectives. About one-third of Boehringer's employees work outside the office. To provide its mobile workforce with up-to-the-minute data, the company deployed software from BackWeb Technologies that allows employees access to current sales information through a Web portal and a custom Web interface, wherever they travel. Boehringer's employees can access and change information presented in the portal when they are offline, and updates are made when they later log on.

By the time Boehringer was finished with its technology makeover, the company had implemented over seven new interconnected computer systems and invested millions in hardware, software, databases, telecommunications, and training. But the investment has paid off. Wherever they may be, employees can now access up-to-date organization-wide information, with the click of a mouse. And decision makers of Boehringer can react as nimbly and quickly to changes as many of its smaller competitors.

Questions

1. In designing its new computer systems, what do you think were Boehringer's most critical goals and considerations?
2. How are hardware, software, databases, telecommunications, people, and procedures used in Boehringer's computer system to provide valuable information?

Sources

1. *Staff. "US drives Boehringer growth,"* Chemist and Druggist, *April 16, 2005, p. 10.*
2. *Songini, M. "Case Study: Boehringer Cures Slow Reporting,"* Computerworld, *July 21, 2003, http://www.computerworld.com.*
3. *"Boehringer Ingelheim Deploys BackWeb's Offline Solution for the Plumtree Corporate Portal," PR Newswire, December 15, 2003.*
4. *Boehringer Ingelheim Web site, accessed January 22, 2004, http://www.boehringer-ingelheim.com/corporate/home/home.asp.*

approaches are used. The steps of systems development may vary from one organization to the next, but most approaches have five common phases: investigation, analysis, design, implementation, and maintenance and review, as shown in Figure 10.1.

The systems development process is an ongoing and cyclic activity. Once a system is developed, it is reviewed and over time often revised and improved. The **systems development life cycle (SDLC)** is the ongoing activities associated with the system development process including investigation, analysis, design, implementation, and maintenance and review. Systems investigation and analysis looks at the existing system and determines if the existing system can

FIGURE 10.1 • The systems development life cycle (SDLC)

The systems development life cycle (SDLC) involves working through a number of steps to go from the initial idea to a finished system. Sometimes information learned in a particular phase requires cycling back to a previous phase.

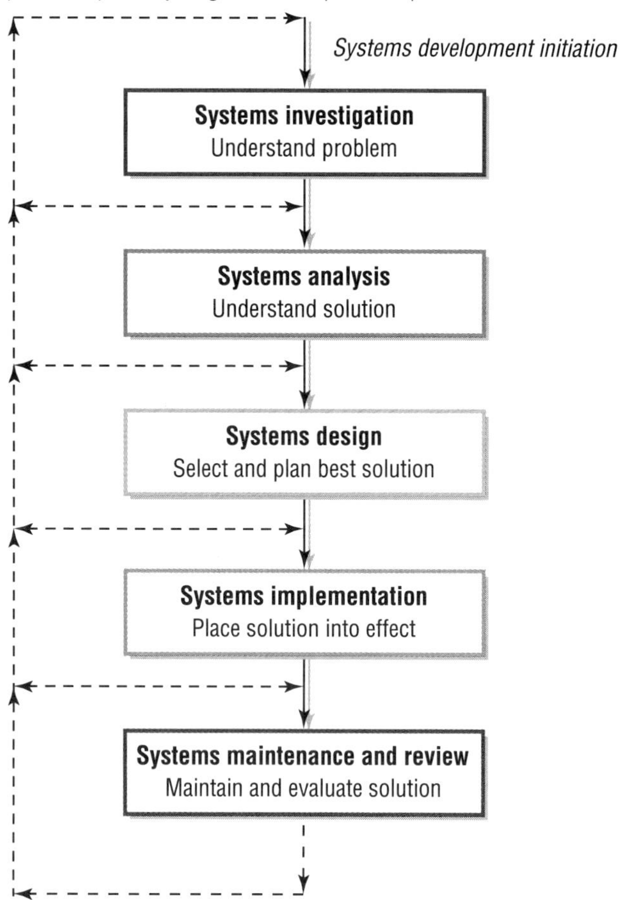

and should be improved. Systems design and implementation involves modifying an existing system or developing a new one and placing it into operation. Finally, maintenance and review makes sure that the new or modified system is operating as intended.

As each system is being built, the project has timelines and deadlines, until at last the system is installed and accepted. The life of the system continues as it is maintained and reviewed. If the system needs significant improvement beyond the scope of maintenance—if it needs to be replaced due to a new generation of technology, or if there is a major change in the organization's requirements of the computer systems—a new systems development project is initiated, and the cycle starts over.

As shown in Figure 10.1, a particular system under development may move from one phase of the SDLC to the next, and then back to a previous phase, and so on. In an ideal world, it is unnecessary to return to a previous phase to correct errors or make adjustments. In reality, however, activities in a later phase of the SDLC may uncover a need to change the results of previous steps. For example, a medical researcher might realize during implementation of a new software system that the current computers are too slow. This might require that hardware and software be upgraded, which could restart the systems design phase. Thus, although it is described as a series of steps, the SDLC is more likely to cycle back and forth between steps to continuously rebuild and refine a system. At each step, there are checkpoints to determine if the step has been successfully completed, if additional work is needed, and if the systems development process should continue.

Even with the steps of systems development, many systems development efforts fail to achieve their goals. Organizations have lost hundreds of millions of dollars on failed systems development efforts. In addition, an organization's clients or customers can also suffer losses. In one classic case, a major stock exchange tried to implement a new computerized trading system. Companies and individuals that interacted with the new system had to spend money to change their systems to work with the new trading system. When the new trading system didn't work and was never implemented, the companies and individuals that interacted with the stock exchange had to change back to their old systems, wasting a tremendous amount of time and money. In another case, a systems development project for the Child Support Agency of the British government was behind schedule and almost $485 million overbudget.[5] The delayed project may have hurt the agency's ability to deliver important services to children.

When Systems Development Projects Fail

Throughout the chapters of this book, you have seen the remarkable use of technology to help people and organizations achieve their goals, advance their careers, and provide entertainment and amusement. Organizations, from small volunteer organizations to the largest corporations, have also used technology to provide better services or increase profits. But success is not guaranteed. Many systems development projects fail, as seen in the cases discussed next.

Because of errors in a US Airways computer system, prices for some flights were advertised to be as low as $1.86. Flights from the United States to Fiji were quoted on an Internet site as $51. Even though the advertised prices were posted on the Internet for less than an hour, the damage was done. In many cases, the airline decided to honor some or all of the flights, even though they were not required to do so in cases where the price was clearly a mistake. Other airlines have had similar computer mistakes. These situations clearly show the importance of good testing before systems are made operational.

The FBI decided to develop a new Virtual Case File system to help its agents identify and fight possible terrorists. The custom system had a projected cost of $170 million. The 4-year project, however, was never completed. The director of the FBI told the U.S. House of Representatives appropriations committee, "I'm disappointed that we did not come through with Virtual Case File." The FBI is now investigating the use of existing off-the-shelf software instead of trying to develop its own programs. The appropriations committee is looking into why the systems development project failed and who is to blame.

Questions:

1. Why do some systems development projects fail?
2. What can be done to prevent these types of systems development failures?

Sources

1. *Johnson, A. "Booking a $51 Flight to Fiji,"* The Wall Street Journal, *April 26, 2005, p. D5.*
2. *Rosencrance, L. "FBI Scuttles $170M System,"* Computerworld, *March 14, 2005, p. 14.*

Participants in Systems Development

Effective systems development requires a team effort. The team usually consists of system stakeholders, users, managers, systems development specialists, and various support personnel. This team, called the development team, is responsible for determining the objectives of the computer system and delivering to the organization a system that meets these objectives. **System stakeholders** are individuals who will ultimately benefit from the systems development project, either directly themselves or through the organization they represent.

Users are a specific type of stakeholder. Users are individuals who will be interacting with the system on a regular basis. They can be employees, managers, customers, suppliers, vendors, and others. Physicians, for example, can be actively involved using developing a 3D system for surgery that allows doctors to practice surgeries using simulations of a patient's internal organs before actually performing the surgery, preventing mistakes and saving lives (see Figure 10.2).

Managers are people within an organization most capable of initiating and maintaining change. For large-scale systems development projects, where the investment in and value of a system can be quite high, it is common to have senior-level managers as part of the development team. The director of a World

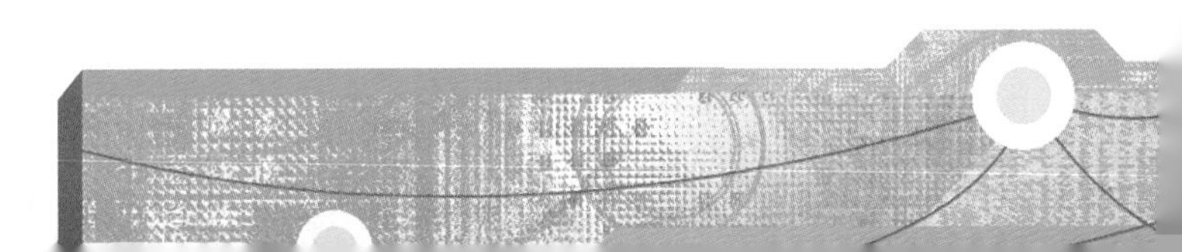

FIGURE 10.2 • A surgical training system

Doctors can use this virtual reality training system to practice new techniques.

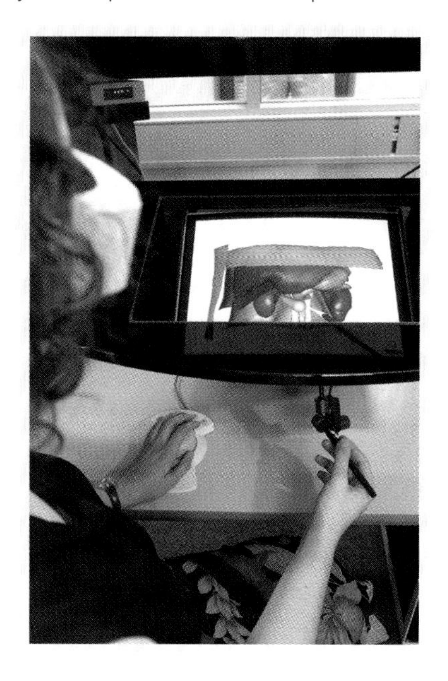

War II museum, for example, may be involved in developing a new museum database system that contains data and information on all of the World War II exhibits and artifacts stored in the museum.

Systems development specialists typically are computer systems personnel. Depending on the nature of the systems project, the development team might include the project leader, systems analysts, and software programmers, among others. A *project leader* is the individual in charge of the systems development effort. This person coordinates all aspects of the systems development effort and is responsible for its success. A **systems analyst** is a professional who specializes in analyzing and designing systems. Systems analysts play important roles while interacting with the system stakeholders and users, management, software programmers, and other computer systems support personnel (see Figure 10.3). Like an architect developing blueprints for a new building, a systems analyst develops detailed plans for the new or modified system. The *software programmer*, sometimes called a computer programmer or software engineer, is responsible for modifying existing programs or developing new programs to satisfy user requirements. Like a contractor constructing a new building or renovating an existing one, the programmer takes the plans from the systems analyst and builds or modifies the necessary software.

The other support personnel on the development team are mostly technical specialists, either computer systems department employees or outside consultants, including database and telecommunications experts, knowledge engineers, and other personnel such as vendor or supplier representatives. Some specialize in building new systems, and others might specialize in implementing computer systems or testing them. Depending on the magnitude of the systems development project and the number of computer systems development specialists on the team, the team may also include one or more computer systems managers. *Computer systems management* can include the chief information officer (CIO) and other computer systems executives.

Why Start a Systems Development Project?

Organizations can start a systems development project for many reasons, including problems with the existing system, mergers, competition, and even pressure from government agencies. Some systems development projects are started to avoid future problems. The U.S. Department of Energy (DOE) and TECSys Development, for example, started a systems development project to protect the electrical infrastructure of the United States.[6] The new DOE system will include software to monitor servers, routers, firewalls, and other components of the U.S. electrical infrastructure.

In the wake of recent financial scandals, the U.S. government is instituting new corporate financial reporting rules under the Sarbanes-Oxley Act. These new reporting regulations have caused many U.S. companies to initiate systems development efforts.[7] Some companies have been overwhelmed by these new, strict reporting requirements.[8] Some are building software from scratch to comply with the act, while others are looking for software they can purchase.[9]

Many organizations initiate systems development projects that they expect will provide a competitive advantage. Software companies are always exploring ways to make new products that people will flock to. Oil companies want to

FIGURE 10.3 • Participants in systems development

The systems analyst plays an important role in the development team, and is often the only person who sees the system in its totality. The arrows in this figure do not mean that there is no direct communication between other team members. Instead, these arrows represent the pivotal role of the systems analyst—an individual who is often called upon to be an interface, facilitator, moderator, negotiator, and interpreter for development activities.

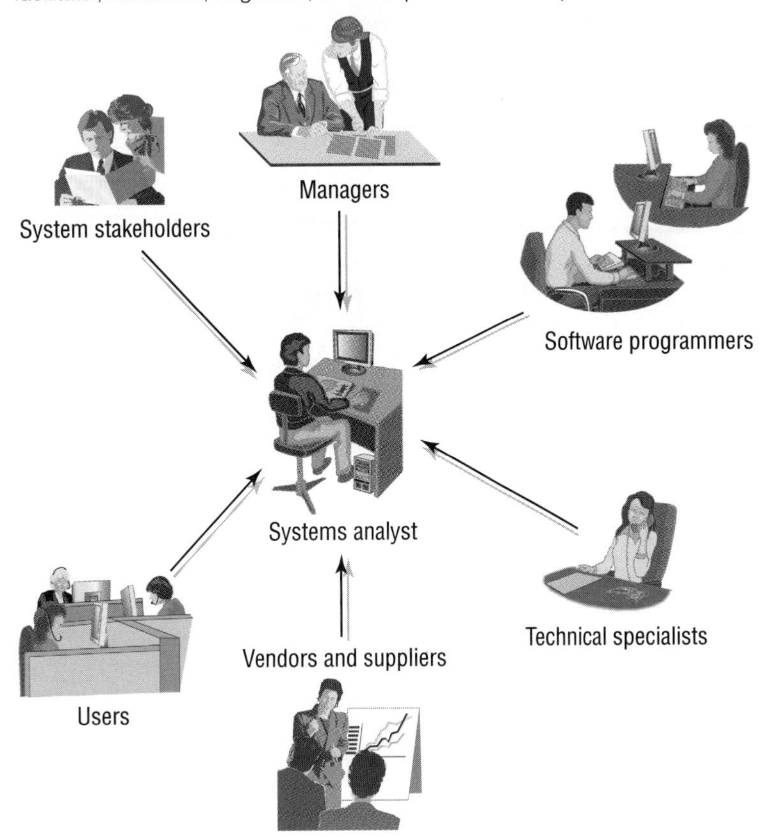

maximize their profits. Figuring out how to gain a competitive advantage usually requires creative and critical analysis.

Creative analysis involves seeking new approaches to existing problems. Some photographic companies, for example, are hoping to achieve a competitive advantage by developing online photo-sharing services. The services are aimed at camera-enabled cell phone users who don't know what to do with the photos after they snap them with their cell phone (see Figure 10.4). Some insurance companies have used creative analysis to develop Web sites to allow customers and sales reps to get information about insurance, retirement, and investment products. This type of Web site can provide a competitive advantage because sales reps can rapidly respond to customer needs. By looking at problems in new or different ways and by introducing innovative methods to solve them, many organizations have gained significant competitive advantage. Typically, these new solutions are inspired by people and things not directly related to the problem.

Critical analysis means being skeptical and doubtful, and requires questioning whether or not the current system is still effective and efficient. Critical analysis can result in finding better ways of doing things or ways of doing business that can result in a competitive advantage. An airline company, for example, can use critical analysis and decide that providing e-mail on many of its flights would attract more business customers, generate additional revenue, and give the company a competitive advantage.

FIGURE 10.4 • Competing creatively

Kodak has created an online photo-sharing service aimed at digital camera users and camera-enabled cell phone users.

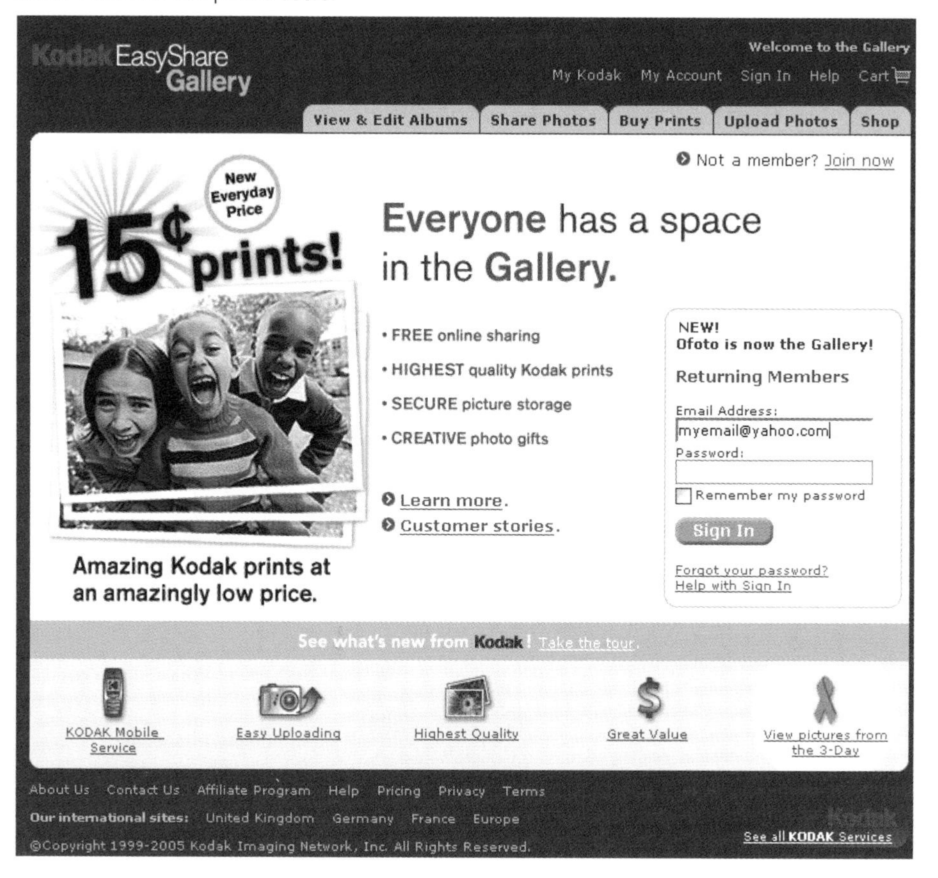

The systems development process often begins with gathering information on users' needs. Questioning users about their needs and being skeptical and doubtful about initial responses can result in better systems and more accurate predictions of how those systems will work. Too often, system stakeholders and users specify certain system requirements because they assume the only way to meet their needs is through those requirements. But often, their needs might best be met through an alternate approach. For example, a movie producer might decide to hire a team of stuntmen for an action scene because this is how he has done it for decades. However, a new computerized imaging system might be able to digitally generate the scenes he wants at less expense. All too often, problem solutions are selected before a complete understanding of the nature of the problem itself is obtained. The understanding can come from systems development planning.

Systems Development Planning

Systems development planning is the translation of organizational or individual goals into systems development initiatives. For example, an organization may identify as organizational goals a doubling of sales revenue within five years, a 20 percent reduction of administrative expenses over three years, acquisition of at least two competing companies within a year, or market leadership in a given product category. An individual might identify the ability to develop beautiful videos or photos as a personal goal. These goals are translated into specific systems development goals, such as acquiring new systems to help slash administrative expenses over three years or acquiring a new personal computer and software to develop digital videos and photos. Although systems development is an important function for any organization, it is increasingly becoming important for end users.

End-User Systems Development

End-user systems development is the development of computer systems by individuals outside of the formal computer systems planning and departmental structure. The increased availability and use of general-purpose information technology and the flexibility of many packaged software programs have allowed noncomputer systems employees to independently develop computer systems that meet their needs. These employees feel that by bypassing the process of going through the computer systems department they can develop systems more quickly. In addition, these individuals often feel that they have better insight into their own needs and can develop systems better suited to their purposes. Macromedia, for example, has an end-user systems development tool called Contribute that is designed to make it easy for people to develop and edit Web pages (see Figure 10.5). Such a Web development tool can be used to give specific people and groups the ability to edit specified Web sites. In addition, many end users are increasingly developing computer systems or solving computer-related problems for others. It should be noted, however, that not all organizations encourage or allow end-user systems development.

FIGURE 10.5 • End-user development

Macromedia Contribute, an end-user systems development tool, is designed to make it easy for people to develop and edit Web pages.

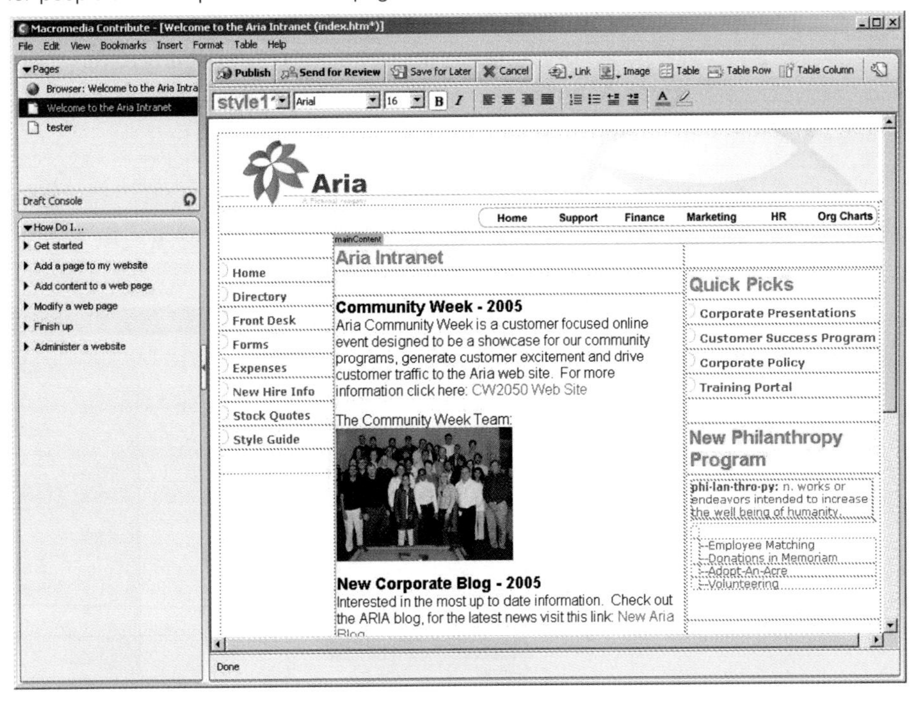

Systems developed by end users range from the very small (a software routine to merge data with form letters) to those of significant organizational value (such as customer contact databases). Like all projects, some systems developed by end users fail and others are successful. Initially, computer systems professionals discounted the value of these projects and basically ignored them. As the number and magnitude of these projects increased, however, computer systems

professionals began to realize that for the good of the entire organization, their involvement with these projects needed to increase. Rather than ignoring these initiatives, astute computer systems professionals encourage them by offering guidance and support. Technical assistance, communication of standards, and the sharing of best practices throughout the organization are just some of the ways computer systems professionals work with motivated managers and employees undertaking their own systems development projects. Computer systems professionals can also help end users apply systems development tools.

TOOLS AND TECHNIQUES FOR SYSTEMS DEVELOPMENT

Just as hammers, screwdrivers, and other tools can be used to make home repairs easier, *systems development tools* can greatly simplify the systems development process. In the strictest sense, a tool can be almost any instrument, from a pencil to a large, complex machine. Systems development tools most often used include computer-aided software engineering tools, flowcharts, decision tables, project management tools, prototyping, outsourcing, and object-oriented systems development.

Computer-Aided Software Engineering

One type of systems development tool consists of software programs that help automate various aspects of the systems development process. *Software engineering,* a formal way of conducting systems development, typically employs systems development software commonly referred to as CASE tools. **Computer-aided software engineering (CASE) tools** automate many of the tasks required in a systems development effort (Figure 10.6).

FIGURE 10.6 • CASE tools

CASE tools can be used to diagram and develop new computer systems.

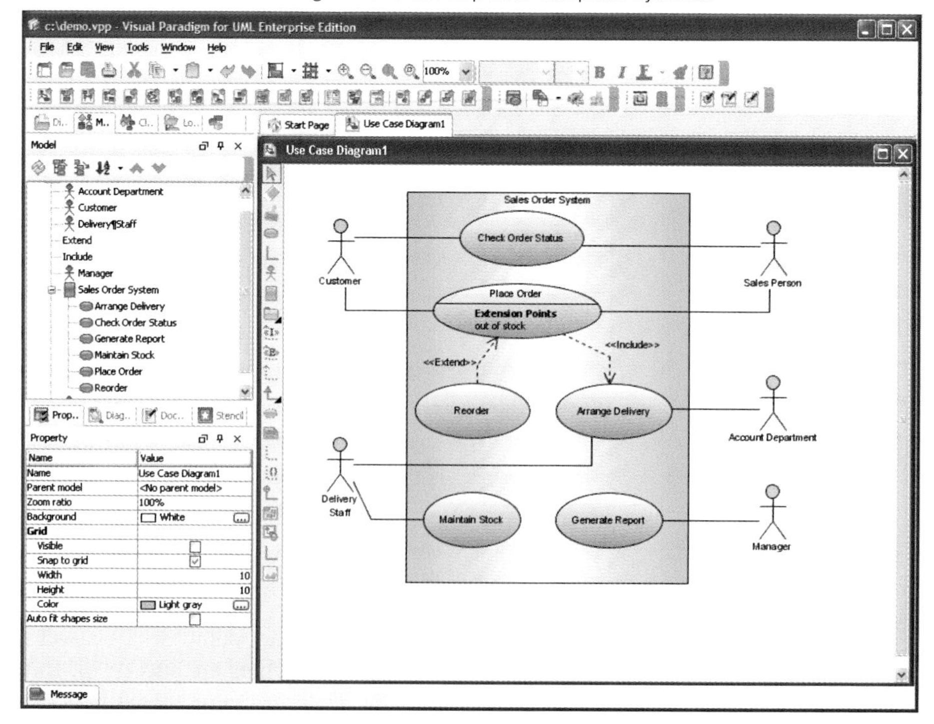

As with any team, coordinating the efforts of members of a systems development team can be a problem. To help address coordination problems, CASE tools allow more than one person to work on the same system at the same time via a multiuser interface. The multiuser interface coordinates and integrates the work performed by all members of the same design team. With this facility, one person working on one aspect of systems development can automatically share results with someone working on another aspect of the same system.

FIGURE 10.7 • Some basic flowcharting symbols

The symbols in a flowchart show the logical relationships between system components. These are some of the more commonly used flowcharting symbols.

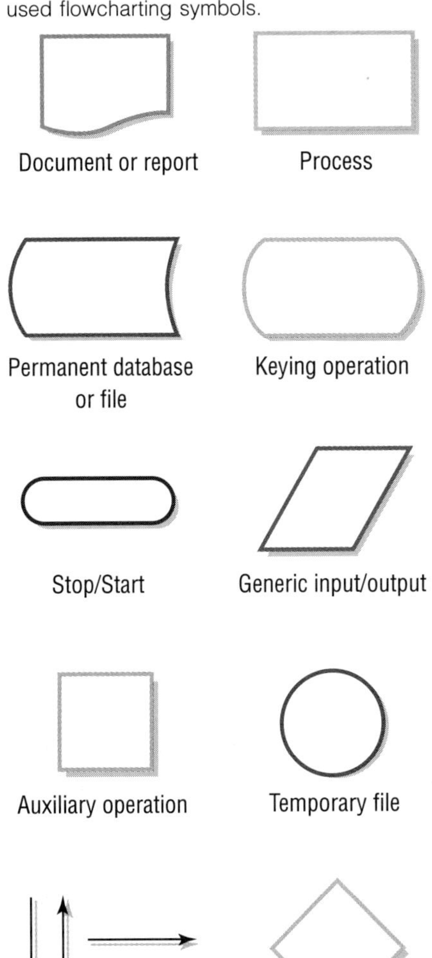

Flowcharts

Like a road map, a **flowchart** is a system design diagram that charts the path from a starting point to the final goal of a system. Flowcharts can display various amounts of detail. Using symbols, flowcharts show the logical relationships between system components. Some common flowcharting symbols are shown in Figure 10.7. Like other systems development tools, flowcharts can be useful in areas other than program design and development. Flowcharts, for example, can be used to display and understand what courses are needed to complete a college or university degree and what activities must be completed to finish a project at work.

When developing a system, a flowchart is used to describe the overall purpose and structure of the system. This is usually called the system flowchart or application flowchart. An *application flowchart* for a simplified payroll application is shown in Figure 10.8. Inputs include an employee file that contains an employee's pay rate and a time file that contains the hours the employee worked during the week. The payroll program multiplies the pay rate times the hours worked and subtracts any deductions to compute the paycheck for the employee. More detailed flowcharts, called *program flowcharts,* are needed to reveal how each software program is to be developed.

Flowcharts have a number of limitations. They were originally developed to help programmers and analysts design and document computer systems and programs. As programs became larger, flowcharts became more difficult to implement. You can imagine how difficult it would be to develop a detailed flowchart for a program containing more than 50,000 program statements. As a result, many organizations are reducing the amount of flowcharting they use. Some rely more on techniques such as computer-aided software engineering (CASE) tools.

FIGURE 10.8 • An application flowchart of a simplified payroll application

Circles in this application flowchart show that inputs include an employee file and a time file. The process square labeled "Payroll program" represents the operations of multiplying the pay rate times the hours worked and subtracting deductions to compute the employee's paycheck.

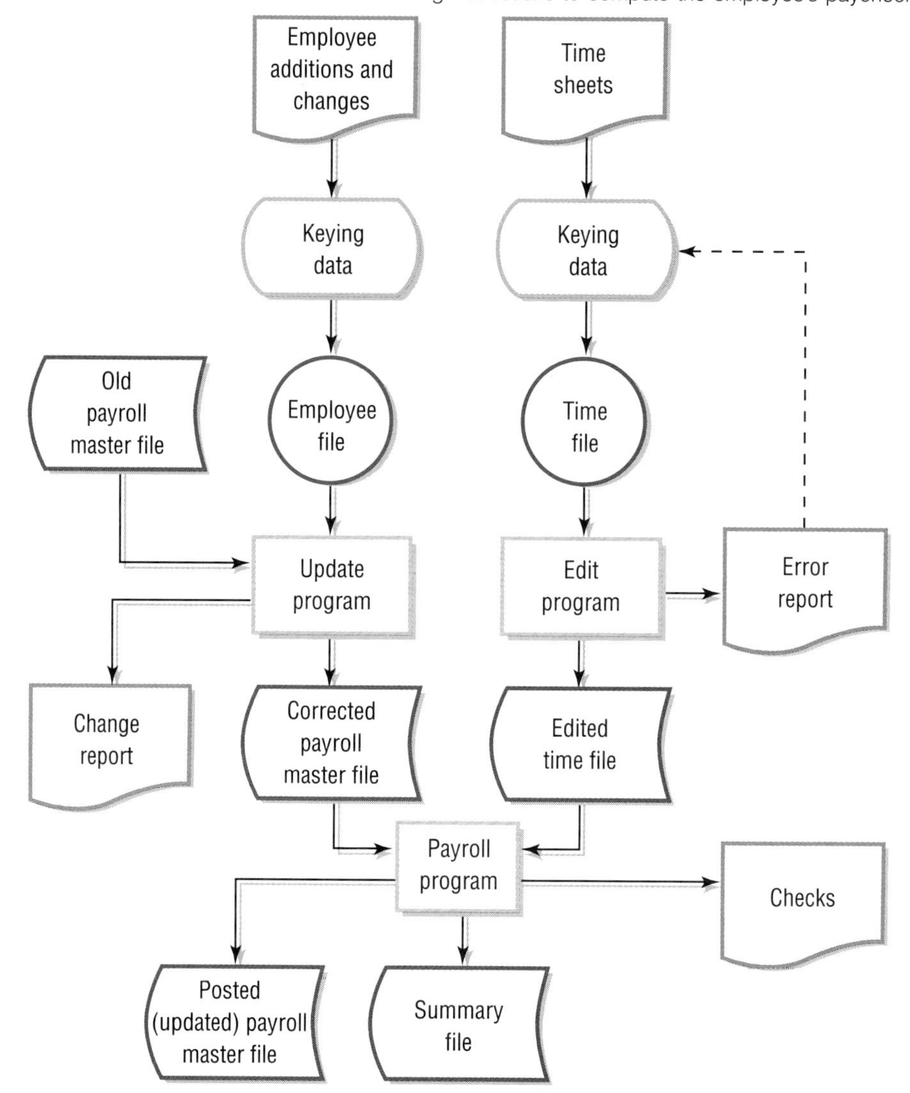

Decision Tables

A **decision table** is a systems development tool that displays the various conditions that could exist in a system and the different actions that the computer should take as a result of these conditions. A decision table can be used as an alternative to or in conjunction with flowcharts. When there are a large number of branches or paths within a software program, decision tables are particularly useful; in fact, in these cases, decision tables are preferable to flowcharts. A decision table that aids decisions regarding airline reservations is shown in Figure 10.9.

FIGURE 10.9 • A decision table for an airline reservation application

A decision table displays the various conditions that can exist and the different actions the computer should take as a result of any one condition.

	Airline reservation application	Rule number				
		1	2	3	4	**Rule numbers**
Name of decision table	*Condition statements*					
	First-class requested	Y	Y	N	N	
Condition statement	First-class available	Y	N	N	N	**Actual conditions**
	Tourist-class requested	N	N	Y	Y	
	Tourist-class available	N	N	Y	N	
	Actions taken					
	First-class ticket issued	X				
Action statement	Tourist-class ticket issued			X		**Action taken**
	First-class wait listed		X			
	Tourist-class wait listed				X	

Project Management Tools

Although the steps of systems development seem straightforward, larger projects can become complex, requiring literally hundreds or thousands of separate activities. For these types of systems development efforts, project management becomes essential. The overall purpose of **project management** is to plan, monitor, and control necessary development activities.

Two techniques frequently used in project management are program evaluation and review technique (PERT) and Gantt charting. PERT is a formalized approach to project management that involves creating three time estimates for an activity: the shortest possible time, the most likely time, and the longest possible time. A formula is then applied to come up with a single PERT time estimate. A Gantt chart is a graphical tool used for planning, monitoring, and coordinating projects. A Gantt chart is essentially a grid that lists activities and deadlines. Each time a task is completed, a darkened line is placed in the proper grid cell to indicate completion of a task (see Figure 10.10).

Both PERT and Gantt techniques can be automated using project management software, such as Microsoft Project. This type of software monitors all project activities and determines if activities and the entire project are on time and within budget. Project management software also has workgroup capabilities, having the ability to handle multiple projects and enabling a team of people to interact with the same software. Some project management software, such as Basecamp.com, is Web-based, which allows easy integration of communications and scheduling among team members wherever they may be located. Project management software helps people determine the best

EXPAND YOUR KNOWLEDGE

To learn more about project management tools, go to www.course.com/swt2/ch10. Click the link "Expand Your Knowledge," and then complete the lab entitled, "Project Management."

ENTER THE MATRIX

Computing has grown so complex that businesses are spending more and more just to manage and update their systems. This trend demands new business models and new systems. Experts in global networking see a solution in a "single, global, and collaborative information technology" that will push toward a decentralized computing system. What's anticipated is a worldwide, self-managing system that directs and self repairs. Sound familiar? It should. It is how the human autonomic nervous system works.

Source:
Supernova 2005: It's a Whole New, Connected World
by Wharton School
CNET News
http://knowledge.wharton.upenn.edu/index.cfm?fa=viewArticle&id=1244&specialId=38July 27, 2005

FIGURE 10.10 • Gantt chart

A Gantt chart shows progress with systems development activities by putting a bar through appropriate cells.

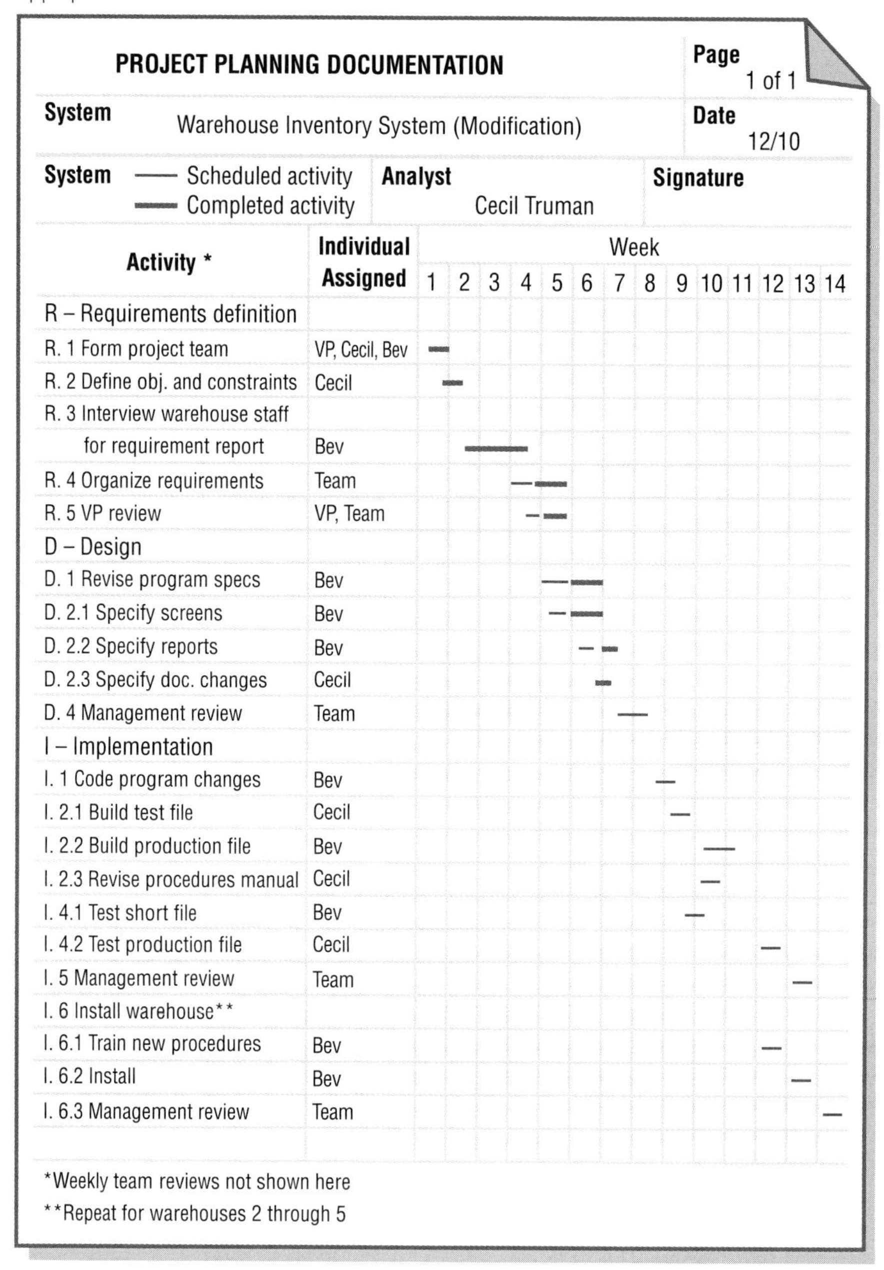

PROJECT PLANNING DOCUMENTATION

Page 1 of 1

System: Warehouse Inventory System (Modification)

Date: 12/10

System: —— Scheduled activity / ▬▬ Completed activity

Analyst: Cecil Truman

Signature

Activity *	Individual Assigned	Week 1	2	3	4	5	6	7	8	9	10	11	12	13	14
R – Requirements definition															
R. 1 Form project team	VP, Cecil, Bev														
R. 2 Define obj. and constraints	Cecil														
R. 3 Interview warehouse staff for requirement report	Bev														
R. 4 Organize requirements	Team														
R. 5 VP review	VP, Team														
D – Design															
D. 1 Revise program specs	Bev														
D. 2.1 Specify screens	Bev														
D. 2.2 Specify reports	Bev														
D. 2.3 Specify doc. changes	Cecil														
D. 4 Management review	Team														
I – Implementation															
I. 1 Code program changes	Bev														
I. 2.1 Build test file	Cecil														
I. 2.2 Build production file	Bev														
I. 2.3 Revise procedures manual	Cecil														
I. 4.1 Test short file	Bev														
I. 4.2 Test production file	Cecil														
I. 5 Management review	Team														
I. 6 Install warehouse**															
I. 6.1 Train new procedures	Bev														
I. 6.2 Install	Bev														
I. 6.3 Management review	Team														

*Weekly team reviews not shown here

**Repeat for warehouses 2 through 5

way to reduce project completion time at the least cost (see Figure 10.11). Reducing project completion time is called *project crashing*. This project management software feature can be very useful if a project starts to fall behind schedule or becomes more expensive than originally planned.

Prototyping

A different technique for systems development uses a phased or *iterative* approach. With the iterative approach to systems development, each phase of the SDLC is repeated several times (iterated). During each iteration, requirements

FIGURE 10.11 • Project crashing in Microsoft Project

Project management software helps people determine the best way to reduce project completion time at the least cost.

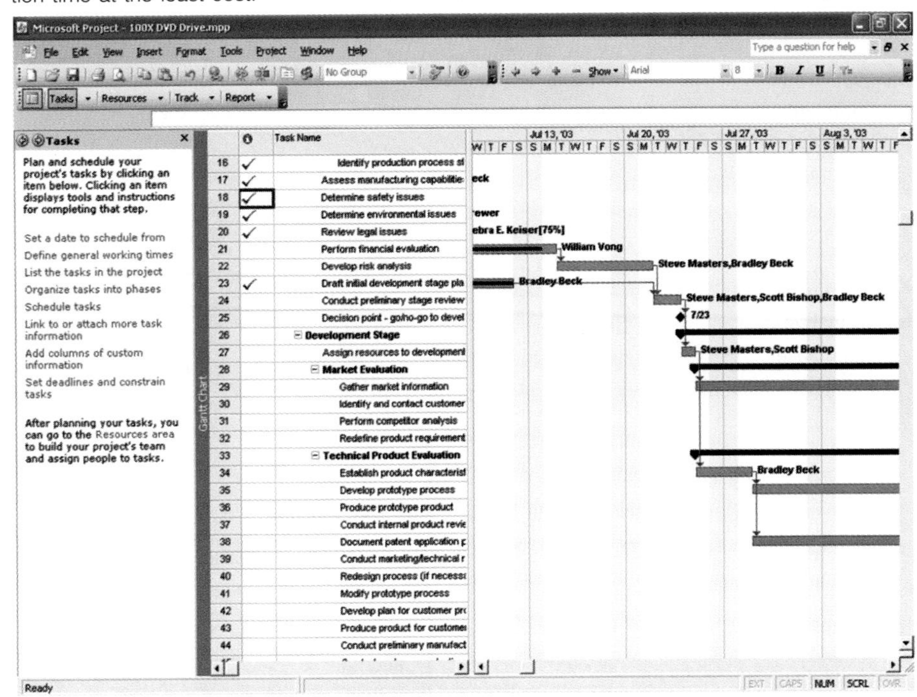

and alternative solutions to the problem are analyzed, solutions are designed, and some portion of the system is implemented and subject to a user review (see Figure 10.12).

A prominent example of an iterative technique for systems development is prototyping. **Prototyping** typically involves creating a preliminary model or version of a major subsystem, or a small or scaled-down version of the entire system. For example, a prototype might be developed to show sample report formats and input screens using a graphics program. Once developed and refined, the prototype reports and input screens developed in the graphics program are used as models for the actual system, which may be developed using a programming language such as C++ or Visual Basic. In many cases, prototyping continues until the complete system is developed.

To gain a better understanding of the use of prototyping, consider a company that sells and services vending equipment in Wisconsin, Illinois, and Indiana. The company wants to develop a new sales application for its five sales representatives to show revenues from their areas. The company decides to use prototyping and starts by developing a program that simply prints sample sales reports. No data is entered into the program at this time. After studying the sales reports for a week, sales representatives and sales management suggest several changes. These changes are made in the program, which is then expanded to accept some data from the sales ordering program and print two sample reports for each representative using actual data. The sales representatives and sales managers evaluate how effective the two actual reports are for their own particular purposes and make further suggestions. The changes are made and additional input data and reports are added to the application. This iterative process, typical of prototyping, continues until the new application is complete and satisfies the needs of both the sales representatives and sales management.

FIGURE 10.12 • An iterative approach to systems development

Using an iterative approach, each phase of the SDLC is repeated several times. System requirements and alternate solutions to the problem are analyzed, solutions are designed, and some portion of the system is implemented and subjected to a user review.

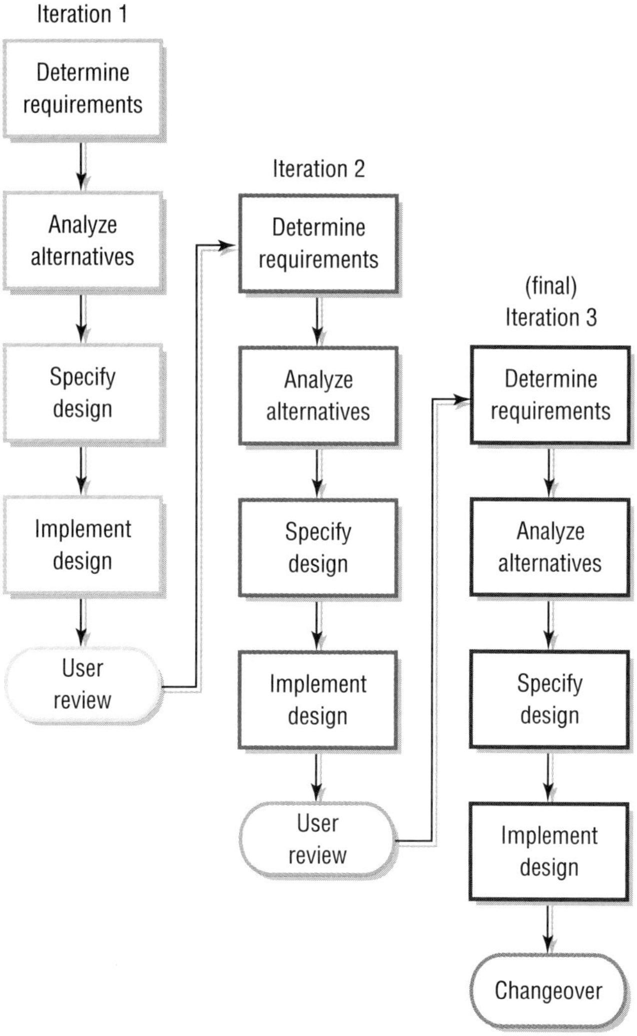

Iteration 1

Determine requirements → Analyze alternatives → Specify design → Implement design → User review

Iteration 2

Determine requirements → Analyze alternatives → Specify design → Implement design → User review

(final) Iteration 3

Determine requirements → Analyze alternatives → Specify design → Implement design → Changeover

Outsourcing

Outsourcing is a business' use of an outside company to take over portions of its workload. Many organizations hire an outside consulting firm that specializes in systems development to take over some or all of its computer systems development activities.[10] Accenture, IBM, and EDS are examples of consulting companies that can be hired to take over some or all computer-related tasks for an organization. Outsourcing is gaining in popularity.[11] See Figure 10.13. Sears, Roebuck, and Co, for example, looked for an outside company to perform many of its systems development and operations activities.[12] Sears is hoping that the outsourcing company it selects will hire some of its 200 people who will be laid off. Outsourcing has also become an important economic and political issue in today's economy for companies that outsource overseas.[13] A group of U.S. companies and organizations have formed the Coalition for Economic Growth and American Jobs to try to appease those that fear job loss in the United States resulting from outsourcing. Companies can spend millions or even billions of dollars to hire other companies to manage Web sites, network servers, data storage devices, and help-desk operations. In one outsourcing deal, a major phone company agreed to pay an outsourcing company $3 billion to provide guidance on cutting costs and improving efficiency. Companies such as Goodyear and Saks have also used outsourcing to increase services and reduce costs.[14]

Much of U.S. outsourcing is being provided by less developed countries where the work can be done less expensively. However, U.S. companies are also getting involved in providing outsourced services.[15] Ciber Inc, for example, is setting up about a half a dozen "Cibersites" containing about 200 technology workers and programmers each to provide outsourcing services and program development for organizations in the United States and around the world.[16] Aelera Corporation spent about 6 months looking for the best outsourcing deal and determined that a company in Savannah, Georgia, was the best.[17] McKesson Corporation saved about $10 million by outsourcing jobs from San Francisco to Dubuque, Iowa. Mattel outsourced to rural Jonesboro, Arkansas. Increasingly, companies are looking to American outsourcing companies to reduce costs and increase services.

FIGURE 10.13 • Outsourcing

Outsourcing can cut costs and improve efficiency, and has become an important economic and political consideration for many companies today.

Reducing costs, obtaining state-of-the-art technology, eliminating staffing and personnel problems, and increasing technological flexibility are reasons that organizations have used outsourcing. One American computer company, for example, estimated that a programmer with three to five years of experience in China would cost $12.50 per hour, while a programmer with similar experience in the United States would cost $56 per hour.[18] Organizations can use a number of guidelines to make outsourcing a success, including:[19]

- Keep tight controls on the outsourcing project. According to Jerry Bartlett of Ameritrade Holding Company, "The primary area where you've got to make a commitment is project management. You've got to have someone in place who has great communications skills, greater problem-resolution skills, and creativity."
- Treat outsourcing companies as partners. According to the CIO of Keystone Automotive Industries, "We were coming out of the dot-com boom, and we really wanted to take a look at some of these outsourcers to see if they understood the concept of what I call corporate hosting."
- Start with smaller outsourcing jobs.
- Create effective communications channels between the organization and the outsourcing company.

Outsourcing can involve a large number of countries and companies in bringing new products and services to market.[20] The idea for a new computer server can originate in Singapore, be approved in Houston, designed in India, engineered in Taiwan, and assembled in Australia. The chain of events can be complex.

There are, however, disadvantages to the outsourcing approach.[21] Internal expertise and loyalty can suffer under an outsourcing arrangement.[22] People who lose their jobs to outsourcing can become very emotional. When an organization uses outsourcing, key computer systems personnel with expertise in technical and organizational functions are no longer needed. Once these computer systems employees leave, the organization loses their experience with the organization and their expertise in computer systems. Outsourcing can also be very costly. Because of a faulty sales ordering system that had been outsourced, one company lost about $70 million in sales and decided to develop its own system for $40 million. The company saw a $400 million increase in sales as a result. For other companies, it can be difficult to achieve a competitive advantage when competitors are using the same outsourcing company. A Gartner, Inc, study estimates that about 80 percent of U.S. companies outsource critical activities to India, Russia, Pakistan, and China, which could jeopardize security.[23] How will important data and trade secrets be guarded?

FIGURE 10.14 • Diagram for a kayak rental program

A kayak rental business has many system objects, including the kayak rental clerk, renting kayaks to customers, and adding new kayaks into the rental program. This figure illustrates one way to diagram these objects.

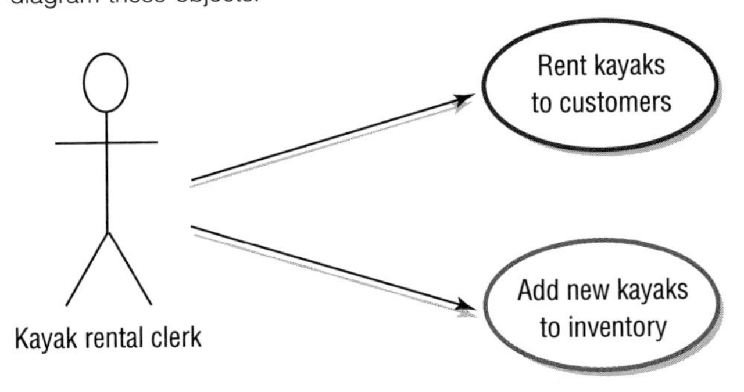

Object-Oriented Systems Development

Object-oriented (OO) systems development is an extension of object-oriented programming. OO development follows a defined system development life cycle, much like the SDLC. The life cycle phases can be, and usually are, completed with lots of iterations.

The object-oriented approach can be used during all phases of systems development, from investigation to maintenance and review. Consider a kayak rental business in Maui, Hawaii, where the owner wants to computerize its operations. This business has many system objects, including the kayak rental clerk, renting kayaks to customers, and adding new kayaks into the rental program. These objects can be diagrammed (see Figure 10.14). As you can see, the kayak rental clerk rents kayaks to customers and adds new kayaks to the current inventory of kayaks available for rent. The stick figure is an example of an *actor,* and the ovals each represent an event, called a *use case.* In our example, the actor (the kayak rental clerk) interacts with two use cases (rent kayaks to customers and add new kayaks to inventory).

Organizations have saved time and money using the object-oriented approach to systems development. AXA Financial Services, for example, was able to save about $55 million in developing a system by using the object-oriented approach.[24] According to the chief technology officer (CTO) for AXA, "Anything we do and have to redo is a negative ROI project because we're just duplicating something we already did and spending money to do it without adding much value." JetBlue Airways used Visual Studio.NET to implement an inventory-tracking system that used the object-oriented approach.[25] The application only took three months to implement and allowed employees to scan shipments using handheld computers. The scanned data was uploaded to the company's database. The application paid for itself in seven months through reduced costs.

SYSTEMS INVESTIGATION

Systems investigation is usually the first step in the development of a new or modified computer system. The overall purpose of **systems investigation** is to determine whether or not the objectives met by the existing system are satisfying the goals of the organization. In systems investigation, potential problems and opportunities are identified. Investigation attempts to reveal the cause and scope of the problem or opportunity. In general, systems investigation attempts to uncover answers to the following types of questions:

- What primary problems might a new or enhanced system solve?
- What opportunities might a new or enhanced system provide?
- What new hardware, software, databases, telecommunications, personnel, or procedures will improve an existing system, or are required in a new system?
- What are the potential costs?
- What are the associated risks?

Conducting a feasibility study is usually an important part of the systems investigation phase.

Feasibility Analysis

A key part of the systems investigation phase is **feasibility analysis**, which investigates the problem to be solved or opportunity to be met. Feasibility analysis involves an investigation into technical, economic, legal, operational, and schedule feasibility (see Table 10.2). *Technical feasibility* is concerned with whether or not hardware, software, and other system components can be acquired or developed to solve the problem. A loan company, for example, can

TABLE 10.2 • Types of feasibility

Type of feasibility	Description
Technical feasibility	Determines whether or not hardware, software, and other system components can be acquired or developed
Economic feasibility	Determines if the project makes financial sense
Legal feasibility	Determines whether laws or regulations may prevent or limit a systems development project
Operational feasibility	Measure of whether or not the project can be put into action or operation
Schedule feasibility	Determines if the project can be completed in a reasonable amount of time

investigate the technical feasibility of using new hardware and software to simplify the loan process. *Economic feasibility* determines if the project makes financial sense and whether predicted benefits offset the cost and time needed to obtain them. For example, a securities firm can investigate the economic feasibility of sending financial research reports to its customers and executives electronically instead of through the mail. Economic feasibility can involve analyzing cash flows and costs. *Legal feasibility* determines whether laws or regulations may prevent or limit a systems development project. For example, an Internet site that allowed users to share music without paying musicians or music producers was sued. Legal feasibility involves an analysis of existing and future laws to determine the likelihood of legal action against the systems development project and the possible consequences.

Operational feasibility is a measure of whether or not the project can be put into action or operation. Operational feasibility includes both physical and motivational considerations. Motivational considerations (acceptance of change) are very important, because new systems affect people and may have unintended consequences. As a result, power and politics may come into play. Some people may resist the new system. Because of deadly hospital errors, some health care organizations have looked into the operational feasibility of developing new computerized physician-order entry systems that check for drug allergies and interactions between drugs. The new system can save lives and prevent lawsuits. Finally, *schedule feasibility* determines if the project can be completed in a reasonable amount of time—a process that involves balancing the time requirements of the project with other projects. In some cases, additional resources can be expended to shorten project completion dates.

If a systems development project is determined to be worthwhile and feasible, systems analysis formally begins.

SYSTEMS ANALYSIS

After a project has been approved for further study during systems investigation, the next step is to perform a detailed analysis of the existing system, whether or not it is currently computer based. **Systems analysis** attempts to understand how the existing system helps solve the problem identified in systems investigation and answers the question, "What must the computer system do to solve the problem?" The process involves understanding the broader aspects of the system that would be required to solve the problem and the limitations of the existing system as identified in systems investigation. The overall emphasis of analysis is to gather data on the existing system and the requirements for the new system, and to consider alternative solutions to the problem within these constraints and the feasibility of these solutions. The primary result of systems analysis is a list of systems requirements and priorities.

General Analysis Considerations

Systems analysis starts by clarifying the overall goals of the individual or organization and determining how the existing or proposed computer system helps meet these goals. A university, for example, might want to develop a fundraising database that contains information on all of the people, trusts, and organizations that have made financial contributions or donations to the university. This goal can be translated into one or more informational needs. One need might be to create and maintain an accurate list of all projects funded by donations made to the university. Another need might be to produce a list of all donors who contributed more than $1000 over the last year. The list can be used to generate personalized thank-you letters.

Analysis of a small organization's computer system can be fairly straightforward. On the other hand, evaluating an existing computer system for a large organization can be a long, tedious process. As a result, large organizations evaluating a major computer system normally follow a formalized analysis procedure involving:

1. Collecting appropriate data
2. Analyzing the data
3. Determining new system requirements and project priorities

Collecting Data

The purpose of data collection is to seek additional information about the problems or needs identified during systems investigation. In many cases, the strengths and weaknesses of the existing system are uncovered.

Data collection involves identifying and locating the various sources of data. In general, there are both internal and external sources (see Table 10.3).

TABLE 10.3 • Internal and external sources of data for systems analysis

Internal sources	External sources
Users, stakeholders, and managers	Customers
Organizational charts	Suppliers
Forms and documents, including input documents from accounting and other transactions	Stockholders
Procedure manuals and written policies	Local, state, and federal government agencies
Financial reports	Competitors
Computer documents, including computer systems manuals	Outside organizations, such as environmental associations
Other measures of existing processes	Trade journals, books, and periodicals related to the organization
	External consultants and other commercial groups

FIGURE 10.15 • The steps in data collection

After data sources have been identified, interviews, direct observation, outputs, questionnaires, and other data collection methods are used to collect the data. Often, it is necessary to clarify what the data means.

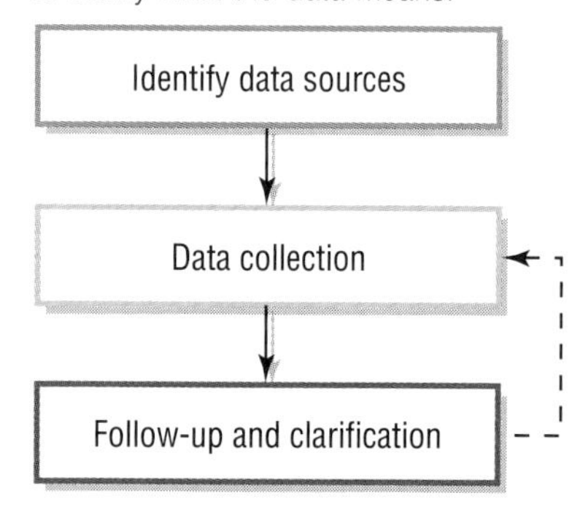

Once data collection sources have been identified, data collection begins. Figure 10.15 shows the steps involved.

Data collection may require a number of tools and techniques, such as:

- Interviews: In a *structured interview,* questions are written in advance. In an *unstructured interview,* the questions are not written in advance; the interviewer relies on experience in asking the best questions to uncover the inherent problems and weaknesses of the existing system.
- Direct observation: With *direct observation,* one or more members of the analysis team directly observe the existing system in action.
- Outputs: Outputs from the existing system, both manual and computerized, are obtained during data collection. See Figure 10.16.
- Questionnaires: When many data sources are spread over a wide geographic area, questionnaires may be the best approach. Like interviews, questionnaires can be either structured or unstructured.
- Other data collection methods: Telephone calls, simulating actual events and activities, and taking a random sample of data are other data collection techniques.

FIGURE 10.16 • Outputs from existing systems

Bills sent out to customers are important output from a computer system.

Data Analysis

Data collected in its raw form is usually not adequate to determine the effectiveness and efficiency of an existing system nor the requirements for a new system. The next step is to use data analysis to put the collected data into a form that is usable by the members of the development team participating in systems analysis. Two commonly used data analysis tools are application flowcharts and CASE tools, which were discussed earlier.

Requirements Analysis

The overall purpose of **requirements analysis** is to determine user, stakeholder, and organizational needs. For an accounts payable application, the stakeholders could include suppliers and members of the Purchasing Department. Questions that should be asked during requirements analysis include the following: Are these stakeholders satisfied with the current accounts payable application? What improvements could be made to satisfy suppliers and help the Purchasing Department?

Numerous tools and techniques can be used to capture systems requirements, including asking directly, critical success factors, joint application development, and rapid application development.

- Asking directly: One of the most basic techniques used in requirements analysis is asking directly. Asking directly is an approach that asks some or

all users and stakeholders what they want and expect from a new or modified system. This approach works best for stable systems in which stakeholders or users have a clear understanding of the system's functions. Unfortunately, many individuals do not know exactly or are unable to adequately articulate what they want or need. The result can be an expensive system that fails to help users or stakeholders. The role of the systems analyst during the analysis phase is to exercise critical and creative thinking skills, questioning statements and assumptions, in order to understand and convey these individual requirements so that the new or modified system best meets users' and stakeholders' requirements. See Figure 10.17.

FIGURE 10.17 • Asking directly

In requirements analysis, a basic technique is to ask users and stakeholders directly what they want and expect from a new system.

- Critical success factors: Another approach uses critical success factors (CSF). Users and stakeholders are asked to list only those factors or items that are critical to the success of their area or the organization. A CSF for a lawyer might be exceptional legal research. A CSF for a sales representative could be a list of customers currently buying a certain type of product. Starting from these CSFs, the system inputs, outputs, performance, and other specific requirements can be determined. One study found that different CSFs might be important during different phases of a systems development project.

- Joint application development: Joint application development (JAD) can be used in place of traditional data collection and requirements analysis procedures. Originally developed by IBM Canada in the 1970s, JAD involves group meetings in which users, stakeholders, and computer systems professionals work together to analyze existing systems, propose possible solutions, and define the requirements of a new or modified system. JAD groups consist of both problem holders and solution providers. A group normally requires one or more top-level executives who initiate the JAD process, a group leader for the meetings, people who will use the system, and one or more individuals who act as secretaries and clerks to record what is accomplished and to provide general support for the sessions (see Figure 10.18). Many organizations have found that groups can develop better requirements than individuals working independently and have assessed JAD as a very successful development technique.

FIGURE 10.18 • Joint application development

Joint application development includes a group leader for the meetings, people who will use the system, and one or more individuals who act as secretaries and clerks to record what is accomplished and to provide general support for the sessions.

- Rapid application development: Another efficient approach to determine and define systems requirements of a group is called rapid application development. Rapid application development (RAD) combines JAD, prototyping, and other techniques to quickly and accurately determine the requirements for a system. RAD involves a process in which a developer first builds a working model, or prototype, of the system to help a group of stakeholders, users, or managers identify how well the system meets their requirements. The prototypes are then refined to more closely align with stated requirements. Rational Software, a division of IBM, has RAD tools to make developing large Java programs and applications easier and faster. Locus Systems, a program developer, used a RAD tool called OptimalJ to generate more than 60 percent of the computer code for three applications it developed.[26] Royal Bank of Canada used OptimalJ to develop a number of customer-based applications. According to David Hewick, group manager of application architecture for the bank, "It was an opportunity to improve the development life cycle, reduce costs, and bring consistency [to our projects]."

Finding Trust in Computer Systems

In June 2004, in what was heralded as "the worst mess since banks put their faith in computers," the Royal Bank of Canada was unable to tell its 10 million Canadian customers exactly how much money was in their accounts. Canada's largest bank had a problem that kept tens of millions of transactions, including every direct payroll deposit it handles, from showing up in accounts.

You can imagine the furor as millions of bank customers discovered that the paycheck they counted on hadn't yet been deposited. While many were just inconvenienced, others were in a panic. Vacations were postponed, bill payments became delinquent, and customers lined up at banks to try to get money that they were due for basic living expenses.

The nightmare began during what was intended to be a routine programming update. Soon afterward, the bank's entire nationwide system failed to register withdrawals and deposits against customer balances for several days. The more days that passed, the larger the backlog, and the worse the chance of the system's being able to recover.

Such computer glitches are not rare in today's computer-dependent society. We have all experienced the frustration of being unable to conduct business because "the system is currently down." Computer system bugs have become a reality that we all endure. They are frustrating for customers and can be devastating to businesses. They plague us in our jobs, when we shop, and at home on our own PCs. As incidents mount, computer users lose trust in the ability of computer systems to work without fail.

Without doubt, we will see increasing efforts from all systems and software developers to deliver trustworthy systems. As our lives become increasingly dependent on computer systems, the future of our society could be jeopardized. It is clear that today's systems, in all of their complexity, require a high degree of automation and artificial intelligence capabilities to successfully manage them. Computer system stability, dependability, and security remains major goals for governments, industries, and technology specialists.

Questions

1. What social factors have arisen to make computer system stability and security an increasing concern?
2. Whose responsibility is it to ensure that computer systems are dependable and secure? Why?

Sources

1. *Staff. "RBC Securities Signs Contract for Brokerage Processing Operations,"* The Asian Banker Journal, *February 15, 2005.*
2. *Saunders, J. & Bloom, R. "Bank's Clients in Limbo,"* The Globe and Mail, *June 4, 2004, www.theglobeandmail.com.*
3. *Walsh, L. "Microsoft's Paradox,"* Information Security, *January 2004, www.lexis-nexis.com.*
4. *Hollands, M. "Microsoft Struggles to Build Trust,"* The Australian, *March 30, 2004, www.lexis-nexis.com.*
5. *Bednarz, A. "Autonomic Authority,"* Network World, *March 22, 2004, www.lexis-nexis.com.*
6. *"Evident Software Partners with IBM to Further Autonomic Computing Initiative,"* Business Wire, *March 8, 2004, www.lexis-nexis.com.*

SYSTEMS DESIGN

The purpose of **systems design** is to select and plan a system that meets the requirements defined in the requirements analysis. This can often involve outside companies and vendors, especially if additional hardware and software are needed. Systems design results in a new or modified system, and thus results in change. If the problems are minor, only small modifications are required. On the other hand, major changes may be suggested by systems analysis. In these cases, major investments in additional hardware, software, and personnel may be necessary. The first step of systems design is to generate systems design alternatives, discussed next.

Generating Systems Design Alternatives

The first step of design is to investigate the various alternatives for all components of the new system. This can include hardware, software, databases, telecommunications, personnel, and more. When additional hardware is not required, alternative designs are often generated without input from vendors. A *vendor* is a company that provides computer hardware, equipment, supplies, and a variety of services. Vendors include hardware companies, such as IBM and Dell; software companies, such as Microsoft; database companies, such as Oracle; and a variety of other companies.

A museum might not need a vendor if it wants to modify one of its databases to include the estimated value of each item in the museum and where each item is located in the museum. This modification likely requires someone on the museum staff working with the database management system without the need for input from outside vendors. However, if the new system is a complex one, the original development team may want to involve additional personnel in generating alternative designs. Florida-based Fidelity National Financial, for example, used IBM to acquire a centralized computer system for its multimillion dollar project to speed processing for the $8 trillion of mortgages and loans it processes for large banks every day.[27] If new hardware and software are to be acquired from an outside vendor, various requests can be made of the vendor.

A *request for information (RFI)* asks a computer systems vendor to provide information about its products or services, and a *request for quotes (RFQ)* asks a computer systems company to give prices for its products or services. The **request for proposal (RFP)** is generated during systems development when an organization wants a computer systems vendor to submit a bid for a new or modified system. It often results in a formal bid that is used to determine who gets a contract for new or modified systems. The RFP specifies, in detail, required resources such as hardware and software. It communicates these needs to one or more vendors, and it provides a way to evaluate whether or not the vendor has delivered what was expected. In some cases, the RFP is made a part of the vendor contract. The table of contents of a typical RFP is shown in Figure 10.19. RFIs, RFQs, and RFPs often ask vendors to provide important information, including:[28]

- A list of the vendor's customers
- Previous experience and financial stability of the vendor
- References on recently completed projects

Evaluating and Selecting a Systems Design

The next step in systems design is to evaluate the various design alternatives and select the design that offers the best solution supporting organizational goals. For a simple design, such as a new graphics program for a commercial artist, one person can complete the system design. A moderate design project can involve a number of people inside the organization. To modify a database or an existing software program at a tax preparation company, programmers from inside the organization can be used.

After the final presentations and demonstrations have been given, the organization makes the final evaluation and selection. Cost comparisons, hardware

FIGURE 10.19 • A typical table of contents for a request for proposal (RFP)

The RFP specifies required resources such as hardware and software. It communicates these needs to one or more vendors, and it provides a way to evaluate whether or not the vendor has delivered what was expected.

JOHNSON & FLORIN, INC.
REQUEST FOR PROPOSAL

Table of Contents

Cover page (with company name and contact person)
Brief description of the company
Overview of the existing computer system
Summary of computer-related needs and/or problems
Objectives of the project
Description of what is needed
Hardware requirements
Software requirements
Personnel requirements
Communications requirements
Procedures to be developed
Training requirements
Maintenance requirements
Evaluation procedures (how vendors will be judged)
Proposal format (description of how vendors should respond)
Important dates (when tasks are to be completed)
Summary

performance, delivery dates, price, modularity, backup facilities, available software training, and maintenance factors are considered. Although it is good to compare computer speeds, storage capacities, and other similar characteristics, it is also necessary to carefully analyze if the characteristics of the proposed systems meet the objectives set for the system and how it will help the organization solve problems and obtain goals. Some journals and magazines evaluate computer vendors and their products and services. J.D. Powers, the company that gives quality awards to car manufacturers, will start evaluating computer systems vendors in terms of their service and support.[29] According to a director at the company, "Typically, every industry evaluated at J.D. Powers has directly or indirectly raised the overall customer satisfaction level of the industry." J.D. Powers is expected to give certifications to those IS vendors that achieve high service and support levels.

The Contract

When large computer systems are purchased, the hardware vendor often requires a contract. Most computer vendors provide standard contracts; however, these contracts are designed to protect the vendor, not necessarily the organization buying the computer equipment. Developing a good contract can be one of the most important steps in systems design if new computer facilities are to be acquired.

More and more organizations are developing their own contracts, stipulating exactly what they expect from the system vendor and what interaction will occur between the vendor and the organization. All equipment specifications, software, training, installation, maintenance, and so on are clearly stated. Furthermore, deadlines for the various stages or milestones of installation and implementation are stipulated, as well as actions to be taken by the vendor in case of delays or problems. Some organizations include penalty clauses in the contract, in case the vendor is unable to meet its obligation by the specified date. Typically, the request for proposal (RFP) becomes part of the contract. This saves a considerable amount of time in developing the contract, because the RFP specifies in detail what is expected from the system vendor or vendors.

SYSTEMS IMPLEMENTATION

After the computer system has been designed, a number of tasks must be completed before the system is installed and ready to operate. This process, called **systems implementation**, includes hardware acquisition, software acquisition or development, user preparation, hiring and training of personnel, site and data preparation, installation, testing, startup, and user acceptance. The typical sequence of these activities is shown in Figure 10.20. In many cases, some of the steps of systems implementation can be performed at the same time. For example, while hardware is being acquired, software can be developed, and new computer personnel can be hired.

At each step shown in Figure 10.20, there are choices and trade-offs to be made that involve analyzing the benefits of the various choices. Unfortunately, many organizations do not take full advantage of these steps or carefully analyze the trade-offs, and hence never realize the full potential of new or modified systems. The carelessness that often causes these steps to be overlooked must be avoided if organizations are to achieve their objectives and get the most from the new or modified computer system.

FIGURE 10.20 • Typical steps in systems implementation

To realize the full potential of new or modified systems, organizations must carefully analyze the trade-offs at each step in the implementation process.

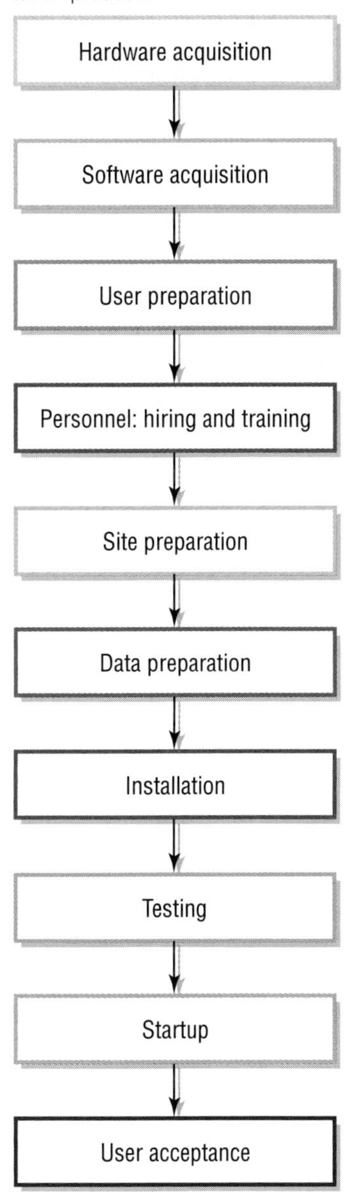

Acquiring Hardware

Although you can build your own computer using commonly available hardware components or have someone build a computer for you, most people and organizations acquire hardware and computers by purchasing, leasing, or renting computer resources from a computer systems vendor.

In addition to buying, leasing, or renting computer hardware, it is also possible for an organization to pay only for the computing that it uses. Called "pay as you go computing" or "utility computing," this approach means that an organization pays only for the computer power it uses. This is similar to paying only for the electricity you use. Many companies, like IBM, offer this service.[30] Hewlett-Packard has a "capacity-on-demand" approach, where organizations pay according to the computer resources actually used, including processors, storage devices, and network facilities. It is also possible to purchase used computer equipment. Every year eBay, a Web auction site, sells millions of dollars of used computer-related equipment. Regardless of how hardware is acquired, it is important to schedule and budget for new hardware acquisitions. Hardware is expensive, and it can become error prone and technologically obsolete over time. As a result, some organizations include hardware replacement costs in their annual budgets.

Selecting and Acquiring Software: Make, Buy, or Rent

As with hardware, software can be acquired in several ways. As previously discussed, software can be purchased from external developers or developed in house. Regis Corporation, an operator of hair salons, spent about $100,000 to purchase software to comply with new federal laws.[31] The District of Columbia decided to purchase three software modules from Hyperion to help its 68 government agencies do a better job of budgeting.[32] The million-dollar deal is part of a larger $71.5 million systems development effort that is expected to save the government more than $60 million annually. Global companies doing business around the world often select a standardized, global software package, such as SAP, a popular enterprise resource planning (ERP) package.

Sometimes developing software in house produces the most effective result. Software engineers, or programmers, are responsible for developing the software that drives the larger information system. The process of developing software involves a series of activities, similar to the systems development life cycle, called the *program development life cycle*. Programmers are provided with a description of what the software must accomplish, called the *program specification*. From the program specification, they use programming logic to develop an *algorithm*, a step-by-step sequence of computer instructions, that when given specified input, yields the desired results. The algorithm is then translated into a program language to create the executable programs that will control the system.

Software is tested thoroughly with every possible input in order to uncover possible bugs. A bug is an error in the program code that causes the software to malfunction. Bugs can be caused by *syntax errors*—errors in the use of the programming language syntax, or bugs can be caused by *logic errors*—errors in the programming logic that governs the action of the software. Once released, software, like systems, is continuously evaluated and periodically revamped and improved with the release of new editions.

FIGURE 10.21 • NASA DAC software

NASA's award-winning DAC software is used to analyze how a spacecraft enters distant environments and atmospheres.

NASA developed a software program called DAC for space exploration to analyze how a spacecraft enters distant environments and atmospheres (see Figure 10.21). This award-winning software has been used for the Mars Global Surveyor and the Mars Odyssey missions. Computer programmers at Sabre Airlines Solutions, a $2 billion airline travel company, developed in-house software that evaluates programming code. Their software, written in Java, helps Saber to eliminate programming errors and shorten program development times. Using their custom-made software, the company manages 13 million lines of programming code in 62 software products.[33]

Some in-house software developers, however, often try to incorporate every conceivable feature and option into the software. The resulting software, often called "bloatware" can be slow, unreliable, and not secure, with too many lines of software code.[34]

In some cases, organizations use a blend of external and internal software development. That is, proprietary software or open-source programs are modified or customized by in-house personnel. Some of the reasons that an organization might purchase or lease externally developed software include lower costs, less risk regarding the features and performance of the package, and ease of installation. The development effort is usually less when programmers use or modify a purchased software product rather than write all of the program code themselves. An *application service provider (ASP)* is a company that provides software and support, such as computer personnel, to run the software. Using an ASP can be faster and less expensive than developing software, but it can be more difficult to get the features that are needed or to make software changes when needed. Software can also be rented.

NOT DEAD YET

A dearth of system developers savvy on mainframes made companies reluctant to invest in them. Current experts were retiring and new grads learned different languages. It looked as if the exit signs were flashing for mainframes. Then IBM introduced its Academic Initiative zSeries program, which has enrolled 150 universities and is aiming for 20,000 mainframe professionals by 2010. Students and professors are discovering to their surprise that mainframe systems are more enjoyable to develop and maintain.

Source:
The Resurgence of Mainframes?
By Clint Boulton
Internetnews.com
http://www.internetnews.com/dev-news/article.php/3520821
July 18, 2005

Hudson River Park Trust Looks for Help in Systems Development

The Hudson River Park Trust (HRPT) was founded in 1998 by the state of New York to create a 5-mile stretch of parkland along Manhattan's West Side. One of the first challenges of the HRPT's staff was to decide how they were going to share information among the hundreds of contractors and suppliers involved in the restoration projects. It was too expensive for the organization to purchase, install, and maintain systems to store the scores of space-intensive project files, such as computer-aided design drawings and blueprints.

With an IT staff of just four people, the HRPT decided that the best approach would be to find outside help. "With so many contractors involved, it would have been a nightmare to manage all of this in house," said Michael Breen, CIO for the organization. "We've saved hundreds of thousands of dollars in cost avoidance in terms of staffing, T1 lines, WANS, LANs, and servers."

The company that saved the HRPT so much money is an applications service provider (ASP) by the name of Constructware. An ASP is a company that "leases" software services to customers, typically over the Internet. Constructware allows the trust and the hundreds of contractors and suppliers it works with to share over the Internet 37,000 blueprints, diagrams, and other documents for various phases of construction. The ASP includes a customized portal that features a personal organizer and tools for project reporting, business development, bid management, project management, file and document management, cost management, and human resources. Constructware covers all the requirements for construction projects.

Under its $65,000 annual licensing agreement, the HRPT can have an unlimited number of people access the system. All Breen does is set up a user account for each contractor and ensures that each user receives one to two hours of training using WebEx software to access the Constructware training modules. Training is done in the HRPT offices or at contractors' offices. Contractors can access the system via 56 KB modems or DSL connections that the trust has established in construction trailers along the waterfront. Currently, 250 people are using the system.

The technology arrangement at HRPT reflects a trend among government agencies to rely more heavily on outsourcing as IT staffs become leaner, said Jim Krouse, an analyst at Input Inc, based in Reston, Virginia. The HRPT is now using the software to manage 165 concurrent projects, including Segment 7, the stretch of land from 46th Street to 59th Street that is known as Clinton Cove. Project work for that section, scheduled to be completed in spring 2005, "is on time and on budget, and not a lot of government organizations can say that," said Breen.

Questions

1. When considering software options, what attributes of a project might lead a project leader to choose an ASP such as Contructware rather than developing something from scratch or hiring an outside company to develop a custom system?
2. What benefits has the HRPT staff experienced by going with an ASP rather than developing and maintaining the system themselves?

Sources

1. *Staff. "RBC Securities Signs Contract For Brokerage Processing Operations,* Constructioneer, *April 4, 2005, p. 32*
2. *Hoffman,T. "ASP Speeds Project Management for NYC Parks Developer,"* Computerworld, *June 17, 2004, www.computerworld.com.*
3. *Constructware Web site, accessed June 21, 2004, www.computerworld.com.*
4. *Hudson River Park Web site, accessed June 21, 2004, www.hudsonriverpark.org.*

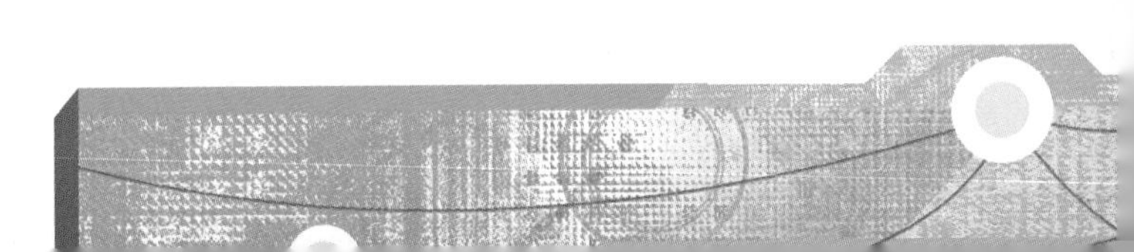

Acquiring Database and Telecommunications Systems

Acquiring or upgrading database systems can be one of the most important parts of a systems development effort. Acquiring a database system can be closely linked to the systems development process, since many systems development projects involve a new or modified database system. Because databases are a blend of hardware and software, many of the approaches discussed for acquiring hardware and software also apply to database systems. For example, an upgraded inventory control system may require database capabilities, including more hard disk storage or a new DBMS. Additional storage hardware would have to be acquired from a computer systems vendor. New or upgraded software could be purchased or developed in house.

Telecommunications is one of the fastest growing applications for today's businesses and individuals. The NASDAQ Stock Market, for example, is investing millions of dollars in a network system to streamline operations and cut costs. Like database systems, telecommunications systems require a blend of hardware and software. Sutter Health, for example, acquired a wireless telecommunications system to help comply with FDA requirements to place bar codes on drugs.[35] The design calls for wireless LAN and IP phones.

As you learned in Chapter 5, for personal computer systems, telecommunications hardware usually includes some type of modem, plus a router or perhaps wireless equipment. For client/server and mainframe systems, the hardware can include multiplexers, concentrators, communications processors, and a variety of network equipment. Communications software also has to be acquired from a software company or developed in house. You acquire telecommunications hardware and software in much the same way you acquire computer system hardware and software.

User Preparation

User preparation is the process of readying managers and decision makers, employees, and other users and stakeholders for the new or modified system. System developers need to provide users with the proper preparation and training to make sure they use the computer system correctly, efficiently, and effectively. User preparation can include marketing, training, documentation, and support.

Without question, training users is an essential part of user preparation, whether they are trained by internal training personnel or by external training firms. In some cases, companies that provide software also provide user training at no charge or at a reasonable price. Training can be negotiated during the selection of new software. Some companies conduct user training throughout the systems development process. Fears and apprehensions about the new system must be eliminated through these training programs. Old and new employees should be acquainted with the system's capabilities and limitations (see Figure 10.22).

Computer Systems Personnel: Hiring and Training

Depending on the size of the new system, a number of computer systems personnel may have to be hired and, in some cases, trained. A computer systems manager, computer programmers, data entry operators, and similar personnel may be needed for the new system. The CEO of PeopleSoft, for example, believes that information services (IS) personnel need training in three areas: business, technology, and an understanding of other countries and people: "If you get a strong background in those three things, you'll be able to participate in the future."[36]

FIGURE 10.22 • User preparation

User preparation is the process of readying managers and decision makers, employees, users, and stakeholders for the new system.

As with users, the eventual success of any system depends on how it is used by computer systems personnel within the organization. This cannot be overemphasized. Training programs should be conducted for the computer systems personnel who will be using or dealing with the computer system. These programs will be similar to those for the users, although they may be more detailed in terms of technical aspects of the systems. Effective training helps computer systems personnel use the new system to perform their jobs and helps them provide support to the other users in the organization.

Site Preparation

The actual location of the new system needs to be prepared in a process called *site preparation*. For a small system, this may simply mean rearranging the furniture in an office to make room for a personal computer. For a larger system, this process is not so easy. Larger systems may require special wiring and air conditioning. One or two rooms may have to be completely renovated. Additional furniture may have to be purchased. A special floor may have to be built, under which the cables connecting the various computer components are placed, and a new security system may have to be installed to protect the equipment. For larger systems, additional electrical circuits may also be required. See Figure 10.23.

Data Preparation

If an organization is about to computerize, all noncomputerized files, such as information on paper in file folders, must be converted into computer files in a process called *data preparation*. For old computerized files, *data conversion* may be required to transform the existing computerized files into the proper format to be used by the new system. All of the permanent data must be placed on a permanent storage device, such as magnetic tape or disk. Usually the organization hires some temporary, part-time data-entry operators or a service company to

FIGURE 10.23 • Site preparation

For larger systems, a secure, climate-controlled room with special wiring and electrical circuits may be needed.

convert manual data. Once the information has been converted into computer files, the data-entry operators or the service company are no longer needed. A computerized database system or other software is used to maintain and update these computer files.

Installation

Installation is the process of physically placing the computer equipment on the site and making it operational. For a small systems development project, this might require making room on top of a desk for a new PC, plugging it into a wall outlet, and following the manufacturer's instructions to turn it on. For a larger project with a mainframe computer system, installation usually involves the computer manufacturer. Although it is normally the responsibility of the manufacturer to install the computer equipment for larger systems development projects, someone from the organization (usually the chief information officer (CIO) or the computer systems manager) should oversee this process, making sure that all of the equipment specified in the contract is installed at the proper location. After the system is installed, the manufacturer performs several tests to ensure that the equipment is operating as it should. See Figure 10.24.

FIGURE 10.24 • Installation

Installation involves placing the computer equipment on the site and making it operational.

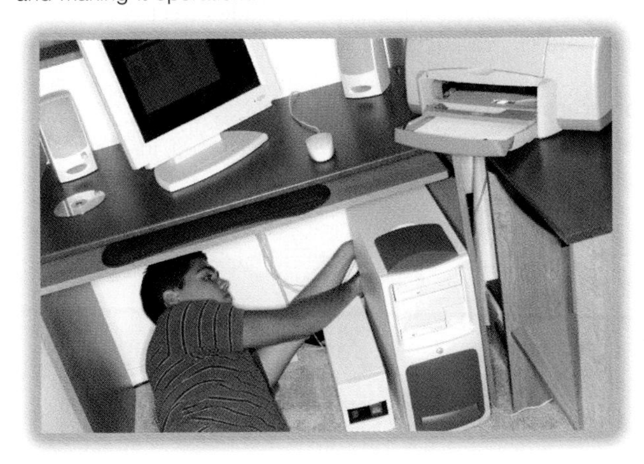

Testing

Testing involves the entire computer system. Millions of dollars or even lives can be lost with inadequate testing. It is always difficult to determine how much testing is needed before a new or modified system is placed into operation.

Testing requires testing each of the individual programs (unit testing), testing the entire system of programs (system testing), testing the application with a large amount of data (volume testing), and testing all related systems together (integration testing), as well as conducting any tests required by the user (acceptance testing).

Unit testing is accomplished by developing test data that forces the computer to execute every statement in the program. In addition, each program is tested with abnormal data to determine how it handles problems with bad data.

System testing requires the testing of all of the programs together. It is not uncommon for the output from one program to become the input for another. In these cases, system testing ensures that the output data from one program can be used as input for another program within the system.

Volume testing is performed to ensure that the entire system can handle a large amount of data under normal operating conditions.

Integration testing ensures that the new program(s) can interact with other major applications. It also makes sure that data flows efficiently and without error to other applications. For example, a new tax preparation program may require data input from a personal accounting program. Integration testing is done to ensure smooth data flow between the new and existing applications. Integration testing is typically done after unit and system testing. Metaserver, a software company for the insurance industry, has developed a tool called iConnect to perform integration testing for different insurance applications and databases.[37] According to Donald Light, a Celenet Communications analyst, "The difference with iConnect is its ability to deliver a return on a relatively small investment. They've taken an area that is critical with independent agents on one hand and providers of external data on the other hand and brought them together."

Finally, *acceptance testing* makes sure that the objectives of the new or modified system are being met. Run times, the amount of memory required, disk access methods, and more can be tested during this phase. Acceptance testing makes sure that this performance objective and all other objectives defined for the system are satisfied. Involving users in acceptance testing may help them better understand and effectively interact with the new system. AON uses acceptance testing to make sure that new software is ready for use.[38] "If we deploy an application now which has 80 percent functionality but contains bugs, we can decide how to move forward," says a company representative. Involving users in acceptance testing may help them understand and effectively interact with the new system. Acceptance testing is the final check of the system before startup.

Startup

Startup begins with the final tested computer system. When startup is finished, the system is fully operational. Different startup approaches include direct conversion, phase-in and pilot conversion (see Figure 10.25).

FIGURE 10.25 • Startup approaches

Startup begins with the final tested computer system. When startup is finished, the system is fully operational.

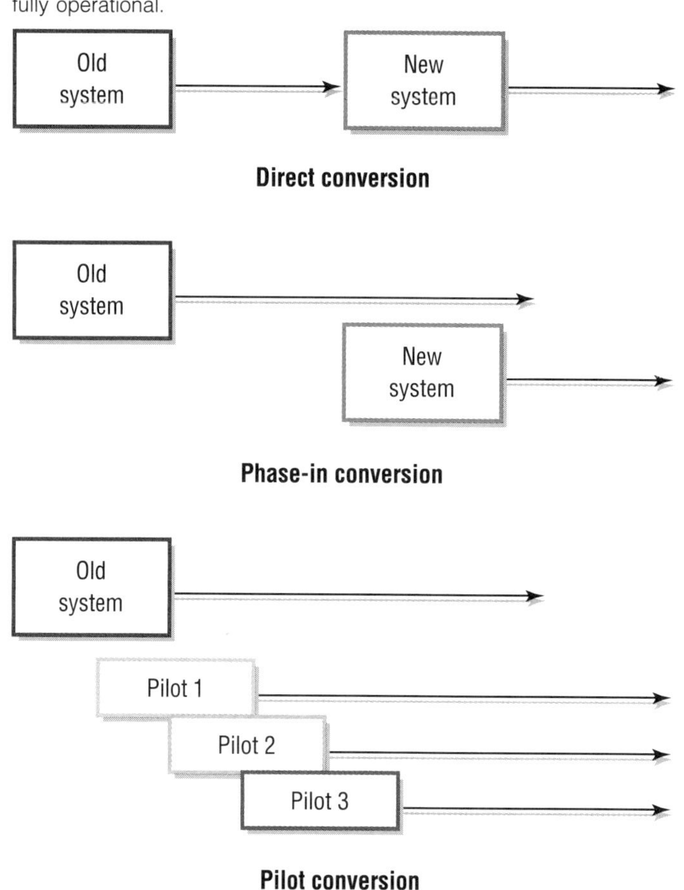

Direct conversion involves stopping the old system and starting the new system on a given date. This is usually the least desirable approach because of the potential for problems and errors when the old system is completely shut off and the new system is turned on at the same instant.

The **phase-in approach** is a popular technique preferred by many organizations. In this approach, the new system is slowly phased in while the old one is slowly phased out. During this process, parts of the old system and new system are running at the same time, in parallel. This is called a *parallel conversion.* When everyone is confident that the new system is performing as expected, the old system is completely phased out. This process is repeated for each application until the new system is running every application.

Pilot startup involves running a pilot or small version of the new system along with the old. After the pilot runs without errors or problems, the old system is stopped and the new system is fully operational. With pilot startup, small pilots can be introduced until the complete new system is operational. For example, a state prison system with a number of correctional facilities throughout the state could use the pilot startup approach and install a new computerized security system at one of the facilities. When this pilot program at the pilot facility runs without errors or problems, the new security system can be implemented at other prisons throughout the state.

User Acceptance and Documentation

User acceptance and documentation is usually done for larger systems development projects that require new computers or servers. Smaller systems development projects, such as a musician acquiring new software to blend several music tracks into a finished song, usually don't require user acceptance and documentation.

A **user acceptance document** is a formal agreement signed by the user that a phase of the installation or the complete system is approved. This is a legal document that usually removes or reduces the vendor's liability or responsibility for problems that occur after the user acceptance document has been signed. Because this document is so important, many organizations get legal assistance before they sign the acceptance document. Stakeholders may also be involved in acceptance to make sure that the benefits to them are indeed realized.

The system should also be fully documented. *Documentation* includes all flowcharts, diagrams, and other written materials that describe the new or modified system. In general, there are two types of documentation. *Systems documentation* describes the technical aspects of the new or modified system. It can be used by the chief information officer (CIO), systems analysts, programmers, and other computer-related staff. *User documentation* describes how the system can be used by noncomputer personnel. A manual on how to use a spreadsheet program or an operating system is an example of user documentation. If a company develops a new game for a PC that simulates a cartoon character, user documentation is needed to describe to people (users) how the new game is to be played. If another company develops new reports for managers, user documentation is needed to describe how managers can use the new reports.

SYSTEMS MAINTENANCE AND REVIEW

The final steps of systems development are systems maintenance and review. **Systems maintenance** involves checking, changing, and enhancing the system to make it more useful in achieving user and organizational goals. In some cases, an organization encounters major problems that involve recycling through the entire systems development process. In other situations, minor modifications are sufficient.

TASTY AND TATTOOED

Hard to put on, devilish to get off, the stickers used on produce have been driving both appliers and removers crazy. National security concerns, however, now mandate food be tracked. The result is a tracing system that keys off a number tattooed on the skin of a fruit or vegetable. Information is added at each stage of the item's journey, identifying its origin, grower, wholesale price, and eventually, what you're going to pay to take it home.

Source:
Tattooed Fruit Is on Way
By Julia Moskin
The New York Times
http://www.nytimes.com/2005/07/19/dining/19fruit.html
July 19, 2005

Reasons for Maintenance

Maintenance can involve all aspects of the system, including hardware, software, databases, telecommunications, personnel, and other system components. Older hardware, for example, may be too slow and without enough storage capacity. Older software can also require maintenance. Once a program is written, it should ideally require little or no maintenance, but old programs require maintenance to make them faster or enhance their capabilities. In addition, new federal regulations or new computer technology may require that computer programs be modified. Experience shows that frequent, minor maintenance to a program, if properly done, can prevent major system failures later on. Today, the maintenance function is becoming more automated. A large home improvement chain, for example, can use new maintenance tools and software that allow the large chain to maintain and upgrade software centrally.

Some of the major reasons for systems maintenance are:

- New requests from stakeholders, users, and managers
- Bugs or errors in the program
- Technical and hardware problems
- Corporate mergers and acquisitions[39]
- Governmental regulations that require changes in programs

When it comes to making necessary changes, most organizations modify their existing programs instead of developing new ones. That is, as new systems needs are identified, most often the burden of fulfilling these needs falls upon the existing system. Old programs are repeatedly modified to meet ever-changing needs. Over time, these modifications tend to interfere with the system's overall structure, reducing its efficiency and making further modifications more burdensome.

The Financial Implications of Maintenance

The cost of maintenance is staggering, including hardware, software, databases, telecommunications, and other computer components. For older software developed in house, for example, the total cost of maintenance can be up to five times greater than the total cost of development. In other words, a program that originally cost $50,000 to develop may cost $250,000 to maintain over its lifetime. The average programmer can spend from 50 percent to over 75 percent of his or her time maintaining existing programs as opposed to developing new ones. Furthermore, as programs get older, total maintenance expenditures in time and money increase, as illustrated in Figure 10.26. With the use of newer programming languages and approaches, including object-oriented programming, maintenance costs are expected to decline. Even so, many organizations have literally millions of dollars invested in applications written in older languages, such as COBOL, that are both expensive and time consuming to maintain.

The financial implications of maintenance make it important to keep track of why systems are maintained, in addition to tracking the cost. For this reason, documentation of maintenance tasks is crucial. A determining factor in the decision to replace a system is the point at which it is costing more to fix it than to enhance or replace it.

DON'T LET THE LIGHTS GO DOWN

Complex system networks run everything in this country from oil refineries and power grids to water treatment plants and transportation networks. In the wrong hands, these systems could be used to paralyze the country and run up billions of dollars in damage. New energy legislation now mandates power companies set up safeguards against hacking with a cyberspace response system. Fueling the fire, cyberattacks on infrastructure systems rose by 50 percent in the first half of 2005.

Source:
New Focus on Cyber-Terrorism. At Risk: Computers that Run Power Grids, Refineries.
By Nathaniel Hoopes
The Christian Science Monitor
http://www.csmonitor.com/2005/0816/p01s02-stct.html
August 16, 2005

FIGURE 10.26 • Maintenance costs as a function of age

As programs get older, total maintenance expenditures in time and money increase.

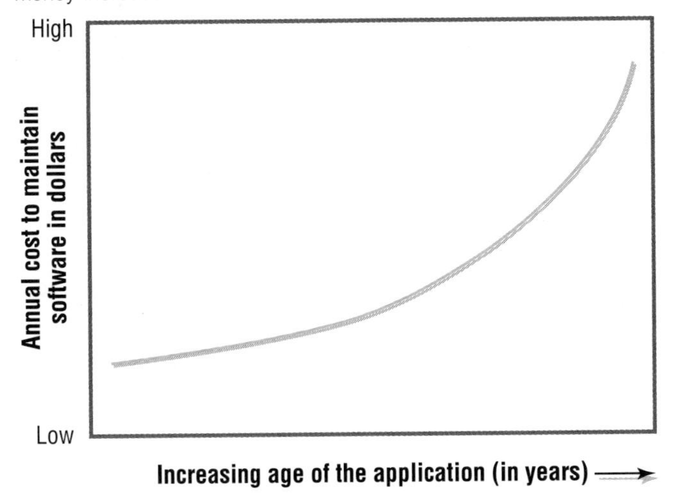

Systems Review

Systems review, the final phase of the systems development life cycle, is the process of analyzing systems to make sure that they are operating as intended.[40] All aspects of the system are reviewed, including hardware, software, database systems, networks and Internet, people, and procedures. Systems review often involves comparing the expected performance and benefits of the system as it was designed with the actual performance and benefits of the system in operation. Increasingly, organizations are using software and the Internet to review existing systems. Wayne State University, for example, uses the Vantage review tool to make sure that its systems are available and running correctly for faculty and students.[41] The City of Boise, Idaho, uses Patrol from BMC to review its computer operations. According to the database administrator for the city, "It puts all the information in one place and makes it easy to drill down into the problem so people at the help desk and other IT managers can see at a glance the root cause of a problem." Hymans Robertson, an actuarial firm, reviewed an existing financial system and discovered that it no longer met its needs.[42] According to the finance director for the firm, "We had been using a finance system that had been implemented three years ago. We soon realized we had outgrown this system as it was not flexible enough to fit our changing needs."

There are two types of review procedures: event driven and time driven. An *event-driven review* is one that is triggered or caused by a problem or opportunity such as an error, a corporate merger, or a new government regulation. In some cases, an individual or organization will wait until a large or important problem or opportunity occurs before a change is made. In this case, minor problems may be ignored. Today, some organizations use a *continuous improvement* approach to systems development. With this approach, an organization makes changes to a system when even small problems or opportunities occur. Although this approach can keep the system current and responsive, doing the repeated design and implementation can be time consuming and expensive. United Parcel Services (UPS), for example, reviewed a new package-flow system that used bar-coded labels and geographic information systems (GISs).[43] The systems development project was expected to save UPS about $700 million a year, but the actual system only saved the company about $100 million. As a result of the review, UPS is retraining many of its employees to take better advantage of the new package-flow system.

A *time-driven review* is one that is started after a specified amount of time. Many application programs are reviewed every 6 months to a year. With this approach, an existing system is monitored on a schedule. If problems or opportunities are uncovered, a new systems development cycle may be initiated. A computer-assisted bicycle design program may be reviewed once a year to make sure that it is still operating as expected. If not, changes are made.

Many organizations use both approaches. A computerized program to choreograph new dance routines for a theater production company, for example, might be reviewed once a year for opportunities to display new dance moves. This is a time-driven approach. In addition, the dance program might be redone if errors or program crashes make the software difficult to use. This is an event-driven approach.

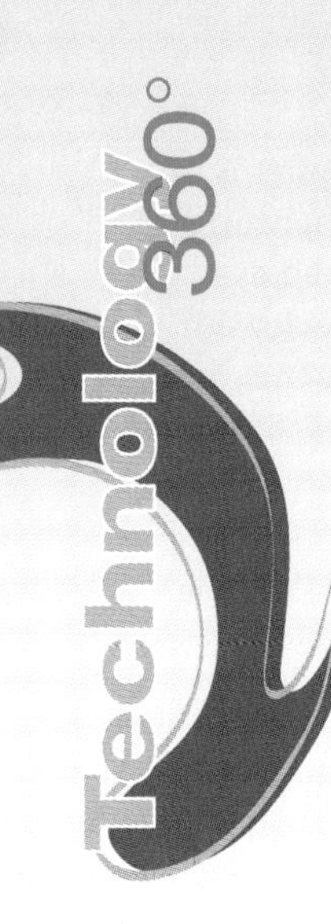

ACTION PLAN

Remember Linda from the beginning of the chapter? Here are answers to her questions.

1. **How should Linda approach her company to get the new program developed?**

   Linda should follow corporate procedures, if they exist. She should demonstrate how such a program can help the company achieve its goals of getting new insurance customers and servicing existing ones.

2. **If the company decides to develop a new insurance program, how should Linda be involved in the process?**

   If the company decides to undertake systems development to create a new computer program, Linda's involvement is critical. She needs to help the people in the Computer Systems Department determine exactly what reports and output can help her and other insurance agents. Linda should also be involved during the process to make sure that her needs are satisfied by the new program.

3. **What role would James and people like him play in developing a new insurance program?**

   As a systems analyst, James could help in developing plans and documents for a new sales program, including CASE documents, flowcharts, and so on. If he is involved in the project, he might even work with Linda in determining her needs and desires for the new program. He would also work with programmers. Working with users and programmers is the classic role of the systems analyst.

Summary

LEARNING OBJECTIVE 1

Describe the systems development life cycle, who participates in it, and why it is important.
The systems development process is called a systems development life cycle (SDLC) because the activities associated with it are ongoing. The five phases of the SDLC are investigation, analysis, design, implementation, and maintenance and review.

The systems development team consists of stakeholders, users, managers, systems development specialists, and various support personnel. The development team is responsible for determining the objectives of the computer system and delivering to the organization a system that meets its objectives.

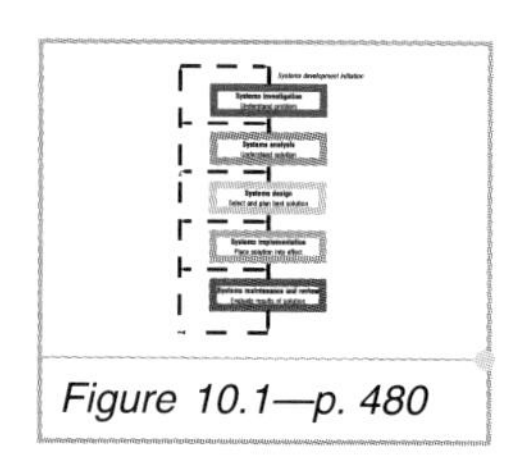
Figure 10.1—p. 480

Many organizations initiate systems development projects to gain a competitive advantage. This usually requires creative and critical analysis. Creative analysis involves the investigation of new approaches to existing problems. Critical analysis means being skeptical and doubtful, and requires questioning whether or not the existing system is effective and efficient.

End-user systems development is a term that was originally used to describe the development of computer systems by individuals outside of the formal computer systems planning and departmental structure. The proliferation of general-purpose information technology and the flexibility of many packaged software programs have allowed noncomputer systems employees to independently develop computer systems that meet their needs.

LEARNING OBJECTIVE 2

Discuss systems development tools.

Some common tools and techniques for systems development include CASE tools, flowcharts, decision tables, project management tools, prototyping, outsourcing, and object-oriented systems development. Some formalized systems development approaches have come to be called *software engineering*. These approaches typically employ the use of software-based systems development tools called computer-aided software engineering (CASE) tools that automate many of the tasks required in a systems development effort. Like a road map, a flowchart reveals the path from a starting point to the final destination. A decision table can be used as an alternative to or in conjunction with flowcharts. When there are a large number of branches or paths within a software program, decision tables are preferable to flowcharts. The overall purpose of project management is to plan, monitor, and control necessary development activities. A prominent technique for systems development is prototyping, which typically involves the creation of some preliminary models or versions of major subsystems or scaled-down versions of the entire system. Many organizations hire an outside consulting firm that specializes in systems development to take over some or all of its computer systems development activities. This approach is called *outsourcing*. Object-oriented (OO) systems development combines a modular approach to structured systems development with the power of object-oriented modeling and programming. OO development follows a defined system development life cycle, much like the SDLC.

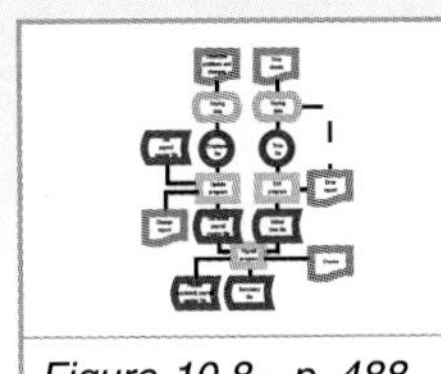

Figure 10.8—p. 488

LEARNING OBJECTIVE 3

Understand how systems development projects are investigated.

Systems investigation is usually the first step in the development of a new or modified computer system. The overall purpose of systems investigation is to determine whether or not the objectives met by the existing system are satisfying the goals of the individual or organization. Systems investigation is designed to assess the feasibility of implementing systems solutions, including technical, economic, legal, operational, and schedule feasibility.

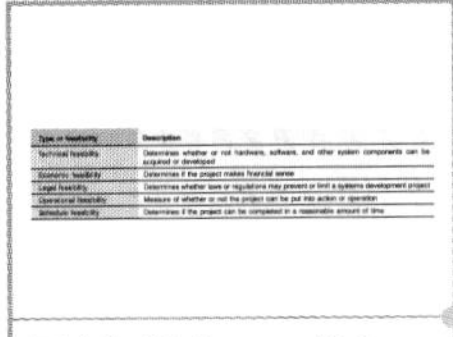

Table10.2—p. 494

LEARNING OBJECTIVE 4

Describe how an existing system can be evaluated.

Systems analysis is the examination of existing systems. This step is undertaken once approval for further study is received. The additional study of a selected system attempts to further understand the systems' weaknesses and potential improvement areas. Data collection methods include observation, interviews, and questionnaires. Data analysis manipulates the data collected. The analysis includes flowcharts, CASE tools, and other approaches. The overall purpose of requirements analysis is to determine user and organizational needs. Asking directly or using critical success factors are two ways to complete requirements analysis. Joint application development (JAD) can be used in place of traditional data collection and requirements analysis procedures. Another efficient approach to determine and define systems requirements from a group is called *rapid application development*. Rapid application development (RAD) combines JAD, prototyping, and other structured techniques in order to quickly and accurately determine the requirements of the system.

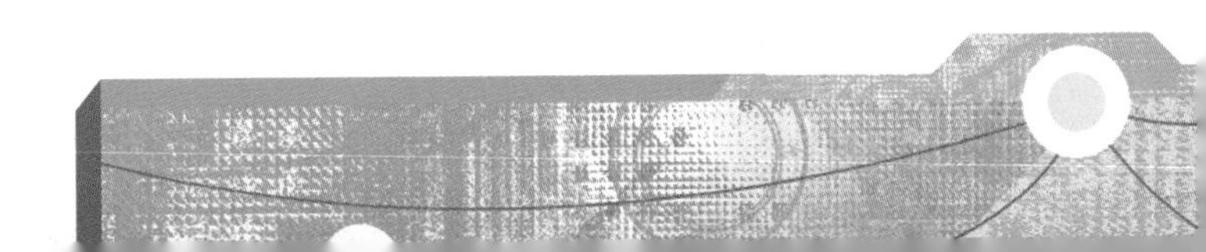

LEARNING OBJECTIVE 5

Discuss what is involved in planning a new system.

The purpose of systems design is to prepare the detailed design needs for a new system or modifications to the existing system. Organizations often develop a request for information (RFI) to get information from computer systems vendors. If new hardware or software will be purchased from a vendor, a formal request for proposal (RFP) is needed. The RFP outlines the company's needs; in response, the vendor provides a written reply. The final phase of system design is evaluation and selection of alternatives. Although most computer systems vendors provide standard contracts for new hardware, software, and systems, organizations today are increasingly developing their own contracts.

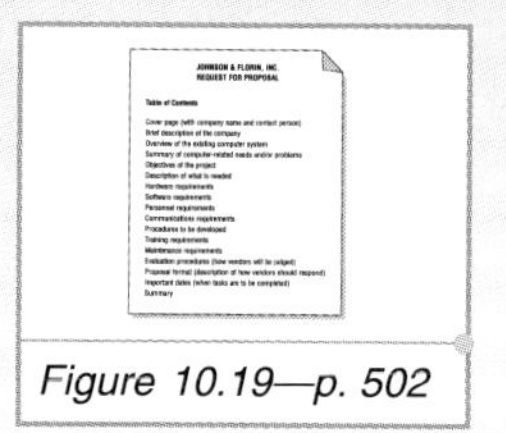

Figure 10.19—p. 502

LEARNING OBJECTIVE 6

List the steps to implement a new or modified system.

Systems implementation includes hardware acquisition, software acquisition or development, user preparation, hiring and training of personnel, site and data preparation, installation, testing, startup, and user acceptance. Hardware acquisition requires purchasing, leasing, or renting computer resources from a vendor. Types of vendors include small and general computer manufacturers, peripheral equipment manufacturers, leasing companies, time-sharing companies, software companies, dealers, distributors, service companies, and others. Software can be purchased from external vendors or developed in house. User preparation involves readying managers, employees, and other users for the new system. New IS personnel may need to be hired, and users must be well trained in the system's functions. Preparation of the physical site of the system must be done, and any existing data to be used in the new system requires conversion to the new format. Hardware installation is done during the implementation step, as is testing. Testing includes program (unit) testing, systems testing, volume testing, integration testing, and acceptance testing. Startup begins with the final tested computer system. Startup approaches include direct conversion, phase in, and pilot startup. Direct conversion involves stopping the old system and starting the new system on a given date. The phase-in approach involves gradually phasing the old system out and the new system in. Pilot startup involves running a pilot or small version of the new software along with the old. Users typically perform an acceptance test to be sure that the capabilities promised were actually delivered.

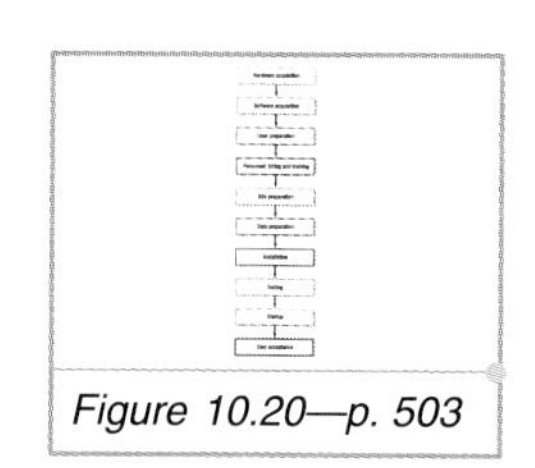

Figure 10.20—p. 503

LEARNING OBJECTIVE 7

Describe the importance of updating and monitoring a system.

Systems maintenance involves checking, changing, and enhancing the system to make it more useful in achieving user and organizational goals. Maintenance is critical for the continued smooth operation of the system. The costs of performing maintenance can well exceed the original cost of acquiring the system. Maintenance can vary from a small change to a large one.

Systems review is the process of analyzing systems to make sure that they are operating as intended. The two types of review procedures are event-driven review and time-driven review. An event-driven review is one that is triggered or caused by a problem or opportunity. A time-driven review is one that is started after a specified amount of time.

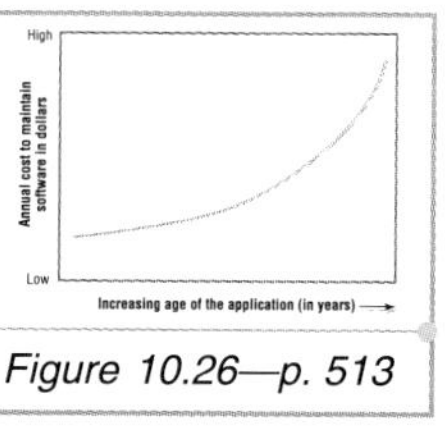

Figure 10.26—p. 513

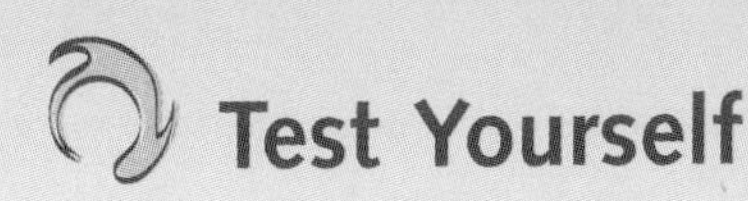

Test Yourself

LEARNING OBJECTIVE 1: Describe the systems development life cycle, who participates in it, and why it is important.

1. True or False: Chief information officers are individuals who ultimately benefit from systems development.
2. Who interacts with the users and others to develop detailed plans for the new or modified systems, like an architect developing blueprints for a new building?
   a. systems analyst
   b. programmer
   c. stakeholder
   d. chief information officer
3. True or False: Developing a competitive advantage typically requires critical and creative analysis.
4. ________________ is a term that was originally used to describe the development of computer systems by individuals outside the formal computer systems department.

LEARNING OBJECTIVE 2: Discuss systems development tools.

5. True or False: CASE tools are used by one person at a time to make sure the results are consistent and accurate.
6. The ________________ is a general flowchart needed to reveal the overall purpose and structure of an application.
7. What systems development tool is appropriate for programs that have a large number of branches or paths?
   a. flowchart
   b. PERT diagram
   c. decision table
   d. Gantt chart
8. The overall purpose of ________________ is to plan, monitor, and control necessary development activities.
9. Use case diagrams are often used with:
   a. flowcharts
   b. object-oriented systems development
   c. PERT
   d. Gantt

LEARNING OBJECTIVE 3: Understand how systems development projects are investigated.

10. ________________ is concerned with whether or not hardware, software, and other systems components can be acquired or developed to solve the problem.
11. True or False: Operational feasibility determines if the project can be completed in a reasonable amount of time.

LEARNING OBJECTIVE 4: Describe how an existing system can be evaluated.

12. The purpose of ________________ is to determine user, stakeholder, and organizational needs.
13. What technique was developed by IBM and uses group meetings in which users, stakeholders, and computer systems personnel work together to analyze the existing system?
    a. joint application development
    b. rapid application development
    c. critical success factors
    d. prototyping
14. True or False: Rapid application development involves a process in which a developer first builds a working model of the system to help a group of stakeholders, users, or managers identify how well the prototype meets their requirements.

LEARNING OBJECTIVE 5: Discuss what is involved in planning a new system.

15. What asks computer systems vendors to provide information about its products or services?
    a. request for information
    b. request for proposal
    c. asking directly
    d. critical success factors

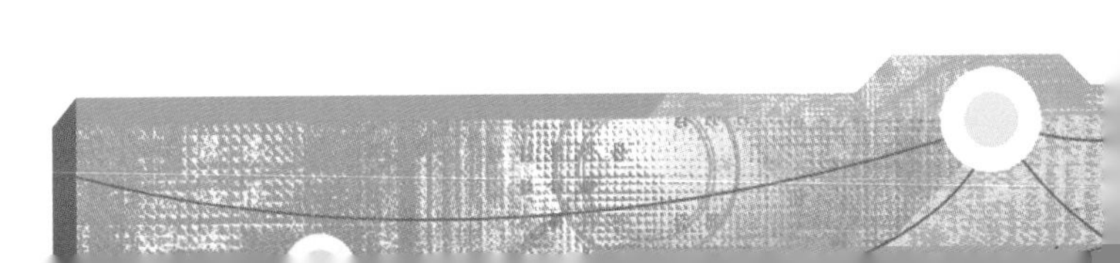

16. Cost comparisons, hardware performance, delivery dates, price, modularity, backup facilities, available software training, and maintenance factors are considered during ______________.

17. True or False: More and more organizations are developing their own contracts, stipulating exactly what they expect from the system vendor and what interaction will occur between the vendor and the organization.

LEARNING OBJECTIVE 6: List the steps to implement a new or modified system.

18. ______________ includes hardware acquisition, software acquisition or development, user preparation, hiring and training of personnel, site and data preparation, installation, testing, startup, and user acceptance.

19. True or False: Site preparation is the process of physically placing the computer equipment on the site and making it operational.

20. What requires the testing of all of the programs together?
    a. unit testing
    b. system testing
    c. integration testing
    d. acceptance testing

LEARNING OBJECTIVE 7: Describe the importance of updating and monitoring a system.

21. ______________ involves checking, changing, and enhancing the system to make it more useful in achieving user and organizational goals.

22. True or False: Systems review can include an event-driven review and a time-driven review.

Test Yourself Solutions: **1.** False, **2.** a. systems analyst, **3.** True, **4.** End-user systems development, **5.** False, **6.** application or system flowchart, **7.** c. decision table, **8.** project management, **9.** b. object-oriented systems development, **10.** Technical feasibility, **11.** False, **12.** requirements analysis, **13.** a. joint application development **14.** True, **15.** a. request for information (RFI), **16.** selection and final evaluation, **17.** True, **18.** Systems implementation, **19.** False, **20.** b. system testing, **21.** systems maintenance, **22.** True.

Key Terms

computer-aided software engineering (CASE) tools, p. 486
decision table, p. 488
direct conversion , p. 510
feasibility analysis, p. 494
flowchart, p. 487
object-oriented (OO) systems development, p. 493
outsourcing, p. 492
phase-in approach, p. 510
pilot startup , p. 510
project management, p. 489
prototyping, p. 491
request for proposal (RFP) , p. 501
requirements analysis , p. 497
system stakeholder, p. 481
systems analysis, p. 495
systems analyst, p. 482
systems design , p. 500
systems development, p. 478
systems development life cycle (SDLC), p. 479
systems implementation, p. 502
systems investigation, p. 494
systems maintenance, p. 511
systems review, p. 513
user acceptance document, p. 511

Questions

Review Questions

1. What are the phases of the systems development life cycle? What tasks are performed in each phase?
2. Who are the participants in systems development?
3. Name three organizations that have benefited from systems development and describe their systems.

4. Give an example of how an organization can use systems development to achieve a competitive advantage.
5. What is end-user systems development?
6. What is an application flowchart?
7. Describe when a decision table should be used.
8. What is prototyping? What are the steps involved in developing a prototype?
9. Describe the object-oriented systems development approach.
10. What is the purpose of systems investigation?
11. What is technical feasibility? What is economic feasibility?
12. What is the difference between operational and schedule feasibility?
13. What is systems analysis? What steps are included in systems analysis?
14. What tools and techniques are used for data collection and analysis?
15. What are joint application development and rapid application development?
16. What is the purpose of systems design?
17. What is a request for proposal (RFP), and why is it important?
18. What are the preliminary and final evaluation steps in systems design?
19. What is systems implementation?
20. List the types of information systems vendors.
21. What steps are involved in testing the information system?
22. What are some of the reasons for program maintenance?
23. What is systems review, and what are the two types of review procedures?

Discussion Questions

24. Why is the term *systems development life cycle* used to describe the process of systems development?
25. For what types of system development projects might prototyping be especially useful? What might be some of the characteristics of a system developed with a prototyping technique?
26. Describe the different types of feasibility. Give an example of each.
27. Describe the various methods used to perform requirements analysis and when each should be used.
28. Describe how you could use systems development tools, such as flowcharts and decision tables, in diagramming tasks and activities you must perform for one of your classes.
29. Why is outsourcing an attractive systems development alternative for many organizations? What are the potential negative aspects of outsourcing? How would you prioritize worker attitudes, the feelings of remaining workers, the lower price for products and services, and other factors in making a decision to outsource systems development projects to other countries?
30. What is the difference between a request for information and a request for proposal? Give examples of each.
31. Your company is developing a new Web site to sell clothes online. The Web site interacts with other programs and several databases. What types of testing are needed? How much testing should be performed before the new Web site is made operational?
32. Assume that you are responsible for the site preparation of the building where a new computer system will be installed. What are some of the equipment, improvement, and other considerations that you would have during this process?
33. You have been put in charge of reviewing the new computer system at your organization. What factors would you consider, and how might you evaluate these factors?

Exercises

Try It Yourself

1. You are developing a new computer system for The Fitness Center, a company that has five fitness centers in your metropolitan area, with about 650 members and 30 employees in each location. Both members and fitness consultants will use the system to keep track of participation in various fitness activities, such as free weights, volleyball, swimming, stair climbers, and aerobic classes. Prepare a brief memo detailing the required participants in the development team for this systems development project. Describe in your memo how you would determine the requirements for the new system. Using a graphics program, develop a chart that shows how participants are organized and their responsibilities in the development team.
2. Using the Internet, investigate several organizations that have used the object-oriented approach to systems development. Develop a presentation on what you found using presentation software such as PowerPoint.
3. Using the Internet, search for fitness centers. Describe these fitness centers. Using your word-processing program, write a brief report describing what you learned about how computer systems are likely used at these fitness centers. What types of systems development tools would be useful for these fitness centers?
4. A consultant has recommended that the management of The Fitness Center not only design a new system, but design it well to avoid maintenance costs. Use your spreadsheet and its graphing capabilities to produce a graph that supports the consultant's recommendation.
5. Use some of the systems development tools discussed in this chapter, including flowcharts and decision tables, to track your progress towards getting your college or university degree. Using a graphics program, diagram important decisions you have to make in the next year.

Virtual Classroom

For the following exercises, do not use face-to-face or telephone communications with your group members. Use only Internet communications.

6. Pick a career area or field that is of interest to your group. Develop a report that describes how a computer system could be developed for this career area or field. Each group member should develop a separate part of the report.
7. Describe a new or modified computer system that can make life easier for students at your college or university. Develop a plan to implement the new system.
8. Your virtual team should investigate three organizations that have recently used systems development successfully. Type a 1-page report that describes the companies and how they benefited from systems development.

Teamwork

9. Effective systems development requires a team effort. The team usually consists of system stakeholders and users, managers, system development specialists, and various support personnel. This team, called the development team, is responsible for determining the objectives of the computer system and delivering to the organization a system that meets these objectives. Have each team member choose a different role to play: chief information officer, systems analyst, senior-level management, or system stakeholder or user. Someone from this group should also be chosen as project leader. Having created the team, develop a profile of the "organization" for which you will develop a system. What is the name of the organization? How many employees does it have? Where is it located? What are its main activities? What are its products or services? Who are its customers and members? Create a document using a word-processing program that can give someone who knows nothing about this organization an understanding of the nature of the organization.
10. Using the organization you created in exercise 9, go through the steps of design for a new computer system.
11. Describe how the organization should implement and maintain the new system.

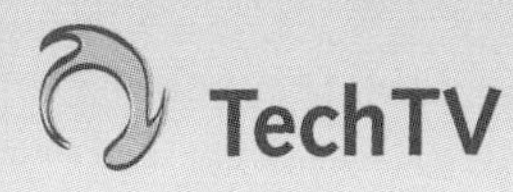

TechTV

Project Management Using Web-Based Tools

Go to www.course.com/swt2/Ch10 and click TechTV. Click the "Project Management Using Web-Based Tools" link to view the TechTV video, and then answer the following questions.

1. What advantages does Web-based project management software have over non Web-based software?
2. What are some possible risks of using Web-based project management software?

Endnotes

[1] Wallace, L. et al. "How Software Project Risk Affects Project Performance," *Decision Sciences,* Spring 2004, p. 289.

[2] Kontzer, T. "Global Appeal," *InformationWeek,* January 19, 2004.

[3] Dwyer, P. "Big Bang at the Big Board," *Business Week,* February 16, 2004, p. 66.

[4] Staff. "Best in Class," *Computerworld,* March 15, 2004, p. 3.

[5] Rhode, L. "Report Details Flaws in UK Case Management IT System," *Computerworld,* April 18, 2005, p. 21.

[6] Hoffman, T. "DOE Works With TECSys on Infrastructure Security," *Computerworld,* January 10, 2005, p. 15.

[7] Kerstetter, J. "Sarbanes-Oxley Sparks a Software Boom," *Business Week,* January 12, 2004, p. 94.

[8] Thurman, M. "Overwhelmed by Sarbanes-Oxley," *Computerworld,* March 1, 2004, p. 33.

[9] Hoffman, T. "Emcor Saves on Sarb-Ox," *Computerworld,* February 2, 2004, p. 7.

[10] Rappa, M.A. "The Utility Business Model and the Future of Computing Services," *IBM Systems Journal,* 2004, 43 (1), p. 32.

[11] Baker, S. et al. "Special Report on Software," *Business Week,* March 1, 2004, p. 84.

[12] Sliwa, C. "Sears Plans to Outsource Part of IT Infrastructure," *Computerworld,* January 19, 2004, p. 1.

[13] Schroeder, M. "Business Coalition Battles Outsourcing Backlash," *The Wall Street Journal,* March 1, 2004, p. A1.

[14] Staff. "On Demand Vision," www.IBM.com, accessed on January 25, 2004.

[15] Fillion, R. "All-American Outsourcing Option," *Rocky Mountain News,* January 14, 2005, p. 1B.

[16] Thurm, S. "Lessons in India," *The Wall Street Journal,* March 3, 2004, p. A1.

[17] King, J. "Home Grown," *Computerworld,* March 28, 2005, p. 45.

[18] Bulkeley, W. "IBM Documents Give Rare Look at Sensitive Plans on Offshoring," *The Wall Street Journal,* January 19, 2004, p. A1.

[19] Vijayan, J. "Outsourcing Savvy," *Computerworld,* January 3, 2005, p. 16.

[20] Buckman, R. "H-P Outsourcing: Beyond China," *The Wall Street Journal,* February 23, 2004, p. A14.

[21] Manes, S. "Electronics Jobs Outsourced – to You," *Forbes,* March 15, 2004, p. 158.

[22] Hamblen, M. "Outsourcing IT Security Functions Can Succeed," *Computerworld,* January 19, 2004, p. 38.

[23] Vijayan, J. "Offshore Outsourcing Poses Privacy Perils," *Computerworld,* February 23, 2004, p. 10.

[24] King, J. "AXA Financial's Blueprint for Software Reuse," *Computerworld,* February 9, 2004, p. 24.

[25] Dragan, R. "Code for the Road," *PC Magazine,* May 18, 2004, p. 124.

[26] Havenstein, H. "Fast-Moving Development," *Computerworld,* January 10, 2005, p. 26.

[27] Mearian, L. "Fidelity National Revamps IT with Single-Vendor Track," *Computerworld,* March 8, 2004, p. 18.

[28] Brandel, M. "Getting to Know You," *Computerworld,* February 21, 2005, p. 36.

[29] Hall, M. "J.D. Powers to Bestow IT Service and Support Certifications," *Computerworld,* March 28, 2005, p. 8.

[30] Crawford, C. et al. "Toward an on Demand Service-Oriented Architecture," *IBM Systems Journal,* 2005, 44 (1), p. 81.

[31] Hoffman, T. "Big Companies Turn to Packaged Sarb-Ox Apps," *Computerworld,* March 1, 2004, p. 6.

[32] Verton, D. "District of Columbia Melds Budgeting for 68 Agencies," *Computerworld,* January 19, 2004, p. 7.

[33] Anthes, G. "Sabre Takes Extreme Measures," *Computerworld,* March 29, 2004, p. 28.

[34] McCracken, H. "How to Build Better Software," *PC World,* February 2005, p. 17.

[35] Brewin, B. "FDA Bar-Code Rule Provides Impetus for More WLANS," *Computerworld,* March 1, 2004, p. 14.

[36] Dunn, D. "There's No Room for Error," *InformationWeek,* February 2, 2004, p. 34.

[37] Ferguson, R. "Server Merges Insurance Data," *eWeek,* January 19, 2004, p. 30.

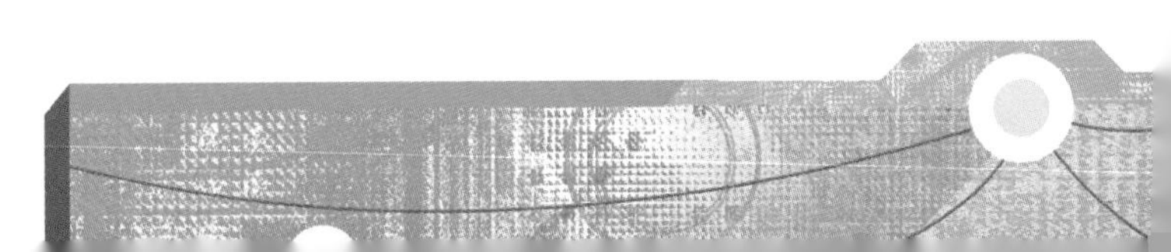

[38] Saran, C. "Insurer Reduces Risk in Rolling Out Software," *Computer Weekly,* January 20, 2004, p. 18.

[39] Thurman, M. "Postmerger Audit Quashes Trust Idea," *Computerworld,* February 9, 2004, p. 34.

[40] Purushothaman, D. et al. "Branch and Price Methods for Prescribing Profitable Upgrades of High Technology Products with Stochastic Demand," *Decision Sciences,* Winter 2004, p. 55.

[41] Hildreth, S. "Application Monitoring Software," *Computerworld,* January 31, 2005, p. 26.

[42] Staff. "Hymans Selects CMS, Net Software," *Pensions Week,* January 26, 2004.

[43] Rosencrance, L. "Planning System Isn't Fully Delivering at UPS," *Computerworld,* February 28, 2005, p. 12.

COMPUTER CRIME AND INFORMATION SECURITY

It was 4:35 p.m. on a Monday when Jan Minski received the call. A polite business-like woman inquired as to whether Jan had recently traveled to Montreal, Canada. "What is this about?" Jan asked. The woman explained that recently there was an unusually large number of purchases made on her credit card in Montreal, and the credit card company found it suspicious so was phoning to investigate.

It so happened that Jan had not been traveling, but someone had been traveling under her identity. She panicked when she found out that $2900.00 had been charged over a 24-hour period. Fortunately, the credit card company insured its customers against identity theft and Jan got off with only a small service fee.

During the investigation of the crime, the company noted that Jan had used her credit card on several occasions to make online purchases and suggested that perhaps Jan's credit card information had been stolen from her PC. Jan's PC was connected to the campus network which was in turn connected to the Internet. She had no idea how someone could steal information from her PC or how to prevent it from happening again.

As you read through this chapter, consider the following:

1. What are the first steps Jan should take in her crime scene investigation to find out if indeed her computer had been broken into?
2. What possible methods could a hacker have used to steal Jan's credit card information?
3. What precautions can Jan take to make sure this doesn't happen again?

Check out Jan's *Action Plan* at the conclusion of this chapter.

11

LEARNING OBJECTIVES

1. Describe the types of information that must be kept secure and the types of threats against them.
2. Describe five methods of keeping a PC safe and secure.
3. Discuss the threats and defenses unique to multiuser networks.
4. Discuss the threats and defenses unique to wireless networks.
5. Describe the threats posed by hackers, viruses, spyware, frauds, and scams, and the methods of defending against them.

CHAPTER CONTENT

Information Security and Vulnerability

Machine-Level Security

Network Security

Wireless Network Security

Internet Security

Introduction

The global information economy depends on computer and information systems to reliably store, process, and transfer information. Threats to system reliability and information integrity undermine our economy and security. Information is money, and thieves are working to divert the flow of information and money to their own pockets, leaving many innocent victims in their wake. In this chapter you will learn about the value of information, the threats to information security, and ways that you can secure information for yourself, your employer, your country, and the world.

You check your e-mail, and discover a happy greeting informing you that an old friend has sent you a virtual postcard with the message, "Click here to access your card." Upon clicking the link, you arrive at the virtual postcard Web site and are asked to enter the claim number provided by the e-mail and then click the Continue button. After doing so, an error message is displayed saying that the claim number is not valid. After several tries you give up and go back to checking your other e-mail, soon forgetting the incident. In the next few days, you notice that your computer isn't behaving like it should. It is sluggish, confusing error messages occasionally pop up, and sometimes the computer locks up or turns itself off.

Chances are that your computer has contracted a virus. Not only is the virus affecting your computer's performance and aggravating you, but it may also be allowing someone on the Internet to access your PC, your files, and personal information. Another real possibility is that your computer has been enslaved and forced to deliver spam or viruses to others on the Internet. Although such news is bound to be disconcerting, you can take some comfort in knowing that you are not alone. Billions of individuals worldwide are dealing with the same problem (see Figure 11.1).

FIGURE 11.1 • Virus infection

On this map from McAfee, areas in dark red had over a billion computers infected by computer viruses in a 30-day period.

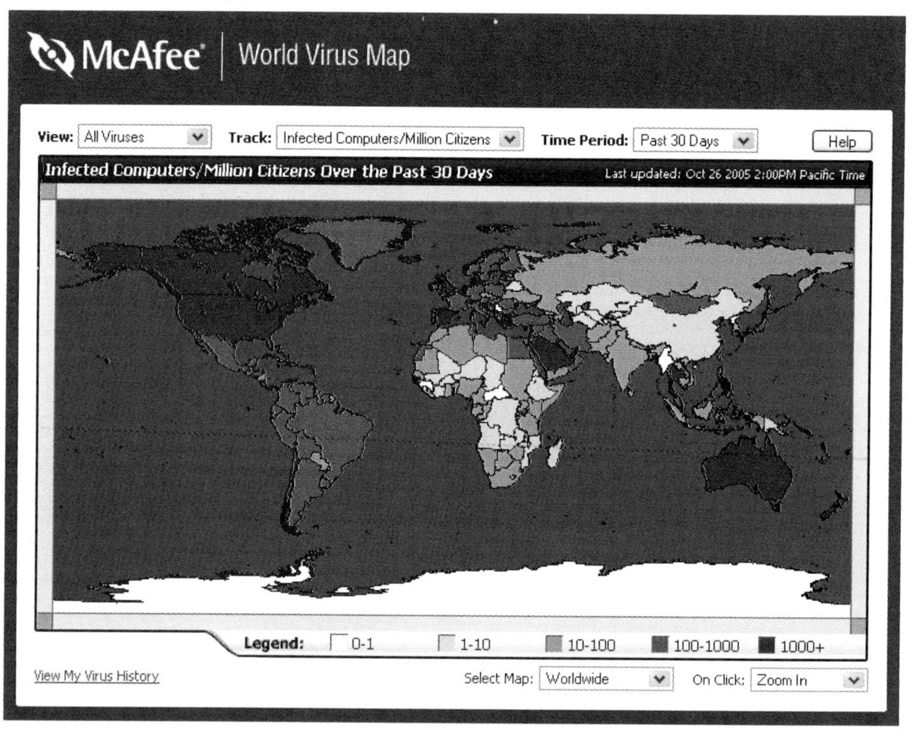

Viruses are one example of the many types of attacks on computer systems that take place every day. Consider these recent news headlines:

- "Electronic account records for some 500,000 banking customers at four different banks were allegedly stolen and sold to collection agencies in a data-theft case that has so far led to criminal charges against nine people, including seven former bank employees."[1]

- "A new Internet attack discovered late Thursday was designed by an infamous group of Russian virus writers to steal credit card and other financial information from Web surfers and send it to Web sites where it can be retrieved by hackers, security experts warned Friday."[2]
- "Microsoft Corp. warned on Tuesday of seven newly found flaws in its software that could allow an attacker to steal data and take over a personal computer running the Windows operating system."[3]
- "Hackers attacked computer servers of a California university and may have gained access to the personal information of 59,000 people affiliated with the school, a university spokesman said on Monday."[4]

FIGURE 11.2 • Jeffrey Lee Parson

Nineteen-year-old Parson received a prison sentence for releasing his version of the Blaster Internet worm.

These are representative of hundreds of similar stories that appear every week. If these headlines give the impression that a war is being waged, it is for good reason. Corporate and government networks and home computers are under attack, and the Internet is the battlefield. Sixty-five percent of businesses admit to having been hacked over a 12-month period.[5] Ninety-five percent admit to having their Web sites hacked. The casualties of this war include your privacy, the theft of valuable data and identities, and heavy financial losses. The total losses reported by U.S. businesses in 2005 from information security breaches were over $130 million.[6] Prisoners in this war include hackers like Jeffrey Lee Parson (see Figure 11.2), a 19-year-old serving time for releasing a version of the Blaster Internet worm that attacked Microsoft's Web site in the summer of 2003 and corrupted an estimated 48,000 computers.[7] "I know I made a huge mistake and that I've hurt a lot of people," Parson said, apologizing to Microsoft and other Blaster victims. "I realize now the effect I was having on a lot of other people."[8]

FIGURE 11.3 • Layers of information security

Computer users are at the heart of information security and are vulnerable to increasing exposure and risks as they move outward through the layers.

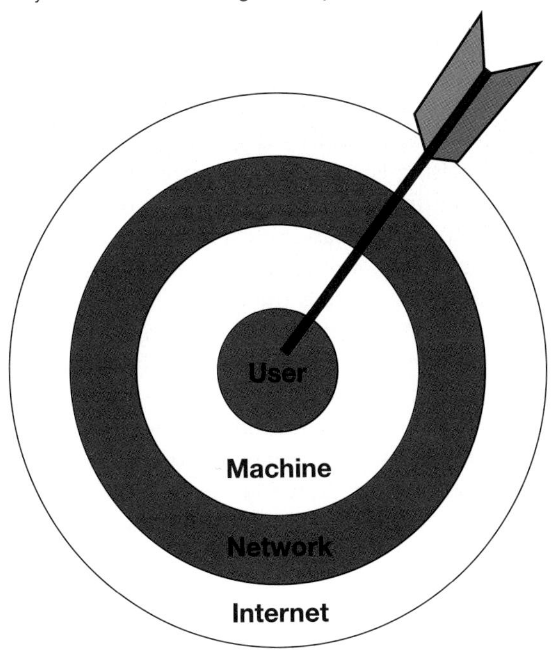

Many believe the battle will only get more intense. Viruses and security attacks increased 50 percent in the first half of 2005. Experts are forecasting serious problems in the near future. Most Internet experts expect a rise in attacks on the fundamental U.S. network infrastructure between now and 2015. Worse, two-thirds surveyed expect a "devastating attack," according to a study by the Pew Internet Project and Elon University on the Future of the Internet.[9]

In Chapter 1, information security was defined as being concerned with three main areas:

- Confidentiality: Information should be available only to those who rightfully have access to it.
- Integrity: Information should be modified only by those who are authorized to do so.
- Availability: Information should be accessible to those who need it when they need it.

This chapter discusses all aspects of information security as it relates to personal computers, organizational systems, and the Internet. It discusses what is at stake, how our computer systems are vulnerable, and how to keep them as safe as possible. The material in this chapter is organized to address the multiple layers of computer security that combine to provide total information security (see Figure 11.3).

Total information security refers to securing all components of the global digital information infrastructure from cell phones, to PCs, to business and government networks, to

Internet routers and communications satellites. You, the computer user, are at the heart of the total information security effort. Your cognizance of security risks and the actions you take to secure the systems with which you interact are the most important component of total information security. It is the user who suffers from the results of attacks against computer systems and who is responsible for safeguarding systems against those attacks. As a computer user, you must learn about security risks at three levels: the machine level, the network level (including wireless networks), and the Internet level. As you move from one level to the next, you face increasing exposure and risks.

INFORMATION SECURITY AND VULNERABILITY

It is clear that information security is an important issue today. To understand its importance, you need to consider exactly what is at stake and the sources of the danger.

What Is at Stake?

The concept of information security is built on the assumption that the information you create, store, maintain, and transfer is valuable, confidential, and worth protecting. This section examines the value of digital information from personal, organizational, national, and international perspectives and the consequences if that information is stolen or made inaccessible.

Personal Information. What would concern you most if a person who wished to do you harm had full control of your personal computer? Unfortunately, this is an all too real possibility. Through any one of many methods discussed later in the chapter, intruders can gain control of home PCs through Internet connections and have a field day with what they find. While at their own computer, intruders can view and manipulate your computer system just as if they were sitting in your home at your computer. The intruder could be in the apartment next door or on the other side of the world. The intruder could steal your Internet access and e-mail passwords, any information that your operating system has stored about you, your Web browser history and cache, your e-mail, and your computer files. If you bank and pay your bills online, it is possible that a hacker could steal your account numbers.

Using key pieces of personal information stolen from a person's personal computer, a business' database, or even discarded paper documents, a thief can steal your identity. **Identity theft** is the criminal act of using stolen information about a person to assume that person's identity. In 2004 the U.S. Federal Trade Commission (FTC) received over 635,000 consumer fraud and identity theft complaints. Consumers reported losses from fraud of more than $547 million (see Figure 11.4).[10] Given a person's social security number, birth date, or other personal identifiers, identity thieves can apply for a new credit card in the victim's name and have it delivered to a post office box. A crime ring

FIGURE 11.4 • Identity theft and fraud

Identity theft is becoming increasingly common with victim complaints rising at an average rate of 25 percent each year, according to the U.S. Federal Trade Commission.

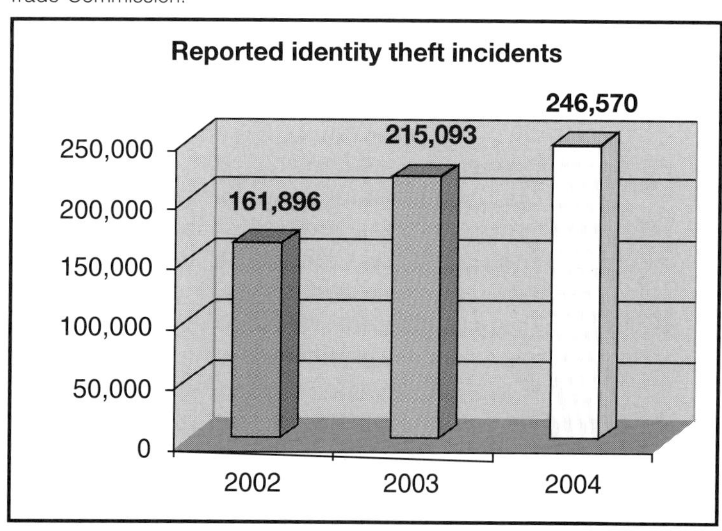

in California, for example, unsuccessfully tried to apply for a credit card using the personal information of one of the authors of this book. Identity thieves can make purchases on stolen credit card numbers, make bank withdrawals, apply for loans, or buy a car. The damage caused can be quite serious. An identity thief can also do damage to a victim's reputation by perpetrating crimes under the victim's identity.

Not only can attackers steal information from your PC, and in some cases steal your identity, but in many instances an attacker or virus can corrupt your system so that you no longer have access to your computer or the data it stores. For many victims, this can be as damaging as having information stolen. Many people collect irreplaceable information on their computers of both practical and sentimental value. People store personal writing, correspondence, music, photos, professional documents, financial records, and much more. To lose computer documents can be like losing a part of your life. The hours and days invested in creating your collection of files are wasted. Moments of your life preserved in bits may be lost forever.

Intellectual Property. **Intellectual property** refers to a product of the mind or intellect over which the owner holds legal entitlement. Intellectual property includes ideas and intangible objects such as poetry, stories, music, and new ways of doing things or making things. Like tangible objects, intellectual property has value and is owned by an individual or organization. Each time you use software, listen to music, read a book or newspaper article, or see a movie, you are enjoying someone else's intellectual property. The most valuable intellectual properties are typically the result of years invested in study and training. The sale of that intellectual property provides an income for the creator and a return on his or her investment.

A respect for the value of intellectual property makes it possible for individuals who create intangible products such as music and software to earn a living. The digitization of many forms of intellectual property including books, movies, artwork, and music has transformed tangible products into intangible products. Unfortunately, people don't always respect the value of these forms of intellectual property or respect intellectual property rights. Consider, for example, the explosion in sales of portable digital music players like the iPod. It is expected that by 2008, sales in the portable music player industry will exceed $52 billion.[11] Although millions of consumers willingly pay anywhere from $80 to $400 for the devices, statistics show that half of them are unwilling to pay for the music they listen to on the device. Instead, they acquire copies of the music for free from friends or strangers over the Internet. While Apple and the other hardware manufacturers are deservedly earning money on the hardware sold to each customer, those who write, record, and sell the music are earning money from only half of those enjoying their products. Which is more important and valuable, the device or the music?

FIGURE 11.5 • Music distribution models

P2P file-sharing networks turn consumers into distributors and detract from the income of the artist.

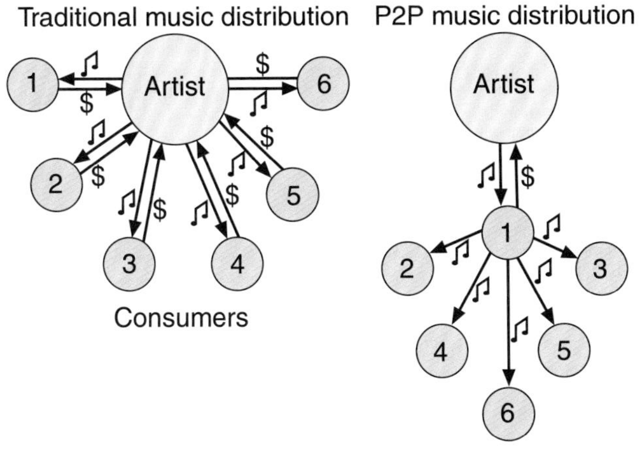

Because software, audio and music recordings, movies, television and other video products, books and other writings are stored as bits, it is easy and convenient to copy and distribute them, sometimes without consideration to the creator. This, in the view of many, is as much theft as if you steal an iPod from someone's backpack. Consider the effects on the artist of the two models illustrated in Figure 11.5. Notice in the P2P model, consumer 1 has taken over as the distributor of someone else's intellectual property. Even though consumer 1 isn't charging people for the music, the effect is a dramatic decrease in the amount of income for the artist. Now consider a situation in which each of the six consumers makes the song available to six other consumers and each of those six, six others, and so on. The result is an exponential increase in free distribution of the music in which nearly everyone has access to the intellectual property for free, and there is very little demand for the product from the artist for a fee. When those who produce intellectual property can no longer earn a living, they will turn to some other career. The P2P model of distribution and a healthy vibrant creative culture are at odds with each other.

This same problem exists in any scenario where customers take over the distribution of intellectual property. Digitization, the Internet, and a general failure to protect intellectual property have combined to create serious issues for the software, music, movie, and publishing industries. An important component of total information security is the protection of intellectual property rights.

Intellectual property rights concern the ownership and use of intellectual property such as software, music, movies, data, and information. Protection of intellectual property can take many forms, including copyrights, trademarks, trade secrets, and patents. See Table 11.1.

TABLE 11.1 • Protecting intellectual property rights

There are a number ways to protect intellectual property, including copyrights, trademarks, trade secrets, and patents.

Protection	Description
Copyright	Protects the words, music, and other expressions for the life of the copyright holder plus 70 years. The *fair use doctrine* describes when and how copyrighted material can be legally used. The *Digital Millennium Copyright Act* provides global copyright protection.
Trademark	Protects a unique symbol or word used by a business to identify a product or service.
Trade secret	Protects secrets or proprietary information of individuals and organizations as long as the trade secret is adequately protected.
Patent	Protects an invention by giving the patent holder a monopoly on the use of the invention for 20 years after the patent application has been applied.

Organizational Information. For a business or nonprofit organization, the information it processes is often highly valuable and key to its success. To have that information compromised can result in a loss of market share and in some cases total business failure. For this reason, businesses and organizations take information security very seriously.

Because businesses typically hold the valuable information, they are the targets of the most attacks. The security threats to businesses, in order of frequency, are:

1. Virus
2. Insider abuse of Internet access
3. Laptop theft
4. Unauthorized access by insiders
5. Denial-of-service attacks
6. System penetration
7. Theft of proprietary information
8. Sabotage
9. Financial fraud
10. Telecommunications fraud
11. Telecommunications eavesdropping
12. Active wiretap

Corporations Urged to Take Action Against Crime

The U.S. government is pushing CEOs to take responsibility for the nation's critical information infrastructure. "In this era of increased cyberattacks and information security breaches, it is essential that all organizations give information security the focus it requires," said Amit Yoran, Director of the National Cyber Security Division, IAIP, within the U.S. Department of Homeland Security. "Addressing these cyber and information security concerns, the private sector will not only strengthen its own security, but help protect the homeland as well."

As noted in a report titled "Information Security Governance: A Call to Action," a group of CEOs observed that information security is too often treated as a technical issue and passed along to the CIO and technical department to handle. But computer systems are so important that top executives and boards of directors must become involved and make security an integral part of core business operations. The report was authored by the Corporate Governance Task Force of the National Cyber Security Partnership. This CEO-led task force is responsible for identifying cybersecurity roles and responsibilities within the corporate management structure.

To coax companies to comply, the Department of Homeland Security is establishing an awards program for companies that meet or exceed the security guidelines. Organizations are asked to state on their Web sites that they intend to use the tools developed by the Corporate Governance Task Force to assess their performance and report the results to their boards of directors. Some believe that if there is not widespread voluntary support of the task force's recommendations, the government may need to apply added incentives.

Questions

1. How might organizations secure sensitive information from cybercriminals? Consider the security concerns and possible solutions regarding data stored in databases and also the important information that is transferred over networks and often printed in reports.
2. How could the U.S. government get 100 percent cooperation from private corporations? Is such a goal attainable? If attained, will it guarantee national security? Why or why not?

Sources

1. *Department of Homeland Security Web page, http://www.dhs.gov, accessed on May 19, 2005.*
2. *"Corporate Governance Task Force of the National Cyber Security Partnership Releases Industry Framework,"* Entrust, *April 12, 2004, www.entrust.com.*
3. *Verton, D. "CEOs Urged to Take Control of Cybersecurity,"* Computerworld, *April 12, 2004, www.computerworld.com.*
4. *National Cyber Security Partnership (NCSP) Web site, accessed April 25, 2004, www.cyberpartnership.org.*

Business intelligence is the process of gathering and analyzing information in the pursuit of business advantage. Companies are continuously gathering and analyzing information about economic indicators, industry statistics, marketing research, public opinion, and anything that can assist in creating a product or service that bests their competition with the lowest amount of investment. *Competitive intelligence* is a form of business intelligence concerned with information about competitors. Sometimes gathering competitive intelligence can become unlawful. Consider the case of the Florida marketing executive who hacked into the computers of competitor marketing company Acxiom Corp, in Little Rock, Arkansas.[12] The hacker accessed the Acxiom databases filled with personal, financial, and company information. The executive hacker was charged with 144 counts of stealing information from Acxiom. *Counterintelligence* is a form of business intelligence concerned with protecting your own information from access by your competitors. The techniques of information security discussed in this chapter provide valuable tools for counterintelligence.

FIGURE 11.6 • HIPAA

The Health Insurance Portability and Accountability Act (HIPAA) privacy requirements have those in the medical profession scrambling to protect patient privacy.

"Take this bag and put it over your head. It's our way of ensuring privacy."

Besides protecting its own proprietary information, a business also has a responsibility to its customers to safeguard their private information. Until recently businesses had little accountability for what they did with customer information. Customer data collected by one company has often been sold to other companies for profit. New laws are attempting to hold companies responsible for maintaining the privacy of their customers' information. For example the Identity Theft Protection Act, introduced in the U.S. Senate Commerce committee by a bipartisan coalition, addresses problems highlighted by numerous high-profile data breaches in 2004 and 2005. The Act requires entities that collect sensitive information such as social security numbers to secure the data physically and technologically and to notify consumers nationwide when data is compromised.[13] In other words, under this Act, companies are responsible for storing sensitive data in a manner that withstands intruder attacks either over the network or by cooperation with insiders. Besides protecting personal identifiers such as social security numbers, the protection of medical and health information has also become a great concern to lawmakers (see Figure 11.6).

Table 11.2 lists several U.S. laws that require companies and other entities to take information security very seriously.

TABLE 11.2 • U.S. laws that protect information and privacy

Law	Description
Consumer Internet Privacy Protection Act of 1997	Data collectors are required to alert people that their personal information is being shared with other organizations
The Children's Online Privacy Protection Act of 2000	Gives parents control over what information is collected from their children online and how such information may be used
Information Protection and Security Act of 2005	Gives the Federal Trade Commission (FTC) the ability to regulate the sale of personal information
Notification of Risk of Personal Data Act of 2003	Businesses have to notify individuals when their personal information is stolen
Identity Theft Protection Act of 2005	Requires businesses to secure sensitive data physically and technologically and to notify consumers nationwide when data is compromised
Health Insurance Portability and Accountability Act (HIPAA) of 1996	Requires those in the health industry to protect the privacy of health information and provides policies and procedures for doing so
Sarbanes-Oxley Act ("Sarbox") of 2002	Fights corporate corruption by imposing stringent reporting requirements and internal controls on electronic financial records and transactions
Gramm-Leach-Bliley Act (GLBA) of 1999	Requires banks and financial institutions to alert customers of their policies and practices in disclosing customer information

Table 11.2 clearly indicates that businesses and other organizations that maintain databases of customer information are under pressure to manage that information responsibly by making sure there is no unauthorized access and that the information is not shared without the customer's knowledge. State laws have an equally strong effect on businesses that operate over the Internet. For

example, California has nearly a dozen privacy laws that must be followed by organizations, wherever they may be located, if they wish to do business with California residents.

Some industries are creating standards of their own to support responsible data management. For example MasterCard and Visa have developed data-protection procedures that are required of companies using the popular credit cards.[14] The Payment Card Industry Data Security Standard has a set of protection rules and procedures, including encryption, logging of credit activities, user access, and monitoring, that merchants and others using MasterCard and Visa credit cards must implement. A certified assessor performs annual tests of compliance.

The United States and other countries are leaning hard on businesses to take up the reins of total information security, as shown in the Job Technology box. It is clear that without the cooperation of businesses, it will not be possible to provide the security that is so important in this information age.

National and Global Security. Just as businesses benefit from digital technologies and the Internet, so do governments and government agencies all over the world. Government computing systems, databases, and networks process, store, and transfer confidential government information, citizen's records, state secrets, national defense strategies, and many other classified documents. Many of these government networks are connected to the Internet and can be open to attack by international hackers. In 2005, a 26-year-old male model from Venezuela, Rafael Nunez-Aponte, known in the hacker world as RaFa, was found guilty of hacking into a U.S. Defense Department computer. According to his plea agreement, Nunez-Aponte was a member of the hacking group "World of Hell."[15] Many such groups exist around the world.

FIGURE 11.7 • Cyber security

The US-CERT Web site provides current information on Internet security threats and how to respond to them.

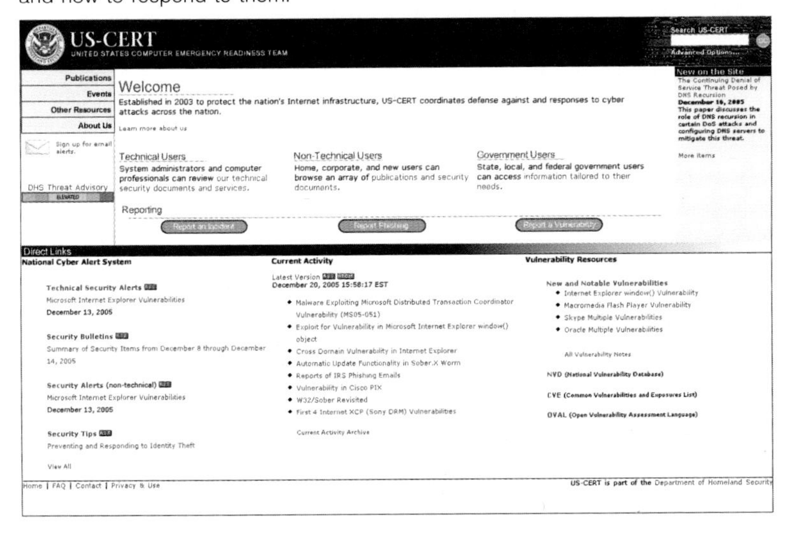

Today's national economies and security depend strongly on technology and the Internet. Technology and the Internet have become tools to protect nations, as well as points of vulnerability. Threats to the Internet and the national information infrastructure that it supports are serious threats to the nation. **Cyberterrorism** is a form of terrorism that uses attacks over the Internet to intimidate and harm a population. The United States Computer Emergency Readiness Team (US-CERT) was established to monitor the security of U.S. networks and the Internet and respond to episodes of cyberterrorism. US-CERT is part of the National Cyber Security Division of the United State's Department of Homeland Security. The US-CERT Web site, *www.us-cert.gov*, gives network administrators and computer users up-to-date information on security threats and defenses. US-CERT is also responsible for supporting and implementing the National Strategy to Secure Cyberspace introduced in Chapter 1. The strategic objectives of the National Strategy to Secure Cyberspace are the following:

- Prevent cyberattacks against America's critical infrastructures.
- Reduce national vulnerability to cyberattacks.
- Minimize damage and recovery time from cyberattacks that do occur.

To prepare for possible cyberattacks, the CIA's Information Operations Center conducted a war game to simulate an unprecedented, catastrophic electronic assault against the United States. The 3-day exercise, known as "Silent Horizon," tested the ability of government and industry to respond to escalating Internet disruptions over many months. About 75 people, mostly from the CIA, along with other U.S. officials, gathered in conference rooms and pretended to react to signs of mock computer attacks. The seldom-publicized Information Operations Center of the CIA is responsible for evaluating threats to U.S. computer systems from foreign governments, criminal organizations, and hackers. The government's assessment of future threats through the year 2020 anticipates many cyberattacks but states that terrorists "will continue to primarily employ conventional weapons."[16] Authorities have expressed concerns about terrorists combining physical attacks such as bombings with hacker attacks to disrupt rescue efforts, known as hybrid or "swarming" attacks.

Many cyberattacks originate overseas. Russia is the home of prolific hacker activity and the source of much of the U.S. credit card theft and fraud.[17] The weakness of the Russian economy combined with the criminal opportunities provided by the Internet and the failure of the Russian government to prosecute hackers have inspired many technically savvy Russians to turn to hacking. "If there were a happy haven for hackers these days, it would be Russia," says Ken Dunham, director of malicious code at U.S.-based iDefense.[18]

Fighting international attacks is particularly difficult as there are currently no global cybercrime laws. The laws of one country cannot be enforced in another without the cooperation of both governments. A computer attack may be designed by hackers in country A and launched in country B in order to attack computers in country C. Some countries are discussing the establishment of a global cybercrime task force, similar to the Interpol international police network. One step in that direction was the Council of Europe's (Figure 11.8) convention on cybercrime. The convention on cybercrime produced a treaty signed in November 2001 that calls on countries to "harmonize their laws on and investigative powers of all illegal behavior, including hacking and child pornography, and to ensure international cooperation in investigations."[19]

FIGURE 11.8 • The Council of Europe

The Council of Europe promotes democracy and human rights as well as developing standards in international cooperation for information security.

Threats to Information Security

To achieve total information security, many diverse threats must be addressed. From the inherent flaws in software, to intentional or unintended acts by law-abiding citizens, to attacks by those wishing to do serious damage, the threats are abundant and complex. This section examines common sources of information security threats: software and network vulnerabilities, user negligence, pirates and plagiarists, and attackers.

Software and Network Vulnerabilities. Perfect software would be impossible to hack. If Microsoft Windows and other operating systems, as well as Web browsers and all Internet software, were perfectly designed no one would be able to access someone else's private computer and data. The fact is that people are not perfect, and their creations such as software are bound to include imperfections.

Some argue that the software industry could do a better job of securing software and computer systems. Some in the software industry argue that the complexity of today's software makes it impossible to guarantee any software to be 100 percent secure. No matter which argument is correct, it is clear that the software and systems that hold and control our data are vulnerable. Security vulnerabilities or **security holes** are software bugs that allow violations of information security.

Many of today's popular operating systems and software were originally created in an era when security wasn't an issue. Users simply didn't have the knowledge or desire to hack a system, and those with the knowledge were, for the most part, trustworthy. Now the software industry is scrambling to secure systems against rapidly expanding attacks. It took a while, but it appears that the big software vendors are finally taking security very seriously.

Microsoft's *Trustworthy Computing* initiative is "a long-term, collaborative effort to provide more secure, private, and reliable computing experiences for everyone."[20] Microsoft claims that Trustworthy Computing is a core company tenet that guides virtually everything they do (see Figure 11.9). Microsoft categorizes Trustworthy Computing into four "pillars": security, privacy, and reliability in their software, services, and products; and integrity in their business practices. One example of Microsoft's efforts to implement security is *Windows Update*, a service that allows patches to be applied to security holes as they are discovered.

FIGURE 11.9 • Bill Gates and Trustworthy Computing

Bill Gates has made information security a top priority for software development at Microsoft with the Trustworthy Computing initiative.

Software patches are corrections to the software bugs that cause security holes. Microsoft releases monthly updates that include patches for minor software flaws as well as *critical patches* that repair flaws that allow for serious breaches of security. Microsoft provides dozens of critical patches through Windows Update every year.

One negative effect of Microsoft's monthly patch announcements is that in addition to informing the general public of security holes, they also serve to inform attackers. Many attackers use Microsoft's monthly announcements to reverse engineer hacking tools that take advantage of the newly found bugs. In many cases code that exploits the security hole is available to hackers on the Web within days of the Microsoft announcement.[21] It becomes a race between hackers who exploit the bug and users applying patches to protect themselves. Microsoft encourages its users to use the automatic update option that applies patches as soon as they are released over the Internet.

User Negligence. There are many situations where innocent human mistakes result in monumental problems. Take for example the Taiwan stock trader who mistakenly bought $251 million worth of shares due to a typing error, causing her company a paper loss of more than $12 million.[22] In another different type of mistake, the U.S. Defense Department, Internal Revenue Service (IRS), and Justice Department have misplaced hundreds of government notebook computers, many of which contain classified government documents. In a more common act of negligence, many users provide hackers with easy access to computer systems by using passwords that are easy to guess, or storing passwords and other private information where others can access them.

People's mistakes can lead to problems with security, accuracy, or reliability of computer systems and information. The major types of human errors include those shown in Table 11.3.

TABLE 11.3 • Types of user negligence

Type of mistake	Example
Data-entry errors	A military commander might enter the wrong GPS position for enemy troops into the computer; this data-entry error might cause friendly troops to be killed
Errors in computer programs	A payroll program might multiply a person's pay rate by 1.5 for overtime instead of 2; this programming error results in a lower paycheck than the employee should receive for overtime work
Improper installation and setup of computer systems	A person may neglect to properly set the necessary security settings to secure a home wireless network
Mishandling of computer output	A medical office might send lab results to the wrong person; the person receiving the results has access to another person's private medical information
Inadequate planning for and control of equipment malfunctions	An individual's hard drive might fail; if the person has not backed up important files recently, he or she will lose access to important information and have to redo work
Inadequate planning for and control of electrical problems, humidity problems, and other environmental difficulties	A power outage in an area could shut down an organization's computer system; without a backup power supply and protection from electrical surges, information and data might be lost and equipment might be ruined

A number of actions can be taken by individuals and organizations to prevent computer-related mistakes. Smart businesses and organizations automate data entry whenever possible to cut down on typing errors. Database management systems and spreadsheets can be programmed to allow only reasonable figures to be entered. Businesses typically have data backup policies and procedures, as well as backup power supplies and surge protection, to assure the safety of computer systems against mechanical failure, natural disaster, and human mistakes. Businesses also employ technology experts to set up and maintain computer systems so that both the hardware and software are safe and secure. It is not uncommon for a business to include policies that outline how employees are to use information systems and handle data in soft and hard copy. For example, a business may have a policy that calls for employees to shred all documents that contain private information. Just as today's best businesses employ smart and thorough information security practices, individuals should do the same in managing their own personal computers and information.

Pirates and Plagiarists. Pirates and plagiarists are two classifications of individuals who violate the laws regarding intellectual property rights. **Piracy** involves the illegal copying, use, and distribution of digital intellectual property such as software, music, and movies. **Plagiarism** involves taking credit for someone else's intellectual property, typically a written idea, by claiming it as your own.

How expensive is piracy? Adams Media Research reports that the annual cost of piracy in the United States is estimated to be $21 billion for DVD sales, $12 billion for CD sales, $4 billion for music sales, and $3 billion for movie sales.[23] According to Thomas Lesinski, president of Paramount, "This is the number one priority at the highest levels. The studios want to have more control over protecting our content."

Pirated software, music, and videos can be accessed through file-sharing networks or from others on homemade CDs and DVDs. Because technology allows media files to be copied, many otherwise law-abiding citizens feel as though it must be okay. Organizations, such as the Recording Industry Association of America (RIAA), the Motion Picture Association of America (MPAA), and the Software & Information Industry Association are making it known that piracy is not okay. The RIAA has sued thousands of individuals involved in illegal MP3 file sharing for significant financial settlements. "The Supreme Court unanimously ruled that businesses that encourage the theft of music [such as KaZaA] can be held accountable for their actions," said RIAA president Cary Sherman. "For businesses and individuals alike, the authority and credibility of the Court's decision could not be more clear: downloading without permission is 'garden-variety theft.' We will continue to send a strong message to the users of these illicit networks that their actions are illegal, they can be identified, and the consequences are real." The RIAA followed this up in July of 2005 with copyright infringement lawsuits against 765 illegal file sharers. This followed 784 lawsuits in June, 725 in April, 753 in February, and 717 in January. Between the pressure of legal action and technology solutions such as digital rights management (discussed in Chapter 6), illegal media distribution is gradually being reined in.

Although the MPAA has not engaged in lawsuits to the same extent as the RIAA, they are gradually ramping up their efforts as movie file sharing is becoming more common. At the close of 2004 the MPAA filed suit on four continents against over 100 individuals providing pirated movies to others through file-sharing services such as BitTorrent, eDonkey, and Direct

VIRTUAL VANDALISM

Founded in 2001, Wikipedia is the Internet encyclopedia written and updated by users. The Web-pedia hosts over 940,000 articles in 105 different languages. Lately, however, attacks by vandals have compromised the system. Most notably, someone replaced a picture of the newly elected pope with one of Darth Vader. In answer, developers plan to freeze pages with stable data and restrict access to others as a way to protect the integrity of their information.

Source:
Web's Wikipedia to tighten editorial rules
Reuters
http://www.washingtonpost.com/wp-dyn/content/article/2005/08/05/AR2005080500771.html
August 5, 2005

Connect. In 2005 the MPAA filed more lawsuits against numerous P2P users. The Software & Information Industry Association's Anti-Piracy Division has taken similar measures in its comprehensive, industry-wide campaign to fight software and content piracy.

International piracy is of great concern to those in the music, motion picture, and software industries. Over 90 percent of motion pictures used in Russia are pirated (see Figure 11.10). In Vietnam, China, and the Ukraine, over 90 percent of software in use is pirated. Forty percent of music acquired by U.S. citizens is pirated.[24]

FIGURE 11.10 • International piracy

In countries such as Romania, street vendors sell pirated copies of software, music, and motion pictures without any fear of legal recrimination.

Plagiarism has been an issue of legal and social concern for as long as people have produced intellectual property. Stealing other's ideas and thoughts and presenting them as your own is a serious breach of ethics matched by serious penalties if you are caught. A student caught submitting the work of another for a grade may face expulsion. A professional journalist, author, or researcher could face an expensive lawsuit and loss of a career. (Social and ethical concerns are discussed in more depth in Chapter 12.)

While plagiarism has been a social issue for a long time, the digitization of the written word, the Internet, and cut-and-paste methods of writing have made it all the more common. Some Web sites provide students with free research papers for class projects. Submitting papers from such Web sites is considered plagiarism and subject to strict and severe penalties by colleges. Software such as Turnitin (*www.turnitin.com*), and iThenticate (*www.ithenticate.com*) are used by many high school and college teachers to find instances of plagiarism in electronically submitted homework files with a high rate of success.

Hackers, Crackers, Intruders, and Attackers. Chapter 1 defined a *hacker* as an individual who subverts computer security without authorization. Security professionals refer to this as *system penetration*. In reality, there are many names used to label those that unlawfully hack into secure computers and networks. The news media generally uses the term *hacker*. Hackers and others who wish to differentiate between law breakers and innocent techies use the term *cracker*, for criminal hacker, claiming that hackers do not necessarily break laws. Those in the information security industry tend to prefer the labels *attacker* and *intruder*.

There are several labels associated with different forms of hacking. Some of the most common include:

- Black-hat hacker: Hackers who take advantage of security vulnerabilities to gain unlawful access to private networks for the purpose of private advantage
- White-hat hacker: Individuals who consider themselves to be working for the common good by hacking into networks in order to call attention to flaws in security so that they can be fixed
- Gray-hat hacker: A hacker of questionable ethics
- Script kiddie: A person with little technical knowledge who follows the instructions of others to hack networks

FIGURE 11.11 • The HOPE conference

Hackers from around the globe gather in July each year for the annual HOPE (hackers on planet earth) conference in New York.

Notice that there are ranges of ethics embraced by varying types of hackers. Not all hackers are considered unethical. Some, such as white hat hackers, consider themselves to have altruistic motivation—even though they do break the letter of the law. Microsoft has recently extended a hand to the hacking community hoping to get white-hat hackers on their side. In 2005 Microsoft invited several hackers to their offices in Redmond for what they dubbed the Blue Hat meeting (Microsoft's corporate color is blue).[25] Microsoft is hoping that altruistic hackers can help them to make Windows and other Microsoft products bullet proof.

Hackers often belong to groups and assist each other with new methods for hacking systems. Hackers often use chat utilities and instant messaging to communicate. The Hacker Quarterly (*www.2600.com*) sponsors a huge hacker convention in New York City biannually called the HOPE conference (see Figure 11.11). HOPE stands for *hackers on planet earth*. They also sponsor monthly meetings in hundreds of cities around the world on the first Friday of every month (*www.2600.com/meetings*).

Law enforcement agencies invest significant amounts of money in tracking criminal hackers and keeping tabs on the hacker population. A new field of law enforcement is computer forensics, also referred to as *digital forensics*. **Computer forensics** is the process of examining computing equipment to determine if it has been used for illegal, unauthorized, or unusual activities. This area of research has become formalized and well respected. There are many educational institutions that offer degrees or certification in computer forensics.

MACHINE-LEVEL SECURITY

At the beginning of this chapter, you learned that information security is implemented in layers. The most basic and fundamental security is implemented at three levels (refer back to Figure 11.3): the individual machine level, the computer network level, and the Internet level. This section examines security from the perspective of the individual machine and discusses security precautions for personal computers that may or may not be connected to a network or the Internet. By learning how to protect stand-alone PCs, you also learn about the first line of defense for the networks to which those PCs may be connected.

FIGURE 11.12 • Authentication

User authentication can be based on something you know, something you carry with you, or some unique personal physical trait.

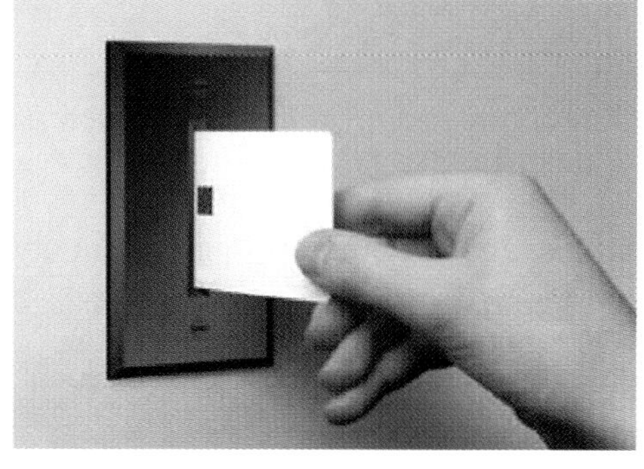

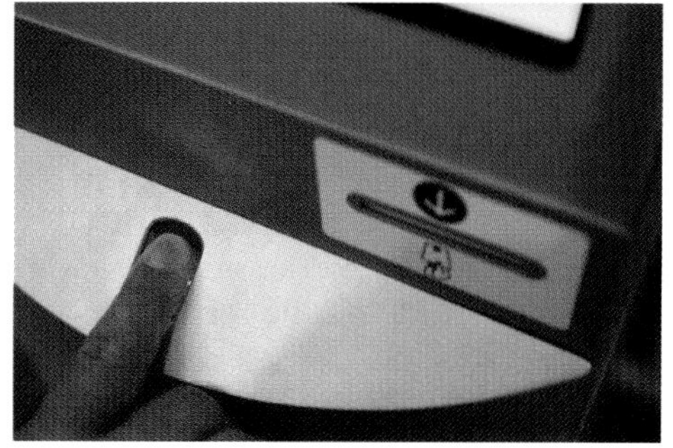

Access to today's PCs is typically guarded with a username and password. In security terms, this is regarded as *authentication*, a manner in which to confirm the identity of a user (see Figure 11.12). There are three common forms of authentication:

- Something you know: Such as a password or personal identification number (PIN)
- Something you have: Such as ID cards, smartcards, badges, keys, and other items designed to be used to authorize access to secure areas, and systems
- Something about you: Unique physical characteristics such as fingerprints, retinal patterns, and facial features can be scanned and used for authentication

This discussion of PC security begins with a short discussion on methods of authentication, then looks at ways to protect the data on your PC. This is followed by some tips for keeping your PC running smoothly and safely.

Passwords

A **username** identifies a user to the computer system. Varying levels of computer access and environmental preferences are associated with a username. For example, the username adkins may be associated with a system administrator who is granted access to the entire system and uses a Starwars wallpaper, while user johnson has access to only his own files, six programs, and uses a plain blue wallpaper. Differing operating systems provide differing levels of access and customization features.

While usernames can act as a form of authentication, they are not very secure. Most systems do not allow access without a valid username, but at the same time, a valid username is typically easy to guess. For example, usernames are often all or part of an e-mail address. For this reason, computers require passwords to be provided with usernames.

A **password** is a combination of characters known only to the user that is used for authentication. Passwords can be an effective form of authentication if they are difficult to guess, kept confidential, and changed regularly. However, passwords are considered the weakest form of authentication because they can be used by others to access systems without your knowledge. For this reason it is important to create passwords that are strong—that is, difficult to guess.

The strongest passwords are a minimum of eight characters in length and do not include any known words or names, especially ones that are related to you or your interests. Strong passwords include upper- and lowercase letters, numbers, and symbols. For example, sk3&KxD$ would be difficult for anyone to guess. Unfortunately, it is also very difficult for the user to remember. The most effective passwords are ones that are difficult for others to guess but easy for you to remember. For this reason, some people combine words they can remember with numbers and symbols, such as bluedoor*45. Another trick is to fabricate words from symbols, numbers, and characters such as L@@K4me or 2gud2btru. This gives the appearance of randomness while providing personal meaning so they can be remembered. Figure 11.13 illustrates passwords created by John Driesdale from New York, across a spectrum from weak to strong. Notice that the most effective passwords are those that include multiple words not closely associated to John or his interests, are a mix of letters numbers and symbols, and are relatively easy to remember.

It is not uncommon for a person to have to remember several passwords for several different systems: a password for your home PC, a password for your school account, another for your e-mail, and several others for Web sites. How can you possibly remember that many complex passwords? The easiest solution, and the most dangerous, is to use the same password for all accounts. This is dangerous because if a hacker discovers your password on one account, all your accounts are vulnerable. For example, many Web sites that require a username and password use your e-mail address as a username and allow you to create a password. It is dangerous to make the assumption that the owners of the Web site will handle your username and password in a responsible manner. In fact, the Web site may have been published with the intent to steal usernames and passwords. If you provide such a Web site with the same password you use for your e-mail account, you have effectively provided all that is required for the hacker to access your e-mail and possibly other private information including bank accounts, school accounts, home PC, and other highly confidential information.

One password strategy is to use different passwords for different resources depending on the level of security required. For example, 80 percent of the passwords you maintain may be for unimportant low-security resources such as online news and information sources. There's no reason why you shouldn't use one common, easy-to-remember password for all of these services. As you move up to resources that include personal accountability, financial access, and access to private information, passwords should become increasingly strong and unique. Table 11.4 illustrates this point.

FIGURE 11.13 • Passwords, weakest to strongest

Strong passwords use words that are unrelated to your interests, and include upper- and lowercase letters, numbers, and symbols.

Weakest
John
Kelley
JohnnieD
yankees
Heresjohnnie
nycoolboy
NYcoolboy#1
Hypertree
nyKOOLB@Y
Hyper#tree9
re@Lpharm#
92Tpo5#cCw
Strongest

Most effective: nyKOOLB@Y, Hyper#tree9, re@Lpharm#

TABLE 11.4 • Passwords

The strength and uniqueness of passwords should be based on the level of importance of the resource being accessed.

Security level	Examples	Password use
High: Resources that include access to finances and private information	• Home PC • School account • E-mail • Bank account • Financial services such as PayPal • E-commerce Web sites that store credit card information for express checkout	Use a unique and strong password for each resource
Medium: Resources that do not contain private or financial data but impact your reputation or require personal accountability	• Online blogs, chat, and discussion • Online support Web sites • Online auctions like eBay	Choose password strength based on level of accountability; do not use any passwords that are used for high-security resources
Low: Resources that do not contain private or financial data and do not impact your reputation in any way	• Free online news sources such as the *New York Times* • Other free information and services provided on the Web	Use one common, easy-to-remember password for all; make sure the password is different from any password used for medium- and high-security resources

Besides choosing passwords wisely and making sure that passwords are unique for important resources, users should also change passwords regularly—typically once or twice a year. It is possible that a hacker could be accessing your account without your knowledge. By changing your password, you can resecure your account.

This discussion about passwords illustrates an important point that will be confirmed throughout this chapter: total information security is inconvenient and difficult. Security and convenience are often diametrically opposed. To achieve total information security, users must be convinced of its importance, and at the same time, security must be made as convenient as possible.

In the case of passwords, there are some very useful tools to help make password management more convenient. Programs such as Password Agent from Moon Software (*www.moonsoftware.com*) act as a secure database for storing passwords. Using Password Agent, you need only remember one password, the one that opens the Password Agent software. Once in the software you can store and retrieve passwords used for all your accounts. This software stores your passwords in an encrypted database, making them impossible for others to read unless they have the one password that accesses the software and database file. Storing passwords in a regular unencrypted file is a big security risk.

Software from Billeo (*www.billeo.com*) also uses encryption to secure password information. Billeo's Password Manager Plus also fills in usernames, passwords, and other data automatically in Web forms. Billeo suggests storing the software on a USB drive so that you can take your usernames and passwords with you everywhere you go. Because the software is password protected, you needn't worry about losing your USB drive and having your database of passwords stolen. Similar products have been designed for use on handheld computers and smart phones.

Apple computers running OS X include software called the Keychain that effectively manages passwords. The Keychain stores all the information required to access password protected resources such as Web servers and local secure resources.

ID Devices and Biometrics

FIGURE 11.14 • Security ID

This security ID token contains a microchip that holds authentication data and is read by pressing the tip of the token into the recessed area of the reader.

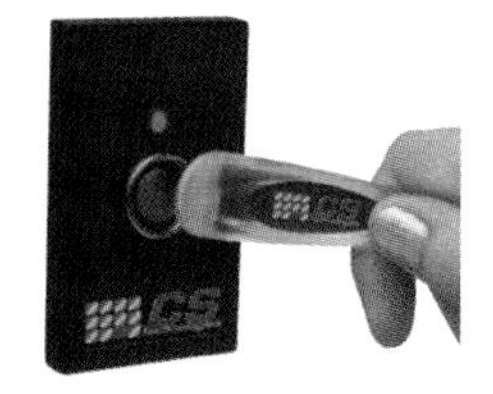

A number of devices are used in corporations and organizations to identify users and provide access to restricted areas and computer systems. These technologies make use of the "something you have" form of authentication. Sometimes they are combined with a password to increase the security level. The most popular of these devices are ID cards and tokens that sometimes take the form of a keychain fob. ID cards may contain forms of identification in a magnetic strip or in a microchip, as is found in a smart card. Tokens contain microchips that hold ID information. Cards with magnetic strips are swiped in card readers, and smart cards may be swiped or simply waved in front of a card reader. ID tokens can also be used with proximity readers or read by silicon key readers as is shown in Figure 11.14.

Although ID cards and tokens are primarily used in businesses today, they will soon be used for accessing services through personal computers. The U.S. federal government is requiring banks to beef up security for Internet customers. Bank Web sites are expected to implement some form of two-factor authentication by the end of 2006.[26] The most likely system to be implemented is one in which customers use the usual username and password, and some form of hardware token that provides a continuously changing numeric access code. Banks may also turn to biometrics or smart cards to positively identify Internet customers.

Biometrics is the science and technology of authentication by scanning and measuring a person's unique physical features such as fingerprints, retinal patterns, and facial characteristics. Fingerprints have been used in areas of security and law enforcement for decades. The digitization of biometric traits and the computer's ability to quickly scan and interpret the data has led to rapid growth in the biometrics field.

FIGURE 11.15 • Facial recognition

By taking measurements of 128 facial features and matching them to the measurements of known faces, biometric software can help identify people.

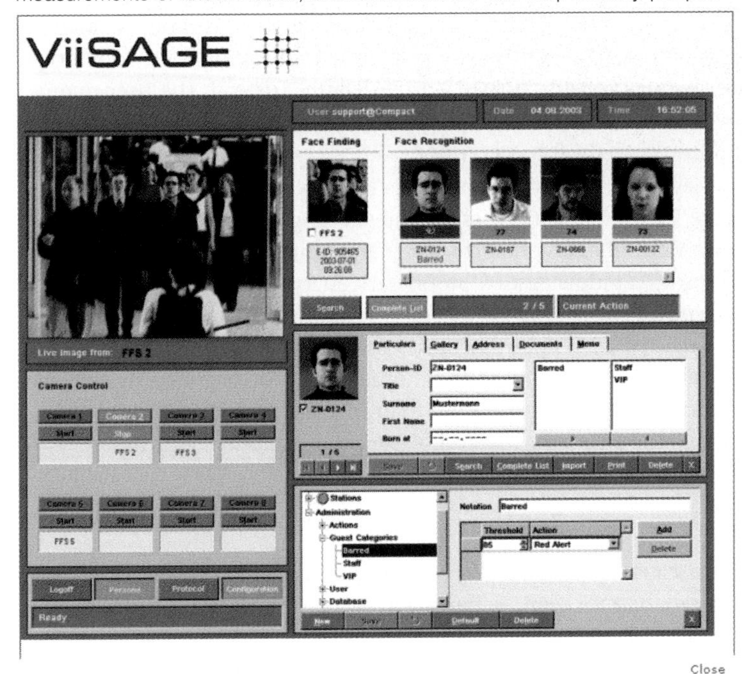

Law enforcement agencies and airport security agents are using facial pattern recognition in efforts to catch criminals and terrorists. *Facial pattern recognition* uses a mathematical technique to measure the distances between 128 points on the face (see Figure 11.15). Taking a weighted sum of these measurements, software can quickly scan a database of known faces and come up with a match if one exists. Cameras posted at airport security gates and on city streets can continuously scan for faces of known criminals. Oki Electric Corporation is marketing face recognition software for camera phones. Once installed, the phone is trained to recognize your face (using the phone's camera), and from that point on will only operate for you—or perhaps your identical twin.[27] Another form of biometrics is the retinal scan. *Retinal scanning* analyzes the pattern of blood vessels at the back of the eye.

Perhaps the easiest form of biometrics is the *fingerprint scan.* Fingerprint scans are an increasingly common method of authentication for access to secure areas, for validating credit card purchases, and for logging on to computers. Using a USB device like the one shown in Figure 11.16 or a built-in fingerprint scanner, a computer user can log in without having to enter a username and password. These devices can also be used to enter your usernames and passwords on Web and other login screens by simply pressing your index finger on the device.

Encrypting Stored Data

With the increase in use of mobile computers such as notebooks, tablets, and handhelds, the possibility of losing a computer or having it stolen is very real. Sometimes the data stored on a portable PC is more valuable than the computer itself. An estimated 2.5 million personal data files were stolen by computer thieves in 2004. For this reason, many businesses have employees store data files on the corporate network servers rather than on their notebook computers. Once in a thief's possession, the hard drive can be removed from a notebook and the files accessed without the need for a username or password. For this reason and others it is a good idea to encrypt any confidential files stored on a computer.

FIGURE 11.16 • Fingerprint ID

Using this Sony USB fingerprint identity device, you simply press your finger to the device every time you are asked for a password.

Encryption has been defined as a security technique that uses high-level mathematical functions and computer algorithms to encode data so that it is unintelligible to all but the intended recipient. Today's Apple and Windows PCs include security tools that can encrypt files stored on disks and flash drives. This is useful in situations where the information you are storing is confidential or valuable, and there is a possibility that your computer can be accessed by others, lost, or stolen.

Files can be encrypted "on the fly" as they are being saved, and decrypted as they are opened. To open encrypted files, you must enter a password. Encryption and decryption does tend to slow a computer down slightly when opening and saving files, so encryption is not typically turned on by default. Instead, the user manually selects the files or folders that contain confidential information and marks them for encryption.

Mac OS X uses a system it calls FileVault for file encryption (see Figure 11.17). FileVault uses the latest U.S. government security standard, AES-128 encryption, to safeguard confidential documents. Microsoft Windows uses a proprietary encryption system called Encrypting File System (EFS). Some Windows PCs include an embedded security subsystem that stores passwords and encryption keys on a dedicated security chip on the motherboard. Storing security information in a dedicated chip rather than in a file on the hard drive offers perhaps the strongest method of personal computer security.

Backing Up Data and Systems

The most common cause of data loss, according to engineers at data recovery vendor Ontrack (*www.ontrack.com*), is hardware failure (56 percent), followed by human error (26 percent), software corruption (9 percent), viruses (4 percent), and natural disasters (2 percent). The only method that provides 100 percent protection for data against all of these disasters is to back it up! Still it is amazing that less than half of computer users have a regular backup procedure in place. Today's backup technologies make backing up important data and system files an effortless task.

System Backup and Restore. When the files that make up an operating system become corrupted, either due to mechanical failure of the hard drive or a virus, your computer may simply not start. To safeguard against operating system failure, operating systems and security software provide a means to create a rescue disk. A *rescue disk*, either a CD or floppy disk, is created while a system is operational and backs up important operating systems files. For example, if the hard drive fails, essential operating system files can be loaded from the rescue disk, which allows you or a computer technician to repair the damage done to the system.

Windows PC users who find their system damaged but are still able to start the computer can benefit from the System

SECURITY AT THE TOUCH OF A HAND

Using passwords to protect personal or professional data is proving cumbersome, even impossible. Passwords can be passed, discovered, or forgotten. Enter biometrics—where sensors read fingerprints, irises of the eye, or other unique physical traits to allow access. New technology and refinements, including a drop in size from a two-inch square fingerprint reader to a thin strip, mean a correspondingly significant drop in production price—from $40 to $50 down to as low as $5. Hewlett-Packard, Toshiba, HP Compaq, and Lenovo are all adding biometric fingerprint scanners to new laptops.

Source:
Biometrics puts security at a user's fingertips: No more jumble of passwords -- personal prints would be the only ID needed
By Benjamin Pimentel
San Francisco Chronicle
http://www.sfgate.com/cgi-bin/article.cgi?file=/chronicle/archive/2005/07/11/BUG7IDL1AH1.DTL&type=tech
July 11, 2005

FIGURE 11.17 • Apple's FileVault

Apple's FileVault system can be used to encrypt files on your hard drive or portable storage device.

Restore utility (See Figure 11.18). Periodically, Windows creates backups of the state of the operating system at what it calls "restore points." If the computer contracts a virus, or newly installed software causes system problems, the system can be "taken back in time" to when the computer was without problems. Data files are left intact, but software and the operating system are reconfigured to match the state they were in at the restore point. So for example, if you recall that your computer was working just fine last week, you can select the restore point created last week to take your system back to its state at that time.

Backing up Data Files. Businesses typically have regular backup procedures that back up corporate data to disk drives, tape, or online storage media. Individuals typically back up their personal data files to CD, DVD, USB drive, or a network drive located on another computer and accessed over a network. Operating systems usually provide a backup utility that can be used to back up data files. Both Apple and Microsoft Windows provide utilities called Backup. Typical backup software collects the files you wish to back up into a compressed file called an archive. Some backup software provides the ability to encrypt the archive and password-protect it. Backup software typically provides the following options:

- Select the files and folders you wish to back up.
- Choose the location to store the archive file.
- Choose whether to back up all files (a *full backup*), or just those that have changed since the last backup (an *incremental backup*).

FIGURE 11.18 • Microsoft Windows System Restore

Microsoft Windows System Restore utility allows users to take their computer back in time to when the computer was functioning correctly.

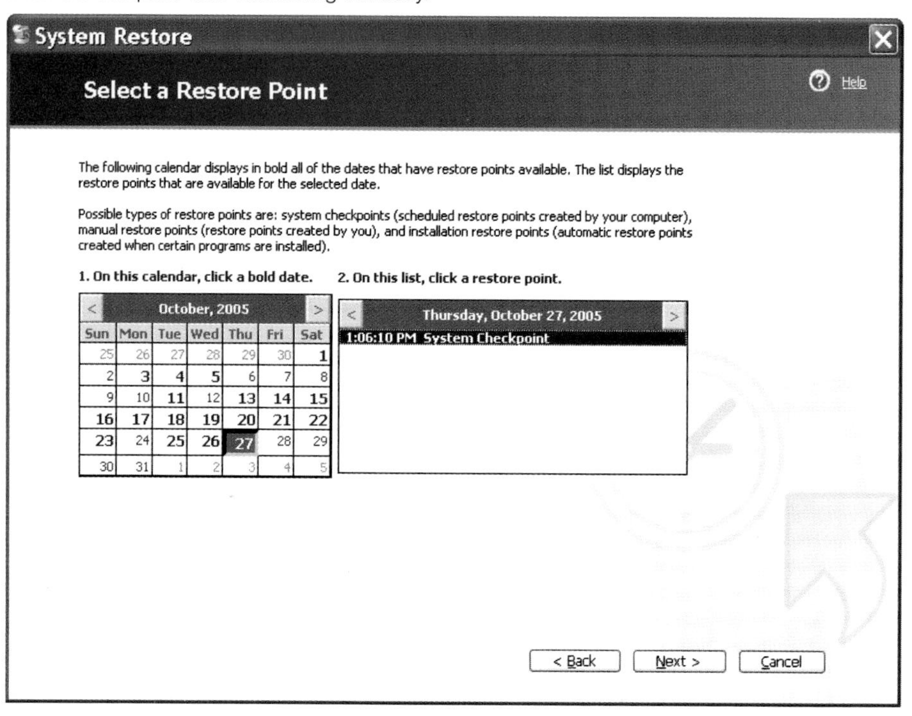

Most backup utilities also provide a scheduling option that allows you to automatically run backup routines at specified dates and times. For example, you might set a backup routine to back up only files on your system that have changed since the last backup and schedule it to run every night at 4:00 a.m.

Some users prefer to have exact copies of their data files on a backup disk rather than building a compressed archive file. Exact copies are useful if you need to occasionally access individual files from the backup. An archive file would not allow this, making you first restore the entire archive. However an archive file takes up much less space than storing an exact copy of a file system.

Creating an exact copy of a system or portion of a system is called *mirroring*. Some software provides real-time mirroring. In *real-time mirroring*, as you save files they are automatically updated in their primary storage space and the mirrored copy. MirrorFolder software from Techsoft (*www.techsoftpl.com*) provides real-time mirroring. It can also be used to keep notebook PCs and desktop PCs synchronized, so that both always retain the most recent version of your files.

The Mirra Personal Server provides real-time mirroring in an external disk drive connected to a home or small business network (Figure 11.19). The Mirra Personal Server automatically and continuously backs up all the computers on your network running Mirra software. What's more, the Mirra Personal Server can be accessed from any Internet-connected computer through *www.mirra.com* using a username and password. This means that you can access your files while away from home, or provide colleagues, friends, and family with access to particular files on your server.

FIGURE 11.19 • Mirra Personal Server

The Mirra Personal Server automatically and continuously backs up computers on a home or small business network and allows network users to access its files from the home network or Web.

Remote Data Backup. Internet-based backup services are becoming increasingly popular as more users connect to the Internet through high-speed connections. For example, Remote Data Backups (*www.remotedatabackups.com*) provides automated backup of your data every night over your Internet connection for a monthly fee. An important feature of this service is that it allows you to back up your data off-site in a different location from your computer. It is smart to keep backups off-site as keeping them on-site may expose them to the same destructive forces that wipe out the original data: fire, flood, power surge, and so on. With Remote Data Backups, every night, changes to your hard drive are collected, compressed, encrypted, and transmitted via your Internet connection to two mirrored data centers.

System Maintenance

In addition to the smart use and management of passwords, encrypting confidential files, and backing up data and systems, there are other steps that can be taken to secure a PC and the data it stores.

System and Software Updates. Most software has an update feature that applies patches to bugs discovered in the software after release. Some updates, such as Windows Update discussed earlier, can be set to download and install automatically when they are available. Other updates require the user to check periodically at the vendor's Web site or click a Check for Updates option in the software's Help menu. Recall that about 9 percent of data loss is attributed to software corruption. Keeping your operating system and software up to date is the best way to avoid problems caused by faulty software. Users also benefit from additional features that are sometimes provided free of charge in software updates.

Computer Housecleaning. *Computer housecleaning* involves organizing the data files and software on your computer. Housecleaning activities can include:

- Deleting unneeded data files
- Organizing the remaining data files logically into folders and subfolders
- Emptying the recycle bin (Windows) or trash can (Mac)
- Deleting unneeded saved e-mail messages
- Cleaning out Web cookie files, and other temporary Internet files
- Uninstalling software that is no longer needed
- Cleaning and reorganizing the computer desktop for quick access to items you use most often
- Organizing Web browser favorites

Your operating system may include a utility that hunts down unnecessary files on your system and allows you to mark them for deletion. After completing your housekeeping, don't be surprised to find several gigabytes of additional space on your hard drive.

While deleting software and files from your system, the remaining files on your hard drive will become fragmented; that is, they will be scattered about the disk in nonadjacent clusters creating slower access times. A *defragmentation utility* provided with your operating system or purchased separately can be used to defragment your hard drive, aligning your files in adjacent clusters and improving your computer's performance.

Windows Cleaners. After a while, Microsoft Windows users may notice problems with system performance: the system may drag, or perhaps error messages pop up during system startup. Although this may indicate the presence of a computer virus or spyware, it may also be that the Windows Registry has become cluttered with data left by low-quality software that has been installed and perhaps uninstalled. Recall that the Windows Registry is a system database file that stores data about the system configuration and the configuration of software running on the system. Utility software referred to as *Windows cleaners*, or *Windows Registry cleaners*, such as CleanMyPC and Registry Mechanic (see Figure 11.20), scan the Windows Registry, correcting incorrect or obsolete information. Some Windows cleaners are available free on the Web, for example at *www.pcpitstop.com*. As the Registry is very important to the proper functioning of Windows, the utmost care must be taken when working with it. Windows cleaners should include a backup/restore function that lets you back up the Registry prior to cleaning so you can restore the Registry to its previous state should you encounter problems.

FIGURE 11.20 • Windows cleaner

Windows cleaners maintain the Windows Registry and other system files to boost the performance of a Windows PC.

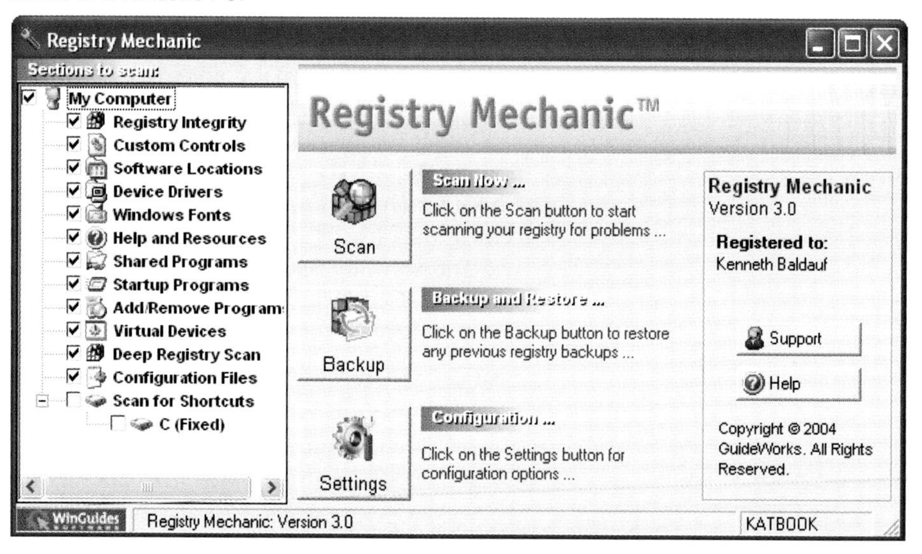

NETWORK SECURITY

When a computer is connected to a network, security risks increase a hundredfold. Connecting to the Internet increases risks a million times that of a stand-alone computer. In fact, for information of highest security, government agencies use computers that are not connected to agency networks or the Internet. So long as there is a network connection, there is danger of unauthorized access from a remote connection.

This section discusses threats, considerations, and security for networked computers in general. A later section deals specifically with security measures for computers connected to the Internet. Networked computers may be in a government agency, a large, medium-sized, or small business, a college campus or campus housing, in an apartment building, or in a home. The considerations are the same in any of these environments, although the value of the information may vary from one environment to the next.

Multiuser System Considerations

A *multiuser system* is a computer system, such as a computer network, where multiple users share access to resources such as file systems (see Figure 11.21). When sharing resources, users are naturally concerned about the privacy and protection of their data files. System administrators are concerned with protecting the system from intentional or unintentional damage. If a corporation owns the network, corporate management is concerned about the security and privacy of corporate information. This section deals with the issue of restricting the access of network users.

FIGURE 11.21 • Multiuser systems

Multiuser systems require special software mechanisms that control user access to shared resources.

User Permissions. *User permissions* refer to the access privileges afforded to each network user. Network operating systems such as Windows XP, Apple OS X, Linux, NetWare, and UNIX provide methods for associating user permissions with each user account. As each user logs on to a workstation connected to the network, the permission policies are applied, and the user can only access the resources defined by those policies.

To control user access to system resources such as files, folders, and disk drives, access policies must be defined for both the resources and the users. For example, a user may be restricted to have access only to files that he or she created. Files and folders on the system must carry information that identifies their creator. This is referred to as *file ownership*. Users are the owners of the files they create.

To facilitate sharing files between users on the network, user groups can be established by the system administrator. For example, you and some of your classmates might be working on a project utilizing the campus network. You might be able to ask the system administrator to create a network group named comp1 that you can use for sharing files with your group. Each user would then be identified as a comp1 member by the system and would be able to access files and folders marked as being accessible to the group. This is referred to as *group ownership*. System resources can also be set so that they are available to everyone on the network. This is sometimes referred as *world ownership*.

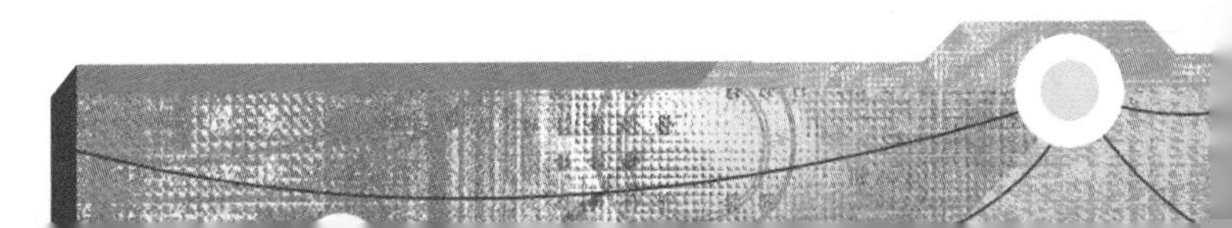

FIGURE 11.22 • User permissions control access to system resources

Bob can access the files he owns and the files owned by his group comp1. Sara can access her own files and comp1 files as well. The system administrator can access everything.

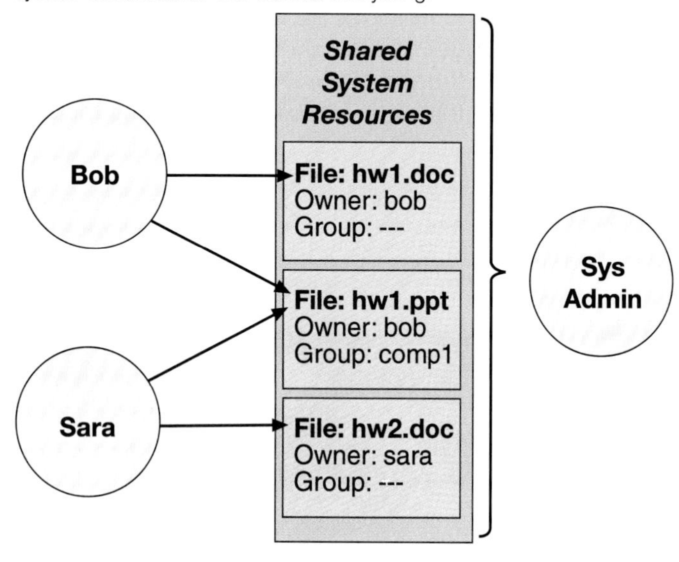

The system administrator is responsible for setting the access rights of users and for setting the permissions on system resources. Users have the power to set permission settings of the files they own. The system administrator is typically the only user who has full access to the system. Figure 11.22 illustrates the effect of user permissions on file access.

The system administrator can also define the type of access allowed. Access to files and folders can be classified as read or write. For example, in Figure 11.22, Bob may have read and write permissions for hw1.ppt, so he can view and edit the file, but Sara may only have read permission so that she can view but not edit the file.

The ability to carry out various system commands can also be restricted through user permissions. For example, Microsoft Windows XP uses two classifications for users: Computer Administrator and Limited. Those with Limited access can only change their own account password and associated icon. They cannot access system settings in the Control Panel or install software. Figure 11.23 illustrates the difference between these account types.

Home users can set permissions on their own files so that others on the network can access them or not. By default, the personal files and folders on both Windows XP and Mac OS X computers are off limits to others on a network.

FIGURE 11.23 • Microsoft Windows XP account types

In Windows XP, users with Limited access do not have the authority to access important and vulnerable system settings.

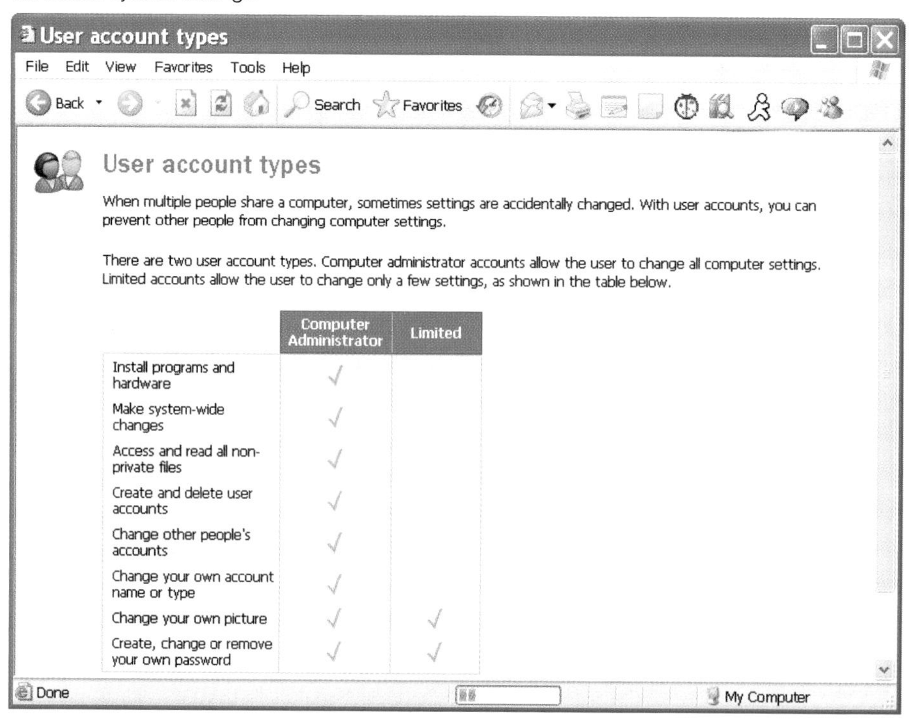

You can adjust file permissions for a specific file or folder by accessing the file properties (in Windows) or info (on Macs). These tools lead you to settings for file permissions.

User permissions provide a second layer of security for computer networks. Usernames and passwords and the login process are the primary way of keeping unregistered users out of the system. User permissions provide registered users with access to the files they need, while restricting access to private resources.

Interior Threats

Threats from within a private network are often referred to as *interior threats*. Interior threats can be intentional or unintentional. Unintentional threats can occur when users make mistakes or exceed their authorization. Intentional threats come from registered users who desire to do the system harm or steal information.

FIGURE 11.24 • Computer labs are fertile ground for viruses

College and student housing computer labs act as breeding grounds for computer viruses as they spread from one computer to the next over the network.

Threats to System Health and Stability. Earlier in this chapter Table 11.3 listed user mistakes that can lead to system instability and data corruption. There are two other common problems that occur on networks, both stemming from allowing network users to introduce software and data files from outside the network.

On most business networks, users are not allowed to install software without the network administrator's authorization. Although such a policy creates inconvenience for users, it is supported by a very good rationale (once again, security at the cost of convenience). There are thousands of software applications available, and many are unstable and dangerous. Although you may be willing to risk the stability of your own PC for the sake of free software, system administrators would be remiss in their duties if they allowed you to jeopardize the entire network. Also, system administrators are responsible for ensuring that all software installed on the network is properly licensed.

In addition, software and data files brought in from outside could contain viruses or spyware. Many viruses are designed to take advantage of network connections and spread to all connected computers. This is the reason that so many college campus computer labs and dorms act as breeding grounds for computer viruses (Figure 11.24), and why you should heed the advice provided in this chapter.

InformationTheft. Many instances of identity theft occur with the assistance of insiders with corporate network access. For example, the personal information, social security numbers, and addresses of 310,000 LexisNexis customers was not stolen by an intruder hacking into the system from the outside, but rather by someone logging in as a registered user.[28] There are a number of instances where data has been mysteriously gone missing. Bank of America Corporation lost a data tape that held credit card information on 1.2 million customers. Ameritrade Holding Corporation was forced to inform 200,000 customers that a tape containing confidential account information had been lost. Time Warner Inc reported 40 tapes containing personal data on 600,000 employees had been lost. Citigroup Inc said that a box of tapes holding personal information on 3.9 million customers had disappeared.[29] All of these instances occurred over a 6-month period in 2005. Were these tapes really lost as claimed, or perhaps stolen?

Businesses are well aware of the rising rate of information theft and many are taking action against it. Companies are further restricting access to physical locations, systems, and databases. Many companies supply their employees with PCs that do not have any external drives or USB ports so that employees are unable to copy data to portable storage media and devices. Even iPods are being scrutinized. A report from the Gartner analyst group suggests that "companies should consider banning portable storage devices such as Apple's iPod from corporate networks as they can be used to introduce malware [malicious software] or steal corporate data."[30]

Security and Usage Policies

To safeguard against threats to a network's health and stability and to prevent information theft, businesses and organizations often implement security and network usage policies. A security and network usage policy is a document, agreement, or contract that defines acceptable and unacceptable uses of computer and network resources. New users are often asked to agree to the conditions of the policy prior to receiving a network account. Users are held liable for upholding the policies and can lose their network account or job if they violate the rules.

Employers are not legally responsible for notifying employees of network usage policies. If policies are not provided at the time of receiving network privileges, it is wise for the new user to ask what is and isn't allowed on the network. Private network administrators have the right to listen in on network communications, read employee e-mail, and electronically monitor employee's Web activities without giving notice to the employee. Employees enjoy very little legal protection in regards to privacy when using employer-owned networks. People have lost their jobs over activities that they thought were okay, but their employer thought was wrong.

Network usage policies typically warn against using the network for illegal activities. They also cover issues that may not be as obvious. For example, some college networks do not allow students to use their network account to run a business. Some businesses do not allow employees to use their business e-mail account for personal correspondence. Table 11.5 lists some corporate security and usage policies and examples.

TABLE 11.5 • Examples of corporate network policies

Corporate policies often address issues regarding the network, e-mail, and Web use.

Policy type	Examples taken from actual company policies
Network and computer use	• Users are responsible for maintaining the security of their password • Users are responsible for using the network facilities in a manner that is ethical, legal, and not to the detriment of others • It is against federal law and corporate policy to violate the copyrights or patents on computer software • Users must request permission of system administration for the installation of software and provide proof of the ownership of the software license
E-mail use	• Employees shall use corporate e-mail systems only for corporate business purposes • E-mail systems shall not be used for transmission or storage of information that promotes discrimination • Employees must use judgment on the type of information sent through e-mail • The use of network systems to send and forward chain letters and other inappropriate messages is prohibited • The office may access an employee's e-mail media
Internet use	• The use of Internet access and the Web should be restricted to corporate business purposes • Users shall request the permission of system administration prior to the installation of Web plug in applications • The use of peer-to-peer networks and file sharing software is strictly forbidden

WIRELESS NETWORK SECURITY

Wireless networks provide wonderful convenience, but as is usually the case, with convenience comes security risks. With wireless technologies, an attacker no longer has to establish a wired connection to a network. Attackers located within the range of the wireless signal, perhaps on the floor above, or in a car parked outside, can attack a wireless network to gain access.

Today the most popular wireless protocol is Wi-Fi. Wi-Fi networks are popping up in offices, homes, on city streets, in airports, coffee shops, even in McDonalds (see Figure 11.25). To make Wi-Fi easy to set up, manufacturers disable all forms of security in new Wi-Fi access points. Several steps are required to secure a Wi-Fi network. Researchers estimate that only 20 percent of Wi-Fi users and network administrators take those steps.[31] For this reason, such networks have become a playground for hackers.

FIGURE 11.25 • Wi-Fi network

Wi-Fi networks are popping up everywhere, causing serious concern over security risks.

Wi-Fi doesn't necessarily have to mean "come and get it" for hackers. There are strong security tools available. This section will examine the threats to Wi-Fi networks and the means with which to thwart them.

Threats to Wireless Networks

Recall that most Wi-Fi networks are centered around a device called an *access point*. The access point sends and receives signals to and from computers on the wireless local area network or *WLAN* (pronounced *W-lan*). By default, access points are set to broadcast their presence. So for instance, if you open up your notebook computer in a coffee shop, a message may pop up on your display letting you know that the Starnet wireless network is within range and asking you if you would like to connect. Starnet is the SSID (service set identifier) of the wireless network. Clicking the Connect icon establishes your wireless connection with the access point. In the case of commercial providers, you may then be asked for a credit card number to pay for the service.

FIGURE 11.26 • War driving

Armed with a notebook, a Wi-Fi antenna, and a power supply, war drivers cruise neighborhoods looking for accessible Wi-Fi networks.

If an access point has no security enabled, clicking the Connect icon puts you on the network, no questions asked. Consider the user in a small apartment who decides to set up a Wi-Fi network. Once the access point is set up, a dialog box appears on the owner's computer with the message, "Network available, would you like to connect?" Easy, right? At the same time, the wireless-enabled computers of neighbors on either side and upstairs and downstairs get a similar message. Now there are five users connected to the "private" home Wi-Fi network without the owner's knowledge. This is the fundamental problem of Wi-Fi networks.

Neighbors may find it hard to resist free wireless network access when a pop-up message offers it. But there are other intruders who go out looking for open wireless networks. *War driving* is the act of driving through neighborhoods with a wireless notebook or handheld computer looking for unsecured Wi-Fi networks (see Figure 11.26). Homemade war driving kits include high-powered antennae attached to the vehicle roof, a long-lasting power supply for the computer, software such as NetStumbler that probes and scans for networks, and sometimes a GPS receiver to mark coordinates that can be shared with others over services on the Web. The legality of war driving is questionable. No one has yet been convicted on charges of war driving. Some feel that if wireless networks are left open, the owner either wishes for others to share the network or takes the responsibility for any problems that ensue.

As wireless networking becomes more established and new wireless technologies emerge, hackers adapt. For example, war nibbling is the activity of hacking into Bluetooth networks. Recall that Bluetooth is the technology behind personal area networks that allows personal digital devices to communicate wirelessly at short distances. It is almost guaranteed that whatever the new technology, somewhere there is a hacker working out plans to alter or attack it.

Securing a Wireless Network

So, how can Wi-Fi network owners keep neighbors and war drivers off their private networks? Access points provide several settings that can all but bullet-proof a wireless network. The access point configuration settings are accessed using a Web browser on a network-connected computer (see Figure 11.27). The Web address and the password are provided in the owner's manual.

FIGURE 11.27 • Wi-Fi access point configuration settings

Wi-Fi access points are configured using a Web browser on a network-connected PC using the Web address and a password.

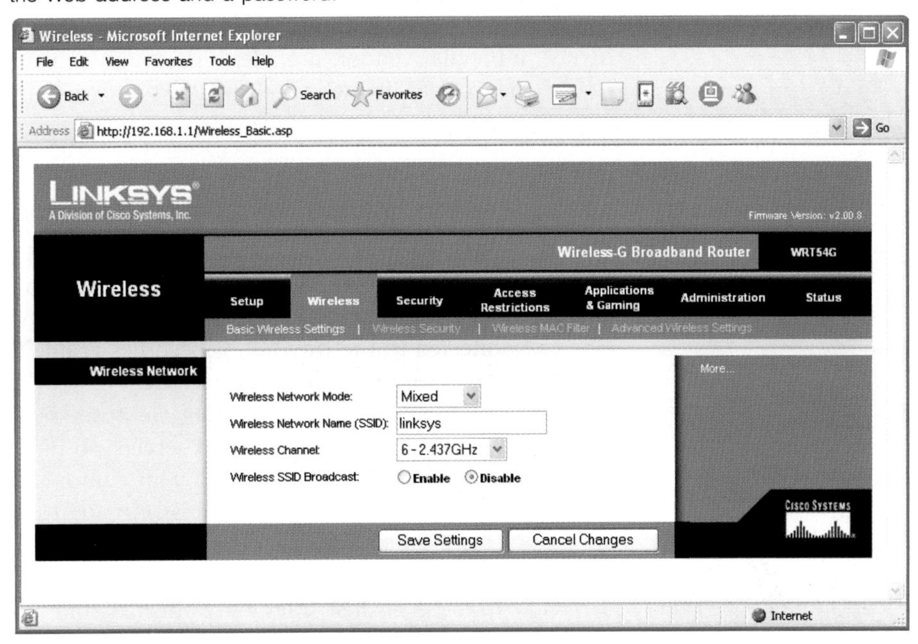

Making a Wireless Network Invisible. Options within the configuration software allow you to disable the access point's broadcasting of the network ID, the SSID. By shutting down the broadcast, neighbors and attackers no longer see a message pop up on their display asking if they want to connect.

Keeping Unwanted Computers off a Wireless Network. Just because the SSID broadcast is disabled doesn't mean that someone who knows the SSID can't still connect. Two other steps are necessary to keep others from connecting to a wireless network.

First, it is very important that you change the password used to connect to the access point. This is easily done using the configuration settings of the access point. The default password is supplied in the owner's manual–and every war driver knows it. You don't want to give attackers the opportunity to log in to your access point and change the settings.

Second, the access point can be set to only allow certain computers to connect. Every Wi-Fi adapter, the hardware used by computers to connect to Wi-Fi networks, has a unique *MAC* (Media Access Control) *address* that is usually printed on the adapter. Notebook computers that come equipped with Wi-Fi capability have a sticker on the bottom of the notebook with the adapter's MAC address. MAC address filtering can be enabled in the access point so that it allows only specified MAC addresses to connect to the network.

Encrypting Data. The above steps are sufficient for keeping others off a WLAN. However attackers may still listen in on the wireless communications. To minimize this possibility, wireless communications can be encrypted. Several wireless encryption protocols exist. *Wired Equivalent Privacy (WEP)* and *Wi-Fi Protected Access (WPA)* are the two most popular. WPA is far more difficult to crack than WEP. Encryption tends to slow down communication, and the better the encryption the slower the communication. Encryption needs to be enabled at both the access point and on all computers using the WLAN.

The above steps take only a short amount of time and very little effort, but the impact on network security is the difference between zero and nearly 100 percent. Detailed instructions for securing a wireless network are typically provided by the manufacturer of the networking equipment. For example, Linksys, a popular vendor, provides a thorough explanation of locking down a wireless network at its Web site (*www.linksys.com*).

INTERNET SECURITY

Expanding from a local area network to the Internet is akin to moving from a small village in Guyana to downtown Manhattan. While your access to information is dramatically increased, so is your exposure to risk. When a computer is connected to the Internet, it becomes visible to billions of Internet users, and a target to millions of various attacks.

Connections to the Internet are not a one-way street. Just as you can request information and services from servers, so too can intruders attempt to access information and services from your computer. On the Internet everyone is just a number—an Internet (IP) address. An address such as 128.186.88.100 could be a Web server, an e-mail server, or your PC. While you may be anonymous, your computer's IP address is registered and known to others.

Attacks against Internet-connected computers can come in the form of direct attacks by hackers (system penetration), or through viruses, worms, or spyware obtained though e-mail, the Web, or from downloaded files. Internet users are also at risk of being manipulated through scams and hoaxes. This section addresses all of these risks and more, and offers practical advice for protecting Internet-connected computers and networks (see Figure 11.28).

FIGURE 11.28 • Internet protection

Although it sometimes feels as though we must take extreme measures to protect our computers and the information they contain, there is no single solution to information security.

Hackers on the Internet

Earlier in this chapter you were introduced to hackers, crackers, attackers, and intruders. This section examines the methods and motivations of these individuals as well as ways to defend against their attacks.

Methods of Attack. Different hacking tools and techniques are used to accomplish different goals. For example, a hacker can remotely install *key-logging software* on a computer to record all keystrokes and commands. The recording is later collected from a remote computer over the Internet and played back in order to spy on the user's actions and sometimes to steal usernames and passwords. A 25-year-old Queens resident was charged with two counts of computer fraud and one charge of unauthorized possession of access codes for a scheme in which he installed key-logging software on computers at 13 Kinko's stores around Manhattan.[32]

Hackers also use *packet-sniffing software* to steal private data being transported over a network. Packet-sniffing software captures and analyzes all packets flowing over a network or the Internet. It can be programmed to look for personal information such as passwords and credit card numbers.

In a targeted attack, an attacker must first find out the IP address of the target computer. Obtaining IP addresses of Web pages is a simple matter using commonly known Internet commands. Once the IP address is discovered, the attacker can attempt attacks based on known software security flaws. The attacker can also scan for open ports on the victim's computer. Recall that a *port* is a logical address used by clients and servers that is associated with a specific service. For example, port 80 sends and receives Web page requests. Ports can act as entry points for attackers. By sending messages to all the ports on a computer, a hacker can establish which ports are open and use known techniques to command the computer through messages to the open port.

For random attacks an attacker may employ *port-scanning software* to search random IP addresses for open ports (see Figure 11.29). Port-scanning software is allowed to run for several hours, after which the attacker can collect a list of IP addresses waiting to be hacked.

Hackers also apply social engineering to acquire information. *Social engineering* exploits the natural human tendency to trust others to acquire private information. For example, a social engineer might phone a person pretending to be a system administrator in order to get that person to provide a password. A social engineer may even go *dumpster diving*; that is, rummaging through trash to steal credit card numbers or other personal information.

Attacks may also be automated in the form of viruses, worms, and spyware. These topics, discussed later, can achieve many of the goals of common attacks without the need for directly manipulating a system.

BAGEL AND NETSKY, OR CRIPS AND BLOODS?

A shrinking pool of unprotected computers has introduced a new twist in the back alley world of computer hacking: turf wars. New viruses not only contain insults aimed at other hacker engineers, but are designed to kill any other competing malware they find on the systems they infect. The focus of the infection has also shifted from mere maliciousness to either hijacking computers for unwholesome aims or stealing a user's identity for fraudulent ambitions.

Source:
Hacker underground erupts in virtual turf wars
By Peter N. Spotts
The Christian Science Monitor
http://www.csmonitor.com/2005/0822/p01s01-stct.html
August 22, 2005

FIGURE 11.29 • Port-scanning software

Port-scanning software such as Nmap from Insecure.org can be used by hackers to probe Internet computers for security holes.

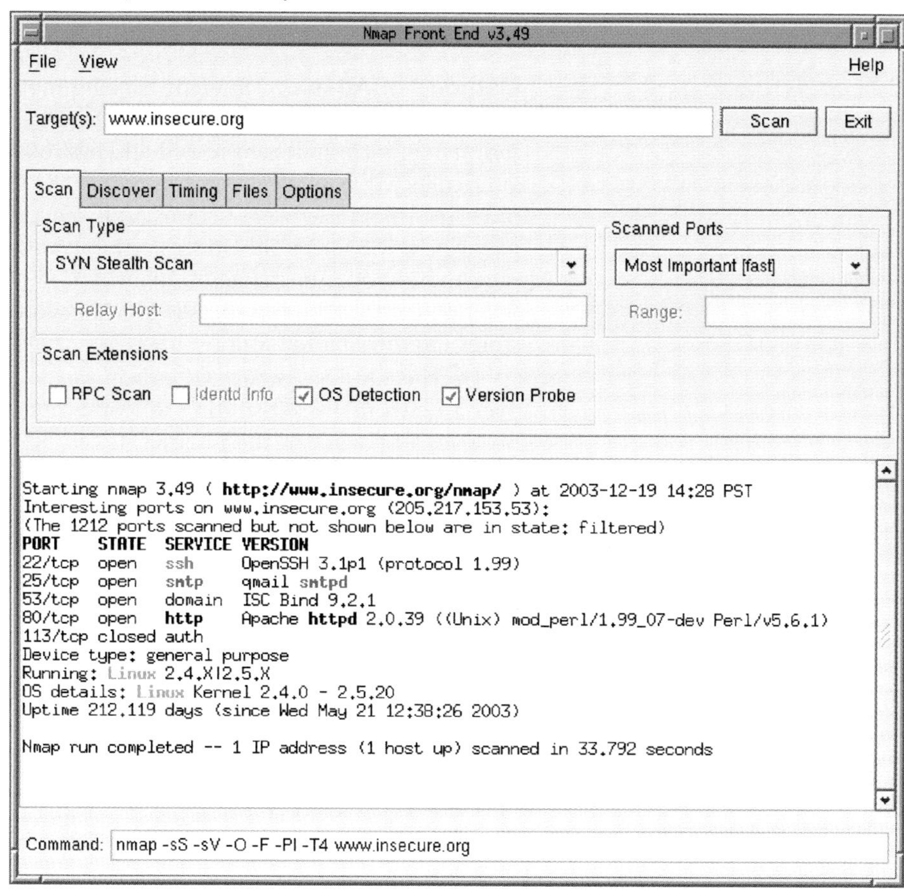

Motivation and Goals for Attacks. Just as there are many different types of attacks, there are many different motivations for Internet attacks. A person or organization may be targeted by an attacker or selected at random. The purpose of the attack may be to vandalize, steal, or spy. Here are some common motivations:

- Hobby and challenge: Many hackers hack for the technical challenge of it. They want to show off their technical prowess by accessing data that is supposedly secure. Being able to access important secure information demonstrats their skill and power. The fact that there is legal risk may provide an additional thrill.
- Malicious vandalism: Some attacks are motivated by hate and resentment. Resentment against the establishment and those with economic or political power has been the cause of many protests and uprisings throughout history. Hacking is a modern tool for protesters. Many attacks have been launched against Microsoft for its capitalistic approach to software development and in resentment of its monopoly. There have been many attacks against Republican political organizations during President Bush's term in office. Random attacks against the general public may simply be an expression of the hacker's resentment of the world.
- Gain a platform for anonymous attacks: Increasingly, computers are being attacked and controlled to be used as a platform to launch attacks on other computers. For example in *distributed denial-of-service (DDoS)*

attacks, many computers are used to launch simultaneous repeated requests at a Web server in order to overwhelm it so that it is unable to function. The Blaster attack, for example, used thousands of remotely controlled PCs to bring down Microsoft's Update Web site by flooding it with requests. DDoS attacks can be costly. The credit card processing firm Authorize.Net was the target of a "large scale" distributed denial-of-service attack that took the network down, making it impossible for it to perform payment processing services for its more than 91,000 small to medium-size e-commerce clients.[33] Attackers can also use hacked computers to distribute viruses, spyware, and spam.

- Steal information and services: In recent years hackers have become increasingly organized and more professional (see Figure 11.30). A higher percentage of hackers are now in it "for the money."[34] Many hire themselves out to businesses that pay them to "take out the competition," steal confidential information, distribute spam, or spy on users. One database expert believes that up to 40 percent of Web sites that connect to corporate databases are susceptible to hackers taking complete control of the database.[35] IBM reported in 2005 that "Hackers have turned toward more criminal and lucrative areas of directing attacks to specific individuals or organizations, often financially, competitively, politically, or socially motivated."[36] Some intruders hack on to computers to steal services such as hacking onto a wireless network to gain Internet access.
- Spying: Through key loggers, packet sniffers, and access to e-mail, Web logs, and file systems, hackers can create a detailed log of everything you do on your computer. Such logs can be very revealing about financial transactions and other private information. An intruder may spy on someone for professional or personal reasons. Consider the 13 high school kids who used their school-issued notebook computers to break into school computer systems. They disabled the remote monitoring software on their own computers and installed it on the network administrator's computer to keep tabs on him.[37]

FIGURE 11.30 • Hacker Kevin Mitnick then and now

Perhaps the most famous hacker ever to serve time, Kevin Mitnick hacked to steal software and expose security holes. After serving nearly five years in U.S. federal prison, Mitnick now has his own information security consulting business.

Defending against Hackers. As mentioned earlier, keeping up to date with operating system and software patches is an important step to protect against intruders. An equally important tool for blocking out hackers is a firewall. A **firewall** is network hardware or software that examines all incoming packets and filters out packets that are potentially dangerous. A firewall protects all the ports of a network or PC from intruders and guards against known methods of attack. Figure 11.31 shows details of many attacks on the author's PC over the course of the day. Notice that most items in the list show the port number that was attacked as well as the IP address of the attacker. Details of the attack are provided at the bottom of the window. Chances are, if your PC is connected to the Internet, it is attacked at least as frequently as is this author's. Fortunately for the author, his firewall was running to catch and stop these attacks.

FIGURE 11.31 • McAfee firewall

A look at McAfee firewall's event log indicates that there have been many attacks on this PC over the course of one day.

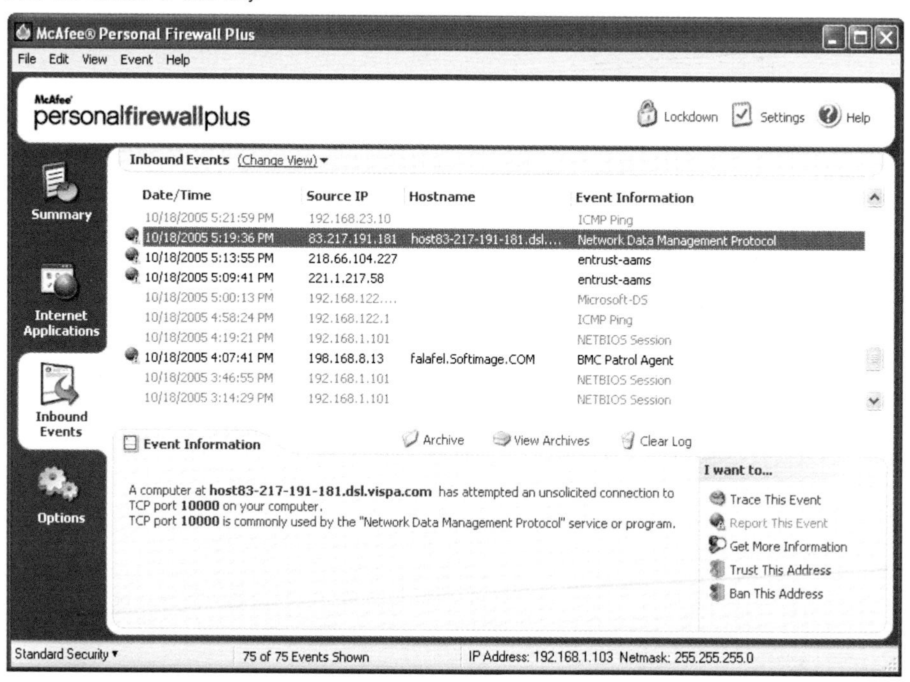

Encrypting confidential data that is stored and transmitted over a wireless network will greatly add to the security of that data. When sending confidential data over the Web, make sure the browser is using a secure connection, indicated by the https:// in the address bar and the closed lock icon at the bottom of the browser screen. Also make sure that any other Internet software that you use applies encryption to sensitive data. For example, traditional FTP (file transfer protocol) software does not use encryption. Newer SFTP (secure FTP) software does. Many colleges now insist that students use SFTP. E-mail servers can also be set to encrypt e-mail that is transferred to your PC. Figure 11.32 shows the security setup options for an e-mail server connection in Microsoft Outlook.

Additional methods of blocking intruders and their tricks are provided in the following sections on viruses and worms, spyware, and other Internet scams, frauds, and hoaxes.

FIGURE 11.32 • Microsoft Outlook security options

E-mail clients provide options for encrypting e-mail that travels between the server and your PC.

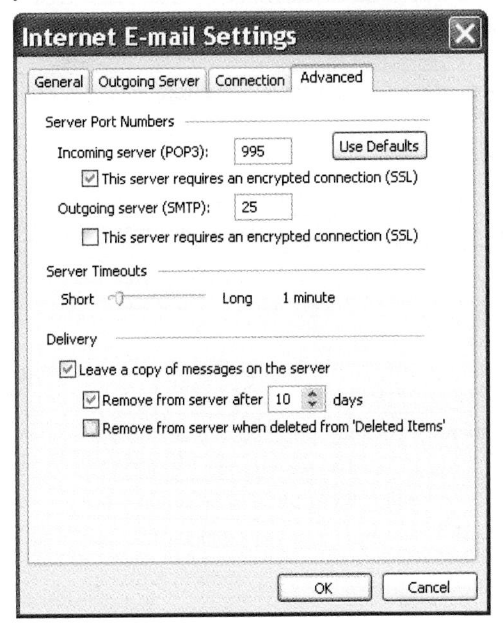

Viruses and Worms

Cyberattacks can take the form of software that is distributed and executed on a computer without the user's knowledge for the purpose of corrupting or disrupting systems. The most common of these types of programs are viruses and worms, which due to their malicious nature are sometimes referred to as *malware.*

A **virus** is a program that attaches itself to a file, spreads to other files, and delivers a destructive action called a *payload.* The payload could be the corruption of computer data files or system files resulting in the loss of data or a malfunctioning computer. A worst-case scenario for damage would be the total loss of data and software. Recovery would involve wiping the hard drive of all files to remove all infection, reinstalling the operating system and all software, and restoring data files from backup. So far, viruses have not been known to damage hardware components of a computer system. During a typical month, 200 or more viruses are on the loose around the world. In 2004 "The count of known viruses broke the 100,000 barrier and the number of new viruses grew by more than 50 percent."[38] They grew by another 50 percent over the first six months of 2005.[39] Table 11.6 provides an overview of the most common types of viruses.

Viruses are sometimes delivered through a technique called a Trojan horse, or just a Trojan. Like the Trojan horse so famous in history books, these *Trojan horses* appear to be harmless programs, but when they run, they install other programs on the computer that can be harmful. A *backdoor Trojan* opens up ports (back doors) on the computer to allow access to intruders.

A **worm** does not attach itself to other programs but rather acts as a free agent, replicating itself numerous times in an effort to overwhelm systems. For example, within 10 minutes of its introduction, the Slammer worm had attacked more than 75,000 computers. Thirty minutes later, some believe the worm had disrupted one in five data packets sent over the Internet.

How Viruses and Worms Spread. Many worms spread through P2P networks and e-mail, as shown in Figure 11.33. Consider this abridged description of the action of the medium-risk Netsky worm from McAfee's virus database:

1. A new variant of W32/Netsky@MM has been received that spreads through e-mail and P2P networks.
2. When run, the worm copies itself to the Windows directory as FVProtect.exe along with six other files.
3. The worm adds itself to the Windows Registry file.
4. The worm sends itself via e-mail—constructing messages using its own e-mail engine.
5. The worm spoofs the From field in the e-mail header and fills it with an address from the infected system.
6. The mailing component harvests addresses from the infected system.
7. The worm copies itself into the folders of file-sharing software on the infected system adopting a name of a media file such as Harry Potter 5.mpg.exe to spread over P2P systems.

There are a few important items to note in this description of an active worm. This worm, like most other worms and viruses, attacks computers running Microsoft Windows, and not Mac or UNIX/Linux machines. There are a few

TABLE 11.6 • Varieties of viruses

Numerous viruses have been designed to infect systems in different ways and avoid detection.

Virus	Description
Boot virus	Infects a computer's startup program so that the virus becomes active as soon as the computer starts up
Direct action virus	Drops payload and spreads when defined conditions are met; can usually be removed without damage to infected files using antivirus software
Directory virus	Changes the paths that indicate the location of files on the computer system
Encrypted virus	Encrypts itself so as to be hidden from scans; before performing its task it decrypts itself
File virus	Attaches to other software so that the virus instructions are processed along with the software instructions
Logic bomb	Delivers its payload when certain system conditions have been met, for example the absence of an employee's name from the corporate database
Macro virus	Infects macros embedded in data files created with other software; infects and spreads to other files viewed by that software
Multipartite virus	Creates multiple types of infections using several techniques, making the virus difficult to detect and remove
Overwrite virus	Deletes information contained in the files that it infects, rendering them partially or totally useless
Polymorphic virus	Encrypts itself in a different way every time it infects a system, making it very difficult to detect
Resident virus	Hides permanently in the system's memory, controlling and intercepting all system operations
Time bomb	Delivers its payload when the system date and clock reach a specified time and date

theories on why worms and viruses target Windows computers. One is that hackers and virus authors hate Microsoft. Another is that hackers and virus writers target the dominant platform in order to do the most damage to the most people. The third is that Microsoft Windows has more security holes than other operating systems and is easier to attack. Perhaps all three theories are correct in part.

FIGURE 11.33 • Worm propagation

Many current worms spread through P2P networks and e-mail.

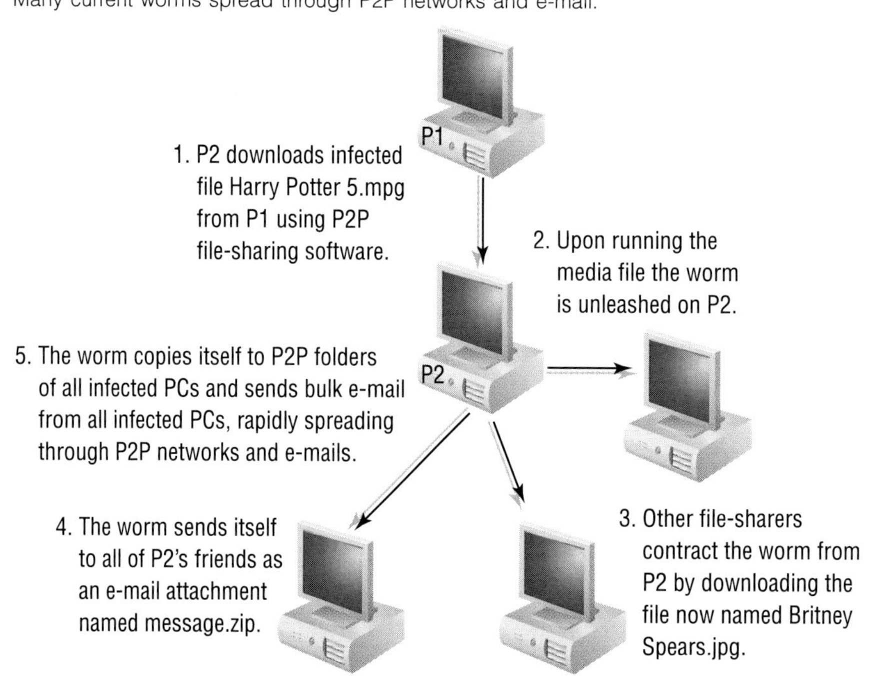
P1
1. P2 downloads infected file Harry Potter 5.mpg from P1 using P2P file-sharing software.
2. Upon running the media file the worm is unleashed on P2.
P2
5. The worm copies itself to P2P folders of all infected PCs and sends bulk e-mail from all infected PCs, rapidly spreading through P2P networks and e-mails.
4. The worm sends itself to all of P2's friends as an e-mail attachment named message.zip.
3. Other file-sharers contract the worm from P2 by downloading the file now named Britney Spears.jpg.

EXPAND YOUR KNOWLEDGE

To learn more about computer viruses, go to www.course.com/swt2/ch11. Click the link "Expand Your Knowledge," and then complete the lab entitled "Keeping Your Computer Virus-Free."

Also note that this worm, and most other recent worms, propagate through e-mail by creating e-mail messages and sending them to every address found on the infected system—in your address book, in your in-box, or in deleted messages. The worm selects one of these addresses and places it in the From field. This means that when you receive a worm through e-mail, it most likely did not come from the person it indicates, but rather, from someone that person knows and has corresponded with. Worms also can include personal data found on the infected PC in the distributed e-mail.

Worms can also propagate through P2P networks. The worm masquerades as a media file in the shared folder of P2P users. It uses a filename with a double extension such as Harry Potter 5.mpg.exe, taking advantage of the fact that Windows, by default, hides file extensions. On most Windows machines this file appears as "Harry Potter 5.mpg," a harmless-looking video file. But when you play it—surprise!—the cycle begins all over from another infected PC.

Viruses and worms also spread through *Web scripts* (small programs that run on Web pages). In 2004 Microsoft's Internet Explorer Web browser was found to have so many security holes that the Department of Homeland Security's U.S. Computer Emergency Readiness Team recommended that for security reasons, citizens should avoid using it.[40] Numerous hackers took advantage of those security holes to spread viruses by way of Web pages. Since then Microsoft has issued patches for Internet Explorer and invested heavily in designing a new, more secure version of the browser.

Microsoft recently took offensive action against Web sites set up to attack Windows PCs. They developed software named HoneyMonkey that goes out on the Web looking for attackers. In its first month, the HoneyMonkey research project located 752 Web addresses linking to 287 sites that could automatically infect unpatched machines. The project also discovered an attack that could penetrate a fully up-to-date Windows XP Service Pack 2 system using a previously unknown vulnerability.[41]

Viruses and worms have found a new home in instant messaging systems. Just like e-mail, instant messages can contain binary attachments, which in turn can hide viruses and worms. There have also been isolated cases of viruses spreading through cell phones. Some believe that by 2008 cell phones will require firewall and antivirus software just as PCs do today.

Computers can also become infected through downloaded software files. Viruses are sometimes hidden in browser plug-in applications. A Web page may inform you that you need a particular viewer in order to view the contents of the Web site and ask you to click OK to download. If the Web site is of questionable origin, chances are the plug-in contains a virus. Viruses also may be included in software available for free on the Web. Typically, if you find a deal that seems too good to be true, there is some hidden agenda at work.

BACK TO ZERO

The Internet developed in an environment of trust as a way to share information, and thus had no defenses against those who would abuse it. That means all security measures are after-the-fact, patchwork additions. The National Science Foundation's solution is a program that will take the Internet back to ground zero and rebuild it, overcoming viruses and other security threats through design. The development team hopes to strike a balance between information safety and privacy.

Source:
NSF Preps New, Improved Internet
By Mark Baard
Wired News
http://www.wirednews.com/news/technology/0,1282,68667,00.html?tw=wn_tophead_3
August 26, 2005

FIGURE 11.34 • Watch out for warnings

Many viruses and worms infect PCs through Web plug-ins. If you get a warning such as this one, consider the source of the Web site carefully before clicking Install.

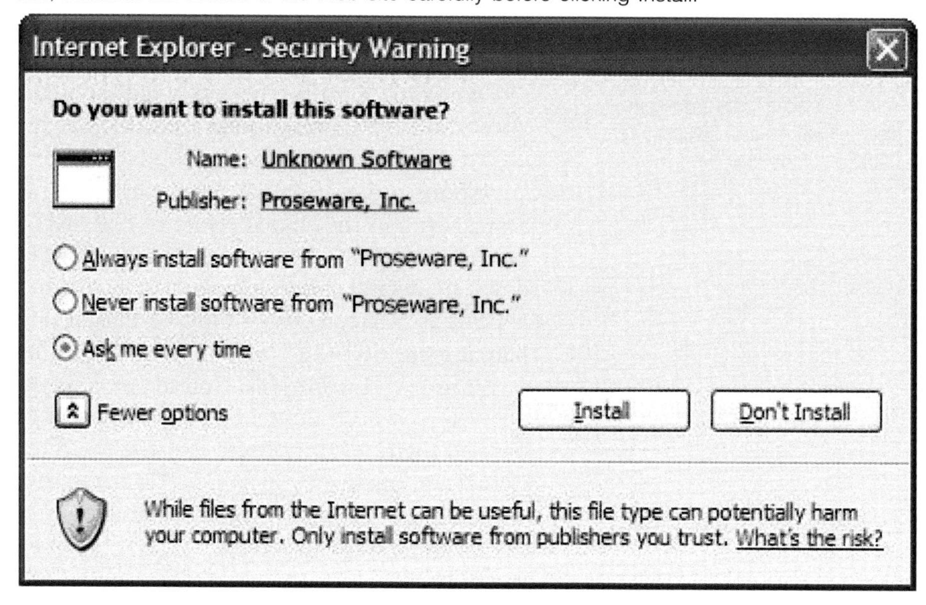

Defending Against Viruses and Worms. The number-one tool against viruses and worms is antivirus software. **Antivirus software**, also known as virus scan software, uses several techniques to find viruses on a computer system, remove them if possible, and keep additional viruses from infecting the system. Antivirus software is only effective if it is updated with new virus information as it becomes known. For this reason, Internet subscription services such as Norton AntiVirus and McAfee Virusscan, both of which update automatically when necessary, are preferred (see Figure 11.35). Antivirus software is often packaged with firewall software for robust Internet security.

FIGURE 11.35 • Norton AntiVirus

Norton AntiVirus scans the computer for known viruses and runs continuously in the background to protect the computer from virus and worm attacks.

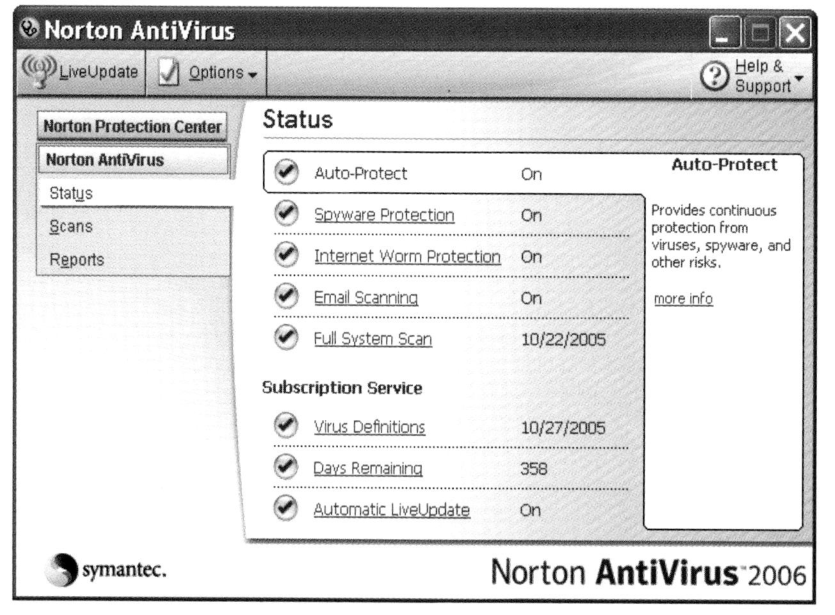

Knowledge and caution play a big part in protecting PCs against viruses and worms:

- Don't open e-mail or IM attachments that come from friends or strangers unless they are expected and inspected by antivirus software.
- Keep up with software patches for your operating system, your Web browser, your e-mail and IM software.
- Use caution when exploring Web sites created and maintained by unknown parties.
- Avoid software from untrusted sources.
- Stay away from file-sharing networks; they do not protect users from dangerous files that are being swapped.

Spyware, Adware, and Zombies

Spyware is software installed on a computer without the user's knowledge to either monitor the user or allow an outside party to control the computer. Spyware is so prolific that any computer that spends time on the Web has probably contracted it. The Internet service provider Earthlink said it uncovered an average of 28 spyware programs on each PC scanned during the first three months of 2004.[42] Spyware differs from viruses and worms in that it does not self-replicate. Spyware is often used for commercial gain by displaying pop-up ads, stealing credit card numbers, distributing spam, monitoring Web activity and delivering it to businesses for marketing purposes, and hijacking the Web browser to show advertising sites. *Adware* is spyware that displays advertisements.

Spyware is distributed by deceiving the user into installing it, by hiding it in shareware, Web plug-ins, and other software acting as a Trojan horse. Spyware can also enter a system through security holes exploited by software on the Web. Once on a system, spyware runs in the background, unknown to the user, carrying out the wishes of its creator. Spyware can communicate with its creator over the Internet.

A computer that carries out actions (often malicious) under the remote control of a hacker either directly or through spyware or a virus is called a **zombie computer**. Experts say hundreds of thousands of computers are added to the ranks of zombies each week.[43] Zombie computers can join together to form zombie networks. Zombie networks apply the power of multiple PCs to overwhelm Web sites with distributed denial-of-service attacks, to crack complicated security codes, or to generate huge batches of spam. It has been estimated that 80 to 90 percent of spam originates from zombie computers.[44] Zombies can also be used for blackmail. In one such case, a small British online payment processing company, Protx, was shut down after being bombarded in a zombie attack and warned that problems would continue unless a $10,000 payment was made.

Defending Against Spyware. Most spyware has so far been directed at Windows PCs, so defenses have also been focused on Windows. **Antispyware** is software that searches a computer for spyware and other software that may violate a user's privacy, allows the user to remove it, and provides continuing protection against future attacks. Spybot Search and Destroy was an early successful spyware tool. More recently, Microsoft has developed its own antispyware tool that is very effective, and antispyware is being incorporated into other security software such as Norton AntiVirus. Spyware that is found by these tools is typically ranked in terms of threat level. For example, low-threat spyware includes commercial software that has the facility to automatically check for updates over the Internet. While this type of spyware provides a service—the automatic update of your software, it may be doing so without your knowledge, and it might be sending information about your computer usage. For example, early versions of Windows Media Player sent information about the music and movies users were accessing. High-threat spyware is spyware that is known to be carrying out illegal activities. Figure 11.36 shows the results page of Microsoft Anti-Spyware illustrating several levels of threat.

FIGURE 11.36 • Microsoft AntiSpyware results page

Spyware that is discovered on the system is classified according to its threat level.

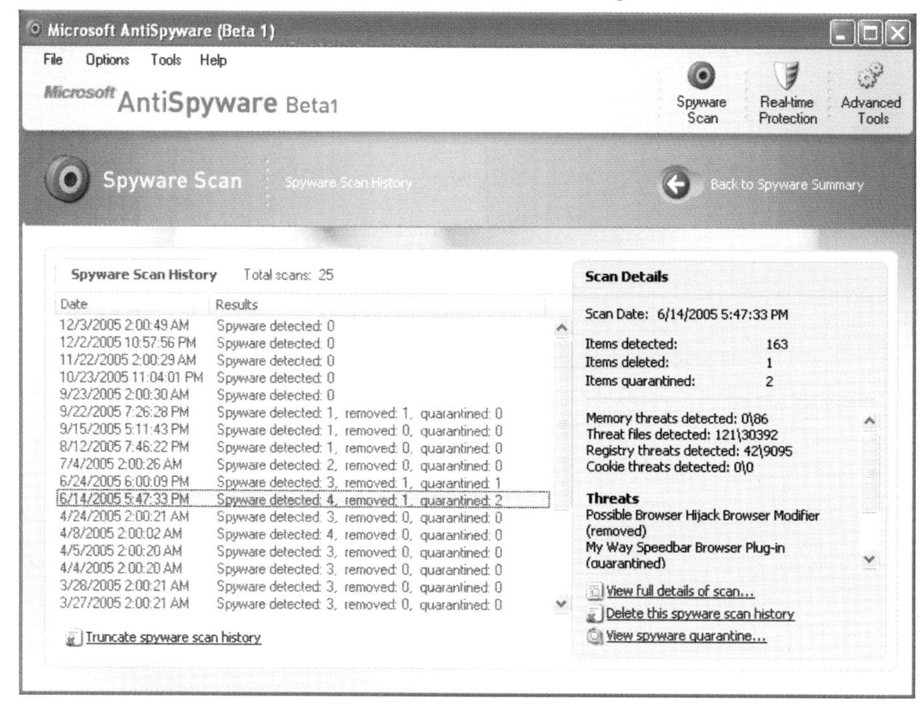

Scams, Spam, Fraud, and Hoaxes

There are many ways in which shysters attempt to take advantage of Internet users and make off with their money. Perhaps you have received e-mail that begins something like this:

- "My name is dr.david konbuna, the manager, credit and foreign bills of bank of africa (boa). I am writing in respect of a foreign customer of my bank with account number 14-255-1004/boa/t who perished in a plane crash with the whole passengers aboard..."
- "This letter may come to you as a surprise due to the fact that we have not yet met. My name is VIJAY SINGH, a merchant in Dubai, in the U.A.E. I have been diagnosed with prostate and esophageal Cancer that was discovered very late due to my laxity in caring for my health. It has defiled all form of medicine and right now, I have only about a few months to live according to medical experts..."

If you have received such an e-mail, it is hoped that you have taken it for what it really is, a scam, and not provided the sender with your bank account information or credit card number as requested.

Internet fraud is the crime of deliberately deceiving a person over the Internet in order to damage them and to obtain property or services from him or her unjustly. Many instances of Internet fraud occur through e-mail, though not all are as obvious as the above examples.

Phishing and Pharming. A **phishing** scam combines both spoofed e-mail and a spoofed Web site in order to trick a person into providing private information. Recall that *spoofing* is the act of assuming the identity of another. In a phishing scam the hacker sends out mass e-mailings (sometimes using zombie networks) that appear to come from a legitimate company such as PayPal, a credit card company, or a bank (see Figure 11.37). The e-mail warns of some trouble with your account and provides a link that should be clicked to address the problem. The link may look safe, such as *https://www.paypal.com/customer-service* but when you click it, it takes you to a different Web address that looks just like the real PayPal login page, so you don't notice a slightly different URL in the address box. The page looks like the PayPal page because the hacker copied the page exactly from PayPal to his own Web server.

FIGURE 11.37 • Phishing

Phishing scams use official-looking e-mail and Web pages to con people into giving private information.

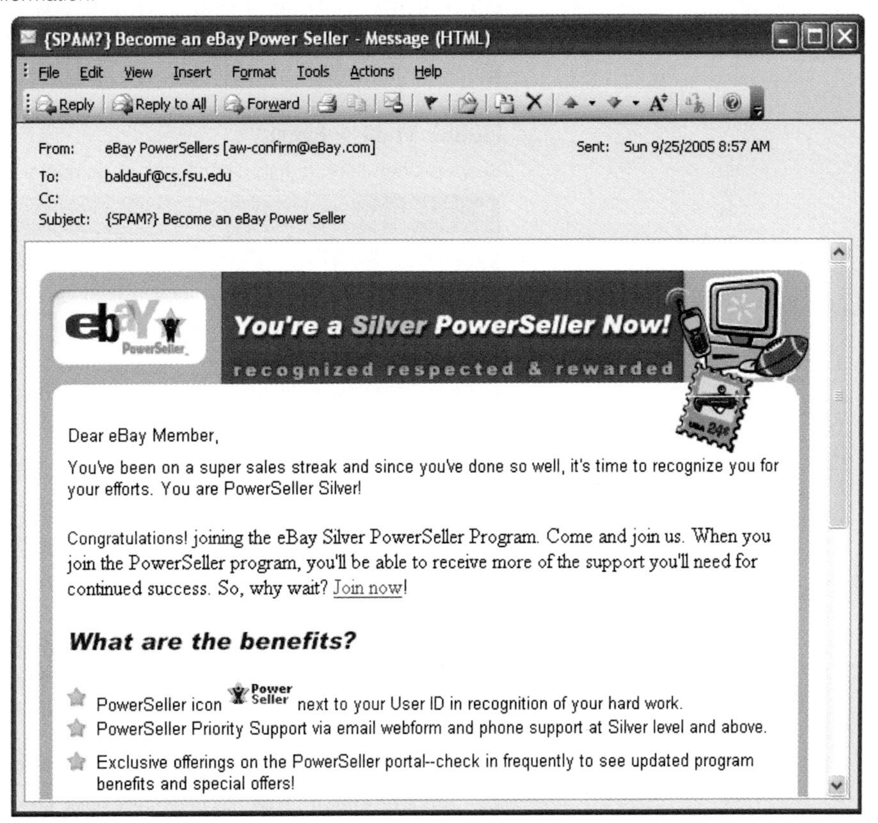

From this point in the phishing scam a number of things can occur. Most commonly the user logs in as requested. The hacker has then obtained his goal of stealing the user's login information. At this point the user gets a message like "Incorrect Password. Click here to try again." When the user clicks to try again, the software may send the user to the real login screen where the user is able to log in without any trouble. The victim is totally unaware of the fraud or identity theft that has occurred.

In addition to stealing login information and perhaps more depending on how long the user can be strung along, phishing pages can also install spyware on a computer. Over 12,000 persons per month reported being taken in by phishing scams in 2005. "The mere threat of these attacks undermines everyone's confidence in the Internet," says U.S. Senator Patrick Leahy, who introduced the Anti-Phishing

Act of 2005.[45] The bill prohibits the creation or procurement of an e-mail or Web site that represents itself as being from a legitimate business but is, in fact, sent with the intent to commit fraud or identity theft.

In the criminal act of *pharming*, hackers hijack a domain name service (DNS) server to automatically redirect users from legitimate Web sites to spoofed Web sites in the effort to steal personal information. In pharming a user may type *www.PayPal.com* into the Web browser address bar, but when the request is received at the DNS server, rather than routing the packets to the PayPal server, they are hijacked to the hacker's Web site. Pharming strikes a serious blow to the underlying architecture of the Web as it is nearly impossible to detect. The ability of hackers to corrupt DNS computers could undermine the public's faith in e-commerce and the Internet altogether.

Spam. Spam has been discussed throughout this book. It is the unsolicited junk mail that makes up more than 60 percent of today's e-mail.[46] There are laws against spam in many states and countries but they have not had much effect on the volume of spam. Because spammers have gone into partnership with hackers they are often well protected behind zombie networks and virus-infected computers (Figure 11.38).

FIGURE 11.38 • Spam

Spam is often tied up with other forms of computer crime such as fraud and spyware.

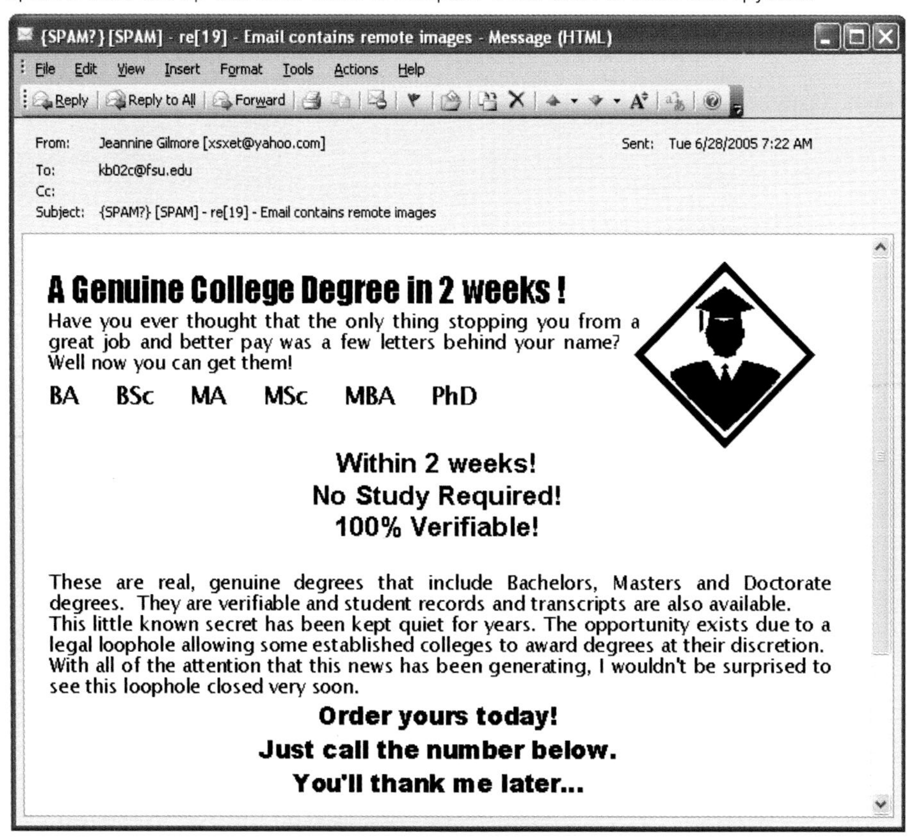

Leaders in the tech industry have teamed up to develop solutions to spam that include:

- Bayesian filters that learn to identify spam and filter it to a spam folder
- Simple authentication technology that verify the sender of a message
- "Trusted sender" technology to identify e-mail senders who can be trusted

- Reputation systems to allow everyone on the Internet to cooperate in identifying good and bad e-mail senders
- Interfaces for client-side tools to allow end users to report spam

The frustration caused by spam presents another threat to the growth of the Internet. It is hoped that a two-pronged approach that attacks spam from technical and legal perspectives can provide a solution that works.

Virus Hoaxes. A **virus hoax** is an e-mail that warns of a virus that doesn't exist. In some cases, a virus hoax is just an inconvenience or nuisance. In other cases, it can cause serious problems. If the hoax tells you to delete a "virus" file that is actually an uninfected and important system file and you delete it, you may not be able to run your computer.

Defending Against Scams, Spam, Fraud, and Hoaxes. The main defense against scams, spam, fraud, and hoaxes is awareness and common sense. If something seems too good to be true, it probably isn't true. Other important safeguards include:

- Do not click links received in e-mails. Hover your mouse pointer over links to expose the real URL.
- Examine Web addresses closely to make sure that they are legitimate and include a https:// for forms.
- Do not believe any virus alert sent through e-mail unless it comes from a verifiable source.

Many spam filters are available that can dramatically reduce the amount of spam in your inbox (Figure 11.39). If you use Web e-mail such as Hotmail, your only option is to use the spam filter of the service provider. If you use an e-mail client such as Microsoft Outlook, you can use a plug-in filter such as Spam Bayes (*http://spambayes.sourceforge.net*).

FIGURE 11.39 • E-mail filter

Web-based e-mail services often offer tools for reporting spam to improve filtering.

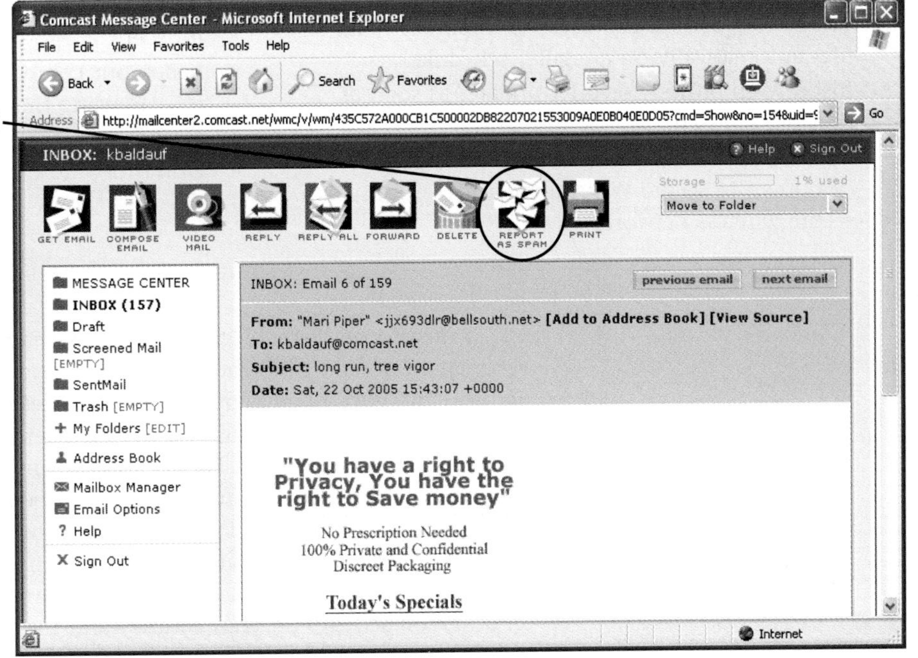

A PC Security Checklist

Keeping your computer and the information it holds safe and secure requires a two-pronged approach: applying software security tools and maintaining safe, vigilant behavior. Throughout this chapter you have been provided with a lot of advice; so much advice that you are likely unable to remember it all. The checklists here are provided in an effort to make personal information security as simple and straightforward as possible. The items in these lists are ordered by level of importance; the most important items listed first. Many of the items at the top of these lists are of equal importance.

Applying Software Security Tools

- Windows users should set Windows to update automatically.
- Install and use antivirus software.
- Install and use firewall software.
- Install and use antispyware software.
- Make sure that all security, antivirus, firewall, and antispyware software is set to update automatically.
- Use backup software to back up important data files automatically on a regular schedule.
- Install Web browser security updates as soon as they are released.
- Use encryption for private files stored on your computer, to secure Web transactions, and to secure wireless networks.
- When connected to a network, make sure that user permissions are set to share only the files you wish to share.
- If you use a home wireless network, disable SSID broadcasting on the access point, change passwords on the access point, and set it to connect with only specific MAC addresses.
- Windows users can consider using Window cleaner software to maintain the Windows registry.
- Use spam filters on your e-mail.

Maintaining Safe and Vigilant Behaviors

- Select passwords carefully and change them regularly.
- Do not open attachments unless expected and scanned for viruses.
- Do not click links received in e-mails.
- Examine Web addresses closely to make sure that they are legitimate and include a https:// for forms .
- Avoid P2P file-sharing networks.
- Avoid Web sites set up for unethical, immoral, or indecent purposes.
- Keep your computer system well organized and up to date using the housecleaning tips provided in this chapter.
- Don't forward e-mail virus alerts.

As mentioned at the beginning of this chapter, to achieve total information security, efforts must be made at multiple levels. Individuals, businesses, organizations, and governments must work together to secure independent computers, private and public networks, and the Internet. Each individual wears different security hats depending on the computing environment. In your personal lives you must be on guard to protect your own personal information and systems. When accessing business or organization networks at work and in your day-to-day lives, you must be on guard to protect those networks and the sensitive and private information they may contain. You must also be aware of the impact of information security on your country and the world.

Information security is often inconvenient. It requires time and effort that most of us would prefer not to invest. However, as this chapter has underscored, your efforts are essential to support this goal.

ACTION PLAN

Remember Jan Minski from the beginning of this chapter? She was the student that had her credit card information stolen. Here are answers to the questions about Jan's situation.

1. **What are the first steps Jan should take in her crime scene investigation to find out if indeed her computer had been broken into?**

   Jan should scan her computer for viruses, worms, and spyware using antivirus software and antispyware. She should also check her data files to see if she has credit card information stored anywhere on her computer where it might be stolen.

2. **What possible methods could a hacker have used to steal Jan's credit card information?**

   A hacker may have a virus or spyware installed on Jan's computer recording her activities. During her last online credit card purchase the illicit software may have recorded her credit card number, expiration date, and code. If Jan stored her credit card info in a data file on her computer, say in her budget spreadsheet, an intruder may have stolen the file to access her info. Another possibility is that Jan was tricked by a phishing scam to supply her credit card information to a fraudulent Web site. Or, her credit card information may have been sniffed off the Internet if she supplied it for a purchase over an unsecured Internet connection.

3. **What precautions can Jan take to make sure this doesn't happen again?**

   Jan should make sure that her operating system and Web browser are patched and up to date. She should install antivirus software, antispyware, and firewall software. She should encrypt her private data files on her computer, and be cautious when supplying credit card information online.

Summary

LEARNING OBJECTIVE 1

Describe the types of information that must be kept secure and the types of threats against them.

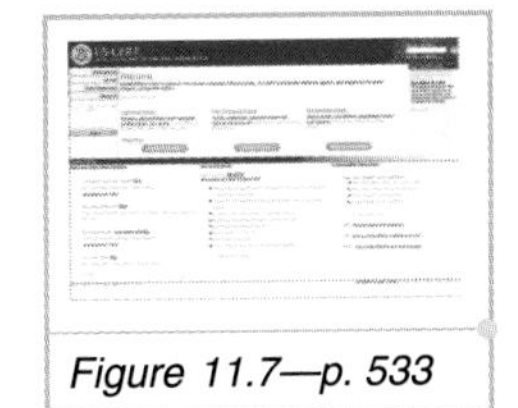
Figure 11.7—p. 533

Total information security refers to securing all components of the global digital information infrastructure from cell phones, to PCs, to business and government networks, to Internet routers and communications satellites. As a computer user, you must learn about security risks at three levels: the machine level, the network level, including wireless networks, and the Internet level. As you move from one level to the next, you face increasing exposure and risks. Personal, organizational, national, and international information is all at risk. Identity theft refers to the criminal act of using stolen information about a person to assume that person's identity. Intellectual property is also the target of crimes. Intellectual property refers to a product of the mind or intellect over which the owner holds legal entitlement. Intellectual property rights concern the ownership and use of intellectual property such as software, music, movies, data, and information. For a business or nonprofit organization, the information it processes is often highly valued and key to its success. For this reason, businesses and organizations take information security very seriously. Besides protecting its own proprietary information, a business also has a responsibility to its customers to safeguard their private information against unauthorized access. New laws are holding companies responsible for maintaining the privacy of their customer's private information. Just as businesses benefit from digital technologies and the Internet, so do governments and government agencies all over the world. Cyberterrorism is a form of terrorism that makes use of attacks over the Internet to intimidate and harm a population.

To achieve total information security, many diverse threats must be addressed. Security vulnerabilities or security holes are software bugs that allow deliberate violations to information security. Software patches are corrections to the software bugs that cause security holes. There are many situations where innocent human mistakes result in monumental problems. Pirates and plagiarists are two classifications of individuals who violate the laws regarding intellectual property rights. A hacker, cracker, attacker, or intruder is an individual who subverts computer security without authorization. Hackers often belong to groups and assist each other with new methods for hacking systems. Computer forensics is the process of examining computing equipment to determine if it has been used for illegal, unauthorized, or unusual activities.

LEARNING OBJECTIVE 2

Describe five methods of keeping a PC safe and secure.

The most basic and fundamental security is implemented at the individual machine level, the point of entry to computers, computer networks, and the Internet. Access to today's PCs is typically guarded with a username and password. In security terms, this is regarded as authentication, a manner in which to confirm the identity of a user. Passwords can be an effective form of authentication if they are difficult to guess, kept confidential, and changed regularly. A number of devices are used in corporations and organizations to provide access to restricted areas and computer systems. The most popular of these devices are ID cards and keychain ID fobs. Biometrics is the science and technology of authentication by scanning and measuring a person's unique physical features such as fingerprints, retinal patterns, and facial characteristics. Today's Apple and Windows PCs include security tools that can encrypt files stored on disks and flash drives. This is useful in situations where the information stored is confidential or valuable, and there is a possibility that your computer can be accessed by others, lost, or stolen. The most effective way to protect data and information is to back it up. Software updates, computer housekeeping, and Windows cleaners are other ways to keep a personal computer running smoothly and securely.

Figure 11.16—p. 544

LEARNING OBJECTIVE 3

Discuss the threats and defenses unique to multiuser networks.

When a computer is connected to a network, security risks increase. A multiuser system is a computer system, such as a computer network, in which multiple users share access to resources such as file systems. User permissions refer to the defined access privileges afforded to each network user. Interior threats, those that come from registered users, can be intentional or unintentional. Many instance of identity theft occur with the assistance of insiders with corporate network access. To safeguard against threats to a network's health and stability and information theft, businesses and organizations often design security and network usage policies.

Figure 11.21—p. 549

LEARNING OBJECTIVE 4

Discuss the threats and defenses unique to wireless networks.

With wireless technologies, an attacker no longer has to establish a wired connection to a network. Attackers located within the range of the wireless signal, perhaps on the floor above or in a car parked outside, can gain access to an unsecured wireless network. To make Wi-Fi easy to set up, manufacturers disable all forms of security in new Wi-Fi access points. Several steps are required to secure a Wi-fi network. Options within the configuration software allow you to disable the broadcasting of the SSID (network ID). Access points can be set to allow only certain computers to connect. Several encryption protocols exist for wireless communications. Wired Equivalent Privacy (WEP) and Wi-Fi Protected Access (WPA) are the two most popular. War driving is the act of driving through neighborhoods with a wireless notebook or handheld computer looking for unsecured Wi-fi networks.

Figure 11.26—p. 554

LEARNING OBJECTIVE 5

Describe the threats posed by hackers, viruses, spyware, frauds, and scams, and the methods of defending against them.

Attacks against Internet-connected computers may come in the form of direct attacks by hackers (system penetration), or through viruses, worms, or spyware obtained though e-mail, the Web, or from downloaded files. Internet users are also at risk of being manipulated through scams and hoaxes. A hacker can remotely install key-logging software on a computer to record all keystrokes and commands. For random attacks an attacker may employ port-scanning software to search IP addresses for ports open to attack. Hackers also apply social engineering to acquire private information. A firewall is network hardware or software that examines all incoming packets and filters out packets that are potentially dangerous. Encrypting confidential data that is stored and transmitted over a network greatly adds to your protection. A virus is a program that attaches itself to a file, spreads to other files, and delivers a destructive action called a payload. A worm does not attach itself to other programs but rather acts as a free agent, replicating itself numerous times in an effort to overwhelm systems. Antivirus software uses several techniques to find viruses on a computer system, remove them if possible, and keep additional viruses from infecting the system. Spyware is software installed on a computer without the user's knowledge to either monitor the user or to allow an outside party to control the computer. Antispyware is software that searches a computer for spyware and any software that may violate a user's privacy, allows the user to remove it, and provides continuing protection against future attacks. Internet fraud is the crime of deliberately deceiving a person over the Internet in order to damage them and to obtain property or services. A phishing scam can be a spoofed e-mail or Web site that tricks a person into providing private information. A virus hoax is e-mail sent to people warning them of a virus that doesn't exist.

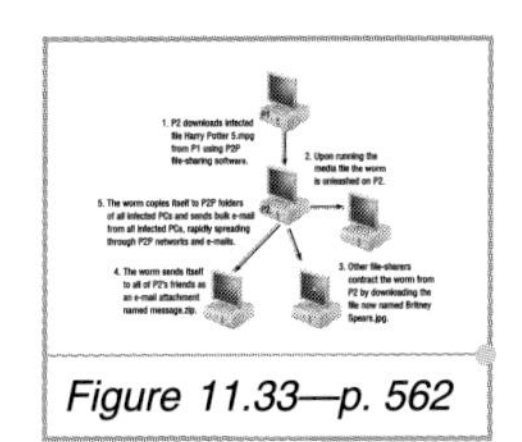

Figure 11.33—p. 562

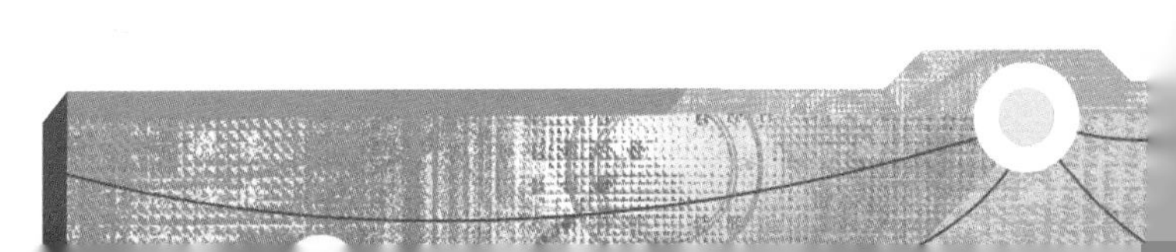

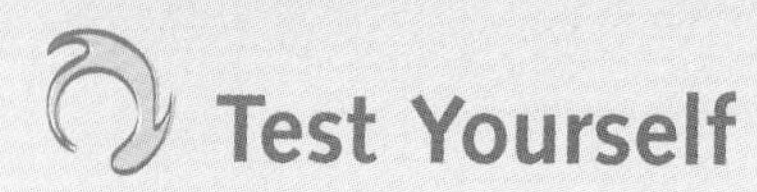

Test Yourself

LEARNING OBJECTIVE 1: Describe the types of information that must be kept secure and the types of threats against them.

1. ________________ refers to the criminal act of using stolen information about a person to assume that person's identity.
2. Illegally copying and distributing music is an issue related to ________________ .
   a. cyberterrorism
   b. intellectual property rights
   c. plagiarism
   d. identity theft
3. True or False: Businesses are fairly unrestricted in their use of information.
4. Software ________________ are bugs that allow deliberate violations to information security.

LEARNING OBJECTIVE 2: Describe five methods of keeping a PC safe and secure.

5. ________________ can be an effective form of authentication if they are difficult to guess, kept confidential, and changed regularly.
6. ________________ is a form of biometrics being used for personal computing.
   a. Facial recognition
   b. Retinal scan
   c. Usernames and passwords
   d. Fingerprint scan
7. True or False: Data files on most PCs are encrypted prior to storage by default.
8. Which of the following protects a computer from hackers, viruses, and hard disk failure?
   a. Hard drive backup
   b. virus detection
   c. firewall
   d. encryption

LEARNING OBJECTIVE 3: Discuss the threats and defenses unique to multiuser networks.

9. User ________________ is/are used to restrict a user's access to computer systems.
   a. names
   b. authentication
   c. permissions
   d. all of the above
10. True or False: Most business networks do not allow users to install software without the network administrator's authorization.
11. True or False: Employers are legally responsible for notifying employees of network usage policies.

LEARNING OBJECTIVE 4: Discuss the threats and defenses unique to wireless networks.

12. By default, Wi-Fi ________________ are set to broadcast their presence.
   a. adapters
   b. access points
   c. routers
   d. computers
13. A technique used by hackers to locate and hack Wi-Fi networks is called ________________ .
14. True or False: Wireless networks are more difficult to hack than wired networks.
15. Which of the following is not a method used to secure a wireless network?
   a. Disable SSID broadcasting.
   b. Encrypt your files.
   c. Set access point to connect to specific MAC addresses.
   d. Place access point in a secure location.

LEARNING OBJECTIVE 5: Describe the threats posed by hackers, viruses, spyware, fraud, and scams, and the methods of defending against them.

16. Hackers often enter a computer through an unprotected ________________ .
   a. user account
   b. port
   c. ISP
   d. firewall
17. True or False: Hackers always have illegal or unethical motivations.
18. A ________________ uses many computers to launch simultaneous repeated requests at a Web server in order to overwhelm it so that it is unable to function.

19. A ______________ is network hardware or software that examines all incoming packets and filters out packets that are potentially dangerous.

20. A ______________ is a program that attaches itself to a file, spreads to other files, and delivers a destructive action called a payload.
    a. virus
    b. worm
    c. Trojan horse
    d. spyware

21. A ______________ scam combines spoofed e-mail and a spoofed Web site in order to trick a person into providing private information.

Test Yourself Solutions: **1.** Identity theft, **2.** b. intellectual property rights, **3.** False, **4.** security holes, **5.** passwords, **6.** d. Fingerprint scan, **7.** False, **8.** Hard drive backup, **9.** d. all of the above, **10.** True, **11.** False, **12.** b. access points, **13.** war driving, **14.** False, **15.** d. access point in a secure location, **16.** b. port, **17.** False, **18.** distributed denial-of-service attack (DDoS), **19.** firewall, **20.** virus, **21.** phishing

Key Terms

antispyware, p. 565
antivirus software, p. 564
biometrics, p. 543
computer forensics, p. 539
cyberterrorism, p. 533
firewall, p. 560
identity theft, p. 528
intellectual property, p. 529
intellectual property rights, p. 530
Internet fraud, p. 566
password, p. 540
phishing, p. 567
piracy, p. 537
plagiarism, p. 537
security holes, p. 535
software patches, p. 536
spyware, p. 565
username, p. 540
virus, p. 561
virus hoax, p. 569
worm, p. 561
zombie computer, p. 565

Questions

Review Questions

1. Information security is concerned with what three main areas?
2. What is "total information security"?
3. What is typically stolen in cases of identity theft?
4. What is competitive intelligence, and how does it differ from counterintelligence?
5. In what ways is cyberterrorism a threat?
6. What is a software patch used for?
7. What is computer forensics?
8. What are three common forms of authentication?
9. Provide some good advice regarding choosing a password.
10. What is biometrics, and how is it used for information security?
11. What is an incremental backup?
12. What is real-time mirroring, and what convenience does it provide?
13. What service does user permissions provide?
14. What software tool is the best defense against hackers?
15. How do viruses and worms differ?
16. In what ways can spyware be acquired?
17. List four actions that can protect a wireless network.
18. What are the best ways to avoid phishing scams?
19. What is a virus hoax?
20. What are the three levels of total information security?

Discussion Questions

21. Could anything less than "total information security" provide society with complete security? Why or why not?
22. In what ways does identity theft impact our economy?
23. What items of intellectual property have you developed? Of what value are these items to you and others?
24. What issues of intellectual property rights are challenging industry, society, and lawmakers today?
25. What is business intelligence, and how is it concerned with information security?
26. Considering the cost to businesses that are passed on to customers, do you think it is necessary to have so many laws about information use and privacy? Why or why not?
27. Who shares the blame for information insecurity?
28. How do piracy and plagiarism differ? What do they have in common?
29. Why is the fingerprint scan the most popular form of biometric authentication for PCs?
30. Explain three ways that encryption can be used to secure information on PCs, networks, and the Internet.
31. What are the risks, costs, and benefits of remote data backup?
32. List several methods of maintaining a personal computer so that it runs smoothly and safely.
33. What are the concerns of users, system administrators, and business managers regarding multiuser systems?
34. What issues are addressed in a typical security and network usage policy?
35. Do you think war driving is ethically wrong? Why or why not?

Exercises

Try It Yourself

1. Use a word-processing application to create a checklist for yourself based on the PC Security Checklist provided in the Home Technology box at the end of the chapter. Check off the items that you feel you have covered adequately. Work towards total information security.
2. If you use Windows, download and install free antispyware software from the Internet (such as from *www.safer-networking.org/en/index.html* or *www.microsoft.com/downloads*) and run the software. Take a screenshot of the results page (Ctrl+PrtSc) and paste it into a word processing document.
3. Create a practice spreadsheet file—it can be empty. Save the file on your computer, then encrypt the file (use the Help feature on your operating system to find out how). Open the encrypted file to witness the password protection.

Virtual Classroom Activities

4. Have an online chat about what each group member feels is the most valuable information stored on his or her computer. Group the items as being either financially or sentimentally valuable, and decide as a group what items are most valuable.
5. Use a discussion board to post your worst computer nightmares—unfortunate events that have occurred on or to your computer that brought loss or destruction of information. If you've never had an unfortunate event, you can share stories you've heard from others.
6. Have a group scavenger hunt to find the usage policies governing your school network. The first one to find it online provides the URL and wins a prize (you determine what it is). Discuss each of the issues on the usage policy and their purpose as well as your feelings about them.

Teamwork

7. Evaluate the top (ranked by Cnet and Users) free antivirus, firewall, and spyware software listed at *www.download.com* and decide as a group which is best. Divide up the research between group members and present your findings in a report.

8. Evaluate Norton and McAfee security software and compare and contrast them against the software evaluated in the previous task. Do you feel that there is significant quality improvement for the paid subscription services over the free services? Divide up the research and present your results in a report.

9. Perform an Internet search on privacy laws for your state and country. Compile a list of the state and federal privacy laws that you find along with a short explanation of each.

TechTV

Understanding Identity, Spoofing, and Internet Attacks

Go to www.course.com/swt2/ch11 and click TechTV. Click the "Understanding Identity, Spoofing, and Internet Attacks" link to view the TechTV video, and then answer the following questions.

1. Which type of attack is most dangerous: ICMP, UDP, or TCP? Why?

2. For which types of attacks does a firewall provide protection?

3. If you have a personal computer, what type of firewall do you use? Do you feel your PC is adequately protected? Why?

Endnotes

[1] Weiss, T.R. "Data Theft Involving Four Banks Could Affect 500,000 Customers 'This thing's getting bigger and bigger,' says one police officer," *Computerworld*, May 18, 2005, http://www.computerworld.com.

[2] Pruitt, S. "Web Attack Aims to Steal Surfers' Financial Details," *ITworld*, June 25, 2004, http://www.itworld.com.

[3] "Microsoft Issues Patches for 7 Software Flaws," *Reuters*, October 12, 2004, http://www.reuters.com.

[4] "Calif. University Says 59,000 Affected by Hackers," Reuters, March 21, 2005, http://www.reuters.com.

[5] 2005 CSI/FBI survey, accessed August 2, 2005, http://www.gocsi.com/forms/fbi/csi_fbi_survey.jhtml.

[6] 2005 CSI/FBI survey, accessed August 2, 2005, http://www.gocsi.com/forms/fbi/csi_fbi_survey.jhtml.

[7] "Microsoft Agrees to Restitution for Teen," *Associated Press*, March 30, 2005, http://www.ap.org.

[8] "Teen Hacker Goes to Jail for Creating Internet Worm," News From Russia, January 31, 2005, http://newsfromrussia.com/ world/2005/01/31/58026.html.

[9] Lewin, J. "Experts Predict Devastating Attack on the Internet," ITworld.com, January 18, 2005, http://www.itworld.com.

[10] "National and State Trends in Fraud & Identity Theft January - December 2004 Report," Accessed August 2, 2005, http://www.consumer.gov/sentinel/pubs/Top10Fraud2004.pdf.

[11] Menta, R. "MP3 Portable Market to Hit $52B by 2008," *MP3 Newswire*, September 21, 2004, http://www.mp3newswire.net/stories/2004/mp3boom.html.

[12] Rhode, L. "Florida Hacker Indicted in Big Online Theft Case," *IDG News*, July 22, 2004, http://security.itworld.com/4368/040722hacker/page_1.html.

[13] Setter, K. "Bill Strives to Protect Privacy ," *Wired*, July 15, 2005, http://www.wired.com/news/privacy/0,1848,68218,00.html.

[14] Vijaya, J. "Merchants Face Deadline for Data Safety," *Computerworld*, April 25, 2005, p. 4.

[15] Abbott, K. "Computer Hacking Brings Guilty Plea," *Rocky Mountain News*, July 29, 2005, http://www.rockymountainnews.com.

[16] "CIA: Take That, Cyberterrorism!," *Wired News*, May 25, 2005, http://www.wired.com/news/politics/0,1283,67644,00.html.

[17] Illet, D. "Russian Hackers 'the Best in the World'," ZDNet UK, April 6, 2005, http://news.zdnet.co.uk/internet/security/0,39020375,39193999,00.htm.

[18] Blau, J. "Russia, - A Happy Haven for Hackers," *Computer Weekly*, May 26, 2004, http://www.computerweekly.com/Article130839.htm.
[19] See note 18 above.
[20] Microsoft's Trustworthy Computing Web site, accessed August 6, 2005, http://www.microsoft.com/mscorp/twc.
[21] "Microsoft Exploit Code Hits the Web," CIOToday, August 12, 2005, http://www.ciotoday.com.
[22] "Bad Keystroke Leads to $251M Stock Buy," *Reuters*, July 1, 2005, http://www.reuters.com.
[23] Pomerantz, D. "Hang the Pirates," *Forbes*, January 31, 2005, p. 96.
[24] "Study: iTunes More Popular than Many P2P sites," *CNET News*, June 7, 2005, http://news.news.com.
[25] Millard, E. "Microsoft Considers Hosting Regular Hacker Conferences," *CIO Today*, August 2, 2005, http://www.cio-today.com/story.xhtml?story_id=37598.
[26] Bergstein, B. "Fed Wants Banks to Strengthen Web Log-ons," *Associated Press*, October 17, 2005, http://news.yahoo.com/s/ap/20051017/ap_on_hi_te/internet_banking_security.
[27] "Cell phones Learn to Recognize Their Owners' Faces," DeviceForge.com, Oct 19, 2005, http://www.deviceforge.com/news/NS2876211743.html.
[28] Silver, C. "LexisNexis Acknowledges More ID Theft," CNN, June 2, 2005, http://money.cnn.com.
[29] Anthes, G.H. "Data: Lost, Stolen or Strayed," *Computerworld*, August 1, 2005, http://www.computerworld.com.
[30] Donoghue, A. "iPods Are Security Risk, Warns Analyst," ZDNet UK, July 05, 2004, http://news.zdnet.co.uk/internet/security.
[31] Kershaw, M. "HERT Interviews Kismet's Author, Mike Kershaw," HERT, *August 19, 2004,* http://hert.org/story.php/46/.
[32] Poulsen, K. "Guilty Plea in Kinko's Keystroke Caper," *The Register*, July 19, 2004, http://www.theregister.co.uk.
[33] Vijayan, J. "Update: Credit Card Firm Hit by DDoS Attack," *Computerworld*, September 22, 2004, http://www.computerworld.com.
[34] Vijayan, J. "Sidebar: Hacking for Profit," *Computerworld*, July 12, 2004, http://www.computerworld.com.
[35] Saran, C. "Code Issue Affects 40% of Websites," *Computer Weekly*, January 13, 2004, p. 5.
[36] "Report Finds Online Attacks Shift Toward Profit," Physorg.com, August 2, 2005, http://www.physorg.com/news5580.html.
[37] Haines, L. "US School Kids Run Amok on Internet," *The Register*, August 10, 2005, http://www.theregister.co.uk.
[38] Ward, M. "Cyber Crime Booms in 2004," *BBCNews*, December 29, 2004, http://news.bbc.co.uk.
[39] "Report Finds Online Attacks Shift Toward Profit," Physorg.com, August 2, 2005, http://www.physorg.com/news5580.html.
[40] Wirbel, L. "U.S. Steers Consumers Away From IE," eetimes, July 1, 2004, http://www.eetimes.com.
[41] Broersma, M. "New Microsoft Security System Scours Web," *TechWorld*, August 10, 2005, http://www.techworld.com.
[42] "PCs 'infested' with spy programs," BBC News, April 16, 2004, http://news.bbc.co.uk/1/hi/technology/3633167.stm.
[43] Labaton, S. "Zombies: An Army of Soulless 1's and 0's," *New York Times*, June 24, 2005 http://www.nytimes.com.
[44] McMillan, R. "Beware of 'Zombies'," *Computerworld*, May 24, 2005, http://www.computerworld.com.
[45] Mark, R. "New Senate Bill Looks to Hook Phishers," InternetNews, March 3, 2005, http://internetnews.com/security/article.php/3487271.
[46] Postini Resource Center—E-mail Stats Web page, http://www.postini.com/stats/, accessed August 12, 2005.

Digital Society, Ethics, and Globalization

John Toh considers himself a naturalist. He has one year to go to finish his degree in Civil and Environmental Engineering. John worries about technology's impact on society and the natural order of the world. He is concerned that people are losing touch with their natural environment and even with each other because of the increasing time everyone spends on computers and cell phones.

John is also concerned that large corporations and government might use technology in ways that negatively impact the environment and citizens' civil rights. He wonders if people might lose their souls and unique identities to technology and become just a list of numbers in a spreadsheet. He also worries about all the people being displaced from their jobs in the advance of technology. He could envision a bleak future where the earth was scoured of nature, where computers, or perhaps tyrant dictators empowered by computers, ran the planet, and people were forced to do their bidding.

As you read through the chapter, consider the following questions:

1. Are computers alienating people from each other and the natural environment?
2. Are computers putting power and control into the hands of a few at the expense of freedom of the many?
3. Are computers reducing the need for a human work force and overshadowing basic human needs?

Check out John's *Action Plan* at the conclusion of this chapter.

LEARNING OBJECTIVES

1. Describe how technology is affecting the definition of community, and list some physical and mental health dangers associated with excessive computer use.
2. Describe the negative and positive impact of technology on freedom of speech, and list forms of speech and expression that are censored on the Web.
3. Explain the ways in which technology is used to invade personal privacy, and provide examples of laws that protect citizens from privacy invasion.
4. List ethical issues related to digital technology that confront individuals in personal and professional life, businesses, and governments.
5. Explain what globalization is, what forces are behind it, and how it is affecting the United States and other nations.

CHAPTER CONTENT

Living Online

Freedom of Speech

Privacy Issues

Ethics and Social Responsibility

Globalization

Introduction

Digital technologies have had a profound impact on most aspects of human life. The rapid pace of technological development has given the current generations one of the most fascinating eras in which to live. Technological advances are leading to life-changing scientific breakthroughs, new business management paradigms, and a smaller, more inclusive and connected global society. The application of digital technologies to accomplish more with fewer resources is turning lives upside down in both negative and positive ways. Much of the social impact of these technologies seems to occur with little or no forethought by those responsible for developing and applying the technology. Governments are scrambling to establish laws to minimize negative impacts, while ethicists struggle to apply traditional ethical standards to brand new modes of human interaction. This chapter examines the impact of digital technologies on one's sense of community, freedom of speech, privacy, ethics, and globalization.

Consider some of the ways that digital technologies have changed an average individual's life over the past 20 years (see Figure 12.1). To make this a manageable task, consider these changes in the areas of communication, information access, commerce, professions, and leisure. Though many examples of the effects of technology have been provided throughout this book, this chapter focuses on the effects of technology from a quality-of-life perspective—what has been lost, and what has been gained.

Popular forms of communications have undergone dramatic changes over the past 20 years. Wireless technologies allow communication with friends and business associates at any time from nearly any place, whether from across town or the other side of the world. Twenty years ago, you would need to be at home, at work, or at a pay phone to chat with a friend or confer with a business associate at some other location. Back then, an executive might dictate a letter to be mailed to an overseas partner. The fax machine was just making its debut as a modern marvel. People spent a lot more time meeting and talking with others around them than those in other locations. Today you can communicate via text, voice, or video, while sharing data and information in any format, instantaneously around the world, making it possible to befriend, communicate with, and collaborate with individuals almost as though they were sitting next to you.

FIGURE 12.1 • Public library, then and now

Today's libraries would be hardly recognizable to students from the 1980s.

Your access to information and the amount of information available to you have vastly expanded over the past 20 years. Back then your primary resources in education were books, journals, textbooks, encyclopedias, and other library reference books. People received their daily news from journalists and reporters through television, radio, and newspapers. Today, in addition to these traditional forms of media and information, you can access an ever-increasing store of public knowledge and opinion on the Internet and Web. Rather than obtaining news and information from the small percentage of the population who are professional journalists, authors, and reporters, you are able to access the news and views from anyone connected to the Internet. More and more news stories are delivered directly from the individuals involved. The Web is a jumble of opinion, information, and misinformation. Blogs, discussion groups, and podcasts have provided a mouthpiece for anyone with something to say, bringing to light both hidden genius and stupidity.

In the area of commerce, technologies have provided access to vast quantities of merchandise and merchants. No longer confined to dealing with local merchants, consumers can use the Web to find the lowest price in the world. Managing money and transferring funds has never been so easy. Nor has there ever been a time when private consumer information was handled by so many individuals and stored in so many databases.

Work environments have been greatly impacted by technologies as well. Many low-level jobs have been automated or outsourced to lower-cost workers in developing countries. Certain unskilled and uneducated workers are finding it increasingly difficult to find work in the United States and other developed countries, while developing countries such as India are experiencing an economic boom. The increased competition in the global market is placing pressure on businesses to innovate like never before. Never has a college education and computer skills been so highly valued.

Leisure activities have also been affected by technology over the past 20 years. More leisure activities are "plugged in," providing people with virtual worlds as their playgrounds. A third grader is more likely to be found designing virtual cities or battling aliens than playing on the backyard swing set. Consider the leisure activities that you enjoy and the role that technology plays in delivering those activities.

FIGURE 12.2 • Distributed workforce

IBM employees coordinate teams of project workers from around the globe over high-speed network connections in the Global Services Network Center in Boulder, Colorado

The power of technology to change individual lives is considerable. Its impact is magnified when you consider its uses in businesses, organizations, and governments. Competition in the marketplace is fierce, and technology is the primary tool for gaining an edge. Businesses are continuously redesigning themselves to take advantage of the latest technological advances. Technology supports an increasingly distributed workforce, where portions of product development may be accomplished in different parts of the world (see Figure 12.2). The goal is always to produce the most attractive product at the lowest possible cost.

Most social change caused by the application of technologies has both benefits and costs. A student benefits from the ability to talk with a friend on a cell phone at the conclusion of class, but may miss out on an opportunity to meet someone new sitting in a neighboring seat. Citizens benefit from the use of technologies in law enforcement by safer living conditions, but might pay a price in terms of personal privacy. Big businesses benefit from outsourcing labor at the expense of the job market at home. The benefits and costs range in scale from personal to global. Balancing benefits against costs and determining which outweighs the other is highly personal and subjective. This chapter examines the ways that technology affects people's lives on many levels so that you can decide for yourself the benefits and costs of technology in your life.

LIVING ONLINE

As people spend increasing amounts of time engaged in virtual space rather than real space, changes occur in social structures and mechanisms. *Virtual space* may be loosely defined as an environment that exists in the mind rather than in physical space. People find virtual space in a daydream, in a book, in a movie, on a cell phone, on the Internet, or when participating in any activity that takes their minds and attentions away from their present physical surroundings. When absorbed in these activities, the here and now of the physical world can become overshadowed by involvement in the alternate space. While books, movies, and daydreams have provided private virtual space for the imagination for decades, today's networking technologies have provided a platform for building virtual space that can be shared by many. Cell phones provide the means to connect with others and, in a manner of speaking, leave the here and now. Other forms of electronic and Internet communications allow you to build virtual communities that foster relationships through electronic communications. The Internet provides an entire virtual world of information, people, and groups, in which you can literally lose yourself.

SLOW LIVING FOR DUMMIES?

Smart phones, handhelds, laptops, Blackberries, and the like have created a 24-hour society. Unlimited choices and information online should be making us happy, but sociologists have noted a rise in decision-making disorders, leading to self-blame, depression, even suicide. Time itself has become the rarest of resources, spawning "slow living" movements worldwide. Giving them momentum in research showing that the torrent of e-mail each day distracts workers to such a degree that their IQs drop by 10 points.

Source:
Time to Switch Off and Slow Down
By Kevin Anderson
BBC News Web site
http://news.bbc.co.uk/1/hi/technology/4682123.stm
July 14, 2005

FIGURE 12.3 • Flash mob

Over 150 flash mobbers spin in circles while crossing busy Market Street in San Francisco. They repeated their behavior for 10 minutes then left.

FIGURE 12.4 • Cell phone etiquette

Nothing can reflect more negatively upon a person than poor cell phone manners.

Computers and Community

On June 3, 2003, more than 100 people converged upon the ninth floor of Macy's New York City department store and gathered around a very expensive rug. When asked by sales assistants what they were doing, each individual answered that they lived together in a warehouse on the outskirts of town and were shopping for a rug. The group dispersed as quickly as it arrived. Later over 200 people flooded the lobby of the New York Hyatt, applauded for 15 seconds, and then disappeared. In both cases, the reality of the matter was that most of the participants had never even met. These episodes marked the beginning of a new fad called flash mobs.

A *flash mob* is a group of people who assemble suddenly in a public place, do something unusual, and then disperse. Flash mobs organize through cell phone text messaging or e-mail. Participants are given precise instructions on where to meet, how to act, and what to say if questioned. Since 2003 many flash mobs have appeared around the world, confused the general public, and disappeared. There have been numerous flash mob pranks in the UK and Europe, Asia, Latin America, and Australia. In India, a large group gathered in a shopping center, talked loudly about stock prices, danced for a few minutes, and dispersed. In December 2004, in Bucharest, Romania, a flash mob assembled to protest censorship in the news media. In 2005 the flash mob concept was applied to a concert tour promotion. Fusion Flash Concerts (*www.fusionflashconcerts.com*) scheduled 10 free concerts in 10 cities across the United States by 10 popular bands with dates, times, and locations to be announced via e-mail or cell phone as the performance approached. Flash mobs serve as an interesting example of the impact of digital technologies on the concept of community. See Figure 12.3.

Increasing amounts of correspondence between friends and acquaintances take place electronically through text messaging, cell phones, e-mail, and other Internet communications. Many of today's relationships are maintained more through electronic communications than face to face. People collaborate on projects, carry out complicated business transactions, meet, and even fall in love in virtual space without ever physically meeting. Communities are increasingly defined by electronic buddy lists, e-mail groups, and blogs, and less by those with whom physical space is shared.

Virtual communities and "anywhere, anytime" communications are also affecting traditional social mechanisms. The new cell phone generation is less concerned with making formal social plans and tends to be more spontaneous than previous generations. Meeting times and locations can change en route. The days of being stranded or stood up due to miscommunication are all but over: just pick up your phone and get an update.

The use of cell phones in public places sometimes offends traditional courtesy and etiquette (see Figure 12.4).

Table 12.1 lists some common dos and don'ts of cell phone etiquette.

TABLE 12.1 • Cell phone etiquette

Ten common courtesies that show respect to those you speak to on the phone as well as those with whom you share physical space.

Topic	Courtesy tip
1. Safety	Do not dial your cell phone while driving or performing other dangerous tasks that require your full attention
2. Volume	Speak softly, at a volume that cannot be overheard by others. Unlike traditional phones, cell phones do not allow you to hear the volume of your own voice on the line. People often speak louder than they need to on cell phones in an effort to be heard.
3. Proximity	Do not intrude on others' personal space. Ten feet (3 m) is the minimum distance that should be maintained between you and others while on the phone.
4. Content	Keep your business private. Typically, others do not wish to hear about the details of your life and find it offensive to have to listen to them. Still others who may be interested in your private affairs should cause you concern.
5. Tone	Keep a civil and pleasant tone. Never lose your temper while on a cell phone in a public place; it does not reflect well on you.
6. Location	Know when to make and accept calls based on your strength of signal. Do not engage in important conversations in areas where the signal is weak. Losing a signal during a conversation can be construed as an insult to the person you phoned.
7. Timing	There is a time and place for cell phone conversations. Most people feel that restaurants, classrooms, concerts, churches, and other public places where personal space is tight are not among them. Turn cell phone ringers off in any such location.
8. Multitasking	Concentrate on one thing at a time. Those you speak to on the phone appreciate your undivided attention while those sharing your public space also deserve your undivided attention
9. Courtesy	Do not interrupt a face-to-face conversation to take a cell phone call. Make sure that your phone call does not inconvenience those around you.
10. Inform	Inform those you call or who phone you that you are on a cell phone and where you are, so they can anticipate distractions or disconnections.

Meeting people has never been easier. Rather than finding friends through chance encounters, the Internet provides a means to meet and get to know many individuals with similar interests. The entire framework of the medium is designed to create connections between people. Social Web sites like *www.facebook.com*, blogs, and other Internet forums make it easy to find others with similar interests, while Internet communications make it quick and easy to meet and get to know friends of friends of friends. If you are looking for romance, you'll find no shortage of online services designed to find Mr. or Ms. Right.

QUIET CAMPUSES

Personal technology has brought startling changes to college campuses and has sparked a debate over whether the wired generation is actually becoming more isolated. Students who once blasted music from their dorm rooms are now wearing iPods; advisors meet with students via instant messaging instead of in person. "It's so noticeable now," said Jerry Rinehart, Vice Provost for student affairs at the University of Minnesota. "Classes come out, and the phones come out. If you see a group of four students, three will be on their cell phones talking to somebody, and it's not each other."

Source:
College students are wired, but can they connect?
By Mary Jane Smetanka
The Minneapolis Star-Tribune
http://www.startribune.com/stories/789/5645266.html
October 2, 2005

Health Issues: Keeping a Balance

Spending more time in virtual communities and in virtual space has its benefits and costs. The benefits have been emphasized throughout this book. There are a number of psychological and physical health concerns that have arisen due to the amount of time people are spending in contact with computers and other devices, and in virtual space. It should be noted that the health problems created by the use of technology pale in comparison to the health solutions that technology provides.

Physical Health Concerns. Working and living with computers and digital technologies can lead to potential physical health problems. As people increasingly use computers at work and at home, more people are suffering from computer-related health problems. Insurance claims relating to repetitive motion disorder, which can be caused by working with computer keyboards and other equipment, have increased greatly in recent years. *Repetitive stress injury (RSI)* is an injury such as tendonitis and tennis elbow, caused by a repetitive motion. The most common RSI for computer users is carpal tunnel syndrome. **Carpal tunnel syndrome (CTS)** is the aggravation of the pathway for nerves that travel through the wrist (the carpal tunnel), typically caused by long hours at the computer keyboard with wrists cocked and fingers typing. CTS can cause wrist pain, a feeling of tingling and numbness, and difficulty in grasping and holding objects. CTS can be difficult to correct. Use of a wrist brace (see Figure 12.5) is typically prescribed, and sometimes surgery is needed.

FIGURE 12.5 • Carpal tunnel syndrome

Long hours of typing in an awkward position can lead to carpal tunnel syndrome, which can sometimes be corrected with a wrist brace.

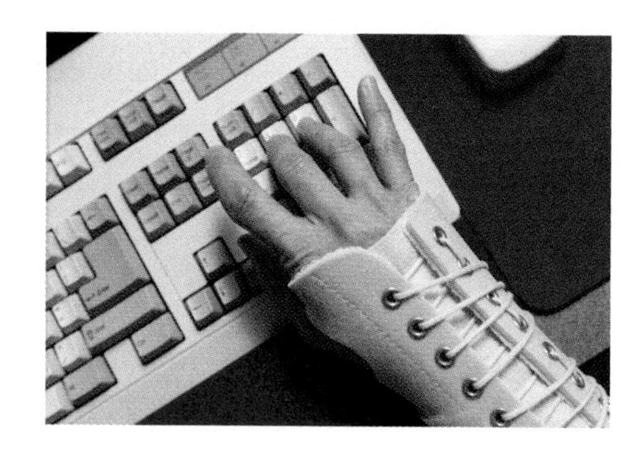

Working long hours staring at a computer screen without proper light can cause a variety of vision problems. In some cases, your eyes get tired, itch, or even burn. In more severe cases, double or blurred vision can result, making it unpleasant to work and reducing your efficiency. In addition to wrists and eyes, you can get a sore back, sore arms, and headaches from long hours working with computer systems without adequate breaks. Even children can get CTS or other physical problems from long sessions with computers.

There are concerns about the effects of radiation emanating from cell phones on the body. One Swedish study suggests that people who use cell phones for more than 10 years may develop a benign (noncancerous) tumor on a nerve in the head where the cell phone is held when used.[1] The Cellular Telecommunications Industry Association (CTIA) defines a metric called the specific absorption rate or SAR to measure the quantity of radiofrequency energy that is absorbed by the body. For a cell phone to pass FCC certification it must have a SAR level less than 1.6 watts per kilogram (W/kg). In Europe, the level is capped at 2 W/kg. Table 12.2 shows the FCC SAR ratings of phones with the highest and lowest radiation metrics.[2]

TABLE 12.2 • SAR ratings of cell phone radiation

Ten highest radiation cell phones		Ten lowest radiation cell phones	
Phone	*SAR*	*Phone*	*SAR*
Motorola V120c	1.55	Audiovox PPC66001	0.12
Motorola V265	1.55	Motorola MPx200	0.20
Motorola V70	1.54	Motorola Timeport L7089	0.22
Motorola P8767	1.53	Qualcomm pdQ-1900	0.26
Motorola ST7868	1.53	T-Mobile Sidekick	0.27
Motorola ST7868W	1.53	Samsung SGH-S100	0.29
Motorola A845	1.51	Samsung SGH-S105	0.29
Panasonic Allure	1.51	Sony Ericsson Z600	0.31
Treo 650 GSM	1.51	Mitsubishi G360	0.32
Sony Ericsson P910	1.50	Siemens S40	0.33

Mental Health and Related Problems. In additional to physical health concerns, the use of computer systems can cause a variety of mental-health issues and related problems. These include gambling, information overload and stress, and Internet addiction and isolation.

Compulsive gambling is a problem for some people using the Internet. Gambling is one of the largest businesses on the Internet. Although Internet gambling is illegal in the United States, it is not illegal in other countries. As a result, Web sites have surfaced in foreign countries that permit gambling on the Internet and have combined to build an $8 billion market.[3] Compulsive gamblers find using the Internet to be much easier to use than driving to a gambling casino. Internet gambling is even becoming available on cell phones.[4]

For some people, working with computers can cause occupational stress. Feelings of job insecurity, loss of control, incompetence, and demotion are just a few of the fears people might experience. These fears could become a serious problem for some employees. Some experts believe this fear might cause threatened workers to sabotage computer systems and equipment. If a manager determines that an employee has this type of fear, training and counseling can often help the employee and avoid potential problems.

The use of the Internet can be addictive. Some people spend most of their time connected to the Internet and staring at the computer screen (see Figure 12.6). When this behavior becomes compulsive, it can interfere with normal daily activities, including work, relationships with others, and other activities. Often, Internet addiction means that a person is isolated and doesn't interact with other people, unless they are also online. *Internet addiction* may exist if people are online for long periods of time, cannot control their online usage, jeopardize their career or family life from excessive Internet usage, and lie to family, friends, and coworkers about excessive Internet usage. Some organizations are investigating Internet addiction prevention and education programs.[5]

FIGURE 12.6 • Internet addiction

Internet addiction can cause otherwise normal people to become withdrawn and isolated.

Avoiding Health Problems. Many computer-related physical health problems are minor and are caused by a poorly designed work environment. The computer screen may be hard to read, with the problems of glare and poor contrast. Desks and chairs may be uncomfortable. Keyboards and computer screens may be fixed and difficult or impossible to move. The hazardous activities associated with these types of unfavorable conditions are collectively referred to as *work stressors*. Although these problems may not be of major concern to casual users of computer systems, continued stressors such as repetitive motion, awkward posture, and eyestrain can cause more serious and long-term injuries. If nothing else, these problems can limit productivity and performance.

The study of designing and positioning work environment and computer equipment in a healthy manner, called **ergonomics**, has suggested a number of approaches to reducing these health problems. The slope of the keyboard, the position and design of display screens, and the placement and design of computer tables and chairs have been carefully studied (see Figure 12.7). Flexibility is a major component of ergonomics and an important feature of computer devices. People of differing sizes and tastes require different positioning of equipment for best results. Some people, for example, want to have the keyboard in their laps; others prefer to place the keyboard on a solid table. Because of these individual differences, computer designers are attempting to develop systems that provide a great deal of flexibility. America Online (AOL) uses ergonomic evaluations and training to avoid injuries and reduce workers compensation claims. AOL's program has helped reduce carpal tunnel-related problems and claims.[6]

FIGURE 12.7 • Ergonomics

People who use computers for long stretches of time should be mindful of good ergonomic practices.

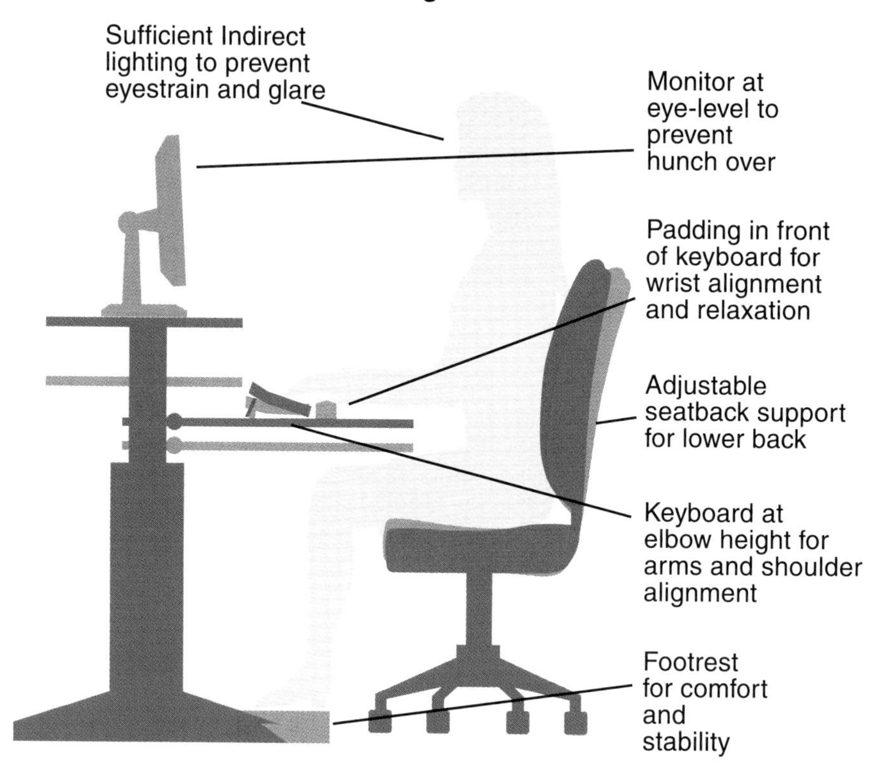
Features of an Ergonomic PC Station
Sufficient Indirect lighting to prevent eyestrain and glare
Monitor at eye-level to prevent hunch over
Padding in front of keyboard for wrist alignment and relaxation
Adjustable seatback support for lower back
Keyboard at elbow height for arms and shoulder alignment
Footrest for comfort and stability

Many feel that the best way to avoid or recover from any computer-related health issue, physical or mental, is to live a balanced life. A balanced life includes technology time and time away from technology; time alone and time spent in the company of others; time on the phone and time in face-to-face interaction; time indoors and time with nature. Some psychologists feel that many young people suffer from nature deficit disorder. They believe that people tend to stay healthier when they stay in touch with nature. The more balance you provide in your life, the more balanced your mental and physical health and development will be.

FREEDOM OF SPEECH

The First Amendment to the U.S. Constitution guarantees citizens the right to free speech. The Internet and Web have provided people with the ability to communicate their views to a greater extent and at a broader reach than has ever before been possible, and many are taking full advantage of this freedom. There are Web sites, blogs, discussion groups, chat rooms, and a growing list of podcasts to cover every conceivable topic and point of view. The Internet is used to espouse the views and beliefs of every religious and political group. Web sites may include points of view that many find offensive and even dangerous, such as Web sites that support or promote suicide or terrorism. This section examines the social implications of freedom of speech on the Internet and how varying societies deal with controlling what is said.

Challenging the Establishment and Traditional Institutions

The Internet has been grasped as a tool to empower those who have traditionally been without a public voice. Consider Web sites such as Complaints.com and the Rip-off Report sponsored by the Bad Business Bureau (*www.badbusinessbureau.com*). Customers who feel they have been ripped off by a business post complaints that act as warnings to others. This Web service stores the complaints registered to create a working history of a particular business and identify habitual bad business practices. Visitors can run searches to see all complaints against a particular business.

Bloggers are becoming an increasingly important component in journalism. This was evident in the case of Dan Rather's 2004 story on the U.S. President's military service records (see Figure 12.8). Bloggers from the conservative forum FreeRepublic.com were quick to disqualify the documents used as evidence in the story as forgeries, causing Rather and CBS a great deal of embarrassment.[7] Rather left his job shortly after the incident. Bloggers also contributed to the abrupt resignation of CNN chief news executive Eason Jordan when they created a "blogswarm" of attention over comments made at what was supposed to be a private World Economic Forum meeting.

FIGURE 12.8 • Dan Rather of CBS News

Professional journalists such as Dan Rather are being challenged by an empowered public who are able to gather facts for themselves and present them online.

Are bloggers "pseudo-journalist lynch mobs" with "no credentials, no sources, no rules, no editors, and no accountability" who interfere with and work to destroy mainstream media to promote their own agenda, as some professional journalists feel? Or, are they a "democratic counterbalance to media arrogance and a much needed call for greater transparency in the media"?[8] Whichever opinion you side with, it is clear that the blogger journalist revolution continues to grow.

The proliferation of digital cameras and camera phones and the ability to transfer digital photos easily has led to many amateur photographers having their photos published. More and more amateur photos are gracing the front pages of newspapers because a member of the public with a digital camera is often at the scene of a breaking story before the press. Twice in one month the biggest Dutch newspaper, *De Telegraaf*, published front-page pictures shot by amateur photographers using their mobile phones.[9] A passerby is often able to get shots that professionals can't, simply by being in the right place at the right time. Much of the film coverage of the Columbia Space Shuttle breaking up over Texas that was presented on television was captured by amateurs on home digital camcorders.

The role of the public in capturing and reporting news, called **consumer-generated media (CGM)**, has vastly increased over the past 10 years. Some are embracing CGM as the way of the future. "Current" is a new television network that uses CGM (Figure 12.9). Viewers contribute a portion of the network's content over the Internet in 2- to 7-minute *vlogs*—video logs—which are woven together into programs. Vlogs, called *pods* by the network, cover topics such as travel, current events, jobs, and technology. Shows can be viewed on television or on the Web site *www.current.tv*. Visitors to the Web site decide which of the

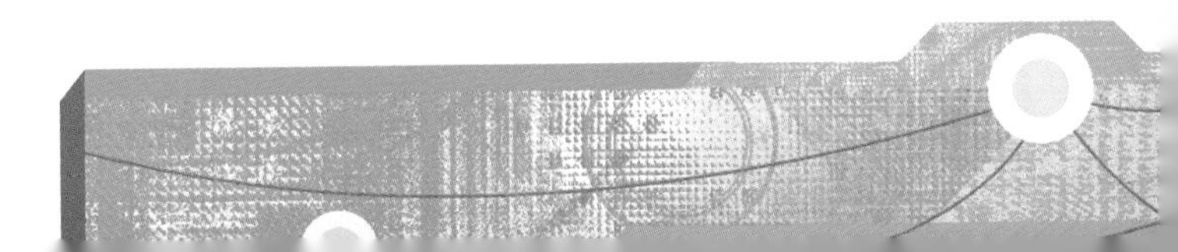

submitted vlogs is broadcast. One of Current's programs is based on whatever is the most popular Google search of the day.

FIGURE 12.9 • Current TV

Current TV empowers the public to create television programming over the Internet to be broadcast on their cable TV network.

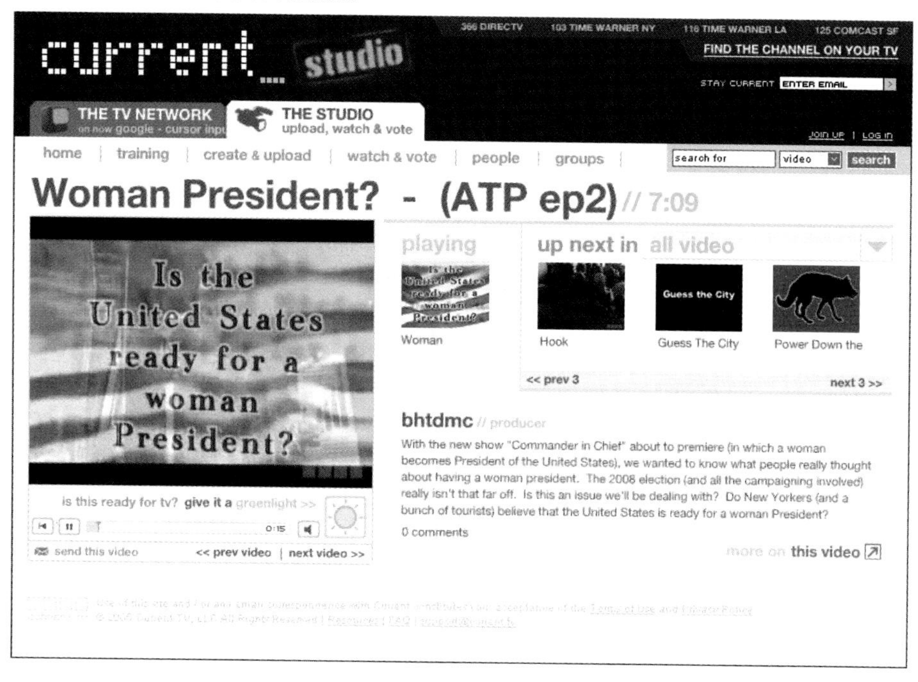

Laws and Censorship

While the First Amendment protects free speech in the United States, there is some restriction placed on speech. *Libel* is the deliberate act of defamation of character by making false statements of fact. Libel and direct, specific threats are not protected under the First Amendment.

Of course, laws regarding speech vary from country to country. When a government or authority controls speech and other forms of expression, it is called **censorship**. Various forms of censorship exist around the world.

Political Freedom. Freedom of speech and Internet technologies are most threatening to oppressive governments whose citizens lack political and social freedom. In 2000 Burma banned the use of the Internet and the creation of Web pages deemed harmful to the government's policies. Vietnam uses Internet filtering technology to block anticommunist communications. The Chinese government channels all Internet traffic through a small number of monitored gateways in order to more easily control what is sent into and out of the country.

It is difficult to control the flow of information over the Internet. Individuals can set up their own satellite dishes to bypass government monitoring. For this reason, some governments make it illegal to have a satellite dish. Cuba does not permit its citizens to own computers.

Pornography and Issues of Decency. Most countries support the Internet's ability to empower its citizens, but struggle with issues regarding the perceived negative aspects. One major concern is keeping indecent content from minors.

Because the Internet does not have a rating system like motion pictures and television, any one who can connect to the Internet can theoretically view any content there. With increasing numbers of very young children making the Internet a part of their daily lives, it is natural for parents and governments to wish to protect them from viewing content that is inappropriate and dangerous.

Australia has a commonwealth law that holds Internet service providers and Internet content hosts responsible for deleting content from their servers that is deemed "objectionable" or "unsuitable for minors" on receipt of a take-down notice from the government regulator, the Australian Broadcasting Authority.[10] Some Australians feel that the law has failed to reduce the availability of pornography as it is still readily available from other countries.

The U.S. government has made similar attempts to eliminate indecent content from the Web with its 1995 Communications and Decency Act. The law was repealed due to the government's inability to define terms such as *indecent*, *obscene*, and *lewd*, on which the law was based. One person's obscene may be another person's work of art. During its brief enactment, the law had a serious effect on legitimate and useful Web sites that may have been considered indecent by terms of the law. Family planning Web sites, medical Web sites, and art and literature Web sites pulled their content in fear of prosecution. For example, it was difficult to find information regarding breast cancer on the Web while the law was in effect.

FIGURE 12.10 • Content-filtering software

Software such as Net Nanny allows parents and administrators to filter out objectionable Web content.

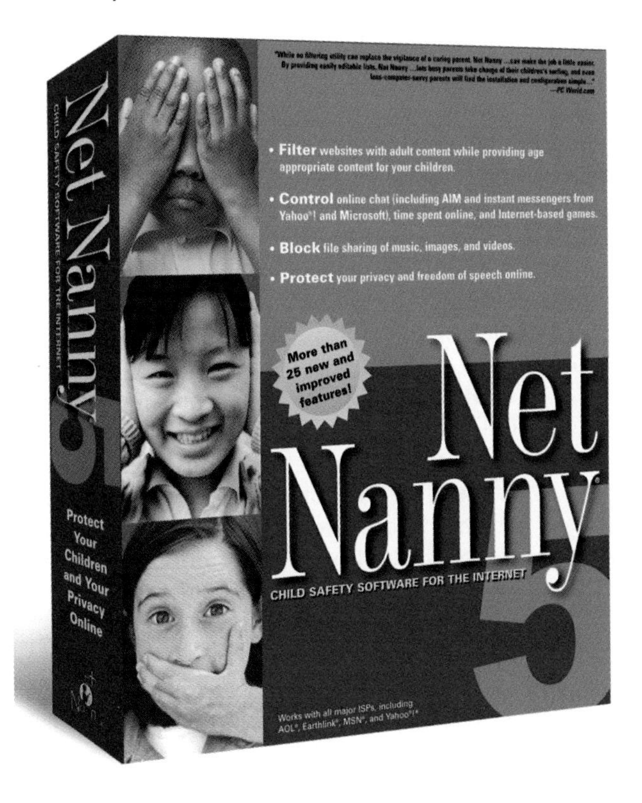

The challenge of censorship is in keeping certain content (perhaps pornography) from a subset of the population (minors), without encroaching upon the freedom of adults. One solution is content-filtering software. **Content-filtering software** works with the Web browser to check each Web site for indecent materials (defined by the installer of the software) and only allows "decent" Web pages to be displayed (see Figure 12.10).

Child pornography goes beyond being indecent and is unlawful in most countries. Many criminals have gone to jail for producing, publishing, and viewing child pornography. British Telecommunications (BT) has applied content-filtering software to the entire British Internet infrastructure to block access to child pornography Web sites. They were shocked to discover that there were over 23,000 attempts each day to access child pornography Web sites.[11]

Content-filtering software is ideal for situations where one person is responsible for setting the rules and defining what is allowable and not. For example, at home parents may use filtering to block out what they consider inappropriate for their children. In the workplace, management may use such software to filter out nonbusiness-related Web sites. Such software becomes problematic in larger democratic situations where definitions of decency may vary. For example, the 2000 Children's Internet Protection Act requires schools and libraries that receive federal funding for technology to implement content filtering. The law created a stir in the public library system when it was discovered that filters block access to many valuable nonpornographic Web sites such as:

- Sites about Middlesex University in London and the University of Essex
- The court decision about the Communications Decency Act
- Student organizations at Carnegie Mellon University

- A Robert Frost poem that includes the phrase "My little horse must think it queer to stop without a farmhouse near."
- The home page of Yale University's Biology Department
- A map of Disney World
- The Heritage Foundation (a conservative think tank)

Libraries bound to their own Library Bill of Rights, which opposes restrictions based on age, were forced to find creative strategies to meet the letter of the law while providing the maximum amount of access to adults.

Dangerous Information. Some information is censored because it is deemed to be dangerous to the public. For example, it is illegal in the United States to make certain encryption technologies available to certain foreign governments. This is in an effort to keep potentially dangerous foreign governments from using U.S. technologies to decrypt national secrets.

After the shootings at Columbine High School in Littleton, Colorado (Figure 12.11), the U.S. Congress passed a law mandating 20 years in prison for anyone distributing bomb making information with the intent to cause violence. Because explosives have numerous industrial uses, there remain many Web sites that contain bomb making instructions. The Internet is not the only method for obtaining such information, however. The Encyclopedia Britannica includes bomb making instructions as does a booklet published by the U.S. Department of Agriculture. The same types of explosives used by farmers to remove tree stumps were used in the Oklahoma City bombing in which 168 people lost their lives. Once again this illustrates the difficulty of censoring information that is valuable for both legal and illegal purposes.

FIGURE 12.11 • Columbine High School massacre

The gunmen responsible for the shootings at Columbine High School created a bomb following instructions found on the Internet and had planned to use it to blow up the school.

Censorship is a hot topic in the scientific research community as an increasing number of scientific publications are being censored on the grounds that they are a threat to national security. Recently the National Academy of Sciences suspended the publication of an article in its journal that described the risk of terrorists poisoning the nation's milk supply using botulinum toxin.[12] The Department of Health and Human Services warned that information in the article provided a "road map for terrorists." The purpose of the article was to better inform those in the dairy industry of the threat so that they could take precautions. While suspending the publication may have kept that information out of the hands of terrorists, it also kept the information away from the people who could prevent the poisoning from taking place.

These examples illustrate the difficulty in censoring public speech. Censorship typically includes an infringement on an individual's rights in exchange for a perceived greater public good. Because concepts such as decency differ in definition from person to person, any government that attempts to define these terms for its citizens risks alienating a percentage of the population. Censorship often contradicts

BAD NEWS/GOOD NEWS

Efficiency, ease, and low cost have made e-mail the weapon of choice in contacting federal lawmakers for constituents and lobbyist alike. Between 2000 and 2004, e-mails to the House doubled (99 million), and tripled to the Senate (83 million), but staff to process them has not. Most go unanswered, and lawmakers send letters when they do because of the political danger of e-mails being altered once they've been received. The upside is it has made lawmakers more responsive.

Source:
On Capitol Hill, the Inboxes Are Overflowing
By Jeffrey H. Birnbaum
The Washington Post
http://www.washingtonpost.com/wp-dyn/content/article/2005/07/10/AR2005071001011.html
July 11, 2005

JOB TECHNOLOGY

Can Businesses Sue or Fire Employees over Blogging?

Freedom for employees has recently become an issue, as big corporations have taken legal action against employees for speaking their minds on the Internet. The first case involved Ken Hamidi, a disgruntled former Intel employee who helped form FACE (Former and Current Employees) of Intel. FACE of Intel bitterly objected to Intel's personnel practices and hoped to act to reform them. Intel sued Hamidi for sending e-mail to thousands of Intel employees containing negative claims about the company, and won an injunction against him. After several years in court, in 2004 Hamidi finally won on the basis of his First Amendment rights.

More recently, Apple Computer was outraged to find its secret plans for new products subject matter for bloggers on three Web sites. Apple took the three Web site operators to court to get them to reveal their sources of information. The case presented an important and precedent-setting question: Should bloggers get the same legal protection as professional journalists when it comes to confidentiality and protecting sources? The Santa Clara County Superior Court ruled in favor of Apple, saying that the three operators must divulge their confidential sources. The ruling is now being appealed. If Apple finds out who leaked the information, they will sue the parties for revealing information protected under trade secret law.

Can employers really fire you or sue you for what you say outside of work? Apparently so. Take for example the flight attendant for Delta Air Lines who lost her job for posting photos on the Internet that featured her almost wearing a Delta flight attendant uniform. A Google employee was fired for discussing the company's financial condition online. The ability of employees to express themselves and broadcast information to millions over the Internet is freedom of speech that makes many employers nervous. While that freedom of speech cannot be denied, employers typically have the right to fire employees who decide to use that freedom against them. If company secrets are divulged, an employee may lose more than just a job; he or she may become the center of an expensive lawsuit.

Questions

1. Should there be a law that protects employees from employer retribution for things employees say or do while off the job site? Why or why not?
2. Why do you think Apple felt as though it had the right to take the bloggers to court and to sue their own employee(s)?
3. What might an employer do to use blogs to their own advantage rather than fearing them?

Sources

1. *Markoff, J. "To Cut Online Chatter, Apple Goes to Court," New York Times, March 20, 2005, http://www.nytimes.com*
2. *Zeller, T. "When the Blogger Blogs, Can the Employer Intervene?"* New York Times, *April 18, 2005, http://www.nytimes.com.*

the basic tenets of societies that value freedom and individual rights. Censorship could backfire if stemming the flow of information impedes people's ability to design effective solutions.

PRIVACY ISSUES

Computer technology provides us with the ability to collect, maintain, process, and transfer much more information than has ever before been possible. This power has given rise to a seemingly endless number of public and private databases that include details about many individuals' private matters. Combined,

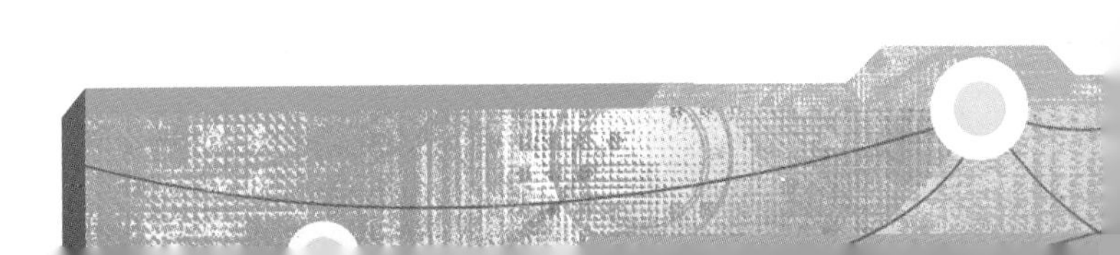

these databases could tell a person's life story in terms of daily activities, personal interests, and affairs. This combined with surveillance technologies that include the increasing use of cameras in public places (see Figure 12.12) has given rise to legitimate concerns over the invasion of privacy in the digital era.

FIGURE 12.12 • Video surveillance

Increasing numbers of video cameras in public places can make you feel as though you are always being watched.

Privacy issues that concern most people include being free from intrusion—the right to be left alone, freedom from surveillance, and control over the information collected and kept about one's self. The previous chapter touched on this last issue of protecting individuals from identity theft. The point was made that security often comes at the cost of some level of convenience and privacy. This section looks at the extent to which your privacy may be sacrificed in order to provide conveniences offered by the digital world and to increase your personal safety and national security.

EXPAND YOUR KNOWLEDGE

To learn more about protecting your privacy, go to www.course.com/swt2/ch12. Click the link "Expand Your Knowledge" and then complete the lab entitled "Protecting Your Privacy Online."

Personal Information Privacy

Much of the information gathered about individuals is done without their knowledge. This invisible information gathering takes many forms. For example, a person might join a discount club at a local grocery store to enjoy special deals on products and to facilitate a faster checkout process. That person might not be aware that the club membership card also allows the store to digitally track his or her buying patterns. The customer may receive special mailings providing information on products that he or she typically buys. Some customers find this to be a valuable service, while others consider it an invasion of privacy.

The Internet acts as a supercharged tool for invisible information gathering. Through the use of cookies, Web companies can accumulate immense amounts of information about customers visiting their Web sites. This type of *computer profiling* is the primary service provided by private information service companies such as

ChoicePoint (see Figure 12.13). ChoicePoint collects and combines information from the three big credit bureaus; public records of numerous local, state, and federal government agencies; telephone records; liens; deeds; and other sources to develop detailed information about individuals, companies, and organizations. Over the years ChoicePoint has purchased many other large personal information services, increasing their database to include drug test records, physician backgrounds, insurance fraud information, and a host of other specialized pieces of valuable personal information. Businesses and organizations contract ChoicePoint to provide information on specific individuals for a variety of uses.

FIGURE 12.13 • ChoicePoint information services

Using ChoicePoint's powerful search capabilities, businesses and government agencies can search over 10 billion records on individuals and businesses.

ChoicePoint has more information about U.S. citizens than the government. As a matter of fact, it has a multimillion-dollar contract with the Justice Department and the IRS. FBI agents consult information supplied by ChoicePoint when involved in criminal investigations.

Privacy and Government

The previous chapter presented some of the laws that govern the privacy of consumer and medical records for businesses and health care companies. Government agencies in the United States and many other democratic countries are regulated far more stringently than businesses and professionals in health care when it comes to the privacy of confidential records.

The Privacy Act of 1974 is the primary law controlling what many government agencies can and cannot do with the information they hold. The primary tenets of the law include the rights of citizens to know what information certain government agencies store about them, and exercise control over the accuracy of that information and how it is used. Other laws that control the U.S. government's handling of information are listed in Table 12.3.

In cases of war, governments can set aside some restrictions over privacy in an effort to capture or defeat the enemy. Such has been the case since the attacks of September 11, 2001, and the beginning of the war on terror. The USA

TABLE 12.3 • U.S. federal privacy laws

Law	Explanation
Educational Privacy Act of 2003	Restricts the collection of data by federally funded schools
USA PATRIOT Act of 2002	Provides the government with the power to view the customer records of Internet service providers and telephone companies without a court order in cases of terrorist investigations
Computer Matching and Privacy Act of 1988	Regulates cross-references of data between federal agencies
Tax Reform Act of 1979	Controls the collection and use of certain information collected by the IRS
Right to Financial Privacy Act of 1978	Restricts the government's access to certain financial records maintained by financial institutions
Freedom of Information Act of 1970	Gives citizens the right to view their own personal records maintained by federal agencies

PATRIOT Act gave the federal government certain liberties regarding access to private information and the treatment of suspected terrorists. It was designed to "deter and punish terrorist acts in the United States and around the world, to enhance law enforcement investigatory tools, and for other purposes."[13] It includes measures to "enhance surveillance procedures."

The easing of government regulations in time of war makes many privacy advocates nervous. They fear that in a state of panic, the government may sacrifice civil liberties and rights. The Fourth Amendment to the U.S. Constitution was created to guarantee a right to privacy: "The right of the people to be secure in their persons, houses, papers, and effects, against unreasonable searches and seizures, shall not be violated, and no Warrants shall issue, but upon probable cause, supported by Oath or affirmation, and particularly describing the place to be searched, and the persons or things to be seized."

When DARPA's Information Awareness Office proposed a new tracking information system called Total Information Awareness (TIA), privacy advocates complained loudly. Total Information Awareness was designed to capture the "information signature" of people so that the government could track potential terrorists and criminals. An information signature is any unique information stored about an individual such as information about property, boats, address history, utility connections, bankruptcies, liens and business filings, as well as a host of other information.[14] Data-mining techniques were to be applied to the database developed by the TIA system in order to "connect the dots" and detect potential terrorist activity.

What outraged privacy advocates was that this system could be used to track all citizens, not just those suspected of crimes. It was essentially a form of the "Big Brother" concept introduced by George Orwell in his book *1984* (see Figure 12.14). In this story the government, Big Brother, watched over everyone in society by using information gathering and video surveillance. In this Orwellian society there was little in the way of crime nor was there any privacy or freedom. It was like living in a prison. In the case of Total Information Awareness, the privacy advocates won out, the concept was dramatically rethought, and the name changed from Total Information Awareness to Terrorism Information Awareness.

Following the fiasco with Total Information Awareness, in 2002 the U.S. government developed a system called Matrix (Multistate Anti-Terrorism Information Exchange). Matrix combined state records and data culled by Seisint, a database and information service provider, to give investigators fast access to

FIGURE 12.14 • Big Brother is watching you

Actor Edmond O'Brien portrayed a citizen in the movie adaptation of George Orwell's, book *1984*, which depicted a totalitarian society where government monitored citizens in all locations, even in their homes.

information on crime and terrorism suspects. Seisint developed a scoring technology that evaluated each citizen and assigned a number used to rate that individual's "terrorist factor." In Florida the system indicated that 120,000 people showed a statistical likelihood of being terrorists, resulting in some investigations and arrests.[15] Because the system included information on innocent people as well as known criminals, like Total Information Awareness, Matrix drew objections from liberal and conservative privacy groups. Many of the states participating in Matrix pulled out due to the controversy, and Matrix was abandoned at the federal level in 2005. However, a few states still use Matrix-like technologies on the state level for finding criminals.[16]

Surveillance Technologies

The personal information stored in databases is just one of several privacy concerns. **Surveillance** is the close monitoring of behavior. Computer-controlled surveillance technologies combined with ubiquitous networks and powerful information processing systems have made it possible to gather huge quantities of video, audio, and communications signals and process them to reveal personal information. While this is mostly done in an effort to curb crime and catch criminals, some people are concerned with the lack of forethought (see Figure 12.15). Who is monitoring those individuals doing the monitoring? The surveillance technologies that cause the most concern are wiretapping, video and audio surveillance, and certain uses of GPS and RFID.

FIGURE 12.15 • USA PATRIOT Act protest

Protestors fear that the USA PATRIOT Act gives government power at the expense of citizen's civil liberties.

Wiretapping. Wiretapping has been around as long as there have been wires to tap. With increased dependence on electronic communications, wiretapping has grown to be an important tool for law enforcement and a major concern for those interested in personal privacy. The act of wiretapping involves secretly listening in on conversations taking place over telecommunications networks including telephone, e-mail, instant messaging, VoIP, and other forms of Internet communications.

Federal laws governing wiretapping are generally the same for all forms of communication. The Federal Wiretap Act, enacted in 1968 and expanded in 1986, sometimes referred to as Title III, sets procedures for court authorization of real-time surveillance of all kinds of electronic communications in criminal investigations. It normally requires a court order issued by a judge who must conclude, based on an affidavit submitted by the government, that there is probable cause to believe that a crime has been, is being, or is about to be committed. The Foreign Intelligence Surveillance Act of 1978 allows wiretapping based on a finding of probable cause to believe that the target is a member of a foreign terrorist group or an agent of a foreign power. Both laws allow the government to carry out wiretaps without a court order in emergency situations involving risk of death or serious bodily injury and in national security cases.

The number of wiretap requests increased by 85 percent in the three years following the September 11 attacks.[17] In 2004, a total of 3464 wiretaps were approved for all U.S. state and federal investigations. Typically, all such requests are granted by the courts.

With cooperation from Internet service providers, an FBI surveillance system called Carnivore has been used to monitor e-mail correspondence. The system has alarmed privacy advocates and some members of Congress because of the manner in which it surveys all e-mail on the system, not just e-mail of suspected criminals. Recently the FBI switched from Carnivore to its own proprietary system, of which little is known.

FBI surveillance grew from telephone networks to broadband cable networks in 2004, and extended to VoIP communications in 2005.[18] The FCC chairman at the time, Michael Powell, emphasized that "law enforcement access to IP-enabled communications is essential" and that police must have "access to communications infrastructure they need to protect our nation." [19] The Communications Assistance for Law Enforcement Act (CALEA) of 1994 requires telecommunications to restructure their networks to support easy wiretapping by police.[20]

As of July 2004, a loophole was found in the Federal Wiretap Act that requires a court order for e-mail surveillance. A federal appeals court in Boston ruled that federal wiretap laws do not apply to e-mail messages if they are stored, even for a millisecond, on the computers of the Internet service providers that process them. Because all e-mail is stored at least momentarily on ISP servers, this implies that almost all e-mail can be snooped without a court order.

With the exception of such loopholes, wiretap laws govern the level to which colleges can snoop on student, staff, and faculty e-mail and electronic correspondence. Without a court order, a university cannot intercept contents of electronic transmissions unless an exception applies.[21] Exceptions include:

- Situations where wiretapping is required to protect the university's rights and property
- Situations where an unauthorized person is using the network
- For the purpose of monitoring network traffic for management purposes

Such laws do not regulate private business networks, however. In May 2005, Proofpoint and Forrester Consulting (a division of leading analyst firm Forrester) conducted an online survey of 332 technology decision makers at large U.S. companies. The study found that more than 63 percent of corporations with 1000 or more employees either employ or plan to hire workers to read outbound e-mail, due to growing concern over sensitive information leaving the enterprise through e-mail.[22]

States also have wiretapping laws to control the use of wiretaps by government agencies. Some nonprofit groups alleged that Google's Gmail service violates California's wiretapping laws.[23] The problem lay in the fact that Gmail provides context-sensitive advertising with each e-mail message it displays. Say, for instance, that a friend e-mailed you about his new Toyota Prius. Gmail might place a link to a Toyota dealership in the margin of the e-mail. Many Gmail users find this acceptable if not useful. While Gmail users may agree to Gmail's "content extraction" system, those who send messages to Gmail users have not given their permission to have their e-mail scanned. Because of this, some feel that Gmail violates the privacy of those users. However, no court case against Gmail has proven successful.

FIGURE 12.16 • Operation Disruption

Chicago police control 80 cameras with microphones mounted on light posts citywide to focus on crime from squad cars or from a central operations center.

Video and Audio Surveillance. Increasingly, cities are turning to networked video surveillance to monitor their streets. Video cameras in public places are assisting in capturing criminals who may otherwise escape. For example, video surveillance cameras in London's King's Cross train station were used to identify suspects in the July 7, 2005, train bombings.

Perhaps the most advanced video surveillance system can be found in Chicago. The multimillion-dollar system dubbed Operation Disruption includes 80 street surveillance cameras in Chicago's most crime-ridden neighborhoods. The cameras include microphones that can detect gunshots, even when a silencer is being used. When a gunshot is detected the camera is electronically triggered to swing around and focus in the direction of the shot. Cameras are perched in bulletproof boxes atop light poles (see Figure 12.16). Video from cameras is delivered wirelessly to squad cars and a central command center, where retired police officers monitor activity. City officials credit the video system for a 12-year low in violent crime rates and a reduction in the number of homicides.[24] Chicago is not alone. Many other cities have installed or are planning to install similar surveillance technology.

For big events that draw thousands of people and may attract terrorists, surveillance becomes essential to security. At the 2004 Democratic National Convention in Boston, an unprecedented surveillance effort was deployed.[25] Besides the more than 100 existing MTBA cameras monitoring subways and bus stations, law enforcement officials also controlled 900 cameras typically operated by the Massachusetts Port Authority, the state Highway Department, and the "Big Dig," a city-wide renovation project. Thirty cameras were installed at the location of the event and dozens of pieces of surveillance equipment mounted on downtown buildings were dedicated to security. All cameras were linked into a surveillance network to monitor crowds for terrorists, unruly demonstrators, and ordinary street crime. In nearby Boston Harbor, the Coast Guard used its new "hawkeye system" to watch area waterways. The network of infrared imaging, radar, and cameras that operate in both day and night conditions provided officials with a real-time picture of the harbor and provided agencies with an early warning if an unexpected vessel entered the harbor.

The increasing numbers of surveillance cameras used for crime detection and prevention by both cities and private businesses leave very few public spaces unmonitored in most cities. The recent addition of audio recording devices with cameras has some concerned about eavesdropping on private conversations. Once again citizens are asking, "Who is monitoring the people doing the monitoring?"

High-resolution cameras attached to satellites and trained on the earth are providing us with amazing new mapping technologies, as illustrated through software like Google Earth (Figure 12.17). These same technologies are used by law enforcement to track criminals and by governments to spy on each other. China is planning to launch more than 100 satellites before 2020 in order to monitor its country and citizens. A large surveying network will be established to monitor water reserves, forests, farmland, city construction, and "various activities of society," a government official said.[26]

FIGURE 12.17 • Satellite photos

Increasing numbers of high-resolution cameras are focused on Earth from satellites, capturing life on the planet, enemy movements, and well-known cities and monuments.

Video and audio digital technologies provide the tools for individuals to practice surveillance. Camera phones are now banned from some businesses, such as bars and health clubs, in order to provide privacy for patrons. A digital video camera was found propped up under a walkway grate on 88th Street near Lexington Avenue in New York City apparently to look up women's skirts.[27]

Some cell phones can be used to bug a remote location. Left hidden in a room with the ringer turned off, the phone can be dialed up to listen in on nearby conversations. Some business executives have been known to use this trick to gain an advantage in negotiations. Leaving their cell phone on the conference table, they excuse themselves to use the restroom then listen to the conversation that takes place in their absence.

Video and audio technologies combine with networking technologies to provide the power of remote presence. As with most technology, such power can be used to the advantage of society or be abused to invade an individual's privacy.

GPS and RFID Surveillance. Global Positioning System (GPS) and radio frequency identification (RFID) technologies are very useful, but they can also be used to invade privacy. Consider the case that was classified by police investigators as a true "21st-century stalking." The case involved Ara Gabrielyan, 32, of Glendale, California. Ara taped a GPS tracking device to the undercarriage of his ex-girlfriend's car to monitor her movements.[28] The device showed the car's location as a blip on a map provided by a Web-based service. Ara's ex-girlfriend became suspicious when he turned up uninvited at various locations including an airport and her brother's grave site. Her suspicions were confirmed when she caught Ara under her car replacing the battery on the device. Ara was picked up by police on suspicion of one count of stalking and three counts of making violent threats; offenses punishable by up to six years in prison.

GPS devices are being considered by some states for tracking ex-cons. Butler County Ohio Commissioner Michael Fox believes that GPS microchips should be implanted in former convicts on parole and probation, so that they can be monitored remotely. "People have these GPS chips put in their pets and—in some cases—in their children, in the event they are lost or kidnapped, I don't see why the same can't be done with probationees."[29] Florida, Missouri, Ohio, and Oklahoma all passed laws in 2005 requiring lifetime GPS monitoring for some sex offenders, even if their sentences would normally have expired.[30] Although this may prevent some ex-cons from backsliding, it also infringes on their privacy rights after they have paid their debt to society.

FIGURE 12.18 • Tracking students with RFID

An elementary school student poses with the RFID badge intended to track her location on school premises.

In a small community in California, parents of Brittan Elementary School students stood up for their children's rights to privacy when the school implemented a new student RFID inventory system. The grade school required seventh- and eighth-grade students to wear RFID badges that tracked their movement around campus (Figure 12.18). Some parents were outraged, fearing it would take away their children's privacy.[31] Similar devices have been used to monitor the whereabouts of youngsters in some parts of Japan. The firm behind the technology decided to abandon the project after parents and the American Civil Liberties Union expressed health and privacy concerns over the deployment of their badges.[32]

Various governments around the world are considering the implementation of RFID in vehicle license plates. The Texas legislature is considering replacing all vehicle inspection stickers with RFID tags. A proposed law contains limited privacy provisions, but does not seem to exclude access to RFID information by law enforcement agencies.[33] Privacy advocates fear that monitoring vehicles electronically can lead to a database of information that can pinpoint the location of any vehicle at any time.

The combination of data mining, consumer and government databases, listening in on electronic communications, video and audio surveillance, satellite surveillance, and GPS and RFID location monitoring adds up to the possibility of serious invasion and abuse of basic privacy rights. Currently, at least in the United States, privacy laws and advocates keep government agencies in check over who can be monitored. A second deterrent to abuse of surveillance technologies by government is a lack of funding and personnel. It takes a considerable investment to monitor video from thousands of cameras. As technology improves and surveillance becomes increasingly automated, funding and personnel will become a nonissue. Once again this matter becomes a case of trust—can those monitoring society be trusted to only use surveillance in the best interest of the public without invading the privacy of law-abiding citizens?

Some experts in this area believe that there are three scenarios regarding the relationship of technology, privacy, and society:

- Full privacy: Citizens should be assured of 100 percent privacy. They should have absolute control over what personal information is maintained in public and private databases, and there should be no surveillance of any kind for any purpose.
- Full trust: Citizens should trust governments to provide surveillance in a safe and secure manner that respects privacy rights.
- Full transparency: All surveillance and information should be accessible to every law-abiding citizen. Governments and law enforcement should not maintain exclusive control over surveillance. Citizens should have the ability to turn the cameras on authority to ensure that power is not being abused.

Some feel that it is too late for full privacy. Technology and its use have progressed past the point of regaining previous levels of privacy. If full privacy is not attainable, some feel it would be a mistake to fully trust those in power to manage privacy responsibly. Power corrupts, and being responsible for all private information of a population would imply absolute power. Such a scenario, some fear, could lead to an Orwellian society.

Full transparency is an intriguing notion. How would society change, if everyone had equal access to all information and surveillance? No group would hold an advantage over others because of the information they controlled. Any person could view surveillance data from any location. Everyone would hold all the cards. While there would be little privacy, there would also be little opportunity for abuse of power.

While fate will probably deal society a mixture of these scenarios, it is an important exercise for you to consider the type of future you desire regarding your personal rights to privacy. This is one area that should not be dictated to you, but rather decided thoughtfully by those that are affected. While the checks and balances of a democratic society may preserve the United States from becoming an Orwellian-type society, bits of freedom and privacy lost over time can add up to significantly affect people's lives.

ETHICS AND SOCIAL RESPONSIBILITY

The field of **ethics** deals with what is generally considered right or wrong. You can imagine how broad and far reaching this topic is. This section focuses on ethical issues that deal with computer use—**computer ethics**. Most authorities on the subject define computer ethics differently. Many constrain the field to computer professionals—those who develop and manage computer systems—and the ethical responsibilities that they shoulder. Rather than focus exclusively on the ethics of computer professionals, this section also discusses governmental computer ethics and personal computer ethics. Each of these three areas has ethical responsibility within different spheres of influence and is important and valuable to examine.

Personal Ethical Considerations

Personal computer ethics involves the responsible use of computers outside of professional environments. Each person has their own sense of what is right and wrong regarding computer use and behavior. So, personal computer ethics are highly subjective except in issues of local, state, and federal law, where ethical issues are clearly defined. Using a computer to create destructive software such as viruses and worms, to steal credit card numbers, or to invade privacy with spyware is illegal, as you learned in Chapter 11. Copyright infringement (P2P file sharing) and plagiarism were covered in that chapter rather than here to emphasize the point that these are crimes and not just ethical points of view. Those who participate in such activities may pay the price in terms of jail time and fines.

However, when a significant portion of a population in a democratic society opposes a law, there arises substantial pressure on the government to change the law or change conditions in society. For example, the MP3 music standard and the availability of P2P file-sharing software created a situation where many law-abiding individuals were actively violating copyright law. In this case, courts

FIGURE 12.19 • Personal computer ethics

Personal computer ethics require an examination of one's own soul.

affirmed that owners of intellectual property deserve compensation. But through a slow process of court battles, technological solutions, and adjustments in the market, a slow but purposeful change is occurring in society's views on intellectual property and the manner in which music and other forms of intellectual property are distributed. It accommodates both the public's need for robust, easy, and inexpensive access and the creators' need to make a living.

This is how governance and law evolve. These institutions should not be rigid and unchanging but flexible and responsive to society. Good laws evaluate a situation from all perspectives and implement a solution that provides the greatest public good. If individuals are able to examine situations intelligently and unselfishly from all perspectives, then personal ethics would also serve the greatest public good.

Personal ethics regarding computer use typically combine legal considerations, what is best for the public good, and what is best for the person in terms of mental and physical well-being. Fear of the law may keep an individual from hacking a computer network. A feeling of social responsibility may guide a computer user to treat others online with respect. A person's own sense of morality may keep that person from becoming involved with Web content that has a negative impact. What that content is may be different for each individual. Personal computer ethics require an examination of one's own soul (Figure 12.19), and weighing benefits against cost in terms of personal, social, and legal considerations.

Professional Ethical Considerations

Professional computer ethics involve the ethical issues faced by professionals in their use of computer systems as part of their jobs. This includes responsibilities toward customers, coworkers, employers, and all others with whom they interact and who are impacted in some way by their computer use on the job. It includes those who produce computers, software, and information systems as well as those who use them at work.

Law can also govern professional ethics. For example the Identity Theft Protection Act of 2005 requires businesses to secure sensitive data physically and technologically and to notify consumers nationwide when such data is compromised. The Children's Online Privacy Protection Act of 2000 prohibits businesses from collecting personal data online from children under the age of 13. These ethics are imposed upon a business by the government.

A number of organizations and associations go beyond what is required by law and establish their own codes of ethics. Codes of ethical conduct can foster ethical behavior in the organization and give confidence to people who interact with the organization, including clients and customers. Some organizations and associations that have developed a code of ethical conduct include:

- Computer Professionals for Social Responsibility (CPSR)
- Association of Information Technology Professionals (AITP)
- The Association for Computing Machinery (ACM)
- The Institute of Electrical and Electronics Engineers (IEEE)
- The British Computer Society (BCS)

Table 12.4 shows the Code of Ethics used by the Association of Computing Machinery (ACM) as listed at *www.acm.org/constitution/code.html*. ACM membership consists of nearly 80,000 computing professionals from industry, academia, and government institutions around the world. Each of these individuals, regardless of whom they work for, is guided by these principles.

TABLE 12.4 • The ACM Code of Ethics and Professional Conduct

1. General Moral Imperatives
   As an ACM member I will...
   1.1. Contribute to society and human well-being.
   1.2. Avoid harm to others.
   1.3. Be honest and trustworthy.
   1.4. Be fair and take action not to discriminate.
   1.5. Honor property rights including copyrights and patents.
   1.6. Give proper credit for intellectual property.
   1.7. Respect the privacy of others.
   1.8. Honor confidentiality.
2. More Specific Professional Responsibilities
   As an ACM member I will...
   2.1. Strive to achieve the highest quality, effectiveness, and dignity in both the process and products of professional work.
   2.2. Acquire and maintain professional competence.
   2.3. Know and respect existing laws pertaining to professional work.
   2.4. Accept and provide appropriate professional review.
   2.5. Give comprehensive and thorough evaluations of computer systems and their impacts, including analysis of possible risks.
   2.6. Honor contracts, agreements, and assigned responsibilities.
   2.7. Improve public understanding of computing and its consequences.
   2.8. Access computing and communication resources only when authorized to do so.
3. Organizational Leadership Imperatives
   As an ACM member I will...
   3.1. Articulate social responsibilities of members of an organizational unit and encourage full acceptance of those responsibilities.
   3.2. Manage personnel and resources to design and build information systems that enhance the quality of working life.
   3.3. Acknowledge and support proper and authorized uses of an organization's computing and communication resources.
   3.4. Ensure that users and those who will be affected by a system have their needs clearly articulated during the assessment and design of requirements; later the system must be validated to meet requirements.
   3.5. Articulate and support policies that protect the dignity of users and others affected by a computing system.
   3.6. Create opportunities for members of the organization to learn the principles and limitations of computer systems.
4. Compliance with the Code
   4.1. Uphold and promote the principles of this code.
   4.2. Treat violations of this code as inconsistent with membership in the ACM.

Governmental Ethical Considerations

Governments face many of the same ethical considerations as businesses in respect to their use of computers and information systems. Governments, however, have the added responsibility of guiding the influence of technology on their population. They create laws that govern the use of technology so that citizens are protected from those who abuse others through the use of technology. Besides keeping information and people safe, government also has a responsibility to make sure that everyone has equal access to technology in order to enjoy the associated benefits.

In Chapter 1, the *digital divide* was defined as the social and economic gap between those who have access to computers and the Internet and those who do not. This is an issue of access, and the difference in opportunities between the "haves" and "have-nots." Most agree that those without access to technology are seriously disenfranchised in today's digital world. The have-nots in this scenario may be unable to access technology due to a physical disability, financial limitations, geographic isolation, or political or social repression. It should be noted that some cultures prefer to do without technology due to their religious or philosophical beliefs. In all but this last category of people, governments and nonprofit organizations can make, and have made, a big difference in broadening access to technology.

The term *digital divide* is used to label the study of technological imbalances amongst many social groups. There are digital divides based on sex, ethnicity, race, age, income, location, and disability. Any group that is not provided equal opportunities in computer use and access can refer to their disadvantage in relation to the rest of the world as a digital divide. This section focuses on digital divides for the disabled and those with socioeconomic disadvantages.

Accessible Computing for the Disabled. There is a growing body of laws and policies in many countries that address accessibility of information and communications technology (ICT), including the Internet and the Web. Laws differ from country to country. Some treat access to ICT as a human or civil right; others control only the purchase of ICT by government, ensuring that government-controlled ICT is accessible to all including the disabled; others specify that ICT sold in a given market must be accessible.

Section 255 of the U.S. Telecommunications Act of 1996 requires telecommunications manufacturers and service providers to make their products and services accessible to people with disabilities, if readily achievable. The World Wide Web Consortium (W3C) has developed Web Content Accessibility Guidelines (*www.w3.org/TR/WAI-WEBCONTENT/*) for businesses and organizations to use in making their Web content accessible to users with disabilities such as those that are unable to see, hear, or move.

The U.S. Department of Education has developed "Requirements for Accessible Software Design" in order to ensure that all software used in schools is accessible to all students, faculty, and staff. The Americans with Disabilities Act of 1990 requires businesses to provide equal access to individuals with disabilities including Web content and services. Many countries including Australia, Canada, Denmark, Finland, France, Germany, Hong Kong, India, Ireland, Italy, Japan, New Zealand, Portugal, Spain, the United Kingdom, and the European Union have instituted similar laws and considerations for the disabled.[34]

There are numerous software and hardware tools available to assist the disabled in using PCs (see Figure 12.20). Individuals may use screen enlargement software to make computers screens easier to read and use. Screen reader software such as JAWS allows blind users to interact with the computer by using text-to-speech technology to read whatever words are displayed on the screen. The user manipulates the software using predefined hotkeys, or shortcut keys. Some phone companies offer a service that uses a Webcam attached to a deaf person's computer. When the person makes a call and uses sign language to communicate, an operator translates the hand signals to a hearing person at the other end of the line.

The Dutch company Secure Internet Machines (SIM) has manufactured a PC designed for the elderly and others who find the complexities of a computer too much to deal with.[35] The simPC (*www.simpc.com*) is small (the size of a video cassette), silent, comes with all the software most people need, and runs on a proprietary operating system guaranteed to boot up in less than 40 seconds (see Figure 12.21). Attach the simPC to a display and a DSL connection to access the Web, e-mail, online banking, digital photos, word processing, movies, games, and Internet phone service. Users are protected with complete security, and backups of data are performed daily by the subscription service. To prevent problems, users are unable to install software, download big files, burn CDs or DVDs, or edit videos. SimPC is currently sold in the Netherlands and Belgium for about $400 plus a $13 monthly subscription.

Socioeconomic Digital Divides. The economic digital divide in the United States, although of significant concern, has been shrinking since the turn of the millennium. The number of Internet users in the low-income range (earning less

FIGURE 12.20 • Stephen Hawking

Computer systems can be liberating to great minds in disabled bodies, or if designed poorly, they can deny the disabled access to valuable information and services.

than $25,000/year) soared, making them the fastest-growing segment of Internet users. According to a U.S. Department of Commerce (USDC) study, more than half of the nation is now online, with 2 million new Internet users per month. Efforts to bring computers and Internet access to all public schools under the No Child Left Behind Act of 2001 have significantly narrowed the digital divide in computer usage rates for children from high- and low-income families.

FIGURE 12.21 • The simPC

The simPC connects to a DSL connection to provide all the popular computer applications with none of the headaches of maintenance and security.

Another contribution that is narrowing the digital divide is provided by public libraries. A study conducted by researchers at Florida State University in 2005 found that 98.9 percent of libraries in the United States offer free public Internet access, up from 21 percent in 1994 and 95 percent in 2002.[36] Such service provides Internet access to citizens without the financial means to purchase their own computers. The study also found that 18 percent of libraries have wireless Internet access and 21 percent plan to get it within the next year. Wireless access allow some people to connect using their own notebook PC, freeing up more desktop PCs for others who don't own a computer.

With numerous projects underway to provide computer access to all citizens, the government is turning its attention to quality. A new emphasis is being placed on access to high-speed Internet services.[37] Because of the significant promise of this technology, President Bush set out a vision, establishing a national goal for "universal, affordable access for broadband technology by the year 2007."[38] Detailing the many benefits of the technology, the president noted that "The spread of broadband will not only help industry, it [will] help the quality of life of our citizens."[39] The graph in Figure 12.22, taken from a thorough study of the topic performed by the U.S. Department of Commerce, shows the rapid diffusion of broadband into the U.S. population.

FIGURE 12.22 • Diffusion of popular technologies in U.S. households

During its first four years of availability, broadband's rate of diffusion outpaced that of many popular technologies in earlier years.

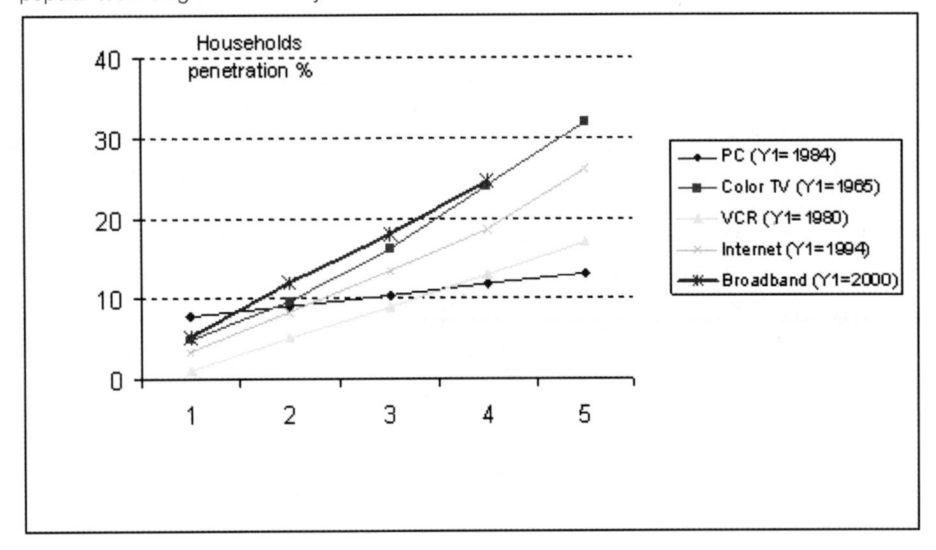

The *global* digital divide provides a greater social and ethical challenge. The World Economic Forum Web site (*www.weforum.org*) states that industrialized countries, with only 15 percent of the world's population, are home to 88 percent of all Internet users. Finland alone has more Internet users than the whole of Latin America. It has been estimated that fewer than 10 percent of the world's population has basic Internet access.

Computer engineers are working on new technologies that can bring computers and the Internet to those that cannot afford them or are in remote locations. A group of not-for-profit developers, called Ndiyo (Swahili for *yes*), has developed an ultrathin client system, which it says could make computing available to billions more people across the planet.[40] Intel is working with the Chinese government to provide Chinese university students with low-priced laptops. Code named Tanggula, after a mountain range in western China, the project is working to provide affordable thin and light laptops based on Intel's Centrino chipset and its Dothan processor to students at more than 300 universities throughout the Asia Pacific region.

An international consortium, including Indian and American companies as well as the World Bank, is building thousands of rural Internet centers in India.[41] Each center connects to the Internet by either land lines or satellite links, and includes 5 to 10 inexpensive thin client PCs to provide access to government, banking, and education services in isolated villages.

FIGURE 12.23 • The digital divide

Urban poverty in Brazil provides a testbed for digital divide solutions developed by the World Economic Forum.

If humans are to utilize the Internet to build a global community, it is clear that the more affluent "neighborhoods" in this community cannot ignore the needs of the less fortunate. Those seeking to assist developing nations believe that societies must move from "divide" to "include" as the central organizing principle of their analysis and actions. Passing out PCs and providing Internet access is the first step, providing education is the next step, and inclusion in the information economy is an important last step.

The World Economic Forum is working to provide all required components to bring developing nations into the digital fold. In 2004, the World Economic Forum launched the IT Access for Everyone (ITAFE) initiative to develop new models of collaboration to accelerate digital access and inclusion.[42] Brazil was selected as a pilot country for prototype solutions to be developed and tested (see Figure 12.23).

A video available at the World Economic Forum Web site (*http://67.104.83.226/WEF/ITAFE_0105.MPG*) provides an excellent overview of the global digital divide and ITAFE's plan to confront it.

Without getting into any more detail regarding ITAFE's program, it should be apparent that bridging global digital divides requires extensive commitment and organization. Many feel that it is an important investment worth the expense and effort. Although the information and communication technology revolution offers genuine potential, if humans do not take a global perspective, there is a risk that a significant portion of the world will lose out. As technologically reaches out to the world with fiber-optic, wireless, and high-speed connections, the rest of the world must be able to respond and participate. The Internet has created a seemingly smaller world and in so doing calls attention to social and economic problems. It is up to the developed nations to recognize these problems and develop solutions to help eliminate them. Only then can a global community be attained.

GLOBALIZATION

Globalization refers to changes in societies and the world economy resulting from dramatically increased international trade and cultural exchange. The largest contributors to globalization are computers, a global telecommunication infrastructure, and the Internet. High-speed fiber-optic global networks make it possible to communicate with, befriend, and collaborate with individuals around the globe as if they were sitting across the table from you. Breaking the traditional barriers of time and space has created a global community and economy (see Figure 12.24).

The technology bubble of the 1990s provided funding for installing transcontinental fiber-optic cables to connect the world in high-speed networks. The bursting of that bubble in the late 1990s dramatically reduced the price for using those connections because there was a huge supply of global bandwidth and a relative lack of demand. Early pioneers of globalization saw business opportunities in these global connections—specifically in the area of outsourcing.

FIGURE 12.24 • Kentucky Fried Chicken in Bangalore, India

A KFC in downtown Bangalore is evidence of globalization and the shrinking world.

Outsourcing

Outsourcing was defined in Chapter 10 as a business' use of an outside company to take over portions of its workload. Remember the Y2K bug? The Y2K bug was the problem faced at the end of the 20th century due to computer systems using only two digits to represent years. When the year switched from 1999 to the year 2000 (Y2K) most computer systems would assume that they had gone back to year zero (00). As a consequence of this error, many anticipated that electronic systems would fail in every industry including vital areas such as banking, medical equipment, manufacturing, and transportation. Because there were not enough computer programmers available to prepare for Y2K, many companies outsourced their Y2K problems to India. When midnight of December 31, 1999, passed without any significant problems, many thought that the entire Y2K scare was all hype. It is more probable that the programmers in India were responsible for saving the day.

Indian programmers earned the respect of many large U.S. corporations, as well as corporations in other developed nations, through their participation in Y2K and other projects. Because Indian programmers worked for pennies on the dollar compared to workers in the United States and other developed nations, it wasn't long before most major companies called upon Indian programmers for many routine programming jobs. Communications lines between the U.S. and India became busy with IT business as teams in India and the United States collaborated. Software to manage distributed project components became an urgent need and software companies responded with sophisticated workflow packages. Workflow software empowered project managers to stay in touch with team members around the world and manage multiple project modules in simultaneous development. Team members dispersed around the world could communicate in real time through text, voice, or video, and pass around project files as though they were sharing a conference table.

Bangalore has become known as the Silicon Valley of India. Many technology companies set up shop in Bangalore, building large corporate offices there (see Figure 12.25). Microsoft, IBM, Texas Instruments, HP, GE, and many others have a major presence in Bangalore. Any service that can be digitized and handled by the Indian work force is being outsourced. Call center operations, U.S. tax returns, medical imaging analysis, news reporting, publishing, you name it—India can do it cheaper than can be done in developed countries. Although much of the work being outsourced is basic, low-skill services, as Indian education levels rise, one can assume that higher level of work will be outsourced. In fact, Microsoft has recently opened a major research facility in Bangalore.

Although India is the biggest destination of outsourced services, Russia, the Philippines, Ireland, Israel, China, and other countries provide outsourced services as well. In fact most nations, including the United States, provide outsourced services. The difference is in the type of services being offered. The United States, with its advanced technology skills, provides many services and products whose quality and price cannot be matched elsewhere.

If you are starting to see the big picture of nations trading services and competing with each other in price and quality on a global scale, you are beginning to understand globalization. Enlarging the marketplace from local to global can have a profound effect on national and international economics and development. For example, outsourcing low-end tech jobs to India has provided

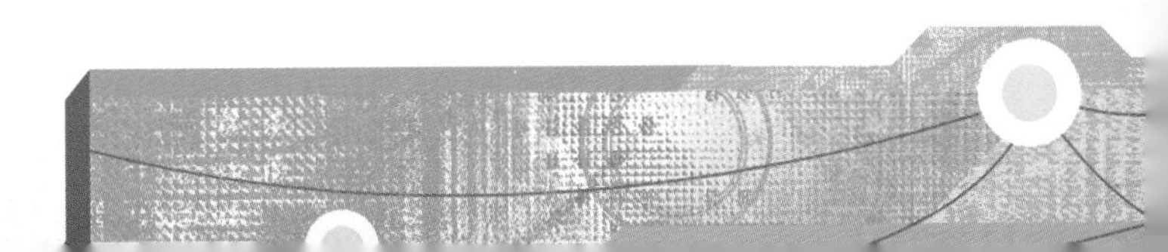

FIGURE 12.25 • Technology companies in Bangalore

Texas Instruments is one of many U.S. corporations that have set up shop on the other side of the world.

a boom in India's economy, while thousands of U.S. workers who used to do the work being outsourced are laid off. The prosperity in India increases the demand for U.S. products in India, which in turn provides growth in some U.S. industries and the need to hire more labor. In the United States there is a general shift in employment from low-skill labor to higher-skill labor and a demand on U.S. companies to be creative innovators in order to remain leaders in the global economy.

Offshoring

Offshoring is somewhat different than outsourcing. With outsourcing, a portion of a work process is hired out. **Offshoring** is a business practice that relocates an entire production line to another location, typically in another country, in order to enjoy cheaper labor, lower taxes, and other forms of lower overhead. If Bangalore, India, is the current global center of outsourcing, then Beijing, China, is the global center of offshoring. At the end of 2001 when China joined the World Trade Organization and agreed to follow international trade laws, it opened a floodgate for trade and innovation. Corporations around the world sought to take advantage of China's large population, cheap labor, and high-quality production to save in manufacturing costs. Textiles, furniture, consumer electronics, and a host of other products are manufactured for global distribution in China for a fraction of the cost of manufacturing in more developed countries.

FIGURE 12.26 • The New World Center in Beijing

The Beijing New World Center, evidence of China's upward-moving economy, includes a 5-story shopping mall, two office towers, deluxe service apartments, an international class hotel, and even a Dairy Queen.

Companies that take advantage of the cost savings found by manufacturing in China and other low-cost manufacturing centers in Eastern Europe and elsewhere in the developing world gain a considerable advantage over competitors that don't. Many business consultants recommend that U.S. companies offshore labor-intensive operations in order to stay competitive. Businesses in other developed nations are following suit. As with outsourcing, offshoring has caused concerns for the labor markets in developed countries, especially in the United States. But as with outsourcing, offshoring is predominantly used for saving money on low-skill, labor-intensive manufacturing. Higher-level manufacturing still takes place in the United States. For example, General Motors may use some parts manufactured in China in vehicles assembled in Detroit.

As with India, a growing and prosperous China means increasing demand for U.S. products (see Figure 12.26). Many U.S. companies have opened branches in China in order to have first crack at the huge market that is developing there. The United States remains the world's largest manufacturer, leading the world in aerospace, pharmaceuticals, automobiles, and other high-tech industries. Through globalization, the field of competition has broadened significantly. As with outsourcing, offshoring is forcing U.S. businesses to stay innovative and competitive.

Outsourcing U.S. Education

Increasing numbers of students are turning to after-school tutoring services for extra help in problem areas. A new form of tutoring is being offered online. Companies such as Sylvan Learning and Growing Stars offer live online instruction. Students are provided with a starter kit that includes a headset/microphone set, a digital writing pad, and digital pencil. They log on to the service over their home broadband Internet connection and work one-on-one with a teacher using a combination of voice communications and a shared electronic whiteboard application. For example, a student may be provided with a paragraph of text on the display and asked to make corrections to the grammar. Voice communication provides instant feedback and assistance to the student.

This digitization of teaching and tutoring provides the prerequisite for it to be outsourced to countries with the lowest wages. Twenty-two-year-old Ms. Salin is employed by Growing Stars and tutors U.S. students in English grammar, comprehension, and writing from her cubicle in Cochin, India. Daniela, an eighth grader at Malibu Middle School, and a student of Ms. Salin, said, "I get Cs in English, and I want to score As." Other than the initial awkwardness of the new situation, she has given no thought to her tutor being 20,000 miles away. There are over a dozen companies like Growing Stars across India that are helping American children improve their academic skills and prepare for tests.

Questions

1. What benefits do U.S. citizens enjoy by using foreign services such as Growing Star that might otherwise not be available?
2. What risks could be involved in children using such a service?
3. Why isn't the need for online tutoring services being fulfilled by more companies in the U.S.?

Sources

1. *Rai, R. "A Tutor Half a World Away, but as Close as a Keyboard,"* New York Times, *September 7, 2005, http://www.nytimes.com.*
2. *Growing Star Web site, http://www.growingstar.com, accessed September 9, 2005.*

Business Challenges in Globalization

There are a number of alternatives for how organizations participate in the global marketplace, depending on where the organization's operations are located and how they are managed. Table 12.5 summarizes these approaches.

TABLE 12.5 • Globalization approaches

Multinational and transnational organizations may enjoy financial benefits by moving some operations to other countries, but they also face significant challenges.

Globalization approach	Management of operations	Location of operations
Importer/exporter	In the home country	In the home country
Multinational	In the home country	Other countries
Transnational	Other countries	Other countries

- Culture: Countries and regional areas have their own cultures and customs that can have a significant impact on individuals and organizations involved in global trade.

- Language: Language differences are another challenge. In some cases, it is difficult to translate exact meanings from one language to another. An exact translation of an advertising slogan, for example, might have a totally different meaning or even be offensive or disgusting to people in other countries.
- Time and distance: Time and distance issues can be difficult to overcome for individuals and organizations involved with global trade in remote locations. In other cases it can be an advantage. For example, outsourcing to India allows a company to work 24 hours a day because it is day in India when it is night in the United States.
- Infrastructure: People and organizations operating in developed countries expect an excellent infrastructure. In some countries, electricity may fluctuate in voltage, damaging machines and computers, or may be off for large periods of the day. Water may be dirty, and phone and Internet service might be problematic.
- Currency: The value of different currencies can vary significantly over time. Sometimes currencies can dramatically fluctuate in a few days or less.
- State, regional, and national laws: Every state, region, and country has a set of laws that must be obeyed by individuals and organizations operating in the country. These laws can deal with a variety of issues, including trade secrets, patents, copyrights, protection of personal data, protection of financial data, privacy, and much more.

Governments can be very helpful in encouraging global trade by implementing trade agreements with other nations. The North American Free Trade Agreement (NAFTA) and the Central American Free Trade Agreement (CAFTA) are examples.[43] The overall objective of NAFTA is to eliminate trade barriers and to facilitate the movement of goods and services between Canada, the United States, and Mexico.[44] It also attempts to protect and enforce intellectual property rights. The trade agreement was signed into law during President Clinton's first term in office. CAFTA extends the same free trade agreements between the United States and various countries in Central America (see Figure 12.27).[45]

FIGURE 12.27 • Signing CAFTA

President Bush and representatives from Costa Rica, El Salvador, Guatemala, Honduras, and Nicaragua sign the Central America Free Trade Agreement (CAFTA) to promote trade between the United States and Central America.

The European Union (EU) is another example of countries with an international trade agreement.[46] The EU is a collection of mostly European countries that have joined together for peace and prosperity. The idea of the EU started in the early 1950s when several European countries signed a trade agreement involving coal and steel. In 1992, a far-reaching treaty formed the EU with a handful of countries. Today, the countries in the EU include Austria, Belgium, Denmark, Finland, France, Germany, Greece, Ireland, Italy, Spain, the United Kingdom, and many other countries. Most of the countries in the EU support the euro, a universal currency that simplifies the purchase and sale of products and services. Like other trade agreements, a primary purpose of the EU was to eliminate trade barriers and obstacles.

CONCLUSION

Throughout this book you have seen how technology assists people to achieve their goals and succeed in life. The positive effects of technology on society can be huge. They include finding cures for deadly diseases, eroding the power of totalitarian governments and dictators, developing alternate environmentally friendly energy sources (see Figure 12.28) and reducing dependence on coal and oil, providing opportunities for developing countries and flattening global economics to spread wealth more evenly, providing people with more engaging and challenging professions, and in general improving the overall quality of life on earth. There is little doubt that technology can be harnessed to improve life.

FIGURE 12.28 • Arklow bank offshore wind power facility, Ireland

Positive uses of technology include GE's giant wind turbines, of which some 1600 are to be installed worldwide in 2005, totaling 2400 megawatts of new wind power capacity.

These last two chapters have focused on some of the challenges of living with technology. The negative effects of technology on society can be significant. Those who wish to oppress and benefit at the expense of others have found a powerful tool in technology. Criminals steal information and identities and cash them in for illegal gains; intellectual property is devalued, copied, and freely distributed at the expense of the creative minds that produced it; scam artists pilfer money from online victims; online vandals corrupt and destroy information systems for the pleasure of it; hate groups, terrorist organizations, and extremists of all kinds use the Web to further their cause and recruit new members (Figure 12.29); and governments could use surveillance and information systems to establish totalitarian rule and a police state that dissolves civil liberties.

Technology can impact the world and society in major ways both positively and negatively. It impacts each of us in small ways on a daily basis too numerous to list. When you weigh the negative against the positive in your life, which way do the scales tip?

As you examine the uses of technology you probably notice that it is a magnifier of human nature—an accelerant to human power. It increases your power to create positive, negative, or indifferent effects. Your feelings toward technology should reflect your feelings about humankind in general. Are humans basically good, with occasional deviant behavior? Are people basically bad and doomed for destruction? Or, is humanity an equal balance of good and evil in continuous flux? Technology is only a tool to be wielded to obtain human goals.

So far, history shows that overall technology has had a more positive than negative effect on the world. Equal access to information for all has undermined those who would otherwise monopolize information to take advantage of others. The Internet and wireless technologies have opened a pipeline of information that is flowing into isolated and oppressed societies to show them that there is a better life.

Political leaders, the news media, and corporate executives have been placed in the spotlight of public awareness and forced to do the right thing or face public humiliation. Widespread access to information is building a more transparent society where everyone who wishes can examine the current state of affairs and take an active role in changing the world for the better. People no longer need to place all their trust in authority because authority no longer has exclusive access to all the facts.

FIGURE 12.29 • July 7, 2005, London bombings

Terrorists can use the Internet to access bomb making information, recruit new members, and communicate with each other to coordinate global action.

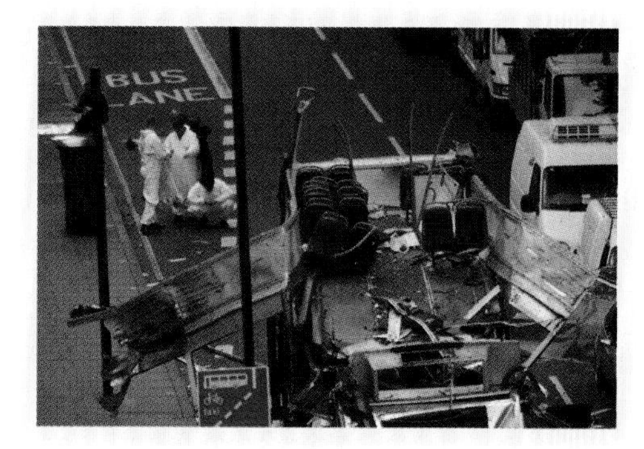

All of this sounds very empowering, but will the citizens of the world bother to take advantage of this power? People taking ownership of their own lives requires time and effort. Consider the effort you are making to educate yourself in order to establish yourself in a career that will support your needs and desires. Taking ownership of your role in society and the world takes even more effort. You can take the time to learn about your local, state, and federal governments and the laws they pass that affect your life. You can learn about international relations. You can examine the practices of big businesses and how they relate to government policies. You can learn about other cultures and other ways of thinking about things. You can follow the news, you can look deeply into issues online, discuss issues with others, and form your own informed opinions. Never before has so much information been so readily available to absorb. And information is power.

The alternative to owning up to your own personal, social, and global responsibilities is to leave it to others to worry about. Those who decide to take this path have no room to complain if the world goes in a direction that they find objectionable.

Thus the torch is handed to you. The torch is technology. The torch is information. The torch is power, freedom, and responsibility (Figure 12.30). You may take up this torch to propel yourself forward to meet your personal and professional objectives, to achieve more than you ever thought possible, to impact those around you, your government, your culture, and the world. You may take up this torch to propagate negativity, hate, and dissonance. Or, you may leave this torch for someone else to pick up. The choice is yours.

FIGURE 12.30 • The torch of freedom and responsibility

Technology provides power, freedom, and responsibility.

ACTION PLAN

Remember John Toh the naturalist, from the beginning of this chapter? John was leery of technology's impact on society and the natural order of the world. Here are some answers to the questions about John's situation.

1. **Are computers alienating the population from each other and the natural environment?**

   Computers can alienate individuals if used excessively. As with any activity, moderation is the key. If used wisely, computers can create and foster healthy relationships between individuals, and place people in environments where they can interact with and impact far more people than they could without the use of technology. Computers and the Internet can be used to learn more about the natural environment and help to preserve it.

2. **Are computers putting power and control into the hands of a few at the expense of freedom of the many?**

   Information and surveillance technology can provide power and control to those who own them. The Internet is a leveling force in that it provides equal access to information for everyone, not just the wealthy and powerful. It is important that society keep an eye on those that are keeping an eye on them. It is also notable that the Internet has provided liberty to many people who were at one time oppressed.

3. **Are computers reducing the need for a human work force and overshadowing basic human needs?**

   Technology is shifting work demands differently in different countries, and almost always elevating the level of work, knowledge, and skills required of the work force. Developing countries are teaching their citizens the basics about technology in order to pull themselves out of poverty, while developed countries are challenged to innovate and take the global society to the next level, whatever that may be. Overall, technology's impact on our global society has been positive, increasing the level of the global economy and providing more people with a worthwhile existence.

Summary

LEARNING OBJECTIVE 1

Describe how technology is affecting the definition of community, and list some physical and mental health dangers associated with excessive computer use.

The Internet provides an entire virtual world of information, people, and groups in which you can literally lose yourself. As people spend increasing amounts of time engaged in virtual space rather than real space, changes occur in social structures and mechanisms. Virtual communities and "anywhere, anytime" communications are affecting traditional social mechanisms. The use of cell phones in public places sometimes offends traditional courtesy and etiquette. Rather than finding friends through chance encounters with strangers that you may happen to meet, the Internet provides means to meet and get to know many individuals with similar interests.

Figure 12.3—p. 584

There are a number of psychological and physical health concerns that have arisen because of the amount of time people are spending on computers and in virtual space. Repetitive stress injury (RSI) can cause physical problems such as tendonitis, the inability to hold objects, and sharp pain in the fingers. Carpal tunnel syndrome (CTS) is the aggravation of the pathway for nerves that travel through the wrist (the carpal tunnel). There are concerns about the effects of radiation emanating from cell phones on the body. In addition to physical health concerns, the use of computer systems can cause a variety of mental health issues and related problems. These include gambling, information overload and stress, and Internet addiction and isolation. The study of designing and positioning the work environment and computer equipment in a healthy manner, called *ergonomics*, has suggested a number of approaches to reducing these health problems. Many feel that the best way to avoid or recover from any computer-related health issue, physical or mental, is to live a balanced life.

LEARNING OBJECTIVE 2

Describe the negative and positive impact of technology on freedom of speech, and list forms of speech and expression that are censored on the Web.

The First Amendment to the U.S. Constitution guarantees citizens the right to free speech. The Internet and Web have been grasped as tools to extend the reach of communications and to empower those who have traditionally been without a public voice. Bloggers are becoming an increasingly important component in journalism, and the role of the public in capturing and reporting news, called consumer-generated media (CGM), has vastly increased over the past 10 years.

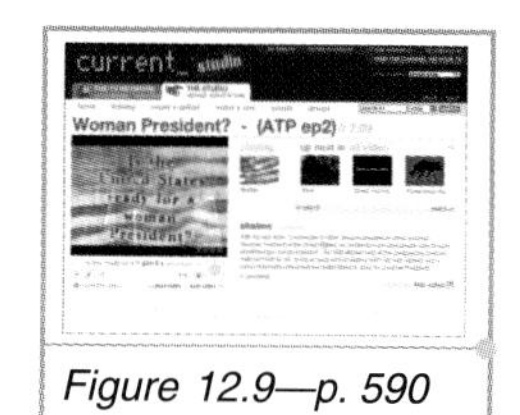

Figure 12.9—p. 590

Libel (the deliberate act of defamation of character by making false statements of fact) and direct, specific threats are not protected under the First Amendment. When a government controls speech and other forms of expression, it is called *censorship*. Various forms of censorship exist around the world. Freedom of speech and Internet technologies are most threatening to oppressive governments whose citizens lack political and social freedom. Most countries support the Internet's ability to empower its citizens, but struggle with issues regarding the perceived negative aspects. One major concern is keeping indecent content from minors. Content-filtering software works with the Web browser to check each Web site for indecent materials (defined by the installer of the software) and only allows "decent" Web pages to be displayed. Some information is censored due to its danger to the public.

LEARNING OBJECTIVE 3

Explain the ways in which technology is used to invade personal privacy, and provide examples of laws that protect citizens from privacy invasion.

Computer technology provides the ability to collect, maintain, process, and transfer much more information than has ever before been possible. This power has given rise to a seemingly endless amount of public and private databases that include details about many individuals. Privacy issues that concern most people include being free from intrusion—the right to be left alone, freedom from surveillance, and control over the information collected and kept about one's self. Much of the information gathered about individuals is done without their knowledge. This invisible information gathering takes many forms. Computer profiling is the primary service provided by private information service companies such as ChoicePoint.

Figure 12.17—p. 600

Government agencies in the United States and many other democratic countries are regulated far more stringently than businesses and professionals when it comes to the privacy of confidential records. The Privacy Act of 1974 is the primary law controlling what many U.S. government agencies can and cannot do with the information they hold. In cases of war, governments sometimes set aside some privacy restrictions in an effort to capture or defeat the enemy.

Computer-controlled surveillance combined with ubiquitous networks and powerful information-processing systems have made it possible to gather huge quantities of video, audio, and communications data and process them to reveal personal information. Although this is mostly done in an effort to curb crime and catch criminals, some people are concerned with the lack of forethought. Who is monitoring the individuals doing the monitoring? Wiretapping involves secretly listening in on conversations taking place over telecommunications networks including telephone, e-mail, instant messaging, VoIP, and other forms of Internet communications. Many cities are using video surveillance to monitor their streets. GPS and RFID are location and tracking technologies that are very useful, but can also be used in manners that invade an individual's privacy.

LEARNING OBJECTIVE 4

List ethical issues related to technology that confront individuals in personal and professional life, businesses, and governments.

The field of ethics deals with what is generally considered right or wrong. Computer ethics deal with ethical issues in regards to computer use. Personal computer ethics involves the responsible use of computers outside of the professional environment. Professional computer ethics involve the ethical issues faced by professionals in their use of computer systems as part of their jobs. This includes responsibilities towards customers, coworkers, employers, and all others with whom they interact and who are impacted in some way by their computer use. A number of organizations and associations go beyond what is required by law and establish their own codes of ethics.

Figure 12.20—p. 606

Governments face many of the same ethical considerations as businesses in respect to their use of information systems. Governments, however, have the added responsibility of guiding the influence of technology on their population. The digital divide is the social and economic gap between those who have access to computers and the Internet and those who do not. There are digital divides based on sex, ethnicity, race, income, location, and disability. There is a growing body of national laws and policies in many countries that address accessibility of information and communications technology (ICT) to users with disabilities such as those that are unable to see, hear, or move.

LEARNING OBJECTIVE 5

Explain what globalization is, what forces are behind it, and how it is affecting the United States and other nations.

Globalization refers to changes in societies and the world economy resulting from dramatically increased international trade and cultural exchange. The largest contributors to globalization are computers, a global telecommunication infrastructure, and the Internet. Outsourcing refers to a business' use of an outside company to take over portions of its workload. Offshoring is somewhat different than outsourcing. With outsourcing a portion of a work process is hired out. Offshoring relocates an entire production line to another location, typically in another country, in order to enjoy cheaper labor, lower taxes, and other forms of lower overhead. There are a number of alternatives for how an organization participates in the global marketplace, depending on where the organization's operations are located and how they are managed. Governments can be very helpful in encouraging global trade by implementing trade agreements with other nations.

Figure 12.25—p. 610

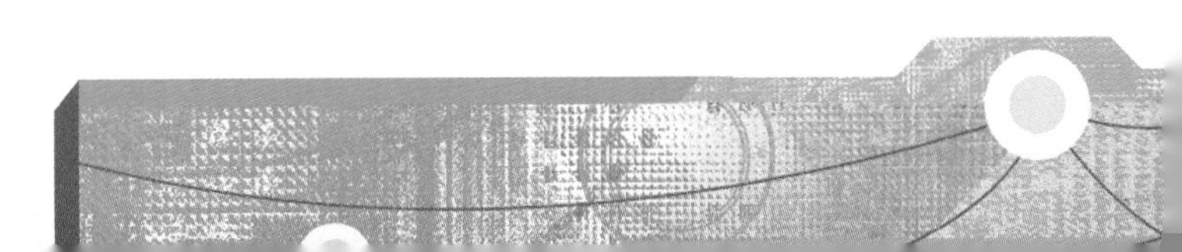

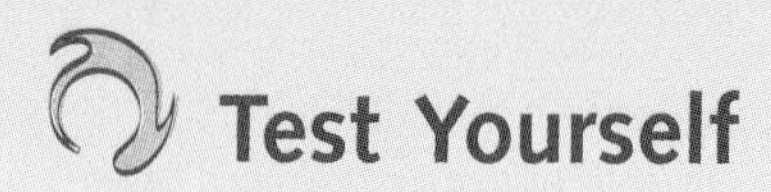

Test Yourself

LEARNING OBJECTIVE 1: Describe how technology is affecting our definition of community, and list some physical and mental health dangers associated with excessive computer use.

1. A ______________ is a spontaneous gathering of a group of people organized through cell phone, text messaging, or e-mail.
2. Which of the following is not included in cell phone etiquette?
   a. Timing: There is a time and place for cell phone conversations.
   b. Content: Keep your business private.
   c. Multitasking: Concentrate on one thing at a time.
   d. Clarity: Speak loudly to be heard.
3. True or False. There are concerns about the effects of radiation emanating from cell phones on the body.

LEARNING OBJECTIVE 2: Describe the negative and positive impact of technology on freedom of speech, and list forms of speech and expression that are censored on the Web.

4. The role of the public in capturing and reporting news is called:
   a. computer-generated media
   b. consumer-generated media
   c. public discourse
   d. blogging
5. When a government controls speech and other forms of expression, it is called ______________.
6. True or False: Content-filtering software has proven to be the ideal solution for libraries who wish to keep pornography off their computers.

LEARNING OBJECTIVE 3: Explain the ways in which technology is used to invade personal privacy, and provide examples of laws that protect citizens from privacy invasion.

7. The practice of using computers to cross-reference and pool electronic customer records into large customer profiles is referred to as:
   a. computer profiling
   b. customer profiling
   c. consumer profiling
   d. e-commerce
8. ______________ is the close monitoring of behavior.
9. True or False: Under no circumstance can the government read private e-mail.

LEARNING OBJECTIVE 4: List ethical issues related to digital technology that confront individuals in personal and professional life, businesses, and governments.

10. ______________ is a field of study that deals with what is generally considered right or wrong.
11. Any group that is not provided equal opportunities in computer use and access can refer to their disadvantage in relation to the rest of the world as a ______________.
12. The ______________ is heading a major effort to provide global access to computers, the Internet, and the associated economic opportunities.
    a. U.S. government
    b. European Union
    c. World Economic Forum
    d. United Nations

LEARNING OBJECTIVE 5: Explain what globalization is, what forces are behind it, and how it is affecting the United States and other nations.

13. ______________ refers to changes in societies and the world economy resulting from dramatically increased international trade and cultural exchange.
14. A business practice that relocates an entire production line to another location, typically in another country, in order to enjoy cheaper labor, lower taxes, and other forms of lower overhead is referred to as:
    a. offshoring
    b. outsourcing
    c. exporting
    d. globalization

15. True or False: Outsourcing and offshoring have a negative effect on the global economy.

Test Yourself Solutions: 1. flash mob, **2.** d. Clarity, **3.** True, **4.** b. consumer-generated media, **5.** censorship, **6.** False, **7.** a. computer profiling, **8.** surveillance, **9.** False, **10.** Ethics, **11.** digital divide, **12.** c. World Economic Forum, **13.** globalization, **14.** a. offshoring, **15.** False.

Key Terms

carpal tunnel syndrome (CTS), p. 586
censorship, p. 590
computer ethics, p. 602
consumer-generated media (CGM), p. 589
content-filtering software, p. 591
ergonomics, p. 587
ethics, p. 602
globalization, p. 608
offshoring, p. 610
surveillance, p. 597

Questions

Review Questions

1. Provide several examples of computer-based virtual space.
2. What is carpal tunnel syndrome? How is it typically caused and treated?
3. What is specific absorption rate (SAR), and why do some people consider it important?
4. List three mental health issues related to computer use.
5. What is the purpose of ergonomics?
6. What legal principle guarantees U.S. citizens the right to free speech?
7. What is the purpose of content-filtering software?
8. What is censorship?
9. What benefit does computer profiling provide for businesses?
10. List four forms of electronic surveillance, and provide examples of each.
11. What are computer ethics, and what are the three areas of computer ethics discussed in this chapter?
12. List six forms of digital divides.

Discussion Questions

13. Provide an example of how the concept of flash mobs can be applied to a useful activity.
14. What is nature deficit disorder? Do you believe that it is an issue for yourself?
15. What is consumer-generated media? Do you think it helps or hurts the public? Why or why not?
16. Do you feel that some form of government censorship is needed on the Web for content that is legal but may be dangerous or unhealthy for society? Why or why not?
17. Should content-filtering software be used in campus computer labs? Why or why not?
18. What are the benefits and short-comings of content-filtering software?
19. Should dangerous information, such as how to build a bomb, be allowed on the Web? Why or why not?
20. Should governments be allowed to use systems such as Total Information Awareness and MATRIX in their efforts to catch terrorists? Why or why not?
21. What categories of Web content do you find most unhealthy for society?
22. Many personal ethical beliefs are not backed by laws. What are some behaviors that you have observed that you wish were prohibited by law?
23. Explain the differences between outsourcing and offshoring.
24. Describe the positive and negative aspects of globalization.

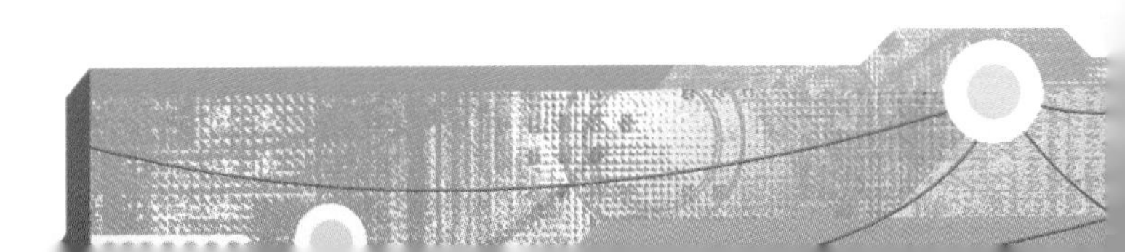

25. Understanding what you do about outsourcing and offshoring, what is the best advice you could provide to a graduating high school student in the United States. Provide a rationale.

26. List several globalization challenges for businesses.

Exercises

Try It Yourself

1. Use a Web search engine to find online dating services. Visit the top six services and write a paragraph review of each. Conclude with a paragraph discussing whether or not you feel that such services are safe and valuable.

2. Do a Web search on ergonomics. Find a list of the most important measurements and angles representing the healthiest computing posture. Use Windows Paint, PowerPoint, or some other graphics tool to draw a stick figure sitting at a computer. Include text labels indicating the proper ergonomic positions and measurements between the person and his or her surroundings.

3. Search for information on *nature deficit disorder* and learn what the experts have to say on the topic. Use Excel to create a week-long hourly schedule for a sixth grader that includes a healthy balance of indoor and outdoor activities. Assume that the child is in school from 8:00 a.m. to 3:00 p.m., Monday through Friday. Include computer time for both home work and fun. Include a reference to the information found on the Web.

Virtual Classroom Activities

4. Divide the class into groups of three to five students. Each group should conduct an online chat providing group members with opportunities to sound off about cell phone etiquette. Use Table 12.1 as a basis for your discussion. Rank the 10 items in order of importance. Decide if any of these items are unrealistic. Summarize your conclusion and make a presentation to your teacher and class on the class discussion board.

5. Learn as much as you can about Google Earth by either downloading and installing free evaluation versions of the software and trying it out, or by reading about it at *http://earth.google.com.* Write a short paper (no less than 400 words) on how this technology affects national security. Conclude with your thoughts on whether such technology should be available to the public.

6. What accommodations does your school provide to assist disabled students in accessing and using computers on campus? Use your school's Web site to find out, and record your findings in a Word document to submit.

Teamwork

7. Have group members scour through your school's policies as listed in the general bulletin, code of conduct, class syllabus, network usage policies, and any other official policies you can find. See if you can find issues of censorship. Build a list of policies that can be considered censorship and vote on whether each is justifiable or not. Provide your results in a document to be submitted to the teacher and/or discussed with the class.

8. Have a surveillance camera scavenger hunt. Use a campus map to divide up the campus between team members. Find as many surveillance cameras as possible. Keep in mind the types of facilities that may use surveillance: entrances to student housing, ATM machines, lecture halls, stores, and so on. Use an electronic version of the campus map (found either on your school Web site, or scan it in yourself) and graphics software to plot the location of all cameras found on the map. Whoever finds the most cameras is the winner! Submit your map electronically for grading.

TechTV

Gmail Privacy

Go to www.course.com/swt2/ch12 and click TechTV. Click the Gmail Privacy link to view the TechTV video, and then answer the following questions.

1. Google now offers more than 2 GB of storage for users of the free Gmail service. Some feel that the narrow column of text ads that run down the right side of the window is a small price to pay for a valuable free service. Others find that the context sensitive ads are actually useful. Still others find the ads invasive and unacceptable. Which sentiment do you agree with and why?
2. Does Gmail's electronic scanning of e-mail for ad placement strike you as a serious invasion of user's privacy? How about the privacy of those that send e-mail to Gmail accounts? Do you feel the Governor of California is over-reacting? Explain.

Endnotes

[1] Staff, Tumor May Be Linked To Cell Phone Use," *CNN Online,* October 14, 2004.

[2] CNet's Quick Guide, Cell Phone Radiation Levels Web site, http://reviews.cnet.com/4520-6602_7-5020357-1.html?tag=nav, accessed on August 27, 2005.

[3] Foroohar, K, "Online Gambling Raises the Ante," *Bloomberg Markets*, October 2005, www.bloomberg.com/media/markets/gambling.pdf.

[4] Staff, "Gaming Corporation Launches Gambling on Mobile Phones," *Investors Chronicle*, April 29, 2005.

[5] Young, K., "Internet Addiction Prevention and Education, *Journal of Employee Assistance*, March 2005, 35 (1), p. 15.

[6] Veysey, S. "AOL Outlines Successes of Ergonomic Strategy," *Business Insurance*, May 2, 2005.

[7] Anderson, K. "American Media vs the Blogs," BBC News, February 22, 2005, http://news.bbc.co.uk/1/hi/world/americas/4279229.stm.

[8] See note 7 above.

[9] van Grinsven, L. "Cell Phones Increasingly Used to Snap the News," Reuters, November 3, 2004, www.reuters.com.

[10] Internet censorship in Australia Web page, http://www.efa.org.au/Issues/Censor/cens1.html, accessed on August 28, 2005.

[11] Warner, B. "BT Child Porn Filter Stopping 23,000 Attempts a Day," Reuters, July 20, 2004, www.reuters.com.

[12] Donohue, L. K. "Censoring Science Won't Make Us Any Safer," *The Washington Post*, June 26, 2005, http://www.washingtonpost.com/wp-dyn/content/article/2005/06/25/AR2005062500077_pf.html.

[13] "The USA PATRIOT Act," http://www.epic.org/privacy/terrorism/hr3162.html, accessed on September 1, 2005.

[14] Electronic Privacy Information Center Web site, http://www.epic.org/privacy/profiling/tia/#introduction, accessed on September 1, 2005.

[15] "Database Tagged 120,000 as Possible Terrorist Suspects, *New York Times/The Associated Press*, May 21, 2004, http://www.nytimes.com/2004/05/21/national/21database.html?th.

[16] Royse, D. "Police Still Using Matrix-Type Database," *AP News Myway*, July 10, 2005, http://apnews.myway.com//article/20050710/D8B8LHQ00.html.

[17] "FBI Said Buried by Security Demands," *The Washington Times/Associated Press*, April 16, 2004, http://www.washingtontimes.com/national/20040415-114936-5908r.htm

[18] Gross, G. "VoIP May Go Under Wiretap Laws," *IDG News Service*, August 5, 2005, http://www.pcworld.com/news/article/0,aid,117270,00.asp.

[19] Charny, B. "Cable Taps into Wiretap Law," *News.com*, March 16 2004, http://news.com.com/Cable+taps+into+wiretap+law/2100-1034_3-5173320.html

[20] The Communications Assistance for Law Enforcement Act (CALEA), http://www.fcc.gov/calea/, accessed on September 3, 2005.

[21] Cal State San Bernardino Web site on Wiretap Laws, http://www.infosec.csusb.edu/policies/wiretap.html, accessed on September 3, 2005.

[22] Mordo, A. "63% of Corporations Plan to Read Outbound E-mail," Aviron's Place Web site, June 7, 2005, http://www.aviransplace.com/index.php/archives/2005/06/07/63-of-corporations-plan-to-read-outbound-email/.

[23] McCullagh, D. "Does Gmail Breach Wiretap Laws?" *ZDNet*, May 4, 2004, http://news.zdnet.com/2100-3513_22-5205554.html

[24] "Chicago to Expand Video Surveillance System," November 30, 2004, *NBC5.com News*, http://www.nbc5.com/news/3959595/detail.html?z=dp&dpswid=2265994&dppid=65193.

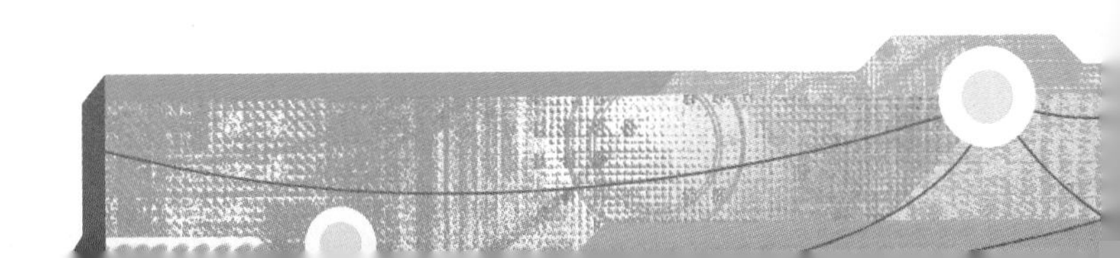

[25] Ranalli, R. and Klein, R. "Surveillance Targeted to Convention," *The Boston Globe*, July 18, 2004, http://www.boston.com/news/politics/conventions/articles/2004/07/18/surveillance_targeted_to_convention?mode=PF

[26] "China Plans to Have Over 100 Eyes in the Sky by 2020," *Reuters*, November 16, 2005, http://www.reuters.com.

[27] Wilson, M. "A Camera Below a Grate? The Police Suspect a Peeper's Work," *The New York Times*, May 19, 2005, http://www.nytimes.com.

[28] Leyden, J. "Suspected GPS Stalker Cuffed," *The Register*, September 6, 2004, http://www.theregister.co.uk/2004/09/06/gps_stalker/

[29] Greene, T.C. "In the Red States, No-One Can Hear You Scream," *The Register*, March 30, 2005, http://www.theregister.co.uk/2005/03/30/gps_chips_for_naughty_people/.

[30] "States Track Sex Offenders by GPS," *Wired News/Associated Press*, July 30, 2005, http://www.wired.com/news/technology/0,1282,68372,00.html?tw=rss.POL.

[31] Leff, L. "Parents Protest Student Computer ID Tags," *Associated Press*, February 10, 2005, http://www.boston.com/business/technology/articles/2005/02/10/parents_protest_student_computer_id_tags/.

[32] Haines, L. "Parent Power Detags US Schoolkids," *The Register*, February 18, 2005, http://www.theregister.co.uk/2005/02/18/schoolkids_detagged/.

[33] Proposed Texas Legislation HB 2893, http://www.capitol.state.tx.us/cgi-bin/tlo/textframe.cmd?LEG=79&SESS=R&CHAMBER=H&BILLTYPE=B&BILLSUFFIX=02893&VERSION=1&TYPE=B, accessed on September 3, 2005.

[34] W3C Web Accessibility Initiative Web site, http://www.w3.org/WAI/Policy/, accessed on September 4, 2005.

[35] Libbenga, J. "New PCs for Old People," *The Register*, January 11, 2005, http://www.theregister.co.uk/2005/01/11/new_pcs_for_old_people/.

[36] Ruethling, G. "Almost All Libraries in U.S. Offer Free Access to Internet," *New York Times*, June 24, 2005, http://www.nytimes.com/2005/06/24/national/24library.html?th&emc=th.

[37] Haley, C.C. "Bush Calls for Universal Broadband by 2007," *Internet News*, March 31, 2004, http://www.internetnews.com/xSP/article.php/3333711.

[38] Remarks by President Bush on homeownership, Expo New Mexico, Albuquerque, New Mexico, March 26, 2004, http://www.whitehouse.gov/news/releases/2004/03/20040326-9.html.

[39] Remarks by President Bush on innovation, U.S. Department of Commerce, Washington, D.C., June 24, 2004, http://www.whitehouse.gov/news/releases/2004/06/20040624-7.html.

[40] Sherriff, L. "Ultra-Thin Client to Close Digital Divide," *The Register*, May 3, 2005, http://www.theregister.co.uk/2005/05/03/ultra_thin_digital_divide/.

[41] Markoff, J. "Plan to Connect Rural India to the Internet," *New York Times*, June 16, 2005, http://www.nytimes.com.

[42] Information Technology Access For Everyone Web site, http://www.weforum.org/site/homepublic.nsf/Content/Information+Technology+Access+For+Everyone, accessed on September 5, 2005.

[43] Smith, G. et al, "Central America is Holding Its Breath," *Business Week*, June 20, 2005, p. 52.

[44] NAFTA Web page, http://www.nafta-sec-alena.org, accessed on May 12, 2005.

[45] CAFTA Web page, http://www.citizen.org/trade/cafta, accessed on May 12, 2005.

[46] European Union Web page, http://www.europa.eu.int, accessed on May 12, 2005.

Glossary

Note: Terms in bold are Key Terms, found in the Key Terms sections at the end of each chapter. Terms in italic are important terms mentioned in the book, but not included in the Key Terms sections at the end of each chapter.

3Cs approach — Emphasizing content, community, and commerce on a B2C e-commerce Web site, a strategy for capturing the interest of the online community.

3D modeling software — Programs that provide graphic tools that allow artists to create pictures of 3-D, realistic models.

3g cellular technology — Current wireless technology that is bringing wireless broadband data services to your mobile phone.

acceptance testing — Conducting tests required by the user, to make sure that the new or modified system is operating as intended.

access points — Wireless network sites, also known as hot spots, distributed around a geographic area that broadcast network traffic using radio frequencies to computers equipped with wireless fidelity (Wi-Fi) cards or adapters.

access time — The amount of time it takes for a request for data to be fulfilled by a storage device.

activate — Make software fully operational, which may require registering the program with the software maker.

ActiveX — Microsoft's alternative to the JavaScript Web programming language.

actor — The object that interacts with events, or use cases, in an object-oriented systems development approach.

adware — Software placed on your computer system without your knowledge or consent, normally through the Internet, to secretly spy on you and collect information, especially for advertising purposes.

agile modeling — A systems development approach that calls for very active participation of customers and other stakeholders in the systems development process.

algorithm — A step-by-step problem-solving process that arrives at a solution in a finite amount of time; sometimes called program logic; a detailed procedure or formula for solving a problem.

alpha testing — The first stage of testing that is implemented by the software developer.

alphanumeric — In database fields, character-type data, including characters or numbers that will not be manipulated or used in calculations.

American Standard Code for Information Interchange (ASCII) — A code for representing text characters that the computer industry agreed upon in the early days of computing.

analog — Signals that vary continuously; the opposite of digital.

analog signal — A signal that continuously fluctuates over time between high and low voltage.

analog-to-digital converter — A device that translates sound and music to digital signals.

animated GIF — Simple drawings, created with simple software tools, that repeat the same motion over and over endlessly; the most basic form of animation.

anomalies — Problems and irregularities in data.

antispyware — Software that searches a computer for spyware and other software that may violate a user's privacy, allows the user to remove it, and provides continuing protection against future attacks.

antivirus software — Software that uses several techniques to find viruses on a computer system, remove them if possible, and keep additional viruses from infecting the system. Also known as virus scan software.

application flowchart — A general flowchart used to describe the overall purpose and structure of a system; also known as the system flowchart.

application layer — The software portion of the three-layer Internet model, which also includes transport (protocol) and physical (hardware) layers.

application programming interfaces (APIs) — Web site development tools that allow software engineers to develop Web-driven programs.

application servers — Computers that store programs, such as word processors and spreadsheets, and deliver them to workstations to run when users click the program icon.

application service provider (ASP) — A company that provides software and support, such as computer personnel, to run the software.

application software — Programs that apply the power of computers to help perform tasks or solve problems for people, groups, and organizations; along with systems software, one of two basic types of software.

arithmetic logic unit (ALU) — One of three primary elements within a central processing unit (CPU), it contains the circuitry to carry out instructions, such as mathematical calculations and logical comparisons.

artificial intelligence (AI) — A term coined in the 1950s to describe computers with the ability to mimic or duplicate the functions of the human brain; computer systems taking on the characteristics of human intelligence.

artificial intelligence language — Programming language used to create artificial intelligence or expert systems applications.

artificial intelligence system — The information system components needed to develop computer systems and machines that demonstrate characteristics of intelligence.

ASCII art — The art of creating drawings with typed characters.

asking directly — A requirements analysis technique that asks users and stakeholders about what they want and expect from a new or modified system.

assembly languages — Second-generation programming languages that replaced the binary digits of first-generation machine languages with symbols that are more easily understood by human programmers.

assigned research — A form of research in which the topic to explore is given, rather than prompted by curiosity, for the purpose of education.

asynchronous communication — A form of electronic communications that allows participants to leave messages for each other to be read, heard, watched, and responded to at the recipient's convenience, such as by using answering machines, voice mail, and e-mail.

attosecond — A billionth of a billionth of a second.

attribute — A characteristic of an entity; in the relational data model, a column of a table; an identifying characteristic, such as a variable, that is associated with an object in object-oriented design.

authentication — A manner in which to confirm the identity of a user. There are three common forms of authentication: something you know (such as a password or PIN), something you have (such as an ID card or badge), and something about you (a unique physical characteristic such as your fingerprint).

automation — Utilizing computers to control otherwise human actions and activities.

avars — Points on a computerized 3-D object that are designed to bend or pivot at specific angles.

avatar — A 3-D representation of a participant in a virtual world environment, which can be used to navigate through the virtual world.

B2B — Business-to-business e-commerce, a system supporting transactions between businesses across private networks, the Internet, and the Web.

B2C — Business-to-consumer e-commerce, the use of the Web to connect individual consumers directly with sellers to purchase products.

back-end application — Software that interacts with other programs or applications, and only indirectly with people or users.

backdoor Trojan — A Trojan horse program that opens up ports (back doors) on the computer to allow access to intruders.

backward-compatible — Capable of supporting previous technology. For example, DVD drives are backward-compatible and thus can play CDs as well as DVDs.

bandwidth — The data transmission rate of a network medium measured in bits per second.

basic input/output system (BIOS) — A set of instructions, activated by one or more computer chips, to perform additional testing and to control various input/output devices, such as keyboards and display screens.

batch processing — A method of processing transactions by collecting them over time and processing them together in batches.

behavioral profiling — Observing customers' cable TV viewing patterns in order to develop an understanding of their interests.

binary data — Data that is intended for a processor to process.

binary number system — A number system that uses only two values, 0 and 1, and is used by computers and digital devices to represent and process numeric data.

bioinformatics — Also called computational biology, a combination of biology and computer science that has been used to help map the human genome and conduct research on biological organisms.

biometric device — A device that can recognize physical traits, such as fingerprints, facial characteristics, or retinal or eye characteristics, and can be used to prevent unauthorized access to a computer system or a security area.

biometrics — The science and technology of authentication by scanning and measuring a person's physical features such as fingerprints, retinal patterns, and facial characteristics.

bit — Short for binary digit, represents data using technologies that can be set to one of two states.

bit-mapped graphics — A representational method that uses bytes to store the color of each pixel in an image. Also known as raster graphics.

bits per second (bps) — A measurement of data transmission speed.

blade computing — A type of enterprise computing that uses stripped-down network PCs called *thin clients* connected to clusters of *blade servers*. This system, paired with a file server that stores user files, is significantly less costly than regular PCs for large enterprises, but offers identical services to users and convenience for system adminstrators.

blade server —PC motherboards that are rack-mounted together in groups of up to 20 to a case.

blog — Short for Web log, Web sites created to express one or more individual's views on a given topic.

Bluetooth — A low-cost, short-range wireless specification for connecting mobile products; Bluetooth technology enables a wide assortment of digital devices to communicate wirelessly over short distances.

booting — When a computer is first turned on, this process is carried out to test the hardware components and load the operating system.

bot — An automated program that scours the Web in an attempt to catalog every Web page by topic; also called a spider or crawler.

brainstorming — A group decision-making approach that often consists of members offering ideas "off the top of their heads."

brick-and-mortar — A term used to refer to a traditional retail store.

broadband — Media advertised by Internet Service Providers as "high-speed"; delivers speeds faster than 200 Kbps.

Broadband over Power Lines (BPL) — A broadband Internet access, not yet widely offered, over the power grids; connecting a computer to the Internet would be as easy as plugging a powerline modem into a wall outlet.

broadband phone— Residential VoIP service that utilizes a high-speed Internet connection to provide telephone services.

buddy lists — Also known as contact lists, names of people who are frequent communication partners in chat rooms or via instant messaging.

buffer — Intermediate storage area between a microprocessor and other hardware or software.

burning — The process of writing to an optical disk.

bugs — Errors in coding or logic that prevent a program from working properly.

bus — A network topology consisting of one main cable or telecommunications line with devices attached to it.

business intelligence — The business use of data mining to help increase efficiency, reduce costs, or increase profits; a term first coined by a consultant at Gartner Group.

business resumption planning (BRP) — The review of every conceivable disaster that could negatively impact a transaction processing system, and the provision of courses of action to minimize their effects.

business to business (B2B) — A type of electronic commerce or transaction that is conducted between two businesses.

business to consumer (B2C) — A type of electronic commerce or transaction that is conducted between a business and a consumer.

business-to-business e-commerce — A system supporting transactions between businesses over private networks, the Internet, and the Web; also known as simply B2B.

business-to-consumer e-commerce — The use of the Web to connect individual consumers directly with sellers to purchase products; also known as simply B2C or e-tailing.

byte — Eight bits combined, the standard unit of storage in digital electronics.

C2C — Consumer-to-consumer e-commerce, the use of the Web to connect individuals who wish to sell their personal belongings with people shopping for used items.

cable modem — A sophisticated signal-conversion device that provides Internet access over a cable television network; it offers faster data transmission rates than a traditional dial-up connection.

cable modem connection — A broadband Internet service, with data transfer rates of around 2 Mbps, provided by cable television providers.

cache memory — A type of high-speed memory that a processor can access more rapidly than RAM, allowing for quick retrieval of program instructions and data.

carpal tunnel syndrome (CTS) — An inflammation of the pathways for nerves that travel through the wrist (the carpal tunnel), that involves wrist pain, a feeling of tingling and numbness, and difficulty in grasping and holding objects; a condition sometimes associated with long hours at the computer keyboard.

CD-RW — Compact disk-rewritable; a type of CD where the data can be written over many times.

CDMA — One of two predominant cell phone networking standards, the other being GSM. CDMA is used more in the United States, and GSM in Europe and Asia, though GSM is rapidly gaining ground in the U.S. as well.

cellular carrier — A company that builds and maintains a cellular network and provides cell phone service to the public.

cellular network — A radio network in which a geographic area is divided into cells with a transceiver antenna (tower) and station at the center of each cell to support mobile communications.

censorship — When a government or authority controls speech and other forms of expression.

central processing unit (CPU) — The group of integrated circuits that work together to perform any system processing, such as arithmetic calculations, logic comparisons, and data access.

certification — The process for testing skills and knowledge resulting in an endorsement by the certifying authority that an individual is capable of performing a particular job.

certification authorities — Businesses, such as VeriSign, that provide digital certificates.

channels — Also known as chat rooms, the various topic related forums on the Internet for synchronous text messaging between two or more participants.

character recognition software — Software that when combined with a scanner can transform document images into editable word-processing documents.

chat — On the Internet, synchronous text messaging between two or more participants.

chat rooms — Also known as channels, the various topic related forums on the Internet for synchronous text messaging between two or more participants.

chip — Another term for microprocessor.

chipset — Set of processors that ties together several bus systems in a computer, sending and receiving bytes from memory, input and output devices, storage, networks, and other motherboard components.

cladding — Thin coating over fiber optic cable that effectively works like a mirror, preventing the light from leaking out of the fiber.

client — The program, in a client/server relationship, that makes a service request.

client/server — A relationship between two computer programs in which one program, the client, makes a service request from another program, the server, which provides the service.

clip art — Drawings and photos, covering a wide variety of objects, offered in presentation graphics or other programs for use in slide presentations and other graphic projects.

clock speed — The rate, typically measured in megahertz or gigahertz, at which a CPU's system clock produces a series of electronic pulses, which is a factor in overall system performance.

clustering — A technique that allows processors from different computers to work together over a network on complex problems. Also called grid computing.

CMOS memory — Short for complementary metal oxide semiconductor, it provides semipermanent storage for system configuration information that may change.

coaxial cable — A type of cable consisting of an inner conductor wire surrounded by insulation, a conductive shield, and a cover; the type of cable provided by cable television services.

cold boot — A computer startup performed by pushing a start button or switch when the computer is not on.

command prompt — A location onscreen where users can enter text-based commands, marked by a prompt symbol or symbols, such as C:\>:.

command-based user interface — Access to and command of a computer system by giving the computer text commands to perform basic activities.

communications medium — Anything that carries a signal between a sender and a receiver.

communications satellite — A microwave station placed in outer space that receives a signal from one point on earth and then rebroadcasts it at a different frequency to a different location.

compact disk read-only memory (CD-ROM) — Commonly referred to as a CD, the first optical media to be mass-marketed to the general public. CDs are read-only—once data has been recorded on one, it cannot be modified.

competitive advantage — A significant and long-term benefit. For example, an athlete might obtain a competitive advantage by using a computer to analyze technique and training methods.

competitive intelligence — A form of business intelligence concerned with information about competitors.

compiler — A language translator that coverts a complete program, such as a COBOL program, into a complete machine-language program.

computational science — An area of computer science that applies the combined power of computer hardware and software to solving difficult problems in various scientific disciplines.

computed — A field determined from other fields, instead of being entered into a database.

computer — A digital electronics device that utilizes hardware to accept the input of data, and software, or a computer program, to process and store the data and produce some useful output.

computer ethics — Ethical issues that deal with computer use; also, the ethical responsibilities of those that work with and develop computer systems.

computer forensics — The process of examining computing equipment to determine if it has been used for illegal, unauthorized, or unusual activities.

computer housecleaning — Involves organizing the data files and software on your computer and removing unneeded files, thus improving computer performance and enhancing your use of the computer.

computer literacy — A working understanding of the fundamentals of computers and their uses.

computer network — A collection of computing devices connected together to share resources such as files, software, processors, storage, and printers; a specific type of telecommunications network that connects computers and computer systems for data communications.

computer platform — The combination of hardware configuration and system software for a particular computer system.

computer profiling — The use of cookies and other Web technologies to gather information about customers visiting Web sites. This information is then used for targeted marketing, and has also raised concerns about privacy.

computer program — Set of instructions or statements to the computer that directs the circuitry within the hardware to operate in a certain fashion.

computer programmers — People who write or create the sets of instructions or statements that become complete programs; software developers, individuals who design and implement software solutions.

computer programming — Software development, the process of developing computer programs to address a specific need or problem.

computer scientist — A person who uses computers to help with the software design process or conducts research into computing topics such as artificial intelligence, robotics, and electrical circuits.

computer system — Any device that supports the activities of input, processing, storage, and output; the hardware, software, the Internet, databases, telecommunications, people, and procedures that comprise the computing experience.

computer systems management — The team of managers that can include the chief information officer (CIO) and other computer systems executives.

computer-aided software engineering (CASE) — Software-based tools that automate many of the tasks required in a systems development effort.

computer-assisted design (CAD) — Software that assists designers, engineers, and architects in designing three-dimensional objects.

computer-based information system (CBIS) — An information system that makes use of computer hardware and software, databases, telecommunications, people, and procedures to manage and distribute digital information.

computerized collaborative work system — The hardware, software, people, databases, and procedures needed to provide effective support in group decision-making settings; also called a group decision support system.

computing platform -- A computer's type, processor, and operating system, for example, Intel Pentium processor running Microsoft Windows operating system, on a tablet PC.

consumer-generated media (CGM) — News or other topical coverage that is captured or reported by non-professionals, that is, the general public.

consumer-to-consumer (C2C) e-commerce — The use of the Web to connect individuals who wish to sell their personal belongings with people shopping for used items; also known as simply C2C.

content-filtering software — Software that works with the Web browser to check each Web site for indecent materials (defined by the installer of the software) and allow only "decent" Web pages to be displayed.

content streaming — Also known as streaming media, streaming video, or streaming audio, a technique to deliver multimedia without a wait, since the media begins playing while the file is still being delivered.

continuous improvement — An approach to systems development in which an organization makes changes to a system when even small problems or opportunities occur.

contract software — A specific software program developed for a particular company or organization.

control unit — One of three primary elements within a central processing unit (CPU), it sequentially accesses program instructions, decodes them, and also coordinates the flow of data in and out of various system components.

cookie — A text file placed on your hard disk by a Web site you visit, for the purpose of storing information about you and your preferences.

coprocessor — Special-purpose processor, typically used in larger workstations, that speeds processing by executing specific types of instructions, while the CPU works on another processing activity.

counterintelligence — A form of business intelligence concerned with protecting your own information from access by your competitors.

cracker — A criminal hacker, a computer savvy person who attempts to gain unauthorized or illegal access to other computer systems to harm the system or to make money illegally.

cradle — *See* docking station.

crash — A program failure that occurs when the program executes commands that cause the computer to malfunction or shut down.

crawling — The process of continually following all Web links in an attempt to catalog every Web page by topic.

creative analysis — The investigation of new approaches to existing problems.

critical analysis — A skeptical and doubtful approach to problems, such as questioning whether or not the current computer system is still effective and efficient.

critical patch — Software update that repairs flaws that allow for serious breaches of security.

critical success factors (CSF) — A requirements analysis approach in which users and stakeholders are asked to list only those factors or items that are critical to the success of their area or the organization.

curiosity-driven — A form of research based on seeking information prompted by a personal thought or question, rather than a topic being assigned, that is responsible for most of the world's great inventions.

custom-designed software — Programs built or developed by individuals and organizations to address specific needs.

cybermall — A Web site that allows visitors to browse through a wide variety of products from varying e-tailers.

cyberterrorism — A form of terrorism that uses attacks over the Internet to intimidate and harm a population.

data — Items stored on a digital electronics device including numbers, characters, sound, music, graphics, anything that can be expressed and recorded.

data analysis — A process that involves developing good, nonredundant, adaptable data, and evaluating data to identify problems with the content of a database.

data communications — A specialized subset of telecommunications, referring to the electronic collection, processing, and distribution of data, typically between computer system hardware devices.

data conversion — The transformation of existing computerized files into the proper format to be used by a new system.

data definition language (DDL) — A collection of instructions and commands used to define and describe data and data relationships in a specific database.

data dictionary — A detailed description of all data used in a database, including information such as the name of the data item, who prepared the data, and the range of values for the data.

data haven — Country that has few restrictions on telecommunications or databases.

data integrity — Ensuring that data stored in the database is accurate and up to date.

data item — The specific value of an attribute, found in the fields of the record describing an entity.

data manipulation language (DML) — A specific language provided with a database management system, that allows people and other database users to access, modify, and make queries about data contained in the database and to generate reports.

data mart — A small data warehouse, often developed for a specific person or purpose.

data mining — The process of extracting information from a data warehouse or a data mart; sifting through the combined information of any one customer or group of customers to recognize trends and tendencies—and ultimately pitch products and services specifically for that customer's interests.

data preparation — The process of converting manual files into computer files.

data redundancy — Duplication of data that occurs when data is copied, stored and used from different locations.

data warehouse — A database that holds important information from a variety of sources.

database — A collection of integrated and related files; a collection of data organized to meet users' needs; an organized collection of facts and information.

database administrator (DBA) — Skilled and trained computer professionals who direct all activities related to an organization's database, including providing security from intruders.

database approach — A data management strategy in which multiple application programs share a pool of related data.

database backup — A copy of all or part of a database, usually made on a regular scheduled basis.

database management software — Personal productivity software that can be used to store large tables of information and produce documents and reports.

database management system (DBMS) — A group of programs that manipulate a database and provide an interface between the database and the user or the database and application programs.

database recovery — The process of returning the database to its original, correct condition if the database has crashed or been corrupted.
database servers — Computers that store organizational databases, and respond to user queries with requested information.
database system — Everything that makes up the database environment, including a database, a database management system, and the application programs that utilize the data in the database.
date field — In a database, field that is a temporal identifier, such as 06/12/06, that can be sorted or used in computations.
decision making — A process that takes place in three stages: intelligence, design, and choice.
decision support system (DSS) — An information system used to support problem-specific decision making.
decision table — A systems development tool, often used as an alternative to or in conjunction with flowcharts, for displaying various conditions that could exist and the different actions the computer should take as a result of these conditions.
dedicated line — A line that leaves the connection open continuously to support a data network connection.
defragmenting — Rearranging files on a hard disk to increase the speed of executing programs and retrieving data on the hard disk.
demand fulfillment — The supply chain management process of getting the product or service to the customer.
demand planning — The supply chain management activity of analyzing buying patterns and forecasting customer demand.
demand report — Report developed to give certain information at a person's request.
desktop computer — One of the most popular types of personal computers, designed to sit on a desktop.
desktop publishing — Software to design page layouts for magazines, newspapers, books, and other publications.
device driver — Software that interfaces with an operating system to control an input or output device, such as a printer.
dial-up connection — A narrowband Internet service, with data transfer rates of around 50 Kbps, provided by a multitude of Internet Service Providers (ISPs).
digital — Signals that exist in one of two possible values; the opposite of analog.
digital art — A new form of art that uses computer software as the brush and the computer display as the canvas.
digital audio — Any type of sound, including voice, music, and sound effects, recorded and stored digitally as a series of 1's and 0's.
digital camcorder — Special-purpose computer device used to take full-length digital video that you can watch on your TV, download to your computer, or transfer to CD, DVD, or VCR tape.
digital camera — An optical recording device or special-purpose computer device that captures images through a lens and stores them digitally rather than on film.
digital certificate — A type of electronic business card that is attached to Internet transaction data to verify the sender of the data.
digital convergence — The trend to merge multiple digital services into one device.
digital divide — The social and economic gap between those that have access to computers and the Internet and those that do not.
digital electronics device — Any device that stores and processes bits electronically.
digital forensics — *See* computer forensics.
digital graphics — Computer-based media applications that support the creation, editing, and viewing of 2-D and 3-D images, animation, and video.
digital graphics animation — The display of digital images in rapid succession to provide the illusion of motion.
digital imaging — Working with photographic images.
digital media — Digital technologies of all kinds that serve and support digital music, video, and graphics.
Digital Millennium Copyright Act — A U.S. law enacted in 1998 that provides global copyright protection.
digital music — A subcategory of digital audio that involves recording and storing music as a series of 1's and 0's.
digital rights management (DRM) — The technology invented to protect intellectual property in digital files.
digital satellite service (DSS) — A wireless broadband Internet service, with data transfer rates of around 400 Kbps, provided by companies such as EarthLink and StarBand, typically used in situations where neither cable broadband nor DSL is available.
digital signal — A signal that at any given time is either high or low, in discrete voltage states.
digital video disk (DVD) — An optical media that stores over 4.7 GB of data in a fashion similar to CDs except that DVDs are able to write and read much smaller pits on the disk surface.
digital voice recorder — A device that stores recordings in standard digital sound formats that can be transferred to a computer for transcription or editing.
digital-to-analog converter — A device that translates sound and music to analog signals.
digitization — The process of transforming non-digital information to bit representation.
direct access — A storage medium feature that allows a computer to go directly to a desired piece of data, by positioning the read-write head over the proper track of a revolving disk.
direct conversion — A startup approach that involves stopping the old system and starting the new system on a given date.
direct observation — A data collection technique in which one or more members of the analysis team directly observe the existing system in action.
display resolution — A measure, typically in terms of width and height, of the number of pixels on the screen, with a larger number of pixels per square inch considered a higher image resolution.
distance education — Conducting classes over the Web with no physical class meetings.
distributed computing — Computing that involves multiple remote computers that work together to solve a computation problem or to perform information processing.

distributed denial of service (DDoS) — A type of computer attack in which many computers are used to launch simultaneous repeated requests at a Web server in order to overwhelm it so that it is unable to function.

distributed database — A database in which the actual data may be spread across several databases at different locations, connected via telecommunications devices.

docking station — A small stand for a handheld device that is used to recharge its battery and to connect to a PC.

documentation — All flowcharts, diagrams, and other written materials that describe the new or modified system.

Domain Name System (DNS) — Used on the Internet to translate domain names into IP addresses.

domain names — The English name associated with a numerical Internet Protocol (IP) address, such as www.fsu.edu.

dot matrix printer — An impact printer that uses a matrix of small printing pins to print characters and graphical images. An older technology.

dot pitch — The measure of space between pixels. The lower the dot pitch, the better the image quality.

dots-per-inch (dpi) — A printer's output resolution; the greater the number of dots printed per inch, the higher the resolution.

downloading — Transferring data or files from a remote computer to the local computer over a network or the Internet.

drawing software — Programs that provide tools to create, arrange, and layer graphical objects on the screen to create pictures; also called vector graphics software.

drill down — Clicking links through a series of Web pages to find additional information.

drum machines — Musical instruments that allow the musician to record drum beat patterns by tapping on pressure-sensitive buttons or pads to produce sampled drum sounds that can be played back in a looping pattern.

DSL (Digital Subscriber Line) connection — A broadband Internet access, with data transfer rates of around 1.5 Mbps, provided by the phone company, or Internet Service Providers working with the phone company.

DSL modem — Digital Subscriber Line modem, for connecting digital devices using a digital signal over telephone lines, for relatively inexpensive high-speed access to the Internet.

dual-core processor — A processor that uses two processors on one chip that work together to provide twice the speed of traditional single-core chips.

dumpster diving — Rummaging through trash to steal credit card numbers or other personal information.

dynamic IP address — An Internet Protocol (IP) address that is assigned to computers as needed.

dynamic Web pages — Web pages that are custom created on-the-fly.

e-book — Digital versions of books and other text-based publications.

e-cash — Electronic cash, a Web service that provides a private and secure method of transferring funds from a bank account or credit card to online vendors or individuals for e-commerce transactions.

e-commerce — Electronic commerce, the process of conducting business or other transactions online, over the Internet, or using other telecommunications and network systems; systems that support electronically executed transactions.

e-commerce host — A company that takes on some or all of the responsibility of setting up and maintaining an e-commerce system for a business or organization.

e-commerce software — Software designed specifically to support e-commerce activities, including catalog management, shopping cart use, and payment.

economic feasibility — The determination of whether a project makes financial sense and whether predicted benefits offset the cost and time needed to obtain them.

electronic cash — A Web service that provides a private and secure method of transferring funds from a bank account or credit card to online vendors or individuals for e-commerce transactions; also known as e-cash or digital cash.

electronic commerce — Also known as e-commerce, the process of conducting business or other transactions online, over the Internet, or using other telecommunications and network systems; systems that support electronically executed transactions.

Electronic Data Interchange (EDI) — The network systems technology, standards, and procedures that allow output from one system to be processed directly as input to other systems, without human intervention.

electronic exchange — An industry-specific Web resource created to provide a convenient centralized platform for B2B e-commerce among manufacturers, suppliers, and customers.

electronic funds transfer (EFT) — Moving money electronically from one institution to another.

electronic wallet — An application that encrypts and stores credit-card information, e-cash information, bank account information—all the personal information required for e-commerce transactions—securely on your computer; also known as e-wallet.

e-mail—Electronic mail; the transmission of messages over a network to support asynchronous text-based communications.

e-mail attachment — Typically a binary file or a formatted text file that travels along with an e-mail message but is not part of the e-mail ASCII text message itself.

e-mail body — The component of an e-mail transmission that contains an ASCII text message written by the sender to the recipient.

e-mail header — The component of an e-mail transmission that contains technical information about the message, such as the destination address, source address, subject, and date and time.

embedded computer — A special-purpose computer that are embedded in and control many electrical devices on which we depend.

embedded operating systems — Operating systems, like those for many small computers and special-purpose devices, that are embedded in a computer chip.

emoticon — Combination of keyboard characters to convey underlying sentiments, such as :-) to create a sideways facial expression meaning happy or smiling.

encoder software — Software used to transfer digital music files from one format to another.

encryption — The use of high-level mathematical functions and computer algorithms to encode data so that it is unintelligible to all but the intended recipient.

end-user systems development — The development of computer systems by individuals outside of the formal computer systems planning and departmental structure.

enterprise resource planning (ERP) software — A set of transaction processing system applications for handling an organization's routine activities, bundled into a unified package.

enterprise service provider (ESP) — A company that provides users, typically large businesses or organizations, server connections and software that allow employees to access a corporate Intranet from outside the office.

enterprises — Large businesses and organizations, which make extensive use of distributed computing.

entity — A generalized class of people, places, or things (objects) for which data is collected, stored, and maintained.

ergonomic keyboard — A set of keys designed so that the user can enter data in a manner that is comfortable and avoids strain on the hands or wrists.

ergonomics — The applied science of designing and arranging the things we work with, such as computer systems, in a manner that promotes health, especially the study of designing and positioning computer equipment to reduce health problems.

e-tailing — B2C e-commerce; a take-off on the term retailing, since it is the electronic equivalent of a brick-and-mortar retail store.

Ethernet — The most widely used network standard for private networks.

ethics — Matters dealing with what is generally considered right or wrong.

event-driven review — A systems review procedure that is triggered or caused by a problem or opportunity such as an error, a corporate merger, or a new government regulation.

e-wallet — Electronic wallet, an application that encrypts and stores credit-card information, e-cash information, bank account information — all the personal information required for e-commerce transactions — securely on your computer.

exception report — Report automatically produced when a situation is unusual or requires action.

expansion board — Also known as an expansion card, a circuit board often packaged with specialized peripheral devices for installation on a computer's motherboard.

expansion card — Also known as an expansion board, a circuit board often packaged with specialized peripheral devices for installation on a computer's motherboard.

expansion slots — Connecting sites on a computer motherboard where circuit boards can be inserted to add additional system capabilities.

expert system (ES) — A computerized system that acts or behaves like a human expert in a field or area; an information system that can make suggestions and reach conclusions in much the same way that a human expert can; a subfield of artificial intelligence.

Extensible Markup Language (XML) — *See* XML.

extension — An addition to a file name, placed after the file name following a period, which identifies the file type. For example, Microsoft Word documents have a .doc extension

extranet — An arrangement whereby Intranet content is extended to specific individuals outside the network, such as customers, partners, or suppliers.

facial pattern recognition — A method of identification that uses a mathematical technique to measure the distances between points on the face. Software can quickly compare these measurements to a database of known faces and come up with a match if one exists.

fair use doctrine — The legal principle that describes when and how copyrighted material can be legally used.

feasibility analysis — A key part of the systems development process that involves investigation into technical, economic, legal, operational, and schedule feasibility of a project or system.

feeds — A blog distribution system that uses RSS technology and XML to deliver Web content that changes on a regular basis.

fiber-optic cable — A type of cable that consists of thousands of extremely thin strands of glass or plastic bound together in sheathing; because it transmits signals via light rather than electricity, fiber-optic cable is extremely fast and reliable.

field — A name, number, or combination of characters that in some way describes an aspect of an object.

field design — The specification of type, size, format, and other aspects of each field.

field name — An identifying label applied to a particular field.

fifth generation languages (5GLs) — High-level programming languages used to create artificial intelligence or expert systems applications.

file — A named collection of instructions or data stored in a computer or computer device; a collection of related records; also called a table in some databases.

file allocation table (FAT) — The file system used by MS-DOS and Windows operating systems prior to Windows XP.

file organizer — A database that goes beyond the capabilities of a flat file to efficiently store and/or retrieve data. File organizers are typically used with personal computers.

file ownership — An identification of the creator of a file or folder. In an operating system, users are the owners of the files they create.

file servers — Computers that store organizational and user files, delivering them to workstations on request.

file system — A way of organizing how data and files are physically stored and how they are logically manipulated; a function of the operating system.

fingerprint scan — A form of biometric authentication in which a user presses a finger on a scanning device to establish his or her identity.

firewall — A device or software that filters the information coming onto a network to protect the network computer from hackers, viruses, and other unwanted network traffic.

FireWire — A standard, and a type of expansion card, for fast video transfer from a camera to the computer.

flash BIOS — A basic input/output system recorded on a flash memory chip rather than a ROM chip. Flash memory can store data permanently, like ROM, but can also be updated with new revisions when they become available.

flash memory card — A chip that, unlike RAM, is nonvolatile and keeps its memory when the power is shut off.

flash mob — A group of people who assemble suddenly in a public place, do something unusual, and then disperse. Flash mobs organize through cell phone text messaging or e-mail.

flat file — A type of database in which there is no relationship between the records; often used to store and manipulate a single table or file.

flat panel display — A flat, space-saving display that uses liquid crystals between two pieces of glass to form characters and graphic images on a backlit screen; also known as liquid crystal display (LCD).

FLOPS — Floating-point operations per second, a more precise measure than MIPS of processor performance.

floppy disk — Portable, low-capacity (1.44 MB) magnetic storage medium. The most popular form of portable storage in the 1990s, floppies are being phased out in favor of higher capacity, less expensive media such as USB drives.

flowchart — A systems development tool or graphical diagram that reveals the path from a starting point to the final destination.

forensic audio — Digital processing to de-noise, enhance, edit, and detect sounds to assist in criminal investigations.

forensic graphics — Art used to create animations and demonstrative exhibits to use in courts of law in order to explain theories and present evidence.

format – A function performed by the operating system on a disk before it can be used.

frames — Series of bit-mapped images that when shown in quick succession create the illusion of movement.

freeware — Software that has been placed in the public domain and is free to use.

frequency — The speed at which an electronic communications signal can change from high to low.

front-end application — Software that directly interacts with people or users.

full backup — Backing up all files. Compare to incremental backup.

fuzzy logic — A specialty research area in computer science developed to deal with ambiguous criteria or probabilities and events that are not mutually exclusive; also known as fuzzy sets; a form of mathematical logic used in artificial intelligence systems.

fuzzy sets — A specialty research area in computer science developed to deal with ambiguous criteria or probabilities and events that are not mutually exclusive; also known as fuzzy logic.

game theory — The study of strategies used by people, organizations, or countries who are competing against each other.

gamepad — A specialized input device used to control and manipulate game characters in a virtual world.

garbage in, garbage out (GIGO) — Inaccurate data being entered into a database, resulting in inaccurate output.

General Public License (GPL) — A legal arrangement that makes software available for free to its users.

general purpose I/O device — Input or output devices designed for a variety of computer environments, such as the standard keyboard and mouse.

general-purpose computer — Computer that can be programmed and used for a wide variety of tasks or purposes, such as mobile and personal computers.

general-purpose database — A database that can be used for a large number of applications.

genetic algorithm — An approach, based on the theory of evolution, to solving large, complex problems where a number of algorithms or models change and evolve until the best one emerges.

geographic information system (GIS) — An application capable of storing, manipulating, and displaying geographic or special information, including maps of locations or regions around the world.

gigaflop — Billions of floating-point operations per second, a measurement for rating the speed of microprocessors.

gigahertz (GHz) — Billions of cycles per second, a measurement used to identify CPU clock speed; for example, a 500 GHz processor runs at 500 billion cycles per second.

global positioning system (GPS) — A sophisticated satellite networking system that is able to pinpoint exact locations on earth; a special-purpose computing device, typically installed in an automobile or boat, that uses satellite and mobile communications technology to pinpoint current location.

global supply management (GSM) — A process that provides methods for businesses to find the best deals on the global market for raw materials and supplies needed to manufacture their products.

globalization — Changes in societies and the world economy resulting from dramatically increased international trade and cultural exchange.

goal-seeking analysis — The process of determining what problem data is required for a given result.

graphical user interface (GUI) — Access to and command of a computer system by using pictures or icons on the screen and menus, which many people find easier to learn and use compared to a command-based user interface.

graphics tablets and pens — Input tools that allow you to draw with a penlike device on a tablet to create drawings on a display.

grid computing — *See* clustering.

group decision support system (GDSS) — The hardware, software, people, databases, and procedures needed to provide effective support in group decision-making settings; also called a computerized collaborative work system.

group ownership — Access to files and folders based on membership in groups established by the system administrator.

groupware — Group decision support system software that helps with joint work group scheduling, communication, and management; software that allows network users to collaborate over the network; also called workgroup software.

GSM — One of two predominant cell phone networking standards, the other being CDMA. CDMA is used more in the United States, and GSM in Europe and Asia, though GSM is rapidly gaining ground in the U.S. as well.

hacker — A person who knows computer technology and spends time learning and using computer systems. More recently used to label individuals who subvert computer security without authorization.

handheld computer — A type of small, easy to use mobile computer.

handheld scanner — Compact data input scanning device that can convert pictures, forms, and text into bit-mapped images.

hard copy — Computer output printed to paper.

hardware — The tangible components of a computer system.

hertz (Hz) — A measurement of signal frequency, in cycles per second.

heuristics — A problem-solving method used in decision support systems, often referred to as "rules of thumb"—commonly accepted guidelines or procedures that usually find a good, though not optimal, solution.

hierarchy of data — The progressive levels of characters, fields, records, files, and databases, such that characters are combined to make a field, fields are combined to make a record, and so forth.

high-capacity diskette — A magnetic storage device, such as the Iomega Zip disk, that stores 69 to 83 times as much data as on a standard floppy disk.

high-definition TV (HDTV) — Television that uses a resolution that is twice that of traditional television displays for sharper, crisper images. HDTV uses a widescreen format, which means it uses the same height and width ratio used in movie theaters.

HomePLC — Home power-line communication, networking that takes advantage of a home's existing power lines and electrical outlets to connect computers; also called power-line networking.

HomePNA — Home phone-line networking alliance, networking that takes advantage of existing phone wiring in a residence; also called phone-line networking.

hotspot — Area within the range of a Wi-Fi network or networks.

HTML tag — Hypertext Markup Language tag, a specific command indicated with angle brackets (<>) that tells a Web browser how to display items on a page.

human-readable data — Data that a person can read and understand.

hydrophone — Underwater microphone.

hyperlink — An element in an electronic document, such as a word, phrase, or an image, that when clicked opens a related document. Hyperlinks are a cornerstone of the World Wide Web.

hypermedia — Pictures or other media that act as a links to related documents.

hypertext — Text that acts as a link to a related document.

Hypertext Markup Language (HTML) — The primary markup language that is used to specify the formatting of a Web page.

Hypertext Transfer Protocol (HTTP) — The protocol used to control communication between Web clients and servers.

identity theft — The criminal act of using stolen information about a person to assume that person's identity.

image-based rendering — A form of computer modeling that allows digital photorealistic actors stored on computer to be manipulated as animated characters.

immediacy — A database characteristic, referring to a measure of how rapidly changes must be made to data.

impact printer — Printer that uses a forcible impact to transfer ink to the paper in order to print characters.

immersive virtual reality — In which the user becomes fully immersed in an artificial, three-dimensional world that is completely generated by a computer.

in-memory database — Uses a computer's memory instead of a hard disk to store and manipulate important data.

incremental backup — Backing up only the files that have changed since the last backup. Compare to full backup.

informatics — The combination of traditional disciplines, like science and medicine, with computer systems and technology.

information — Data organized and presented in a manner that has additional value beyond the value of the data itself.

information overload — The inability to find the information you need on the ever-growing Web due to an overabundance of unrelated information.

information security — The protection of information systems and the information they manage against unauthorized access, use, manipulation, or destruction, and against the denial of service to authorized users.

information system — A computer system that makes use of hardware, software, databases, telecommunications, people, and procedures to manage and distribute digital information.

information technology — Issues related to the components of an information system.

infrared transmission — The sending of signals through the air via light waves, a type of wireless media.

ink-jet printer — A popular type of printer, providing economical but relatively low-print-quality black and white or color output.

input — As a verb, the capturing and gathering of raw data. As a noun, the data to be entered into a computer system.

input device — A device that helps to capture and enter raw data into a computer system, such as a keyboard, mouse, or touch screen.

installation — The process of physically placing computer equipment on the site and making it operational.

instant messaging — Synchronous one-to-one text-based communication over the Internet.

instruction set — The specific set of instructions that a processor is engineered to carry out.

integrated circuit — A module, or chip, consisting of multiple electronic components and used to store and process bits and bytes in today's computers.

integrated digital studio — High-tech electronic system that packages many digital recording devices in one unit for convenient home recording.

integrated software package — An application that contains several basic programs, such as word processing, spreadsheet, and calendar, offering a range of capabilities, but with less power and for less money than the standalone software included in software suites.

integration testing — Testing all related systems together, to ensure that the new program(s) can interact with other major applications.

intellectual property — A product of the mind or intellect over which the owner holds legal entitlement. Intellectual property includes ideas and intangible objects such as poetry, stories, music, and new ways of doing things or making things.

intellectual property rights — Rights associated with the ownership and use of software, music, movies, data, and information.

intelligent agent — An intelligent robot, or bot, consisting of programs and a knowledge base used to perform a specific task for a person, a process, or another program.

interactive media — Multimedia presentations that involve user interaction for education, training, or entertainment, typically by combining both digital audio and digital video for a full multimedia experience.

interactive TV — A digital television service that includes one or more of the following: video on demand, personal video recorder, local information on TV, purchase over TV, Internet access over TV, and video games over TV.

interior threats — Threats from within a private network that can be intentional or unintentional. Unintentional threats can occur when users make mistakes or exceed their authorization. Intentional threats come from registered users who desire to do the system harm or steal information.

Internet — The world's largest public computer network; a network of networks that provides a vast array of services to individuals, businesses, and organizations around the world.

Internet2 — A research and development consortium led by U.S. universities and supported by industry and government to develop and deploy advanced network applications and technologies for tomorrow's Internet.

Internet addiction — A condition that can exist when people are online for long periods of time, cannot control their online usage, jeopardize their career or family life from excessive Internet usage, and lie to family, friends, and coworkers about excessive Internet use.

Internet access over TV — Services, such as WebTV, that allow viewers to navigate the Web on their television sets.

Internet backbone — The collection of the many national and international communications networks owned by major telecommunication companies, such as Verizon and Sprint, that provides the hardware over which Internet traffic travels.

Internet fraud — The crime of deliberately deceiving a person over the Internet in order to damage them and to obtain property or services from him or her unjustly.

Internet hosts — The more than 200 million computers joined together to create the Internet, the world's largest network.

Internet Protocol — Along with Transmission Control Protocol (TCP), one of the sets of policies, procedures, and standards used on the Internet to enable communications between two devices; defines the format and addressing scheme used for packets.

Internet radio — Radio programming that is similar to local AM and FM radio, except that it is digitally delivered to a computer over the Internet, and there are more choices of stations.

Internet service providers (ISPs) — Companies that provide users access to the Internet through points of presence.

internetwork — Networks joined together to make larger networks, such as today's Internet, so that users on different networks can communicate and share data.

interpreter — A language translator that converts each statement in a programming language into a machine language and executes the statement.

intranet — A private network that uses the protocols of the Internet and the Web, TCP/IP and HTTP, along with Internet services such as Web browsers.

invisible information gathering — Collecting and storing information about individuals without their knowledge or explicit consent.

IP address — A unique Internet Protocol address consisting of a series of four numbers (0 to 255) separated by periods, assigned to all devices connected to the Internet.

IrDA ports — Infrared Data Association ports, for using infrared transmission to connect most handheld and notebook computers to desktop computers and other digital devices.

iterative approach — An approach to systems development in which each phase of the SDLC is repeated several times (iterated).

Java — An object-oriented programming language developed by Sun Microsystems that can be used to create programs that run on any operating system and on the Internet.

Java applets — Small applications developed using the Java programming language.

JavaScript — A programming language developed specifically for the Web and more limited in nature than Java and other high-level programming languages.

joining — A basic data manipulation that involves combining two or more tables.

joystick — A specialized input device, in the form of a swiveling stick, used to control and manipulate game characters in a virtual world.

jukebox software — Software that allows computer users to categorize and organize digital music files for easy access.

key — A field in a record that is used to identify the record.

key-indicator report — A special type of scheduled report that summarizes the previous day's critical activities, and is typically available at the beginning of each workday.

key-logging software — Software that is installed on a computer to record all keystrokes and commands. The recording is later collected from a remote computer over the Internet and played back in order to spy on the user's actions and sometimes to steal usernames and passwords.

keywords — Words that are specified to a search engine to find information on a topic of interest.

kiosk — Special-purpose computing station, often equipped with touch-sensitive screens, made available where the public can access location-relevant information.

knowledge worker — A professional who makes use of information and knowledge as a significant part of his or her work.

language translator — System software that converts a statement from a high-level programming language into machine language to be executed by the CPU.

laptop computer — Also called a notebook, a type of personal computer designed for portability.

laser printer — A popular type of printer that provides the cleanest output and is typically used when professional-quality documents are required.

LCD projector — Small, portable device used to project presentations from a computer onto a larger screen.

learning system — A combination of software and hardware that allows the computer to change how it functions or that reacts to situations based on feedback it receives.

legal feasibility — The determination of whether laws or regulations may prevent or limit a systems development project.

libel — The deliberate act of defamation of character by making false statements of fact.

license agreement — A contract that software companies may require users to accept that specifies how the software can be used.

line-of-sight — A medium in which the straight-line view between sender and receiver must be and unobstructed.

Linux — An open-source, GUI-based operating system developed in 1991 by Linus Torvalds, which runs on computer systems ranging from small personal computers to large mainframe systems.

liquid crystal display (LCD) — A flat, space-saving panel display that uses liquid crystals between two pieces of glass to form characters and graphic images on a backlit screen; also known as flat panel display.

listservs — Special interest groups that create online communities for discussing topic-related issues via e-mail.

load balancing — The sharing of system processing among servers so that if one goes down, the others pick up the slack.

local area network (LAN) — A network that connects computer systems and devices within the same geographical area.

Local information on TV — A feature of interactive TV that provides local community news and information.

local resource — Along with network, one of two types of resources that workstations typically have access to: the files, drives and printers or other peripheral devices that are accessible to the workstation on or off the network.

localization — The process of creating multiple versions of a Web site, each in a different language and catering to a different cultural bias.

location-based m-commerce — Making use of global positioning system (GPS) or the cell network to track your current location in order to provide location-related services such as weather reports, road maps, lists of nearby merchants, lists of nearby friends, and traffic reports.

logic error — The result of a poor algorithm that contains a flaw in reasoning, causing a program to crash, behave in an unexpected fashion, or not effectively solve the problem for which it was designed.

logical field — Database field limited to certain operators, such as "yes" or "no."

logical view —The way a programmer or user thinks about data. With a logical view, the programmer or user doesn't have to know where the data is physically stored in the computer system. Compare to physical view.

logical versus physical access — Two ways of viewing memory access, exemplified by how memory-management programs convert a logical request for data or instructions into the physical location where the data or instructions are stored.

lossless compression — A type of file compression that allows the original data to be reconstructed without loss.

lossy compression — A type of file compression that accepts some loss of data to achieve higher rates of compression. Savings in file size can be considerable and is essential for fast-loading Web pages.

MAC address — Media Access Control address, a unique hardware code printed by the manufacturer on network interface cards and wireless adapters, and is how the devices are identified on the network. Notebook computers with Wi-Fi capability have a sticker on the bottom with the adapter's MAC address. MAC address filtering can be enabled in routers and access points so that they allow only specified MAC addresses to connect to the network.

machine cycle — The combination of the instruction phase and the execution phase, which together are required to execute an instruction.

machine languages — First-generation programming languages with instructions written in binary code, telling the CPU exactly which circuits to switch on and off.

machine-readable data — Information that is usually stored as bits and bytes and can be read and understood by a computer system.

magnetic disk — A thin steel platter (hard disk) or piece of Mylar film (floppy disk) used to store data, with widely varying capacity and portability.

magnetic ink character recognition (MICR) — Use of a special-purpose optical reading device to read magnetic-ink characters, such as those written on the bottom of a check.

magnetic storage — Technology that uses magnetic properties of iron oxide particles to store data more permanently than RAM.

magnetic tape — Mylar film coated with an iron oxide; used as a sequential access storage medium.

mail server — A server that handles sending and receiving e-mail messages.

mainframe server — Computer system typically used by universities and large organizations, which requires plenty of processing power and speed.

management information system (MIS) — An information system used to provide useful information to decision makers usually in the form of a report.

managers — The people within an organization who are most capable of initiating and maintaining change.

market segmentation — A method of market research in which customer opinions are divided into categories of race, gender, and age to determine which segment a product appeals to most.

markup language — Used to describe how information is to be displayed on a Web browser. It typically combines the information, such as text and images, along with additional instructions for formatting. HTML is an example.

massively parallel processing — A form of multiprocessing, used in supercomputers, that works by linking a large number of powerful processors to operate together.

master files — Permanent files that are updated over time.

m-commerce — Mobile commerce, a form of e-commerce that takes place over wireless mobile devices such as handheld computers and cell phones.

media cards — Flash memory cards used in media devices such as digital cameras, camcorders, and portable MP3 players.

media player software — Media software programs that combine digital music functions with video support; popular media players such as Windows Media Player from Microsoft and QuickTime Player from Apple are free downloads.

medical informatics — Combining traditional medical research with computer systems and technology to reduce medical errors and improve health care delivery.

megahertz (MHz) — Millions of cycles per second, a measurement used to identify CPU clock speed; for example, a 500 MHz processor runs at 500 million cycles per second.

megapixels — 1 megapixel = 1 million pixels, a measure of picture resolution in digital cameras. An inexpensive digital camera can capture around 2 megapixels, and the best can capture over 8 megapixels.

meta search engine — A tool that allows you to run keyword searches on several search engines at once.

meta tags — Information tags, containing terms such as business-related keywords, that are read by search engines and Web servers, but not displayed on the page by a Web browser.

methods — Instructions to perform a specific task in an object-oriented program.

metropolitan area network (MAN) — A large, high-speed network connecting a series of smaller networks within a city or metropolitan size area.

microcontroller — Special-purpose computer (typically an entire computer on one chip) that is embedded in electrical and mechanical devices in order to control them. Also called embedded computer.

microdrive — Tiny hard drives that can store gigabytes of data on a disk one or two inches in size, usually found in handheld computers.

microprocessor — A single module, smaller than a fingernail, that holds all of a computer's central processing unit (CPU) circuits and performs systemprocessing.

microwave transmission — Also known as terrestrial microwave, the line-of-sight sending of high-frequency radio signals through the air.

MIDI card — Musical instrument digital interface expansion board, available for Windows PCs and included as standard equipment in most Apple computers, that allows computers to be connected to digital music devices.

midrange server — Also called mini-computer, computer system that is more powerful than personal computers and is used by small businesses and organizations to perform business functions, scientific research, and more.

MIPS — Millions of instructions per second, indicating the amount of time it takes a processor to execute an instruction, a measure of processor performance.

mirroring — Creating an exact copy of a system or portion of a system. Some software provides real-time mirroring, in which files are automatically updated in their primary storage space and the mirrored copy whenever they are created or changed.

mixing board — A large panel with many dials, buttons, and sliders that a sound engineer uses to adjust the sound quality of each instrument separately.

mobile commerce (m-commerce) — A form of e-commerce that takes place over wireless mobile devices such as handheld computers and cell phones; also known as m-commerce.

modem — An external or internal device that converts analog and digital signals from one form to the other.

Moore's Law — A trend, first predicted by Intel cofounder Gordon Moore in 1965 and since proven true, that technological innovations would be capable of doubling the transistor densities in an integrated circuit every 18 months, resulting in increased processor speeds.

motherboard — Also called the system board, the main circuit board of the computer where many of the hardware components are placed, such as the central processing unit, memory, storage, and other supportive chips.

MP3 — A digital music file format that compresses music files to less than 10 percent of their original size.

MP3 players — Special-purpose computer devices used for listening to digital music files.

multicore technology — The latest technique in chip design, housing more than one processor on a chip.

multifunction printer — A printer that combines the functionality of a printer, fax machine, copy machine, and digital scanner in one device.

multimedia — Digital devices of all kinds that serve and support digital media such as music, video, and graphics; the computer's ability to present and manipulate visual and audio media such as graphics, animation, video, sound, and music.

multimedia messaging service (MMS) — A service for mobile cell phone users to send pictures, voice recordings, and video clips to other cell phones or e-mail accounts.

multiprocessing — The simultaneous operation of more than one processing unit.

Multipurpose Internet Mail Extensions (MIME) — The protocol that e-mail servers follow to govern the transportation of e-mail attachments.

multitasking — Running more than one application at the same time.

multithreading — Running several parts of an application at the same time.

multiuser system — A computer system, such as a computer network, where multiple users share access to resources such as file systems.

musical instrument digital interface (MIDI) — A protocol, implemented in 1983, that provides a standard language for digital music devices to use in communicating with each other.

narrowband — A category of bandwidth; narrowband is slower than broadband and is more restricted in its applications and uses.

National Lambda Rail (NLR) — A U.S. cross-country, high-speed, fiber-optic network dedicated to research in high-speed networking applications.

natural language processing — Often referred to as speech recognition, the ability that allows a computer to understand and react to statements and commands made in a "natural" language, such as English.

netiquette — Network etiquette, the informal set of conduct rules for communicating online.

network — Interlinked system that can connect computers and computer equipment in a building, across the country, or around the world.

network access server (NAS) — A computer that large businesses or organizations use to access the network.

network adapter — A computer circuit board, PC Card, or USB device installed in a computing device so that it can be connected to a network.

network administrator — Person responsible for setting up and maintaining a network, implementing policies, and assigning user access permissions. Also known as system administrator.

network interface card (NIC) — A circuit board or PC card that, when installed, provides a port for the device to connect to a wired network with traditional network cables.

network operating system (NOS) — Operating system software, such as Windows 2000 and Windows 2003, that controls the computer systems and devices on a network and allows them to communicate with each other; includes security and network management features.

network service provider (NSP) — Major telecom companies, such as Sprint and Verizon, that agree to connect their networks so that users on all the networks can share information over the Internet.

networking devices — Hardware components that together with networking software enable and control communications signals between communications and computer devices.

networking media — Anything that carries an electronic signal and creates an interface between a sending device and a receiving device.

networking software — Software components that work together with hardware components to enable and control communications signals between communications and computer devices.

neural network — A computer system that can act like or simulate the functioning of a human brain, process many pieces of data at once, and learn to recognize patterns; a branch of artificial intelligence.

newsletter — Subscription-based broadcast e-mail communication, such as the *New York Times* newsletter.

node — Device attached to a network.

nonprogrammed decision — Decision that deals with unusual or exceptional situations, and in many cases are difficult to look at as a matter of a rule, procedure, or quantitative method.

nonvolatile — A characteristic of permanent or secondary memory storage, such that a loss of power to the computer does not cause a loss of data and programs.

normalization — The process of correcting data problems or anomalies to insure that the database contains good data.

notebook computer — Also called a laptop, a popular type of personal computer designed for its portability.

numeric field — A database field that contains numbers that can be used in making calculations.

object — An element of the object-oriented design approach that is composed of attributes and methods.

object code — The machine-language code necessary for a computer to execute programming instructions.

object linking and embedding (OLE) — A technology that allows Microsoft Office users to copy and paste data between office applications and link the data so that when it is changed in the source document it is automatically updated in the target document.

object-oriented (OO) systems development — A systems development process that follows a defined life cycle, much like the SDLC, and typically involves defining requirements, designing the system, implementation and programming, evaluation, and operation.

object-oriented database — A database that stores both data and the processing instructions for it.

object-oriented database management system (OODBMS) — A group of programs that manipulates an object-oriented database, providing an interface between the database and the user or the database and application programs.

object-oriented programming — An approach to programming that derives the solution to the program specification from the interaction of objects.

object-oriented programming languages — High-level programming languages, such as Visual Basic .NET, C++, and Java, that group together data, instructions, and other programming procedures.

object-relational database — Modern database management systems such as Oracle and DB2 that incorporate object-oriented capabilities in a relational database.

off-the-shelf software — An existing software program developed for the general market.

offshoring — A business practice that relocates an entire production line to another location, typically in another country, in order to enjoy cheaper labor, lower taxes, and other forms of lower overhead.

on-board synthesizers — Synthesizers located on most personal computers' sound cards that include all the standard synthesizer keyboard sounds.

on-demand media — The ability to view or listen to programming or music at any time rather than at a time dictated by television and radio schedules.

online clearinghouses — E-commerce sites that provide a method for manufacturers to liquidate stock and consumers to find a good deal.

online transaction processing — A method of processing transactions at the point of sale, which is critical for time-sensitive transactions such as making flight reservations.

Open Mobile Alliance (OMA) — An organization comprised of hundreds of the world's leading mobile operators, device and network suppliers, information technology companies, and content providers, which have joined together to create standards and ensure interoperability between mobile devices.

Open System Interconnection (OSI) model — A detailed, seven-layer model for networks, such as the Internet, that provides network technicians and administrators with a deeper understanding of the technology to design and troubleshoot networks.

open-source software — Software programs that make the source or machine code available to the public, allowing users to make changes to the software or develop new software that integrates with the open-source software.

operating system (OS) — A set of computer programs that runs or controls the computer hardware, acts like a buffer between hardware and application software, and acts as an interface between application programs and users.

operational feasibility — The determination, affected by both physical and motivational considerations, of whether or not a project can be put into action or operation.

optical character recognition (OCR) — Use of a special-purpose reading device to read hand-printed characters.

optical mark recognition (OMR) — Use of a special-purpose reading device to read "bubbled-in" forms, such as those found on exams and ballots.

optical processor — Microprocessors that use light waves instead of electrical current. It has been estimated that optical processors have the potential of being 500 times faster than traditional electronic circuits.

optical storage — Technology that uses an optical laser to burn pits into the surface of a highly reflective disk, such as a CD or DVD, to store data. Such disks hold significantly more data than a magnetic storage device.

optimization — A feature of many spreadsheet programs that allows the spreadsheet to maximize or minimize a quantity subject to certain constraints.

optimization model — A popular problem-solving method, used in decision support systems, that identifies the best solution, usually the one that best helps individuals or organizations meet their goals.

order processing system — A type of transaction processing system that supports the sales of goods or services to customers and arranges for shipment of products.

outboard device — Large racks of interconnected digital audio devices used in sound recording studios to process digital music and audio signals.

output — The results of the processing produced in a manner that is discernable to human senses or used as input into another system.

output device — A piece of hardware that allows us to observe the results of computer processing with one or more of our senses, such as a display monitor or printer.

outsourcing — A business' use of an outside company to take over portions of its workload.

packet — Data transported over the Internet as a small group of bytes, which includes the data being sent as well as a header containing information about the data, such as its destination, origin, size, and identification number.

packet-sniffing software — Software that captures and analyzes all packets flowing over a network or the Internet, often used by criminals to steal private data such as passwords and credit card numbers.

packet switching — The dividing of information into small groups of bytes in order to make efficient use of the network.

page scanner — Data input scanning device that can convert pictures, forms, and text into bit-mapped images.

pager — Small lightweight device that receives signals from transmitters.

pages printed per minute (ppm) — A measurement used to compare the speed of printers.

parallel conversion — A process during the phase-in startup approach in which parts of the old system and new system are running at the same time.

parallel processing — A form of multiprocessing that works by linking several microprocessors to operate at the same time, or in parallel.

parental controls — Applications, such as Net Nanny, that filter out adult content to make Web browsing safe for young users.

partitioning — Dividing a hard disk into two or more sections, usually to contain different operating systems.

passwords — A secret combination of characters, known only to the user, that is used for authentication

payload — The destructive action of a computer virus, such as the corruption of computer data files or system files resulting in the loss of data or a malfunctioning computer.

payment software — E-commerce software to facilitate payment for merchandise and arrange shipping.

PC Cards — Short for PCMCIA cards, small expansion cards typically inserted into notebook computers to support network adapters, modems, and additional storage devices.

PCMCIA cards — Usually called PC Cards, small expansion cards typically inserted into notebook computers to support network adapters, modems, and additional storage devices. PCMCIA is an acronym for the standard-developing Personal Computer Memory Card International Association.

PCMCIA slots — The area on notebook computers where PCMCIA cards can be inserted.

PDA (personal digital assistant) — A type of small, easy to use mobile computer.

peer-to-peer (P2P) — A network architecture that does not utilize a central server, but facilitate communications directly between clients.

performance tuning — Adjusting the database to enhance overall performance.

personal area network (PAN) — The interconnection of personal information technology devices within the range of an individual, usually around 32 feet.

personal computer (PC) — A general-purpose computer designed to accommodate the needs of an individual, such as a desktop, notebook, tablet, or handheld computer.

personal information managers (PIMs) — Software that individuals, groups, and organizations can use to store information, such as a list of tasks to complete or a list of names and addresses; a special-purpose database for storing personal information.

personal video recorder (PVR) — Electronic device, such as Tivo and Replay TV, which provides large hard drive storage to record dozens of movies and programs to be watched at your leisure.

pervasive communications — The ability to communicate with anyone, anywhere, anytime, through a variety of formats, resulting from advances in wireless and Internet communications.

pervasive computing — The growing spread of computer devices to the point where everything and anything can be used for input and output.

peta — A prefix that represents 2 to the 50th power (roughly a quadrillion, or a thousand million).

pharming — A criminal act in which hackers hijack a domain name service (DNS) server to automatically redirect users from legitimate Web sites to spoofed Web sites in order to steal personal information.

phase-in approach — A popular startup technique preferred by many organizations, in which the new system is slowly phased in, while the old one is slowly phased out.

phishing — A scam in which e-mail and Web sites are used to impersonate an authentic business in an effort to get unsuspecting customers to provide personal and private information.

photo printer — Printer, often used in conjunction with digital cameras, that produces photo-quality images on special photo-quality paper.

photo-editing software — Programs with special tools and effects designed for improving or manipulating bitmapped photograph images.

physical layer — The hardware portion of the three-layer Internet model, which also includes application (software) and transport (protocol) layers.

physical view — A view of data which includes the specific location of the data in storage or memory and the techniques needed to access the data.

pilot startup — A startup approach that involves running a pilot or small version of the new system along with the old.

piracy — The illegal copying, use, and distribution of digital intellectual property such as software, music, and movies.

pixels — Short for picture elements, small points of light or dots of ink that make up digital images.

pixilation — Fuzziness that occurs when bit-mapped images are made larger than the size at which they are captured.

plagiarism — Taking credit for someone else's intellectual property, typically a written idea, by claiming it as your own.

plotter — Output device that produces hard copy for general design work, such as blueprints and schematics.

plasma display — A flat panel display that uses plasma gas between two flat panels to excite phosphors and create light.

plug and play (PnP) — An operating system feature that allows users to attach a new hardware device and have it automatically installed and configured by the operating system.

plug-in — An application, such as Macromedia's Flash, that works with a Web browser to offer extended services, such as the ability to view audio, animations, or video.

podcast — An MP3 audio file that contains a recorded broadcast distributed over the Internet.

podcast aggregator — Software such as iPodder (http://ipodder.sourceforge.net), Doppler (www.dopplerradio.net), and PlayPod for Mac (www.iggsoftware.com/playpod) that allows you to subscribe to your favorite podcasts.

podcasting — PC-based home recording that is distributed over the Internet.

point of presence (PoP) — Utility station that enables Internet users to connect to network service providers, including networking hardware for dial-up connections.

point of sale (POS) device — Terminal or other I/O device connected to a larger system, with scanners that read codes on retail items and enter the item number into a computer system.

portable media center — Portable devices that play not only digital music but also play digital movies and sometimes TV.

portable operating system — Sometimes called generic, an operating system that can function with different hardware configurations.

port — Logical addresses used by clients and servers a logical addresses used by clients and servers that is associated with a specific servicethat are associated with a specific service.

port-scanning software — Software that searches random IP addresses for open ports. The software is allowed to run for several hours, after which the attacker can collect a list of IP addresses waiting to be hacked.

Post Office Protocol (POP) — The standard used to transfer e-mails from an e-mail server to a PC.

postproduction sound engineering — The addition of sound tracks, sound effects, and voice-overs to a movie after the movie has been recorded.

power-on self test (POST) — A test that is performed during the booting process to make sure that components are working correctly.

presentation graphics — Personal productivity software programs, such as Microsoft PowerPoint, that enable people to create slideshow presentations using built-in features ranging from developing charts and drawings, to formatting text, to inserting movies and sound clips.

primary key — A database field that uniquely identifies a record.

primary storage — Another term for RAM or volatile computer memory.

print server — A server that manages the printing requests for a printer shared by multiple users on a network.

proactive problem-solving approach — An approach to solving problems in which the problem solver seeks out potential problems before they become serious. Compare to reactive problem-solving approach.

problem solving — A process that combines the three phases of decision making—intelligence, design, and choice—with implementation and monitoring.

procedure — In structured design, a code module or subroutine, which is designed to solve a subproblem.

processing — An action a computer takes to convert or transform data into useful outputs.

processor — *See* microprocessor.

production studios — Sound production studios that work with motion pictures.

productivity software — Software designed to help individuals be more productive, such as word processors, spreadsheets, database-management systems, presentation graphics software, and personal information management software.

program code — The set of instructions that signal the CPU to perform circuit-switching operations.

program development life cycle — A five-step sequence of activities for developing and maintaining programming code that includes problem analysis, program design, program implementation, testing and debugging, and maintenance.

program flowcharts — Detailed charts that reveal how each software program in a system is to be developed.

program specification — A document, resulting from problem analysis, that defines the requirements of the program in terms of input, processing, and output.

programmed decisions — Decisions that are made using a rule, procedure, or quantitative method.

programmer — An individual who writes or codes the instructions that make up a computer program.

programming — The writing or coding of instructions to create a computer program.

programming language — The primary tool of computer programmers, provides English-like commands for writing software that is translated to the detailed step-by-step instructions executed by the processor.

programming language standard — A set of rules that describes how programming statements and commands should be written.

project crashing — Reducing project development time.

project leader — The individual in charge of the systems development effort, who coordinates all aspects and is responsible for its success.

project management — The planning, monitoring, and controlling of necessary systems development activities.

proprietary — Services or products that are protected by exclusive legal rights and thus, for example, often do not communicate or easily interconnect with each other.

proprietary operating system — A vendor developed operating system intended for use with specific computer hardware.

proprietary software — An program developed or customized by an individual, group, or organization for a specific application.

protocols — Rules that ensure that devices participating in a network are communicating in a uniform and manageable manner; an agreed-upon format for transmitting data between two devices.

prototyping — An iterative technique for systems development that typically involves the creation of some preliminary model or version of a major subsystem, or a small or "scaled-down" version of the entire system.

proximity payment system — A mobile commerce transaction method using devices that allow customers to transfer funds wirelessly between their mobile device and a point-of-sale terminal.

public domain software — Software that is not protected by copyright laws and can be freely copied and used.

purchase over TV — A feature of interactive TV that allows viewers to make purchases over their cable TV connection, much as computer users make purchases on the Web; sometimes called t-commerce.

purchasing system — A type of transaction processing system that supports the purchase of goods and raw materials from suppliers for the manufacturing of products.

quality — The ability of a product, including services, to meet or exceed customer expectations.

qubit — A quantum bit. A qubit displays properties in adherence to the laws of quantum mechanics, which differ radically from the laws of classical physics.

query by example (QBE) — An easy and fast way to make queries about data contained in a database; many databases use this approach to give users ideas and examples of how queries can be made.

radio frequency identification (RFID) — A tiny microprocessor combined with an antenna that is able to broadcast identifying information to an RFID reader, primarily used to track merchandise from supplier to retailer to customer.

radio wave — An electromagnetic wave transmitted through an antenna at different frequencies.

RAID — Redundant array of independent disks, a system of magnetic disks used to maintain a backup copy of the data stored on the primary disks. If the original drives or data become damaged the secondary disks can take over with little loss of time or work.

random access memory (RAM) — Temporary, or *volatile*, memory that stores bytes of data and program instructions in addressed cells for the processor to access.

raster graphics — Bit-mapped graphics, in which bytes are used to store the color of each pixel in an image.

ray tracing — Creating computerized 3-D models by adding shadows and light, which 3-D modeling software does by tracing beams of light as they would interact with the models in the real world.

reactive problem-solving approach — An approach to solving problems in which the problem solver waits until a problem surfaces or becomes apparent before any action is taken. Compare to proactive problem-solving approach.

readme file — File included with software that often contains last minute updates or disclosures, including how to deal with bugs.

read- only memory (ROM) — Permanent storage for data and instructions that do not change, like programs and data from the computer manufacturer including the boot process used to start the computer.

real-time mirroring — Making an exact copy of files in a system in real time as they are changed or saved.

record — A collection of related fields that describe some object or activity.

recording studio — Sound production studio that specializes in recording music.

recovery disk — A disk that stores an operating system for starting the computer when there is a problem with the hard disk and the computer is not booting correctly.

registers — One of three primary elements within a central processing unit (CPU), they hold the bytes that are currently being processed.

Registry — A database in Windows operating systems that stores important information on hardware devices, software settings, and user preferences.

relational model — A database model in which all data elements are placed in two-dimensional tables called relations that are the logical equivalent of files.

relations — Two-dimensional tables that are the logical equivalent of files.

remote resource — Along with local, one of two types of resources that workstations typically have access to: those that the workstation can access only while connected to the network; also called network resources.

rendering — The process of calculating the light interaction with the virtual 3-D models in a scene and presenting the final drawing in two dimensions to be viewed on the screen or printed.

repetitive stress injury (RSI) — A potentially computer-related health problem characterized by conditions such as tendonitis, tennis elbow, the inability to hold objects, and sharp pain in the fingers, caused by long hours at the computer keyboard; also known as repetitive motion disorder.

replicated database — A database that holds a duplicate set of frequently used data.

request for information (RFI) — A request directed to a computer systems vendor to provide information about its products or services.

request for proposal (RFP) — A request directed to a computer systems vendor to submit a bid for a new or modified system.

request for quotes (RFQ) — A request directed to a computer systems vendor to give prices for its products or services.

requirements analysis — The process of determining user, stakeholder, and organizational needs.

rescue disk — Software on a floppy disk, CD ROM, or other disk used to start or boot your computer in case you can't start or boot your computer from your hard disk.

restart — A procedure or button on a computer that when clicked performs a warm boot.

retinal scanning — A form of biometrics that analyzes the pattern of blood vessels at the back of the eye.

rich content — Web content that contains more than just text and simple images, such as video and interactive media.

rich media — When different digital media types are combined, such as animation or video and audio, Also called multimedia.

ripper software — Programs that can be used to translate music CDs to MP3 files on your hard drive.

ripping — The process of transferring music from CD to MP3.

roaming fee — Fees incurred when you use your cell phone outside your carrier's network.

robotics — The development of mechanical or computer devices to perform tasks that require a high degree of precision or that are tedious or hazardous for humans.

rollover — A cell phone plan feature that allows you to apply unused minutes from one month to the next.

router — Special-purpose computing device, typically small to large box with network ports, that manages network traffic by evaluating messages and routing them to their destination.

RSS — Really Simple Syndication, a Web technology that enables subscribers to receive daily or periodic updates of their favorite blogs. RSS uses XML to deliver Web content that changes on a regular basis.

sampling — Measuring a sound wave's amplitude at regular timed intervals.

sampler — A digital music instrument that digitally records real musical instrument sounds and allows them to be played back at various pitches using an electronic keyboard.

satellite radio — A form of digital radio that receives broadcast signals via a communications satellite. Satellite radio services charge a monthly subscription fee and offer stations featuring commercial-free music, comedy, news, talk, and sports programming.

schedule feasibility — The determination of whether a project can be completed in a reasonable amount of time.

scheduled report — Report produced periodically or on a schedule, such as daily, weekly, or monthly.

schema — A description of the logical and physical structure of the data and relationships among the data in a database; a description of the entire database.

scientific visualization — The use of computer graphics to provide visual representations that improve our understanding of some phenomenon.

search engine — A tool that enables a user to find information on the Web by specifying keywords.

secondary storage — Nonvolatile devices that are used to store data and programs more permanently than RAM, while the computer is turned off.

Secure Sockets Layer (SSL) — Technology that encrypts data sent over the Web and verifies the identity of the Web server; in combination with digital certificates it allows for encrypted communications between Web browser and Web server. A newer version is called Transport Layer Security (TLS).

security holes — Software bugs that allow violations of information security.

selecting — A basic data manipulation, involving eliminating rows according to certain criteria.

Semantic Web — The seamless integration of traditional databases with the Internet, allowing people to access and manipulate a number of traditional databases at the same time.

sequencer — A device that allows musicians to create multitrack recordings with a minimal investment in equipment.

sequential access — A storage medium feature that makes a computer that needs to read data from, for example, the middle of a reel of magnetic tape, sequentially pass over all of the tape before reaching the desired piece of data. It is one disadvantage of magnetic tape.

server — A computer system with hardware and software operating over a network or the Internet, and often sharing common resources, such as disks and printers; in a client/server relationship, the program that receives a service request and provides the service.

shareware — Software, usually for personal computers, that is inexpensive and can often be tried before purchase from the software developer.

Short Message Service (SMS) — A method for sending short messages, no longer than 160 characters, between cell phones. Also called text messaging or texting.

signal — The communications element, containing a message comprised of data and information, that is transmitted by way of a medium from a sender to a receiver.

simulation — Mimicking or modeling an event or situation that could happen in the future.

single in-line memory module (SIMM) — A circuit board that holds a group of memory chips, or RAM.

site preparation — The process of preparing the actual location so it is ready for a new computer system, which may mean simply rearranging furniture in an office or may require installing special wiring and air conditioning.

smart cards — Credit cards with embedded microchips, which are playing an increasing role in e-commerce payment methods.

smart home — Technology that allows residents to open and close curtains, turn on sprinkler systems, control media throughout the house, and adjust environmental controls from any Internet-connected computer or wall-mounted display.

smart phone — Handheld computers that include cell phone capabilities. A high-end smart phone can offer the benefits of several digital devices in one: a PDA, a cell phone, a digital music player, and a portable video player.

SMS text messaging — A method of sending short messages, no longer than 160 characters, between cell phones, also called texting or text messaging.

social engineering — A method used by hackers acquire information such as usernamesa nd passwords that will gain them access to a computer system or database. Social engineering exploits the natural human tendency to trust others to acquire private information.

software — Programs and instructions given to the computer to execute or run; computer programs that control the workings of the computer hardware.

software engineer — A software developer, an individual who designs and implements software solutions.

software engineering — Software development, the process of developing computer programs to address a specific need or problem.

software patches — Corrections to the software bugs that cause security holes.

software programmer — The person in the systems development process who is responsible for modifying existing programs or developing new programs to satisfy user requirements.

software suite — A collection of application software bundled together into one package, often including word processors, spreadsheets, presentation graphics, and more.

sound compression — The ability to remove those frequencies that are beyond the range of human hearing and in so doing reduce the size of a digital music file.

sound production studios — Facilities that use a wide variety of audio hardware and software to record and produce music and sound recordings.

source code — The high-level program code that a language translator converts into object code.

source data automation — The process of automating the entry of data close to where it is created, thus ensuring accuracy and timeliness.

special purpose I/O device — Input or output device designed for a unique purpose, such as a pill-sized camera that can be swallowed to record images of the digestive system.

special-purpose computers — Computers developed and used for primarily one task or function, such as MP3 players for listening to digital music files.

special-purpose database — A database designed for one purpose or a limited number of applications.

speech recognition software — Software that translates human speech into text or commands.

spider — Automated search program that follows all Web links in an attempt to catalog every Web page by topic.

spoofing — A technique used to impersonate others on the Internet.

spooling — An operating system's method of storing data temporarily in a buffer or queue area before transferring it to, for example, a slower printer, in order to free the processor for other tasks.

spreadsheet software — Personal productivity software that stores rows and columns of data and is used for making calculations, analyzing data, and generating graphs.

spyware — Software placed on your computer system without your knowledge or consent, normally through the Internet, to secretly spy on you and collect information.

standard — An agreed-upon way of doing something within an industry.

startup — The next to last step in systems implementation, beginning with the final tested computer system and finishing with the fully operational system.

static — When referring to an Internet Protocol (IP) address, one that is permanently assigned to a particular computer. When referring to a Web page, a page whose content does not change.

storage — The ability to maintain data within the system temporarily or permanently.

storage area network (SAN) — A relatively new technology that links together many storage devices over a network and treats them as one large disk.

storage capacity — The maximum number of bytes a storage medium can hold.

storage device — The drive that read, writes, and stores data.

storage media — Objects that hold data, such as disks.

storyline — A feature of video-editing software that allows the videographer to arrange video scenes sequentially and specify the transition effects between each scene.

streaming media — An Internet technology that plays audio or video files as they are in the process of being delivered.

structured interview — A data collection technique in which an interviewer relies on questions written in advance.

structured problem — A problem that is routine and where the relationships are well defined.

Structured Query Language (SQL) — A popular standardized data manipulation language.

stylus — A short, penlike device, without ink, used to select items on a touch screen.

subject directory — A catalog of sites collected and organized by human beings, such as the directory found at Yahoo.com.

supercomputer — The most powerful and advanced type of computer, often used for sophisticated and complex calculations.

supply chain management (SCM) — The process of producing and selling goods, involving demand planning, supply planning, and demand fulfillment.

supply planning — The supply chain management activity of producing and making logistical arrangements to ensure that a company is able to meet the forecasted demand.

surround sound — Audio that makes use of special sound recording techniques and multiple speakers placed around the audience, so that sound-producing objects can be heard from all directions; also known as 3-D audio.

surveillance — The close monitoring of behavior. Computer-controlled surveillance technologies combined with ubiquitous networks and powerful information processing systems have made it possible to gather huge quantities of video, audio, and communications signals and process them to reveal personal information.

switched line — A line that maintains a network connection only as long as the receiver is active.

synchronous communication — Along with asynchronous, one of two forms of electronic communications, allowing participants to communicate in real time as phrases are transmitted, whether spoken or typed, such as by using telephones, online chat, and instant messaging.

syntax — The set of rules each programming language has that dictates how the symbols should be combined into statements capable of conveying meaningful instructions to the CPU.

syntax error — An error in the form of a program's coding, such as a missing semicolon at the end of a command.

synthesizer — A digital music instrument that electronically produces sounds designed to be similar to real instruments or produces new sounds unlike any that a traditional instrument could produce.

system administrator — Person responsible for setting up and maintaining the network, implementing network policies, and assigning user access permissions.

system board — Also called the motherboard, the main circuit board in the computer where many of the hardware components are placed, such as the central processing unit, memory, and storage.

system bus — A collection of parallel pathways between the CPU and RAM that supports the transporting of several bytes at a time.

system clock — The CPU component that determines the speed at which the processor can carry out an instruction.

system penetration — Unlawful access hack into computers and networks.

system requirements — The storage, processor, and memory requirements to run software.

system software — A collection of programs that interact with the computer hardware and application programs.

system stakeholders — The individuals who, either themselves or through the area of the organization which they represent, will ultimately benefit from a systems development project.

system storage — Storage that is used by a computer system for standard operations.

system testing — Testing an entire system of programs together.

systems analysis — A process that attempts to understand how the existing system helps solve the problem identified in systems investigation and answer the question, "What must the computer system do to solve the problem?"

systems analyst — A professional who specializes in analyzing and designing computer systems.

systems design — The selection and planning of a system to meet the requirements outlined during systems analysis, which are needed to deliver a problem solution.

systems development — The activity or process of creating new or modifying existing computer systems.

systems development life cycle (SDLC) — The ongoing activities associated with the systems development process, including investigation, analysis, design, implementation, and maintenance and review.

systems development specialists — Computer systems personnel who might include a project leader, systems analysts, and software programmers.

systems development tools — Instruments such as computer-aided software engineering tools, flowcharts, and decision tables that can greatly simplify the systems development process.

systems documentation — Written materials that describe the technical aspects of a new or modified system.

systems implementation — A process that includes hardware acquisition, software acquisition or development, user preparation, hiring and training of personnel, site and data preparation, installation, testing, startup, and user acceptance, according to the systems design.

systems investigation — The first step in the development of a new or modified computer system, the purpose of which is to determine whether the objectives met by the existing system are satisfying the goals of the organization; the activity of exploring potential problems or opportunities in an existing system or situation.

systems maintenance — One of the final systems development steps, involving the checking, changing, and enhancing of the system to make it more useful in achieving user and organizational goals.

systems review — The process of analyzing systems to make sure that they are operating as intended, the final phase of the systems development life cycle.

T1 line — A network line that supports high data transmission rates by carrying twenty-four signals on one line.

T3 lines — Network lines that carry 672 signals on one line and are used by telecommunications companies; some act as the Internet's backbone.

table — A collection of related records; also called a file.

tablet PC — A type of small, easy to use mobile computer. Tablet PCs come in slate models that have a shape similar to a writing tablet or slate, and convertible models, which can convert from a slate style to a more standard laptop style, and include a keyboard.

t-commerce — A feature of interactive TV that allows viewers to make purchases over their cable TV connection, much as computer users make purchases on the Web; also known as purchase over TV.

TCP/IP — Transmission Control Protocol/Internet Protocol, the two Internet protocols.

technical feasibility — The determination of whether or not hardware, software, and other system components can be acquired or developed to solve the problem.

technology — Tools, materials, and processes that help solve human problems. Many of today's technologies fall under the classification of digital electronics.

telecommunications — The electronic transmission of signals for communications.

telecommunications network — A network that connects communications and computing devices.

telecommuting — The ability to work away from the office using network connections.

teraflop — Trillions of floating-point operations per second, a measurement for rating the speed of microprocessors.

testing — Conducting tests on the entire computer system, including each of the individual programs, the entire system of programs, the application with a large amount of data, and all related systems together.

text messaging — Using a cell phone to send short text messages to other cell phone users. Also known as *Short Message Service (SMS)* and *texting.*

texting — *See* text messaging.

thin client — Stripped-down network PCs, which include a keyboard, mouse, display, and a small system unit that supplies only enough computing power to connect the device to a server over the network.

time-driven review — A systems review procedure that is started after a specified amount of time.

time sharing — More than one person using a computer system at the same time.

token ring — A network standard using unique hardware and software that does not work with Ethernet.

top-level domain (TLD) — The final portion, such as .com or .edu, of a domain name, classifying Internet locations by type or, in the case of international Web sites, by location.

topology — The physical arrangement of a local area network, such as bus, star, or ring.

total information security — The goal of securing all components of the global digital information infrastructure from cell phones, to PCs, to business and government networks, to Internet routers and communications satellites.

touch pad — A touch-sensitive input pad below the spacebar on notebook computers that allows you to control the mouse pointer.

touch screen — An input device that allows users to select screen items by touching them on the display.

trackball — An input device that allows you to control the mouse pointer by rolling a stationary, mounted ball.

TrackPoint — An input nub in the center of the keyboard on notebook computers that allows you to control the mouse pointer.

transaction — An exchange involving goods or services, such as buying medical supplies at a hospital or downloading music on the Internet.

transaction file — Temporary file that contains data representing transactions or actions that must be taken.

transaction processing cycle — The common set of activities, including the collection, editing, correction, manipulation, and storage of data, that e-commerce and all other forms of transaction processing systems share.

transaction processing system (TPS) — An information system used to support and record transactions.

transborder data flow — Specific national and international laws regulating the electronic flow of data across international boundaries.

transistor — An electronics component composed of semiconducting material, typically silicon, that opens or closes a circuit to alter the flow of electricity to store and manipulate bits.

Transmission Control Protocol(TCP) — Along with Internet Protocol (IP), one of the sets of policies, procedures, and standards used on the Internet to enable communications between two devices.

transparency — A feature of network management, in which the underlying network structure is hidden from the user, promoting ease of use.

transport layer — The protocol portion of the three-layer Internet model, which also includes application (software) and physical (hardware) layers.

Transport Layer Security (TLS) — A more recent implementation of Secure Sockets Layer (SSL).

Trojan horse — Program that appears to be useful but actually masks a destructive program.

Trustworthy Computing — A long-term Microsoft initiative to provide more reliable, secure, and private computing experiences for their users.

tunneling — A technology used by virtual private networks to securely send private network data over the Internet.

Turing test — A proposal by British mathematician Alan Turing that says that a computer exhibits "intelligent" behavior if the responses from the computer are indistinguishable from responses from a human.

twisted-pair cable — A cable consisting of pairs of insulated twisted wires bound together in a sheath; the type of cable that brings telephone service to homes.

ubiquitous computing — A vision of a future so completely saturated with computer technology that we no longer even notice it.

Unified Modeling Language (UML) — A language that provides tools for creating object-oriented models for software solutions.

Uniform Resource Locator (URL) — The unique string of characters, such as http://www.course.com, that indicates where a particular Web page resides on the Internet; a Web address.

unit testing — Testing each of the individual programs in the computer system, which is accomplished by developing test data that will force the computer to execute every statement in the program.

Universal Product Code (UPC) — An identification code placed on products that can be read by scanners.

Universal Serial Bus (USB) — A standard that allows a wide array of devices to connect to a computer through a common port.

UNIX — A powerful command-based operating system developed in the 1970s by AT&T for minicomputers.

unstructured database — Database that contains data that is difficult to place in a traditional database system, such as notes, drawings, fingerprints, medical abstracts, or sound recordings

unstructured interview — A data collection technique in which an interviewer relies on experience, rather than on questions written in advance, to formulate questions designed to uncover the inherent problems and weaknesses of an existing system.

unstructured problem — A problem that is not routine and does not have well-defined rules and relationships.

uploading — Transferring data or files from a local computer over a network or the Internet to a remote computer.

USB storage — Small flash memory modules that plug into the universal serial bus port found on many computers, digital cameras, and MP3 players. Also called a thumb drive or USB drive.

use case — The event than an actor interacts with in an object-oriented systems development approach.

user acceptance document — A formal agreement signed by the user that a phase of the installation or the complete system is approved.

user documentation — Written materials that describe how the computer system can be used by noncomputer personnel.

user permissions — The access privileges afforded to each network user. Network operating systems such as Windows XP, Apple OS X, Linux, NetWare, and UNIX provide methods for associating user permissions with each user account.

user preparation — The process of readying managers and decision makers, employees, and other users and stakeholders for a new systems.

username — A name entered at a computer that identifies a user to the computer system.

users — A specific type of stakeholder who will be interacting with the system on a regular basis; people who use computers to their benefit.

utility programs — System software that is used to perform important routine tasks, such as to merge and sort sets of data or to keep track of computer jobs being run.

vector graphics — A representational method that uses bytes to store mathematical formulas that define all the shapes in an image.

vector graphics software — Programs that provide tools to create, arrange, and layer graphical objects on the screen to create pictures; also called drawing software.

vendor — A company that provides computer hardware, equipment, supplies, and a variety of services.

video conferencing — Technology that combines video and phone call capabilities along with shared data and document access.

video game consoles — High-powered multiprocessor computers designed to support 3-D interactive multimedia.

video games over TV — A feature of interactive TV that provides access to video games.

video on demand (VoD) — Technology that allows digital cable customers to select from hundreds of movies and programs to watch at anytime they choose.

video RAM (VRAM) — A buffer, sometimes referred to as a frame buffer, that stores image data after they are read from RAM and before they are written to the display.

video-editing software — Programs that allow professional and amateur videographers to edit bad footage out of digital video and rearrange the good footage to produce a professional video production.

virtual memory — Also known as virtual storage, an operating system feature that allows users to store and retrieve more data without physically increasing the actual storage capacity of memory.

virtual office — A substitute office, as in workers' homes and cars, or at remote job sites, that can be worked out of using cell phones, pagers, and portable computers.

virtual private network (VPN) — A network technology that uses a technique called tunneling to securely send private network data over the Internet.

virtual reality — A computer-simulated environment or event.

virtual reality headset — A gogglelike device, with spatial sensors as input devices, that projects output in the form of three-dimensional color images.

virtual space — An environment that exists in the mind rather than in physical space.

virtual storage — Also known as virtual memory, an operating system feature that allows users to store and retrieve more data without physically increasing the actual storage capacity of memory.

virus — A program that attaches itself to a file, spreads to other files, and delivers a destructive action called a *payload*.

virus hoax — E-mail sent to people warning them of a virus that doesn't actually exist.

virus-scanning software — Programs that detect viruses and worms on a personal computer.

vision systems — The hardware and software that permit computers to capture, store, and manipulate visual images and pictures.

visual languages — Programming languages that use a graphical or visual interface, allowing programmers to "drag and drop" programming objects onto the computer screen.

vlog — Video log. Similar to a blog , but using video, covering topics such as travel, current events, jobs, and technology.

Voice over Internet Protocol (VoIP) — A popular technology that allows phone conversations to travel over the Internet or other data network.

voice recognition — A technology, similar to speech recognition, used by security systems to allow only authorized personnel into restricted areas.

volatile — A characteristic of primary memory storage, such that a loss of power to the computer means that the contents of memory is also lost or eliminated; temporary memory storage that is cleared each time the computer is shut down.

volatility — A database characteristic, referring to a measure of the changes, such as additions, deletions, or modifications, typically required in a given period of time.

volume testing — Testing an application with a large amount of data, to ensure that the entire system can handle it under normal operating conditions.

war driving — The act of driving through neighborhoods with a wireless notebook or handheld computer looking for unsecured Wi-Fi networks.

warm boot — A computer startup performed while the computer is currently running, as by clicking the restart button.

wearable PC — Small system units that can clip to a belt or fit in a pack, head-mounted displays that only partially obscure vision, and hands-free or one-handed input devices. They are used in industry for individuals who need to have access to data while doing physical work.

Web - Short for World Wide Web, an application that makes use of the Internet to deliver information and services through a convenient interface utilizing hyperlinks.

Web auctions — E-commerce sites that provide a virtual auction block where users can place bids on items.

Web authoring software — Software that allows users to create HTML documents using word-processor-like programs.

Web browser — Web client software such as Internet Explorer, and Firefox, that is used to request Web pages from Web servers.

Web conferencing — Online video conferencing, allowing groups to see, hear, text chat, present, and share information in a collaborative manner.

Web portals — Web pages that serve as entry points to the Web.

Web script — Small programs that run on Web pages. One avenue for the spread of viruses and worms.

Web server — Server that stores and delivers Web pages and other Web services such as interactive Web content.

Web server software — Software whose primary purpose is to respond to requests for Web pages from browsers.

Web server utility programs — Software programs that provide statistical information about server usage and Web site traffic patterns.

Web Services — Programs that automate tasks by communicating with each other over the Web.

Web-based e-mail — Messages stored on a Web server, and viewed using a Web browser.

Webcams — Typically low-priced video cameras, often used for video conferencing over the Internet.

Webcasting — A technology that that provides television-style delivery of information using steaming video and high-speed Internet connections.

Web-driven programs — Programs that allow users to interact with Web sites to access useful information and services.

what-if analysis — The process of making hypothetical changes to problem data and observing the impact on the results.

wide area network (WAN) — A network connecting local area networks between cities, cross country, and around the world using microwave and satellite transmission or telephone lines.

Wi-Fi — Short for *wireless fidelity*, is a popular networking technology based on the IEEE 802.11 standards that connects computers wirelessly to other computers and to computer networks.

Wi-Fi Protected Access (WPA) — The most secure wireless encryption protocol available at present, more secure than Wired Equivalent Privacy (WEP).

WiMAX — Also known as IEEE 802.16, the next-generation wireless broadband technology that is both faster and has a longer range than Wi-Fi. WiMAX is built on Wi-Fi standards and is able to interoperate with Wi-Fi networks. A WiMAX access point has a 31-mile (50-kilometer) range.

Windows cleaners — Utility software that scans the Windows Registry, correcting incorrect or obsolete information.

Windows Update — A Microsoft service that allows patches to be applied to security holes in Windows as they are discovered. Windows Update can be accessed from the Microsoft Web site, or you can set your Windows operating system to automatically run Windows Update and install the patches.

Wintel — Also known as IBM-compatible, one of two popular personal computing platforms, along with Apple. These computers typically use the Microsoft Windows operating system and Intel or Intel-compatible processors.

Wired Equivalent Privacy (WEP) — An early wireless encryption protocol, less secure than Wi-Fi Protected Access (WPA).

wireless access point — A site that is connected to a wired network and receives data from and transmits data to wireless adapters installed in computers.

wireless adapter — A circuit board, PC card, or an external USB device that provides an external antenna to send and receive network radio signals

wireless fidelity (Wi-Fi) — Wireless networking technology that makes use of access points to wirelessly connect users to networks within a range of 250–1000 feet (75–300 meters). The Wi-Fi standards, also known as the 802.11 family of standards, were developed by the Institute of Electrical and Electronics Engineers (IEEE).

wireless networking — Uses radio signals or infrared rather than cables to connect computers and digital devices to computer networks and through those networks to the Internet.

WLAN — Wireless local area network, usually based on the IEEE 802.11 standards.

word processing software — Text and document manipulation software, such as Microsoft Word, that is perhaps the most highly used application software for individuals.

wordlength — The number of bits that a CPU can process at one time; the larger the wordlength, the more powerful the computer.

work stressors — Hazardous activities associated with unfavorable work conditions, such as hard-to-read computer screens, uncomfortable desks and chairs, and fixed keyboards, possibly leading to serious and long-term injuries.

workgroup software — Software that allows network users to collaborate over the network; also called groupware.

workstation — A powerful desktop computer used to make sophisticated calculations or graphic manipulations; a personal computer attached to a network.

world ownership — Files, folders, or other resources on a network that are available to all users.

World Wide Web — Also known as the Web, a client/server Internet application that links together related documents from diverse sources providing an easy navigation system with which to find information; an Internet-access application that uses a graphical interface to ease Internet navigation.

worm — A program that, rather than attaching itself to another program, acts as a free agent, placing copies of itself into other systems, destroying programs, and interrupting the operation of networks and computer systems.

wysiwyg — Pronounced wizzie-wig, and short for "what you see is what you get." It is a feature of Web-site editing programs, such as Dreamweaver, that allows users to design Web pages that will look the same when published on the Web.

XHTML — A successor to Hypertext Markup Language that embodies the best of HTML and XML in one markup language.

XML — Extensible Markup Language, a markup language for designing data classification to organize the content of Web pages and other documents.

zombie computer — Computer on the Internet that is either hacked into or under the influence of a virus or worm, and made to carry out Internet activities on the hacker's behalf.

Subject Index

Bold page numbers indicate where a key terms is defined in the text.

A

Academic Initiative zSeries program (IBM), 504
acceptance testing, 509
Access Grid, 221
access, Internet, 179–183
Access (Microsoft), 358
access points, 241, 256, 553
access time, hardware, 74
accessibility of information and communications technology (ICT), 605
accessibility options, Windows, 128
accessible computing for disabled, 605, 606
Accordant Health Services, 457
accuracy of DBMS, 364
acronyms for common phrases, 201
activating software, 159, 162
ActiveX, 193
Add/Remove Program feature, 161
Advanced Research Projects Agency (ARPA), 177
advertising, 406, 409
adware, 565
Aelera Corporation, 492
Agronow, Dan, 162
AI. *See* artificial intelligence (AI)
AirPort Express, 272
Akimbo service, 324
algorithms, 458–460, 504
alphanumeric data, 362
ALU (arithmetic/logic unit), **63**
Amazon.com, 192, 390, 391, 393, 395–396, 401, 409, 419, 425, 459
America Online Instant Messenger (AIM), 203
American National Standards Institute (ANSI), 368
American Standard Code for Information Interchange (ASCII), 59
Americans with Disabilities Act of 1990, 605
Ameritrade, 410, 551
analog signal, **234**, 235, 285
analog to digital conversion (ADC), 285–286
animated GIF, 312
animation
- computer 2D, 312–314
- computer 3D, 315–316
- film, 34
- films, 312, 313, 315, 316
- Web browser, 194

anomalies, 353
Anti-Phishing Act of 2005, 567–568
antispyware, **565**
antivirus software, **564**
AOL, 182, 199
AOL Chat, 204
Apollo space missions, 33
Apple computers
- computer platforms, 12
- development of OS, 130
- display monitors, 88
- file name conventions, 127
- iPod. *See* iPod (Apple)
- Keychain password manager, 542
- Macintosh GUI, 123
- online purchases, 97
- processors, 65, 66
- trade secrets revealed in blogs, 593
- Wi-Fi capability, 256

application flowchart, 487, 488
application layer, 186–187
application programming interfaces (APIs), 418
application servers, 263
application service provider (ASP), 504–505
application software. *See also* software
- acquiring, 155–158
- described, **112**–113, 139–140
- for e-commerce implementation, 417–419
- for groups and organizations, 152–153
- for individuals, 150–152
- for information, decision support and specialized purposes, 153–155
- productivity software, **140**–150

Arcadia, 263
architecture, 66
Ariba, 406
arithmetic/logic unit (ALU), **63**
Armour, Tom, 464
ARPANET, 177, 178
art. *See* digital graphics
Art and Antique Organizer Deluxe, 359
ArticWays.com, 390
artificial intelligence (AI)
- compared to natural intelligence, 454–455
- conceptual model of, 455
- described, **22**, 453
- expert systems, **460**–461. *See also* expert system (ES)
- fraud detection, 459
- fuzzy logic, **458**
- genetic algorithms, **458**–460
- intelligent agents, **460**
- languages, 116
- learning systems, **457**–458
- natural language processing, 22, 457
- neural networks, **458**
- in perspective, 454
- predicting human behavior, 464
- software, 155
- systems, 454
- timeline in movies, 454
- Turing Test, 454
- vision systems, 22, **456**

ASCII (American Standard Code for Information Interchange), 59
ASCII art, 200–201
ASCII text message, 201
asking directly for requirements analysis, 497–498
assembly language, 116
assigned research, 218
Associated Press, 209
asynchronous communications, **200**
AT&T, 130, 178, 237
Atlanta Veterans Administration, 445
ATMs, 14, 393
attachment, e-mail, **201**
attackers, 43, 538–539
attoseconds, 92
attributes, 346
audio databases, 375
audio file formats, 294–295
audio players, 296–297
audio storage media, 295–296
Australian Marine Mammal Research Center, 286

authentication, 540
Authorize.net, 559
automated retrieval system (ARS), 22, 23
automation
  bots (automated programs), 196
  described, 22–23
  source data, **83**
automobile industry, 258–259, 261, 325, 328, 329, 406, 407, 463
automobile shopping, 407–408
AutoTradeCenter, Inc., 358
avars, 315
avatar, 204
AVG Anti-Virus, 162
AXA Financial Services, 494
AXS-One Email, 378

B

"back door" access, 43
backbone, Internet, **178**
backdoor Trojan, 561
back-end applications, 363–364
backup
  data and systems, 544–547
  data files, 545–546
  databases, 369
  remote, 547
  utility programs for, 137
backward compatibility, 77, 148
Bad Business Bureau, 589
Bailey, Charles, 460
bandwidth, **235**
Bangalore, India, 609, 610
Bank of America Corporation, 551
banking online, 409–410, 426, 427
banner ads, 406, 409
Bartlett, Jerry, 493
batch processing, 393
Baudenbacher, Beat, 9
Bear Stearns, 195
behavioral profiling, 331
Beijing, China, 610
Ben and Jerry's, 372
Berners-Lee, Tim, 187, 188
Best Buy, 449
Beth Israel Medical Center, 327
"Big Dig" construction project, 162, 164
bin Laden, Osama, 287
binary data, 201
binary number systems, **59**
bioinformatics, 465
biometrics, 86, **543**, 544
BIOS (basic input/output system), 72, 124
bitmap (.bmp) format, 302
bit-mapped graphics, **301**
bits (binary digits)
  described, **6**, 59
  prefixes for, 7
bits per second (bps), 235
.biz TLD, 189
Blackboard educational software, 152, 210
blackjack tracking, 158
blade computing, 263–264
Blaster attack, 527, 559
bloatware, 504
blocking spam and pop-ups, 137
blogs
  business communities, 422
  citizen journalist, 199, 589
  described, 23, **205**–206
  freedom of speech issues, 593
Bluetooth, 252, **258**–259, 264–265
Bluetooth Special Interest Group (BSIG), 258
Blu-ray disks (BDs), 77
.bmp files, 302
Boehringer Ingelheim, 479
Boeing, 9–10, 325
Boingo, 183, 212, 256
Bombardier Flexjet, 440
booting, **124**
bots (automated programs), 196
bots (intelligent robots), 460
Bower, David, 445
brainstorming session, 200
Breen, Michael, 505
brick-and-mortar retail store, 396
bridges, 241
British Broadcasting System, 152
British Phonographic Industry (BPI), 185
British Telecommunications (BT), 178, 409, 591
broadband, **235**–236, 606–607
broadband over power line (BPL), 182
broadband phones, 24
brokerages, 410
browser plug-ins, **194**
browsers, Internet, 128
buffer, 71
bugs, software, 160–161, 504
Burlington Northern and Santa Fe Railway Company, 465
burning CDs/DVDs, 77
buses, 62
Bush, George W., 36, 606, 612
business application software, 152–153
business intelligence, 372, 531
business resumption planning, 428
business-to-business e-commerce (B2B), **396**–397, 399, 400, 405–406, 415, 423. *See also* e-commerce
business-to-consumer e-commerce (B2C), **396**–397, 400, 415, 416, 420, 423. *See also* e-commerce
bytes
  described, **6**, 59–60
  prefixes for, 7

C

C++ programming language, 117
cable connections, 235
cable modem connections, **180**
cable modems, 239–240
cables, physical
  coaxial cable, **237**
  described, 237
  fiber-optic cable, **238**
  twisted pair cable, **237**
cache memory, 66–67
CAD (computer-aided design), 32
Caldera OpenLinux, 133
Callahan, Kevin, 211
CAM (computer-aided manufacturing), 32
camcorders, digital, 86, 322
cameras, digital, 85–86, 316–318
career information on Web, 213–214
Carnivore (FBI) system, 598
carpal tunnel syndrome (CTS), **586**
CaseMap, 359
catalog management software, 417
CBS News, 406, 589
CDMA cell phones, 248
CD-ROM (compact disk read-only memory), **76**
CD-RW format, **77**
CDs, 76–78, 80–81
Ceiva Digital Photo Receiver, 320
Cell chip (Sony), 67
cell phone etiquette, 584, 585
cell phone service providers, 412
cell phones
  candy bar style, 250
  capabilities of, 16, 246–247
  CDMA standard, 248
  features and services, 251–253
  flip phones, 250
  games, 329
  GSM standard, 248
  handsets, 250–251
  Internet access, 181
  Lexus and Bluetooth, 258–259
  m-commerce. *See* mobile commerce
  operating systems, 135
  roaming fees, 249
  rollover plans, 249
  SAR ratings, 586
  service plans, 248–250
  SIM cards, 248
  text messaging, 16, 205, 252, 412–413
  texting, 16, 205
  third generation (3G), 16, 247
  TV on, 324
cellular carriers, **247**–248
cellular networks, **246**, 248
Cellular Telecommunications Industry Association (CTIA), 586
censorship, **590**–593
Census data, 21, 35, 36, 75
CenterPoint Energy, 182
Centra Software, 452
Central American Free Trade Agreement (CAFTA), 612
Central Intelligence Agency (CIA), 461, 533–534
Central Missouri State University, 222
central processing unit (CPU)
  characteristics, 64–68
  components, 63
  described, **62**
  RAM, **63**–64
CERN (European Organization for Nuclear Research), 197, 355
certification authorities, 427
Chabot, Christian, 359
Championchip USA, 261
Chandler, Mark, 115
channels, chat, 204
character recognition software, 86
chat, 203, **204**–205
chat rooms, 204
ChatBlazer, 203
CheckPoint, 46

Chicago Police Department, 38
Chicago Tribune (newspaper), 12
child pornography, 591
Children's Internet Protection Act of 2000, 591
Children's Online Privacy Protection Act of 2000, 532, 603
chips
  described, 6, 62
  multicore technology, 67
chipset, 63
ChoicePoint database, 379, 595
Chomsky, Noam, 205
Cingular, 248–250
circuits, integrated, **62**
Cisco Systems, Inc., 115, 142, 269
Citigroup, 551
citizen journalist, 199, 589
cladding, 238
Clarissa (voice-activated computer), 87
ClearForest Intelligence Database, 372
client, 184
client/server relationship, **184**, 263
Clinton, Bill, 612
clip art, 144
clock speed, 66
closing programs, 113
clustering, 68
CMOS memory, 73
cnet.com, 96, 182
Coalition for Economic Growth and American Jobs, 492
coaxial cable, **237**
code. *See* program code
codes of ethical conduct, 603, 604
Cohen, Jay, 378
cold boot, 124
collaboration, 199, 221, 452
college students and technology, 585
collegeclub.com, 223
ColorNet (Polygon) database, 370
.com TLD, 189
Comcast, 325, 326
command-based user interface, **121**, 122
command prompt, 121
comma-separated values (CSV), 356
Commercial Internet Exchange (CIX) Association, 178
communications
  asynchronous, **200**
  data. *See* telecommunications
  digital sound in, 287–289
  enhancing, 23–24
  evaluating forms of, 200
  mobile, 16
  pervasive, 199
  protocols and standards, 245
  synchronous, **200**
  text, 200–201
  video, 207, 321–326
  voice, 24, 206–207
  Web applications for, 199
Communications Assistance for Law Enforcement Act (CALEA), 598
communications medium, 232
communications processor, 243
communications satellite, **242**
community and computers, 584–585. *See also* society and technology
compatibility, backward, 77, 148
competitive intelligence, 531
compiler, **119**
compressed files, 136–137
computational science, 21
compute, 21
computed fields, 362
computer animation, 312–316
computer competency, 5
computer crime
  adopting security guidelines, 531
  cyberterrorism, **533**
  hackers, crackers, intruders, and attackers, 43, 538–539
  headlines describing, 526–527
  identity theft, **528**–529
  laws protecting information and privacy, 532
  pharming, 568
  phishing, **567**–568
  projects for, 527
  scams, spam, fraud, and hoaxes, 566–570
  spyware, adware, and zombies, 43, 565–566
  viruses and worms, 43, 136, 526, 561–565
Computer Discount Warehouse, 98
computer ethics, **602**
computer fluency, 5
computer forensics, **539**
computer housekeeping, 547
computer literacy, **4**–5
computer networks, **18**, 177, 233
computer profiling, 594
computer programming. *See* programming
computer programs, **112**
Computer Shopper, 96
computer systems
  described, 27
  management, 482
  trusting, 500
computer vision, 22
computer-aided instruction (CAI), 38
computer-aided software engineering (CASE) tools, **486**–487
computer-assisted design (CAD), 152, **307**–308
computer-based information system (CBIS), **26**
computer-based professionals, 31–32
computerized collaborative work system, 452
computerized physician order entry (CPOE), 36
computers
  applications for, 20–28
  careers using, 30–38
  described, **5**, 9, 27
  digital technology, 6–7
  functions of, 7–8
  general-purpose, 8
  hardware. *See* hardware
  mobile digital devices, 15–17
  personal uses for, 39–42
  reasons to study, 4–5
  selecting and purchasing, 71, 95–100
  special-purpose, 8, 14–15
  types of, 8–15
computing
  described, 21
  history of, 58
  pervasive, 44
  ubiquitous, 44
comScore Networks, 217
connecting to Internet, 179–183
Connexion service (Boeing), 325
consumer complaints, 589
Consumer Price Index, 33
Consumer Reports, 246, 248
consumer-generated media (CGM), **589**
ConsumerReports.org, 407, 408
consumer-to-consumer e-commerce (C2C), **396**, 423. *See also* e-commerce
content streaming, **194**
content-filtering software, 210, **591**
Continental Airlines, 361
continuous improvement, 513
contract, 502
contract software, 156
Contribute (Macromedia), 485
control unit, **63**
convertible tablet PCs, 10
cookies (files), **192**
Coolpix (Nikon), 317
coprocessors, 68
copyrights
  described, 161–162, 530
  on shared files, 184–186
Cornell University, 448
corporate network policies, 552
Costco, 320
Council of Europe, 534
counterintelligence, 531
Covisint, 406, 407
CP/M (Control Program for Microcomputers), 128
CPU. *See* central processing unit (CPU)
crackers, 43, 538–539
cradles, 10
Creative, 16
Creative Suite (Adobe), 34, 35
credit card transactions, 423
crime detection and prevention, 37, 38
crime surveillance systems, 599
critical patches, 536
critical success factors (CSF), 498
Cross, Wayne, 415
curiosity-driven research, 218
Current TV channel, 589–590
customized software, 27, 155–156, 158
customizing products, 404
Cyberkinetics, 465
cybermall, 404
cyberterrorism, **533**

D

D'Amico, Bob, 94
dangerous information and censorship, 592
Dartmouth College, 318, 453

data
alphanumeric, 362
conversion, 507
described, **6**
hierarchy of, 344–345
human-readable, 83
machine-readable, 83
organizing in databases, 350. *See also* databases
physical vs. logical views of, 125
preparation, 507–508
source, automation of, 83
data analysis
for databases, **352**–353
for systems analysis, 497
data definition language (DDL), 365
data dictionary, **365**–367
data havens, 267
data integrity, **364**
data items, 346
data management
basic concepts of, 342
database approach to, 348–350
for individuals and organizations, 343–344
simple approach to, 348
data manipulation language (DML), 367
data marts, 371–372
data mining, 222, **371**–372
data packets, 183
data redundancy, 349
data warehouses, **371**–372
database(s)
anomalies, 353
approach to data management, 348–350
attributes, 346
characteristics, 355
data analysis, **352**–353
data entities, 346
described, **26**, 345
distributed, **373**–374
government regulations and costs, 378
hierarchy of data, 344–345
immediacy, 355
Internet and network access to, 374–375, 376
joining data, 351
keys, 346–348
normalization, 354
object-oriented, **354**–355
organizing data in, 350–355
primary key, **346**–348
relating data, 352
relational model, **351**–354
relations, 351
replicated, 374
research, 219–220
security, 559
selecting data, 351
servers, 263
size, 355
use, policies, and security, 379
visual, audio, and unstructured, 375–376
volatility, 355
database administrators (DBAs), **377**
Database Lite 10g (Oracle), 358
database management system (DBMS)
accuracy of, 364
administration of, 377
advantages of, 350
approach to data management, 349
back-end applications, 363–364
backup and recovery, 369
comma-separated values, 356
creating and modifying databases, 365–367
data integrity, **364**
data marts, 371–372
data mining, **371**–372
data warehouses, **371**–372
database design, 361–363
described, **342**
field design, 361
file organizers, 357
flat files, 356
front-end applications, 363–364
general-purpose databases, 359
for information and decision support, 370–371
input and output interface design, 362–363
list of popular, 360
manipulating data and generating reports, 367–369
multiuser, 358–359
open-source, 359–361
record and table design, 362
for routine processing, 370
single user, 358
software, 145, 146
special-purpose databases, 359
updating databases, 367
uses for, 145
database system
described, 342
for systems implementation, 506
date fields, 362
dating services, online, 29
Davis, Miles, 300
De Telegraaf (newspaper), 589
decision(s)
nonprogrammed, **440**
programmed, **439**
decision making, **438**
decision support software, 153–155
decision support system (DSS)
characteristics of, 448–451
data sources, 448
described, **29**–30, 447–448
goal-seeking analysis, **450**
nonprogrammed decisions, 440
optimization and heuristic approaches to, 440–443, 450
range of data handled, 448
report and presentation flexibility, 449
simulations, 450–451
software, 155, 450
what-if analysis, **450**
decision tables, **488**–489
dedicated lines, 243
Deere, 441
Defense Advanced Research Projects Agency (DARPA), 464
Defense Evaluation Research Agency, 446
defragmenting disks, 137, 138, 547
Dell computers, 66, 88, 300
dell.com, 96
Delphi, 325
Delta Airlines, 260
demand fulfillment, 400
demand reports, **445**–446
Denver Art Museum, 462
Department of Commerce (USDC), 606
Department of Defense (DoD), 177
Department of Education, 605
Department of Energy (DOE), 482
Department of Health and Human Services, 592
Department of Homeland Security (DHS), 35, 43, 531, 563
Department of Human and Health Services (HHS), 448
design. *See* systems design
desktop computers
clock speeds, 66
described, 9
expansion, 93–94
Intel processors for, 65
platforms for, 11–12
desktop publishing software, 152, 305
desktop search utilities, 138
Deutsche Bahn, 13
development. *See* systems development
DeviantArt.com, 214
device drivers, 124
devices, digital. *See also specific device*
DIADs (delivery information access devices), 268, 269
dial-up connections, **179**–180
digital art, 304
digital audio
described, **285**
digital sound for professionals, 286–288
digitizing audio, 285–286
entertainment and communications, 287–289
law enforcement, 286–287
scientific research, 286
digital camcorders, 86

digital cameras, 85–86, 316–318
digital certificate, **427**
digital convergence, **60**
digital data representation
binary number system, **59**–60
look-up tables, 59
digital devices, 58. *See also specific device*
digital divide
described, **44**
global, 607–608
U.S., 44, 45, 178, 604–606
digital electronics device, **6**
digital forensics, 539
digital graphics
communicating ideas, 306
computer animation, 312–316
computer-assisted design (CAD), **307**
creative expression, 304–305
described, **301**
digitizing graphics, 301–302
documenting life, 308
entertainment, 306
exploring new ideas, 306
file formats, 302–304
independent film productions, 303
lossless compression, 303
lossy compression, 303
presenting information, 305
3D modeling software, **310**–311
vector graphics software, 308–309
digital images, 316
Digital Library project (Google), 179
digital media
described, 25, **284**
digitizing music, 285–286
Digital Millennium Copyright Act (DMCA), 185
digital music
audio file formats, 294–295
audio players, 296–297
audio storage media, 295–296
described, **285**
digitizing music and audio, 285–286
distribution, 214–215, 297, 300, 529–530
home recording studios, 290–291
media player software, **296**
online music services, 297–299
podcasting, 292–294
professional music production, 289–290
satellite radio, **300**
software, 150, 151, 296–297
digital photography
creating digital photos, 316–318
described, 316
digital cameras, 85–86, 316–318
photo-editing software, **318**–319
printing photos, 90, 91, 320–321
storage for photos, 317–318
viewing and sharing photos, 320
digital rights management (DRM), 298, 299
digital satellite service (DSS) connections, 181
digital signal, **234**, 235
digital society. *See* society and technology
digital sound. *See* digital audio
digital subscriber line. *See* DSL
digital technology
computers, 6–7
world-wide impact of, 44–46
digital to analog conversion, 285–286
digital video
conferencing, **207**
creating, 322
described, 321–322
Internet, 207
personal video recorders (PVRs), 331
video logs, 323
video recorders, 86, 322
video, television, and movie services, 324–326
video-editing software, 152, **322**–324
digital video recorders (DVRs), 78
digital voice recorders, 286
digitization
described, **6**
of graphics, 301–302, 312
of music, 285–286
smell and taste, 8
Dinero, Steven, 390
direct access, 74
direct conversion, **510**
directories, file, 127
directories, Web, 197, 199
directoryofschools.com, 211
disabled, accessible computing for, 605, 606
disk drives, 74–75, 78, 80
disk partitioning, 120
disk utilities, 137
diskettes, high-capacity, 75
Disney, 13, 325
display monitors, 88–90
display resolution, **88**
distance education, 210–211
distributed computing, 263, 418
distributed databases, **373**–374
distributed denial-of-service attacks, 558–559
distribution of digital music, 214–215, 297, 300, 529–530
DNA computing, 70
DNA database, 346
docking stations, 10, 146
DoCoMo, 265
documentation, 511
dogpile.com, 197, 198
Domain Name System (DNS), 184
domain names, 184, 188, 420
DOS (Disk Operating System), 128–129
dot pitch, 88
dot-matrix printers, 91
DoubleClick, 409
download.com, 206, 216
download-free-games.com, 162
downloading, 379
dpi (dot-per-inch) resolution, 91
Drexel University, 82
drill down method, 208
drivers, device, 124
drum machines, 289
DSL (digital subscriber line)
connections, **181**, 235
modem, 240
dual-core processors, 67
Duke University, 82, 374
dumb surf, 400
dumpster diving, 557
Dunham, Ken, 534
DuPont, 115
DVD (digital video disk), **77**–78
DVD movies, 34, 216
dynamic IP address, 183–184
dynamic Web pages, 192

E

E. W. Scripps Company, 195
Earth Simulator, 13
Earthlink, 565
eBay, 98, 153, 396, 405, 423
e-books, 10
e-cash, **423**–424
EcoBot, 456
e-commerce. *See also* e-commerce implementation
applications, 403–410
automobile shopping, 407–408
banking, 409–410, 426, 427
benefits and challenges of, 400–403
from buyer's perspective, 397–399
customizing products, 404
described, **28**, 212, **390**
electronic exchange, 406
finance and investment, 410
global supply management, 405–406
history of, 392–393
host, **414**, 415
market research, 408–409
marketing, 406, 422
marketplaces, 404–405
milestones, 391
online clearinghouses, 404–405
product information, 407–408
retail and online shopping, 403–404
sales projections, 391
from seller's perspective, 399–400
transaction processing, 28, 153, 393–394
transaction processing cycle, 394
types of, 395–397
Web auctions, 404–405
e-commerce implementation
3Cs approach (content, community, and commerce), 420, 421
building traffic, 419–422
business resumption planning, 428
catalog management software, 417
e-commerce host, **414**, 415
electronic payment systems, 423–425
electronic shopping cart software, 417
graphics applications, 417

graphics development, 417
hardware and networking, 417
identity verification, 426–427
infrastructure, 415–416
international markets, 425–426
keywords and search engines, 420–422
marketing, 406, 422
payment software, 417
purchasing domain names, 420
securing data in transit, 427
security issues, 426–428
software, 417–419
Web server software, 417
Web server utility programs, 417
Web services, **418**–419
Web site design tools, 417
Web site development tools, 417
economic feasibility, 494, 495
Edge legal software, 115
education
digital divide. *See* digital divide
online, 210–212
software, 152
U.S., outsourcing, 611
.edu TLD, 189
eHarmony, 41
eHobbies, 422
Eisenhower, Dwight, 177
electronic cash, **423**–424
electronic commerce. *See* e-commerce
electronic data interchange (EDI), 268–269, 392
electronic exchange, 406
electronic funds transfer (EFT), 32
electronic health records (EHRs), 36, 374
electronic shopping cart software, 417
electronic wallet, 424
Elliott, Frank, 76
e-mail
attachment, **201**, 202
body, 201, 202
described, **201**–202
header, 201, 202
to lawmakers, 592
scams, spam, fraud, and hoaxes, 566–570
servers, 12
spam blocking, 137
spam filters, 569
surveillance, 598
viruses. *See* viruses
embedded computers, **14**–15
embedded operating systems, 135
emoticons, 200
Empire Online, 217
employee blogs, 593
employment, seeking via Web, 213–214
encoder software, 296
encryption, **427**, 544, 555, 560
encryption devices, 244
end-user computing, 379
end-user systems development, 485–486
enterprise information and management system (EIMS), 347
enterprise resource planning (ERP), 153
enterprise service provider (ESP), 266
Enterprise Systems Architecture/370 (ESA/370), 134
enterprises, 263
entertainment
digital graphics in, 306
digital sound in, 287–289
digital technology, 25–26
Web applications for, 214–217
entity, 346
eProcrates, 448
ergonomic keyboards, 84, 588
ergonomics, **587**–588
errors and user negligence, 536–537, 551
establishment, challenging, 589–590
e-tailing, 396
Ethernet, **245**
ethical issues. *See also* society and technology
accessible computing for disabled, 605, 606
computer access, 45, 402, 606
computer ethics, **602**
digital divide. *See* digital divide
file sharing, 185, 529
governmental considerations, 604–608
hackers, 539
intellectual property rights, **530**
personal considerations, 602–603
professional considerations, 603–604
ethics, **602**
European Union (EU), 605, 612
event-driven review, 513
e-wallet, 424
Excel (Microsoft), 142, 143, 441, 442, 451
exception reports, **446**
exchange rates, 424
executable files, 113, 119
execution phase of machine cycle, 64
executive support software, 155
expansion boards/cards, 93
expansion slots, 94
expedia.com, 213
expert system (ES)
described, 22, **30**, **460**
heuristics, **442**–443, 450, 461
nonprogrammed decisions, 440
software, 36, 155
uses for, 461
Extensible Markup Language (XML), **189**–191, 418
extensions, file name, 113
Exxon, 242

F

FACE (Former and Current Employees) of Intel, 593
face identification software, 153
facebook.com, 41, 585
facial pattern recognition, 37, 543
Farid, Haney, 318
Farmyard Nurseries, 400, 401
feasibility analysis, **494**–495
Feder, Bart, 195
Federal Bureau of Investigation (FBI), 213, 372, 481
Federal Communications Commission (FCC), 238, 239
Federal Trade Commission, 213
Federal Wiretap Act, 597, 598
FedEx Global Trade Manager, 426
FedEx Kinko's, 208
FeedRoom, 195
feeds, RSS (Really Simple Syndication), 206
Ferion game network, 216, 217
fiber-optic cable, **238**
field name, 344
fields, **344**
fifth-generation languages (5GLs), 116
file allocation table (FAT), 127
file compression, 136–137
file names
extensions, 113
operating system conventions, 127
file organizers, 357
file ownership, 549
file servers, 12, 262, 263
file systems, 127
files
backing up, 545–546
database, **345**
described, **6**
operating system management, 127
file-sharing software, 184–186, 297
FileVault (Apple), 544, 545
finance and e-commerce, 410
finance, personal, 39
financial management software, 151
finger scanner in automobiles, 43
fingerprint databases, 342, 375
fingerprint scanning, 86, 543, 544
firewalls, 241, 266, 267, **560**
FireWire, 94
First Financial Credit Union, 86
flash BIOS, 72
flash drives, 78–80
Flash (Macromedia), 194, 313–314
flash memory card, **78**
flash mobs, 584
flat files, 356
flat panel display, **89**
floppy disks, 75, 80–81
FLOPS (floating-point operations per second), 67
flowcharts, **487**
FocalPoint (TriPath Imaging), 461
folders, file, 127
Food Allergen and Consumer Protection Act, 370
Food and Drug Administration (FDA), 378, 461, 506
food tracking, 511
Foreign Intelligence Surveillance Act of 1978, 597
forensic audio, 286
forensic graphics, 321
Forex Trader, 327
formatting disks, 124
Forrester Research, 128
fourth-generation languages, 116
Fox, Michael, 601
frames, 302
fraud, Internet, 213, 459

Fraunhofer Institute for Telecommunications, 90
Free Software Foundation, 148, 162
Free2C 3D display, 89, 90
freedom of speech, 590–593
freeware, **162**
frequency, signal, 235
front side bus (FSB) performance, 66
front-end applications, 363–364
Frontline Management Training Program (FLM), 157
FSB (front side bus) performance, 66
fsf.org, 162
FTP (file transfer protocol), 138, 560
Fusion Flash Concerts, 584
fuzzy logic, **458**
fuzzy sets, 458

G

Gabrielyan, Ara, 600
Gaeta, John, 25
Gaim, 162
gambling, online, 217, 587
game consoles, 217
game theory, **464**
GameCube (Nintendo), 328
gamepad, **85**, 86
games
- interactive video, 328–330
- online, 216–217
- software, 128, 152, 162
- training using, 327
- video, over TV, 331

gaming, 85
Gantt charting, 489, 490
Gap. Inc. Direct, 65
GarageBand, 211, 291, 293
garbage in, garbage out (GIGO), **364**
Gartner research, 361, 376, 426, 493, 552
Gates, Bill, 535
gateway.com, 96
gateways, 241
Gelenbe, Pamir, 413
Gemini observatories, 222
General Electric, 374
General Nutrition Web site, 420, 421
General Public License (GPL), 133
genetic algorithms, **458**–460
Genoa II program, 464
geographic information system, **463**
geographic information system (GIS), 38
Georgia College & State University, 82
German University of Teubingen, 90
.gif files, 302
gigaflop (billions of floating-point operations per second), **67**
gigahertz (GHz), **66**
Given Imaging, 84
global digital divide, 607–608
global information security, 533–534
global networks, 267–268, 489
Global Positioning System (GPS)
- applications for, 113, 254–255, 261, 306, 412
- described, 95, **253**–255
- surveillance, 600–602

global supply management (GSM), 405–406
globalization
- business challenges, 611–612
- described, **608**
- offshoring, **610**
- outsourcing, 419, 420, **492**–493, 609–610, 611

Gmail, 598
GMC, 195
Gnutella file-sharing system, 186
goal-seeking analysis, **450**
goggles, swimming, 72
Google, 179, 196–197, 209, 325, 422
Google Earth, 213, 312, 313, 599–600
Google Maps, 212
Google Picasa, 320
government regulations, 378
.gov TLD, 189
GPS. *See* Global Positioning System (GPS)
Graffiti (Palm) system, 84, 85
Grand, Steve, 454
Grand Theft Auto, San Andreas, 330
graphic files formats, 302–304
graphical information systems (GISs), 33
graphical user interface (GUI)
- advantages of, 123
- Apple computers, 130
- described, **122**
- Linux, 133
- UNIX, 130–132
- Windows (Microsoft), 129–130

graphics. *See* digital graphics
graphics applications, 417
Graphics Interchange Format (.gif), 302
graphics, presentation, 142–144
graphics tablets, 84
Greenberg, Seth, 422
grid computing, 68
Grigsby, Paul, 376
group decision support system (GDSS)
- characteristics of, 452
- described, 29, **451**
- software, 155, 452

group ownership, 549
groupware, 155, **452**
Groves, Kursty, 255
Growing Stars, 611
GSM cell phones, 248
GUI. *See* graphical user interface (GUI)

H

Hackensack University Medical Center, 75–76
hackers, 538–539
- defending against, 560
- described, **43**, 538
- groups, 539
- methods of attack, 557–558
- motivation and goals, 558–559
- tracking by law enforcement, 539
- turf wars, 557
- types of, 539

Hadjikhani, Nouchine, 61
Hamidi, Ken, 593
handheld computers
- applications, 10, 82, 83, 96
- clock speeds, 66
- described, 10
- digital convergence, **60**
- docking stations, 146
- expansion, 95
- input devices, 84
- mobile software packages, 149–150
- operating systems, 135
- platforms for, 11–12
- ROM, 73
- smart phones. *See* smart phones
- Web access, 188

handheld scanners, 86
handwriting recognition
- Graffiti (Palm) system, 84, 85
- handheld computers, 10
- processing power for, 61, 62

Hannaford Brothers, 133
hard copy, 90
hard drives, 74–75, 78, 80
hardware. *See also specific hardware*
- CPU, **62**–70
- described, **7**
- digital representation, 58–60
- for e-commerce implementation, 417
- expansion, 93–95
- input/output devices, 81–95
- integrated circuits, 61–**62**
- Internet, 178–183
- networking, 233
- operating system control of, 124–126
- selecting and purchasing computers, 71, 95–100
- storage, 70–81
- for systems implementation, 503
- telecommunications, 233, 236–244
- utility programs, 137

Hawking, Stephen, 606
HaxMat:Hotzone, 327
Hazmat database, 359
health concerns
- avoiding health problems, 587–588
- mental health and related problems, 587
- physical health problems, 586

Health Record Network Foundation, 374
Healthcare Informatics journal, 465
Hein, Gary, 361
Help feature of software, 112
Herbruck, David, 9
Herndon, Hillary, 221
hersheys.com, 421
hertz (Hz), 235
heuristics, **442**–443, 450, 461
Hewick, David, 499
Hewlett-Packard (HP), 263, 503
high definition TV (HDTV), 89
high-level languages, 116

High-Performance Storage System (HPSS), 73, 74
high-speed Internet, 235
Holiday Inn, 242
Holland American Cruise Lines, 376
hologram interface to surgical instruments, 81
home networks
  benefits of, 269
  phone-line networking, 270
  power-line networking, 270
  setting up, 270
  uses for, 271–272
  wireless networking, 271
Home Phoneline Networking Alliance, 270
home recording studios, 290–291
Honda robot (Asimo), 454, 457
Honey Monkey (Microsoft), 563
HOPE (hackers on planet earth) conference, 539
horizontal Web portals, 199
Hostmann, Bill, 376
hosts, 177
hotjobs.com, 214
hotspots, Internet, 183, 256–257, 329
HotWired RGB Gallery, 304
House of Blues Entertainment, Inc., 236, 237
HTML (Hypertext Markup Language), **189**
hubs, 240
Hudson River Park Trust (HRPT), 505
human-readable data, 83
Hutchinson Port Holdings, 441
hydrophones, 286
hyperlinks, 19, **188**
Hypertext Transfer Protocol (HTTP), 187, **188**
Hyuandai Motot Company, 347

I

IBM, 65, 76, 133, 153, 155, 164, 182, 269, 414, 503
IBM ASCI White computer, 67
IBM BladeCenter systems, 263
IBM Blue Gene/L computer, 13, 21, 68
Ice Age, 312, 313
iChat (Apple), 207
ICQ, 203
ID devices, 542–543
IDC, 73, 412
identity theft, **528**–529
Identity Theft Protection Act of 2005, 532, 603
identity verification, 426–427
IEEE (Institute of Electrical and Electronics Engineers) standards
  FireWire (1394), 94
  Wi-Fi (802.11), 256
  WiMAX (802.16), **257**
iManage, 36
immersive virtual reality, 462
impact printers, 91
India, affordable computers, 96
industrial telecommunications, 242–244
industry news, 209–210
InfoPath (Microsoft), 358
informatics, **464**–465
Informatics for Integrating Biology and the Bedside (I2B2), 465
information
  described, **6**
  managing, 26
  online resources, 217–220
  overload, 26, 196
information and communications technology literacy, 5
information security. *See also* network security; security
  described, **42**–43
  intellectual property, **529**–530
  intellectual property rights, **530**
  national and global security, 533–534
  organizational information, 530–534
  personal information, 528–529, 594–595
  piracy, **537**–538
  plagiarism, **537**, 538
  software and network vulnerabilities, 535–536, 551
  threats to, 535–539
  total, 527–528
  user negligence, 536–537, 551
information systems
  described, 27–28
  types of, 28–30
information technology (IT), **27**
information technology literacy, 5
information theft, 551–552
infrared transmission, 260
ING Direct, 449
Initiative, 164
ink-jet printers, 90
input
  concepts, 81–83
  described, 7
  to management information system, 443–444
input devices
  data readability, 83
  described, **81**
  digital cameras, 85–86, 316–318
  gaming, 85
  hologram interface to surgical instruments, 81
  microphone, 85
  mobile, 84
  operating system management, 127
  personal computer, 84
  scanning, 83, 86–87
  source data automation, **83**
  speed and functionality, 82
installation, 159–160, 508, 509
instant messaging (IM), **203**–204
instruction phase of machine cycle, 64
instruction set, 63
integrated circuits, 61–**62**
integrated software packages, 149
integrated studios, 290
integration testing, 509
Intel, 257, 607
Intel processors
  Centrino, 256, 607
  code numbers, 66
  Microsoft Windows support, 11, 130
  Pentium, 63, 64, 65, 66, 67, 69, 256
  types of, 65
Intel Research, 46
intellectual property, **529**–530
intellectual property rights, **530**
Intellifit, 370
intelligent agents, **460**
interactive collaboration, 221
interactive media, 284
  commercial applications of, 328
  described, **326**–327
  for education and training, 327–328
  TV, **331**–332
  video games, 328–330
Intermountain Health Care, 374
Internal Revenue Service, 35, 378, 536
international markets, 425–426
international networks, 267–268
Internet. *See also* Web
  access over TV, 331
  access to databases, 374–375, 376
  accessing (connecting to), 179–183
  backbone, **178**
  blogs, 23, 199, **205**–206, 422, 589, 593
  chat, 203, **204**–205
  client/server relationship, **184**
  communication and collaboration, 199
  connection speeds, 181
  data mining, 222
  described, **18**–19
  e-mail, 12, 137, **201**–202, 561
  freedom of speech on, 588–593
  future of, 220–222
  groupware on, 452
  hardware, 178–179
  high-speed applications, 221
  history of, 18, 177–178, 195
  hosts, 177
  instant messaging, **203**–204
  interactive collaboration, 221
  large-scale projects, 222
  as layered system, 186
  multisite computation, 222
  proposed re-design, 563
  protocols, 178, 183–184
  radio, 214
  real-time access to remote resources, 222
  relationship with Web, 176, 196
  shared virtual reality, 222
  SMS text messaging, **205**, 412–413
  software, 184–186
  telecommuting, **208**
  text communication, 200–201
  video communications, 207
  visualizing tools, 178, 179
  voice communications, 24, 206–207
  wireless access, 182–183, 256–257, 271

Internet addiction, 587
Internet Archive project, 73
Internet cafes, 256
Internet Corporation for Assigned Names and Numbers (ICANN), 184
Internet fraud, **566**
Internet Fraud Complaint Center, 213
Internet Protocol (IP), 183
Internet Protocol Television (IPTV), 128
Internet Relay Chat (IRC), 204
Internet security
  adware, 565
  hacker attacks, 43, 538–539, 557–560
  need for, 556
  pharming, 568
  phishing, **567**–568
  scams, spam, fraud, and hoaxes, 566–570
  spyware, **565**–566
  viruses, 43, 136, 526, **561**–565
  worms, **561**–565
  zombie computers, 43, 559
Internet service providers (ISPs)
  described, **179**
  selecting, 182
Internet2 and beyond, 220–221
Internet3, 221
internetwork, 177
internetworldstats.com, 178
interpreted HTML code, 189
interpreter, **118**
interviews, 496
intranets, **265**–266
intruders, 43, 538–539
investing online, 410
invisible information gathering, 45
I/O devices. *See also* input devices; output devices
  general-purpose, 84
  special-purpose, 84, 92–93
IP addresses, **183**–184
iPAQ Pocket PC (HP), 60
iPod (Apple), 16, 78, 82, 84, 123, 325
irDA ports, 260
Irfan-View freeware, 162
Israeli Holocaust Database, 359
IT Access for Everyone (ITAFE) initiative, 608
iThenticate, 538
iTunes, 137, 215, 297–299
iTunes Originals, 300
Izumiya Co. Ltd., 416

J

Jacobi Medical Center, 260
Java, 117, 193
Java Applets, 117
Java Web Services Developer Pack (Java WSDP), 419
JavaScript, 193
JAWS (Freedom Scientific Corporation), 92, 605
J.D. Edwards, 354
J.D. Powers, 502
Jet Blue Airways, 494
job hunting strategies, 214
jobweb.com, 214
joining data, 351
joint application development (JAD), 499
Joint Photographic Experts Group (.jpg or .jpeg), 302
Jordan, Easton, 589
joystick, 85
J.P. Morgan Chase & Co., 359
.jpg files, 302
Jukebox software, 296
Justice Department, 456

K

Kamen, Dean, 465
KaZaA, 184, 186, 214–215, 267
Keesal, Young & Logan, 10
Kemp, Liam, 311
Kennametal, 377
keyboards, 84
key-indicator report, 445
key-logging software, 557
keywords, 196
keywords and search engines, 420–422
Kezon, Joe, 38
Kiku-Masamune, 447
Kilby, Jack, 62
kiosks, **14**, 86, 320
Kismet robot (MIT), 456
knowledge worker, 4, 6
Kodak, 320, 401–402
Kolhatkar, Jaya, 459
Krouse, Jim, 505

L

Laibson, Dr., 438
Lander, Eric, 454
Land's End Web sites, 404, 425
language support, 142
languages. *See* programming languages
laptops. *See* notebook computers
large-scale projects, 222
laser printers, 90
Laurel Pub Company, 445
laws protecting information and privacy (U.S.), 532, 596
LCD projectors, 89–90
Leahy, Patrick, 567
leased lines, 243
legal feasibility, 494, 495
Levelle, Ed, 444
LexisNexis Academic Universe, 36, 219–220, 551
Lexus and Bluetooth, 258–259
libel, 590
Library Bill of Rights, 592
library Internet access, 606
library resources, 219–220, 364
licenses, 161–162
LifeDrive handheld (PalmOne), 75
Light, Donald, 509
Lindows, 133, 163
line-of-sight medium, 242
Linux
  computer platforms, 12
  described, 133
  file name conventions, 127
  open-source suites, 148, 162
  proprietary software vs., 163
liquid crystal display (LCD), **89**
listservs, 202
LiveNote, 359
living.com, 195
load-balancing, 417
local area network (LAN), **265**–266
local information on TV, 331
local vendors, 98
localization, 425
Location Free TV (Sony), 325
location-based m-commerce applications, 412
locators, GPS, 254–255
Locus Systems, 499
Loebner Prize, 454
logic errors, 504
logical fields, 362
logical view of data, 125
logos, 214
look-up tables, 59
LotusNotes (IBM), 155, 452
Loyalkaspar, 9
Lucile Packard Children's Hospital, 260
Lufthansa, 183
lycos.com, 199
Lynch, William, 82
Lyons Bakeries, 153

M

MAC (Media Access Control) address, 555
machine cycle, 64
machine language, 116, 118–119
machine-level security
  authentication, 540
  backing up data and systems, 544–547
  biometrics, **543**
  encrypting stored data, 543–544
  ID devices, 542–543
  passwords, **540**–542
  system maintenance, 547–548
  username, **540**
machine-readable data, 83
MacWorld, 96
mad cow disease database, 343
magnetic disks, 74–75
magnetic ink character recognition (MICR) devices, 87
magnetic resonance imaging (MRI), 36
magnetic storage, **74**
magnetic tape, 75–76, 80
mainframe servers, 12
mainframes, 12, 134, 504
maintenance. *See* systems maintenance
Mallis, Laurie, 154
malware, 561
management information systems (MIS)
  described, **28**–29, **443**
  inputs to, 443–444
  outputs of, 444–447
  programmed decisions, 439
management reporting software, 155
managers and system development, 481–482
mapping using GPS, 254–255, 306
Mapquest, 212
maps, 212–213
market research, 408–409
market segmentation, 408
marketing, 406, 422
marketplaces, 404–405
markup languages, 189–190

massively parallel processing (MPP), 68
master files, 345
MasterCard credit cards, 425
Matchmaker System, 446
Matrix (Multistate Anti-Terrorism Information Exchange), 596–597
Matrix Reloaded, The, 25
Matthews, Dave, 300
McAfee software, 136, 137, 526, 560, 564
McCarthy, John, 453
McClellan, Mark, 343
MCI, 18, 178
McLachlan, Sarah, 300
m-commerce, **410**. *See also* mobile commerce
media cards, 79
media center, personal, 41–42
media, mobile, 16
Media Player (Microsoft), 128
media player software, **296**
media, storage, 73
medical informatics, 465
Medicare, 343
Medicare database software, 362
megahertz (MHz), **66**
megapixels, 316
memory
  cache, 66–67
  CMOS, 73
  flash card, **78**–80
  operating systems and management of, 125
  RAM, **63**–64
  RAM SIMMs, 70–71
  ROM, **72**–73
  virtual, 125
mental health concerns, 587
Mercedes, 328
messaging. *See* Internet
meta search engine, 197, 198
meta tags, 421
metropolitan area network (MAN), **267**
microcontrollers, 14
microdrives, 74
microphone input devices, 85
microprocessors, **6**, 62–68
Microsoft. *See also specific products*
  Blue Hat meeting, 539
  device drivers, 124
  research, 152
  software, vs. open-source, 163
  surface computing project, 83
  video download service, 324
microwave transmissions, **242**
microwaves, 238
MIDI (musical instrument digital interface), 34, 151
MIDI cards, 290
midrange servers, 12
Milan, AC, 447
MindManager, 359
Minneapolis MAN, 267
MIPS (millions of instructions per second), 67
Mirra Personal Server, 546, 547
MirrorFolder (Techsoft), 546
mirroring, 546
Missing Child Act, 37
MIT Media Lab, 93
Mitchell, Melanie, 464
Mitnick, Kevin, 43, 559
mixing board, 289
Mobile, 242
mobile commerce
  described, **410**
  over mobile Web, 412
  technology, 411
  through cell phone service providers, 412
  through short-range wireless data communications, 413–414
  through SMS text messaging, 412–413
mobile communications, 16
mobile computers
  expansion of, 95
  types of, 16
mobile computing, 16
mobile digital devices, 15–17. *See also specific device*
mobile input devices, 84
mobile media, 16
Mobile Plate Hunter 900, 88
modeling, 68
modems, 180–182, **239**–241
monitors, baby, 46
Moore, Gordon, 69
Moore's Law, **69**
Mosaic Web browser, 178, 187
motherboard, **62**, 63, 71
Motion Picture Association of America (MPAA), 537–538
Motion Picture Industry of America, 215
Motion Pictures Experts Group-Layer 3 standard. *See* MP3 standard
Motley Fool, The, 39, 410
Motorola processors, 130
mouse, PC, 84
Move2Mac, 138
movie services, 324
movielink.com, 215–216
Moving Picture software, 152
MP3 players, 16
MP3 standard, 59, 136–137, 151, **294**
MPE/iX (Hewlett Packard), 134
MPP(massively parallel processing), 68
MS-DOS (Microsoft Disk Operating System), 128
MSN, 182, 197, 199, 305
MSN Messenger, 203–204
MSN Music, 297–299
multicore technology, 67
multifunction printers, 90–91
multimedia
  described, 284
  Web applications for, 214–217
multiplayer games, 216–217
Multiple Virtual Storage/Enterprise Systems Architecture (MVS/ESA), 134
multiplexer, 243
multiprocessing, 68
multisite computation, 222
multitasking, 126
multitrack recording, 289
music. *See* digital music
music lessons, 211, 221
musical instrument digital interface (MIDI), **290**
Muzzy Lane Software Company, 327
MyCourse, 210
MySQL, 359, 360, 361
Mystic Aquarium, 222
MyTraffic service, 412

N

.name TLD, 188, 189
Napster, 186, 297–299, 300
narrowband, 235
NASA Columbia supercomputer, 68–69
NASA DAC software, 504
NASDAQ Stock Market, 506
Nash, John, 464
National Academy of Sciences, 592
National Aeronautics and Space Administration (NASA), 177
National Educational Technology Standards, 44
National Energy Research Scientific Computing Center (NERSC), 73, 74
National Football League, 35
national information security, 533–534
National Insurance Crime Bureau, 32
National LambdaRail (NLR), 221
National Nuclear Security Agency, 13
National Scalable Cluster Project (NSCP), 222
National Science Foundation, 390, 563
National Security Entry-Exit Registration System, 342
National Strategy to Secure Cyberspace, 34
National Strategy to Secure Cyberspace, 533
National White Collar Crime Center (NW3C), 213
natural intelligence, 454–455
natural language processing, 22, 457
natural language programming, 116
NaturallySpeaking (ScanSoft), 457
nature deficit disorder, 588
NaviTag, 465
Ndiyo development group, 607
Net Nanny, 591
.net TLD, 189
.NET Web services development (Microsoft), 419
Netflix, 216
NETg, 211–212
network(s). *See also* network security; telecommunications
  access to databases, 374–375
  adapters, **240**
  administrators, 244
  computer, **18**, 177, 233
  device software, 245
  digital convergence, **60**
  distributed, 262, 263
  for e-commerce implementation, 417
  electronic data interchange, 268–269, 392
  global, 267–268
  home, 269–272
  interface card, 240

internetworks, 18. *See also* Internet
intranets, **265**–266
local area network, **265**–266
local/remote resources, 262
management software, 244–245
metropolitan area network, **267**
networking software, 233, 244–245
nodes, 262
operating systems, 127, 134
packet-switching, 183
peer-to-peer, **186**, 562
personal area network, **264**–265
private, 262–272
SANs, 75–76, 369
types of, 264–269
virtual private network, **266**
vulnerabilities, 535–536
wide area network, **267**
workstations, 262
network access server (NAS), 266
network operating systems (NOS), 129, 134, 244
network security
information theft, 551–552
interior threats, 551–552
multiuser system considerations, 549
need for, 548
security and usage policies, 552
threats to system health and stability, 551
user permissions, 549–550
vulnerabilities, 551
network service providers (NSPs), 178
networking devices
described, 233
modems, 239–240
network adapters, 240
network control devices, 240–241
networking media
considerations, 236–237
described, **233**
microwave and satellite transmission, 242–244
physical cables, 237–238
radio signals and light, 238–239
neural networks, **458**
neuroimaging, 61
Never Lost service (Hertz), 254
New Technology File System (NTFS), 127
New York Stock Exchange, 117, 445
New York Times newsletter, 202
New York Times online, 24, 409
NewLeads software (Xerox), 152–153
news media, impact of technology on, 199, 589
news, online, 208–210
newsletters, 202
newspapers, online, 209
N-Gage cell phone/video game (Nokia), 62
NIC (network interface card), 240
nic.name, 188
Nike.com, 404
1984 (Orwell), 596–597
Nintendo DS, 329
No Child Left Behind Act, 44, 606
Nokia, 257, 425
nonprogrammed decisions, **440**
normalization, 354
North American Free Trade Agreement (NAFTA), 612
Norton software, 161, 564, 565
notebook computers
clock speeds, 66
described, 9
docking stations, 146
expansion, 95
input devices, 84
Intel processors for, 65
platforms for, 11–12
Wi-Fi capability, 256
Novell, 373
Noyce, Robert, 62
number systems, 59
numeric fields, 361–362
Nuovo home robot, 457
nytimes.com, 202

O

object code, 118
object linking and embedding (OLE), **149**
object-oriented database management system (OODBMS), 354
object-oriented databases, **354**–355
object-oriented (OO) systems development, **493**–494
object-oriented programming, **117**–118
object-oriented programming languages, 117
objects, **117**
O'Brien, Edmond, 597
observation, direct, 496
Octane2 workstation, 68
Office Depot, 414, 415
Office Enterprise Edition (Microsoft), 147
Office (Microsoft), 147, 149, 165, 190
offshoring, **610**
off-the-shelf software, 27, 155, 157–158
on-board synthesizers, 290
on-demand media, 284
OneNote (Microsoft), 149–150, 357
OneReach, 113
online brokerages, 410
online clearinghouses, 404–405
online music services, 297–299
online shopping, 370, 403–404
online training, 157, 210–212
online transaction processing, 393–394
online tutoring, 611
online vendors, 97–98
OnQ (Hilton), 371
on-the-fly Web pages, 192
Ontrack, 544
Open Mobile Alliance (OMA), 411
Open System Interconnection (OSI) model, 186
OpenOffice.org, 148, 162
open-source DBMS, 359–361
open-source software, **133**, 162–164
opensource.org, 164
operating system (OS)
Apple computers, 130
described, **120**
embedded, 135
evolution of, 128–133
file management, 127
functions of, 121
for handheld computers, 135
hardware control, 124–126
I/O, storage, and peripheral management, 127
Linux, 133
market share by type, 130, 135
memory management, 125
miscellaneous functions of, 127–128
network, 129, 134, 244
network management, 127
for personal computers, 128–133
portable, 123
processor management, 125–126
proprietary, 123
for servers, networks, and mainframes, 134
for special-purpose devices, 135
TinyOS, 466
UNIX, 130–132
user interfaces, 121–123
Windows (Microsoft), 129–130
operational feasibility, 494, 495
optical character recognition (OCR) readers, 87, 88
optical mark recognition (OMR) readers, 87
optical processing, 70
optical storage, **76**–78, 80–81
OptimalJ, 499
optimization in spreadsheets, 142, 143, 441, 442
optimization model, **440**–441, 450
order processing system, 394–395
organizational information security, 530–534
.org TLD, 189
Orwell, George, 596
outboard devices, 290
Outlook, Microsoft, 146, 560, 561
output
concepts, 81–83
data collection, 496
described, 7
of management information system, 444–447
output devices
described, **81**, 87
display monitors, 88–90
operating system management, 127
printers and plotters, 90–91
sound systems, 91–92
outsourcing, 419, 420, **492**–493, 609–610, 611
Overstock.com, 400

P

packet-sniffing software, 557
packet-switching network, 183
page scanners, 86
pagers, 253
Palm Graffiti system, 84, 85
Palm OS, 12, 123, 135
PalmOne, 135
PalmSource, 135

Panzano restaurant, 342
parallel conversion, 510
parallel processing, 68
parental control, 210
parking meters, wireless, 246
Parson, Jeffrey Lee, 527
partitioning drives, 120
PartyGaming, 217
Password Agent (Moon Software), 542
Password Manager Plus (Billeo), 542
passwords, **540**–542
patches, software, **536**
patent application software, 115
patents, 530
Patriot Act, 378, 596
Patrol (BMC), 513
Payment Card Industry Data Security Standard, 533
payment options, 99
payment software, 417
payment systems, 413–414, 423–425
PayPal, 423–424, 426, 427
PayPass, 425
PC Cards, 95
PC Relocator, 138
PC security checklist, 570
PCs. *See* personal computers (PCs)
PCMCIA cards, 95
PCMCIA slots, 95
peer-to-peer (P2P) networks, **186**, 562
Pentagon, 35, 466
performance
  cache, 66–67
  clock speed comparisons, 66
  factors, 66
  Intel processors, 65
peripherals, 127. *See also* input devices; output devices
permissions, 549–551
Personal Area Network (Microsoft), 265
personal area network (PAN), **264**–265
personal computers (PCs)
  clock speeds, 66
  described, **9**
  input devices, 84
  Intel processors for, 65
  operating systems, 128–133
  types of, 9–11
personal digital assistant (PDA), 10, 135. *See also* handheld computers
personal information
  privacy, 594–595
  security, 528–529
personal information management (PIM) software, 10, 40, 145–147
personal video recorders (PVRs), 331
personnel hiring and training, 506
pervasive communications, 199
pervasive computing, 44
Pfizer, 113
pharming, 568
phase-in approach, **510**
Philadelphia University, 390
Philips, 152
phishing, **567**–568
phone-line networking (HomePNA), 270
photo printers, 90, 91, 320–321
photo software, 152
photo-editing software, **318**–319
photography. *See* digital photography
physical health problems, 586
physical layer, 186–187
physical view of data, 125
pilot startup, **510**
piracy, **537**–538
Pixar, 313, 316
pixels, 85, 88, **301**
pixilation, 301
plagiarism, **537**, 538
plasma display, 89
platform, computer
  described, **11**, 112
  selecting, 99
  types of, 11–12
platters, **75**
PlayStation 2 (Sony), 217, 328
PlayStation Portable (Sony), 16, 17, 329, 330
plotters, 91
plug and play (PnP), 127
plug-ins, **194**
.png files, 302
Pocket PC (Microsoft), 135
podcast, **292**
podcasting, 292–294
pods, 589
point-of-sale (POS) devices, 87, 393, 413, 414
points of presence (PoPs), 178–179
political freedom, 590
pop-up ads, 406, 409
pop-up ads, blocking, 137
pornography and issues of decency, 590–592
portable media center, 16
Portable Network Graphics (.png), 302
portable operating systems, 123
portals, 199, 218, 404
ports, 93, **184**, 260, 557
port-scanning software, 557
PostgreSQL, 359, 360
Powell, Michael, 598
power company safeguards, 512
power-line networking (HomePLC), 270
power-on self test (POST), 124
PowerPC processors, 65, 67
PowerPoint (Microsoft), 143–144
ppm (pages printed per minute), 91
Premiere (Adobe), 152
presentation graphics software, 142–144
price quotes, 407
priceline.com, 213
primary key, **346**–348
primary storage, 70
Primerica Life Insurance, 152
print servers, 12, 262
printer driver, 124
printers, 90–91
Prism Visualization system (Silicon Graphics), 68
privacy, 45–46
Privacy Act of 1974, 595
privacy issues
  described, 593–594
  future scenarios, 601–602
  personal information privacy, 594–595
  surveillance, **597**–602
  video surveillance, 594
private networks, 262–272
problem solving
  described, **438**
  heuristics, **442**–443, 450, 461
  optimization model, **440**–441, 450
  proactive approach to, 439
  reactive approach to, 439
  stages in, 438, 439
processing, 63, 68
processors
  coprocessors, 68
  CPU. *See* central processing unit (CPU)
  described, 6
  dual-core, 67
  Intel, types of, 65
  operating systems and management of, 125–126
  optical, 70
  performance factors, 67
  silicon-based, and alternatives, 69–70
  Wi-Fi capability, 256
product information, 407–408
production studios, 287–289
productivity software, **140**
program code
  described, 114–115
  object code, 118
  source code, 118
program development life cycle, 503
program evaluation and review technique (PERT), 489
program flowchart, 487
program specifications, 503
programmed decisions, **439**
programming
  application programming interfaces, 418
  natural language, 116
  object-oriented, **117**–118
  for Web, 192–193
programming language standard, 115
programming languages. *See also* software
  described, **114**
  evolution of, 116
  machine, 116
  newest types of, 116–118
  syntax, 115
  translators, 118–119
  for Web pages, 193
programming, Web, 192–193
programs, computer, **112**
project crashing, 490
project leaders, 482
project management, **489**

project management software, 151, 489
project management tools, 489–490
Project (Microsoft), 489, 491
Proofpoint and Forrester Consulting, 598
proprietary operating systems, 123
proprietary software, 155, 163, 203
protocols. *See also specific protocols*
  described, **18**, 245
  Internet, 178, 183–184
  wireless encryption, 555
prototyping, **491**
Protx, 565
Providence Washington Insurance Company, 444
proximity payment system, 413–414
.psd files, 302
Psion, 135
public domain, 162
public libraries and Internet access, 606
purchase over TV, 331
purchasing computers
  needs analysis, 95–96, 99
  payment options, 99
  research, 71, 96–97
  shopping strategies, 99
  vendors, 97–98
purchasing system, 394–395

Q

Quadrani, Alexia, 195
quantum computing, 70
Quantum View Manage software, 154
qubits, 70
query by example (QBE), 367–368
questionnaires, 496
QuickBooks, 359
Quicken, 39, 358
quintcareers.com, 214

R

R (recordable) media, 77
radio frequency identification (RFID), **260**–261, 371, 600–602
radio, Internet, 214
radio spectrum, 238, 239
radio wave, **238**
RAID (redundant array of independent disks), 76, 369
RAM. *See* random access memory (RAM)
random access memory (RAM)
  described, **63**–64
  SIMMs, 70–71
  types of, 71
  video, 71–72
rapid application development (RAD), 499
raster graphics, 301
Rather, Dan, 589
Rational Software, 499
ray tracing, 310
Reader (Adobe), 162
readme files, 159, 160
read-only memory (ROM), **72**–73
Real Time Crime Center, 353
RealNetworks, 305
real-time access to remote resources, 222
real-time mirroring, 546
Recording Industry Association of America (RIAA), 185, 186, 297, 537
records, database, **345**, 362
recovery, database, 369
recovery disk, 124
Red Hat Linux, 133, 373
Reeves, Keanu, 25
reference software, 152
registering software, 159, 160, 162
registers, 63
Registry Mechanic, 548
registry, Windows, 127–128, 138, 548
relating data, 352
relational database model, **351**–354
relations, personal, 41
remote data backup, 547
Remote Data Backups, 547
removing software, 161
rendering, 311
repeaters, 241
repetitive stress injury (RSI), 586
replicated databases, 374
ReportNet (Cognos), 444
reports
  demand, **445**–446
  exception, **446**
  flexibility of, 449
  key-indicator, 445
  management information systems, 444
  scheduled, **445**
  uses for, 447
request for information (RFI), 501
request for proposal (RFP), **501**, 502
requirements analysis, **497**–499
rescue disk, 544
research
  assigned, 218
  curiosity-driven, 218
  databases, 219–220
  library resources, 219–220
  personal, 40
  sources of online information, 218–220
  topics, 218
  Wikipedia.com, 219
restart procedure, 124
Restoration Hardware, 457
restore points, 137, 545
restore utility, 545
retinal scanning, 543
Reuters, 209
review, systems, **513**
Reviewing in Microsoft Word, 140
Rhapsody, 297–299
rich content, 194
rich media, 284
Rights Management Services, Windows, 138
ripper software, 296
Robertson, Hymans, 513
robotics, 22, **456**
robots, uses for, 454, 456, 457
Rodriguez, Robert, 303
ROM (read-only memory), **72**–73
Rosser, Jr., James, 327
routers, **179**, 180, 241
Royal Bank of Canada, 499
RSS (Really Simple Syndication) feeds, 206
"rules of thumb", 442
running programs, 113
RW (rewritable) media, 77, 80–81

S

Sabre Airlines Solutions, 504
sampler, **289**
sampling, 285
Samsung, 16, 86
Sandia Laboratories, 458
Sarbanes-Oxley Act, 378
Sarbanes-Oxley Act of 2002, 482, 532
SAS Institute, 459
satellite images, 212
satellite Internet, 181
satellite radio, **300**
satellite transmissions, 242–243
scams, 566
scanning devices
  fingerprint, 86, 543, 544
  source data automation, **83**
  types of, 86–87
Scent Dome (Trisenx), 8
schedule feasibility, 494, 495
scheduled reports, **445**
schema, **365**
scientific visualization, **306**, 307
Scottish Life, 154
screensavers, 138
ScripTalk, 261
search engines, **196**–197, 198, 420–422
secondary storage
  described, 73–74
  types of, 74–76
Secure Internet Machines (SIM), 605
Secure Sockets Layer (SSL), 427, 428
Securities and Exchange Commission (SEC), 378
security. *See also* computer crime; information security; Internet security; machine-level security; network security
  database, 379
  guidelines, 531
  national information, 533–534
  PC security checklist, 570
  personal information, 528–529
  wireless networking, 553–556
security holes, **535**
security levels by resource, 541
Segway personal transporter, 465
Seisint, 596–597
selecting data, 351
Semantic Web, 189
sequencer, **289**
sequential access, 75
servers
  application servers, 263
  client/server relationship, **184**, 263
  database servers, 263
  described, **12**–13
  file servers, 12, 262, 263, 546
  Intel processors for, 65
  network access server, 266
  operating systems, 134
  print servers, 12, 262

on private networks, 263
Web, 12, 184, **188**
Service Patch 2, Windows XP (Microsoft), 160
SETI (Search for Extra Terrestrial Intelligence), 68, 286
setup program, 113
Severino, Victoria, 372
SFTP (secure file transfer protocol), 138, 560
shared virtual reality, 222
shareware, **162**
Shepard's Citations, 220
shipping management software, 154
shopping for computers. *See* purchasing computers
Short Message Service (SMS) text messaging, 16, **205**, 252, 412–413
short-range wireless data communications, 413–414
Sibelius software, 151
Sidekick II (T-Mobile), 123
signals, telecommunications, 232, 234, 238
Silicon Graphics, Inc., 68
silicon-based chips, 69–70
Simeone, Anna, 221
SIMM (single in-line memory module), 70–71
Simpay, 425
simPC, 605, 606
simulated society, 460
simulations, 450–451
Sirius Satellite Radio, 325
site preparation, 507, 508
Slammer worm, 561
slate model tablet PCs, 10
slides and presentation graphics, 143
smart cards (credit cards), **424**–425
smart containers, 465
smart dust, 466
smart homes, 23, 24
Smart Personal Objects Technology (SPOT), 466
smart phones
described, 10–11
digital convergence, **60**
features of, 113, 329
m-commerce, 411–413
PayPass payment system, 425
platforms for, 11–12, 99
SmartCar, 261
SmartForce, 211–212
SmartSuite (Lotus), 148, 149
social engineering, 557
Social Fabric software, 247
Social Security numbers, 346, 377
society and technology. *See also* ethical issues
cell phone etiquette, 584, 585
challenging establishment and traditional institutions, 589–590
computers and community, 584–585
decision-making disorders, 583
ethics and social responsibility, 602–608
freedom of speech, 588–593
health issues, 585–588
impact of technology on, 582–583
living online, 583–588
nature deficit disorder, 588
privacy issues, 593–602
software
antivirus, **564**
artificial intelligence, 155
backward compatibility, 77, 148
bugs, 160–161, 504
computer-aided design, 152
contract, 156
customized, 27, 155–156, 158
database. *See* database management system (DBMS)
decision support, 153–155, 450
described, **7**, 112
desktop publishing, 152
digital music, 296–297
for e-commerce implementation, 417–419
educational and reference, 152
entertainment, games, and leisure, 152
expert systems, 155
file-sharing, 184–186, 297
financial management and tax preparation, 151
freeware, **162**
functions of, 113
for groups and organizations, 152–153
importance of, 112
installing new, 159–160
integrated packages, 149
Internet, 184–186
management reporting, 155
media player, **296**
mobile packages, 149–150
music, 150, 151
network management, 244–245
networking, 233, 244–245
object linking and embedding, **149**
off-the-shelf, 27, 155, 157–158
open-source, **133**, 162–164
patches, **536**
photo-editing, 152, **318**–319
PIM, 40, 145–147
presentation graphics, 142–144
programming languages. *See* programming languages
project management, 151
proprietary, 155, 163, 203
removing (uninstalling), 161
security holes, **535**
shareware, **162**
spreadsheet applications, 140–142
statistical, 152
system. *See* system software
transaction processing, 153
updating, 159, 547
upgrading, 158–161
used in schools, 605
vector graphics, 308–309
video editing, 152, 322–324
virus detecting, 136, 564–565
vulnerabilities, 535–536, 551
Web authoring, 151, **191**–192
word-processing applications, 113–114, 140
Software & Information Industry Association, 537–538
software engineering, 486
software network devices, 245
software programmer, 482
software suites, **147**–149
Solaris (Sun Microsystems), 134
Solver, Microsoft Excel, 142, 143
Sonoma State University, 22
Sonos, 296
Sony DVRs, 78
sound compression, 294
sound production studios, 287–289
sound systems, 91–92
SoundBlaster wireless music system, 271, 272
source code, 118
source data automation, **83**
spam, 412–413, 565, 566, 568–569
spam blocking, 137, 442, 569
special-purpose systems
game theory systems, **464**
geographic information system, **463**
informatics, 464–465
other specialized systems, 464–466
virtual reality systems, **462**–463
specific absorption rate (SAR), 586
speech recognition
applications for, 85, 257
natural language processing, **457**
Speedpass, 242
spiders and Web crawling, 196
spoofing, 561, 567, 568
spooling, 124
spreadsheet software, 140–142, 441, 442
spreadsheets, 140–141
Springs Retreat, 415
Sprint, 18, 24, 178, 248, 326, 414
Sputnik, 177
spying methods, 559
spyware
defending against, 565–566
described, **565**
SSID (service set identifier), 553, 555
standards, 59, 245
Stanford University, 177, 465
Starbucks, 183, 300
Starclass software, 151
StarOffice (Sun Microsystems), 148, 149, 163
startup, 510–511
STATCare, 327
static IP address, 183
static Web pages, 192

statistical software, 152
Stillman College, 82
stop-sign.com, 162
storage, 70–81
 for audio media, 295–296
 capacity examples, 7
 described, 70
 digital photos, 317–318
 evaluating devices and media, 80–81
 flash drives and cards, 78–80
 operating system management, 127
 optical, **76**–78, 80–81
 secondary, 73–76
 system, 70–73
 virtual, 125
storage area network (SAN), 75–76, 369
storage capacity, 73–74
storage devices, 73
storage media, 73
storyline, 323
streaming audio, 298
streaming media, 194
stress, 587
structured interviews, 496
structured problems, 439
structured query language (SQL), **368**–369
student.com, 223
stylus, 84, 85
subject directories, 197, 199, 218
subscriber identity module (SIM) cards, 248
Sun Microsystems, 117, 263, 269
supercomputers, **13**–14, 21, 67–69, 73
Superdisk (Imation), 75
supply chain management, 400
supply planning, 400
surgery, robotic, 456
surgery training system, 481, 482
surgery, Webcasting, 210
surveillance
 described, **597**
 GPS and RFID, 600–602
 video and audio, 594, 599–600
 wiretapping, 597–598
Swift, Tom, 152
switched lines, 243
switches, network, 240
Sybase, 359
Sylvan Learning, 611
Symantec, 136, 137
Symbian operating system, 12, 135
synchronization, 10
synchronous communications, **200**
syntax errors, 504
syntax, programming language, 115
synthesizer, **289**
system administrators, 244, 550
system backup and restore, 544–545
system bus, 63
system clock, 66
system penetration, 538
system requirements, 159
System Restore, 137
system software. *See also* software
 described, 112, **120**
 operating systems, 120–135
 utility programs, **135**–139
system stakeholders, **481**
system storage, 70–73
system testing, 509
system X supercomputer, 21
systems analysis
 collecting data, 496–497
 data analysis, 497
 described, **495**
 general considerations, 495–496
 requirements analysis, **497**–499
systems analysts, **482**, 483
systems design
 contract, 502
 described, **500**
 evaluating and selecting, 501–502
 generating alternatives, 501
systems development. *See also* systems implementation
 creative analysis, 483
 critical analysis, 483
 described, **28**, **478**
 end-user development, 485–486
 failed projects, 481
 iterative approach, 490, 492
 needs analysis, 484
 participants in, 481–482
 planning, 484–485
 reasons for starting, 482–484
 specialists, 482
 successful projects, 478
 technology makeover with, 479
 tools. *See* systems development tools
systems development life cycle (SDLC), **479**–480
systems development tools
 computer-aided software engineering tools, **486**–487
 decision tables, **488**–489
 flowcharts, **487**
 object-oriented systems development, **493**–494
 outsourcing, 419, 420, **492**–493, 609–610, 611
 project management tools, 489–490
 prototyping, **491**
systems documentation, 511
systems implementation. *See also* systems development
 acquiring database and telecommunications systems, 506
 acquiring hardware, 503
 data preparation, 507–508
 described, **502**
 installation, 508, 509
 personnel hiring and training, 506
 selecting and acquiring software, 503–504
 site preparation, 507, 508
 startup, 510–511
 testing, 509
 typical steps in, 503
 user acceptance and documentation, 511
 user preparation for, 506, 507
systems investigation, **494**–495
systems maintenance
 described, **511**
 financial implications of, 512, 513
 reasons for starting, 511–512
systems review, **513**

T

T1 lines, 243
T3 lines, 243
Tableau, 359
tables, database, 345, 362
tablet PCs
 described, 9–10
 platforms for, 11–12
 software, 149–150
tablets, graphics, 84
Tagged Image File Format (.tiff or .tif), 302
TalkToAliens.com, 243
Tanggula project, 607
Tarantino, Quentin, 303
Target, 320, 321
TaskTracker, 357
tax preparation software, 151
TaylorMade, 441
t-commerce, 331
TCP/IP, **183**
Techno Bra, 255
technology, **6**
telecom (telecommunications) industry, 18
telecommunications
 bandwidth, **235**
 Bluetooth, 252, **258**–259
 broadband, **235**–236, 606–607
 cell phone technologies, 246–253. *See also* cell phones
 characteristics of, 234–236
 control devices, 240–241
 data communications, 233
 described, **17**–18, **233**
 fundamentals of, 232
 GPS, 95, 113, **253**–255
 industrial media and devices, 242–244
 infrared transmission, 260
 interpreting information, 232
 microwave and satellite transmission, 242–244
 modems, 239–240
 network adapters, 240

networking hardware devices, 233, 239–241
networking media, **233**, 236–239
networking software, 233, 244–245
pagers, 253
physical cables, 237–238
radio frequency identification, **260**–261
radio signals and light, 238–239
signals, types of, 234
surveillance, 598
for systems implementation, 506
transmission capacities, 234–236
Wi-Fi, **19**–20, **256**–258, 271, 553–555
WiMAX, 182, 256–**257**
Telecommunications Act of 1996, 605
telecommuting, **208**, 269
telecoms, 178
telescopes, real-time access to, 222
television, 324–326
television displays, 89
Telstra, 157
teraflop (trillions of floating-point operations per second), 67
terminals, 13
terrestrial microwave, 242
Terrorism Information Awareness Program, 372
testing systems implementation, 509
text communication, Internet, 200–201
text messaging, 16, 205, 252, 412–413
Thawte, 427
thin client, 263–264
Thompson IV, William, 291
3D computer animation, 315–316
3D displays, 89, 90
3D modeling software, **310**–311
thumb drives, 79
Thunderbird, 162
.tif files, 302
time sharing, 126
Time Warner Inc., 551
time-driven review, 513
TiVo, 78, 136
TiVoToGo, 325
T-Mobile, 248–250
token ring, 245
top-level domain (TLD), 188
topologies, 265, 266
Toronto Airport, 233
Torvalds, Linus, 133
Toshiba hard drives, 78
Total Information Awareness (TIA), 596
touch pad, 84
touch screen, **84**–85
Toy Story, 315
Toyota, 32, 328, 329, 450
Track Changes in Microsoft Word, 140
trackballs, 84
tracking devices, 260–261
TrackPoint, 84
trade secrets, 530, 593
trademarks, 530
training, online, 157, 210–212
transaction files, 345
transaction processing
batch processing, 393
business resumption planning, 428
cycle, 394
for different needs, 394–395
online processing, 393–394
software, 153
transaction processing system (TPS)
described, **28**
for e-commerce, 393
input to management information system, 443–444
software, 153
transborder data flow, 267
transistors, **62**, 69
translators, language, 118–119
transmission capacities, 234–236
Transmission Control Protocol (TCP), 183
transparent activity, 124, 262
transport layer, 186–187
Transport Layer Security (TLS), 427
travelocity.com, 213
Trend Micro, 442
Trojan horses, 561
Troublemaker Digital Studios, 303
Trustworthy Computing (Microsoft), 535
TUI, 448
tunneling, 266
Turing, Alan, 454
Turing Test, 454
Turnitin, 538
TV
consumer-generated media on, 589–590
interactive, **331**–332
Twenty Questions, 457
twisted pair cable, **237**
2D computer animation, 312–314

U

ubid.com, 404
ubiquitous computing, 44
Uniform Resource Locator (URL), **188**–189
uninstalling software, 161
unit testing, 509
United Airlines, 183
United Nations, 43
United Parcel Service (UPS), 154, 513
United States Computer Emergency Readiness Team (US-CERT), 533, 563
United States Postal Service (USPS), 35
Universal Product Codes (UPCs), 32
Universal Serial Bus (USB)
described, **79**
drives, 79–80
University of California at Berkeley, 466
University of California at Los Angeles, 177
University of California at Santa Barbara, 177
University of California at Santa Cruz, 222
University of Connecticut, 222
University of Denver, 37–38
University of Edinburgh in Scotland, 465
University of Missouri, 327
University of Missouri-Columbia, 222
University of Utah, 177
University of Washington, 300
University of West England, 456
UNIX, 130–132
unstructured databases, 375–376
unstructured interviews, 496
unstructured problems, 440
updates, software, 159, 547
upgrading software, 158–161
uploading, 379
UPS (United Parcel Service)
database, 342, 343
signature device, 82–83
UPSnet, 268
US Airways, 481
U.S. Army, 456, 460
U.S. laws protecting information and privacy, 532, 596
USB fingerprint ID device, 544
USB keychain drives, 10
USB ports, 93–95
user acceptance document, **511**
user documentation, 511
user interface
command-based, **121**, 122
GUI environment, **122**–123
operating systems, 121–123
user permissions, 549–551
user preparation, 506, 507
username, **540**
users and systems development, 481
utility programs
backup utilities, 137
described, **135**–136
file compression utilities, 136–137
file transfer utilities, 138
hardware and disk utilities, 137, 138, 547
search utilities, 138
spam and pop-up guards, 137
virus detection and recovery, 136, 564–565
Web server, 417

V

Vaio Type X (Sony), 78
value-added software vendors, 156
ValueWeb, 414
Vanderbilt University, 249
Vantage review tool, 513
Vcast (Verizon), 325
vector graphics, **301**
vector graphics software, 308–309
vendors, 97–98, 501
VeriSign, 427
Verizon, 248–250
Verizon Internet Services, 185
vertical Web portals, 199
video conferencing, **207**
video, digital. *See* digital video
video game consoles, **328**
video logs, 589
video on demand (VoD), 331
Video RAM (VRAM), 71–72
video surveillance, 594
violence, video game, 330
Virginia Tech, 21
Virtual Case File system, 481
Virtual Harlem, 222
virtual memory (VM), 125

virtual office telecommuting centers, 208
Virtual PC for Mac, 139
virtual private network (VPN), **266**
virtual reality
described, 326, **462**
headsets, 93, 462
shared, 222
systems, **462**–463
virtual space, **583**
virtual storage, 125
Virtual U, 327
virtualized databases, 373
Virtually Human (Sony), 152
virus hoax, **569**
viruses
defending against, 136, 564–565
described, 43, **561**
detection and recovery, 136, 237
spreading, 561–564
symptom of, 526
varieties of, 562
Visa credit cards, 424–425
vision systems, 22, **456**
Visual Basic, 116–117
Visual Basic.NET, 117
Visual C++.NET, 117
visual databases, 375
visual languages, 116
Visual Studio.NET, 494
visualizing tools, Internet, 178, 179
Visualware software, 180
Vivo, 413, 414
vlogs, 323, 589
Vocera Communications, 257
voice communications, 24, 206–207
Voice over Internet Protocol (VoIP), **24**, 206–207, 452
voice recognition, 85
voice recorders, 286
voice-print identification, 287
voiceprints, 37
volatile memory, 63
volume testing, 509

W

Wal-Mart, 242, 260, 261, 266, 396
Walrus, 178
war driving, 554
Warehouse Management (Oracle) software, 371
warm boot, 124
Washington Post online, 409
Wayne State University, 513
wearable PCs, 93
Weather Channel, The, 425
weather.com, 162
Web. *See also* e-commerce; Internet
auctions, 404–405
authoring software, 151, **191**–192
browser plug-ins, **194**
browsers, **188**
comparison shopping, 398, 399, 403
components of, 188–189
content streaming, **194**
creating pages for, 190
described, **19**, 187
designing pages for, 191
e-commerce. *See* e-commerce
education and training, 210–212
employment and careers, 213–214
games, 216–217
history of, 19, 187, 195
information resources, 217–220
logs, 205
markup languages, 189–190
movies, 215–216
multimedia and entertainment, 214–217
music, 214–215
news, 208–210
portals, 199, 218, 404
programming, 192–193
relationship with Internet, 176, 196
search engines, **196**–197, 198, 420–422
subject directories, 197, 199, 218
travel, 212–213
Web conferencing, 207
Web Content Accessibility Guidelines, 605
Web crawling, 196
Web scripts, 563
Web servers
described, 12, 184, **188**
software, 417
utility programs, 417
Web services, 191, **418**–419
Web Services Interoperability Organization (WS-I), 419
Web site
design tools, 417
development tools, 417
Web site design tools, 417
Webcams, 86, 87
Webcasting, 209, 210
WebCT educational software, 152, 210
Web-driven programs, 418
WebFocus, 376
WebSphere (IBM), 419
"what if" analysis, 140
what-if analysis, **450**
Wherify GPS locator, 255
wide area network (WAN), **267**
Widgit Software, 138
Wi-Fi Protected Access (WPA), 555
Wi-Fi (wireless fidelity), **19**–20, **256**–258, 271, 553–555
Wiki projects, 219
Wikipedia, 219, 537
Wild Brain, 419
WiMAX, 182, **257**
Windows Automotive (Microsoft), 135
Windows CE (Microsoft), 123, 135
Windows cleaners, 138, 548
Windows Embedded (Microsoft), 135
Windows (Microsoft)
computer platforms, 11–12
computing platform, 11
development of OS, 129–130
file name conventions, 127
GUI environment, 123
Windows Mobile (Microsoft), 135
Windows Registry cleaners, 548
Windows Server 2003 (Microsoft), 134
Windows Update (Microsoft), 535, 536, 547, 559
Windows Vista (Microsoft), 130
Windows XP Media Center Edition (Microsoft), 42, 60, 320, 332
Windows XP (Microsoft), 129
Windows XP, Service Pack 2 (Microsoft), 160
Windows XP Tablet PC Edition (Microsoft), 149
WinPatrol, 162
winzip.com, 136
Wired Equivalent Privacy (WEP), 555
wireless access points, 241, 256, 553
wireless adapters, 240
Wireless Application Protocol (WAP), 411
wireless citywide network, 18
wireless Internet access. *See also* wireless networks
in classrooms, 249
for homes, 271
locations, 212
types of, 182–183
Wireless Markup Language (WML), 411, 412
wireless networking, **19**–20
wireless networks
controlling access to, 555
encrypting data, 555
invisible, 555
need for security, 553
securing, 554–556
threats to, 553–554
wireless telecommunications
Bluetooth, 252, **258**–259
cell phone technologies, 246–253. *See also* cell phones
GPS, 95, 113, **253**–255
home networks, 271
infrared transmission, 260
pagers, 253
radio frequency identification, **260**–261
Wi-Fi, **19**–20, **256**–258, 271, 553–555
WiMAX, 182, 256–**257**
wiretapping, 597–598
Word (Microsoft), 140, 141
wordlength, 67
WordPerfect Office (Corel), 148, 149
word-processing software, 113–114, 140, 141
work stressors, 587
Works (Microsoft), 149
Works4you software, 154
WorkSite, 36
workstations, 262
World Economic Forum, 607, 608
world ownership, 549
World Trade Organization, 610
World Wide Web. *See* Web
World Wide Web Consortium (W3C), 190, 605
worlds.com, 204
WORM (write-once, read many) technology, 378
worms
defending against, 136, 564–565
described, **561**
spreading, 561–564

Wyndham International, 452
Wyndham Resorts, 96
WYSIWYG (what-you-see-is-what-you-get) editor, 191–192

X

X Window System, 131, 132
Xbox Live (Microsoft), 217
Xbox (Microsoft), 217, 328
XHTML, 190
XML (Extensible Markup Language), **189**–191
.xml files, 190

Y

Y2K preparation, 609
Yahoo!, 26, 197, 199, 209, 218, 422
Yahoo! Chat, 204
Yahoo! Messenger, 203–204
Yahoo! Mobile Photos, 320
Yahoo! Photos, 320
Yoran, Amit, 531
Yoshihiko Handa, 447

Z

Zaman, Quazi, 164
zdnet.com, 96
Zimmerman, Thomas, 265
Zip disk (Iomega), 75, 80–81
zip files, 136–137
zombie computers, 43, **565**
ZOOP, 260
z/OS (IBM), 134

Career Index

A

accountant, 61, 365
accounting, 32, 36
advertising, 32
advisor, 30
aeronautical engineer, 33
agents, 236
air traffic control, 451
air traffic controllers, 451
airlines, 234
animators, 68
architect, 90, 93, 152, 284, 463
astronomy, 33
athletics, 35, 321, 442, 443, 447

B

banking, 32
bioinformatics, 465
biology, 21, 33, 465
biomechanics, 35
business, 27, 28, 32–33, 141, 143, 438, 451, 476

C

call center operator, 364
cardiologist, 306
chemical engineer, 33
chemistry, 33, 92
chief information officer (CIO), 31, 444, 482, 508, 511
chief technology officer (CTO), 31
civil engineering, 33, 362
clerk, 364
communications, 32–33
computer engineer, 607
computer operator, 31
computer professional, 31–32
computer programmer, 31, 112, 115, 116, 125, 165, 366, 368, 482, 503, 609
computer scientist, 31, 93, 136
computer systems manager, 482, 508
computer technician, 128
construction, 234
criminology, 37, 438

D

dance, 34, 513
data entry operator, 506
database administrator (DBA), 377
database expert, 369
delivery driver, 82–83, 268
dentist, 36
designer, 152, 284
detective, 37

E

economics, 33
educator, 38, 307, 327, 364
electrical engineer, 33
emergency (911) operator, 463
emergency medical technician, 36
engineering, 33, 90, 114, 141, 152, 284, 438
entertainment artist, 306
entertainment artists, 236
environmental engineer, 33, 580
environmental sciences, 33
executive, 365
exercise science, 34–35, 36

F

family therapy, 36
farmer, 451
FBI agent, 439, 595
film studies, 34
finance, 32, 409–410
financial analyst, 61
financial manager, 450
fine arts, 34
firefighters, 327
forensic audio specialist, 286
forensic graphics expert, 321
forest-service consultant, 143

G

geography, 33
government, 35, 36
graphic artist, 9, 34, 56, 61, 82, 304–315

H

health care, 36
hospitality, 32, 96, 370
human resources, 32

I

illustrator, 306, 307
industrial engineer, 33
information systems, 32
insurance, 32, 36, 75, 152, 154, 483
IT (information technology) professional, 96, 513

J

journalist, 286, 582

K

knowledge engineer, 482

L

laboratory technician, 36
law, 35–36, 110, 115
law enforcement, 32, 37, 286–287, 439, 464
lawyer, 10, 286, 498
leisure, 32
librarians, 83
literature, 34

M

managers, 155, 157, 162, 364, 441, 444, 481
manufacturing, 32, 85, 142, 284, 463
marketing, 32
mathematics, 33
mechanical engineer, 33
medical infomatics, 36, 465
medical records, 36, 42, 361, 595
medical research, 33, 465, 480
medicine, 28, 36–37, 61, 257, 258, 327, 458, 462
meteorology, 21, 33, 306
military, 33, 35, 451, 458, 463
motion picture industry, 45, 70, 215, 306, 324
movie producer, 484
music, 34, 70, 93, 94, 285, 289–292

N

national defense, 42, 533
network administrator, 244
network engineer, 267
neurobiologists, 92
news correspondent, 230
nursing, 15, 36, 440
nutrition, 35, 36, 142, 420

O

oceanography, 33
organizational behavior, 32
ornithologist, 367

P

paleontologist, 367
payroll, 32, 118, 349, 393, 487–488
pharmacist, 36
photographer, 27, 34, 94, 305, 320, 401, 589
physician, 15, 27, 143, 154, 191, 286, 356, 446, 448, 465, 482
political science, 33
project manager, 609
psychology, 33, 36
publisher, 141, 190, 209, 284, 305

R
real estate professional, 27, 436, 463
reporter, 82
researcher, 36, 68, 133, 176, 364, 365, 454
retail, 32, 284

S
sales, 32, 234–235, 340, 436, 441, 463, 476
science, 33, 141
scientist, 21, 27, 92, 93, 321, 344, 454
sculptor, 305
security expert, 15, 37–38
social sciences, 33–34
social work, 36, 463
sociology, 33
software developer, 11, 162, 164, 504
software engineer, 31, 112, 125, 193, 418, 482, 486, 503
software programmer, 482, 483
sound engineer, 287, 288, 289
space exploration, 504
special effects specialists, 68
sports science, 33, 34–35
statistics, 33, 141, 306, 318, 343
surgeon, 36, 81, 90, 93, 327, 456, 481
system administrator, 244, 550
systems analyst, 28, 31, 482, 511
systems development specialist, 481, 482

T
technical artist, 307
theater, 34
traffic control, 321, 412
training, 38
travel, 234
TV producer, 447

U
urban planning, 33

V
videographer, 94, 305, 322, 323
visual arts, 34

W
Web developer, 138, 190, 196, 412, 417, 425

Photo Credits

Chapter 1

Figure 1.1: © David R. Frazier Photolibrary, Inc. / Alamy
Figure 1.2: © Mark Scott/ Getty Images
Figure 1.4: Courtesy of Apple Computer, Inc.
Figure 1.5: Courtesy of Apple Computer, Inc.
Figure 1.6: © Don Mason/CORBIS
Figure 1.7: Courtesy of Palm, Inc.
Figure 1.8: Courtesy of Symbian, Ltd. and Microsoft Corporation
Figure 1.9, Figure 1.10: Courtesy of IBM Corporation
Figure 1.11: Courtesy of Eastman Kodak Company
Figure 1.12: Courtesy of Brookhaven National Laboratory
Figure 1.13: © Royalty-Free/Corbis
Figure 1.14: Courtesy of Apple Computer, Inc., Creative Technology, Ltd., Sony Computer Entertainment, Inc.
Figure 1.16: Courtesy of Christina Micek
Figure 1.17: © *BananaStock / SuperStock*
Figure 1.19: Courtesy of Visual & Broadcast Communications Virginia Tech
Figure 1.20: Linnea Mullins/ Sonoma State University
Figure 1.21: Courtesy of Insteon
Figure 1.22: Courtesy of Microsoft Corporation
Figure 1.24: © Jim Craigmyle/Corbis
Figure 1.25: © Tim Boyle/ Getty Images
Figure 1.26: Courtesy of NASA
Figure 1.27: Courtesy of SRA Touchstone Consulting Group, Inc.
Figure 1.28: Courtesy of Intel Corporation
Figure 1.29: Ryan McVay/ Getty Images, Kim Steele/ Getty Images, Daniel Allen/ Getty Images
Figure 1.30: Courtesy of Computer Science Corporation
Figure 1.31: David Sailors/ Corbis
Figure 1.32: Courtesy of Adobe Systems, Inc.
Figure 1.33: Courtesy of Vicon Motion Systems Ltd.
Figure 1.34: Courtesy of U.S. Census Bureau
Figure 1.35: © Steve Berg/ INSIGHT PHOTOGRAPHY
Figure 1.36: Courtesy of Intuit Inc.
Figure 1.37: Courtesy of The Knot Inc.
Figure 1.38: Courtesy of EHARMONY. COM, Inc.
Figure 1.39: Courtesy of Microsoft Corporation
Figure 1.41: © Marc Romanelli/ Getty Images
Figure 1.42: ©LWA- JDC/CORBIS

Chapter 2

Tech360 © Hemera Photo Objects
Figure 2.1: Courtesy of Nokia; Courtesy of Siemens AG, Munich/ Berlin; Courtesy of RCA/Thomson Consumer Electronics
Figure 2.3: © AP Photo/Paul Sakuma; Courtesy of Apple Computer, Inc.; Courtesy of Hewlett Packard Development Company
Figure 2.4: Courtesy of Hewlett Packard Development Company
Figure 2.5: Courtesy of Dustin Driver and Apple Computers, Inc.
Figure 2.6: Courtesy of Microsoft Corporation
Figure 2.7: Courtesy of Nokia
Figure 2.8: Courtesy of Intel Corporation; Ron Chapple/ Getty Images
Figure 2.9: Courtesy of Intel Corporation
Figure 2.11: Courtesy of Intel Corporation
Figure 2.12: Courtesy of Silicon Graphics
Figure 2.13: Courtesy of NASA
Figure 2.15: © Goodwin Photography
Figure 2.16: Courtesy of CompUSA Management Company
Figure 2.17: Courtesy of ATI Technologies Inc.
Figure 2.18: Courtesy of Roy Kaltschmidt -LBL photographer
Figure 2.20: Courtesy of Toshiba
Figure 2.21: Courtesy of IBM
Figure 2.23: Courtesy of Iomega
Figure 2.25: Courtesy of Symbol Technologies, Inc.
Figure 2.26: © AP photo/Steve Shelton
Figure 2.27: Courtesy of Apple Computer, Inc.
Figure 2.28: Courtesy of Wacom Technology Co.
Figure 2.29: Courtesy of Nokia
Figure 2.30: Courtesy of Palm, Inc.
Figure 2.31: Courtesy of Logitech, Inc.
Figure 2.32: Courtesy of First Virtual Communications, Inc.
Figure 2.33: Courtesy of Apple Computer, Inc.
Figure 2.34: Courtesy of Frauenhofer Institute for Telecommunications
Figure 2.35: Courtesy of Hewlett Packard
Figure 2.36: Courtesy of Logitech
Figure 2.37: Courtesy of WetPC Pty Ltd
Figure 2.38: Courtesy of Fujitsu Siemens Computers
Figure 2.39: © Goodwin Photography
Figure 2.40: Courtesy of SanDisk Corporation
Figure 2.41: Courtesy of Symbol Technologies, Inc.
Figure 2.42: Courtesy of Apple Computers, Inc.
Figure 2.43: Courtesy of CDW Corporation

Chapter 3

Tech360 © John Kelley/ Getty Images
Figure 3.2 Courtesy of Microsoft Corporation

Figure 3.3: Courtesy of Microsoft Corporation
Figure 3.5: Courtesy of Microsoft Corporation
Figure 3.10: Courtesy of Microsoft Corporation
Figure 3.11: Courtesy of Microsoft Corporation
Figure 3.14: Courtesy of Apple Computer, Inc
Figure 3.16: Courtesy of Microsoft Corporation
Figure 3.18: Courtesy of Apple Computer, Inc
Figure 3.19: Copyright 2003 Course Technology
Figure 3.20: Courtesy of Tim Menzies
Figure 3.21: Courtesy of Lindows.com
Figure 3.22: Courtesy of Microsoft
Figure 3.24: © 2005 TiVo Inc. All Rights Reserved.
Figure 3.25: Courtesy of Symantec
Figure 3.26: Courtesy of Symantec
Figure 3.27: Courtesy of Google
Figure 3.28: Courtesy of Corel Corporation
Figure 3.30: Courtesy of Microsoft Corporation
Figure 3.31: Courtesy of Microsoft Corporation
Figure 3.34: Courtesy of Microsoft Corporation
Figure 3.35: Courtesy of Microsoft Corporation
Figure 3.36: Courtesy of Adobe Systems, Inc.
Figure 3.37: Courtesy of Course Technology
Figure 3.39: Courtesy of Microsoft Corporation
Figure 3.40: Courtesy of Microsoft Corporation

Chapter 4

Figure 4.1: © Philippe Galvez, Caltech VRVS Project at *www.vrvs.org*
Figure 4.2: © Lucent Technologies
Figure 4.3: Courtesy of William Decker/ University of California San Diego
Figure 4.11: Courtesy of iAnywhere Solutions
Figure 4.15: Courtesy of Microsoft Corporation
Figure 4.16: Courtesy of Macromedia
Figure 4.19: Courtesy of Mini USA
Figure 4.22: Courtesy of ivillage.com and Courtesy of IGN Entertainment
Figure 4.27: Courtesy of Creative Technology, Ltd.
Figure 4.30: Courtesy of Worlds.com
Figure 4.31: Courtesy of Apple Computer, Inc.
Figure 4.35: Courtesy of Google, Inc.
Figure 4.36: Courtesy of Apple Computer, Inc.
Figure 4.37: Courtesy of Movielink
Figure 4.38: Courtesy of Ferion
Figure 4.39: Courtesy of Yahoo
Figure 4.40: Courtesy of Lexis Nexis
Figure 4.41: Courtesy of National Lambda Rail
Figure 4.42: Courtesy of Gemini Observatory

Chapter 5

Figure 5.1: © Powerstock/ Superstock
Figure 5.5: © *Richard Cummins / SuperStock*
Figure 5.11: Courtesy of Dlink
Figure 5.13: © Michael Dunning/ The Image Bank
Figure 5.16: Courtesy of Nokia
Figure 5.18: Courtesy of Cingular
Figure 5.19: Courtesy of Samsung and Courtesy of Motorola
Figure 5.20: Courtesy of Jabra
Figure 5.21: Courtesy of Thales Navigation, Inc.
Figure 5.22: Courtesy of Wherify, Inc.
Figure 5.24: Courtesy of Vocera
Figure 5.25 Courtesy of PCTEL Antenna Products Group, Inc.
Figure 5.26: Courtesy of Toyota
Figure 5.27: Courtesy of Walmart; Courtesy of the Kennedy Group
Figure 5.29: Courtesy of Fujitsu Siemens Computers
Figure 5.33: Courtesy of UPS

Chapter 6

Figure 6.1: © Mark Scheuern / Alamy; Noel Hendrickson/ Getty Images
Figure 6.3: Courtesy of Sony Corporation of America
Figure 6.4: Courtesy of Creative Forensic Services
Figure 6.5: Courtesy of Westlake Audio
Figure 6.6: Courtesy of Yamaha Motif 6 ES Music Production Synthesizer Workstation. Photo courtesy Yamaha Corporation of America
Figure 6.7: Courtesy of Apple Computer, Inc.
Figure 6.8: Courtesy of Podcast Networks
Figure 6.9: Courtesy of Apple Computer, Inc.
Figure 6.10: Courtesy of Apple Computer, Inc.
Figure 6.11: Courtesy of Samsung
Figure 6.12 Courtesy of Oakley, Inc.
Figure 6.13: Courtesy of RealNetworks
Figure 6.14: Courtesy of Delphi Corporation
Figure 6.15: Courtesy of Kenneth J. Baldauf
Figure 6.17: Courtesy of Microsoft Corporation
Figure 6.18 Courtesy of Mindjet
Figure 6.19: Reprinted with permission of Quark, Inc., and its affiliates.
Figure 6.20: © Sharl Heller/ Resonance Fine Art
Figure 6.21: Courtesy of Daka Design
Figure 6.24: © Liam Kemp
Figure 6.29: Courtesy of Google, Inc.
Figure 6.27: 20TH CENTURY FOX / THE KOBAL COLLECTION
Figure 6.28: Courtesy of Macromedia
Figure 6.30: Courtesy of Nikon
Figure 6.31: Courtesy of Epson
Figure 6.32: Image Courtesy of Andrew Hall *www.worth1000.com*
Figure 6.34: Courtesy of Yahoo
Figure 6.35: Courtesy of AEC-Michael Bloomenfeld
Figure 6.37: Courtesy of Samsung
Figure 6.38: Courtesy of Sony Computer Entertainment, Inc.
Figure 6.39: Courtesy of Lucas Arts
Figure 6.40: Courtesy of Toyota
Figure 6.41: AKIO SUGA/EPA /Landov
Figure 6.42: 2003 Take-Two Interactive Software, Inc.
Figure 6.43: Courtesy of Hewlett Packard

Chapter 7

Tech360 © Stockdisc/ Getty Images
Figure 7.1: Courtesy of UPS
Figure 7.2: © AP Photo/Bob Child
Figure 7.5: Courtesy of IBM Corporation
Figure 7.11: Courtesy of Versant Corporation
Figure 7.13: Courtesy of Microsoft Corporation
Figure 7.14: Courtesy of Intuit Inc.
Figure 7.15: Courtesy of Sybase
Figure 7.16: Courtesy of Mindjet Corporation
Figure 7.17: Courtesy of Microsoft Corporation
Figure 7.20: © AP Photo/Dennis Cook
Figure 7.22: Courtesy of BirdLife International
Figure 7.23: Courtesy of IBM
Figure 7.24: Courtesy of Intellifit
Figure 7.25: Courtesy of Hilton Hospitality, Inc.

Figure 7.27: Courtesy of Business Objects
Figure 7.29: Courtesy of Shopping.com
Figure 7.30: © AP Photo/Denis Poroy

Chapter 8

Figure 8.1: Courtesy of Articways.com
Figure 8.7: Courtesy of Peapod, LLC.; Michael Keller/ CORBIS
Figure 8.8: Courtesy of S.E.O Technologies Pty Ltd.
Figure 8.10: Courtesy of NexTag, Inc.
Figure 8.12: © Frank Siteman/ Getty Images
Figure 8.13: Courtesy of Eastman Kodak Company
Figure 8.14: Courtesy of CNET Networks, Inc.
Figure 8.15: Courtesy of eBay Inc.
Figure 8.16: Courtesy of Compuware Corporation
Figure 8.17: Courtesy of Consumers Union of U.S., Inc.
Figure 8.19: Courtesy of Ameritrade, Inc.
Figure 8.20: Courtesy of Verizon
Figure 8.21: Courtesy of AT&T Wireless; Photographer's Choice/ Getty Images
Figure 8.22: Courtesy of Vivotech.com
Figure 8.25: Courtesy of Izumiya, Inc.
Figure 8.26: Courtesy of Amazon.com, Inc.
Figure 8.28: Courtesy of International Computer Science Institute
Figure 8.29: Courtesy of General Nutrition Center, Inc.
Figure 8.31: Courtesy of Paypal
Figure 8.32: Courtesy of Cherry Corporation

Chapter 9

Figure 9.1: © PhotoAlto/ Superstock
Figure 9.3: © AP Photo/Bebeto Matthews
Figure 9.4 Courtesy of RyTech Software; reprinted with permission of MatheMEDics, Inc.
Figure 9.6: © AP Photo/Gerry Broome; AP Photo/Amy Sancetta
Figure 9.8: Courtesy of Cognos, Inc.
Figure 9.12: Courtesy of ePocrates, Inc.; Jim Craigmyle/Corbis
Figure 9.13: Courtesy of ING Direct Worldwide
Figure 9.14: © Mario Tama/Getty Images
Figure 9.15: Courtesy of National Severe Storms Laboratory
Figure 9.16: © AP Photo/ The Oklahoman, Paul B. Southerland
Figure 9.17: Courtesy of Touchstone Consulting Group, Inc.
Figure 9.18: Courtesy of GoToMetting.com
Figure 9.22: © AP Photo/Shizuo Kambayashi
Figure 9.23: © Peter Dazeley/ Getty Images
Figure 9.25: Courtesy of Agentland.com
Figure 9.26: © AP Photo/Neil Brake
Figure 9.27: © The Cover Story/ CORBIS
Figure 9.28: Courtesy of GeoSim Systems
Figure 9.29: Courtesy of Segway LLC

Chapter 10

Tech360 © Stockdisc/ Getty Images
Figure 10.2: Courtesy of Duncan Stevenson at CSIRO.AU
Figure 10.4: Courtesy of Eastman Kodak Company
Figure 10.5: Courtesy of Macromedia
Figure 10.6: Courtesy of Visual-Paradigm
Figure 10.11: Courtesy of Microsoft Corporation
Figure 10.13: © Brian Lee/Corbis
Figure 10.16: Courtesy of Comcast
Figure 10.17: © Alamy
Figure 10.18: © Steve Chen/Corbis
Figure 10.22: © Dennis MacDonald / Alamy
Figure 10.23: © Paul Eekhoff/ Masterfile
Figure 10.24: © ThinkStock / SuperStock

Chapter 11

Tech360 © John Kelley/ Getty Images
Figure 11.2: © AP Photo/Elaine Thompson
Figure 11.6: © 2003. Nursing Spectrum Nurse Wire (www.nursingspectrum.com). All rights reserved. Used with permission.
Figure 11.8: © JANEK SKARZYNSKI/ AFP/Getty Images
Figure 11.9: © Spencer Platt/Getty Images
Figure 11.10: © ROBYN BECK/AFP/ Getty Images
Figure 11.11: Courtesy of 2600.com
Figure 11.12: Courtesy of International Barcode; Eric Miller/Getty Images
Figure 11.14: Courtesy of CS Technologies
Figure 11.15: Courtesy of Viisage
Figure 11.16: Courtesy of Sony Corporation of America
Figure 11.17: Courtesy of Apple Computer, Inc.
Figure 11.18: Courtesy of Microsoft Corporation
Figure 11.19: Courtesy of Mirra
Figure 11.21: © Visions of America, LLC / Alamy
Figure 11.24: © *Jiang Jin / SuperStock*
Figure 11.25: © Royalty-Free/Corbis
Figure 11.26: ©Kenny@Hektik.org
Figure 11.28: Courtesy of Sandia Corporation
Figure 11.29: Courtesy of Source Forge

Chapter 12

Figure 12.1: Courtesy of the University of Wisconsin; Photodisc/ Getty Images
Figure 12.2: © Kevin Moloney/Getty Images
Figure 12.3: © Sean Savage
Figure 12.4: © Chuck Savage/CORBIS
Figure 12.5: © Will & Deni McIntyre/ Photo Researchers, Inc.
Figure 12.6: © Tom Grill/ Corbis
Figure 12.8: © Evan Kafka/Liaison
Figure 12.9: Courtesy of Current
Figure 12.10: Courtesy of Net Nanny
Figure 12.11: © Najlah Feanny/CORBIS SABA
Figure 12.12: Courtesy of RG Networks
Figure 12.13: © Erik S. Lesser/Getty Images
Figure 12.14: © COLUMBIA / THE KOBAL COLLECTION
Figure 12.15: © Randy Emmitt
Figure 12.16: © Tim Boyle/Getty Images
Figure 12.17: Courtesy of Google, Inc.
Figure 12.18: © AP Photo/Max Whittaker
Figure 12.19: © The M.C. Escher Company BV
Figure 12.20: © Jim Sugar/CORBIS
Figure 12.21: Courtesy of Secure Internet Machines
Figure 12.23: © Dave G. Houser/Post-Houserstock/Corbis

Figure 12.24: © SIMONPIETRI CHRISTIAN/CORBIS SYGMA
Figure 12.25: Courtesy of Texas Instruments
Figure 12.26: Courtesy of Champigny Triathlon
Figure 12.27: © AP Photo/Charles Dharapak
Figure 12.28: Courtesy of General Electric
Figure 12.29: © Sion Touhig/ /Corbis
Figure 12.30: © Index Open